Quantum Heterostructures

Quantum Heterostructures provides a detailed description of the key physical and engineering principles of quantum semiconductor heterostructures. Blending important concepts from physics, materials science, and electrical engineering, it also explains clearly the behavior and the operating features of modern microelectronic and optoelectronic devices.

The authors begin by outlining the trends that have driven development in this field, most importantly the need for high-performance devices in computer and communications technologies. They then describe the basics of quantum nanoelectronics, including various transport mechanisms. In the latter part of the book, they cover novel microelectronic devices and optical devices based on quantum heterostructures.

The book contains many homework problems and is suitable as a textbook for undergraduate and graduate courses in electrical engineering, physics, or materials science. It will also be of great interest to those involved in research or development in microelectronic or optoelectronic devices.

Vladimir Mitin is Professor of Electrical and Computer Engineering at Wayne State University. He obtained his Ph.D. from the Institute for Semiconductor Physics in Kiev, Ukraine. He is the author of over 280 scientific papers and co-author of five books, including *Introduction to Solid-State Electronics* and *The Physics of Instabilities in Solid-State Electron Devices*.

Viatcheslav Kochelap is Professor of Theoretical Physics and Head of the Theoretical Physics Department in the National Academy of Sciences of Ukraine. He obtained his Ph.D. from the Institute for Semiconductor Physics in Kiev. He is the author of over 200 scientific papers and co-author of two books.

Michael Stroscio is Professor of Physics at Duke University. He obtained his Ph.D. from Yale University. He has served as a policy analyst for the White House Office of Science and Technology and as Vice Chairman of the White House Panel on Scientific Communication. Professor Stroscio is the author of over 400 papers, presentations, and patents spanning a broad spectrum of topics in the physical sciences and electrical engineering. He is a Fellow of the IEEE.

Quantum Heterostructures

Microelectronics and Optoelectronics

VLADIMIR V. MITIN
Wayne State University

VIATCHESLAV A. KOCHELAP
National Academy of Sciences of Ukraine

MICHAEL A. STROSCIO
Duke University

PUBLISHED BY THE PRESS SYNDICATE OF THE UNIVERSITY OF CAMBRIDGE
The Pitt Building, Trumpington Street, Cambridge, United Kingdom

CAMBRIDGE UNIVERSITY PRESS
The Edinburgh Building, Cambridge CB2 2RU, UK   http://www.cup.cam.ac.uk
40 West 20th Street, New York, NY 10011-4211, USA   http://www.cup.org
10 Stamford Road, Oakleigh, Melbourne 3166, Australia

First published 1999

Printed in the United States of America

Typeset in Sabon 10/13pt. and Antique Olive in LaTeX 2$_\varepsilon$ [TB]

*A catalog record for this book is available from
the British Library.*

Library of Congress Cataloging-in-Publication Data
Mitin, V. V. (Vladimir Vasilévich)
    Quantum heterostructures : microelectronics and optoelectronics /
Vladimir V. Mitin, Viatcheslav A. Kochelap, Michael A. Stroscio.
        p.   cm.
    ISBN 0-521-63177-7. – ISBN 0-521-63635-3 (pbk.)
    1. Semiconductors – Materials.   2. Heterostructures.
3. Optoelectronics.   4. Microelectronics.   I. Kochelap, V. A.
(Viacheslav Aleksandrovich)   II. Stroscio, Michael A., 1949–   .
III. Title.
QC611.M665   1999
    621.381–dc21   98-40658
                                                           CIP
ISBN 0 521 63177 7 hardback
ISBN 0 521 63635 3 paperback

Contents

Preface   *page* xiii

Notation   xv

Chapter 1
Trends in Microelectronics and Optoelectronics   1

Chapter 2
Theoretical Basis of Nanoelectronics   12

  2.1 Introduction   12
  2.2 Wave–Particle Duality in Quantum Physics   13
  2.3 Time and Length Scales   15
     2.3.1 Electron Fundamental Lengths in Solids   15
     2.3.2 Quantum and Classical Regimes of Electron Transport   20
     2.3.3 Time Scales and Temporal (Frequency) Regimes   21
  2.4 Schrödinger Equation   23
     2.4.1 Average Values of Physical Quantities   29
     2.4.2 Separation of Variables   30
     2.4.3 Variational Method   32
     2.4.4 Perturbation Methods   33
  2.5 Many-Particle Systems   37
     2.5.1 Spin and Electron Statistics   37
     2.5.2 Many-Particle Schrödinger Equation: Self-Consistent
           Potential   40
  2.6 An Electron in Crystalline Potential: Effective-Mass
     Approximation   44
     2.6.1 Bloch Functions   45
     2.6.2 Occupation of Energy Bands: Metals, Semiconductors,
           and Dielectrics   47
     2.6.3 Effective Masses   50
     2.6.4 The Envelope Function   53
  2.7 Quantum Electron Transport: Landauer Formula   55
  2.8 Boltzmann Transport Equation   61
  2.9 Local Approximations for Electron Transport   64
     2.9.1 Electron Mobility and Diffusion Coefficient   64
     2.9.2 Electron Temperature (Hydrodynamic) Approximation   67
     2.9.3 Drift-Diffusion Model   70

2.10	Closing Remarks	70
	PROBLEMS	71

Chapter 3
Electrons in Quantum Structures   73

3.1	Introduction	73
3.2	Quantum Wells	75
	3.2.1 Density of States of a Two-Dimensional Electron Gas	81
	3.2.2 Quantum Effects in a Continuum-Electron Spectrum	83
	3.2.3 Two-Dimensional Electron Motion in a Smooth Potential	85
3.3	Quantum Wires	87
	3.3.1 Wave Functions and Energy Subbands	88
	3.3.2 Density of States for a One-Dimensional Electron Gas	90
3.4	Quantum Dots	91
	3.4.1 Wave Functions and Energy Levels	91
	3.4.2 Density of States for Zero-Dimensional Electrons	94
3.5	Coupling between Quantum Wells	96
3.6	Superlattices	99
	3.6.1 Density of States	104
3.7	Excitons in Quantum Structures	106
	3.7.1 Excitons	106
	3.7.2 Excitons in Quantum Wells	110
3.8	Coulomb Bound States and Defects in Quantum Structures	115
	3.8.1 Coulomb Bound States of Impurities	116
	3.8.2 Interfacial Defects	119
3.9	Closing Remarks	123
	PROBLEMS	125

Chapter 4
Properties of Particular Quantum Structures   126

4.1	Introduction	126
4.2	Energy Spectra of Some Semiconductor Materials	127
	4.2.1 Symmetry of Crystals and General Properties of Electron Spectra	127
	4.2.2 Band Structures of Semiconductor Alloys	132
	4.2.3 Electron Affinities: Energy-Band Discontinuities of Heterostructures	134
4.3	Lattice-Matched and Pseudomorphic Heterostructures	136
	4.3.1 Lattice-Matched and Lattice-Mismatched Materials	138
	4.3.2 Lattice-Matched Heterostructures	141
	4.3.3 Strained Pseudomorphic Structures	143
	4.3.4 Si/Ge Strained Heterostructures	146
4.4	Single-Heterojunction Devices: Selective Doping	151

	4.4.1	Introduction	151
	4.4.2	Metal-Oxide-Semiconductor Structures	153
	4.4.3	Single Modulation-Doped Heterojunction	155
	4.4.4	Basic Equations and Quantitative Results for a Single Heterostructure	156
	4.4.5	Simple Analytical Estimates for a Selectively Doped Single Heterostructure	161
	4.4.6	Numerical Analysis of a Modulation-Doped Single Heterojunction	163
	4.4.7	Control of Charge Transfer	167
4.5	Modulation-Doped Quantum Structures		171
	4.5.1	Modulation-Doped Quantum Wells	171
	4.5.2	n–i–p Structures	176
	4.5.3	Delta Doping	180
4.6	Closing Remarks		182
	PROBLEMS		184

Chapter 5
Lattice Vibrations in Quantum Structures
185

5.1	Introduction		185
5.2	Vibrations of Atomic Linear Chains		186
	5.2.1	Monoatomic Chain	186
	5.2.2	Diatomic Chain	189
5.3	Normal Coordinates. Three-Dimensional Case		192
5.4	Phonons		196
5.5	Acoustic Vibrations in Quantum Structures		199
	5.5.1	Acoustic Modes in the Long-Wavelength Limit	199
	5.5.2	Acoustic-Mode Localization in Heterostructures with Quantum Wells and Wires	202
5.6	Short-Wavelength and Optical Vibrations in Quantum Structures		218
	5.6.1	Qualitative Analysis of Short-Wavelength Vibrations	218
	5.6.2	Optical Vibrations in Polar Crystals	223
5.7	Closing Remarks		231
	PROBLEMS		233

Chapter 6
Electron Scattering in Quantum Structures
235

6.1	Introduction	235
6.2	Elastic Scattering in Two-Dimensional Electron Systems	239
6.3	Screening of a Two-Dimensional Electron Gas	242
6.4	Scattering by Remote Ionized Impurities	246
6.5	Scattering by Interface Roughness	249
6.6	Electron–Phonon Interaction	253

6.6.1   Transitions due to Electron–Phonon Interactions   253
6.6.2   Short-Range and Long-Range Electron–Phonon
          Interactions   255
6.6.3   The Interaction with Different Phonon Modes   256
6.7   Interaction with Acoustic Phonons   260
6.8   Interaction with Optical Phonons   267
6.9   Scattering of Electrons by Acoustic Phonons in Quantum
       Wells and Wires   273
6.9.1   Quantum Wells   276
6.9.2   Quantum Wires   281
6.9.3   Free-Standing Quantum Structures   281
6.10 Scattering of Electrons by Optical Phonons in Quantum
       Wells and Wires   284
6.10.1 Quantum Wells   284
6.10.2 Quantum Wires   288
6.11 Closing Remarks   290
PROBLEMS   292

Chapter 7
Parallel Transport in Quantum Structures   295

7.1   Introduction   295
7.2   Linear Electron Transport   297
7.3   High-Field Electron Transport   303
7.3.1   Hot, Warm, and Cold Electrons   304
7.3.2   Velocity Saturation   307
7.3.3   Transient Overshoot Effect   311
7.3.4   Gunn Effect   312
7.3.5   Nonequilibrium Phonons   315
7.3.6   Hot-Electron Size Effect   316
7.4   Hot Electrons in Quantum Structures   317
7.4.1   Nonlinear Transport in Two-Dimensional Electron
          Gases   317
7.4.2   Nonlinear Electron Transport in Quantum Wires   319
7.4.3   Real-Space Transfer of Hot Electrons   324
7.4.4   Other Effects of High-Field Electron Transport in
          Quantum Structures   326
7.5   Closing Remarks   328
PROBLEMS   330

Chapter 8
Perpendicular Transport in Quantum Structures   333

8.1   Introduction   333
8.2   Double-Barrier Resonant-Tunneling Structures   333
8.2.1   Coherent Tunneling   336

8.2.2 Sequential Tunneling 340
8.2.3 Comparison of Two Mechanisms of Resonant Tunneling 342
8.2.4 Negative Differential Resistance under Resonant Tunneling 343
8.3 Superlattices and Ballistic-Injection Devices 346
8.3.1 Negative Differential Resistance and Transconductance of Devices with Ballistic Superlattices 346
8.3.2 Bloch Oscillations 350
8.3.3 Wannier–Stark Energy Ladder 352
8.3.4 Negative Differential Resistance in Superlattices for Electron-Collision Regimes 355
8.3.5 Ballistic-Injection Devices 356
8.4 Single-Electron Transfer and Coulomb Blockade 357
8.5 Closing Remarks 359
PROBLEMS 361

Chapter 9
Electronic Devices Based on Quantum Heterostructures 362

9.1 Introduction 362
9.2 Field-Effect Transistors 367
9.2.1 Principle of Field-Effect-Transistor Operation 367
9.2.2 Amplification and Switching 372
9.2.3 Heterostructure Field-Effect Transistors 377
9.3 Velocity-Modulation and Quantum-Interference Transistors 385
9.3.1 Velocity-Modulation Transistors 385
9.3.2 Quantum-Interference Transistors 389
9.4 Bipolar Heterostructure Transistors 394
9.4.1 p–n Junctions and Homostructure Bipolar Transistors 394
9.4.2 Heterostructure Bipolar Transistors 408
9.5 Si/SiGe Heterostructure Bipolar Transistors 412
9.6 Hot-Electron Transistors 416
9.6.1 Ballistic-Injection Devices 416
9.6.2 Real-Space Transfer Devices 420
9.7 Applications of the Resonant-Tunneling Effect 425
9.7.1 Resonant-Tunneling Oscillators 425
9.7.2 Resonant-Tunneling Diode as Frequency Multiplier 431
9.7.3 Resonant-Tunneling Transistors 435
9.8 Circuit Applications of Resonant-Tunneling Transistors 446
9.8.1 Multipeak Current-Voltage Characteristics and Multivalued Logic Applications 448
9.9 Closing Remarks 452
PROBLEMS 455

Chapter 10
Optics of Quantum Structures 456

10.1 Introduction 456
10.2 Electromagnetic Waves and Photons 457
    10.2.1 Electromagnetic Fields, Modes, and Photons in Free
        Space 457
    10.2.2 Photons in Nonuniform Dielectric Media 460
    10.2.3 Optical Resonators 461
    10.2.4 Photon Statistics 464
10.3 Light Interaction with Matter: Phototransitions 464
    10.3.1 Photon Absorption and Emission 464
    10.3.2 Calculation of Phototransition Probabilities 467
10.4 Optical Properties of Bulk Semiconductors 470
    10.4.1 Interband Emission and Absorption in Bulk
        Semiconductors 471
    10.4.2 Spectral Density of Spontaneous Emission 479
    10.4.3 Phototransitions in III–V Compounds 480
    10.4.4 Excitonic Effects 484
    10.4.5 Refractive Index 484
10.5 Optical Properties of Quantum Structures 487
    10.5.1 Electrodynamics of Heterostructures 487
    10.5.2 Light Absorption by Confined Electrons 490
    10.5.3 Effects of Complex Valence Bands of III–V
        Compounds 494
    10.5.4 Other Factors Affecting the Interband Optical Spectra 496
    10.5.5 Polarization Effects 499
10.6 Intraband Transitions in Quantum Structures 500
    10.6.1 Intraband Absorption and Conservation Laws 500
    10.6.2 Probability of Intersubband Phototransitions 502
    10.6.3 Probability of Phototransitions to Extended States 508
10.7 Closing Remarks 511
PROBLEMS 512

Chapter 11
Electro-Optics and Nonlinear Optics 513

11.1 Introduction 513
11.2 Electro-Optics in Semiconductors 513
    11.2.1 Electro-Optical Effect in Conventional Materials 513
    11.2.2 Electro-Optical Effect in Quantum Wells 518
    11.2.3 Electro-Optical Effect in Superlattices 524
    11.2.4 Terahertz Coherent Oscillations of Electrons
        in an Electric Field 529

11.3 Nonlinear Optics in Heterostructures   532
      11.3.1 Linear and Nonlinear Optics   532
      11.3.2 Optical Nonlinearities in Quantum Wells   534
      11.3.3 Virtual Field-Induced Mechanism of Nonlinear
              Optical Effects   535
      11.3.4 Nonlinear Optical Effects due to Generation
              of Excitons and Electron–Hole Plasma   536
      11.3.5 Nonlinear Effects Induced by Nonthermalized
              Electron–Hole Plasma   542
      11.3.6 Optical Bistability   543
      11.3.7 Applications of Nonlinear Optical Effects in Quantum
              Wells   548
11.4 Closing Remarks   553
      PROBLEMS   555

Chapter 12
Optical Devices Based on Quantum Structures   556

12.1 Introduction   556
12.2 Light Amplification in Semiconductors   556
      12.2.1 Criteria for Light Amplification   557
      12.2.2 Estimates of Light Gain   560
      12.2.3 Methods of Pumping   562
      12.2.4 Motivations for Using Heterostructures for Light
              Amplifications   566
      12.2.5 Light Amplification in Quantum Wells and Quantum
              Wires   570
12.3 Light-Emitting Diodes and Lasers   574
      12.3.1 Light-Emitting Diodes   574
      12.3.2 Amplification, Feedback, and Laser Oscillations   579
      12.3.3 Laser Output Power and Emission Spectra   581
      12.3.4 Modulation of the Laser Output   584
      12.3.5 Quantum-Well Lasers   587
      12.3.6 Surface-Emitting Lasers   592
      12.3.7 Quantum-Wire Lasers   594
      12.3.8 Blue Quantum-Well Lasers   599
      12.3.9 Unipolar Intersubband Quantum-Cascade
              Laser   602
12.4 Self-Electro-Optic-Effect Devices   609
      12.4.1 Resistor SEED   609
      12.4.2 Symmetric SEED   611
12.5 Photodetectors on Intraband Phototransitions   614
      12.5.1 Photoconductive Detectors   615

12.5.2 Intraband Phototransitions and Electron Transport
in Multiple-Quantum-Well Structures   618

12.5.3 Simple Model of Multiple-Quantum-Well
Photodetectors   624

12.6 Closing Remarks   631

PROBLEMS   633

Index   635

Preface

Welcome to the world of quantum-based devices! In this book you will find the fundamental physics and engineering principles underlying quantum semiconductor heterostructures as well as new concepts in the field of quantum-based microelectronic and optoelectronic devices. We ask the readers to disregard the thickness of this volume. Just follow us from simple and basic definitions into this exciting field.

The recent and diverse trends in semiconductor and device technologies as well as in novel device concepts are driving the establishment of new subdisciplines of electronics and optoelectronics based on quantum structures, as reviewed in Chapter 1. These trends make it important for the electrical engineer, the materials engineer, the materials scientist, and the solid-state physicist alike to understand the fundamentals underlying devices based on quantum structures. Our goal in writing this book is to help you in this endeavor.

The book grew out of our research and teaching experience in these subjects. Many of the ideas and the achievements in the field can be explained in a relatively simple setting, if the necessary underlying fundamentals are presented properly. Our book provides a unifying framework for the basic ideas needed to understand recent developments and culminates with treatments of electronic and optoelectronic devices based on quantum structures.

With this purpose in mind, we present the material on quantum-heterostructure physics and engineering in Chapters 2–6. The foundational topics covered in these early chapters set the stage for the study of a variety of important electron transport regimes in Chapters 7 and 8 and the optics of quantum structures in Chapter 10. Chapters 9, 11, and 12 are built on the preceding chapters to describe successively the subdisciplines that are emerging from the theory underlying quantum-based structures; these are electronic, electro-optical, and optical devices based on quantum structures.

The scope and the organization of this book facilitate instruction on both the undergraduate and the graduate levels. In our experience, the material can be covered in a one-semester course with sixty contacts and with a few topics from the book assigned to students for self-study. For courses that are two quarters in duration, instructors may find it convenient to present the material of Chapters 2–6 during the first quarter and that of Chapters 7–12 during the second quarter; this method of presentation separates the fundamental background from the subject matter on specific quantum structures and devices. The problems included in each chapter of this book focus on amplifying the understanding of the key concepts underlying quantum-based devices. Many of the problems contain extensive discussions that lead the readers to the frontiers of electronics and optoelectronics. Every chapter also has a list of recommended literature, which typically includes the most useful reviews and books on the subjects.

The authors have many professional colleagues and friends from different countries who must be acknowledged. Without their contributions and sacrifices this work would not have been completed. These people include Professors Michael P. Polis and Frank Westervelt – the former and the current Chairs of the Electrical and Computer Engineering of the Wayne State University; Professors Z. Gribnikov, S. Svechnikov, V. Sokolov, and V. Pipa of the Institute of Semiconductor Physics, Kiev, Ukraine; Professor Gerald J. Iafrate of Notre Dame University; Professors Ki Wook Kim, M. A. Littlejohn, and Mitra Dutta of the North Carolina State University; Professors Pallab K. Bhattacharya, George I. Haddad, and Jasprit Singh of the University of Michigan; Professors Karl Hess and J.-P. Leburton of the University of Illinois; and Professors H. Craig Casey and Steven Teitsworth of Duke University. The help of Dr. R. Mickevičius in preparing Chapters 5 and 6 is especially appreciated. We also thank the students who took this course with us. Their feedback helped us in the choice and presentation of the material.

VM acknowledges former graduate students and associates for their critical reading of the manuscript and their constructive comments. Especially appreciated are the comments of Dr. N. Vagidov (Chapters 1–12), B. Glavin (Chapters 2–6), Dr. D. Romanov (Chapters 7 and 8), Dr. A. Korshak (Chapter 9), and Professor V. Pipa (Chapters 10–12). G. Paulavičius helped in preparing the final versions of most of the figures. S. Rolnik, B. Glavin, A. Brailovsky, I. Gordion, and G. Paulavičius helped with some editorial work. VM is also thankful to his family and friends for their understanding and forgiving that he did not devote enough time to them while working on the book, and especially to his mother Vera Mitina and granddaughter Christina whom he missed the most.

VK is deeply thankful to all his family members who have waited for him during long periods of lecturing and working on this book in Detroit; these include his father and mother, wife and daughter, and, finally, granddaughter Asya.

MAS thanks family members who have been attentive during the period when this book was being written; these include Anthony and Norma Sidbury Stroscio, Mitra Dutta, and Gautam and Marshall Stroscio.

Vladimir Mitin, *Detroit, MI*
Viatcheslav Kochelap, *Detroit, MI; Kiev, Ukraine*
Michael Stroscio, *Raleigh, NC*

Notation

Symbols

Other symbols are self-evident and are given as they occur within specific analyses.

A	cross-sectional area
a	lattice constant
a_B	Bohr radius
$\vec{a}_i$	basic vectors of the Bravais lattice
a_0^{ex}	radius of bulk exciton
a_0	Bohr radius for an impurity atom in semiconductors
c	velocity of light
c_{ij}	elastic moduli of the crystal
$c_{\vec{q},s}$	phonon annihilation operator
$c_{\vec{q},s}^{+}$	phonon creation operator
D_{ij}	tensor of the deformation potential
D	diffusion coefficient
d	distance between wells; superlattice period
d_{sp}	width of spacer
E	photon energy; electron energy
$E_{A,D}$	ionization energy of acceptors and donors, respectively
$E_{c,v}(\vec{k})$	energy of the electron in the conduction and the valence bands, respectively
E_{el}	elastic energy
E_{ex}	exciton energy
E_{im}	energy associated with the misfit dislocations or misfit defects of a crystal
E_F	Fermi level
E_{Fn}	electron quasi-Fermi level
E_{Fp}	hole quasi-Fermi level
E_g	bandgap
E_h	energy of hole
$\mathcal{E}$	electric field of electromagnetic wave; energy functional
F	electric-field intensity; restoring force
$F_\alpha(\vec{r})$	envelope function
$\mathcal{F}$	finesse of resonator
$\mathcal{F}_F$	Fermi distribution function
$\mathcal{F}(\vec{k}, \vec{r}, t)$	electron distribution function
$\vec{f}$	external force applied to the surface of a crystal

f_T	cutoff frequency of a field-effect transistor
G	conductance; total gain
$G_0 = 39.6\,\mu S$	quantum of conductance
g_m	transconductance
g_0	conductance of a field-effect transistor for a completely opened channel
$\mathcal{H}$	Hamiltonian of a system
$h = 6.62 \times 10^{-34}\,\text{Js}$	Planck's constant
hh	heavy-hole band
lh	light-hole band
sh	split-off band
$\hbar = h/2\pi$	reduced Planck's constant
I	current; collision integral; total intensity of waveguide mode
$\vec{i}$	density of particle flow
j	electron current density
k	wave number, wave vector of the electron in the crystal
$\vec{k}$	wave vector
$k_\parallel$	parallel component of three-dimensional vector $\vec{k}$
k_B	Boltzmann constant
k_F	Fermi wave vector
L	length, thickness of a well
L_E	inelastic-scattering length
L_R	ambipolar length
L_T	thermal diffusion length
L_i	geometrical size of a sample in the direction i
$L_{n,p}$	electron L_n and hole L_p recombination lengths
l	angular-momentum quantum number
l_e	electron mean free path
l_ϕ	coherence length
M	electron effective mass along the superlattice axis
M^{ex}	exciton effective mass
m	free-electron mass
m^*	effective mass
$(1/m^*)_{ik}$	reciprocal effective-mass tensor
m_e	electron effective mass
m_h	hole effective mass
m_l	longitudinal effective mass
m_{lh}	effective mass of the light hole
m_{hh}	effective mass of the heavy hole
m_r	reduced mass of the electron–hole pair
m_t	transverse effective mass
N	total number of phonons, electrons, atoms
$\vec{N}$	vector perpendicular to the surface of a crystal

$N_{A,D}$	concentration per unit volume of ionized acceptors and donors, respectively
$N_{c,v}$	effective density of states for electrons and holes
$\mathcal{N}$	number of atoms per primitive crystal cell; total number of photons
n	electron concentration per unit volume; principal quantum number; refractive index
n_e	electron concentration per unit volume; area of length
n_i	intrinsic electron concentration
n_S	surface concentration of electrons
n_{3D}	electron concentration per unit volume
n_{2D}	electron concentration per unit area
n_{1D}	electron concentration per unit length
$P_{i \to f}$	probability of electron transition from state i to state f
$\vec{\mathcal{P}}$	operator of momentum; macroscopic polarization vector
P_b	probability of finding the electron in the barrier layer
$\vec{p}$	momentum vector
p_i	intrinsic hole concentration
q	wave vector
R	reflection coefficient
R_y	Rydberg constant; ionization energy of the hydrogen atom
R_y^{ex}	ionization energy of bulk exciton
R_y^*	effective Rydberg constant
r	space coordinate; amplitude of a reflected wave
S	cross-sectional area
s	spin number; sound velocity
T	temperature; transmission coefficient
T_e	electron temperature
T_t	tunneling matrix element
$T_{\vec{d}}$	translational operator
t	amplitude of a transmitted wave; time
t_{lw}	characteristic time for changes in the light-wave amplitude
t_{tr}	transit time
U	elastic energy per unit volume; power gain
$\vec{u}(\vec{r})$	vector of relative displacement of point $\vec{r}$
u_{ij}	strain tensor
$u_{\vec{k}}(\vec{r})$	Bloch function
V	potential energy, volume of the system
V_{FB}	voltage on the metal required for reaching the flat-band condition in a metal-oxide-semiconductor structure
V_b	depth of a potential well; potential-barrier height
V_0	volume of the primitive cell
v	electron velocity; group velocity
v_p	phase velocity

v_g	group velocity
$W(\vec{r})$	crystalline potential
$W(\vec{k}', \vec{k})$	probability of transition from an initial state $\vec{k}'$ to a final state $\vec{k}$
W_{km}	probability of transition from an initial state k to a final state m per unit time
$Y_{l,m}(\theta, \varphi)$	spherical functions
α	Frölich coupling constant
β	quasi-elastic force coefficient
Δ_{so}	spin-orbital splitting of valence bands
∇	dell operator
$\delta(x)$	Dirac delta function
δ_{ij}	Kronecker delta
ϵ	energy of a transverse motion of an electron in a well or a wire
$\Theta(x)$	Heaviside step function
θ	angle in spherical coordinates
κ	dielectric permitivity of the medium
κ_{el}	dielectric permitivity of an electron gas
λ	wave vector; de Broglie wavelength; elastic modulus
μ	electron mobility
ν	frequency
$\vec{\xi}$	vector of polarization of electromagnetic wave
ρ	linear density of the string; density of a semiconductor; density of states
$\vec{\rho}$	two-dimensional vector of space coordinate
σ_{ij}	stress tensor
τ_E	mean time between two inelastic collisions; energy-relaxation time
τ_R	lifetime of a charge carrier
τ_T	thermal diffusion spreading time
τ_e	mean time between two elastic-scattering events; mean free time
τ_{ee}	time of electron–electron interaction
τ_{ep}	time of electron interaction with lattice
$\tau_{n,p}$	electron–hole recombination time for p-type material τ_n, and n-type material τ_p
τ_p	momentum-relaxation time
τ_s	duration of scattering
Υ	total photon flux
Υ_0	photon flux in steady state
Φ	voltage applied to a device; many-electron time-dependent wave function
Φ_p	internal pinch-off voltage
ϕ	wave phase

$\phi(\vec{r})$	electrostatic potential
$\phi(z)$	electrostatic potential
φ	many-electron time-independent wave function
χ	electron wave function; electric susceptibility
$\Psi(\vec{r}, t)$	one-electron time-dependent wave function
Ψ_{tr}	true wave function
$\psi(\vec{r}, t)$	one-electron time-dependent wave function
Ω	angular frequency of the de Broglie wave
ω	angular frequency of a wave
ω_{IF}	frequency of interface modes
ω_{LO}	frequency of longitudinal-optical modes
ω_{TO}	frequency of transverse-optical modes

ACRONYMS

BT	bipolar transistor
CHINT	charge-injection transistor
CMOS	complementary MOS, i.e., NMOS and PMOS on the same chip
FET	field-effect transistor
JBT	homojunction BT
JFET	junction FET
HBT	heterojunction BT
HEMT	high-electron-mobility transistor
HFET	heterostructure FET
MES	metal-semiconductor
MESFET	metal-semiconductor FET
MODFET	modulation-doped FET
MOS	metal-oxide-semiconductor
MOSFET	metal-oxide-semiconductor FET
NERFET	negative-resistance FET
NMOS	n-channel MOSFET
PMOS	p-channel MOSFET
QUIT	quantum-interference transistor
RHET	resonant hot-electron transistor
RST	real-space transfer
VMT	velocity-modulation transistor
SEED	self-electro-optic-effect device

1

Trends in Microelectronics and Optoelectronics

This book is intended to provide the foundations of the physics and engineering of quantum heterostructures and the physics of novel microelectronics and optoelectronics devices.

A wide variety of quantum heterostructures and devices has become possible because of dramatic improvements in semiconductor materials and technology as well as because of a deeper understanding of the underlying physics and the new device concepts. Progress in each of these areas has been stimulated, in part, by the enormous demands for information and communication technologies as well as by numerous special applications. The continuous demands for steady growth in memory and computational capabilities and for increasing processing and transmission speeds of signals appear to be insatiable. They have determined the dominant trends of contemporary microelectronics and optoelectronics – two components of the rapidly growing information industry. In this chapter we briefly analyze some of these trends and discuss the use of quantum heterostructures and quantum physics to realize electronic devices with greatly enhanced performance. Then we discuss new quantum effects discovered in heterostructures as well as their impact on the operation of conventional devices and on new device concepts.

To trace how the dominant trends are evolving, Fig. 1.1 illustrates the relationships among end-use technologies, physics of materials and devices, and new device concepts. The upper level of this schematic presents end-use technologies. As is well known, these information and communication technologies are essential to the functions and progress of society. There are also other special applications based on microelectronics and optoelectronics; these applications underlie many high-technology industries including those supporting aeronautics, space, and the military. These end-use technologies are based on supporting technical capabilities.

The next level of Fig. 1.1 focuses on the general demands for technical capabilities and systems. Modern information technology depends heavily on the systems that are highly integrated with great numbers of devices per unit area or on a single chip; moreover, there are increasing demands for high-speed operation and low power consumption. Communication technology relies on microwave and optical-fiber transmission and is based on systems operating with both high-frequency electrical and optical signals. Special applications result in additional

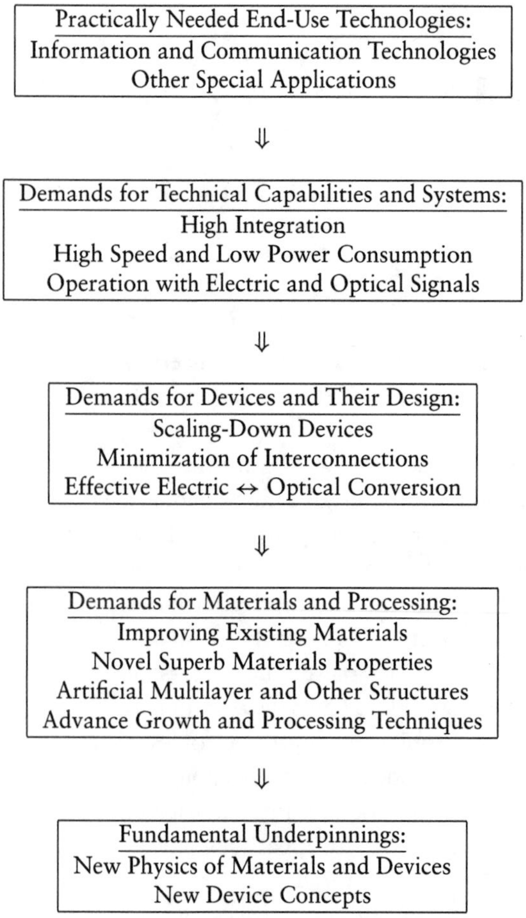

Practically Needed End-Use Technologies:
Information and Communication Technologies
Other Special Applications

⇓

Demands for Technical Capabilities and Systems:
High Integration
High Speed and Low Power Consumption
Operation with Electric and Optical Signals

⇓

Demands for Devices and Their Design:
Scaling-Down Devices
Minimization of Interconnections
Effective Electric ↔ Optical Conversion

Figure 1.1. This schematic illustrates the relationships among end-use technologies, technical systems, single devices, material science and engineering, physics, and new device concepts and systems.

⇓

Demands for Materials and Processing:
Improving Existing Materials
Novel Superb Materials Properties
Artificial Multilayer and Other Structures
Advance Growth and Processing Techniques

⇓

Fundamental Underpinnings:
New Physics of Materials and Devices
New Device Concepts

demands such as high-temperature operation and the handling of high-power signals. The technical systems consist of active devices, passive elements, a number of interdevice connections, and so on. The next level of Fig. 1.1 highlights the demands on single devices. For high-performance systems, single devices have to be as small as possible, and it is highly desirable to minimize interconnections on chip. Clearly, efficient conversion from electric to optical signals and vice versa are necessary. It is possible to achieve many of these goals through advances in materials and processing, as presented in the next level of Fig. 1.1. This includes improving existing semiconductor and optical materials, growing new materials with superb properties and perfectness, and the fabrication of artificial structures like multilayered structures and other semiconductor heterostructures. Techniques for processing materials and heterostructures are essential for these advances. Currently, processes such as patterning, etching, implantation, metallization, and others are carried out with nanoscale control. Finally, the basis for future progress in all these technologies is the physics of new materials, structures, and devices. A principal task for researchers in these fields is to establish the fundamental properties of materials, to model processes in devices, and to find ultimate regimes and operation limits for devices. Equally important, scientists and engineers must generate new conceptual ideas of devices.

It is instructive to illustrate these trends and achievements through the example of Si-based electronics. Contemporary microelectronics is based almost entirely on Si technology because of the unique properties of Si. This semiconductor material has high mechanical stability as well as good electrical isolation and thermal conductivity. Furthermore, the thin and stable high-resistance oxide, SiO_2, is capable of withstanding high voltage and can be patterned and processed by numerous methods. Si technology also enjoys the advantage in that one can grow larger-area Si substrates (wafers) than for other semiconductor materials. The high level of device integration realizable with Si-based electronics technology is illustrated for the important integrated circuit element of any computer, controller, etc. – the dynamic random-access memory (DRAM). We shall consider DRAMs based on the Si complementary metal-oxide-semiconductor (CMOS) technology. The main elements of such DRAMs are metal-oxide-semiconductor (MOS) field-effect transistors (FETs) and passive capacitors. For Si MOSFETs, current channels are created in the Si substrate between the source and the drain contacts, and the currents are controlled by electrodes – metal gates – that are isolated electrically by very thin SiO_2 layers, typically ~100 Å thick. In a CMOS, the storage element uses two MOSFETs with p and n channels. Figure 1.2 illustrates the evolution of such DRAMs: the minimum MOSFET feature size is presented as a function of time for DRAMs with different integration levels and capabilities. The chip also contains, besides MOSFETs, capacitors, metallic line connections, etc. The number of such devices and elements increases progressively with increasing information capability of the DRAM chip. Figure 1.2 illustrates the steady scaling down of the devices as well as increasing levels of integration. For example, the modern 64-Mbit DRAM chip contains approximately 10^8 MOSFETs/cm^2 and these MOSFETs have feature sizes of the order of 0.3 μm. The MOSFETs in this DRAM as well as those of the more highly integrated 256-Mbit chip operate as conventional devices and obey the laws of classical physics. The next generation of devices is going to be in a transition regime and further in a quantum regime of operation.

One of the factors driving the huge production and wide use of microelectronic systems is the relatively low cost of their fabrication. Moreover, despite an increase in complexity, microelectronic systems continue to be produced at lower costs. In Table 1.1 the costs per bit and costs per chip as well as the associated performance levels are given as functions of the integration level. One can see that every 3 years the bit number per chip increases by a factor of 4 and the cost per bit decreases by a factor of 2 or more.

In Table 1.1, the integration levels of logic circuits and microprocessors are forecast. We see that for this case, device integration is also large but it will increase slightly slower than for DRAMs. The cost of the principal elements of logic circuits – transistors – is sufficiently greater, but it also tends to decrease.

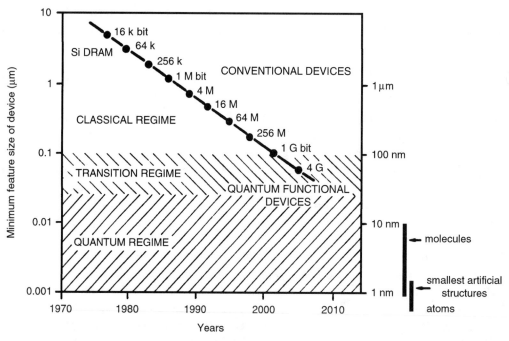

Figure 1.2. Evolution of Si-based DRAM chips as a function of time. The ordinate presents the minimal feature sizes of MOSFETs. The number of bits per chip is indicated as a function of time. Regimes of electron transport in FETs, classical, transitional, and quantum-mechanical devices are identified in terms of relevant feature sizes. For comparison, the sizes of large molecules, the smallest artificial systems, and atoms are presented. [After H. Lüth, "Nanostructures and semiconductor electronics," Phys. Status Solidi B **192**, pp. 287–299 (1995).]

In the table, a forecast for the necessary power supply is presented that shows a slow, but persistent, decrease. Thus one can expect favorable trends for the power consumption of microelectronic systems.

At the bottom of the table the necessary technological parameters for ultra-high integration are presented; these include minimum feature sizes, diameters of wafers, and electrical defect densities. It is anticipated that by 2001 feature sizes of ~0.18 μm will be realized; at this dimensional scale, the lithographic technique must be changed. For larger sizes, optical (ultraviolet) lithography is applicable, while for small dimensions, innovative tools such as x-ray and electron-beam lithography will have to be used. As is well known, wafers cut from monocrystalline Si ingots are used as substrate materials for subsequent processing and device fabrication; the larger the wafer size, the greater the number of devices that can be fabricated on chip. Table 1.1 forecasts that wafer diameters will increase by a factor of 2 and the defect density will decrease by a factor of 6.

Such ultra-large-scale integration is possible only for Si-based electronics because of the unique properties of Si. Other advanced semiconductor technologies based on III–V compounds also facilitate high integration, but they cannot match Si technology in terms of applicability to ultra-large-scale integration.

Table 1.1.
Roadmap for Si-Based Microelectronics: Predictions of the Semiconductor
Industry Association

Years	1995	1998	2001	2004	2007	2010
Memories, DRAM						
Bits per chip	64 M	256 M	1 G	4 G	16 G	64 G
Cost per bit (millicent)	0.017	0.007	0.003	0.001	0.0005	0.0002
Cost per chip (US$)	11	18	30	40	80	130
Logic, microprocessors						
Transistors per cm^2	4 M	7 M	13 M	25 M	50 M	90 M
Cost per transistor (millicent)	1	0.5	0.2	0.1	0.05	0.02
Power supply (V)	3.3	2.5	1.8	1.5	1.2	0.9
Parameters						
Minimum feature size (μm)	0.35	0.25	0.18	0.13	0.10	0.07
Wafer size (in.)	8	8	12	12	16	16
Electric defect density ($1/m^2$)	240	160	140	120	100	25

The data are taken from U. König, Phys. Scripta **T68**, 90 (1996). M, $\times 10^6$; G, $\times 10^9$.

Concerning speed of operation and lower power consumption for a single device, the scaling of III–V devices to small dimensions appears to offer potential advantages compared with the figures of merit for Si-based electronics. The potential advantages of III–V devices stem from higher mobilities and hence higher electron velocities.

It is worth emphasizing that quantum effects appear in III–V-based devices for larger device features than for their Si-based counterparts. Achieving high-speed operation of integrated circuits also requires the minimization and the optimization of interconnections on chips. An alternative way to satisfy this demand is to use optically active semiconductor materials, as is discussed below.

In the processes of achieving minimum device sizes and ultrahigh levels of integration it is necessary to identify the limiting and critical parameters for improved performance. In reality, these parameters depend on the integrated elements of each individual material system. For example, for transistors, two parameters of the host material are of special importance: the ultimate electron velocity – the so-called saturated velocity – and the limiting electric field that does not induce electric breakdown – the so-called dielectric breakdown strength. Further improvements in the parameters are possible to achieve through materials engineering; examples include growing alloys and fabricating high-quality artificial multilayered structures.

The simplest multilayered structure has a single heterojunction, that is, a single-heterojunction structure is made of two different materials. At the interface of such a heterojunction, the electronic properties can be changed to improve selected physical characteristics. In particular, electrons can be confined in a thin

---

Table 1.2.
Advances in Growth, Characterization, and Processing of
Quantum Heterostructures

1960s–1970s	Molecular-beam epitaxy
	Ultra-thin-layer fabrication
	Superlattice fabrication
	Qualitative electron-beam and x-ray microscopies
	Lithographic microstructuring
1980s–1990s	Metal-organic vapor-phase epitaxy
	Metal-organic molecular-beam epitaxy
	Atomic layer accuracy fabrication
	δ doping
	Controlled strained layers
	Quantitative electron-beam and x-ray macroscopies
	Scanning tunneling microscopy
	Atomic force microscopy
	Lithography and etching for nanostructuring
	Picosecond and femtosecond spectroscopy

---

layer near the interface and spatially separated from their parent impurities; this so-called modulation doping greatly enhances electron mobility. In fact, the layers with confined electrons are so thin that electrons become quantized, that is, they obey the laws of quantum physics. The same is valid for different multilayered structures, which can be grown with high quality. Two- and one-dimensional electron channels, including quantum wells and quantum wires, as well as cells for electrons known as quantum boxes or dots, are currently fabricated in many laboratories throughout the world. Such structures are known as quantum semiconductor heterostructures. The progress in heterostructure technology has been made possible largely as a result of new advances in fabrication techniques. In Table 1.2, we give a brief summary of some important steps now used in the growth, characterization, and processing of heterostructures.

In the 1960s and the 1970s, molecular-beam epitaxy was invented, developed, and used to fabricate high-quality and ultrathin layers and superlattices. Qualitative electron-beam and x-ray microscope technologies were used to characterize the perfectness of structures, including interface disorder. During this period, lithographic and etching methods suitable for microscale devices were proposed and realized. In the 1980s and later, new epitaxial techniques were developed; these included metal-organic vapor-phase epitaxy, metal-organic molecular-beam epitaxy, and others. These innovations made possible the fabrication of layers with atomic-level accuracy. Desirable spatial-modulation doping by impurities has become possible, including δ doping: doping of one or a few atomic monolayers. The thin-layer fabrication technique facilitated atomic-scale control and the use of materials with quite different lattice parameters. Such layers are strained, but in many cases they can be almost perfect. New tools – scanning-tunneling microscopy (STM) and atomic force microscopy (AFM) – emerged that portend numerous applications in high-precision fabrication. Lithography and etching methods were improved to the point that they can be used for nanoscale structuring. Finally, picosecond and femtosecond spectroscopy progressed substantially, and they were applied to characterize the electronic and lattice properties of heterostructures.

Perfect heterostructures can have properties superior to those of bulk materials and alloys. Currently, they have become key elements of microelectronics technology. Initially, quantum heterostructures were developed on the basis of III–V materials. Heterostructure FETs, which are also known as high-electron-mobility transistors (HEMTs), heterostructure bipolar transistors (HBTs), double-barrier resonant-tunneling diodes (DBRTDs) for microwave applications, real-space transfer (RST) transistors, hot-electron transistors (HETs), and other microelectronic devices have been fabricated with high-performance characteristics. For example, these devices can operate in the high-frequency range up to hundreds of gigahertz. Furthermore, heterostructures based on Si/Ge have significant potential for being compatible with Si technology. Many properties of these structures portend devices with advantages over Si devices. In particular, Si/Ge heterostructure bipolar transistors can operate at frequencies up to 100 GHz, which substantially exceeds the frequencies of operation of homostructure Si bipolar transistors. A variety of different devices may be fabricated with these Si/Ge heterostructures. These developments may lead to major extensions of current Si-based electronics technology.

Fundamental questions arise when conventional principles of device operation fail as a result of entering the small-scale domain of quantum physics. In Fig. 1.2 the feature sizes corresponding to classical (conventional), transitional, and quantum-mechanical regimes of operation are indicated. One can see that integration above 250 Mbits on a single chip makes it necessary to take into account new regimes and even to modify the principles underlying device operation. Further device downscaling and higher integration densities of Si-based chips lead to information capacities exceeding 1 Gbit/chip and imply the need to explore by use of quantum regimes of operation in future years. Besides quantum effects, reducing device dimensions results in a decrease in the number of electrons participating in the transfer of an electric signal. As a result, nanoscale devices may operate on the basis of single-electron transfer. A variety of novel single-electron devices has been proposed and demonstrated. These devices are fabricated on the basis of III–V and Si/Ge heterostructures. When the sizes of quantum dots are reduced to 100 Å or less, it is possible to operate with single electrons at temperatures near or close to room temperature.

In conclusion, the current and the projected trends in microelectronics lead to the use of quantum heterostructures and to the reliance on novel quantum effects as an avenue to realizing further progress. A number of such relevant quantum effects have been discovered. In Table 1.3 we present the list of some of these

Table 1.3.
Physical Effects in Quantum Structures and Associated Microelectronic
(ME) and Optoelectronic (OE) Devices

	Physics		Devices
1970	Proposal for multilayered structures (superlattices) (Chapters 3, 4)	1979	(OE) Injection quantum well laser (Chapter 12)
1974	Resonant-tunneling Effect (Chapter 8)	1980	(ME) HEMT and other transistors (Chapter 9)
1974	Quantum-well-size quantization effect (Chapters 3, 4)	1983	(ME) Microwave DBRTD oscillator (Chapters 8, 9)
1978	Modulation doping effect (Chapters 4, 6, 7)	1984	(OE) Self-electron-optic-effect device (Chapters 11, 12)
1981	Quantum-confined Stark effect (Chapters 11, 12)	1984	(ME) Resonant HET (Chapters 8, 9)
1986	Quantum wires and waveguides (Chapters 3, 6, 7)	1984	(OE) Quantum-well, nonlinear (conventional) etalon (Chapters 11, 12)
1987	Quantum point contact (Chapter 2)	1987	(OE) Quantum-well infrared photodetector (Chapters 10, 12)
1987	Stark ladder transitions in superlattices (Chapter 11)	1989	(OE) Surface-emitting laser (Chapters 10, 12)
1990	Single-electron tunneling (Chapter 8)	1991	(OE) II–VI blue laser (Chapter 12)
1991	Quantum microcavities (Chapters 10, 12)	1994	(OE) Quantum-cascade laser (Chapter 12)
		1995	(OE) Quantum-box laser (Chapter 12)
		1995	(OE) GaN blue laser

effects and identify novel devices that use them; we include those effects that are analyzed in this book. Some other important phenomena observed in quantum structures such as the quantum Hall effect, the suppression of shot noise, and so on, are not presented.

In this book we study conditions associated with the transition between classical and quantum regimes of operation as well as quantum physics and new microelectronic devices and concepts based on the underlying physics. In Table 1.3, we identify the chapters in which these effects and devices are analyzed.

Now we briefly consider optoelectronic systems that are frequently referred to as photonic systems. Optoelectronics complements microelectronics in many applications and systems. First, optoelectronics provides means to make electronic systems compatible with light-wave communication technologies. Furthermore, optoelectronics can be used to accomplish the tasks of acquisition, storage, and processing of information. Advances in optoelectronics make significant contributions to the transmission of information by means of optical fibers (including communication between processing machines as well as within them), to the high-capacity mass storage of information in laser disks, and to a number of other specific applications.

The principal components of optoelectronic systems are light sources, sensitive optical detectors, and properly designed light waveguides, for example, optical fibers. These devices and passive optical elements are fabricated with optically active semiconductor materials. The III–V, IV–VI, and II–VI compounds belong to this group; most of these compounds have direct bandgaps, which make them suitable materials for optoelectronic devices. For comparison, Si is an indirect-energy-gap material, and it is not suitable for conventional optical applications. With direct-bandgap semiconductor materials, two main types of light sources have been developed: light-emitting diodes (LEDs) that produce spontaneous incoherent emission and lasers that emit stimulated coherent light. In both cases, electrical energy is converted into light energy. General goals for these devices include electric control, high-speed optical tuning, and achieving operation in the desired optical spectral range.

Optoelectronic devices and systems use a variety of different optical and electro-optical effects. Quantum heterostructures provide a means to enhance many of the effects known in bulklike materials, such as excitonic effects and optical nonlinearities near the fundamental edge of optical absorption. Quantum heterostructures also exhibit the new optical effects listed in Table 1.3.

Semiconductor junction lasers are quantum devices. They have superior properties compared with those of LEDs and are preferable for many technologies. These lasers are quite compact and are highly compatible with semiconductor electronic circuits. The original semiconductor lasers were homojunctions, i.e., they were made of one material, usually GaAs, doped to form a p–n junction. For these lasers, the injection of electrons and holes from both sides of the junction into the active region provided the population inversion necessary for lasing, and stimulated emission occurred because of radiative recombination of highly nonequilibrium electrons and holes in the active region. The next generation of the lasers used two heterojunctions. These double-heterojunction structures served both to confine electrons and holes in a precisely defined active region and to provide a waveguide for the stimulated emitted light. These lasers were designed successfully for different spectral ranges. For example, AlGaAs/GaAs double-heterostructure lasers operated in the 0.75–0.9-μm range, while GaInAsP/InP lasers covered the 1.2–1.6-μm range, which is ideally suited for low-attenuation optical-fiber transmission. There are several different critical parameters of semiconductor lasers: the threshold current, temperature sensitivity, modulation bandwidth, speed of modulation, coherence, and so on. All these demands can be met if nonequilibrium electrons and holes are squeezed together in such a narrow active region; accordingly, quantum effects in electron transport become significant. The demand for advanced semiconductor lasers promotes the reliance on various heterostructure materials. Devices with low costs and long life are required as well. In Table 1.3 several new types of heterostructure lasers are highlighted: quantum-well injection lasers, surface-emitting lasers, quantum-wire and quantum-dot lasers, quantum-cascade lasers, and short-wavelength injection lasers.

As for the previously studied case of microelectronics, the trends in optoelectronics are scaling down the sizes of these devices and achieving high levels of integration in systems such as arrays of light diodes, laser arrays, and integrated systems with other electronic elements on the same chip. Note that there is a fundamental limit to the size scaling of optical devices: light can not be spatially confined below λ/n_{ref}, where λ is the wavelength of light in vacuum and n_{ref} is the refractive index of optical material. Light confinement on scales of the order of λ/n_{ref} is possible in waveguides or specially designed optical microcavities such as a Fabry–Perot resonator with highly reflective multilayered mirrors.

Generally, there are two approaches to device operation with optical signals. The first, currently the most widely used, is optical-to-electrical signal conversion and subsequent processing by electronic means; these systems are referred to as hybrid optoelectronic systems. To achieve this goal, one needs optical detectors; in addition, optical modulators, optical gates, and other electrically controlled devices are to be used. The essential performance requirements are fast response, high sensitivity, and high quantum efficiency. Electro-optical devices based on heterostructures satisfy these conditions and play a leading role in optoelectronics. A particular example is provided by the so-called self-electro-optic-effect devices (SEEDs). These quantum heterostructure optoelectronic devices serve as multifunctional devices and facilitate the realization of optical set–reset latches, differential logic gates, differential modulators or detectors, and so on. Another approach to optical signal processing is based on nonlinear optical elements that make light-by-light control possible and exhibit all necessary functions for all-optical addressed systems. The latter should include such devices as optical switches, dynamic optical memory, and optical transistors. A digital hybrid optoelectronic or all-optical addressed system should use two-dimensional arrays of lasers, optical switches (optical gates), and optical detectors, as well as such optic elements as lenses and light splitters. The benefits of this approach are the substantial decrease in the number of optical-to-electrical and electrical-to-optical conversions, the dramatically decreased number of interdevice connections, and powerful parallel processing of signals.

Special techniques for the growth of optical semiconductor materials and their processing were developed to fabricate these optoelectronic devices with sizes close to the above-mentioned fundamental optical limit and with sizes that led to confining electrons and holes in the quantum limit. Large arrays of emitting diodes or lasers, nonlinear elements, and optical detectors have been fabricated for this purpose. Their fabrication is based on the heterostructure manufacturing and processing techniques presented in Table 1.2 and tends to provide devices with long lifetimes for low costs. In Table 1.3, we indicate several heterostructure devices used for optical signal processing: SEEDs, conventional nonlinear multilayered optical photodetectors, and quantum-well photodetectors.

Thus we can conclude that optoelectronics benefits substantially through the use of quantum heterostructures and becomes competitive with its microelectronic counterpart.

More information on the progress made on the fabrication of scaled-down microelectronic and optoelectronic devices, quantum structures, and their general properties and perspectives can be found in the following reviews:

The National Technology Roadmap for Semiconductors (Semiconductor Industry Association, San Jose, CA, 1994).

H. Lüth, "Nanostructures and semiconductor electronics," Phys. Status Solidi B **192**, pp. 287–299 (1995).

M. A. Reed and J. F. Sleight, "Fabrication of nanoscale device," in *Quantum Transport in Ultrasmall Devices*, D. K. Ferry et al., eds. (Plenum, New York, 1995) pp. 111–132.

L. Geppert, "Semiconductor lithography for the next millennium," IEEE Spectrum **33**, pp. 33–38 (1996).

C. Weisbuch, "The future of physics of heterostructures: a glance into the crystal (quantum) ball," Phys. Scripta **T68**, pp. 102–112 (1996).

U. König, "Future applications of heterostructures," Phys. Scripta **T68**, 90–101 (1996).

These recent and diverse trends in semiconductor and device technologies as well as in novel device concepts are driving the establishment of new subdisciplines of electronics and optoelectronics based on quantum structures. These subdisciplines and their foundations will be studied in the following chapters.

The rest of this book is organized as follows. In Chapter 2 we review the essential fundamentals of quantum physics and solid-state physics. Chapters 3–5 are introductory chapters on the physics of quantum heterostructures. In Chapter 3 we present a detailed analysis of the electronic properties in ideal quantum heterostructures. Materials and heterostructures of special interest are studied in Chapter 4, in which we consider typical structures such as single heterojunctions, quantum wells, and superlattices. Chapter 5 is devoted to the lattice properties of heterostructures; there we provide a detailed introduction to confined optical and acoustical phonons in quantum wells, quantum wires, and free-standing structures. In Chapter 6, an essential step toward analyzing practical structures is made: we analyze scattering of electrons by impurities and phonons, as well as screening effects. In Chapters 7 and 8 the reader is introduced to two kinds of electron transport in heterostructures; these two types of transport are distinguished according to whether the predominant direction of current flow is parallel or perpendicular to the heterointerfaces. The results obtained in these chapters facilitate the study of microelectronic devices in Chapter 9. The optical properties of heterostructures are studied in Chapter 10. In Chapter 11 the reader is introduced to nonlinear optical properties and the electro-optical effect in heterostructures. Finally, Chapter 12 is devoted to the description of various optoelectronic applications of quantum structures.

2

Theoretical Basis of Nanoelectronics

2.1 Introduction

The evolution of microelectronics toward reduced device dimensional scales has proceeded to a degree that renders conventional models, approaches, and theories inapplicable. Indeed, for nanoelectronics it is frequently the case that the length scales of fundamental physical processes are comparable with the geometrical size of the device; as well, the fundamental time scales are of the order of the time parameters of device operation. Therefore, on the nanoscale, the theories and the models underlying modern nanoelectronics become more complicated, more mathematical, but, at the same time, more physical. Quantum mechanics, equilibrium and nonequilibrium statistics, physical kinetics, and electrodynamics are necessary theories in understanding physical phenomena on the nanoscale. Thus nanoelectronics is a discipline that relies more and more on basic science.

Any scientific solution to a problem including a problem in nanoelectronics must be based on a clear formulation of the problem and the selection of a specific theoretical approach to solving it. Any real problem is too complicated to be solved without certain simplifications. The choice of theoretical approach depends on the level of simplification one can permit. The same problem may require different levels of simplification and different theoretical approaches: one level for simple transparent results determining the general trends and for obtaining rough estimates and another level for more elaborate theories and more accurate estimates. In any case the simplifications cannot go beyond common sense; the limits of applicability of all theoretical approaches have to be thoroughly checked before one attempts to solve the problem.

In this textbook we intentionally use different theoretical approaches for various problems of nanoelectronics. It is not our goal to force students to memorize these approaches but rather to teach them the reasons for choosing one or another theoretical approach, i.e., our primary purpose is to provide a rigorous scientific basis for understanding how to attack and solve technical problems. In this chapter we discuss basic equations and approaches that are used to solve various problems of nanoelectronics. In the following chapters we make extensive use of these equations and various simplifications and methods of their solution.

2.2 Wave–Particle Duality in Quantum Physics

An underlying principle of central importance in this book is the fundamental quantum-mechanical concept that all matter, including electromagnetic fields, electrons, nuclei, etc., behaves as both waves and particles, that is, wave–particle duality is a basic characteristic of all matter.

We are used to treating radiation, say in the radio-frequency or the visible spectral regions, as waves, which can be characterized by a wavelength λ, a wave number $k = 2\pi/\lambda$, the corresponding frequency $\nu = c/\lambda$, where c is the velocity of light, and the phase $\phi = \vec{k}\vec{r} - \omega t$, where $\vec{k}$ is the wave vector with the direction of wave propagation and magnitude equal to the wave number $k = |\vec{k}|$, $\vec{r}$ is the space coordinate, $\omega = 2\pi\nu$, and t is time. According to Planck and Einstein, electromagnetic radiation is emitted and absorbed as discrete-energy quanta – photons. The energy of a photon E is proportional to the frequency of the wave:

$$E = h\nu = \hbar\omega, \tag{2.1}$$

where $h = 6.62 \times 10^{-34}$ Js is Planck's constant and $\hbar = h/2\pi$ is the reduced Planck's constant. Each photon, like a particle, has a momentum

$$p = E/c = \hbar k. \tag{2.2}$$

The direction of the momentum vector $\vec{p}$ coincides with the direction of propagation of the wave, $\vec{p} = \hbar\vec{k}$. For each of the two possible polarizations of the radiation, one introduces appropriate characteristics of photons: each polarization of light is related to a certain photon. Instead of the wave amplitude (or intensity of the radiation), in quantum theory one introduces the number of photons N. This number is subject to the uncertainty principle, that is, it is not possible to know the number of photons and the phase of the wave ϕ at the same time. Uncertainties in the photon number ΔN and phase $\Delta\phi$ are related by the inequality

$$\Delta N \Delta\phi > 1. \tag{2.3}$$

Uncertainty relationships like that of inequality (2.3), which limit both the particlelike and the wavelike quantities, are included among the basic principles of quantum mechanics.

Considering a particle, say an electron, on the basis of classical mechanics, one can attempt to characterize it by a mass m and by a vector representing the momentum $\vec{p}$. In classical physics we can know with certainty that the center of mass of a particle is at a certain position in space $\vec{r}$. From a quantum-mechanical point of view, a particle is characterized by the wave function $\psi(\vec{r})$ such that the value $|\psi(\vec{r})|^2 \, d\vec{r}$ gives the probability of finding a particle inside a small volume $d\vec{r}$ around point $\vec{r}$. Thus the wave function ψ may be interpreted as the probability amplitude corresponding to probability density for finding a particle at a particular point of space $\vec{r}$.

Probabilistic behavior is one of the key features of quantum mechanics; thus a word of explanation is necessary to define what is meant by probability in this context. To understand probability in a quantum-mechanical context, it is convenient to have in mind the following situation. Imagine an ensemble of similarly prepared systems. By similarly prepared we mean identical systems as far as any physical measurement is concerned. Now if a measurement is made on one of the systems to determine whether a particle is in a particular volume element, the result will be definite: either a particle is there or it is not. When the same measurements are made on a large number of similarly prepared systems, the number of times a particle is found in the fixed volume is taken as a measure of the probability of finding a particle in the elementary volume.

For the simplest case of a particle in free space, the wave function has a plane waveform,

$$\Psi(\vec{r}, t) = Ae^{i\phi} = Ae^{i(\vec{k}\vec{r} - \Omega t)}, \tag{2.4}$$

where $\vec{k}$ is the wave vector of the particle, A is the constant amplitude of the wave, Ω is the angular frequency associated with the energy of the particle, and $\phi = \vec{k}\vec{r} - \Omega t$ is its phase.[†] The wave vector, or more precisely its magnitude, called the wave number $k = |\vec{k}|$, is related to the wavelength of a particle λ:

$$\lambda = 2\pi/k. \tag{2.5}$$

According to the de Broglie relationship, the momentum of a particle is related to a wavelength associated with the particle through the equation

$$\lambda = h/p. \tag{2.6}$$

From Eqs. (2.5) and (2.6) we obtain a relation between the wave number and the momentum of a particle: $p = \hbar k$, or, in the vector form, $\vec{p} = \hbar \vec{k}$. This simplest case demonstrates that one may describe a particle in terms of both particlelike and wavelike pictures. Within the plane-wave description of a free particle, it follows from Eq. (2.4) that there is equal probability of finding a particle in any point of space:

$$|\psi(\vec{r})|^2 = |A|^2 = \text{const.}$$

This result is in complete contradiction with the classical description of a particle. Just as for the above case of the radiation, there is a relationship determining the range of position values and the range of momentum values. People use the

---

[†] In this textbook both ω and Ω are used to denote angular frequency; in general, ω is used for conventional wavelike phenomena and Ω is used to describe de Broglie waves.

word simultaneous but it is not really essential. Uncertainties in the quantities Δp and Δr have to satisfy the following inequalities:

$$\Delta p_x \, \Delta x > \hbar/2, \qquad \Delta p_y \, \Delta y > \hbar/2, \qquad \Delta p_z \, \Delta z > \hbar/2. \tag{2.7}$$

Thus, if a particle is localized in a space region of width Δx, the uncertainty in the x component of its momentum will be more than or equal to $\hbar/(2\Delta x)$.

Note that the phase of the wave ϕ in Eq. (2.4) depends on time. The angular frequency of the oscillations of this phase is related to the energy of the particle E through $\Omega \equiv E/\hbar$. This kind of phase temporal dependence remains valid for any complex system under stationary conditions, including those of constant external fields.

Let us estimate the consequences of the uncertainty principle for a free electron moving with the velocity of $\sim 10^7$ cm/s. The mass of a free electron is $m = 9.11 \times 10^{-28}$ g, which gives $p = mv = 9 \times 10^{-21}$ g cm s^{-1}, $k = p/\hbar = 8.7 \times 10^6$ cm^{-1}, and the de Broglie wavelength of free electron, $\lambda_0 = 2\pi/k = 7.3 \times 10^{-7}$ cm $= 73$ Å. If we need to measure both the position and the momentum of the electron and we impose a limit of 10% accuracy on the value of its momentum, we cannot indicate the position of this electron with an accuracy greater than $\Delta x = 6 \times 10^{-7}$ cm $= 60$ Å. This last value is of the order of the wavelength of the electron.

2.3 Time and Length Scales

2.3.1 Electron Fundamental Lengths in Solids

The de Broglie wavelength of an electron in a semiconductor: An electron in a semiconductor is characterized by the so-called effective mass m^*, which is usually less than the free-electron mass. Thus the de Broglie wavelength of an electron in a semiconductor λ is greater than that of a free electron λ_0:

$$\lambda = \frac{h}{p} = \frac{h}{\sqrt{2m^* E}} = \lambda_0 \sqrt{\frac{m}{m^*}}. \tag{2.8}$$

In Fig. 2.1 the value λ is depicted as a function of m^*/m. Points 1–4 on the curve indicate wavelengths for electrons in InSb, GaAs, GaN, and SiC, respectively. We have used effective masses m^*/m equal to 0.014, 0.067, 0.172, and 0.41, respectively, for these materials. Absolute values are shown for the electrons with thermal energy $E = k_B T$, where $T = 300$ K is the ambient temperature and k_B is Boltzmann's constant: $k_B = 1.38062 \times 10^{-23}$ J K^{-1}. We see that the de Broglie wavelength of an electron in typical semiconductors with m^* in the range $(0.01–1)m$ is of the order of 730–73 Å. As a temperature decreases to 3 K, the de Broglie wavelength increases by 1 order of magnitude. This wavelength becomes comparable with the sizes of semiconductor structures and devices fabricated by modern nanofabrication technology. This is one of the major reasons

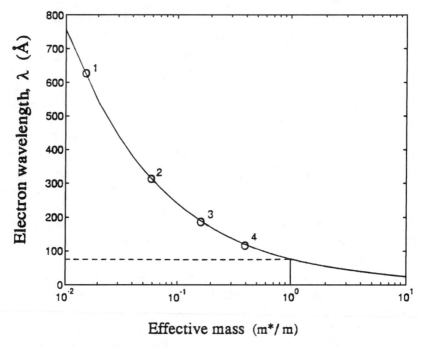

Figure 2.1. Electron wavelength versus the electron effective mass for room temperature ($E = k_B T$, $T = 300$ K). Points 1–4 correspond to InSb ($m^*/m = 0.014$), GaAs (0.067), GaN (0.172), and SiC (0.41), respectively.

why quantum-mechanical phenomena must be considered in order to understand the physics of nanostructures.

Size of a device and electron spectrum quantization: Let us introduce a geometrical size of semiconductor sample $L_x \times L_y \times L_z$, as shown schematically in Fig. 2.2. Without loss of generality we assume that $L_x < L_y < L_z$. If the system is free of randomness and other scattering mechanisms are sufficiently weak, the electron motion is quasi-ballistic and the only length that the geometrical sizes need to be compared with is the electron de Broglie wavelength λ. Since only an integer number of half-waves of the electrons can be put in any finite system, instead of a continuous-energy spectrum and a continuous number of the electron states, one obtains a set of discrete electron states and energy levels, each of which is characterized by the corresponding number of half-wavelengths. This

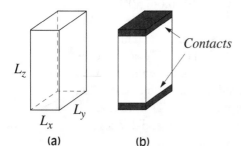

Figure 2.2. (a) Geometric sizes of a semiconductor sample, (b) a sample with contacts: the electron transport occurs along the z direction.

is frequently referred to as quantization of electron motion. Depending on the system dimensions, one can distinguish different cases:

- The three-dimensional or bulklike case, in which the electron spectrum quantization is not important at all,[‡]

$$\lambda \ll L_x, L_y, L_z, \tag{2.9}$$

and an electron behaves like a free particle characterized by the effective mass m^*.

- The two-dimensional or quantum-well case, in which the quantization of the electron motion occurs in one direction while in the two other directions electron motion is free:

$$\lambda \simeq L_x \ll L_y, L_z. \tag{2.10}$$

- The one-dimensional or quantum-wire case, in which the quantization occurs in two directions, so that the electron moves freely only along the wire:

$$L_x \simeq L_y \simeq \lambda \ll L_z. \tag{2.11}$$

- The zero-dimensional or quantum-box (dot) case, in which the quantization occurs in all three directions and the electron cannot move freely in any direction:

$$L_x \simeq L_y \simeq L_z \simeq \lambda. \tag{2.12}$$

The last three cases also illustrate the quantum size effects in one, two, or three dimensions, respectively.

If at least one geometrical size of a device is comparable with the electron wavelength, a quantum-mechanical treatment of the problem is strictly required.

The mean free path and the coherence length: Let us analyze the conditions under which and the reasons why carriers lose their wavelike behavior so that they can be considered as classical particles. There are two major reasons. The first is nonideality of the system, which leads to electron scattering. The second is related to finite temperature and electron statistics.

Electrons in solid-state devices are subjected to scattering[§] by crystal imperfections, impurities, lattice vibrations, interfaces, etc. These scattering processes are divided into two groups: elastic and inelastic. It is well known that in classical physics an elastic collision leads to a change in the particle momentum, while in an inelastic collision both the momentum and the energy change. Let us discuss how different types of scattering affect the quantum behavior of electrons.

[‡] Although in principle spatial quantization exists in any finite system, its role in bulk systems is negligible and is usually ignored. This is why when we consider quantization we imply that the effect of the quantization is appreciable.

[§] Sometimes we say collisions, although this classical term loses its meaning when dealing with waves.

Consider, for example, elastic scattering by an impurity. It is easy to imagine that an electron plane wave $\Psi_i(\vec{r}, t)$ with the initial wave vector $\vec{k}_i = \vec{p}_i/\hbar$ after scattering on the impurity potential is transformed into another wave $\Psi_{sc}(\vec{r}, t)$ with a wave vector $\vec{k} = \vec{p}/\hbar$ and the same energy $E(p) = E(p_i) = \hbar\Omega$:

$$\Psi_i(\vec{r}, t) = e^{-i\Omega t} e^{i\frac{\vec{p}_i \vec{r}}{\hbar}} \rightarrow \Psi_{sc}(\vec{r}, t) = e^{-i\Omega t} \sum_{\vec{p}, |\vec{p}|=p_i} A_{\vec{p}} e^{i\frac{\vec{p}\vec{r}}{\hbar}} = e^{-i\Omega t} \psi(\vec{r}). \quad (2.13)$$

Thus there is a probability $|A_{\vec{p}}|^2$ of finding the electron with any momentum direction $\vec{p}$ after scattering. For elastic scattering, $p = p_i$. The incident and the scattered waves produce a complex wave pattern, but an essential property of such an elastic collision is that it does not destroy the phase of the electron. The spatial distribution of $|\Psi_{sc}(\vec{r}, t)|^2 = |\psi(\vec{r})|^2$ remains independent of time. In other words, elastic scattering does not destroy the coherence of electron motion. The same is true for the case of two or more impurities: the spatial wave pattern is complex, but it remains coherent.

If τ_e refers to the mean time between two elastic scattering events, we can define the mean free path of the electrons between elastic scattering events as $l_e = v\tau_e$, where v is the average electron velocity[||] and τ_e is time between two scattering events. Therefore, even for distances exceeding l_e, the wavelike properties of electrons are coherent.

Inelastic scattering leads to a new result. This scattering produces electron waves with different energies and, according to Eq. (2.4), with different time dependences:

$$\Psi_i(\vec{r}, t) = A e^{i\frac{\vec{p}_i \vec{r}}{\hbar}} e^{-i\Omega(p_i)t} \rightarrow \Psi_{sc}(\vec{r}, t) = \sum_{\vec{p}, p \neq p_i} A_{\vec{p}} e^{i\frac{\vec{p}\vec{r}}{\hbar}} e^{-i\Omega(p)t}. \quad (2.14)$$

Now the resulting wave function $\Psi(\vec{r}, t)$ has a complex dependence on both position and time; the beating of different wave components in time washes out the coherence effects and the wavelike behavior of electrons. Indeed, $|\Psi_{sc}|^2$ now is a function of not only the space coordinates but also of the time. Let τ_E be the mean time between two inelastic collisions. The distance the electron propagates between these collisions can be called inelastic-scattering length L_E. The electron preserves its quantum coherence for distances of less than L_E and it loses coherence for larger distances. Generally, $L_E > \lambda$ unless extremely nonequilibrium conditions exist. Often L_E far exceeds the mean free path l_e, as shown in Fig. 2.3. In this case, the electron undergoes many collisions before losing its energy. As a result, its motion is chaotic and its displacement during τ_E is

$$L_E = \sqrt{D\tau_E} \quad (\tau_E \gg \tau_e), \quad (2.15)$$

where the diffusion coefficient D is given by $D = v^2\tau_e/\alpha$, where $\alpha = 3$ for a three-dimensional electron gas, $\alpha = 2$ for a two-dimensional electron gas, and $\alpha = 1$ for

---

[||] In quantum mechanics one usually deals with the momentum p and the group velocity v, which in our discussion will be taken to be equal to p/m.

Figure 2.3. Intervals for the characteristic lengths: λ, de Broglie wavelength; l_e, mean free path; L_E, energy relaxation length; and l_ϕ, coherence length in semiconductor materials. As an example, the lengths are marked by circles for Si at $T = 77\,\mathrm{K}$ assuming that the electron mobility equals $10^4\,\mathrm{cm^2/V\,s}$.

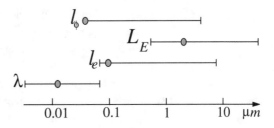

a one-dimensional electron gas. Usually τ_E and L_E decrease as the temperature of the system increases.

Thermal diffusion length: The second reason for averaging out the quantum behavior is the temperature effect on the electron statistics. This effect occurs even for elastic scattering. As is well known, electrons obey Fermi statistics, for which at zero temperature all electron states with energy less than the so-called Fermi energy E_F are occupied. This Fermi energy is determined by the total electron concentration. If the temperature of the system is equal to zero the electrons with energy equal to the Fermi energy can undergo transitions from one state to another since there are empty states in the vicinity of E_F. As the temperature is increased, the number of empty electron states around the Fermi energy and the number of electrons participating in transport both increase as $k_B T$. These trends reflect a broadening of the distribution function. Let $k_B T \ll E_F$; then each electronic state has almost the same amplitude but a slightly different phase. If this difference is small, averaging over the states within the temperature-broadened electron distribution does not break the phase and destroy the coherence of the electrons. But at high temperatures, when electrons with significantly different energies take part in kinetic processes, the coherence in the system is destroyed as a result of a large spreading of the wave function phases. One can estimate the characteristic length L_T associated with such a temperature smearing of the coherence. Temperature broadening of the electron distribution $k_B T$ leads to a spreading of the phases with time that goes as $t(k_B T/\hbar)$. Thus the time of spreading can be estimated as $\tau_T = \hbar/k_B T$. If the only scattering is elastic scattering, an electron diffuses in space over a distance of approximately $\sqrt{Dt}$ during the time t exceeding the mean free-flight time τ_e. Therefore, during the time interval τ_T, one obtains the thermal diffusion length $L_T = \sqrt{D\tau_T} = \sqrt{D\hbar/k_B T}$. At distances exceeding L_T, the coherence of the electron will be lost.

In fact, the dephasing effects caused by inelastic collisions and the temperature spreading of phases exist simultaneously. The spatial scale associated with the loss of quantum-mechanical coherence should be determined by the smaller of these two lengths:

$$l_\phi \simeq \min(L_E, L_T). \tag{2.16}$$

Electron transport is determined by the superposition of the electron wave functions. From the above consideration we can conclude that the coherence length

l_ϕ defines the limit below which electron transport has a quantum character. Devices with geometrical sizes of the order of the coherence length are no longer characterized by macroscopic material parameters such as conductivity, drift velocity, etc. Such systems are called mesoscopic systems. The proper theory for the description of mesoscopic devices is therefore quantum theory; the properties of such mesoscopic systems are determined by wavelike phenomena and are thus strongly dependent on geometry of the sample, contacts, positions of scatterers, etc.

For the case in which the transport distance L_z is long compared with l_ϕ, the device can be described within the framework of classical physics. In this case, besides the scales l_e and L_E, other kinetic lengths can be defined. If some process in the device is characterized by a time τ_{ch} such that $\tau_{ch} > \tau_e$, one can introduce a corresponding length $L_{ch} = \sqrt{D\tau_{ch}}$. For example, in nonequilibrium bipolar systems containing electrons and holes, $\tau_{ch} = \tau_R$, where τ_R is the lifetime of carriers in the system. One can introduce the ambipolar length $L_R = \sqrt{D\tau_R}$ that defines the distance during which nonequilibrium carriers diffuse before they recombine.

2.3.2 Quantum and Classical Regimes of Electron Transport

Let us compare the previous discussion of fundamental lengths with characteristic device sizes to illustrate and explain possible electron-transport regimes. For simplicity we suppose that transport occurs along one dimension, say, the z direction. The total current in each of the other two directions is zero, but these transverse sizes of the device can be important.

Quantum and mesoscopic regimes of transport: We can define two nonclassical regimes. If the de Broglie wavelength is about the length of a device L_z,

$$\lambda \sim L_z, \tag{2.17}$$

and $l_e \gg \lambda$, electron transport is described in terms of the quantum-ballistic transport regime. If the coherence length (often referred to also as the dephasing length) exceeds L_z and λ,

$$l_\phi > L_z, \lambda, \tag{2.18}$$

electron transport is described in terms of the mesoscopic transport regime.

Combining these results with conditions (2.9)–(2.11), we conclude that the transverse device dimensions define one-, two-, or three-dimensional quantum-ballistic regimes. The same is valid for the mesoscopic transport.

Classical transport regime: In the case in which size L_z exceeds the dephasing length,

$$L_z > l_\phi, \tag{2.19}$$

electron transport is described in terms of the classical regime. If the dimension L_z is less than the mean free path,

$$l_e > L_z, \tag{2.20}$$

electron transport is described in terms of the classical ballistic regime, which means that electrons can move through the device along classical trajectories without collisions.

If the dimension L_z is greater than the mean free path,

$$L_z \gg l_e, \tag{2.21}$$

electron transport is of a diffusive nature. If $L_z \sim l_E \gg l_e$, electrons do not lose their energy in moving across the device. Such transport is called quasi-ballistic transport. In the absence of an electric field, the electrons preserve their energy under the quasi-ballistic regime. By combining inequalities (2.17)–(2.21) with inequalities (2.9)–(2.11), one can see that there are three classical transport regimes for one-, two-, and three-dimensional electrons.

If the transverse dimensions L_x and L_y are both greater than the de Broglie wavelength but are comparable with one of the characteristic classical lengths, the transport regime is characterized by transverse classical size effects. In this case, collisions with the device boundaries affect electron transport through the device. For example, if one or both of the transverse dimensions are of the order of the mean free path,

$$L_x, L_y \sim l_e, \tag{2.22}$$

the resistance of the device depends strongly on the properties of the side boundaries. Roughness of the boundaries increases the resistance and entirely controls it if $L_x, L_y \ll l_e$.

If the transverse dimensions become comparable with one of the diffusion lengths, we deal with another kind of classical size effect, namely, diffusive classical size effects. For example, if L_x or L_y is of the order of the energy-relaxation length L_E, the device boundaries provide an additional energy-relaxation channel. This diffusive size effect controls the mean energy of nonequilibrium electrons.

For convenience the classification of the possible transport regimes is presented in Table 2.1.

2.3.3 Time Scales and Temporal (Frequency) Regimes

The time scales that characterize transport phenomena determine the temporal and the frequency properties of materials and devices.

There are two fundamental times that define the character of electron-transport behavior: the time between two successive scattering events or the free-flight (scattering) time τ_e and the time that characterizes the duration of a

Table 2.1.
Classification of Transport Regimes

Quantum regime	Intercontact distance L_z is comparable with the electron wavelength, $L_z \sim \lambda$
Mesoscopic regime	Intercontact distance is less than the dephasing length, $L_z \leq l_\phi$
Classical regime (one-, two-, and three-dimensional electron transport)	Intercontact distance exceeds the dephasing length, $L_z > l_\phi$: • classical ballistic regime, $l_e \geq L_z$ • quasi-ballistic regime (energy conserving): $L_E \geq L_z \geq l_e,\ l_\phi$ • transverse size effects: effect related to the mean free path, $L_x, L_y \sim l_e$ diffusion effects, $L_x, L_y \sim L_E, L_R$

scattering event τ_s. Under ordinary conditions $\tau_e \gg \tau_s$. In fact, it is usually assumed that the scattering event is instantaneous, i.e., $\tau_s \to 0$). In this case either classical or quantum theory can be applied for the description of electron behavior, depending on the length scales. If, however, τ_e is comparable with or smaller than τ_s, which may happen under extremely strong scattering of nonequilibrium electrons, the quantum description of electron behavior is required, regardless the size of the system.

In classical transport regimes the characteristic times and their relationships to the device sizes are of critical importance. They determine temporal and frequency regimes of device operation. For example, the transit time $t_{\text{tr}} = L_z/v$ determines the speed of signal propagation through a device, where v is the electron-drift velocity. Therefore t_{tr} defines the ultimate speed limit of the device: The device cannot effectively operate in the time range less than t_{tr} or at frequencies greater than t_{tr}^{-1}. This explains one of the trends of modern electronics: downscaling of device sizes.

The times related to transverse dimensions, $t_b = L_{x,y}/v$ (near the ballistic regimes) or $t_D = l_{x,y}^2/D$ (for diffusive size effects), determine features of electron transport at frequencies of the order of t_b^{-1} or t_D^{-1}. For example, the spectral densities of the electron fluctuations and current noise in devices depend on their sizes.

In quantum physics, as described later in Section 2.4, if external fields are steady, the electrons are in stationary states. In this case, despite possible complex dependence of the wave function on position, the temporal evolution of a stationary state is always determined by an exponential factor $\exp[i(E/\hbar)t]$. If the alternating external field of an angular frequency ω is applied to the stationary electron system, the response of the electron system may be referred to one of the three different regimes, depending on the frequency of the external field.

Ultrahigh (quantum) frequencies: If $\hbar\omega$ is comparable with the characteristic steady-state electron energy E, the nature of the electron response will be essentially quantum mechanical. Only transitions between the states with energy difference $\Delta E = \hbar\omega$ are allowed. If E is quantized, the interaction is possible only at resonant frequencies. By varying the device size, one can vary the energy spectra and, as a result, change the frequency properties over a wide range. The kinetic times τ_e, τ_E, etc., lead to a broadening of these resonances. If this broadening exceeds the energy separations between quantized levels, the discrete quantum behavior changes to a continuumlike behavior reminiscent of classical mechanics.

If $\hbar\omega \ll E$ the electron response to an alternating field is classical (quantization of transitions can be neglected). In the classical picture, the external alternating field will cause periodic electron acceleration and deceleration. The scattering interrupts these accelerations and decelerations. Depending on how many scattering events occur during the one period, we distinguish two different regimes of electron behavior.

High (classical) frequencies: If $\omega\tau_e \gg 1$ the electron motion during one period is not interrupted by scattering. In accordance with classical mechanics, the electron momentum oscillates with a phase opposite that of the field.

Low frequencies: If $\omega\tau_e \ll 1$ the electron undergoes many scattering events during one period of the external field. Multiple scattering during the period brings the electron into a quasi-stationary state that follows the oscillations of the external field. In other words, the electron momentum oscillates in phase with the field.

In closing this section we conclude that, depending on the device dimensions, the temperature, and other conditions, there is a variety of transport regimes. Each of these regimes requires a physical description suited to the relevant conditions. Furthermore we consider the basic equations for most of the regimes mentioned above.

2.4 Schrödinger Equation

From Section 2.3 we conclude that nanostructures are quantum-mechanical systems inasmuch as their sizes are comparable with the typical de Broglie wavelength of electrons in solids. In dealing with quantum-mechanical systems one aims at determining the wave function of a single electron or of the whole system. As is demonstrated in the subsequent discussion, the wave function in quantum mechanics is sufficient to describe a particle or even a system of particles. In other words, if we know the wave function of an electron system in a solid, we can, in principle, calculate all macroscopic parameters that define the electronic performance of the device.

The wave function Ψ of an electron or an electron system satisfies the principal equation of quantum mechanics, the Schrödinger equation,

$$i\hbar\frac{\partial\Psi}{\partial t} - \mathcal{H}\Psi = 0, \tag{2.23}$$

where $\mathcal{H}$ is the Hamiltonian operator of the system,

$$\mathcal{H} = -\frac{\hbar^2\nabla^2}{2m} + V(\vec{r}), \tag{2.24}$$

$V(\vec{r})$ is the potential energy [note that $V(\vec{r})$ might be a function of time as well], and the first term is the operator of the kinetic energy, with

$$\nabla^2 = \frac{\partial^2}{\partial x^2} + \frac{\partial^2}{\partial y^2} + \frac{\partial^2}{\partial z^2} \qquad (2.25)$$

being the divergence of the gradient or the Laplacian operator. The form of Eq. (2.23) is that of a wave equation and its solutions are expected to be wavelike in nature. If $V(\vec{r})$ is time independent, one may separate the dependences on the time and spatial coordinates:

$$\Psi(\vec{r}, t) = e^{-iEt/\hbar}\psi(\vec{r}), \qquad (2.26)$$

where ψ is a function of only spatial coordinates. Substituting Eq. (2.26) into Eq. (2.23), one gets the time-independent Schrödinger equation:

$$\left[-\frac{\hbar^2 \nabla^2}{2m} + V(\vec{r}) \right] \psi(\vec{r}) = E\psi(\vec{r}). \qquad (2.27)$$

In Eqs. (2.26) and (2.27) E is the energy of a particle, or, as we shall explain, of a system of particles. The major task of quantum mechanics is to solve the Schrödinger equation.

Conditionally, applications of the Schrödinger equation to solid-state electronics can be divided into two groups. The first is related to electron bound states in heterostructures (impurities, defects, quantum dots, etc.). For these problems, the major goals are to find the discrete energies of bound states, to calculate a relaxation of excited states that is due to interactions with free electrons, phonons, and other defects, and to understand the results of interactions with electromagnetic radiation. For the same group of problems we may define stationary-state problems of a free electron, i.e., calculation of the electron-energy spectra and electron interactions with crystal vibrations and defects as well as with other electrons and with electromagnetic waves. This group of problems requires a solution of the stationary Schrödinger equation (2.27) and, in general, the application of time dependent perturbation theory for Eq. 12.23.

The second group of problems concerns dynamics of free carriers in time-dependent strong fields. In general, this group of problems requires a solution of the time-dependent Schrödinger equation (2.23). The time-dependent as well as time-independent Schrödinger equations may be solved exactly only for a few very simple Hamiltonians. A whole arsenal of approximate methods for solving the Schrödinger equations is presented in the many textbooks on quantum mechanics. We cite and use these results as necessary to analyze particular problems of interest.

As we mentioned in Section 2.2, the wave function of a particle in free, $V(\vec{r}) = 0$, space has a plane waveform, Eq. (2.4). By substituting Eq. (2.4) into the Schrödinger equation (2.23), one gets the relationship between the electron's wave vector and its energy:

$$E = \hbar\Omega = \frac{\hbar^2 k^2}{2m} = \frac{p^2}{2m}, \tag{2.28}$$

which coincides with the classical relationship between the particle momentum and its energy.

To describe localized electrons, one may construct a wave packet by considering of a group of plane waves with finite amplitudes over a certain range of wave vectors:

$$\Psi(\vec{r}, t) = \int A(\vec{k}) e^{i(\vec{k}\vec{r} - \Omega t)} \, d\vec{k}, \tag{2.29}$$

where $A(\vec{k})$ is nonzero in a range of wave vector $\Delta\vec{k}$ around $\vec{k}_0$. Using the Taylor series expansion of Ω about $\vec{k}_0$, i.e.,

$$\Omega = \Omega_0 + (\vec{k} - \vec{k}_0)\left(\frac{d\Omega}{d\vec{k}}\right)_0 + \cdots, \tag{2.30}$$

one can write

$$\begin{aligned}
\Psi(\vec{r}, t) &= \exp[i(\vec{k}_0\vec{r} - \Omega_0 t)] \int A(\vec{k}) \exp\left\{ i(\vec{k} - \vec{k}_0)\left[\vec{r} - t\left(\frac{d\Omega}{d\vec{k}}\right)_0\right]\right\} d\vec{k} \\
&= B\left[\vec{r} - t\left(\frac{d\Omega}{d\vec{k}}\right)_0\right] \exp[i(\vec{k}_0\vec{r} - \Omega_0 t)],
\end{aligned} \tag{2.31}$$

where $B[\vec{r} - t(d\Omega/d\vec{k})_0]$ is the envelope function, which modulates the plane wave. The momentum-coordinate uncertainty principle follows immediately from the description of this wave packet. Indeed, if the electron wave vector $\vec{k}$, or equivalently its momentum, is defined exactly, the amplitude $A(\vec{k})$ in Eq. (2.29) is

$$A(\vec{k}) = \delta(\vec{k} - \vec{k}_0).$$

As a result, the envelope function in Eq. (2.31) becomes independent of coordinate $\vec{r}$: $B =$ const., i.e., the wave packet extends infinitely in space and is, in fact, a plane wave. In the opposite case, when electron momentum is not defined, i.e., $A(\vec{k}) =$ const., the envelope function B collapses to a delta function:

$$B\left[\vec{r} - t\left(\frac{d\Omega}{d\vec{k}}\right)_0\right] = \delta\left[\vec{r} - t\left(\frac{d\Omega}{d\vec{k}}\right)_0\right].$$

This means that the position of the particle in the latter case is defined at $\vec{r} = t[(d\Omega/d\vec{k})_0]$.

In an intermediate case that is, in fact, the case corresponding to reality, the function $A(\vec{k})$ has a pronounced maximum at $\vec{k} = \vec{k}_0$. This leads to a pronounced maximum of function B at $\vec{r} = t[(d\Omega/d\vec{k})_0]$. This means that the group velocity of the wave packet, $\vec{v}_g = [(d\Omega/d\vec{k})_0]$, is just equal to the velocity of a particle with momentum p_0:

$$\vec{v}_g = \left(\frac{d\Omega}{d\vec{k}}\right)_0 = \frac{\vec{p}_0}{m}. \tag{2.32}$$

The phase velocity of the plane wave is

$$\vec{v}_p = \frac{\vec{k}_0 \Omega_0}{k_0^2}. \tag{2.33}$$

This result shows that a superposition of plane waves can be used to characterize the localization of a particle. The assumption of a well-pronounced maximum of $A(\vec{k})$ means that in the $\vec{k}$-space expansion, $\Delta \vec{k}$ of this function around $\vec{k}$ is small, that is, $|\Delta \vec{k}| \ll |\vec{k}|$. A range of $\vec{k}$ values, $\Delta \vec{k}$, corresponds to a spreading of the function B in real space with $\Delta x \approx \pi/\Delta k_x$, $\Delta y \approx \pi/\Delta k_y$, $\Delta z \approx \pi/\Delta k_z$. Using the assumed inequality between $|\Delta \vec{k}|$ and $|\vec{k}|$, one can conclude that the uncertainty relations (2.7) are satisfied as strong inequalities. This explains why one can describe a particle in terms of a localized entity.

The exact value of the energy E characterizes the system only for the stationary-state case, in which the potential energy and the Hamiltonian do not depend on time. If they depend on time, the Schrödinger equation becomes time dependent. There is one important case in which the time-dependent part of the Hamiltonian constitutes just a small addition to the stationary Hamiltonian. In this case, the small addition can be treated as a perturbation and the change in the wave function due to the perturbation may be described in terms of transitions between the stationary (unperturbed) states. Perturbation theory describes such electron transitions due to the perturbation of the stationary Hamiltonian by lattice vibrations, other electrons, crystal defects, electromagnetic radiation, etc. We use this theory when considering electron scattering and interactions with light.

In addition to the stationary situation, a determination of energy requires an infinite time of observation (measurement). If this time Δt is finite, an accuracy of measurement of the energy ΔE should satisfy the following inequality:

$$\Delta E \, \Delta t \geq \hbar/2. \tag{2.34}$$

This is the uncertainty relation between the energy and the time. Inequalities (2.7) and (2.34) comprise the set of fundamental uncertainty relations.

Equation (2.27) has the form known as an eigenvalue equation. The energy E is its eigenvalue and the wave function $\psi_E(\vec{r})$ is its eigenfunction. The eigenvalue E may run over discrete or continuum values, depending on the shape of the potential function $V(\vec{r})$ and the boundary conditions.

To illustrate both possible types of solutions of the Schrödinger equation and energy states, as well as to clarify tasks that arise, let us consider the one-dimensional problem for a system with potential energy $V(\vec{r}) = V(x)$, as shown in Fig. 2.4. Here the vertical axis depicts the energy E and x represents only one

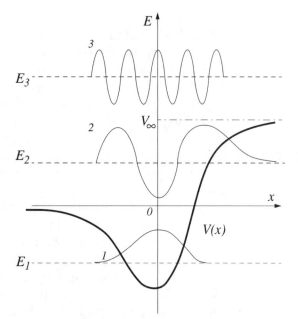

Figure 2.4. One-dimensional potential well $V(x)$ and three types of solutions (1,2,3) of the Schrödinger equation corresponding to the bound state (energy E_1) and unbound states (energies E_2 and E_3).

space coordinate. The potential has a minimum at $x = 0$, vanishes at $x \to -\infty$, and saturates to a finite value V_∞ at $x \to \infty$. This potential is the most general form of a potential well.

At this point a short qualitative discussion serves to emphasize that the boundary conditions define the type of solution. These solutions will be obtained and discussed in more detail in Chapter 3. Among the possible solutions of the Schrödinger equation (2.27) with the chosen potential there can exist solutions with negative energy $E = E_i < 0$. The wave function corresponding to energy $E = E_1 < 0$ is sketched in Fig. 2.4 by curve 1. One of the peculiarities of solutions with negative energy is that the spatial region with classically allowed motion, where kinetic energy $p^2/2m = E - V(x) > 0$, is certainly restricted. In the classically forbidden regions, where $V(x) > E$, the ψ function decays for $|x| \to \infty$. The states of particles, like those just described, are so-called bound states and they are characterized by a discrete-energy spectrum. Consider next the possible solutions for the energy range $0 \le E \le V_\infty$, as shown by line E_2 in Fig. 2.4. These solutions exist for any values of E; they are finite at $x \to -\infty$ and penetrate slightly into the barrier region $U(x) > E$, as shown by curve 2 in Fig. 2.4. In the limit $x \to -\infty$ these solutions may be represented as a sum of two waves traveling in opposite directions: One wave is incident upon and the other is reflected from the barrier. For each energy in the range $0 \le E \le V_\infty$ there is only one solution satisfying the physical requirements. For any energy $E > V_\infty$ there are two independent solutions; see curve 3 corresponding to the energy E_3. One can be chosen in the form of the wave propagating from left to right. At $x \to \infty$ it has only one component, namely the wave overcoming the barrier, and at $x \to -\infty$ there is a superposition of both incident and reflected waves. It is emphasized that the wave reflected from the barrier when its energy exceeds the height of the barrier arises only in quantum physics. The other wave function can be chosen in the form of waves propagating from right to left. Examples of continuous-energy spectra and nonbound states are given by consideration of cases with energies $E > V_\infty$ and wave functions that are finite far away from the potential relief.

The previous considerations demonstrate the importance of boundary conditions for the Schrödinger problem: If a decay of wave functions (i.e., $\Psi \to 0$ at $x \to \pm\infty$) away from the potential well is required, the discrete energies and the bound states can be found and they are of principal interest; if the boundary conditions correspond to the incident particle, continuous-energy spectra will be obtained.

To conclude the discussion of boundary conditions, we often use potential energies with discontinuities. In this case, at the point of discontinuity of the potential we should require both continuity of the wave function and continuity of the derivative of the wave function with respect to coordinates.

Since the Schrödinger equation is linear, it is clear that if a function Ψ is a solution of the equation, then any function of the form of const. $\times \Psi$ is also a solution of the same equation. To eliminate this ambiguity, we have to take into account the probabilistic character of the wave function. Indeed, if a physical system is enclosed in finite volume, the actual probability of finding a particle in this volume must equal 1, i.e.,

$$\int |\Psi(\vec{r}, t)|^2 \, d\vec{r} = 1. \tag{2.35}$$

Equation (2.35) is called the normalization condition. It provides the condition needed to determine the constant multiplier of the wave function for the case of a system of finite extent. If an infinite volume is under consideration and the integral of Eq. (2.35) does not exist, instead of Eq. (2.35) there are other normalization conditions. Consider, for example, the so-called scattering problem, in which electrons come from infinity and are scattered by a local potential. For this problem one can assume that the incident wave is a plane wave with a given amplitude A: $\Psi(x, t) = Ae^{ikx}e^{-iEt/\hbar}$. Then the scattered waves will be proportional to the amplitude A because of the linearity of the Schrödinger equation. Very often the last condition is referred to as an initial condition instead of a boundary condition because we are dealing with the state before the scattering and the states after the scattering.

Eigenvalues and eigenstates: To determine measurable physical quantities we focus attention on the eigenstate problem. The stationary Schrödinger equation may be written as an operator equation of the form $\mathcal{H}\psi = E\psi$. Thus the energy is an eigenvalue for the Hamiltonian operator $\mathcal{H}$. In quantum mechanics the same is true for any physical observable a: One introduces an operator $\mathcal{A}$ associated with this observable a. The operator has eigenfunctions ψ_a, so that

$$\mathcal{A}\psi_a = a\psi_a. \tag{2.36}$$

Eigenfunctions corresponding to different eigenvalues are orthogonal. For example, the momentum of a particle $\vec{p}$ is associated with the operator $\vec{\mathcal{P}} = -i\hbar(\partial/\partial\vec{r}) = -i\hbar\nabla$, and the plane wave of Eq. (2.4) is an eigenfunction of this operator:

$$\vec{\mathcal{P}}e^{i\vec{p}\vec{r}/\hbar} = \vec{p}e^{i\vec{p}\vec{r}/\hbar}. \tag{2.37}$$

In the same manner one may introduce quantum-mechanical operators for angular momentum, spin, current, etc.

Flux in quantum mechanics: In an overview of quantum mechanics it is very useful to introduce the density of particle flow $\vec{i}$. In classical physics the density of flow is a vector, which specifies the direction of particle flow and has a modulus equal to the number of particles crossing a unit area per unit time. In quantum mechanics this quantity is given by the following formula:

$$\vec{i} = -\frac{i\hbar}{2m}\left(\psi^*\nabla\psi - \psi\nabla\psi^*\right). \tag{2.38}$$

For example, the density of flow of particles described by the plane wave of Eq. (2.4) is straightforwardly found to be $\vec{i} = (\vec{p}/m)|A|^2 = \vec{v}|A|^2$.

2.4.1 Average Values of Physical Quantities

Because of the probabilistic character of the description of quantum-mechanical systems we have to clarify how to determine the average values of quantities that characterize such systems. The simplest case is the calculation of the average coordinate of a particle. Indeed, the absolute square of the normalized wave function gives the actual probability per unit volume of finding a particle at a particular point in space. Hence the average value of a particular coordinate, say x, is given by

$$\langle x \rangle = \int x|\psi|^2\,d\vec{r} = \int \psi^*x\psi\,d\vec{r}. \tag{2.39}$$

Thus the integral over space gives the mean, or expectation, value of coordinate x. It must be stressed again that the meaning of the expectation value is the average over the number of measurements of the coordinate x carried out over an ensemble of identical particles.

The last expression, Eq. (2.39), represents a more general form for the calculation of the expectation value of any observable a:

$$\langle a \rangle = \langle A \rangle = \int \psi^*A\psi\,d\vec{r}, \tag{2.40}$$

where the operator A is associated with the observable a. From the definition of the expectation value, Eq. (2.40), one can see that if ψ_a is an eigenfunction of the operator A and corresponds to a certain eigenvalue a, the expectation of the physical observable a coincides with the eigenvalue: $\langle a \rangle = a$. For example, if the wave function is the solution of the Schrödinger equation (2.27), one may easily calculate the mean energy:

$$\langle E \rangle = \langle \mathcal{H} \rangle = \int \psi_E^* \mathcal{H} \psi_E \, d\vec{r} = E. \tag{2.41}$$

But if a particle is in the state with no well-defined energy, say a state that is characterized by the superposition of the solutions ψ_{E_i},

$$\psi = \sum_i C_i \psi_{E_i}, \tag{2.42}$$

one would obtain the average energy in the form

$$\langle E \rangle = \sum_i |C_i|^2 E_i. \tag{2.43}$$

Here we take into account the orthonormality condition:

$$\int \psi_{E_i} \psi_{E_j} \, d\vec{r} = \delta_{ij}, \tag{2.44}$$

that is, wave functions of different energy states are orthogonal to each other, while each function ψ_{E_i} is normalized by the condition of Eq. (2.35).

It is timely to highlight the differences between the two cases given by Eqs. (2.41) and (2.43). The first case is related to a system characterized by a wave function, which is an eigenfunction of the operator $\vec{A}$; in this particular case $\vec{A} = \vec{H}$. The second case corresponds to the situation described by a super-position of the eigenfunctions of the same operator. Measurements of the value of the energy for the first case would reproducibly give the same result, E. In the second case the measurements would give us different probabilistic results, and only their average $\langle E \rangle$ remains the same.

Thus the Schrödinger equation describes the evolution of the wave function of the quantum-mechanical system of particles. Its solution with proper boundary or initial conditions gives us all information necessary to calculate macroscopic parameters of device operation. Now we review several of the most common methods of solution of the Schrödinger equation.

2.4.2 Separation of Variables

The Schrödinger equation (2.27) is written for the three-dimensional case. It can be reduced to lower dimensions for some particular, but rather frequent, situations. For example, let the potential energy $V(\vec{r}) \equiv V(x, y, z)$ in Eq. (2.27) be of the form

$$V(x, y, z) = V_1(x) + V_2(y) + V_3(z). \tag{2.45}$$

Then one can seek the solution as the product of wave functions:

$$\psi(x, y, z) = \psi_1(x)\psi_2(y)\psi_3(z). \tag{2.46}$$

Substituting Eq. (2.46) into Eq. (2.27) and dividing by $\psi_1\psi_2\psi_3$, one gets

$$\frac{1}{\psi_1(x)}\left[-\frac{\hbar^2}{2m}\frac{d^2}{dx^2} + V_1(x)\right]\psi_1(x) + \frac{1}{\psi_2(y)}\left[-\frac{\hbar^2}{2m}\frac{d^2}{dy^2} + V_2(y)\right]\psi_2(y)$$

$$+ \frac{1}{\psi_3(z)}\left[-\frac{\hbar^2}{2m}\frac{d^2}{dz^2} + V_3(z)\right]\psi_3(z) = E. \tag{2.47}$$

Each term on the left-hand side of this last equation is an arbitrary function of one of the coordinates x, y, or z. The right-hand side is the coordinate-independent function E. The only way to satisfy this equation is to have each of these three terms be equal to three arbitrary coordinate-independent constants E_1, E_2, and E_3, which satisfy the condition

$$E_1 + E_2 + E_3 = E. \tag{2.48}$$

As a result, the separation of variables leads to three one-dimensional equations:

$$\left[-\frac{\hbar^2}{2m}\frac{d^2}{dx^2} + V_1(x)\right]\psi_1(x) = E_1\psi_1(x),$$

$$\left[-\frac{\hbar^2}{2m}\frac{d^2}{dy^2} + V_2(y)\right]\psi_2(y) = E_2\psi_2(y),$$

$$\left[-\frac{\hbar^2}{2m}\frac{d^2}{dz^2} + V_3(z)\right]\psi_3(z) = E_3\psi_3(z). \tag{2.49}$$

Equations (2.49) are usually much easier to solve than Eq. (2.27).

Let $\psi_{1,n}(x)$, $\psi_{2,m}(y)$, and $\psi_{3,l}(z)$ be the solutions of these equations corresponding to the eigenvalues $E_{1,n}$, $E_{2,m}$, and $E_{3,l}$. Then the total wave function and the energy are given by

$$\psi_{n,m,l}(x, y, z) = \psi_{1,n}(x)\psi_{2,m}(y)\psi_{3,l}(z), \qquad E_{n,m,l} = E_{1,n} + E_{2,m} + E_{3,l}. \tag{2.50}$$

This procedure is frequently referred to as the method of separation of variables.

Important particular cases of the form of Eq. (2.45) are those in which the potential energy V is independent of one or two coordinates. In these cases one or two equations, respectively, may be solved immediately. For example, let $V(x, y, z) = V_1(x)$, i.e., there is no dependence of V on y and z so that it is permissible to take $V_2 = V_3 \equiv 0$. Then the initial equation easily reduces to a single one-dimensional Schrödinger equation. The solutions for the wave function and the energies for the other two dimensions, the y and the z directions, are just the solutions of Schrödinger equation for free motion in the y and the z directions. The result is then

$$\psi(x, y, z) = \psi_1(x) A_y A_z e^{i(k_y y + k_z z)}, \qquad E = E_1 + \frac{\hbar^2 (k_y^2 + k_z^2)}{2m},$$

where A_y and A_z are the normalization constants for the solutions of the plane-wave type in the y and the z directions. This method of separation of variables will be used widely in due course.

2.4.3 Variational Method

The Schrödinger equation in the general form of Eq. (2.27) may be obtained from the so-called variational principle. Consider the following functional:

$$\int \psi^* (\mathcal{H} - E) \psi \, d\vec{r}. \tag{2.51}$$

Let us require that the variation of this functional with respect to ψ or ψ^* equal zero:

$$\delta \int \psi^* (\mathcal{H} - E) \psi \, d\vec{r} = 0. \tag{2.52}$$

Varying ψ^*, we get

$$\int \delta \psi^* (\mathcal{H} - E) \psi \, d\vec{r} = 0. \tag{2.53}$$

Since $\delta \psi^*$ is an arbitrary function, the last integral is equal to zero if and only if

$$(\mathcal{H} - E) \psi = 0. \tag{2.54}$$

Varying Eq. (2.52) with respect to ψ, we obtain the complex-conjugate equation

$$(\mathcal{H}^* - E)\psi^* = 0,$$

that is, the wave function, which results in vanishing variations of Eq. (2.52), satisfies the Schrödinger equation. If the wave function is normalized, from Eq. (2.35) we find

$$\delta \int \psi^* \psi \, d\vec{r} = 0. \tag{2.55}$$

Combining Eq. (2.55) with Eq. (2.52), one obtains the variational principle in the form

$$\delta \mathcal{E}\{\psi\} = \delta \int \psi^* \mathcal{H} \psi \, d\vec{r} = 0. \tag{2.56}$$

Both Eqs. (2.52) and (2.56) lead to identical results.

Equation (2.56) means that the true wave function ψ_{tr} has to result in an extremum of the functional

$$\mathcal{E}\{\psi\} = \int \psi^* \mathcal{H} \psi \, d\vec{r}. \tag{2.57}$$

The value of this functional calculated with the true function ψ_{tr} coincides with the corresponding eigenvalue E_{tr}:

$$\mathcal{E}\{\psi_{tr}\} = E_{tr}. \tag{2.58}$$

This variational principle gives a powerful method for the calculation of eigenfunctions and eigenvalues. One may select a particular shape of $\psi(\vec{r}, \alpha, \beta, \dots,)$, with any number of the parameters $\alpha, \beta, \dots$. Usually the shape of the wave function is selected from physical considerations (expected shape of ψ, symmetry, etc.). Then one calculates the functional and obtains it as a function of the parameters: $\mathcal{E}\{\psi\} = \mathcal{E}(\alpha, \beta, \dots,)$. Whereas the true function should extremize the value of $\mathcal{E}$, the parameters $\alpha, \beta, \dots$, can be determined from the extremum conditions

$$\frac{\partial \mathcal{E}}{\partial \alpha} = 0, \qquad \frac{\partial \mathcal{E}}{\partial \beta} = 0, \dots \tag{2.59}$$

Such a procedure gives the best approximation to the wave function of the selected shape.

In fact, only the wave function of the lowest energy (ground state) results in the lowest minimum of $\mathcal{E}$. Wave functions of the excited states correspond only to extrema. The most straightforward procedure for finding eigenstates is the following. First, one determines the function of the ground state, say ψ_0. Looking for the next eigenstate, one restricts the choice to those functions $\psi_1(\vec{r}, \alpha', \beta', \dots,)$ that are orthogonal to ψ_0, that is, apart from the normalization conditions of Eq. (2.35), one imposes the additional restriction

$$\int \psi_0 \psi_1 \, d\vec{r} = 0. \tag{2.60}$$

By calculating $\mathcal{E}\{\psi_1\} = \mathcal{E}(\alpha', \beta', \dots,)$ and extremum conditions together with the restriction of Eq. (2.60), one determines ψ_1. After finding ψ_1, one can select ψ_2 to be orthogonal to ψ_0, ψ_1, etc.

2.4.4 Perturbation Methods

The majority of problems in quantum mechanics cannot be solved exactly. The fact that different physical interactions have different orders of magnitude can be extremely useful for approximate solutions by use of various perturbation methods. Common for all perturbation methods is that in the first step of the calculations one neglects the weaker interaction and simplifies the problem. The second step takes into account the previously neglected interaction and gives corrections to the initial results. Such a procedure can be repeated to achieve the necessary accuracy.

Stationary Problem

Let us suppose that the Hamiltonian of the system under consideration is

$$\mathcal{H} = \mathcal{H}_0 + V, \tag{2.61}$$

where V is a small correction to the unperturbed operator $\mathcal{H}_0$. Suppose that both $\mathcal{H}_0$ and V are independent of time, i.e., we consider a stationary problem. Let us assume that the Schrödinger equation with the Hamiltonian $\mathcal{H}_0$ is solved, that is, we know the eigenfunctions ψ_{0n} and the eigenvalues E_{0n}:

$$\mathcal{H}_0 \psi_{0n} = E_{0n} \psi_{0n}.$$

Let these eigenstates be nondegenerate and correspond to a discrete spectrum. One can represent the required wave functions in terms of a complete set of the wave functions ψ_{0n}:

$$\psi = \sum_n C_n \psi_{0n}. \tag{2.62}$$

Substituting ψ into the Schrödinger equation with the Hamiltonian of Eq. (2.61), we obtain

$$\sum_m C_m (E_{0m} + V) \psi_{0m} = \sum_m C_m E \psi_{0m}.$$

Multiplying both sides by ψ_{0n}^* and integrating over space coordinates, we get

$$(E - E_{0n}) C_n = \sum_m V_{nm} C_m, \tag{2.63}$$

where we introduce the matrix elements

$$V_{nm} = \langle \psi_{0n} \mid V \mid \psi_{0m} \rangle = \int \psi_{0n}^* V \psi_{0m} \, d\vec{r}. \tag{2.64}$$

Now we can represent E and C_m as a series:

$$E = E_0 + E_1 + \cdots,$$
$$C_m = C_{0m} + C_{1m} + \cdots,$$

where E_1 and C_{1m} are supposed to be of the same order of magnitude as the perturbation potential V.

Let us find the corrections to a certain eigenvalue E_{0k} and eigenfunction ψ_{0k}. It is obvious that in the lowest order of approximation, $C_{0k} = 1$ and $C_{0m \neq k} = 0$. Now from Eq. (2.63) we obtain

$$E_{1k} = V_{kk}, \tag{2.65}$$

$$C_{1m} = \frac{V_{mk}}{E_{0k} - E_{0m}}, \quad m \neq k. \tag{2.66}$$

Here we introduce the matrix elements $V_{mk} = \int \psi_{0m}^* V \psi_{0k} \, d\vec{r}$. As for C_{1k}, it can be chosen so that the function $\psi_k = \psi_{0k} + \psi_{1k}$ is normalized to unity, as in Eq. (2.35). This requirement gives $C_{1k} = 0$. As a result, we get the correction to the wave function:

$$\psi_{1k} = \sum_m \frac{V_{km}}{E_{0k} - E_{0m}} \psi_{0m}, \quad m \neq k. \tag{2.67}$$

Correction ψ_{1k} should be small in comparison with ψ_{0k}. Hence we obtain the criterion of applicability of such an approach: $V_{km} \ll |E_{0k} - E_{0m}|$.

Thus we obtain corrections to both the energy level of Eq. (2.65) and the wave function of Eq. (2.67).

Time-Dependent Perturbations: Fermi's Golden Rule

In this book we deal frequently with time-dependent external fields. Generally in this case there is no reason to speak about the electron stationary states because the electron energy is no longer conserved at all. But if the interaction with external fields is weak and can be considered as a small perturbation to the stationary Hamiltonian, one can think in terms of transitions between the stationary states as a result of such a weak perturbation. The goal of the perturbation theory in this case is to calculate the probability of such transitions per unit time.

Let us study the Hamiltonian of Eq. (2.61), in which the potential is now time dependent: $V = V(t)$. Again we can represent the required wave functions as a series with respect to the wave functions of the stationary states, $\Psi_{0n}(\vec{r}, t) = e^{-i\frac{E_{0n}}{\hbar}t} \psi_{0n}(\vec{r})$, i.e.,

$$\Psi(\vec{r}, t) = \sum_n C_n(t) \Psi_{0n}(\vec{r}, t) = \sum_n C_n(t) e^{-i\frac{E_{0n}}{\hbar}t} \psi_{0n}(\vec{r}). \tag{2.68}$$

Each Ψ_{0n} satisfies the Schrödinger equation (2.23) at $V = 0$:

$$i\hbar \frac{\partial \Psi_{0n}}{\partial t} = \mathcal{H}_0 \Psi_{0n} = E_{0n} \Psi_{0n}.$$

Substituting Eq. (2.68) into the Schrödinger equation (2.23), multiplying it by Ψ_{0m}^*, and integrating over the space coordinates, we get the system of differential equations for the coefficients $C_m(t)$:

$$i\hbar \frac{dC_m(t)}{dt} = \sum_n \mathcal{V}_{mn}(t) C_n(t) \tag{2.69}$$

with

$$\mathcal{V}_{mn}(t) = V_{mn} e^{i\frac{(E_{0m} - E_{0n})t}{\hbar}} \equiv V_{mn} e^{i\omega_{mn}t}.$$

We define the resonant frequency $\omega_{mn} = [(E_{0m} - E_{0n})/\hbar]$, while V_{mn} is the same as in Eq. (2.64).

Let the initial stationary state be state k at $t = 0$. Thus

$$C_k(t = 0) = 1, \qquad C_{m \neq k} = 0.$$

Using these initial conditions for Eqs. (2.69), we can find in the lowest-order approximation,

$$i\hbar \frac{dC_{1m}}{dt} = V_{mk}(t), \qquad m \neq k. \tag{2.70}$$

The solution of Eq. (2.70) with the above boundary conditions is

$$C_{1m}(t) = -\frac{i}{\hbar} \int_0^t V_{mk} e^{i\omega_{mk}t} \, dt. \tag{2.71}$$

Using $C_{1m}(t)$, we can calculate the function $\Psi(\vec{r}, t)$. The coefficients $C_{1m}(t)$ determine the probability $\mathcal{P}_m(t)$ of finding the system in the state m at any time t:

$$\mathcal{P}_m(t) = |C_{1m}|^2.$$

For the most important cases the perturbation has a harmonic time dependence:

$$V(t) = F e^{i\omega t} + F^* e^{-i\omega t}, \tag{2.72}$$

where F is independent of time. Substituting Eq. (2.72) into Eq. (2.71), we get

$$C_{1m}(t) = -\frac{F_{mk} \left[e^{i(\omega_{mk} - \omega)t} - 1 \right]}{\hbar(\omega_{mk} - \omega)} - \frac{F_{mk}^* \left[e^{i(\omega_{mk} + \omega)t} - 1 \right]}{\hbar(\omega_{mk} + \omega)},$$

where matrix elements F_{mk} are introduced in accordance with Eq. (2.64). It is easy to see that only the first term is important as it is much larger than the second term when $\omega \to \omega_{mk}$. Accordingly, the second nonresonant term may be dropped. Then the probability $\mathcal{P}_m(t)$ for this perturbation gives

$$\mathcal{P}_m = 4|F_{mk}|^2 \frac{\sin^2 \left(\frac{\omega_{mk} - \omega}{2} t \right)}{\hbar^2 (\omega_{mk} - \omega)^2}. \tag{2.73}$$

From mathematics we know that

$$\lim_{t \to \infty} \frac{\sin^2 \alpha t}{\pi \alpha^2 t} = \delta(\alpha),$$

where δ is the Dirac delta function. Then Eq. (2.73) at large t may be written as

$$\mathcal{P}_m(t) = \frac{2\pi}{\hbar^2} |F_{mk}|^2 \delta(\omega_{mk} - \omega) t. \tag{2.74}$$

The linear dependence on t allows one to introduce the probability of transition from an initial state k to any other state m per unit time:

$$W_{km} = \frac{2\pi}{\hbar} |F_{mk}|^2 \delta(\hbar\omega_{mk} - \hbar\omega), \tag{2.75}$$

where we use the well-known relation $\delta(ax) = \delta(x)/a$ at $a \neq 0$.

Remarkably, this probability is independent of time. This formula is known as Fermi's golden rule. Fermi's golden rule is an extremely important and useful result in quantum mechanics and it finds extensive applications in solid-state theory. In due course we will make frequent use of Fermi's golden rule.

In this section we have considered a single-particle Schrödinger equation. In general the wave function and the Hamiltonian represents the system of all interacting particles as a whole. It is a tremendously difficult task to solve the Schrödinger equation for such a complex system. The complete approach to such systems involves additional quantum-mechanical concepts, such as the Pauli exclusion principle. In Section 2.5 we discuss such many-particle quantum-mechanical systems.

2.5 Many-Particle Systems

Although any semiconductor material consists of a vast number of electrons, ions, atoms, etc., the major properties of semiconductor devices are generally determined by the electron subsystem. It is reasonable to separate the electron subsystem and the lattice vibrations and consider them to be weakly coupled. In the spirit of this approach we study a many-electron subsystem and then take into account its interaction with the lattice vibrations.

For many-electron systems, quantum physics brings additional features that are absent in the classical description. These features are associated with the fact that elementary particles, including electrons, are identical and the impossibility, in principle, of specifying the coordinates and tracing to a given electron. In addition, an internal characteristic of a particle, the spin, plays an important role in many-particle physics.

2.5.1 Spin and Electron Statistics

So far we have studied how an electron's wave function evolves in space as revealed by quantum mechanics. In addition to the parameters that characterize this spatial evolution, quantum-mechanical particles possess an additional internal degree of freedom, which is referred to as spin. Although one can compare spin with classical rotation, in fact, spin is a strictly quantum-mechanical quantity and differs substantially from its classical analog. The principal quantitative characteristic of spin is a dimensionless quantity called the spin number s. It is well established experimentally that an electron has a spin number equal to 1/2. If one fixes an axis in space, the projection of the electron spin on this axis can be either $+1/2$ or $-1/2$. Thus a complete description of an electron state requires at least two quantum numbers: one corresponding to the spatial wave function, say n, and another corresponding to the spin number s.

In the absence of a magnetic field, one may neglect the weak interaction between electron spin and electron translational (orbital) motion, which is known as the spin-orbital interaction. In this case, the electron spin does not affect the electron's spatial properties. Hence an electron occupying any energy level may have two possible spin orientations. This case corresponds to a twofold degeneracy for each energy level.

Although for many actual cases the electron spin is not important in altering the energy spectra or the spatial dependence of wave functions, etc., there is one crucially important consequence of the fact that the electron spin number is a half-integer. The point is that particles with half-integer spin numbers obey the Pauli exclusion principle: any quantum state $\{n, s\}$ can be occupied by only a single particle. In other words, two electrons in a system cannot be simultaneously in the same quantum state. Stated another way, two electrons may be in the same energy state if their spin quantum numbers are different; if one spin quantum number is $+1/2$, the other must be $-1/2$. It is clear that this restriction leads to a new, nonclassical statistics of electrons. Such statistics are called the Fermi statistics. Under equilibrium the occupation of the energy levels is described by the Fermi distribution function,

$$\mathcal{F}_F(E_{n,s}) = \frac{1}{1 + e^{\frac{E_{n,s} - E_F}{k_B T}}}, \tag{2.76}$$

where $k_B T$ is the temperature of the system, E_F is the so-called Fermi energy or Fermi level, and $E_{n,s}$ is the energy of the quantum state characterized by two quantum numbers n and s. The Fermi function $\mathcal{F}_F(E)$ is shown in Fig. 2.5 for different temperatures. The Fermi energy depends on the electron concentration and may be found from the normalization condition:

$$\sum_{n,s} \mathcal{F}_F(E_{n,s}) = N. \tag{2.77}$$

Here N is the total number of the electrons in the system.

If the positions of the energy levels are independent of spin, $E_{n,s} \equiv E_n$, the occupation of these levels is simply equal to $2\mathcal{F}_F(E_n)$, and the normalization condition may be written as

$$2 \sum_n \mathcal{F}_F(E_n) = N. \tag{2.78}$$

One can see that in accordance with the Pauli principle the occupation of any energy state, $\{n, s\}$, defined by Eq. (2.76), is always less than or equal to 1.

At high temperatures the Fermi distribution is

Figure 2.5. Fermi distribution function for different crystal temperatures.

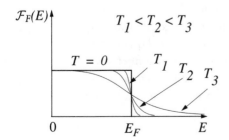

close to the Boltzmann distribution:

$$\mathcal{F}_F(E) \approx e^{\frac{E_F-E}{k_BT}}. \tag{2.79}$$

The corresponding curve is shown schematically in Fig. 2.5 at $T = T_3$. The evolution of $\mathcal{F}_F(E)$ with the temperature is illustrated by curves corresponding to temperatures $T = 0, T_1, T_2, T_3$. In the limit of low temperatures, $T \to 0$, the function $\mathcal{F}_F$ is transformed into a step function: $\mathcal{F}_F(E) = 1$ for the energy levels below the Fermi energy E_F since all levels with $E < E_F$ are occupied and $\mathcal{F}_F(E) = 0$ for energies above E_F since these levels are empty. In this limit, the electron subsystem is frequently referred to as a highly degenerate electron gas.

For a three-dimensional electron gas, the index n may be replaced by the wave vector $\vec{k}$, and the summation over $\vec{k}$ in Eq. (2.78) may be converted to an integration:

$$\sum_{\vec{k}} = \frac{V}{(2\pi)^3} \int d\vec{k}, \tag{2.80}$$

where V is the volume of the system under consideration. Performing the integration in the spherical coordinate system and using the dispersion relation of Eq. (2.28), one obtains from Eq. (2.78) the following expression for the Fermi energy of the degenerate electrons:

$$E_F = (3\pi^2)^{2/3} \frac{\hbar^2 n_{3D}^{2/3}}{2m^*}, \quad T \to 0, \tag{2.81}$$

i.e., it increases as the 2/3 power of electron concentration n_{3D}. Here m^* is the mass of electrons and $n_{3D} = N/V$ is the concentration per unit volume. Since the Fermi function contains the exponential factor, the low-temperature limit corresponds to

$$E_F \gg k_BT. \tag{2.82}$$

In metals and heavily doped semiconductors the electron gas remains degenerate up to room temperature.

In Chapter 3 it will be shown that in the limit defined by Eq. (2.82) the Fermi energy of two-dimensional electrons is

$$E_F = \frac{\pi\hbar^2}{m^*} n_{2D}, \quad T \to 0, \tag{2.83}$$

i.e., it increases as the first power of electron concentration n_{2D}, where $n_{2D} = N/S$ is concentration per unit area. For one-dimensional electrons the Fermi energy is given by

$$E_F = \frac{\pi^2\hbar^2}{8m^*} n_{1D}^2, \quad T \to 0, \tag{2.84}$$

where $n_{1D} = N/L$ is concentration per unit length of the one-dimensional electron gas.

All three formulas, (2.81), (2.83), and (2.84), may be unified by introduction of the Fermi wave vector k_F, i.e., the wave vector of electrons with the Fermi energy:

$$E_F = \frac{\hbar^2 k_F^2}{2m^*}.$$ (2.85)

Comparison of these three formulas gives k_F for all three dimensionalities:

$$k_{F,3D} = (3\pi^2 n_{3D})^{1/3}, \qquad k_{F,2D} = (2\pi n_{2D})^{1/2}, \qquad k_{F,1D} = \frac{1}{2}\pi n_{1D}.$$ (2.86)

From these results one can conclude that as the dimensionality of the electron gas becomes lower, the effects of degeneracy become more pronounced. Another important conclusion is that for a light mass m^*, the effects of degeneracy are always stronger; cf. Eqs. (2.81)–(2.84).

The concept of Fermi statistics is one of the fundamental ideas of modern solid-state physics and electronics. We widely apply Fermi statistics to the description of electronic and photonic devices.

2.5.2 Many-Particle Schrödinger Equation: Self-Consistent Potential

The basic equations of quantum mechanics can be generalized for the many-particle case. Instead of the one-particle wave function $\Psi(\vec{r}, t)$, one has to introduce the many-electron wave function,

$$\Phi(\vec{r}_1, \vec{r}_2, \vec{r}_3, \ldots, \vec{r}_N, t),$$ (2.87)

where $\vec{r}_1, \ldots, \vec{r}_N$ is the set of coordinates of the N electron subsystem. For simplicity, we drop spin indices. The physical meaning of the function Φ is the same as for the single-electron wave function: $|\Phi|^2$ is the probability of finding N electrons at the N points $\vec{r}_1, \ldots, \vec{r}_N$. The normalization of this function is given by

$$\int |\Phi|^2 d\vec{r}_1 \cdots d\vec{r}_N = 1.$$ (2.88)

The wave function Φ satisfies the many-particle Schrödinger equation,

$$i\hbar \frac{\partial \Phi}{\partial t} - \mathcal{H}^{mp}\Phi = 0.$$ (2.89)

Here, the many-particle Hamiltonian $\mathcal{H}^{mp}$ consists of the sum over single-particle Hamiltonians and the potential energy of the Coulomb interaction among these particles (electrons):

$$\mathcal{H}^{\mathrm{mp}} = \sum_{i=1}^{N} \mathcal{H}(\vec{r}_i) + \frac{e^2}{2\kappa} \sum_{i,j} \frac{1}{|\vec{r}_i - \vec{r}_j|}, \tag{2.90}$$

where e is the charge of the electron and κ is the dielectric permittivity of the medium.

In the stationary case, the time dependence of the many-particle wave function is given by a simple exponential prefactor,

$$\Phi = e^{-iE_\nu t/\hbar} \, \varphi_\nu(\vec{r}_1, \ldots, \vec{r}_N), \tag{2.91}$$

where E_ν is the total energy of the many-particle system in the stationary state ν. The time-independent function φ_ν is defined by the stationary many-particle Schrödinger equation:

$$H^{\mathrm{mp}}\varphi_\nu(\vec{r}_1, \ldots, \vec{r}_N) = E_\nu\varphi_\nu(\vec{r}_1, \ldots, \vec{r}_N); \tag{2.92}$$

ν represents all quantum numbers of the system including the spins.

These many-particle equations are straightforward generalizations of the corresponding one-particle equations. They do not take into account any specific statistics. To include the statistics, let us consider the fact that electrons are all identical, i.e., the fact that no experiment can distinguish one electron from another because they have the same mass, spin, and charge. Consequently, there would be no change in the properties of an electron system if any two electrons are interchanged. In other words, the square of the modulus of the many-electron wave function should not change if the electron coordinates are permuted. The Pauli exclusion principle imposes additional restrictions on the form of the many-electron wave function. The wave function should be antisymmetric with respect to the permutation of any pair of electron coordinates:

$$\varphi_\nu(\vec{r}_1, \ldots, \vec{r}_i, \ldots, \vec{r}_j, \ldots, \vec{r}_N) = -\varphi_\nu(\vec{r}_1, \ldots, \vec{r}_j, \ldots, \vec{r}_i, \ldots, \vec{r}_N). \tag{2.93}$$

Note that the latter restriction leads to correlation between electrons even when no direct interaction is present. For example, for two noninteracting electrons the wave function is

$$\Phi_\nu(\vec{r}_1, \vec{r}_2) = \frac{1}{\sqrt{2}} e^{-i(E_{\nu_1} + E_{\nu_2})t/\hbar} [\psi_{\nu_1}(\vec{r}_1)\psi_{\nu_2}(\vec{r}_2) - \psi_{\nu_1}(\vec{r}_2)\psi_{\nu_2}(\vec{r}_1)], \tag{2.94}$$

where ψ are the normalized one-electron wave functions and E_{ν_i} are the eigenenergies of a single electron, $i = 1, 2$. The total energy of this system is an arithmetic sum of two single-electron energies $E_\nu = E_{\nu_1} + E_{\nu_2}$. As can be seen from Eq. (2.94), the wave function takes into account the spatial correlation of the two electrons. For example, states ν_1 and ν_2 must be different, otherwise the wave function is identically zero. The wave function is also zero if $\vec{r}_1 = \vec{r}_2$, i.e., the probability of finding two electrons with the same spin at the same point of space is equal to zero in agreement with the Pauli exclusion principle.

For interacting particles the above antisymmetrization of the wave function yields an additional contribution to the energy of a many-particle system. This contribution is called the exchange-correlation energy.

Generally, for interacting particles the wave function cannot be factorized into the form of Eq. (2.94). This form is only an approximation to the real nonfactorized many-electron wave function. For an N-electron system this approximation can be written as the determinant:

$$\varphi(\vec{r}_1, \ldots, \vec{r}_N) = \frac{1}{\sqrt{N!}} \begin{vmatrix} \psi_{v_1}(\vec{r}_1), \psi_{v_2}(\vec{r}_1), \ldots, \psi_{v_N}(\vec{r}_1) \\ \psi_{v_1}(\vec{r}_2), \psi_{v_2}(\vec{r}_2), \ldots, \psi_{v_N}(\vec{r}_2) \\ \cdots\cdots\cdots\cdots\cdots\cdots\cdots\cdots \\ \psi_{v_1}(\vec{r}_N), \psi_{v_2}(\vec{r}_N), \ldots, \psi_{v_N}(\vec{r}_N) \end{vmatrix} . \tag{2.95}$$

Here it is supposed that each of the ψ_{v_i} is normalized to 1. It is easy to see that the wave function of Eq. (2.95) is antisymmetric with respect to any permutations of the pairs of coordinates. All single-electron wave functions ψ_{v_i} must be different, otherwise the determinant of Eq. (2.95) is equal to zero, i.e., electrons must occupy different states.

In contrast to Eq. (2.94), the single-electron wave functions ψ_{v_i} in Eq. (2.95) are wave functions of interacting electrons. Substituting Eq. (2.95) into Eq. (2.92), multiplying by wave functions $\psi_{v_{j\neq i}}$, and integrating over all coordinates, one obtains a system of coupled equations for the single-electron wave functions:

$$H\psi_{v_i}(\vec{r}) + \frac{e^2}{\kappa} \sum_{j\neq i} \left[\int d\vec{r}' \frac{|\psi_{v_j}(\vec{r}')|^2}{|\vec{r} - \vec{r}'|} \right] \psi_{v_i}(\vec{r})$$

$$- \frac{e^2}{\kappa} \sum_{j\neq i} \left[\int d\vec{r}' \frac{|\psi_{v_j}^*(\vec{r}')\psi_{v_i}(\vec{r}')|}{|\vec{r} - \vec{r}'|} \right] \psi_{v_j}(\vec{r}) = E_{v_i}\psi_{v_i}. \tag{2.96}$$

This is a system of N equations ($i = 1, \ldots, N$) that are called the Fock equations. The Fock equations can be interpreted easily. Let us represent each equation as

$$(H + V_H + V_{EC})\psi_{v_i}(\vec{r}) = E_{v_i}\psi_{v_i}(\vec{r}), \tag{2.97}$$

where the first term in the parentheses on the left-hand side is the single-particle Hamiltonian containing the kinetic energy and the external potential. In the second term,

$$V_H = \frac{e^2}{\kappa} \sum_{j\neq i} \int d\vec{r}' \frac{|\psi_{v_j}(\vec{r}')|^2}{|\vec{r} - \vec{r}'|} \tag{2.98}$$

is the electrostatic potential created by all electrons except the ith one; V_H is called the Hartree potential. The term

$$V_{EC}\psi_{v_i}(\vec{r}) = \frac{e^2}{\kappa} \sum_{j \neq i} \left[\int d\vec{r}' \frac{|\psi_{v_j}^*(\vec{r}')\psi_{v_i}(\vec{r}')|}{|\vec{r} - \vec{r}'|} \right] \psi_{v_j}(\vec{r}) \tag{2.99}$$

represents the electron exchange correlation and V_{EC} is known as the exchange-correlation potential. Thus the Fock representation describes the motion of a single electron in the effective potential created by all other electrons of the system. The motion of all electrons is strongly correlated: if one of them changes its state, all other electrons will be immediately affected by this change. The total energy of the electron system is now

$$E = \sum_{i=1}^{N} E_{v_i}. \tag{2.100}$$

The Fock equations form a system of nonlinear integrodifferential equations. They are used widely in molecular and atomic calculations, but remain difficult to apply to solid-state problems because of their extreme complexity. However, Eq. (2.97) can be simplified if we compare the characteristic kinetic energy of electrons and the Coulomb interaction energy. According to Eq. (2.90), the interaction energy between two electrons can be estimated as

$$\langle E_{e-e} \rangle = \frac{e^2}{2\kappa \langle r_{12} \rangle}, \tag{2.101}$$

where $\langle r_{12} \rangle$ is the average distance between two electrons. In a bulk sample with a three-dimensional electron gas, $\langle r_{12} \rangle \approx 1/n_{3D}^{1/3}$, where n_{3D} is the number of electrons per unit volume. On the other hand, the energy of the electrons, participating in transport, optical transitions, etc., is of the order of the Fermi energy defined by Eq. (2.81) for a degenerate electron gas. Comparing these energies, we find that

$$\frac{\langle E_{e-e} \rangle}{E_F} = \frac{1}{(3\pi^2)^{2/3} a_B n_{3D}^{1/3}}, \tag{2.102}$$

where $a_B = \kappa \hbar^2 / 2m^* e^2$ is the Bohr radius. Hence, if $a_B^3 n_{3D} \gg 1$, the Coulomb interaction energy is small and may be considered as a perturbation. This assumption holds for metals and most semiconductors. Using the same estimates for two-dimensional ($\langle r_{12} \rangle = 1/\sqrt{n_{2D}}$) and one-dimensional ($\langle r_{12} \rangle = 1/n_{1D}$) electrons, one can come to the general conclusion that at high electron concentrations the electron interaction may be considered as a small perturbation. This is due to the fact that for greater electron concentrations the Fermi energy, cf. Eqs. (2.81)–(2.84), increases faster than the electron–electron interaction energy.

In this case, the exchange-correlation potential can be neglected. Instead of the Fock equations, one gets the so-called Hartree approximation for the one-electron state in a many-electron system:

$$(H + V_H)\psi_{v_i} = E_{v_i}\psi_{v_i}, \tag{2.103}$$

where the Hartree potential V_H is determined by Eq. (2.98). The Hartree potential V_H is the solution of the Poisson equation:

$$\nabla^2 V_H = -\frac{4\pi e^2}{\kappa}\left[\sum_j |\psi_{v_j}(\vec{r})|^2 - N_D\right], \tag{2.104}$$

where N_D is the ionized donor concentration that contributes positive charge to the crystal. The Hartree equation (2.103) is still a nonlinear integrodifferential system of equations, but it can be solved and analyzed for some particular cases.

The variational method studied in Section 2.4 is the most suitable method of the solution of the Hartree equations. In applying this method it is important to keep in mind that the self-consistent Hartree potential is now also a functional of the unknown functions ψ_{v_i}. Therefore the variational functional in the explicit form reads as

$$\mathcal{E}\{\psi\} = \int d\vec{r}\left[\frac{-\hbar^2}{2m^*}|\nabla\psi|^2 + V_{\text{ext}}(\vec{r})|\psi|^2 + \frac{1}{2}V_H\{\psi\}|\psi|^2\right]. \tag{2.105}$$

Here V_{ext} is the external potential and the factor of $1/2$ before the last term appears because one has to vary both multipliers $V\{\psi\}$ and $|\psi|^2$.

Minimization of $\mathcal{E}\{\psi\}$ with respect to ψ gives the one-particle wave function in the Hartree approximation ψ_H. The corresponding one-particle energy E_H is equal to

$$E_H = \int d\vec{r}\left[\frac{-\hbar^2}{2m^*}|\nabla\psi|^2 + V_{\text{ext}}(\vec{r})|\psi|^2 + V_H\{\psi\}|\psi|^2\right]. \tag{2.106}$$

Of physical importance is the fact that the Hartree energy does not coincide with the value of the functional $\mathcal{E}\{\psi_H\}$.

The one-particle wave functions ψ_H and energies E_H allow one not only to describe different states of a complex many-particle system, but also to calculate the response of the system to external perturbations. For example, if a many-particle system is subjected to illumination by light, the Hartree approximation allows one to find a polarization of the system, as well as to calculate the absorption and the emission of light in terms of phototransitions between one-particle states.

2.6 An Electron in Crystalline Potential: Effective-Mass Approximation

Actually, in a crystal, an electron moves in the field created by positively charged ions and all other electrons. This field is frequently referred to as the crystalline potential. In this section we consider one-particle states for electrons in an ideal crystal. We present a classification of these states and find a general form for the wave functions and energies.

2.6.1 Bloch Functions

For an ideal crystal, the crystalline potential is periodic with the period of the crystalline lattice. Let $\vec{a}_i$, $i = 1, 2, 3$, be three basic vectors of the Bravais lattice that define the three primitive translations. The periodicity of the crystalline potential $W(\vec{r})$ implies that

$$W\left(\vec{r} + \sum_i n_i \vec{a}_i\right) = W(\vec{r}),
\tag{2.107}$$

where $\vec{r}$ is an arbitrary point of the crystal and n_i are any integers. The one-particle wave function should satisfy the Schrödinger equation,

$$H_{cr}\psi(\vec{r}) = \left[-\frac{\hbar^2}{2m}\nabla^2 + W(\vec{r})\right]\psi(\vec{r}) = E\psi(\vec{r}),
\tag{2.108}$$

where m is the free-electron mass and H_{cr} is the crystalline Hamiltonian.

Because of the potential periodicity (2.107), the wave functions $\psi(\vec{r})$ can be classified and presented in a special form. To find this form we introduce the translation operator $T_{\vec{d}}$, that acts on the coordinate vector $\vec{r}$ as

$$T_{\vec{d}}\vec{r} = \vec{r} + \vec{d}, \qquad \vec{d} = \sum_i n_i \vec{a}_i.
\tag{2.109}$$

Applying this operator to the wave function we find that the function $T_{\vec{d}}\psi(\vec{r}) \equiv \psi(\vec{r} + \vec{d})$ is also a solution of Eq. (2.108) at the same energy E. Let us assume that the electron state with energy E is not degenerate. Then we conclude that both wave functions $\psi(\vec{r})$ and $\psi(\vec{r} + \vec{d})$ can differ by only a multiplier:

$$\psi(\vec{r} + \vec{d}) = C_{\vec{d}}\,\psi(\vec{r}).
\tag{2.110}$$

From the normalization condition we obtain $|C_{\vec{d}}|^2 = 1$. Two different translations $\vec{d}_1$ and $\vec{d}_2$ should lead to the same result as the single translation $\vec{d} = \vec{d}_1 + \vec{d}_2$, i.e., $C_{\vec{d}_1} C_{\vec{d}_2} = C_{\vec{d}_1 + \vec{d}_2}$. From this last result it follows that $C_{\vec{d}}$ may be represented in an exponential form:

$$C_{\vec{d}} = e^{i\vec{k}\vec{d}} = \exp\left(i\vec{k}\sum_i n_i \vec{a}_i\right),
\tag{2.111}$$

where $\vec{k}$ is a constant vector. Thus from Eq. (2.110) we get the wave function in the Bloch form:

$$\psi(\vec{r}) = e^{-i\vec{k}\vec{d}}\psi(\vec{r} + \vec{d}) = e^{i\vec{k}\vec{r}}u_{\vec{k}}(\vec{r}),
\tag{2.112}$$

where

$$u_{\vec{k}}(\vec{r}) = e^{-i\vec{k}(\vec{r}+\vec{d})}\psi(\vec{r}+\vec{d}). \tag{2.113}$$

One can easily check that the so-called Bloch function $u_{\vec{k}}(\vec{r})$ is a periodic function:

$$u_{\vec{k}}(\vec{r}+\vec{d}) = u_{\vec{k}}(\vec{r}).$$

Therefore the stationary one-particle wave function in a crystalline potential has the form of a plane wave modulated by the Bloch function with the lattice periodicity. The wave vector $\vec{k}$ is called the wave vector of the electron in the crystal. This wave vector is one of the quantum numbers of electron states in crystals.

Applying cyclic boundary conditions for the crystal with a number of periods N_i along the direction $\vec{a}_i$,

$$\psi(\vec{r}+N_i\vec{a}_i) = \psi(\vec{r}), \quad N_i \to \infty, \tag{2.114}$$

we find for $\vec{k}$,

$$\vec{k}\vec{a}_i N_i = 2\pi n_i, \quad n_i = 1, 2, 3, \ldots, N_i. \tag{2.115}$$

These allowed quasi-continuum values of $\vec{k}$ form the so-called first Brillouin zone of the crystal. It is important that the symmetry of the Brillouin zone in $\vec{k}$ space be determined by the crystal symmetry.

Let the one-particle energy corresponding to the wave vector $\vec{k}$ be $E = E(\vec{k})$. If the wave vector changes within the Brillouin zone, one gets a continuum-energy band, i.e., an electron-energy band. At fixed $\vec{k}$, the Schrödinger equation (2.108) has a number of solutions in the Bloch form:

$$\psi_{\alpha,\vec{k}}(\vec{r}) = \frac{1}{\sqrt{V}}e^{i\vec{k}\vec{r}}u_{\alpha,\vec{k}}, \tag{2.116}$$

where α numerates energy bands. Here we normalize the wave function $\psi_{\alpha,\vec{k}}$ for the crystal volume $V = NV_0$; $N = N_1 N_2 N_3$ and V_0 are the number and the volume of the primitive crystal cells, respectively. For the chosen normalization condition, one can easily get

$$\frac{1}{V_0}\int_{V_0}|u_{\alpha,\vec{k}}|^2\,\mathrm{d}\vec{r} = 1,$$

where the integral is calculated over the primitive cell. The latter formula allows one to estimate the order of value of $u_{\alpha,\vec{k}}$: $|u_{\alpha,\vec{k}}| \approx 1$. Note that for the Bloch functions with different α and $\vec{k}$, the orthogonality conditions are valid:

$$\frac{1}{V_0}\int_{V_0}u^*_{\alpha,\vec{k}}u_{\alpha',\vec{k}'}\,\mathrm{d}\vec{r} = \delta_{\alpha\alpha'}\delta_{\vec{k}\vec{k}'}. \tag{2.117}$$

The Bloch functions $u_{\alpha,\vec{k}}(\vec{r})$ satisfy the following equation:

$$\left[-\frac{\hbar^2}{2m}\nabla^2 - i\frac{\hbar^2}{m}(\vec{k}\nabla) + W(\vec{r})\right]u_{\alpha,\vec{k}} = \left[E_\alpha(\vec{k}) - \frac{\hbar^2 k^2}{2m}\right]u_{\alpha,\vec{k}}. \qquad (2.118)$$

Because of the crystal periodicity, the Bloch function can be calculated within a single primitive cell.

2.6.2 Occupation of Energy Bands: Metals, Semiconductors, and Dielectrics

From the previous analysis we can conclude that the energy spectrum of an ideal crystal consists of a series of energy bands $E_\alpha(\vec{k})$. The energy bands are separated by bandgaps; no allowed energy states exist in bandgaps. In Fig. 2.6 such a series is sketched. According to Eq. (2.115) the total number of wave vectors $\vec{k}$ within any band is $N_1 N_2 N_3 = N$, i.e., it equals the total number of primitive cells in the crystal. Because of the electron spin, the number of energy states for any band is $2N$. The number of crystalline electrons per the primitive cell depends on atoms composing the crystal. Let us use the Pauli exclusion principle (see Subsection 2.5.1) to find the band occupation. Each band state can be occupied by only one electron; therefore for different numbers of electrons per cell we get the following results at $T = 0$:

1. one atom per cell and one electron per atom: the lowest energy band is half-full;
2. one atom per cell and two electrons per atom: the lowest band is full, but next band is empty;
3. two atoms per cell and each atom gives at least one electron: the lowest band is full, but the next is empty and, in general, is separated by an energy gap from the first, etc.

Figure 2.6. An energy-band structure of (a) semiconductors and dielectrics, (b) metals. Four occupied bands are shown for a semiconductor, while for a metal only one half-filled conduction band is presented.

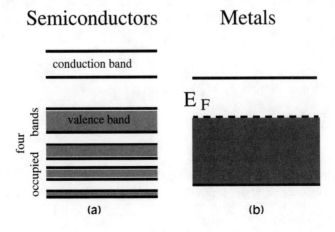

For crystals of type 1, the partly occupied band is the conduction band. Such materials possess a conductivity even at $T = 0$ and are metals. For crystals of types 2 and 3 the lowest full bands are the so-called valence bands. The empty energy band is the conduction band. There is no conductivity in such crystals at low temperatures until the conduction band is empty. Thus they are semiconductors or dielectrics.

For atoms, which consist of many electrons, only valence atomic electrons determine the band filling and the type of crystal. If there is an odd number of such electrons per cell we always get a metal. If this number is even we usually have a semiconductor or a dielectric. Consider III–V compounds, for example, the GaAs crystal. In each primitive cell there are two atoms, Ga and As, which supply three and five valence electrons, respectively. Thus eight valence electrons per primitive cell provide the semiconductor type of the crystals.

Note that, in some crystals, such a scheme fails because different energy bands can overlap and materials satisfying the above-discussed semiconductor classification can actually be metallic.

At finite temperatures for an evaluation of the band occupation, one should use the Fermi distribution function (2.76), which for electrons from the energy band $E_\alpha(\vec{k})$ has the form

$$\mathcal{F}_F[E_\alpha(\vec{k})] = \frac{1}{1 + e^{\frac{E_\alpha(\vec{k}) - E_F}{k_B T}}}.$$

Let us designate $\alpha = v$ for the valence band and $\alpha = c$ for the conduction band. Then the concentration of electrons in the conduction band can be calculated as

$$n = \frac{2}{V} \sum_{\vec{k}} \frac{1}{1 + e^{\frac{E_c(\vec{k}) - E_F}{k_B T}}}, \tag{2.119}$$

where V is the crystal volume and $\vec{k}$ runs over the first Brillouin zone. As for the valence band, it is full at $T = 0$, but at finite temperatures some of the electrons leave this band and transfer to the conduction band, that is, the valence band contains some unoccupied states or, in other words, it contains holes. Using the distribution function of electrons in the valence band $\mathcal{F}_v(\vec{k})$, we can find the distribution function $\mathcal{F}_h(\vec{k})$ of these holes over energy in the valence band:

$$\mathcal{F}_h(\vec{k}) = 1 - \mathcal{F}_v(\vec{k}). \tag{2.120}$$

For equilibrium conditions $\mathcal{F}_v = \mathcal{F}_F$ and we get

$$\mathcal{F}_h(\vec{k}) = \frac{1}{1 + e^{\frac{E_F - E_v(\vec{k})}{k_B T}}}. \tag{2.121}$$

Now we can find the concentration of the holes – the number of unoccupied states in the valence band per unit volume – by calculating the sum over all states in this band:

$$p = \frac{2}{V} \sum_{\vec{k}} \frac{1}{1 + e^{\frac{E_F - E_v(\vec{k})}{k_B T}}}. \tag{2.122}$$

The Fermi level E_F is determined by the total number of electrons in the crystal. For metals, the Fermi level is situated in the conduction band. For undoped semiconductors, the Fermi energy lies within the energy gap between the valence and the conduction bands. To utilize these results fully, we need to specify the energy dependence on the wave vector for both $E_v(\vec{k})$ and $E_c(\vec{k})$. This can be done in the effective-mass approximation, which we analyze in Section 2.6.3.

From analysis of the occupation of the valence bands, it is clear that one may usefully adopt the concept of holes as quasi-particles instead of electrons missing from these bands. These quasi-particles can be introduced by the following simple consideration. If the valence bands are full, the total wave vector of valence electrons is zero:

$$\vec{K}_v = \sum_i \vec{k}_i = 0, \tag{2.123}$$

where the sum accounts for all occupied valence states. Let us assume that one of the electrons with wave vector $\vec{k}_e$ is removed from the valence band. The total wave vector of the valence electrons becomes

$$\vec{K}_v = \sum_i \vec{k}_i = -\vec{k}_e. \tag{2.124}$$

On the other hand, removing this electron is identical to the creation of a hole. One can attribute the wave vector of Eq. (2.124) to this hole:

$$\vec{k}_h = -\vec{k}_e.$$

Then the energy of the valence electrons decreases by the factor $E_v(\vec{k}_e)$; thus one can also attribute the energy

$$E_h(\vec{k}_h) = -E_v(\vec{k}_e) \tag{2.125}$$

to this hole. If the energy band is symmetric and $E_v(\vec{k}) = E_v(-\vec{k})$, we can write

$$E_h(\vec{k}_h) = -E_v(\vec{k}_h) \tag{2.126}$$

for the hole energy. Thus we can characterize the hole by a wave vector $\vec{k}_h$ and energy $E_h(\vec{k}_h)$ and consider the hole as a new quasi-particle created when the electron is removed from the valence band. In the conduction band the electron energy $E_c(\vec{k})$ increases as the wave vector increases. In the valence band, near the maximum energy of the band, the electron energy $E_v(\vec{k})$ decreases as $\vec{k}$ increases. But, according to Eq. (2.126), the hole energy increases with the hole wave vector, that is, the hole behaves as a usual particle. Thus one can introduce the velocity of the hole

$$v_h = \frac{1}{\hbar}\nabla_{\vec{k}_h} E_h(\vec{k}_h),$$

employ the Newton equations, etc. The absence of a negative charge in the valence band when an electron is removed makes it possible to characterize a hole by a positive elementary charge, that is, the holes carry the positive electric charge. The distribution function of Eq. (2.121) can also be rewritten in terms of the hole energies:

$$\mathcal{F}_h(\vec{k}) = \frac{1}{1 + e^{\frac{E_F + E_h(\vec{k})}{k_B T}}}. \tag{2.127}$$

2.6.3 Effective Masses

To find $u_{\alpha,\vec{k}}$, and $E_\alpha(\vec{k})$, one should know the crystalline potential $W(\vec{r})$ in detail. However, most aspects of electron properties in a crystal require knowledge of these quantities in a small range of the $\vec{k}$ space around the energy-band extrema. Let an extremum of $E_\alpha(\vec{k})$ be at the point $\vec{k} = 0$. We can rewrite Eq. (2.118) as

$$[H(\vec{k} = 0) + w(\vec{k})]u_{\alpha,\vec{k}} = E_\alpha(\vec{k})u_{\alpha,\vec{k}}, \tag{2.128}$$

where

$$H(\vec{k} = 0) = -\frac{\hbar^2}{2m}\nabla^2 + W(\vec{r})$$

is the Hamiltonian, whose eigenfunctions are $u_{\alpha,0}$:

$$H(\vec{k} = 0)u_{\alpha,0} = E_\alpha(0)u_{\alpha,0},$$

$$w(\vec{k}) = -i\frac{\hbar^2}{m}(\vec{k}\nabla) + \frac{\hbar^2 k^2}{2m} \to 0, \quad \text{as } \vec{k} \to 0.$$

So at small $\vec{k}$ we may consider $w(\vec{k})$ as a small perturbation.

The Bloch function with a finite wave vector can be expanded in a series:

$$u_{\alpha,\vec{k}}(\vec{r}) = \sum_{\alpha'} C_{\alpha'}(\vec{k})u_{\alpha',0}(\vec{r}). \tag{2.129}$$

Inserting the expansion of Eq. (2.129) into Eq. (2.118), multiplying by a complex-conjugate quantity $u_{\alpha,0}^*(\vec{r})$, and integrating over the primitive cell, we obtain

$$\sum_{\alpha'}\left\{\left[E_{\alpha'}(0) - E_\alpha(\vec{k}) + \frac{\hbar^2 \vec{k}^2}{2m}\right]\delta_{\alpha\alpha'} + \frac{\hbar\vec{k}}{m}\vec{\pi}_{\alpha\alpha'}\right\}C_{\alpha'}(\vec{k}) = 0, \tag{2.130}$$

where

$$\vec{\pi}_{\alpha\alpha'} = \int_{V_0} u^*_{\alpha,0}(\vec{r}) \left(-i\hbar\nabla\right) u_{\alpha',0}(\vec{r}). \tag{2.131}$$

Let us consider the terms containing $\vec{k}$ as small ones. Then we can find corrections to the eigenvalue $E_\alpha(0)$ by applying the perturbation theory of Section 2.4. According to this method, we can set $C_{\alpha'}(0) = \delta_{\alpha\alpha'}$. Then we get

$$C_{\alpha' \neq \alpha} = \frac{\hbar\vec{k}\vec{\pi}_{\alpha\alpha'}}{m} \frac{1}{E_\alpha(0) - E_{\alpha'}(0)}, \tag{2.132}$$

$$E_\alpha(\vec{k}) = E_\alpha(0) + \frac{\hbar^2 k^2}{2m} + \frac{\hbar^2}{m^2} \sum_{\alpha'} \frac{|\vec{k}\vec{\pi}_{\alpha\alpha'}|^2}{E_\alpha(0) - E_{\alpha'}(0)}. \tag{2.133}$$

Equation (2.133) can be rewritten as

$$E_\alpha(\vec{k}) = E_\alpha(0) + \frac{\hbar^2}{2} \left(\frac{1}{m^*}\right)_{ik} k_i k_k, \tag{2.134}$$

where we introduce the reciprocal effective-mass tensor $(1/m^*)_{ik}$, which describes properties of the electron band α in the vicinity of the point $\vec{k} = 0$. This tensor depends on the crystal symmetry. For the cubic crystal it reduces to a scalar:

$$\left(\frac{1}{m^*}\right)_{ik} = \frac{1}{m^*}\delta_{ik}, \qquad \frac{1}{m^*} = \frac{1}{m} + \frac{2}{m^2} \sum_{\alpha'} \frac{|\vec{\pi}_{\alpha\alpha'}|^2}{E_\alpha(0) - E_{\alpha'}(0)}, \tag{2.135}$$

that is, we get as a quadratic dispersion for the electron-energy vicinity the extremum of $E_\alpha(\vec{k})$ at $\vec{k} = 0$.

The effective mass m^* is determined by matrix elements (2.131) and by distances between a chosen and all other electron bands. As a rule, the dominant contribution to the effective mass of a chosen extremum comes from the nearest band. Particularly, for a minimum of a given energy band α, the nearest band α' lies below $[E_{\alpha'}(0) < E_\alpha(0)]$. From Eq. (2.135), it follows that the dominant contribution of the band α' and hence the effective mass are positive. Thus for the conduction band with nondegenerate minimum at $\vec{k} = 0$, we get the electron energy in the conventional form:

$$E_e = E_c(0) + \frac{\hbar^2\vec{k}^2}{2m_e}, \tag{2.136}$$

where m_e is the effective mass for electrons. On the other hand, for a maximum of $E_\alpha(\vec{k})$, the contribution comes from the upper band and is negative. The effective mass of an electron near the top of a band is also negative.

Using the definition of the hole energy given by Eq. (2.126), one can find this energy in the form

$$E_h = E_v(0) + \frac{\hbar^2\vec{k}^2}{2m_h}, \tag{2.137}$$

where a simple nondegenerate parabolic dependence on $\vec{k}$ is supposed for $E_v(\vec{k})$. Note that the effective mass for holes m_h is positive.

It is worth emphasizing that we have obtained results of great importance in solid-state physics: electron motion in a crystalline potential can be described by a quasi-continuum wave vector $\vec{k}$. According to Eq. (2.2), one can introduce a quasi-momentum of the electron $\vec{p} = \hbar\vec{k}$, that is, an electron in a crystal can be described as a free electron in the vacuum, but with an effective mass instead of the free-electron mass m. The effective mass differs from m. Indeed, it can be anisotropic and even negative.

If an external potential $U(\vec{r})$ is applied to the crystal, an electron moves in a complicated potential field composed of the crystalline potentials $W(\vec{r})$ and $U(\vec{r})$:

$$W(\vec{r}) + U(\vec{r}). \tag{2.138}$$

If $U(\vec{r})$ is slowly varying with a coordinate, we can use the previous results of this section to simplify the description of the electron motion significantly.

Consider first semiclassical electron motion. In Section 2.4 we have shown that semiclassical motion of a free electron is described by a wave packet built of plane waves, as in Eq. (2.29). For a crystalline electron occupying an energy band $E_\alpha(\vec{k})$, we should use a wave packet built of the Bloch wave functions with wave vectors near a certain value of k_0:

$$\Phi_\alpha(\vec{r}) = \int A(\vec{k}) e^{i[\vec{k}\vec{r} - \frac{E_\alpha(\vec{k})}{\hbar}t]} u_{\alpha,\vec{k}}(\vec{r}) \, d\vec{k}. \tag{2.139}$$

Following Eqs. (2.28) and (2.32), we get the expression for the group velocity of a crystalline electron:

$$\vec{v}_g = \frac{1}{\hbar} \frac{dE_\alpha(\vec{k})}{d\vec{k}}. \tag{2.140}$$

In the presence of an external force $\vec{f} = [-(dU/d\vec{r})]$ we can write the rate of the change of the electron energy as

$$\frac{dE_\alpha(\vec{k})}{dt} = \vec{f}\vec{v}_g, \tag{2.141}$$

where the right-hand side is the work done on the electron per unit time. The left-hand side is

$$\frac{dE_\alpha(\vec{k})}{dt} = \frac{dE}{d\vec{k}} \frac{d\vec{k}}{dt} = \hbar\vec{v}_g \frac{d\vec{k}}{dt}.$$

Comparing the last result and Eq. (2.141), we get the equation for the time rate of change of the wave vector of a classical crystalline electron:

$$\hbar \frac{d\vec{k}}{dt} = \frac{d\vec{p}}{dt} = \vec{f}. \tag{2.142}$$

Thus we see that although the electron is moving in the composite potential of expression (2.138), for its wave vector $\vec{k}$ and quasi-momentum $\vec{p} = \hbar \vec{k}$ we get a Newton-type equation (2.142) containing only the external force. Exactly the same equation can be derived for the holes with the dispersion relation of Eq. (2.137).

2.6.4 The Envelope Function

The effective-mass approximation can be applied also to describe the quantum-mechanical behavior of an electron in the potential $U(\vec{r})$. The Schrödinger equation is

$$[H_{cr} + U(\vec{r})]\,\psi(\vec{r}) = E\psi(\vec{r}). \tag{2.143}$$

Let us expand the wave function $\psi(\vec{r})$ in the series

$$\psi(\vec{r}) = \sum_{\alpha,\vec{k}} C_{\alpha,\vec{k}} \psi_{\alpha,\vec{k}}, \tag{2.144}$$

where $\psi_{\alpha,\vec{k}}(\vec{r})$ are solutions of unperturbed problem of Eq. (2.108). By the standard method we can transform Eq. (2.143) to the following form:

$$E_\alpha(\vec{k})C_{\alpha,\vec{k}} + \sum_{\alpha',\vec{k}'} \langle \alpha, \vec{k} \mid U \mid \alpha', \vec{k}' \rangle C_{\alpha',\vec{k}'} = EC_{\alpha,\vec{k}}, \tag{2.145}$$

where the matrix elements are

$$\langle \alpha, \vec{k} \mid U \mid \alpha', \vec{k}' \rangle = \frac{1}{NV_0} \int_{V_0} \mathrm{d}\vec{r}\, e^{i(\vec{k}'-\vec{k})\vec{r}} u^*_{\alpha,\vec{k}}(\vec{r}) U(\vec{r}) u_{\alpha',\vec{k}'}(\vec{r}). \tag{2.146}$$

Next we expand the potential $U(\vec{r})$ in a Fourier series:

$$U(\vec{r}) = \sum_{\vec{K}} U_{\vec{K}} e^{i(\vec{K}\vec{r})}. \tag{2.147}$$

Since $U(\vec{r})$ is a slowly varying function, in the expansion of Eq. (2.147) only terms with small $\vec{K}$ are essential; others are almost negligible.

Equation (2.146) can be rewritten as

$$\langle \alpha, \vec{k} \mid U \mid \alpha', \vec{k}' \rangle = \frac{1}{NV_0} \sum_{\vec{K}} \int \mathrm{d}\vec{r}\, e^{i(\vec{k}'-\vec{k}+\vec{K})\vec{r}} u^*_{\alpha,\vec{k}}(\vec{r}) U_{\vec{K}} u_{\alpha',\vec{k}'}(\vec{r})$$

$$\approx \frac{1}{NV_0} \sum_{\vec{K}} U_{\vec{K}} \sum_{\vec{n}} e^{i(\vec{k}'-\vec{k}+\vec{K})\vec{n}} \int_{V_0} \mathrm{d}\vec{r}\, u_{\alpha,\vec{k}}(\vec{r}) u_{\alpha',\vec{k}'}(\vec{r}). \tag{2.148}$$

Here $\vec{n}$ is the position of the nth primitive cell. We take into account that the Bloch functions are periodic and only small $\vec{K}$ contribute to the result. The sum $\frac{1}{N}\sum_{\vec{n}}(\cdots)$ vanishes unless

$$\vec{k}' - \vec{k} + \vec{K} = 0. \tag{2.149}$$

Thus we can represent Eq. (2.145) as

$$E_\alpha(\vec{k})C_{\alpha,\vec{k}} + \sum_{\alpha',k-\vec{K}} U_{\vec{K}} C_{\alpha',\vec{k}-\vec{K}} \Delta_{\vec{k}\vec{k}-\vec{K}}^{\alpha\alpha'} = EC_{\alpha,\vec{k}}, \tag{2.150}$$

where

$$\Delta_{\vec{k}\vec{k}-\vec{K}}^{\alpha\alpha'} = \frac{1}{V_0} \int_{V_0} d\vec{r} u_{\alpha,\vec{k}}^*(\vec{r}) u_{\alpha',\vec{k}-\vec{K}}(\vec{r}). \tag{2.151}$$

For small $\vec{K}$, which is realistic for our situation, we can make the approximation

$$\Delta_{\vec{k}\vec{k}-\vec{K}}^{\alpha\alpha'} \approx \Delta_{\vec{k}\vec{k}}^{\alpha\alpha'} = \delta_{\alpha\alpha'}. \tag{2.152}$$

The latter equality follows from Eq. (2.117). The physical meaning of the approximation of Eq. (2.152) is that different energy bands are entirely independent, i.e., the smooth potential $U(\vec{r})$ does not lead to interband transitions. Now we can simplify Eq. (2.150) to the form

$$E_\alpha(\vec{k})C_{\alpha,\vec{k}} + \sum_{\vec{K}} U_{\vec{K}} C_{\alpha,\vec{k}-\vec{K}} = EC_{\alpha,\vec{k}}. \tag{2.153}$$

Returning to the wave function $\psi(\vec{r})$ and taking in the series of Eq. (2.144) terms with fixed α, we find

$$\psi(\vec{r}) = \sum_{\vec{k}} C_{\alpha,\vec{k}} \psi_{\alpha,\vec{k}} = \frac{1}{\sqrt{V}} \sum_{\vec{k}} C_{\alpha,\vec{k}} e^{i\vec{k}\vec{r}} u_{\alpha,\vec{k}}(\vec{r}). \tag{2.154}$$

To calculate $\psi(\vec{r})$ we introduce the function

$$F_\alpha(\vec{r}) = \frac{1}{\sqrt{V}} \sum_{\vec{k}} e^{i\vec{k}\vec{r}} C_{\alpha,\vec{k}}. \tag{2.155}$$

From Eq. (2.153) we obtain

$$\sum_{\vec{k}} E_\alpha(\vec{k}) C_{\alpha,\vec{k}} e^{i\vec{k}\vec{r}} + \sum_{\vec{k},\vec{K}} U_{\vec{K}} C_{\alpha,\vec{k}-\vec{K}} e^{i\vec{k}\vec{r}} = E \sum_{\vec{k}} C_{\alpha,\vec{k}} e^{i\vec{k}\vec{r}}.$$

The first term can be transformed as follows:

$$\sum_{\vec{k}} E_\alpha(\vec{k}) C_{\alpha,\vec{k}} e^{i\vec{k}\vec{r}} \equiv \sum_{\vec{k}} E_\alpha(-i\nabla) C_{\alpha,\vec{k}} e^{i\vec{k}\vec{r}}$$

$$= E_\alpha(-i\nabla) \sum_{\vec{k}} C_{\alpha,\vec{k}} e^{i\vec{k}\vec{r}} = E_\alpha(-i\nabla)\sqrt{V} F_\alpha(\vec{r}),$$

where $E_\alpha(-i\nabla)$ is the energy dispersion with the wave vector $\vec{k}$ replaced with the operator $-i\nabla$. The second term is identical to

$$\sum_{\vec{K}} U_{\vec{K}} e^{i\vec{K}\vec{r}} \sum_{\vec{k}} e^{i(\vec{k}-\vec{K})\vec{r}} C_{\alpha,\vec{k}-\vec{K}}.$$

At fixed $\vec{K}$, one can show for the internal sum that

$$\sum_{\vec{k}} e^{i(\vec{k}-\vec{K})\vec{r}} C_{\alpha,\vec{k}-\vec{K}} = \sum_{\vec{k}-\vec{K}} e^{i(\vec{k}-\vec{K})\vec{r}} C_{\alpha,\vec{k}-\vec{K}} = \sqrt{V} F_\alpha(\vec{r}).$$

This result is independent of $\vec{K}$. Obviously the sum over $\vec{K}$ gives the potential $U(\vec{r})$. Therefore we get the equation for $F_\alpha(\vec{r})$:

$$[E_\alpha(-i\nabla) + U(\vec{r})] F_\alpha(\vec{r}) = E F_\alpha(\vec{r}). \tag{2.156}$$

In the effective-mass approximation, $E_\alpha(\vec{k}) = E(0) + \hbar^2 k^2/(2m^*)$, so instead of Eq. (2.156) we obtain

$$\left[-\frac{\hbar^2}{2m^*}\nabla^2 + U(\vec{r}) \right] F_\alpha(\vec{r}) = [E - E(0)] F_\alpha(\vec{r}). \tag{2.157}$$

The latter has the form of the Schrödinger equation for an electron with an effective mass m^* moving in the external potential $U(\vec{r})$. In the same approximation we can represent the wave function of Eq. (2.154) as

$$\psi(\vec{r}) \approx \frac{1}{\sqrt{V}} \sum_{\vec{k}} C_{\alpha,\vec{k}} e^{i\vec{k}\vec{r}} u_{\alpha,0}(\vec{r}) = F_\alpha(\vec{r}) u_{\alpha,0}(\vec{r}). \tag{2.158}$$

Thus we have found that in a slowly varying external potential the crystalline electron can be described in term of a slow modulation on the Bloch function $u_{\alpha,0}$. The function $F_\alpha(\vec{r})$ is called the envelope function and satisfies the Schrödinger equation with the effective mass used in Eq. (2.157). The normalization condition for $F_\alpha(\vec{r})$ is $\int d\vec{r} |F_\alpha(\vec{r})|^2 = 1$.

2.7 Quantum Electron Transport: Landauer Formula

We have seen from the previous sections that the Schrödinger equation is a sufficient theoretical tool for dealing with single-electron mechanics or mechanics of noninteracting electrons. It is obvious that a complete theory of electron kinetics in solids requires a broader spectrum of approaches than the cases in which the Schrödinger equation is reducible to a single-particle problem. We have discussed in Sections 2.2 and 2.3 several regimes of electron transport that belong to two major groups: quantum and classical transport. Now we consider more detailed theoretical approaches to both regimes of transport. We begin our analysis by considering the quantum-transport regime.

The general theoretical description of the various quantum-transport regimes is too complex a problem to be presented in this book. We consider the simplest limiting case of time-independent transport, in which inelastic processes are negligible. As was pointed out previously, transport through nanostructure devices depends on both the geometry of the nanostructure and the leads (electrodes, contacts, wires, interconnections, etc.), which connect the device to an external

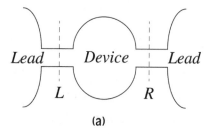

Lead | Device | Lead

L | R

(a)

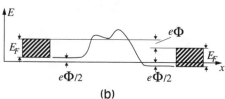

(b)

Figure 2.7. (a) Sketch of a nanostructure consisting of a nanodevice connected with leads, (b) energy diagram for an electron in the nanostructure. Potential profile in the region of the nanodevice is shown conditionally.

electric circuit. Thus we should consider the whole system: the device and the leads.

Figure 2.7 illustrates essential features of the simplified model. To avoid a detailed description of the leads, we assume that the leads are reservoirs of electrons in which energy- and momentum-relaxation processes are so effective that the electron system remains in equilibrium even under a given applied voltage bias. Hence the boundary condition at the interface between the lead and the device is assumed to be determined by the equilibrium Fermi distribution. The electron concentration in the leads is so high that the electrostatic potential in each lead is taken to be constant, as for the case of a metal. Let E_F be the Fermi energy of the electrons in the leads in the absence of a bias. When a voltage bias Φ is applied, the Fermi level in one of the leads becomes $E_F - e\Phi/2$, while that of the second lead becomes $E_F + e\Phi/2$, as depicted in Fig. 2.7(b). Thus the electron distribution functions are

$$\mathcal{F}_F\left(E - \frac{1}{2}e\Phi - E_F\right), \qquad \mathcal{F}_F\left(E + \frac{1}{2}e\Phi - E_F\right), \tag{2.159}$$

at the left and at the right leads, respectively. Here E is the kinetic energy of the electrons. We suppose that the system under consideration may be analyzed with sufficient accuracy by the method of separation of variables, i.e., the potential energy is taken to be

$$V(x, y, z) = V_1(x) + V_2(y, z).$$

According to the results of Section 2.4.1, in this case the wave function can be written as

$$\psi(x, y, z) = \psi_{\parallel}(x)\psi_T(y, z). \tag{2.160}$$

As far as the transverse solution is concerned, we denote the corresponding states by $\psi_T(y, z) = \psi_{n,m}(y, z)$ and their energies by $E_T = E_{n,m}$. Since electron transport occurs along the x direction, we must analyze the wave function $\psi_\parallel$. Let us write the boundary conditions for the wave function of the electrons $\psi_\parallel$ at the interfaces between the leads and the device. These interfaces are indicated in Fig. 2.7 by the L and the R cross sections. The wave function of the electrons incoming from the left lead is

$$\psi_{\parallel,l} = \begin{cases} e^{ik_l(x-x_l)} + r_l e^{-ik_l(x-x_l)} & x \sim x_l \\ t_r e^{ik_r(x-x_r)} & x \sim x_r \end{cases}, \tag{2.161}$$

that is, at the cross section L the wave function consists of incoming and reflected waves, while at the cross section R there is only the wave that has passed through the device. The coefficients t_r and r_l are amplitudes of these transmitted and reflected waves. These coefficients depend on the potential profile in the device, its geometry, etc. In a similar manner, one can write the wave function of the electrons incoming from the right lead as

$$\psi_{\parallel,r} = \begin{cases} t_l e^{-ik_l(x-x_l)} & x \sim x_l \\ e^{-ik_r(x-x_r)} + r_r e^{ik_r(x-x_r)} & x \sim x_r \end{cases}. \tag{2.162}$$

In Eqs. (2.161) and (2.162) values k_l and k_r correspond to the parallel components of wave vectors $k_\parallel$ in the left and the right cross sections, respectively; these vectors are, of course, the x-directed components of the wave vectors.

The distribution of external voltage across the device can also be taken into account; in this case t_r and r_l become dependent on Φ.

Among the coefficients t_r, r_l, t_l, and r_r there are several fundamental relationships, which do not depend on a specific design of the device. The requirement of the current continuity for both wave functions $\psi_{\parallel,l}$ and $\psi_{\parallel,r}$ gives

$$k_l(1 - |r_l|^2) = k_r|t_r|^2, \qquad k_r(1 - |r_r|^2) = k_l|t_l|^2. \tag{2.163}$$

Other relationships can be found as follows. Both functions $\psi_{\parallel,l}$ and $\psi_{\parallel,r}$ are solutions of the Schrödinger equation at the same energy E. The equation is real, and therefore $\psi_{\parallel,l}^*$ and $\psi_{\parallel,r}^*$ are also solutions. Since a differential equation of second order has only two independent solutions, one can represent the latter functions in terms of the former:

$$\psi_{\parallel,l}^* = A\psi_{\parallel,l} + B\psi_{\parallel,r}, \tag{2.164}$$

$$\psi_{\parallel,r}^* = C\psi_{\parallel,l} + D\psi_{\parallel,r}, \tag{2.165}$$

where A, B, C, and D are constants. Substituting Eqs. (2.161) and (2.162) into Eqs. (2.164) and (2.165) and matching the coefficients of the functions $\exp[\pm ik_l(x - x_l)]$ and $\exp[\pm ik_r(x - x_r)]$ for both of the modified versions of equations (2.164) and (2.165), we find eight equations, which facilitate the elimination of A, B, C, and D; we find four fundamental relationships:

$$1 = r_r r_r^* + t_r t_l^*, \qquad 1 = r_l r_l^* + t_l t_r^*, \tag{2.166}$$

$$r_l t_l^* + r_r^* t_l = 0, \qquad r_r t_r^* + r_l^* t_r = 0. \tag{2.167}$$

From these equations it follows that

$$|r_l|^2 = |r_r|^2, \qquad t_r^* t_l = t_r t_l^*. \tag{2.168}$$

Then, by using Eqs. (2.163) we get

$$k_r^2 |t_r|^2 = k_l^2 |t_l|^2. \tag{2.169}$$

Substituting the wave functions of Eqs. (2.161) and (2.162) into Eq. (2.38), we calculate both the incoming i_{in} and outgoing i_{out} electron flows:

$$i_{\text{in}} = v_l, \qquad i_{\text{out}} = v_r |t_r|^2, \tag{2.170}$$

where $v_l = \hbar k_l / m^*$ and $v_r = \hbar k_r / m^*$ are the velocities at the L and the R cross-sections, respectively. The ratio of these quantities defines the transmission coefficient $T_{l \to r}(E)$ through the device:

$$T_{l \to r}(E) = \frac{i_{\text{out}}}{i_{\text{in}}} = \frac{k_r}{k_l} |t_r|^2. \tag{2.171}$$

The transmission coefficient corresponding to the wave function $\psi_{\parallel, r}$ is

$$T_{r \to l}(E) = \frac{k_l}{k_r} |t_l|^2. \tag{2.172}$$

From the above relationships between the coefficients t_r and t_l, one can conclude that

$$T_{l \to r}(E) = T_{r \to l}(E) = T(E_{\parallel}), \tag{2.173}$$

where $E_{\parallel} = \hbar^2 k_{\parallel}^2 / (2m^*)$ is the kinetic energy corresponding to the longitudinal component of the electron's momentum.

Thus the transmission coefficients are the same for both directions of incoming electrons. The ratio of reflected and incoming electron flows defines the reflection coefficient:

$$R(E) = \frac{i_r}{i_{\text{in}}} = |r_l|^2 = |r_r|^2. \tag{2.174}$$

It is obvious that

$$T(E) + R(E) = 1.$$

Now we can take into account contributions to the total current from all electrons entering the device from both leads. Consider a state of the electrons, say, in the left lead with quantum numbers $k_{\parallel}$, n, and m. The number of electrons in this state is given by $2\mathcal{F}_F[E(k_{\parallel}, n, m) + \frac{1}{2}e\Phi - E_F]$, where the factor 2 comes from spin degeneracy. If the connection length is L_c, the number of electrons per unit length of the connection is $2\mathcal{F}/L_c$. The total contribution to the electric current from the electrons entering from the left is

$$I_l = -\frac{2e}{L_c} \sum_{n,m} \sum_{k_\parallel > 0} v_\parallel T(E_\parallel) \mathcal{F}_F \left[E(k_\parallel, n, m) - \frac{1}{2}e\Phi - E_F \right]. \qquad (2.175)$$

Note that I_l is a part of the total electric current, while Eq. (2.38) defines the electron flow density i and the current density $j = I/S$, $j = -ei$. Similarly, for the electrons from the right lead one gets

$$I_r = \frac{2e}{L_c} \sum_{n,m} \sum_{k_\parallel < 0} v_\parallel T(E_\parallel) \mathcal{F}_F \left[E(k_\parallel, n, m) + \frac{1}{2}e\Phi - E_F \right]. \qquad (2.176)$$

Therefore the total current flowing through the device is

$$I = I_l + I_r = \frac{2e}{L_c} \sum_{n,m} \sum_{k_\parallel > 0} v_\parallel T(E_\parallel) \left\{ \mathcal{F}_F \left[E(k_\parallel, n, m) + \frac{1}{2}e\Phi - E_F \right] \right.$$
$$\left. - \mathcal{F}_F \left[E(k_\parallel, n, m) - \frac{1}{2}e\Phi - E_F \right] \right\}. \qquad (2.177)$$

Since the electron velocity $v_\parallel$ and the transmission T are independent of the transverse quantum numbers n and m, we can calculate the sum over n and m. When the explicit form of the Fermi function is taken into account, it is convenient to introduce the function

$$\mathcal{F}(E) = 2 \sum_{n,m} \frac{1}{1 + e^{\frac{E + E_{n,m}}{k_B T}}}. \qquad (2.178)$$

Next, the summation over $k_\parallel$ in Eq. (2.177) may be replaced by an integration:

$$\sum_{k_\parallel} \{\cdots\} \rightarrow L_c \int \frac{dk_\parallel}{2\pi} \{\cdots\} = L_c \int \frac{dE_\parallel}{2\pi \hbar v_\parallel} \{\cdots\}. \qquad (2.179)$$

Finally, we obtain the following expression for the total current:

$$I = e \int \frac{dE_\parallel}{2\pi \hbar} T(E_\parallel) \left[\mathcal{F}\left(E_\parallel + \frac{1}{2}e\Phi - E_F \right) - \mathcal{F}\left(E_\parallel - \frac{1}{2}e\Phi - E_F \right) \right], \qquad (2.180)$$

where the integration range runs over the kinetic energy of the longitudinal motion $E_\parallel$. Note that the electron velocity does not appear in the final expression for the total current.

This general result may be applied to a variety of different cases. Let us consider two of them.

Device macroscopically large in the transverse directions: In this case the transverse quantum numbers are the wave vectors: $n = k_y$, $m = k_z$, and $E_{n,m} = \hbar^2(k_y^2 + k_z^2)/(2m^*)$. The function of Eq. (2.178) can be calculated explicitly:

$$\mathcal{F}(E) = S\frac{m^* k_B T}{\pi \hbar^2} \ln\left(1 + e^{\frac{-E}{k_B T}}\right), \tag{2.181}$$

where S is the cross-sectional area. The $\mathcal{F}(E)$ has the meaning of the number of electrons with energy E. Then the total current density $j = I/S$ is

$$j = -e\frac{m^* k_B T}{\pi \hbar^2} \int \frac{dE_\parallel}{2\pi\hbar} T(E_\parallel) \ln\left(\frac{1 + e^{\frac{E_F - E_\parallel + e\Phi/2}{k_B T}}}{1 + e^{\frac{E_F - E_\parallel - e\Phi/2}{k_B T}}}\right). \tag{2.182}$$

This is indeed a useful result since Eq. (2.182) allows one to calculate the current-voltage characteristics of a nanostructure device and its dependence on the electron concentration, temperature, etc.

Device conductance at low temperatures: Landauer formula. Let us turn to the general result of Eq. (2.177). It can be simplified significantly for near-equilibrium transport at low temperatures. We know that in the limit of zero temperature the Fermi distribution becomes a step function:

$$\lim_{T\to 0} \mathcal{F}_F(E - E_F) = \Theta(E_F - E).$$

If the applied voltage Φ is small, the difference in distribution functions in the sum of Eq. (2.177) is

$$\mathcal{F}_F\left[E(k_\parallel, n, m) + \frac{1}{2}e\Phi - E_F\right] - \mathcal{F}_F\left[E(k_\parallel, n, m) - \frac{1}{2}e\Phi - E_F\right]$$
$$= -e\Phi\delta(E_F - E), \tag{2.183}$$

where we have taken into account the fact that the derivative of the steplike Fermi distribution function is a δ function. Hence the current is proportional to the voltage bias. One can introduce a conductance of a nanostructure device:

$$G = \frac{I}{\Phi}. \tag{2.184}$$

From Eqs. (2.177), (2.179), (2.183), and (2.184) we obtain the conductance at low temperatures in the form

$$G = \frac{e^2}{h} \sum_{n,m,s} T(E_F, n, m) = 2\frac{e^2}{h} \sum_{n,m} T(E_F, n, m), \tag{2.185}$$

where the sum extends only over electron states (n, m) with energy $E < E_F$. The coefficient in front of the sum of Eq. (2.185),

$$G_0 = \frac{e^2}{h}, \tag{2.186}$$

is called the quantum of conductance. The quantum of conductance is equal to 39.6 μS; its inverse value is equal to 25.2 kΩ.

Within the approximation of separable variables [Eq. (2.160)] electrons with fixed spin propagate through the device, conserving the pair of transverse quantum numbers (n, m). Sometimes it is convenient to consider electron states that correspond to different quantum numbers (n, m) and spins in terms of separate electron-conduction channels. In this latter formulation Eq. (2.185) may be rewritten in the form

$$G = 2G_0 \sum_{n,m} T_{n,m}; \qquad T_{n,m} = T(E_F, n, m). \qquad (2.187)$$

Here each channel (n, m) contributes the value $G_0 T_{n,m}$ to the conductance. If the channel corresponding to the (n, m) state is transparent to electrons, $T_{n,m} = 1$, and the contribution of this channel is equal to the quantum of conductance G_0. We will see in due course that Eq. (2.187) explains the quantization of the conductance in quantum structures at low temperatures.

In conclusion, the Landauer formula describes quantum transport in mesoscopic devices. It is valid at low temperatures and small voltage biases.

2.8 Boltzmann Transport Equation

In Section 2.7 we overviewed some approaches to the quantum transport of electrons in nanostructures. It must be noted that there are a great number of other quantum-transport theories, which remain beyond the scope of this textbook. As for classical electron transport, there is no such diversity of approaches and we will deal with the most general of them.

We have established in Subsection 2.3.2 that electron transport is classical in nature if the time between two subsequent scattering events is much larger than the duration of the scattering event and if the size of the system in the direction of transport is much larger than the typical de Broglie wavelength of electrons in this system. In this case the electron dynamics can be separated into classical electron motion in an external smoothly varying field and instantaneous electron scattering. The scattering is usually considered within Fermi's golden rule approximation, which was introduced in Subsection 2.4.4. Keeping in mind the preceding analysis of the length and time scales, one can derive the transport equation that describes electron behavior for the following length L and time T scales:

$$\lambda, \, l_\varphi \ll L, \qquad (2.188)$$

$$\frac{\hbar}{E} \ll T. \qquad (2.189)$$

When dealing with low-dimensional semiconductor structures one has to consider the fact that at least in directions of electron confinement the electron spectrum is quantized and quantum mechanics must be considered for modeling transport in these directions even if the transport in the unconfined dimensions is describable in terms of classical physics. In other words, the longitudinal motion of the electron remains classical but its transverse motion is quantum mechanical when there is significant dimensional confinement in the transverse directions. Electron transport in long nanostructures is sometimes referred to as semiclassical electron transport.

From the theory of the bulk materials it is well known that the most useful and sufficiently rigorous approach to the classical transport problem is based on the Boltzmann transport equation for the distribution function of electrons, $\mathcal{F}(\vec{k}, \vec{r}, t)$. The meaning of the distribution function is as follows: the quantity

$$\mathcal{F}(\vec{k}, \vec{r}, t) \, d\vec{k} \, d\vec{r} \qquad (2.190)$$

characterizes the probability of finding an electron in a wave-vector range $\vec{k}$ to $\vec{k} + d\vec{k}$ and in a range $\vec{r}$ to $\vec{r} + d\vec{r}$ around the spatial location $\vec{r}$.

Under equilibrium conditions the function $\mathcal{F}(\vec{k}, \vec{r}, t)$ coincides with the Fermi distribution. There are equal numbers of the particles with oppositely directed wave vectors [$\mathcal{F}(\vec{k}, \vec{r}, t)$ is an even function of $\vec{k}$] and the net electron current is zero. Hence, for transport induced by an electric field $\vec{F}$, the function $\mathcal{F}(\vec{k}, \vec{r}, t)$ has to be modified by the applied field to become asymmetric with respect to $\vec{k} = 0$. (The same is true for any other transport process: diffusion, thermal transport, etc.) The change in $\mathcal{F}(\vec{k}, \vec{r}, t)$ caused by an external field $d\mathcal{F}/dt$ is balanced by scattering processes. Introducing the rate of change of $\mathcal{F}(\vec{k}, \vec{r}, t)$ that is due to scattering as

$$\left[\frac{d\mathcal{F}(\vec{k}, \vec{r}, t)}{dt} \right]_{sc},$$

one can write

$$\frac{d}{dt} \mathcal{F}(\vec{k}, \vec{r}, t) = \left[\frac{d\mathcal{F}(\vec{k}, \vec{r}, t)}{dt} \right]_{sc}.$$

Calculation of the derivative with respect to time gives

$$\frac{\partial \mathcal{F}(\vec{k}, \vec{r}, t)}{\partial t} + \vec{v} \nabla_{\vec{r}} \mathcal{F}(\vec{k}, \vec{r}, t) - \frac{e}{\hbar} \vec{F} \nabla_{\vec{k}} \mathcal{F}(\vec{k}, \vec{r}, t) = \left[\frac{\partial \mathcal{F}(\vec{k}, \vec{r}, t)}{\partial t} \right]_{sc}, \qquad (2.191)$$

where $\vec{v} = d\vec{r}/dt$ is the electron group velocity and $-e\vec{F} = \hbar d\vec{k}/dt$ is the force on the electron caused by an electric field $\vec{F}$. Note that in this section we use two differential operators: $\nabla_{\vec{r}} \equiv \partial/\partial\vec{r}$ and $\nabla_{\vec{k}} \equiv \partial/\partial\vec{k}$.

The Boltzmann transport equation (2.191) is actually the rate equation for a distribution function. Without specification of the scattering term (collision integral), $(\partial \mathcal{F}/\partial t)_{sc}$, it is clear that there are two different types of scattering: one type involves single-electron scattering on impurities, defects, lattice vibrations, etc., and the other type involves at least two electron interactions such as electron–electron, electron–hole scattering, impact ionization, etc. In the case of single-electron scattering, the collision integral is a linear functional of $\mathcal{F}$:

$$\left(\frac{\partial \mathcal{F}}{\partial t}\right)_{sc} = I_{e-l}\{\mathcal{F}\}, \tag{2.192}$$

while in the case of two-electron processes the scattering term is a quadratic functional of $\mathcal{F}$:

$$\left(\frac{\partial \mathcal{F}}{\partial t}\right)_{sc} = I_{e-e}\{\mathcal{F}, \mathcal{F}\}. \tag{2.193}$$

Unless specifically mentioned, electron–electron scattering is neglected in the remainder of this book, and we deal with collision integrals in the form of Eq. (2.192). Let $W(\vec{k}', \vec{k})$ be the probability of a transition from the electron state with wave vector $\vec{k}'$ to the state with the wave vector $\vec{k}$ per unit time. Then, taking into consideration all possible scattering processes leading to the population and depopulation of the state $\vec{k}$ in the vicinity of space point $\vec{r}$, one can obtain the Boltzmann equation in the form

$$\frac{\partial \mathcal{F}(\vec{k}, \vec{r}, t)}{\partial t} + \vec{v}\nabla_{\vec{r}}\mathcal{F}(\vec{k}, \vec{r}, t) - \frac{e}{\hbar}\vec{F}\nabla_{\vec{k}}\mathcal{F}(\vec{k}, \vec{r}, t)$$
$$= \sum_{\vec{k}'}[W(\vec{k}', \vec{k})\mathcal{F}(\vec{k}', \vec{r}, t) - W(\vec{k}, \vec{k}')\mathcal{F}(\vec{k}, \vec{r}, t)]. \tag{2.194}$$

We assume conservation of the electron spin during collisions. Equation (2.194) is written for nondegenerate statistics when $\mathcal{F} \ll 1$.

For an elastic scattering and the equilibrium case it follows from the statistical principle of detailed balance that

$$W(\vec{k}', \vec{k}) = W(\vec{k}, \vec{k}'). \tag{2.195}$$

The solution of the Boltzmann transport equation is the central problem of classical physical kinetics. Indeed, the concept of the distribution function is critical since it provides all the necessary information about the electron ensemble. Any transport parameter and characteristics can be expressed in terms of the distribution function. For example, the electric current density can be written as

$$\vec{j}(\vec{r}, t) = -2\frac{e}{V}\sum_{\vec{k}}\vec{v}\mathcal{F}(\vec{k}, \vec{r}, t), \tag{2.196}$$

where V is the volume of the conductive structure. The factor 2 in Eq. (2.196) accounts for the spin degeneracy of the electrons.

Because of the integrodifferential nature of the Boltzmann transport equation and complex probabilities of scattering, the solution of this equation is quite complicated. We review some simplified methods of solving the Boltzmann transport equation in Section 2.9. However, for most practical cases the Boltzmann transport equation can be solved only numerically.

An important group of problems deals with stationary and homogeneous cases in which the distribution function does not depend on time and the position vector. In this case the distribution function depends only on $\vec{k}$ [$\mathcal{F} = \mathcal{F}(\vec{k})$], the first two terms can be dropped from the left-hand part of the Boltzmann transport equation (2.194), and the resultant equation is commonly known as the stationary and homogeneous Boltzmann equation.

Since we are dealing primarily with low-dimensional structures, we pay more attention to the Boltzmann transport equation for low-dimensional electron systems. Let us generalize the distribution function for low-dimensional systems. When electron transport along the quantum-well layers or quantum wires – so-called parallel transport – can be described semiclassically, it is possible to introduce the distribution function $\mathcal{F}_n(\vec{k})$, where a quantum number n is the label for a two- or a one-dimensional subband (which will be introduced in Chapter 3), and $\vec{k}$ is a two- or one-dimensional wave vector. Let us consider a two-dimensional electron gas and introduce a two-dimensional vector in the plane of the two-dimensional electron gas, $\vec{r} = z, \vec{\rho}$. Then $\mathcal{F}_n(\vec{k}, \vec{\rho})$ is the density of electrons with fixed $\vec{k}, \vec{\rho}$ and subband number n per unit area. The forces acting on low-dimensional electrons are derived in Subsection 3.2.2. The scattering probabilities are, of course, dependent on quantum numbers of the initial and the final subbands n and m. Thus for a stationary problem and a homogeneous electron gas in the x, y plane, we obtain the equation analogous to Eq. (2.194):

$$-\frac{e}{\hbar}\vec{F}\vec{\nabla}_{\vec{k}}\mathcal{F}_n(\vec{k}) = \sum_{\vec{k}',m}[W(m, \vec{k}', n, \vec{k})\mathcal{F}_m(k') - W(n, \vec{k}, m, \vec{k}')\mathcal{F}_n(k)]. \qquad (2.197)$$

This system of equations for $\mathcal{F}_n(\vec{k})$ could be solved either numerically or analytically by use of approximations that are introduced in Section 2.9 for bulk semiconductors.

2.9 Local Approximations for Electron Transport

In this section we return to the general form of the Boltzmann equation (2.191) and we simplify it to a form that is often used for an evaluation of electron transport in devices. We discuss physical cases for which one can use the so-called local approximations for electron transport when the basic characteristics such as the electron current, electron energy flux, etc., can be represented in terms of the electron concentration, effective temperature, and their derivatives with respect to coordinates.

2.9.1 Electron Mobility and Diffusion Coefficient

We can always represent the distribution function as a sum of symmetric and asymmetric parts:

$$\mathcal{F}(\vec{k}, \vec{r}) = \mathcal{F}_s(\vec{k}, \vec{r}) + \mathcal{F}_a(\vec{k}, \vec{r});$$
$$\mathcal{F}_s(\vec{k}, \vec{r}) = \mathcal{F}_s(-\vec{k}, \vec{r}); \qquad \mathcal{F}_a(\vec{k}, \vec{r}) = -\mathcal{F}_a(-\vec{k}, \vec{r}). \qquad (2.198)$$

Substituting $\mathcal{F}$ of the form of Eqs. (2.198) into Eq. (2.196), we see that the summation containing the symmetric part $\mathcal{F}_s$ gives zero and the electric current of Eq. (2.196) is determined by the asymmetric part:

$$\vec{j}(\vec{r}) = -2\frac{e}{V}\sum_{\vec{k}}\vec{v}\mathcal{F}_a(\vec{k},\vec{r}).$$ (2.199)

(Remember that, throughout this book, we take the electron charge to equal $-e$, $e > 0$.)

In Chapter 7 we will see that there are physical situations in which the symmetric and asymmetric parts of the distribution function may be of the same order of magnitude. However, we restrict our consideration in this section to the case in which

$$|\mathcal{F}_s| \gg |\mathcal{F}_a|.$$ (2.200)

Equation (2.191) may be easily separated into two equations[†]: one is symmetric in $\vec{k}$ and the other is asymmetric in $\vec{k}$. We can linearize the nonlinear collision integral in the asymmetric equation with respect to $\mathcal{F}_a$:

$$I_{sc}\{\mathcal{F}_a\} \approx -\frac{\mathcal{F}_a}{\tau_p},$$ (2.201)

where τ_p is a function of electron energy and is called the momentum-relaxation time; for more details see Chapter 7. Now the antisymmetric equation can be rewritten as

$$\mathcal{F}_a = -\tau_p\left(\vec{v}\nabla_{\vec{r}}\mathcal{F}_s - \frac{e}{\hbar}\vec{F}\nabla_{\vec{k}}\mathcal{F}_s\right).$$ (2.202)

The two terms on the right-hand side of Eq. (2.202) give two different contributions to the electric current density:

$$\vec{j} = \vec{j}_1 + \vec{j}_2,$$ (2.203)

where

$$\vec{j}_1 = -\frac{2e^2}{\hbar V}\sum_{\vec{k}}\vec{v}(\vec{F}\nabla_{\vec{k}}\mathcal{F}_s)\tau_p$$ (2.204)

is proportional to the electric field and is called the drift current, and

$$\vec{j}_2 = \frac{2e^2}{V}\sum_{\vec{k}}\vec{v}(\vec{v}\nabla_{\vec{r}}\mathcal{F}_s)\tau_p$$ (2.205)

is associated with the nonuniformity of the electron distribution and is called the diffusion current.

[†] We need to do this in more detail, writing both symmetric and antisymmetric equations.

If we assume that τ_p is a function of the electron energy E and that the electric field is applied along the z axis, after the summation is replaced by integration, as in Eq. (2.80), we can find the drift current density:

$$j_{1z} = -e^2 \frac{2\sqrt{2m^*}}{\pi^2 \hbar^3} F_z \int_0^\infty dE \; E^{3/2} \tau_p(E) \frac{\partial \mathcal{F}_s}{\partial E}. \tag{2.206}$$

The electron concentration is given by Eq. (2.78):

$$n(\vec{r}) = \frac{2}{(2\pi)^3} \int d\vec{k} \; \mathcal{F}_s(\vec{k}, \vec{r}) = \frac{m^*\sqrt{2m^*}}{\pi^2 \hbar^3} \int_0^\infty dE \; E^{1/2} \mathcal{F}_s. \tag{2.207}$$

Hence the current density of Eq. (2.206) can be rewritten as

$$j_{1z} = e\mu n_e F_z, \tag{2.208}$$

where we have introduced the electron mobility, defined as

$$\mu \equiv \frac{e\langle \tau_p \rangle}{m^*}, \qquad \langle \tau_p \rangle = -\frac{2}{3} \frac{\int_0^\infty dE \; E^{3/2} \tau_p(E) \frac{\partial \mathcal{F}_s}{\partial E}}{\int_0^\infty dE \; E^{1/2} \mathcal{F}_s(E)}. \tag{2.209}$$

Since in the general case $\mathcal{F}_s$ depends on the electric field, so does μ. Therefore the drift velocity $v_d = \mu(F)F$ is in general a nonlinear function of the electric field.

In the same way, the second contribution to the current can be written as

$$\vec{j}_2 = e\nabla_{\vec{r}}(D\,n_e), \tag{2.210}$$

where D is the diffusion coefficient defined as

$$D = \frac{2}{3} \frac{1}{m^*} \frac{\int_0^\infty dE \; E^{3/2} \tau_p(E)\mathcal{F}_s(E)}{\int_0^\infty dE \; E^{1/2} \mathcal{F}_s(E)}. \tag{2.211}$$

It is worth emphasizing that the symmetric function $\mathcal{F}_s$ is still not specified. It can depend on the field. Thus the forms of Eqs. (2.211) and (2.209) suggest that in general μ and D depend on the electric field by means of $\mathcal{F}_s$.

For weak electric fields, at nearly equilibrium conditions, when $\mathcal{F}_s$ can be chosen equal to the equilibrium Boltzmann distribution function of Eq. (2.79), from Eqs. (2.209) and (2.211) it follows that

$$\frac{D}{\mu} = \frac{k_B T}{e}. \tag{2.212}$$

This is the well-known Einstein relation between the mobility and the diffusion coefficient of a nondegenerate electron gas.

Finally, we obtain the electric current density in the form

$$\vec{j} = e\mu n_e \vec{F} + e\nabla_{\vec{r}}(D\,n_e). \tag{2.213}$$

These results have been obtained for a three-dimensional electron gas. It can be easily shown that the same equation (2.213) is valid for two-dimensional electrons when the current is in the plane and for one-dimensional electrons when the current is along the wire axis. Care must be taken in using the expressions for μ and D in low-dimensional cases, since the expressions we have derived contain the density of states.

2.9.2 Electron Temperature (Hydrodynamic) Approximation

Further calculations of the symmetric function $\mathcal{F}_s$, mobility μ, and diffusion coefficient D are possible only when specific scattering mechanisms are defined. Such cases will be analyzed in Chapter 7 in more detail. Here we briefly consider a common approximation that defines the shape of the symmetric part of the distribution function. It is called the electron-temperature approximation.

Let us evaluate the collision terms as

$$I_{e-l} = \frac{\mathcal{F}}{\tau_{ep}}, \qquad I_{e-e} = \frac{\mathcal{F}}{\tau_{ee}}, \tag{2.214}$$

where τ_{ep} is the time of the electron interaction with the lattice and τ_{ee} is the scattering time for the electron–electron interaction. If the electron concentration is high, so that

$$\tau_{ep} \gg \tau_{ee}, \tag{2.215}$$

one may neglect all terms in the transport equation and, as a first approximation, find

$$I_{e-e}\{\mathcal{F}, \mathcal{F}\} = 0. \tag{2.216}$$

To analyze Eq. (2.216) we note that during a two-electron interaction without participation of the foreign body, the momentum conservation implies that

$$\vec{k}_1 + \vec{k}_2 = \vec{k}'_1 + \vec{k}'_2.$$

Thus the probability of an electron–electron scattering event W satisfies the proportionality

$$W(\vec{k}_1, \vec{k}_2, \vec{k}'_1, \vec{k}'_2) \propto \delta(\vec{k}_1 + \vec{k}_2 - \vec{k}'_1 - \vec{k}'_2). \tag{2.217}$$

The specific form of the probability is not important in this case. One can represent Eq. (2.216) for an electron with an arbitrary wave vector $\vec{k}_1$ as

$$I_{e-e}\{\mathcal{F}, \mathcal{F}\} \equiv - \sum_{\vec{k}_2, \vec{k}'_1, \vec{k}'_2} W[\mathcal{F}(\vec{k}_1)\mathcal{F}(\vec{k}_2) - \mathcal{F}(\vec{k}'_1)\mathcal{F}(\vec{k}'_2)] = 0, \tag{2.218}$$

where we have assumed a nondegenerate electron gas. Equation (2.218) is a nonlinear integral equation. We can find the solution of this equation by keeping in mind that under equilibrium conditions the Boltzmann distribution function satisfies this equation. Generalizing, we use a Boltzmann-like distribution for the symmetric part of the distribution function:

$$\mathcal{F}_s(k) = n_e e^{-\hbar^2 k^2 / (2m^* k_B T_e)}.\tag{2.219}$$

The physical meanings of T_e and n_e are quite transparent: they are the electron temperature and the concentration, respectively. These quantities can be obtained by the solution of the Boltzmann transport equation. The length scale L required for applying this method is

$$L \gg \bar{v}\tau_{ee},\tag{2.220}$$

where

$$\bar{v} = \sqrt{\frac{3k_B T_e}{m^*}}.$$

A straightforward method for deriving the equations for n_e and T_e from the Boltzmann transport equation is the following. Let us take a sum (integral) over all possible k in the Boltzmann transport equation (2.191). The first term gives the time derivative of n_e. The second term gives the divergence of the total current density, div$\vec{j}$. The third term equals zero because of the boundary conditions $[\mathcal{F}(\vec{k}) \to 0$ for $|\vec{k}| \to \infty]$, and the summation of all possible scattering processes also yields zero as a result of the detailed balance principle. Accordingly, we obtain the result

$$-e\frac{\partial n_e}{\partial t} + \mathrm{div}\,\vec{j} = 0.\tag{2.221}$$

Equation (2.221) is called the continuity equation. For the current density j one should use Eq. (2.213) with the mobility and diffusion coefficient dependent on the electron temperature T_e.

Now if we multiply the Boltzmann equation (2.191) by the electron energy $\hbar^2 k^2 / 2m^*$ and perform the sum over all k we can find the balance equation for the electron temperature:

$$\frac{v}{2}\frac{\partial}{\partial t}(n_e k_B T_e) + \mathrm{div}\,\vec{Q} - \vec{j}\vec{F} + n_e P(T_e) = 0.\tag{2.222}$$

Here v is the dimensionality of the electron gas, $v = 3, 2,$ and 1 for the three-, two-, and one-dimensional cases, respectively, $v k_B T_e / 2$ is the average energy of the electron gas, $\vec{j}\vec{F}$ is the Joule power transferred to the electron system, and $\vec{Q}$ is the electron-energy flux:

$$\vec{Q} = -\frac{1}{e}\theta\vec{j} - \chi(T_e)\,\nabla T_e.\tag{2.223}$$

The first term in Eq. (2.223) is the convective flux, where θ is the energy transferred by one electron, and the second term is the heat conduction flux. The last term in Eq. (2.222),

$$n_e P \equiv 2 \sum_k \frac{\hbar^2 k^2}{2m^*} I_{sc}\{\mathcal{F}\} \tag{2.224}$$

is in fact the rate of electron-energy dissipation. It is obvious that only inelastic-scattering mechanisms contribute to the energy dissipation. Within the simplest approximation,

$$P(T_e) = \frac{T_e - T}{\tau_E}, \tag{2.225}$$

where τ_E is the energy relaxation time. In Chapter 6 we shall discuss the value of τ_E.

To close the system of equations comprising the so-called hydrodynamic approximation, one must add the equations describing the self-consistent electric field:

$$\text{div}\,\vec{F} = -\frac{4\pi e}{\kappa}(n_e - N_D), \qquad \text{rot}\,\vec{F} = 0. \tag{2.226}$$

The second equation of Eqs. (2.226) states that the electric field is a potential field. Note that when Poisson's equations (2.226) are used for time-dependent problems, the continuity equation can be transformed to a form expressing the conservation of the total electric current:

$$\text{div}\left(\frac{\kappa}{4\pi}\frac{\partial \vec{F}}{\partial t} + \vec{j}\right) = 0, \tag{2.227}$$

where the first contribution is the so-called displacement current, while the second is the electron current defined by Eq. (2.213). If proper boundary conditions for electron temperature, concentration, and current density are specified, the system of differential equations (2.221), (2.222), and (2.226) describes electronic processes in devices of sizes conforming with inequality (2.220).

The above-discussed electron-temperature approximation can be applied to low-dimensional electron systems as well. The generic form we have used accommodates any dimensionality.

It is important to note that the above considerations are based on the assumption of intensive electron–electron scattering, which is responsible for the formation of the electron distribution function (2.219). For a two-dimensional electron gas this assumption is as good as in the three-dimensional case, but there is a problem with a one-dimensional electron gas. As follows directly from the identities for momentum and electron-number conservation, there is no momentum transfer during a two-electron collision in a one-dimensional system. The electrons simply exchange their momenta. To make this approximation work it is necessary that three and more electron collisions be included in the analysis of the one-dimensional system.

2.9.3 Drift-Diffusion Model

The results derived thus far in this chapter comprise the basis of the most applicable drift-diffusion approximation for modeling conventional semiconductor devices operating in the classical regimes of the electron transport; see Table 2.1. Collecting Eq. (2.213) for the electric current, the continuity equation (2.221), and the Poisson equations (2.226), we get the basic system of self-consistent equations for this modeling:

$$\vec{j} = e\mu(\vec{F})n_e\vec{F} + e\nabla_{\vec{r}}[D(\vec{F})\,n_e],$$

$$-e\frac{\partial n_e}{\partial t} + \text{div}\,\vec{j} = 0,$$

$$\text{div}\,\vec{F} = -\frac{4\pi e}{\kappa}(n_e - N_D). \tag{2.228}$$

Here the kinetic coefficients μ and D are regarded as functions of the local electric field $\vec{F}(\vec{r})$. These functions are to be taken either from a microscopic calculation of the distribution function or from simple and reasonable approximations that reflect the main features of the electrons in high electric fields; such features include saturation electron velocity and velocity overshoot. The proper boundary conditions for the electron concentration n_e (or the current) and the electric field $\vec{F}$ (or the electrostatic potential) are imposed at the contacts and boundaries of a device.

2.10 Closing Remarks

In this chapter we have discussed the basic physical concepts necessary to study nanoscale devices. The goal of this treatment is not to give a comprehensive review of these concepts but to give readers an idea of the major quantum and classical regimes of heterostructure devices as well as the conditions for crossover between the different regimes of electron transport.

By analyzing characteristic lengths and time scales of physical processes and comparing them with device dimensions and frequencies of external fields interacting with devices, we have shown how one can choose an adequate approach for describing a particular nanoscale device. We have presented different theoretical approaches for describing and modeling nanoscale devices. We exploit all these approaches in subsequent chapters.

In this chapter we have focused on the behavior of electrons in nanoscale devices; much of our emphasis has been on quantum-mechanical behavior. However, the quantization of both lattice vibrations (phonons) and the light (photons) in nanostructures can be important in the operation of these devices. Many of the basic quantum-mechanical concepts introduced in this chapter are applicable for phonons and photons, and they are studied in subsequent Chapters 5 and 10.

For those who want to explore the basic principles of quantum physics in more depth we recommend the following textbooks:

L. Schiff, *Quantum Mechanics* (McGraw-Hill, New York, 1968).

D. Saxon, *Elementary Quantum Mechanics* (Holden-Day, San Francisco, 1968).

R. P. Feynman, R. B. Leighton, and M. Sands, *Lectures on Physics* (Addison-Wesley, New York, 1964), Vol. 3.

Additional information on basic solid-state physics can be found in

C. Kittel, *Introduction in Solid-State Physics* (Wiley, New York, London, 1986).

I. Ipatova and V. Mitin, *Introduction to Solid-State Electronics* (Addison-Wesley, New York, 1996).

J. M. Ziman, *Principles of the Theory of Solids* (Cambridge U. Press, London, 1972).

Detailed discussions of different applications of the Boltzmann transport equation to semiconductors and methods of device modeling can be found in the following references:

K. Hess, *Advanced Theory of Semiconductor Devices* (Prentice-Hall, Englewood Cliffs, NJ, 1988).

C. M. Snowden and R. E. Miles, eds., *Compound Semiconductor Device Modeling* (Springer-Verlag, Berlin, 1989).

K. Tomizawa, *Numerical Simulation of Semiconductor Devices* (Artech House, Boston, 1992).

For much more information on electron quantum transport and its applications to the device physics the reader can consult the following references:

R. Landauer, "Electrical resistance of disordered one-dimensional lattices," Philos. Mag. **21**, pp. 863–867 (1970).

D. K. Ferry, *Semiconductors* (Macmillan, New York, 1991).

H. Grubin, D. K. Ferry, and C. Jacoboni, eds., *The Physics of Submicron Semiconductor Devices* (Plenum, New York, 1989).

C. Jacoboni, L. Reggiani, and D. K. Ferry, eds., *Quantum Transport in Semiconductors* (Plenum, New York, 1992).

M. A. Stroscio, "Quantum mechanical corrections to classical transport in submicron/ultrasubmicron dimensions," in *Introduction to Semiconductor Technology*, Cheng T. Wang, ed. (Wiley, New York, 1990), pp. 551–593.

PROBLEMS

1. Derive the amplitude reflection coefficient r and amplitude transmission coefficient t for a particle of wave number k incident upon a δ-function potential energy $V(x) = A\delta(x)$. Plot $R = |r|^2$ for both positive and negative values of A as a function of $\hbar^2 k^2/(2m|A|)$.

2. For the nth state of a quantum-mechanical harmonic oscillator, it is known that the uncertainty relation takes the form $\Delta x \Delta p \approx (n + 1/2)\hbar$. Is this consistent with the Heisenberg uncertainty relation?

3. By application of the uncertainty relation, estimate the binding energy of a particle of mass m^* in a one-dimensional quantum well of depth V and width d for very small values of the dimensionless parameter $m^* V d^2/\hbar^2$.

4. For an electron of effective mass m^* in the one-dimensional potential $V(x) = V d^2 \delta(x^2 - d^2)$ plot the relationship between V and d so there are just barely two bound states for the case in which $V < 0$.

5. For initial and final wave functions of the form $\Psi_i = [(1/\sqrt{V})e^{i\vec{k}\vec{r}}]$ and $\Psi_f = [(1/\sqrt{V})e^{i\vec{k}\vec{r}}]$, respectively, calculate the matrix element of the potential energy of interaction: (a) $V(\vec{r}) = A\delta(\vec{r})$, (b) $V(\vec{r}) = B/(|\vec{r}|^2)$, and (c) $V(\vec{r}) = C\exp(-|\vec{r}|/a)$.

6. Consider one-dimensional coherent electron transport of a free electron with the de Broglie wavelength λ over a linear distance interval $0 \leq x \leq d$. If the probability amplitude of the electron wave function has a maximum at $x = 0$, how are d and λ related if the electron's probability amplitude is equal to zero at $x = d$?

3

Electrons in Quantum Structures

3.1 Introduction

In conventional semiconductor devices with one or two types of carriers, different operational functions are achieved by the creation of a junction within the same material, for example, p–n, n^+–n junctions, etc. These kinds of junctions are commonly referred to as homojunctions, and devices fabricated from such structures are called homostructure devices. If more than one material is used in a device, this device is sometimes called a heterostructure device. In this chapter we study the behavior of carriers in different semiconductor heterostructures.

A heterojunction is formed when two different materials are joined together at a junction. It is now possible to make such heterojunctions extremely abrupt through modern materials growth processes; in fact, heterojunction interfacial thicknesses may be fabricated on a dimensional scale approaching only one atomic monolayer. The simplest heterostructure is a structure that consists of one heterojunction of two different semiconductor materials. There is a rich selection of various materials (semiconductors, metals, and insulators) that leads to high-quality interfaces and permits the fabrication of different heterostructures that manifest desired electron characteristics as functions of applied potentials. The most well known and applicable example is the SiO_2/Si heterojunction with a very small density of defects at the interface. The following pairs of semiconductor III–V compounds have great importance in technological applications and in studies of the properties of heterostructures: GaAs/AlGaAs, GaInAs/InP, GaInAs/AlInAs, GaSb/AlSb, GaN/AlN, InN/GaN, etc. Another class of actively studied and applied heterostructures is based on the II–VI compounds: CdZnSe/ZnSe, ZnSTeSe/ZnSSe, etc.

One of the major goals of the fabrication of heterostructures is the controllable modification of the energy bands of carriers. Energy-band diagrams for several semiconductor heterostructures are shown in Fig. 3.1. This figure illustrates the following: (a), (b) two types of single-junction heterostructures; (c), (e) different double-junction heterostructures; (f) a quaternary-junction heterostructure; and (d), (g), (h), (i) several multiple-junction structures. All these heterostructures are studied in this textbook. It is important that the energy positions of conduction and valence bands on different sides of heterojunctions differ by an amount of the order of a few hundred millielectron volts or more. These energies exceed by a considerable amount the thermal energy of carriers under typical

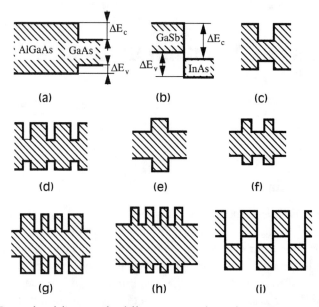

Figure 3.1. Energy-band diagrams for different semiconductor heterostructures. Depending on band discontinuities they are classified as (a) type-I or (b) type-II. The following variants are represented: (a), (b) single heterojunctions; (c) single, (d) multiple quantum wells; (e) single , (f) double, (g) multiple barrier structures; (h) type-I, (i) type-II superlattices. [After C. Weisbuch and B. Vinter, *Quantum Semiconductor Structures* (Academic, New York, 1991).]

operational conditions. This is why carriers can be confined easily inside the electrically active regions of a heterostructure bounded by potential barriers. These electrically active regions are referred to frequently as potential wells. In two special limiting cases, potential wells may be classified straightforwardly: classical and quantum wells. The width of the classical well is large compared with the de Broglie wavelength of an electron or a hole in the well. In the other limit, for a quantum well, Fig. 3.1(c), the width is comparable with or smaller than the de Broglie wavelength. For this case, the classical treatment of carrier motion breaks down and quantum behavior is manifested clearly in quantum wells.

In this chapter we study an idealized picture of electron states for three types of quantum structures: quantum wells, wires, and dots. For each of these structures our goals are to study quantum-mechanical features and to reveal pure effects of quantization. We also consider exciton effects and electron states on defects in these idealized quantum structures. By idealized we mean that the following set of simplifications applies:

1. We describe an electron by an isotropic effective mass m^* that is independent of both the position and the energy of the electron. As we have shown in Section 2.6, the effective mass m^* characterizes the motion of an electron in the periodic potential of a crystal and differs from the free-electron mass. In general, m^* changes when the electron passes through a heterojunction, but, in our idealized approach we neglect this effect.

2. The second simplification is related to the profile of the potential energy in a heterostructure. The idealized approach is based on the steplike potential profile, which can be analyzed easily.

Besides discontinuities of energy bands, there are other factors that control the profile of the potential and other properties of electrons. In real devices these factors include a space-dependent composition of compounds, a dependence on spatial-coordinate of doping, an electrostatic potential, different effective masses in layers composing a heterostructure, possible mechanical strain of layers, etc. Properties of real heterostructures will be analyzed in Chapter 4.

3.2 Quantum Wells

A quantum well can be fabricated when a single layer of one material is grown between two other layers of different material(s), which are characterized by wider bandgaps than the central layer, as shown in Fig. 3.2. The band discontinuity provides for confinement of carriers inside the well. We therefore accept the following idealized form of the potential, as illustrated in Fig. 3.3:

$$V(z) = \begin{cases} 0 & \text{for } |z| \leq L/2 \\ V_b & \text{for } |z| \geq L/2 \end{cases}, \tag{3.1}$$

where V_b and L are the depth and the thickness of the well, respectively.

The potential $V(z)$ is a function of coordinate z only. In this case, according to the method of separation of variables as discussed in Chapter 2, the electron motion in the two other directions, x and y, is free and can be described by a plane wave. Thus we can write for the wave function of the electron

$$\Psi(x, y, z) = e^{ik_x x + ik_y y} \chi(z). \tag{3.2}$$

After substitution of this wave function into the Schrödinger equation (2.27), we obtain the following equation for $\chi(z)$:

$$\left[-\frac{\hbar^2}{2m^*} \frac{\partial^2}{\partial z^2} + V(z) \right] \chi = \left(E - \frac{\hbar^2 \vec{k}_\parallel^2}{2m^*} \right) \chi, \tag{3.3}$$

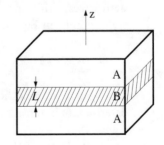

Figure 3.2. A three-layer structure providing electron confinement in layer B. Layers A are of a wide-bandgap material; layer B is of a narrow-bandgap material.

where $\vec{k}_\parallel = (k_x, k_y)$. The quantity $(\hbar^2 \vec{k}_\parallel^2)/(2m^*)$ is the kinetic energy of motion of the electron in the x, y plane. Let us define ϵ by

$$\epsilon = E - \frac{\hbar^2 k_\parallel^2}{2m^*}. \tag{3.4}$$

Since E is total energy, ϵ is the energy of the motion transverse to the x, y plane. Thus we have separated the in-plane and the transverse variables, and we

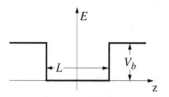

Figure 3.3. The potential energy of a quantum well.

have reduced the three-dimensional problem to the one-dimensional Schrödinger problem:

$$\left[-\frac{\hbar^2}{2m^*}\frac{\partial^2}{\partial z^2} + V(z)\right]\chi = \epsilon\chi. \tag{3.5}$$

According to Chapter 2, we can expect two types of solutions to Eq. (3.5). The first is characterized by a discrete-energy spectrum for $\epsilon < V_b$ of bound states of electrons. The second corresponds to a continuous spectrum for $\epsilon > V_b$ of unbound states. In spite of the discontinuity in the potential at $z = \pm L/2$, χ and $d\chi/dz$ are continuous everywhere, and χ is finite for unbound states or vanishes for bound states at $z \to \pm\infty$.

For $\epsilon < V_b$ outside the well, the solution has the form

$$\chi(z) = \begin{cases} Ae^{-\kappa_b(z-L/2)} & \text{for } z \geq L/2 \\ Be^{\kappa_b(z+L/2)} & \text{for } z \leq -L/2 \end{cases}, \tag{3.6}$$

where $\kappa_b = \sqrt{-2m^*(\epsilon - V_b)/\hbar^2}$. The solution inside the well is a simple combination of plane waves,

$$\chi(z) = C\cos k_w z + D\sin k_w z, \quad \text{for } |z| \leq L/2, \tag{3.7}$$

where $k_w = \sqrt{2m^*\epsilon/\hbar^2}$; A, B, C, and D are arbitrary constants. As a result of the symmetry of the problem, we can choose either even or odd combinations in Eqs. (3.6) and (3.7). Consequently, continuity of the wave function implies that $A = B$ for the even solution and $A = -B$ for the odd solution. Now we have two constants for odd solutions and two for even solutions in our problem. For example, for even solutions we find

$$\chi(z) = \begin{cases} C\cos k_w z & \text{for } |z| \leq L/2 \\ Ae^{\mp\kappa_b(z\mp L/2)} & \text{for } |z| \geq L/2 \end{cases}, \tag{3.8}$$

where the signs $-$ and $+$ correspond to positive and negative values of z, respectively.

The next step in finding the solution is to match the functions and their derivatives with respect to z at points $z = \pm L/2$. For example, for even solutions we obtain from Eq. (3.8) the system of algebraic equations:

$$C\cos k_w L/2 = A, \quad Ck_w\sin k_w L/2 = A\kappa_b. \tag{3.9}$$

This system of linear homogeneous equations has solutions if the corresponding determinant is zero:

$$\tan k_w L/2 = \kappa_b/k_w. \tag{3.10}$$

An analogous relationship can be found for odd solutions:

$$\cot k_w L/2 = -\kappa_b/k_w. \tag{3.11}$$

The algebraic equations (3.10) and (3.11) can be solved numerically, but it is more instructive to analyze them graphically. We transform Eqs. (3.10) and (3.11) into the following results:

$$\cos k_w L/2 = \pm k_w/\kappa_0, \quad \text{for } \tan k_w L/2 > 0, \tag{3.12}$$

$$\sin k_w L/2 = \pm k_w/\kappa_0, \quad \text{for } \cot k_w L/2 < 0, \tag{3.13}$$

where $\kappa_0 = \sqrt{2m^* V_b/\hbar^2}$. The signs $+$ and $-$ in Eq. (3.12) are to be chosen when $\cos k_w L/2$ is positive or negative, respectively. The same is valid for $+$, $-$, and $\sin k_w L/2$ in Eq. (3.13) for the odd solutions.

The left-hand $\mathcal{L}$ and the right-hand $\mathcal{R}$ sides of Eqs. (3.12) and (3.13) can be displayed on the same plot as the functions of k_w; see Fig. 3.4. The portions of the curve in the intervals $[2\pi(l-1)/L, 2\pi(l-1/2)/L]$ correspond to even solutions and in the intervals $[2\pi(l-1/2)/L, 2\pi l/L]$ to odd solutions; here $l = 1, 2, 3, \ldots$. The right-hand sides of Eqs. (3.12) and (3.13) are linear functions with a slope equal to κ_0^{-1}. The left-hand side is a cosine or sine function. Intersections of these two curves give us values $k_{w,n}$, for which our problem has solutions satisfying all necessary conditions. All solutions are represented in the first quadrant since positive and negative values of $\mathcal{L}$ and $\mathcal{R}$ are both plotted in the first quadrant.

To analyze the results we note that the problem is characterized by two independent parameters: the depth of the well V_b and the width L. We can fix one of these parameters and vary the other. Let us vary the depth of the well V_b, i.e., the parameter κ_0. In this case the left-hand side of Eq. (3.12) and the corresponding curves $\mathcal{L}$ in Fig. 3.4 do not change, but the slope of the line, k_w/κ_0, is controlled by κ_0. We can see that at small κ_0, when the slope is large, as is shown by the dashed curves in Fig. 3.4, there is only one solution $k_{w,1}$ corresponding to small k_0. This first solution exists at any κ_0 and gives the first energy level

Figure 3.4. Graphical solution for Eqs. (3.12) and (3.13). Solutions correspond to the intersections of the curves. The dashed curves correspond to small κ_0, resulting in only one solution of Eq. (3.12). The dotted–dashed lines correspond to a critical value of κ_0 when the second level appears in the well as a result of solution of Eq. (3.13). The solid line corresponds to an intermediate value of κ_0, leading to five solutions.

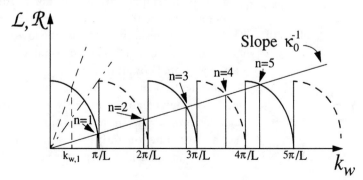

$\epsilon_1 = \hbar^2 k_{w,1}^2 / 2m^*$. As κ_0 increases, a new energy level occurs at $k_w = \kappa_0 = \pi/L$, as is shown by the dotted–dashed lines in Fig. 3.4, with energy slightly below V_b ($\epsilon_2 \approx V_b$). When κ_0 increases further, the first and the second levels become deeper and a third level occurs in the well, and so on. Indeed, new levels occur when the parameter $\sqrt{(2m^* V_b L^2)/(\pi^2 \hbar^2)}$ becomes an integer, so that the number of levels with energy $\epsilon < V_b$ is

$$1 + \mathrm{Int}\left[\sqrt{\frac{2m^* V_b L^2}{\pi^2 \hbar^2}}\right], \tag{3.14}$$

where $\mathrm{Int}[x]$ indicates the integer part of x. Figure 3.5 depicts the number of bound states as a function of well thickness for two specific values of V_b and for effective masses close to those of electrons and holes in AlGaAs/GaAs structures.

The explicit expressions for the energies of bound states can be found easily for two extreme cases. The first corresponds to a very shallow well, when only one level exists: $\epsilon_1 \approx m^* L^2 V_b^2 / 2\hbar^2$. The second case is related to an infinitely deep well, when $\kappa_0 \to \infty$; in this case, the slope of the linear function in Fig. 3.4 tends to zero and solutions are $k_w = \pi n/L$, or

$$\epsilon_n = \frac{\hbar^2 \pi^2 n^2}{2m^* L^2}, \quad n = 1, 2, 3 \dots . \tag{3.15}$$

We can see that for the latter case the separations between levels increase when n increases. Let us apply the last expression to a square quantum well with the effective mass $m^* = 0.067\,\mathrm{m}$ (like GaAs) and $L = 125\,\text{Å}$. We obtain $\epsilon_1 \approx 35\,\mathrm{meV}$, $\epsilon_2 \approx 140\,\mathrm{meV}$, etc.

One can see that the energy spacing between levels for the quantum wells with thicknesses of the order of 100 Å are tens to hundreds of millielectron volts. These values have to be compared with the Fermi energy of a degenerate two-dimensional electron gas as given by Eq. (2.83) at a low temperature or with the thermal (average) energy of the electrons at high temperature $k_B T/e$. For example, at room temperature the thermal energy is $\sim 26\,\mathrm{meV}$. This comparison shows that only the lowest energy levels can be occupied by electrons under typical device operational conditions.

Another frequently encountered well-like potential is that of an idealized single heterojunction. As a rule, the most important portion of this potential is near a heterojunction and, as illustrated in Fig. 3.6, it has a triangularlike shape formed by both a

Figure 3.5. The number of the bound states of a square well plotted as a function of well thickness: (a) $V_b = 224\,\mathrm{meV}, m^* = 0.067\,\mathrm{m}$; (b) $V_b = 150\,\mathrm{meV}, m^* = 0.4\,\mathrm{m}$. [After G. Bastard, *Wave Mechanics Applied to Semiconductor Heterostructures* (Halsted, New York, 1988).]

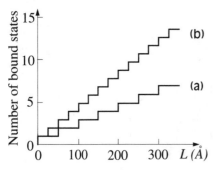

discontinuity of the electron (hole) band and an electrostatic field of electrons or remote ionized impurities. If the potential is characterized by high barriers, the lower energy levels may be studied with some accuracy by application of the following approximation:

$$V(z) = \begin{cases} eFz & \text{for } z > 0 \\ \infty & \text{for } z \leq 0 \end{cases}. \tag{3.16}$$

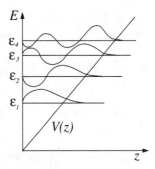

Figure 3.6. Triangular shape of the well potential, subband energies, and wavefunctions.

Here F is the electrostatic field. For this triangular model of the quantum well, the Schrödinger equation for the transverse component of the electronic wave function has the following form inside the well:

$$\left(-\frac{\hbar^2}{2m^*} \frac{\partial^2}{\partial z^2} + eFz \right) \chi = \epsilon \chi. \tag{3.17}$$

Outside the well, at $z \leq 0$ we set $\chi(z) = 0$ because the high barrier prevents significant penetration of electrons into barrier region. Equation (3.17) is known as the Airy equation and its solutions are given by a special form of the Airy functions, $\mathcal{A}i$:

$$\chi(z) = \text{const} \times \mathcal{A}i \left[\left(\frac{2m^*}{\hbar^2 e^2 F^2} \right)^{1/3} (eFz - \epsilon) \right]. \tag{3.18}$$

The special function $\mathcal{A}i(p)$ is presented in Fig. 3.7. Here

$$p = \left(\frac{2m^*}{\hbar^2 e^2 F^2} \right)^{1/3} (eFz - \epsilon)$$

Figure 3.7. The Airy function $\mathcal{A}i(z)$.

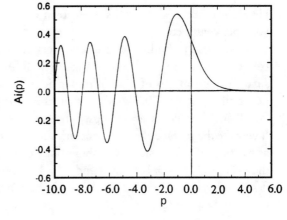

is the argument of the Airy function. One can present the Airy function in the analytical form for two extreme cases: one is for $p \gg 1$,

$$Ai(p) = \frac{1}{2p^{1/4}} \exp\left(-\frac{2}{3}p^{3/2}\right), \tag{3.19}$$

and the second is for $p < 0, |p| \gg 1$,

$$Ai(p) = \frac{1}{|p|^{1/4}} \sin\left(\frac{2}{3}p^{3/2} + \frac{\pi}{4}\right). \tag{3.20}$$

Accordingly, at $z \to \infty$, the solutions (3.18) vanish for any ϵ, but the requirement for the wave function to vanish at $z \leq 0$ gives us the following condition: The argument of the function of Eq. (3.18) at $z = 0$ must coincide with one of the zeros of the Airy function, p_n. As a result we obtain the following quantization of the energy ϵ:

$$\epsilon_n = -\left(\frac{e^2\hbar^2 F^2}{2m^*}\right)^{1/3} p_n. \tag{3.21}$$

The first root of the Airy function is $p_1 \approx -2.35$. For $n \gg 1$ it is possible to use the approximation of Eq. (3.20) to find the zeros,

$$p_n = -\left[\frac{3\pi}{2}(n + 3/4)\right]^{2/3}, \tag{3.22}$$

so that at large n the energy ϵ is proportional to $n^{2/3}$. This last result implies that, in contrast with the previous case, the distance between levels decreases when the energy increases.

Using Eqs. (3.21) and (3.22), one can evaluate the positions of the lowest energy levels for the AlGaAs/GaAs junction:

$$\epsilon_1 = 3.94 \times 10^{-5} F^{2/3} \text{ (eV)}, \qquad \epsilon_2 = 6.96 \times 10^{-5} F^{2/3} \text{ (eV)}. \tag{3.23}$$

Here, the energies are in units of electron volts and the electric field is in units of volts per centimeter.

It is simple to find the relation between the electrostatic field F and the electron concentration inside the triangular well. Actually, in the limit of negligibly small concentrations of ionized impurities in a narrow-bandgap semiconductor, i.e., in the region $z > 0$, Gauss's law gives $F = 4\pi e N_s/\kappa$, where N_s is the electron area (surface) concentration and κ is the dielectric constant of the semiconductor. If we take the surface electron concentration N_s to be 10^{12} cm^{-2} and take $\kappa = 13$, as for the AlGaAs system, we obtain an electric field $F = 1.39 \times 10^5$ V/cm and the energies are $\epsilon_1 = 106$ meV and $\epsilon_2 = 190$ meV. The separation between the first two levels is almost 100 meV. So, even at room temperature, only the lowest level will be populated.

Thus, in contrast to the case of motion of three-dimensional electrons in bulk semiconductors, the confinement of electrons in one dimension results in the creation of energy subbands ϵ_n, which contribute to the energy spectrum:

$$E_{n,\vec{k}_\parallel} = \epsilon_n + \frac{\hbar^2}{2m^*}\left(k_x^2 + k_y^2\right). \qquad (3.24)$$

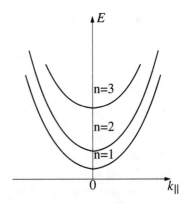

Here ϵ_n is the quantized energy associated with the transverse (perpendicular to the heterostructure) confinement. Thus two quantum numbers, one discrete n and another continuous $\vec{k}_\parallel$, are now associated with each electron subband. At fixed n, the continuum range of $\vec{k}_\parallel$ spans the energy band, which is usually referred to as a two-dimensional subband; see Fig. 3.8. One can interpret the subbands of

Figure 3.8. Schematic of the energy spectrum of the two-dimensional electrons in the first three subbands.

Eq. (3.24) to be a set of minima in the electron energies. The existence of such a set of minima drastically changes some characteristics of the electron system. For example, an impurity creates a series of energy levels under the electron band in the three-dimensional case. In a quantum well, each subband generates a series of impurity levels.

If the temperature and the electron concentration are such that only the lowest energy level is filled, free motion of the electron is possible only in the x, y plane, i.e., in two directions. This system is referred to frequently as a two-dimensional electron gas. The behavior of a two-dimensional electron gas differs strongly from that of a bulk crystal and will be studied in the Chapters 4, 6–12. We now give an analysis of the density of states for electrons in a quantum well.

3.2.1 Density of States of a Two-Dimensional Electron Gas

From the results presented in the previous subsection we see that the electron-energy spectrum in a quantum well is complex and consists of a series of subbands. The distances between subbands are determined by the profile of the potential, while inside each subband the spectrum is continuous and these continuous spectra overlap. For these complex spectra it is convenient to introduce a special function known as the density of states $\varrho(E)$ that gives the number of quantum states $d\mathcal{N}(E)$ in a small interval dE around energy E:

$$d\mathcal{N} = \varrho(E)dE. \qquad (3.25)$$

If the set of quantum numbers corresponding to a certain quantum state is designated as ν, the general expression for the density of states is defined by

$$\varrho(E) = \sum_{\nu} \delta(E - E_\nu), \qquad (3.26)$$

where E_ν is the energy associated with the quantum state ν. In our case the set of quantum numbers includes a spin quantum number s, a quantum number n characterizing the transverse quantization of the electron states, and a continuous two-dimensional vector $\vec{k}_\parallel$. Hence $\nu \equiv \{s, n, \vec{k}_\parallel\}$. There is a twofold spin degeneracy of each state ($s = \pm 1/2$) so that

$$\varrho(E) = 2 \sum_{n, k_x, k_y} \delta \left[E - \epsilon_n - \frac{\hbar^2(k_x^2 + k_y^2)}{2m^*} \right]. \tag{3.27}$$

To calculate the sum over k_x and k_y we can define the area of the surface of the quantum well: $S = L_x \times L_y$, where L_x and L_y are the sizes of the quantum well in the x and the y directions, respectively. If cyclic boundary conditions are assumed in the x and the y directions, the possible values of k_x and k_y are

$$k_x = 2\pi l_x / L_x, \qquad k_y = 2\pi l_y / L_y, \quad l_x, l_y = 0, 1, 2, \ldots. \tag{3.28}$$

It is well known that these results are independent of the assumption of cyclic boundary conditions.

Thus $\Delta k_x = 2\pi / L_x$, $\Delta k_y = 2\pi / L_y$, and we can write

$$\sum_{k_x, k_y} (\cdots) = \frac{L_x L_y}{(2\pi)^2} \int\int dk_x \, dk_y (\cdots). \tag{3.29}$$

By using Eq. (3.29), we transform Eq. (3.27) into

$$\varrho = \frac{m^* L_x L_y}{\pi \hbar^2} \sum_n \int_0^\infty d\epsilon_\parallel \, \delta(E - \epsilon_n - \epsilon_\parallel) = \frac{S m^*}{\pi \hbar^2} \sum_n \Theta(E - \epsilon_n), \tag{3.30}$$

where $\epsilon_\parallel = [(\hbar^2 k_\parallel^2)/(2m^*)]$, $S = L_x L_y$, and $\Theta(x)$ is the Heaviside step function: $\Theta(x) = 1$ for $x > 0$ and $\Theta(x) = 0$ for $x < 0$. Often the density of states per unit area, ρ/S, is used to eliminate the size of a sample. Each term in the sum of Eq. (3.30) corresponds to the contribution from one subband. The contributions of all subbands are equal and independent of energy. As a result, the density of states of two-dimensional electrons exhibits a staircase-shaped energy dependence, with each step being associated with one of the energy states ϵ_n. Figure 3.9 depicts the two-dimensional density of states.

It is instructive to compare these results with the density of states of electrons in bulk crystals:

Figure 3.9. Density of states for two-dimensional (2D) electrons in an infinitely deep potential well.

$$\varrho^{3D} = \left(\frac{m^*}{\hbar^2}\right)^{3/2} \frac{V}{\pi^2} \sqrt{2E}, \tag{3.31}$$

where V is the crystal volume: $V = SL$.

It is seen from Fig. 3.9 that the differences between the two- and the three-dimensional cases are most pronounced in the energy regions of the lowest subbands. For large n the staircase function lies very close to the bulk curve $\varrho^{3D}(E)$ and coincides with it asymptotically.

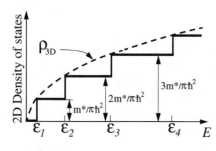

The dramatic changes in the electron density of states caused by dimensionally confining crystals to produce low-dimensional systems are manifest in a variety of major modifications in conductivity, optical properties, etc. Indeed, as we shall see, these modifications in the density of states also lead to new physical phenomena.

3.2.2 Quantum Effects in a Continuum-Electron Spectrum

So far, in studying the energy states of idealized quantum wells, we have concentrated on the discrete spectra generated by the potential profiles of heterostructures. Let us now consider another important case; specifically, we consider continuous-electron spectra.

We return to the case of the simplest quantum well with the rectangular potential profile of Eq. (3.1), as in Fig. 3.3, and consider the electron states in the quantum well, as well as the reflection of wavelike solutions from the heterointerfaces. Indeed, the quantum-well potential discontinuities affect not only the electron solutions in the well but also those above the well, although to a lesser extent. Analyzing the Schrödinger equation (3.2) for energy $\epsilon > V_b$, one can conclude that for regions far away from the quantum well, there are two plane-wave solutions propagating in opposite directions. The first wave corresponds to the propagation of an incident electron from left to right; then at $z \to -\infty$ there are incident and reflected plane waves, and at $z \to +\infty$ there are only transmitted plane waves. The second situation is related to the opposite direction of the electron propagation. Both cases can occur at the same energy. In addition, there is a twofold degeneracy of these states related to electron spin. Since both directions are identical, we need to study only one of them, say the case in which an electron propagates from left to right.

We introduce a wave vector in the regions of the barrier as

$$k_b = \sqrt{\frac{2m^*}{\hbar^2}(\epsilon - V_b)}. \tag{3.32}$$

Note that for $\epsilon > V_b$ the wave vector k_b is a real quantity. Then the wave function is

$$\chi(z) = \begin{cases} e^{i[k_b(z+L/2)]} + re^{-i[k_b(z+L/2)]} & z \leq -L/2 \\ ae^{ik_w z} + be^{-ik_w z} & -L/2 \leq z \leq L/2 \, . \\ te^{i[k_b(z-L/2)]} & z \geq L/2 \end{cases} \tag{3.33}$$

Here r, a, b, and t are still arbitrary constants, which we find by matching these solutions and their derivatives at the walls of the well, $z = \pm L/2$. This procedure is almost the same as the one we used in the previous subsection; however, there is one significant difference. Specifically, in this case the set of four linear algebraic equations is inhomogeneous and the conditions for the existence of a solution do not require any special restriction on the energy. As a result, we find that a continuum-energy spectrum and all constants r, a, b, and t can be expressed in terms of the coefficient of the incident plane wave, which we choose to equal 1.

The solution is straightforward with the following results for the constants t and r, which determine the transmitted and the reflected electron waves, respectively:

$$t(\epsilon) = \frac{1}{\cos k_w L - (i/2)(k_w/k_b + k_b/k_w)\sin k_w L}, \tag{3.34}$$

$$r(\epsilon) = \frac{i}{2}\left(\frac{k_w}{k_b} - \frac{k_b}{k_w}\right) t(\epsilon)\, \sin k_w L. \tag{3.35}$$

Now we can calculate two important characteristics: the transmission and the reflection coefficients $T(\epsilon)$ and $R(\epsilon)$, respectively. See Chapter 2 for a related discussion of these coefficients. If j_{in}, j_{tr}, and j_r are current densities for the incident, transmitted, and reflected electron flows, respectively, the definitions of $T(\epsilon)$ and $R(\epsilon)$ are

$$T(\epsilon) = \frac{j_{\text{tr}}}{j_{\text{in}}}, \qquad R(\epsilon) = \frac{j_r}{j_{\text{in}}}. \tag{3.36}$$

The current densities can be calculated with the wave function of Eq. (3.33) in the manner presented in Chapter 2. The results are

$$T(\epsilon) = tt^* = \frac{1}{1 + 1/4\,(k_w/k_b - k_b/k_w)^2 \sin^2 k_w L}, \tag{3.37}$$

$$R(\epsilon) = rr^* = 1 - T(\epsilon). \tag{3.38}$$

Figure 3.10. Energy dependence of the transmission coefficient T in a square quantum well of thickness $L = 250$ Å: (a) $V_b = 224$ meV, $m^* = 0.067$ m; (b) $V_b = 150$ meV, $m^* = 0.48$ m. [After G. Bastard, *Wave Mechanics Applied to Semiconductor Heterostructures* (Halsted, New York, 1988).]

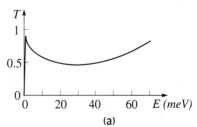

(a)

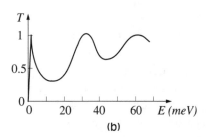

(b)

The last relationship is evident because the total probability of transmission and reflection is equal to 1. For a more general case, this relation was derived in Chapter 2; see Eqs. (2.173) and (2.174).

The formulas that we have obtained are interesting because they illustrate purely quantum-mechanical features. First, one can see that for $\epsilon > V_b$ the R coefficient is not zero, which means there is a purely quantum effect of reflections from the boundaries of the potential well, which vanishes in the classical limit. Second, the denominator of Eq. (3.37) oscillates as a function of energy. Hence the transmission coefficient becomes a nonmonotonous function of the energy. At energies satisfying the condition $k_w L = \pi n$, the transmission coefficient equals 1, and consequently the reflection coefficient R equals zero; for other energies the transmission coefficient is less than 1. At small energies $T(\epsilon) \propto \epsilon$, and $T(\epsilon)$ vanishes at $\epsilon = 0$. In Fig. 3.10, numerical results are presented for typical depths of potential wells in AlGaAs/GaAs structures for both electrons [Fig. 3.10(a)] and holes [Fig. 3.10(b)]. Quantum effects may be enhanced in structures with multiple quantum wells. For the case of superlattices, this is illustrated in Section 3.6.

Other interesting quantum-mechanical effects are related to the behavior of the electron wave function. The probability of finding a propagating particle in the region of a quantum well oscillates as a function of the energy. This behavior affects the technologically important processes of electron capture of and escape from the quantum wells. These processes are important for numerous applications of quantum structures including the quantum-well semiconductor laser.

It is instructive to mention an illuminating analogy between the foregoing effect and the transmission of a light through a dielectric slab structure known as the Fabry–Perot interferometer. In optics a nonmonotonous dependence in the transmission of light is observed because of the occurrence of multiple reflections from boundaries of the dielectric slab and the vacuum. This analogy has a fundamental basis, which can be seen from the similar forms of the one-dimensional Schrödinger equation and the electromagnetic wave equation for a medium with the refractive index that depends on one coordinate. This analogy is especially pronounced in the problem of continuous spectrum of electrons in the so-called resonant-tunneling structures; see Chapter 8 for additional information on such continuous spectra.

3.2.3 Two-Dimensional Electron Motion in a Smooth Potential

The concept of two-dimensional electron motion has more general applicability than for the strictly one-dimensional dependence of the potential V, as was introduced above. The concept can be extended to the case of an external potential $U(x, y, z)$ if the x, y dependences are associated with slow variations on scales compared with the quantum-well thickness L. Consider the following approximation for the wave function of the electron in the well defined by Eq. (3.1)

$$\Psi(x, y, z) = \sum_n \chi_n(z \,|\, x, y) \, \psi_n(x, y). \tag{3.39}$$

In contrast to Eq. (3.2), a general form for the wave function describing in-plane motion and a wave function χ_n for the transverse dependence that varies slowly with x and y are still assumed. The equation for $\chi_n(z \,|\, x, y)$ is derived as follows. Substitute Eq. (3.39) into the general Schrödinger equation (2.27) with the potential $V(z) + U(x, y, z)$ and drop all derivatives with respect to the variables x, y since these dependences are slowly varying. Then $\psi_n(x, y)$ can be cancelled on both sides of our equation to yield

$$\left[-\frac{\hbar^2}{2m} \frac{\partial^2}{\partial z^2} + V(z) + U(x, y, z) \right] \chi_n(z \,|\, x, y) = \epsilon_n(x, y) \chi_n(z \,|\, x, y). \tag{3.40}$$

Equation (3.40) is a generalization of Eq. (3.5) with important differences: χ_n and ϵ depend parametrically on x and y.

Suppose we can find the total set χ_n, ϵ_n of solutions of Eq. (3.40). Next, substituting Eq. (3.39) into the Schrödinger equation (2.27) and retaining all derivatives, we obtain

$$\sum_n \left\{ -\frac{\hbar^2}{2m} \nabla_2^2 \psi_n + [\epsilon_n(x, y) - E]\psi_n \right\} \chi_n$$

$$= \sum_n \left(\frac{\hbar^2}{2m} \right) \left[\psi_n \nabla_2^2 \chi_n + 2(\nabla_2 \psi_n \nabla_2 \chi_n) \right]. \tag{3.41}$$

Here $\nabla_2^2 = \partial^2/\partial x^2 + \partial^2/\partial y^2$ is a two-dimensional operator. Since χ_n is the total set of functions, one can multiply by $\chi_n(z \mid x, y)$ and then integrate both sides of Eq. (3.41) over z. The result is

$$-\frac{\hbar^2}{2m} \nabla_2^2 \psi_n + [\epsilon_n(x, y) - E]\psi_n$$

$$= \frac{\hbar^2}{2m} \sum_l \int dz \{ \chi_n(z \mid x, y)\psi_l(x, y)\nabla_2^2 \chi_l(z \mid x, y)$$

$$+ 2\chi_n [\nabla_2 \psi_l(x, y)\nabla_2 \chi_l(z \mid x, y)] \}. \tag{3.42}$$

The right-hand side of Eq. (3.42) is proportional to derivatives with respect to the in-plane coordinates. Since we have assumed a smooth external potential $U(x, y, z)$, these derivatives may be dropped in the first approximation. This method is called the adiabatic approximation: dependences on x, y are adiabatically slow as compared with z dependences. In this approximation we get the following two-dimensional Schrödinger equation:

$$-\frac{\hbar^2}{2m} \nabla_2^2 \psi_n + \epsilon_n(x, y)\psi_n = E\psi_n. \tag{3.43}$$

In the absence of an external potential, Eq. (3.43) immediately gives the wave function of Eq. (3.2) and energy of Eq. (3.4).

In the general case, Eq. (3.43) can be thought of as the Schrödinger equation for a particle with only 2 degrees of freedom, i.e., a strictly two-dimensional electron of the nth subband with a potential energy given by the variation of the bottom of this subband in the x, y plane. Since the potential U is smooth in the x, y coordinates, one can neglect the right-hand side of Eq. (3.42) and the electron remains in the same subband during its in-plane motion. The corrections coming from the right-hand side of Eq. (3.42) lead to less frequent transitions between the subbands; these transitions are the so-called intersubband transitions.

The above derivation of the two-dimensional Schrödinger equation for electrons in quantum wells can be supplemented by simple but important examples. Let U be small in magnitude. In this case one can use the perturbation theory of

Subsection 2.4.4 to obtain an explicit formula for $\epsilon_n(x, y)$:

$$\epsilon_n(x, y) \approx \epsilon_n + \int dz\, \chi_n^2(z) U(x, y, z), \tag{3.44}$$

where $\chi_n(z)$ and ϵ_n are solutions of Eq. (3.5).

Another useful benchmark case corresponds to the case in which the potential U is a smooth function of all three coordinates x, y, z but without restriction on the magnitude. The latter condition implies that U can be considered as an almost constant function across the quantum well, so that

$$\epsilon_n(x, y) \approx \epsilon_n + U(x, y, z=0). \tag{3.45}$$

Returning to Eq. (3.43) one can conclude that an external potential that is smooth in the x and the y coordinates modifies the in-plane electron motion. For example, such a potential can provide supplementary localization in one or two additional dimensions. It can create the barriers for the electrons, accelerate them, etc.

Equation (3.43) allows one to find the condition for the important case of semiclassical in-plane motion of the electrons. According to the analysis given in Chapter 2, one should compare the local de Broglie wavelength, $\lambda_n(x, y) = h/\sqrt{2m[E - \epsilon_n(x, y)]}$, with the characteristic length scale for the variation of the subband energy, $L(x, y) = \epsilon_n(x, y)/|\nabla_2 \epsilon_n(x, y)|$. Thus the condition of semiclassical motion of two-dimensional electrons is

$$\frac{h|\nabla_2 \epsilon_n(x, y)|}{\epsilon_n(x, y)\sqrt{2m[E - \epsilon_n(x, y)]}} \ll 1. \tag{3.46}$$

If inequality (3.46) holds, one can use the Newton equations,

$$\frac{d^2 x}{dt^2} = -\frac{\partial \epsilon_n(x, y)}{\partial x}, \qquad \frac{d^2 y}{dt^2} = -\frac{\partial \epsilon_n(x, y)}{\partial y}, \tag{3.47}$$

to study the in-plane motion of the electrons, that is, we have found that derivatives of $\epsilon_n(x, y)$ with respect to the in-plane coordinates give the force acting on the two-dimensional electrons $\vec{F}_{2D,n} = -\nabla_2 \epsilon_n(x, y)$. Generally this force is drastically different from that associated with the initial external potential $U(x, y, z)$: $\vec{F}_{3D} = -\nabla_3 U(x, y, z)$.

3.3 Quantum Wires

We have seen that the transition from a three-dimensional electron gas to a two-dimensional electron gas imposes quantization of the electron motion in one direction. As a result, the electron is characterized by 2 degrees of freedom. The creation of a two-dimensional electron gas is possible by the imposition of a one-dimensional confining potential $V(z)$, which can be formed by one plane heterojunction or by layered heterostructures.

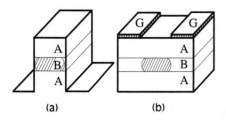

Figure 3.11. Two of the simplest examples of structures providing electron confinement in two dimensions: case (a) uses an etching technique, while case (b) is based on the split-gate technique.

To make the transition from a two-dimensional electron gas to a one-dimensional electron gas, the electrons should be confined in two directions and only 1 degree of freedom should remain, that is, one should design a two-dimensional confining potential $V(y, z)$. The x direction remains the only one for free-electron propagation.

There are several modern technological methods for the formation of such structures. The simplest two are illustrated by Fig. 3.11: case (a) (compare with Fig. 3.2) corresponds to etching of the top layer A, the confining material B, and part of bottom layer A; in case (b) two gates are deposited on the top layer A. A negative potential applied to these gates controls the lateral electron confinement.

3.3.1 Wave Functions and Energy Subbands

Let us consider general features of electrons in heterostructures with a confining potential $V(y, z)$. According to the method of the separation of variables, solutions of the Schrödinger equation (2.27) can be sought in the form

$$\psi(x, y, z) = e^{ik_x x}\chi(y, z),\tag{3.48}$$

so that the equation for the transverse electron wave function $\chi(y, z)$ is

$$\left[-\frac{\hbar^2}{2m^*}\left(\frac{\partial^2}{\partial y^2} + \frac{\partial^2}{\partial z^2}\right) + V(y, z)\right]\chi(y, z) = \epsilon\chi(y, z),\tag{3.49}$$

where $\epsilon = E - \hbar^2 k_x^2/(2m^*)$ is the electron energy in two transverse directions. If one can find a solution $\chi_i(y, z)$ of Eq. (3.49) corresponding to the discrete energy ϵ_i, one can obtain the total energy of the electrons in a form analogous to that of Eq. (3.24):

$$E = \epsilon_i + \frac{\hbar^2 k_x^2}{2m^*},\tag{3.50}$$

where k_x is now a one-dimensional vector. A comparison of Eqs. (3.50) and (3.24) reveals important properties of one-dimensional subbands. The wave function $\chi_i(y, z)$ corresponding to the discrete-energy level ϵ_i is localized in some area of the y, z plane. This means that the electrons of this quantum state i are confined in y and z directions around the minimum V_0 of the potential $V(y, z)$ and they are able to propagate along the x axis only. Such artificial systems in which electrons are guided along only one direction are called quantum wires.

The simplest case, in which the two-dimensional Schrödinger problem can be solved, is given by an infinitely deep rectangular potential,

$$V(y, z) = \begin{cases} 0 & \text{for } 0 \le y \le L_y, 0 \le z \le L_z \\ \infty & \text{for } y \le 0, z \le 0, y \ge L_y, z \ge L_z \end{cases}. \tag{3.51}$$

Here L_y and L_z are the transverse dimensions of the wires. For this case the electron wave function $\chi(y, z)$ can be represented as a product of functions depending separately on y and z:

$$\chi(y, z) = \chi(y)_{n_1} \chi(z)_{n_2}. \tag{3.52}$$

For each of the directions, solutions of the one-dimensional Schrödinger problem have the same form:

$$\chi_{n_1}(y) = \sqrt{\frac{2}{L_y}} \sin \frac{\pi y n_1}{L_y}, \qquad \chi_{n_2}(z) = \sqrt{\frac{2}{L_z}} \sin \frac{\pi z n_2}{L_z}, \quad n_1, n_2 = 1, 2, 3, \dots, \tag{3.53}$$

and the quantized energy ϵ_i of the transverse motion of the electrons is

$$\epsilon_{n_1, n_2} = \frac{\hbar^2 \pi^2}{2m^*} \left(\frac{n_1^2}{L_y^2} + \frac{n_2^2}{L_z^2} \right). \tag{3.54}$$

These results are applicable for quantum wires that can be fabricated from two-dimensional electron systems by the patterning of one-dimensional features in the plane of the electron gas through the use of nanoscale lithographic techniques with subsequent processing by etching, etc.; see Fig. 3.11(a).

Several other methods provide for the supplemental confinement of two-dimensional electrons in the second direction; as examples, such one-dimensional confinement of a two-dimensional electron gas can be accomplished by the application of an external voltage, a strain-induced potential, etc. The case, illustrated by Fig. 3.11(b), is known as the split-gate technique and is used frequently. For this case the confining external potential can be approximated by a parabolic dependence, so that

$$V(y, z) = V_1(y) + V_2(z), \qquad V_1(y) = \frac{1}{2} \left(\frac{\partial^2 V_1}{\partial y^2} \right)_{y=0} y^2. \tag{3.55}$$

If the potential $V_2(z)$ provides high barriers at $|z| = L_z/2$, the solution of the problem is given by following expressions:

$$\chi_{n_1, n_2}(y, z) = \text{const} \times e^{-\alpha^2 y^2} \mathcal{H}_{n_1}(\alpha y) \sin \frac{\pi z n_2}{L_z}. \tag{3.56}$$

$$\epsilon_{n_1, n_2} = \hbar\omega(n_1 + 1/2) + \frac{\hbar^2 \pi^2 n_2^2}{2m^* L_z^2}. \tag{3.57}$$

Here we introduce the notations

$$\omega = \sqrt{\frac{2}{m^*}\left(\frac{\partial^2 V_1}{\partial y^2}\right)_{y=0}}, \qquad \alpha = \frac{m^*\omega}{\hbar}, \qquad \mathcal{H}_n(y) = (-1)^n e^{y^2}\frac{d^n}{dy^n}e^{-y^2},$$

where $\mathcal{H}_n$ are the Hermite polynomials.

The examples presented for the potentials defined by Eqs. (3.51) and (3.55) demonstrate that the energy levels, which arise in quantum wires, are strongly dependent on the form of the confining potentials. However, the following important conclusions are quite general: additional confinement of electrons leads to an increase of the lowest energy level and there is a twofold set of the levels for quantum wires. Two quantum numbers, n_1 and n_2, now characterize any energy level corresponding to the transverse directions. The total energy of electrons, defined by Eq. (3.50), also includes the kinetic energy of one-dimensional electron propagation characterized by the one-dimensional wave vector k_x.

3.3.2 Density of States for a One-Dimensional Electron Gas

Let us calculate the density of states of a one-dimensional electron gas. In the definition of Eq. (3.26), ν numerates all possible quantum states of the system. In the case of quantum wires, $\nu = \{s, n_1, n_2, k_x\}$. Rewriting Eq. (3.26) in the form

$$\varrho(E) = \sum_{n_1, n_2} \varrho_{n_1, n_2}(E), \tag{3.58}$$

one can calculate the contribution to the density of states from a single subband:

$$\varrho_{n_1 n_2}(E) = 2\sum_{k_x} \delta\left(E - \epsilon_{n_1, n_2} - \frac{\hbar^2 k_x^2}{2m^*}\right), \tag{3.59}$$

where the factor of 2 is due to the spin s. The sum has to be calculated in the same way as for Eq. (3.27):

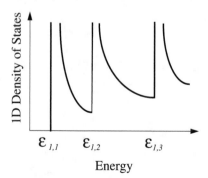

Figure 3.12. Density of states for one-dimensional (1D) electrons.

1D Density of States

$\mathcal{E}_{1,1}$ $\mathcal{E}_{1,2}$ $\mathcal{E}_{1,3}$

Energy

$$\varrho_{n_1 n_2}(E) = \frac{2L_x}{\pi}\int_0^\infty dk_x \delta\left(E - \epsilon_{n_1, n_2} - \frac{\hbar^2 k_x^2}{2m^*}\right)$$

$$= \frac{L_x}{\pi}\sqrt{\frac{2m^*}{\hbar^2}}\frac{1}{\sqrt{E - \epsilon_{n_1, n_2}}}\Theta(E - \epsilon_{n_1, n_2}). \tag{3.60}$$

Here L_x is the length of the wire, and the factor 2 comes from the fact that the summation in Eq. (3.59) is from $-\infty$ to $+\infty$ and we replace it by the integration from 0 to ∞. Schematically, $\varrho(E)$ for one-dimensional electrons is shown in Fig. 3.12.

Let us compare the density of states of the one-dimensional electron gas of Eq. (3.60) with the density of states of the two-dimensional gas of Eq. (3.30). The characteristics of these two densities of states are very different. Instead of the step-function behavior of the two-dimensional case, the one-dimensional density of states $\varrho(E)$ diverges at the bottom of each subband and then decreases as the kinetic energy increases. This behavior is remarkable because it leads to a whole set of electrical and optical effects peculiar to quantum wires.

3.4 Quantum Dots

So far we have considered semiconductor heterostructures in which an electron is confined in one or two directions. This leads to quantization of the electron spectrum, resulting in two- or one-dimensional energy subbands. This drastically changes the density of states. But still there is at least one direction for free propagation of the electron along the structure parallel to the barriers associated with the confining potentials; these structures can be used for electronic devices based on this transport.

The advances in semiconductor technology allow one to go further and fabricate heterostructures in which all existing degrees of freedom of electron propagation are quantized. These so-called quantum-dot, or quantum-box, systems are like artificial atoms and they demonstrate extremely interesting behavior.

Before we begin our study of such zero-dimensional systems, let us emphasize briefly that the trend toward further lowering of dimensionality of the electron gas portends useful applications in both electronics and optoelectronics. Let us imagine that one may quantize the electron states in all three possible directions and obtain a new physical object – a macroatom. Many questions concerning the usefulness of these objects for applications would naturally arise. And these doubts are understandable, because most electronic systems use electric voltages and electric currents. A fundamental question is What is the current through a macroatom? Valid answers are as follows. First, there is the possibility of passing an electric current through an artificial atom because of the tunneling of electrons through quantum levels of the atom. Second, electrical methods of control of functions are not the only methods possible. The control can also be realized by light, sound waves, etc. Hence there are several possible ways to achieve a necessary control of useful functions of zero-dimensional devices.

3.4.1 Wave Functions and Energy Levels

When considering the energy spectrum of a zero-dimensional system, we have to study the Schrödinger equation (2.27) with the confining potential, which is a function of all three coordinates and confines the electron in all three directions. The simplest potential $V(x, y, z)$ of this type is

$$V(x, y, z) = \begin{cases} 0 & \text{inside the box} \\ +\infty & \text{outside the box} \end{cases}, \tag{3.61}$$

where the box is restricted by the conditions $0 \leq x \leq L_x$, $0 \leq y \leq L_y$, $0 \leq z \leq L_z$; see Fig. 3.13. For this case, one can write down the solutions of the Schrödinger equation (2.27) immediately:

$$\psi_{n_1,n_2,n_3}(x,y,z) = \sqrt{\frac{8}{L_x L_y L_z}} \sin\frac{\pi x n_1}{L_x} \sin\frac{\pi y n_2}{L_y} \sin\frac{\pi z n_3}{L_z}, \qquad (3.62)$$

$$E_{n_1,n_2,n_3} = \frac{\hbar^2 \pi^2}{2m^*}\left(\frac{n_1^2}{L_x^2} + \frac{n_2^2}{L_y^2} + \frac{n_3^2}{L_z^2}\right), \qquad (3.63)$$

where $n_1, n_2, n_3 = 1, 2, 3, \ldots$. Of fundamental importance is the fact that E_{n_1,n_2,n_3} is the total electron energy, in contrast with previous cases, in which the solution for the bound state in a quantum well and wire gave us only the energy spectrum associated with transverse confinement. Another unique feature is the presence of three discrete quantum numbers resulting straightforwardly from the existence of three directions of quantization. Thus we obtain threefold discrete-energy levels and wave functions localized in all three dimensions of the quantum box.

Generally, all energies are different, i.e., the levels are not degenerated. However, if two or all dimensions of the box are equal or their ratios are integers, some levels with different quantum numbers coincide. Such a situation results in degeneracy: twofold degeneracy if two dimensions are equal and sixfold degeneracy for a cube. This discrete spectrum in a quantum box and the lack of free-electron propagation are the main features distinguishing quantum boxes from quantum wells and wires. As is well known, these features are typical for atomic systems.

The similarity with atoms is easily seen from another example of potential: the case of spherical dot. In this case, the potential is

$$V(r) = \begin{cases} 0 & \text{for } r \leq R \\ V_b & \text{for } r \geq R \end{cases}, \qquad (3.64)$$

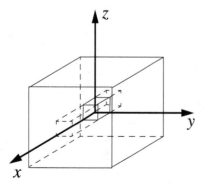

Figure 3.13. Sketch of a quantum box embedded in a matrix.

Here r is the magnitude of the radius vector, or coordinate, and R is the radius of the spherical dot.

From quantum mechanics it is known that under spherical symmetry the solutions of the Schrödinger equation can be expressed by the separation of angular and radial dependences in the form

$$\psi(r,\theta,\phi) = \mathcal{R}(r)Y_{l,m}(\theta,\phi).$$

Here r, θ, and ϕ are the spherical coordinates, $Y_{l,m}(\theta,\phi)$ are the well-known spherical functions, and l and m are quantum numbers corresponding to the angular momentum and its projection on axis z, respectively. For the radial function $\mathcal{R}(r)$, the

Schrödinger equation is

$$\left[-\frac{\hbar^2}{2m^*} \frac{\partial^2 \chi(r)}{\partial r^2} + V_{\text{eff}}(r) \right] \chi(r) = E\chi(r), \tag{3.65}$$

where

$$\chi(r) = r\mathcal{R}(r), \qquad V_{\text{eff}}(r) = V(r) + \frac{\hbar^2 l(l-1)}{r^2}. \tag{3.66}$$

Thus, as a result of the spherical symmetry, the problem reduces formally to a one-dimensional equation. The effective potential $V_{\text{eff}}(r)$ depends on the quantum number l, but does not depend on m. Accordingly, the energy levels must be degenerate with respect to the quantum number m and the number of degenerated states is equal to $2l + 1$. The energy is a function of two quantum numbers: the principal quantum number n, which comes from Eq. (3.65), and the angular-momentum quantum number l. One can find the general analysis of the problem with the potential $V_{\text{eff}}(r)$ in most textbooks on quantum mechanics. Below we present results for only the simplest cases.

If $l = 0$, one can obtain easily the solutions of Eq. (3.65) with the potential of Eq. (3.64):

$$\psi(r) = \begin{cases} A \sin k_w r / r, \; k_w = \sqrt{2m^* E} / \hbar & \text{if } r < R \\ B e^{-\kappa_b r} / r, \; \kappa_b = \sqrt{2m^* (V_b - E)} / \hbar & \text{if } r > R \end{cases}. \tag{3.67}$$

The matching of functions (3.67) and their derivatives at $r = R$ gives us equations similar to those studied for the one-dimensional problem. As a result, one can find an equation for the energy similar to Eq. (3.11):

$$\sin k_w R = \pm \sqrt{\frac{\hbar^2}{2m^* V_b}} k_w. \tag{3.68}$$

Only the roots of this equation satisfying the condition $\cot k_w R < 0$ can be chosen. We can repeat the analysis given in the previous subsections for the rectangular one-dimensional potential well of finite depth. The solutions of Eq. (3.68) are identical to the even states of the previous case represented by Eq. (3.13). Thus we may apply the same qualitative analysis by using Fig. 3.4. Only the portions of the curve in the intervals $[2\pi(i - 1/2)/L, 2\pi i/L]$, $i = 1, 2, 3, \ldots$, in Fig. 3.4 should be taken into consideration. The analysis shows that a level exists inside the spherical well if

$$V_b \geq \frac{\pi^2 \hbar^2}{8m^* R^2}. \tag{3.69}$$

Thus, Eq. (3.69) quantifies that a potential well must be large enough or deep enough to confine the electron. If the potential well is very deep, one can obtain solutions of Eq. (3.68):

$$k_w R = \pi n; \qquad E_{n,l=0} = -V_b + \frac{\hbar^2 \pi^2 n^2}{2m^* R^2}, \tag{3.70}$$

where we shifted the origin of energy coordinate so that $E_{n,l=0} < 0$ corresponds to bound states. The series of the levels with $l = 0$ is similar to that for the quantum-well problem.

If the angular momentum $l > 0$, the problem can be analyzed for a large depth of the well, $|V_b| \to \infty$. In this case the radial wave functions have the form

$$\mathcal{R}\{r\} = \sqrt{\frac{2\pi k}{r}} J_{l+1/2}(k_w r), \tag{3.71}$$

where $J_\nu(r)$ are the Bessel functions. The roots of the equation $J_{l+1/2}(k_w R) = 0$ give the energy levels. Now they are dependent on the angular momentum l. In the theory of atomic spectra one refers to the spherical states with $l = 0$ as s states and the states with $l = 1, 2, 3, \ldots$, as $p, d, f, \ldots$, states, respectively. An analysis of the roots of the Bessel functions gives us the following series of the energy levels in spherical quantum dots:

$$1s(2), \ 1p(6), \ 1d(10), \ 2s(2), \ 1f(14), \ 2p(6), \ldots. \tag{3.72}$$

The number in each set of parentheses shows the degeneracy of the states (electron spin is taken into account). One can compare this series with that of the electron in the Coulomb potential $V = -e^2/r$:

$$1s(2), \ 2s(2), \ 2p(8), \ 3s(2), \ 3p(6), \ 3d(18). \tag{3.73}$$

Thus there is a similarity between the spectra of atoms and the spherical quantum dots. Certainly the high degree of degeneracy will normally be broken in any real situation and many of the degenerate energy levels will split. The actual picture is thus quite complicated but the same is true for the atomic spectra of all but the simplest atoms.

3.4.2 Density of States for Zero-Dimensional Electrons

According to the definition of Eq. (3.26), in the case of quantum boxes or dots the spectra are discrete; accordingly, the density of states is simply a set of δ-shaped peaks:

$$\varrho(E) = \sum_\nu \delta(E - E_\nu), \tag{3.74}$$

where $\nu = (n_1, n_2, n_3)$. For an idealized system, the peaks are very narrow and infinitely high, as illustrated in Fig. 3.14. In fact, interactions between electrons and impurities as well as collisions with phonons bring about a broadening of the discrete levels and, as a result, the peaks for physically realizable systems have finite amplitudes and widths. Nevertheless, the major trend of sharpening the spectral density dependences as a result of lowering the system dimensionality is a dominant effect for near-perfect structures at low temperatures.

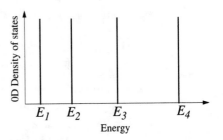

Figure 3.14. Density of states for zero-dimensional (0D) electrons.

From the previous examples of a potential energy $V(x, y, z)$, which creates the quantum box, it is evident that the structure of the energy spectra of the boxes (dots) is complex, especially for large quantum numbers. For each particular case the spectrum can be calculated numerically. However, it is usually straightforward to make reasonable estimates for the total number of energy levels inside a quantum box or dot if this number is large.

Several times in the previous discussions we have exploited the number of states of one-dimensional free motion in the space region of length ΔL in the interval of wave vectors Δk:

$$\Delta \varrho^{1D} = 2\frac{\Delta L \Delta k}{2\pi}. \tag{3.75}$$

One can generalize this expression to the three-dimensional case:

$$\Delta \varrho^{3D} = 2\frac{\Delta x \Delta k_x \Delta y \Delta k_y \Delta z \Delta k_z}{(2\pi)^3}. \tag{3.76}$$

Let us consider an arbitrary potential $V(x, y, z) < 0$. If one fixes some point of space $\vec{r}$ and calculates the number of states corresponding to the small volume $\Delta x \Delta y \Delta z$ around $\vec{r}$, one has to apply Eq. (3.76) and take into account all possible values of the electron wave vector within the region of confinement:

$$0 \le k(\vec{r}) \le k_{\max}(\vec{r}). \tag{3.77}$$

Here $k_{\max}(\vec{r})$ can be defined from the semi-classical approach, for which the total energy is

$$E = \frac{\hbar^2 k^2}{2m^*} + V(\vec{r}). \tag{3.78}$$

For bound states, $E \le 0$, so that $k_{\max}(\vec{r}) = \sqrt{2m^* |V(\vec{r})|/\hbar^2}$. The integration over all possible k that satisfy relation (3.77) gives the number of states in the volume $\Delta x \Delta y \Delta z$:

$$\Delta \rho^{3D} = \frac{k_{\max}^3 \Delta x \Delta y \Delta z}{3\pi^2}. \tag{3.79}$$

Now, by the summing over of all classically allowed electron coordinates, the total number of the energy states inside a quantum box is found to be

$$N_t = \frac{2\sqrt{2}m^{*3/2}}{3\pi^2\hbar^3} \int dx \, dy \, dz \, |V(\vec{r})|^{3/2}. \tag{3.80}$$

For example, for a box with potential given by Eq. (3.61) and with a finite potential depth V_b, one obtains

$$N_t = \frac{2\sqrt{2}m^{*3/2}}{3\pi^2\hbar^3}|V_b|^{3/2}L_xL_yL_z. \tag{3.81}$$

As a numerical example, with $L_x = L_y = L_z = 100$ Å, $V_b = 0.2$ eV, and $m^* = 0.067\,m$, it follows that the total number of the energy levels inside this box is $N_t = 75$. The actual number of the electrons inside the box is less then N_t; the amount of this reduction is determined by the level of impurity doping. It is possible to control the number of the localized carriers by application of an external voltage. Indeed, it is possible to change this number from a few electrons to tens of electrons.

3.5 Coupling between Quantum Wells

When we studied a single quantum well, we found that the electron wave function is not zero in the barrier region. Accordingly, the electron has some finite probability of penetrating into the barrier layer of a heterostructure. This effect may be illustrated by calculation of the total probability of finding the electron in the barrier layer:

$$P_b = \int_{|z|>L/2} dz\,|\chi(z)|^2, \tag{3.82}$$

Figure 3.15. Probability of finding the electrons in the barrier layer as function of the well thickness L. The electron state is assumed to be the ground state of the well. (a) $V_b = 224$ meV, $m^* = 0.067\,m$; (b) $V_b = 150$ meV, $m^* = 0.4\,m$. [After G. Bastard, *Wave Mechanics Applied to Semiconductor Heterostructures* (Halsted, New York, 1988).]

where the integration is performed over the barrier layer. The results are shown in Fig. 3.15. This tunneling behavior is one of the important manifestations of quantum mechanics and it leads to a range of physical phenomena that are exploited in various electronic and optoelectronic devices. Next we study the role of tunneling as a means of providing the coupling between quantum wells.

Let us consider the double-quantum-well structure for the case in which both wells, for simplicity, have identical rectangular shapes and are characterized by the same depth V_b and width L. Suppose the distance between wells is d, as in Fig. 3.16, and the wells are situated symmetrically with respect to the plane $z = 0$. Let us study the z-directed, one-dimensional propagation of the electron across the barrier. For such a double-well system the Schrödinger equation (2.27) of the transverse motion has the simple form:

$$\left[-\frac{\hbar^2}{2m^*}\frac{\partial^2}{\partial z^2} + V(z+d/2) + V(z-d/2)\right]\chi(z)$$

$$= \epsilon\chi(z). \tag{3.83}$$

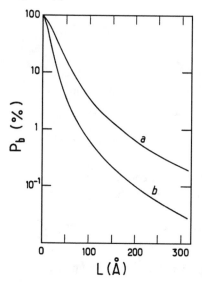

The term $V(z + d/2)$ is the potential of the left well centered at $z = -d/2$, while the term $V(z - d/2)$ corresponds to the right well. Here $V = V_b > 0$ in the barrier and V is zero inside the wells.

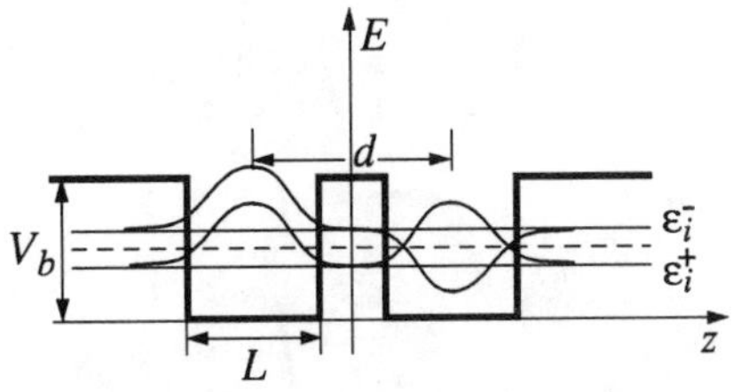

One can represent the wave functions of the double-well system through superposition of the solutions of the single-well problem $\chi_\nu(z)$:

$$\chi(z) = \sum_\nu [A_\nu \chi_\nu(z - d/2) + B_\nu \chi_\nu(z + d/2)],$$

$$(3.84)$$

Figure 3.16. Energy levels and wave functions for a double-well heterostructure. The dashed line corresponds to the position of the energy level ϵ_i in separated quantum wells. For coupled wells the splitting of this level is marked by $\epsilon_i^{\pm}$. Wave functions of these levels are presented schematically.

where ν runs over all possible bound and continuum states of the single-well problem. However, if the separation between wells is large enough so that the expected tunneling from one well to another has a small probability, we can restrict ourselves to only one level, say i, and take from Eq. (3.84) the combination

$$\chi(z) = A_i \chi_i(z - d/2) + B_i \chi_i(z + d/2). \tag{3.85}$$

This is the wave function for the electron states with energy near one of the bound states ϵ_i of the separated wells. Our goal now is to find the coefficients A_i and B_i. Because the distance d is supposed to be large, in the first approximation one can conclude that electron state ϵ_i is twofold degenerate. The electron in state i may occupy either the left well or the right well with equal probability. To correct the energies and find the coefficients A_i and B_i, we use the following procedure of the perturbation theory for degenerate states. We substitute the wave function of Eq. (3.85) into Eq. (3.83). Then we multiply this equation first by $\chi_i^*(z - d/2)$ to obtain one equation and then by $\chi_i^*(z + d/2)$ to obtain another equation. Next we integrate left-hand and right-hand sides of both of these equations and, as a result, we find two homogeneous linear equations for A_i and B_i:

$$(\epsilon_i + s_i - \epsilon) A_i + [(\epsilon_i - \epsilon)r_i + t_i]B_i = 0,$$

$$[(\epsilon_i - \epsilon)r_i + t_i]A_i + (\epsilon_i + s_i - \epsilon)B_i = 0,$$

where r_i, s_i, and t_i are constants that will be defined precisely in the following discussion. This system has nonzero solutions for only the following two energies $\epsilon = \epsilon_i^{\pm}$:

$$\epsilon_i^{\pm} = \epsilon_i + \frac{s_i}{1 \pm r_i} \pm \frac{t_i}{1 \pm r_i}. \tag{3.86}$$

The quantities r_i, s_i, and t_i represent three integrals that characterize electron tunneling in a double-well structure:

$$r_i \equiv \int_{-\infty}^{\infty} dz\, \chi_i^*(z - d/2)\chi_i(z + d/2),$$

$$s_i \equiv -\int_{-\infty}^{\infty} dz\, \chi_i^*(z - d/2)V(z + d/2)\chi_i(z - d/2),$$

$$t_i \equiv -\int_{-\infty}^{\infty} dz\, \chi_i^*(z - d/2)V(z - d/2)\chi_i(z + d/2). \tag{3.87}$$

The first integral is called an overlap integral, the second a shift integral, and the last a transfer integral. Because of weak penetration of the electron wave function into the barrier region, the overlap integral is small compared with 1, so that we can neglect the corresponding terms in Eq. (3.86) and simplify the equations for the energies to yield the form

$$\epsilon_i^{\pm} = \epsilon_i + s_i \pm t_i. \tag{3.88}$$

The diagram in Fig. 3.17 depicts the shift of the energy levels and clarifies the meaning of the shift and the transfer integrals. Since s_i and t_i are always less than zero, the first is responsible for shifting down the levels and leads to the energy decrease down from the initial energy position of the single well. Likewise, the second determines the splitting of the initial twofold degenerate level. For the lowest level, which is designated as ϵ_i^+ in Eq. (3.86), the combination in Eq. (3.85) is symmetric, i.e., $A_i = B_i$. The upper level ϵ_i^- is antisymmetric with $A_i = -B_i$. From Eqs. (3.87) one can see that both shifting and splitting are caused by the tunneling phenomenon. The wave functions of these two-well states are sketched in Fig. 3.16. The same consideration is valid for any electron states i localized in coupled quantum wells.

Although the approximation of Eq. (3.85) is easy to interpret, its applicability is limited to the case of thick and high barriers. Meanwhile, our simple model of two identical rectangular wells can be solved exactly without any approximations. The method of solution is similar to that for the rectangular well. Here we present only the expression that defines the energies of the system:

$$2\cos k_w L + \left(\frac{\kappa_b}{k_w} - \frac{k_w}{\kappa_b}\right)\sin k_w L \pm \left(\frac{\kappa_b}{k_w} + \frac{k_w}{\kappa_b}\right)e^{-\kappa_b d}\sin k_w L = 0, \tag{3.89}$$

where k_w and κ_b are the same as in Eqs. (3.6) and (3.8). The minus and the plus signs in Eq. (3.89) correspond, respectively, to symmetric and antisymmetric solutions, which were discussed in Section 3.2. For thick barriers between wells, when the last term of Eq. (3.89) can be dropped, the equation becomes equivalent to Eqs. (3.12) and (3.13), that is, to the equations of the single-well problem. If the distance between the wells, d, is finite, the last term of Eq. (3.89) shows

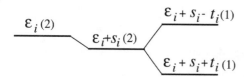

Figure 3.17. Diagram of an energy-level modification in a double-well heterostructure when shift and transfer integrals are taken into account. The degeneracy of levels is indicated in parentheses.

the splitting of each single-well level and reconstruction of the spectrum. When d is zero, a new set of levels coincides with that of the single-well problem for the case in which the well thickness is $2L$.

Note that Eq. (3.89) gives the energies of the transverse electron motion in a double-well structure. The total energies including the in-plane kinetic energy are

$$E^{(\pm)}_{i,\vec{k}_\parallel} = \epsilon^{(\pm)}_i + \frac{\hbar^2}{2m^*}\left(k_x^2 + k_y^2\right).$$

The corresponding wave functions are

$$\Psi^{(\pm)}_{i,\vec{k}_\parallel} = \frac{1}{\sqrt{2(1 \pm r_i)}} e^{i(k_x x + k_y y)}\left[\chi_i(z - d/2) \pm \chi_i(z + d/2)\right].$$

The major result of this section is the existence of a special type of interaction between quantum wells, that is, there is a coupling associated with tunneling between adjacent quantum wells. This interaction is caused by the quantum nature of electron propagation and shows that localized electrons can be found at the same time inside both wells. It also shows that the electrons are propagating in plane x, y, being in both layers.

Let us discuss briefly this in-plane motion and a crossover to the classical case. If at some initial instant of time we localize the electron inside one of the wells by constructing some wave packet, this wave packet will not be an eigenfunction of the stationary equation (3.83). As a result, the wave packet initially localized in the left well will tunnel into the right well; next, the packet will tunnel back into the left well. This cyclic process will be repeated over and over until some inelastic-scattering process causes collapse of the wave packet. The characteristic time of the tunneling process τ_t can be calculated with the transfer integral t_i from Eq. (3.87). According to the uncertainty relation for time and energy, Eq. (2.34), we can estimate τ_t as $\tau_t \approx \hbar/\Delta\epsilon \approx \hbar/t_i$. On the other hand, as just discussed, these quantum effects associated with free-particle propagation disappear if there is some scattering process. Let us characterize the scattering process by a relaxation time τ_r. Thus we can say that if the tunneling time τ_t is considerably less than the relaxation time τ_r, then the electron belongs to both wells, which guide it. In the opposite case, $\tau_t \gg \tau_r$, quantum effects vanish and the electron propagates through one of the wells with infrequent transitions to another well.

The results obtained for the double-quantum-well problem will be exploited further to study the case of multiple quantum wells. But before we close this section, we point out that double-quantum-well structures attract much attention because of interesting phenomena in electron parallel transport and various electronic and optoelectronic device applications, such as the velocity-modulated-transistor, the generation of high-frequency microwave radiation, unipolar cascade laser and others, which will be discussed in Chapters 9, 11 and 12.

3.6 Superlattices

Modern semiconductor technology facilitates the fabrication of not only single- or double-quantum-well structures, but also perfectly regular periodic systems

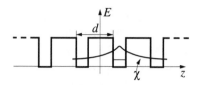

Figure 3.18. Schematic presentation of the potential energy of a superlattice with the period d. An energy level and a wave function are shown for a single well.

of layers. Each layer can act as either a potential well or a barrier for electrons. If these layers alternate and the number of them is large enough, the system can be considered as an artificial one-dimensional lattice with a period exceeding the period of the bulk crystal lattice. Such heterostructures are called superlattices. Superlattices have many interesting features and are used in different types of electronic and optoelectronic devices. In this section, we study the basic properties of superlattices.

An example of a superlattice is shown in Fig. 3.18, in which the wells and the barriers have simple rectangular forms. The theory of real superlattices is not simple, because the real wells and the barriers have complex profiles. In addition, the effective masses can be different in the wells and the barriers. Moreover, as the thicknesses of wells and barriers are reduced to a few atomic monolayers, the commonly used effective-mass approximation is not applicable.

To understand the basic behavior of superlattice structures, we should use the experience accumulated in the problem of periodic structures in crystals and then illustrate the major properties of the structure presented in Fig. 3.18.

This structure can be described by the Kronig–Penney model, which is a principal model for one-dimensional crystals in solid-state physics. The only essential difference is that we need to modify the Kronig–Penney model by using the effective mass instead of the free-electron mass. According to the Kronig–Penney model, any energy level of the single well ϵ_i splits into series of N levels, where N is the number of the wells in the superlattice, as depicted in Fig. 3.19(a). The physical reason of this splitting may be viewed as being caused by electron tunneling among the wells, as was described in detail in Section 3.5. For an N-well system with high and thick barriers between the wells, in the first approximation the energy levels coincide with their single-well positions, but they are N-fold degenerate. Tunneling between the wells breaks the degeneracy and results in a splitting into N levels with the wave functions common for all the wells. This implies that in any of these states one can find electrons inside any well. It is worth noting that a width of the energy region occupied by these levels is independent of the number of wells N if N is large. This energy region is the so-called energy band and it depends on only the tunneling characteristics of the system. Generally the energy bands appear for any system that exhibits this tunneling-induced coupling and does not require periodicity. Indeed, disordered condensed materials are also characterized by energy bands. However, the periodicity brings about additional nontrivial behavior of the system, as was studied in Section 2.6 for an electron in an ideal crystal. This periodicity leads to the condition

Figure 3.19. Formation of superlattice minibands: (a) splitting of a single level of a separated quantum well into an energy band because of tunneling coupling, (b) minibands originating from different energy levels of separated wells.

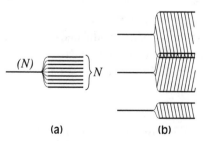

(a)   (b)

that the physical characteristics of the system do not change when the electron is shifted exactly by one period or an integer number of periods. Applying this requirement to the wave function, we get a direct analog of the Bloch theorem [Eq. (2.112)]:

$$\chi(z + nd) = C_n \chi(z) = e^{iqnd} \chi(z),\tag{3.90}$$

where d is the period of the superlattice. Equation (3.90) introduces a new parameter in the wave function, namely, the one-dimensional wave vector q. Wave functions with different values of q obey different transformations in accordance with Eq. (3.90). The wave functions corresponding to different states of the same energy band are characterized by different values of q. Thus q is a new quantum number of the system. In addition, we can introduce a hypothetical length of the superlattice Nd and exploit the cyclic boundary conditions:

$$\chi(z + Nd) = \chi(z).\tag{3.91}$$

This yields the relationship $e^{iqNd} = 1$ and $q = [(2\pi/Nd)p]$, $p = 1, 2, \ldots, N$. The latter relationship implies that the total number of quantum states arising as a result of the coupling of the N identical, uncoupled single-well levels is the same as the number of the wells in the superlattice (the spin is not considered). The spacing between neighboring values of q is equal to $2\pi/Nd$. It is customary to introduce the following interval for q:

$$\left\{ -\frac{\pi}{d}, +\frac{\pi}{d} \right\},\tag{3.92}$$

which is the first Brillouin zone for the electrons in a superlattice.

As just discussed, the energy band originates from the N isolated levels of the N single quantum wells as the well separation is decreased and the tunneling coupling among the wells is thereby increased. Now it is clear that each energy level of an isolated single quantum well evolves into an energy band containing N levels when N such isolated quantum wells are coupled through tunneling effects to form a superlattice; see Fig. 3.19(a). Because of the periodicity, each band is characterized by a one-dimensional wave vector q, so that for the ith level we obtain the band $\epsilon_i(q)$. Since the tunneling probability increases with increasing level index i, the width of the bands also increases with i. The narrowest band is the lowest one. Frequently the higher bands in superlattices overlap. This situation is sketched in Fig. 3.19(b).

Important features of superlattices are the bandgaps and the new man-made dispersion curves $\epsilon_i(q)$ that characterize the energy states of the superlattice. The bandgaps, of course, define the energy intervals in which propagating states, the Bloch waves, do not exist. In these intervals electrons cannot propagate in the superlattice.

The dispersion relation $\epsilon_i(q)$ for the model presented in Fig. 3.18 can be calculated in the so-called tight-binding approximation when the Bloch wave functions are constructed from the wave functions of the single-well problem with the potential of Eq. (3.1). Thus the wave function for the superlattice band i that originates from the energy level i can be written as

$$\psi_{i,q}(z) = \frac{1}{\sqrt{N}} \sum_{n=1} e^{iqnd} \chi_i(z - nd). \tag{3.93}$$

When the wave function of Eq. (3.93) is substitued into the Schrödinger equation with the superlattice potential $V(z) = \sum_n V(z - nd)$ and the nearest-neighbor wells are taken into account, it follows that the energy of the ith subband is

$$\epsilon_i(q) = \epsilon_i + s_i + 2t_i \cos qd, \tag{3.94}$$

where s_i and t_i are given by Eqs. (3.87). Relation $\epsilon_i(q)$ is shown schematically in Fig. 3.20(a) (note that t_i is negative). One can compare this result with those of the double-well problem studied in the previous section. The dispersion curve $\epsilon_i(q)$ reveals the downshift in energies, which is twice that for the double-well problem; in the present case, two neighboring wells are taken into account. Instead of the previous splitting into two levels with energies separated by $2|t_i|$, there is an entire continuum-energy band of width $\Delta\epsilon_i = 4|t_i|$. One can see that the width does not depend on the number of wells in the superlattice. The last term of Eq. (3.94) gives us the dispersion relation for the ith energy band. These superlattice bands are frequently referred to as minibands. The bandwidths of minibands are finite and determined by the transfer integral and depend strongly on the superlattice parameters, as illustrated in Fig. 3.20(b).

So far, we have studied only the transverse propagation of the electron, i.e., the propagation along the axis of the superlattice. The total electron energy including the energy associated with the in-plane propagation is

$$E_i(\vec{k}, q) = \epsilon_i + s_i + \frac{\hbar^2 k_\parallel^2}{2m^*} - 2|t_i| \cos qd, \tag{3.95}$$

where $\vec{k}_\parallel$ is the two-dimensional wave vector in the x, y plane. This formula demonstrates how dramatically the electron spectrum is modified and controlled by an artificial lattice.

Near the bottom and the top of the minibands it is possible to simplify further the energy spectrum of Eq. (3.95). Using the series expansion of $\cos qd$ leads to the following approximation around the point $q = 0$:

$$E_i(\vec{k}, q) = \epsilon_i + s_i - 2|t_i| + \frac{\hbar^2 q^2}{2M_i} + \frac{\hbar^2 k_\parallel^2}{2m^*}, \tag{3.96}$$

where $M_i = (\hbar^2)/(2d^2|t_i|)$. The effective mass M_i in the z direction is determined by the transfer integral, in accordance with the tunneling character of the transverse propagation. Equation (3.96) reveals that the electron propagation is

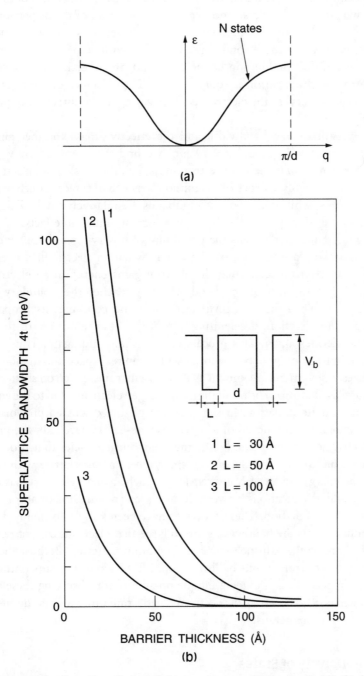

Figure 3.20. (a) The dispersion relation of electrons in a superlattice, Eq. (3.94), (b) the bandwidth of the lowest miniband for the GaAs/Al$_x$Ga$_{1-x}$As superlattice as a function of the barrier thickness ($x = 0.2$, $V_b = 212$ meV). [(b) is after G. Bastard, "Band structure, impurity binding energies and Stark effect in superlattices," Acta Electron **25**, pp. 147–161 (1993).]

extremely anisotropic. The anisotropy is controlled by the thickness of the layers and the height of the barriers. Another interesting aspect of the dispersion curve of Eq. (3.95) is the existence of a portion of the dispersion curve with a negative effective mass where the second derivative of ϵ_i with respect to q changes its sign at $q = \pi/2d$. Near the top of the miniband, the effective mass associated with transverse propagation is equal to $-M_i$. The existence of the negative effective mass means that the electron is propagating in the direction opposite to the applied force.

Let us estimate the bandwidth and the effective mass for the miniband. Assuming that $L = 5$ nm and $d = 10$ nm, we find $\Delta\epsilon_i = 4t \approx 15$ meV. Hence $M_i = 4.4\,m$, where m is the free-electron mass, i.e., the mass of an electron in a superlattice is roughly 1 order of magnitude larger than typical effective masses of bulk crystals, while widths of the minibands are relatively small. This is why low temperatures are preferable for applications of miniband effects.

The finite bandwidth opens the possibility of observing and applying novel effects that take place in any periodic system when an electric field is applied. If there is no electron scattering, the electron gains energy in an electric field until it reaches the top of the band where q reaches the boundary of the Brillouin zone. Then the electron reflects back and continues its propagation in the opposite direction, decelerating until it reaches the bottom of the band. Then it accelerates again. This process repeats again and again. Thus an electron in a miniband of finite bandwidth oscillates in real space and in momentum space. These oscillations are known as Bloch oscillations. Electron scattering destroys these oscillations. The oscillations of charged electrons lead to the emission of electromagnetic radiation. In the quantum picture discussed previously, the energy miniband is composed of a set of discrete-energy levels. In the presence of an electric field, band bending lifts the energy degeneracies that produce the superlattice minibands and the energy spectrum of a superlattice with large N becomes an energy ladder of N discrete levels. Such an energy spectrum is known as a Stark ladder. Transitions between the levels lead to emission or absorption of electromagnetic radiation. These processes have been known for many decades, but in bulk crystals the bandwidths are so large that electrons cannot reach the top of the band in the ballistic regime, i.e., without scattering. Because the bandwidth can be controlled and ballistic propagation along the superlattice axis can occur, superlattices provide a unique possibility for observing these effects. This principle of generation of electromagnetic radiation will be discussed in Chapter 11 in more details.

3.6.1 Density of States

Let us calculate the density of states for the superlattices $\varrho(E)$, where E is the total electron energy; we shall apply Eq. (3.26). The set of quantum numbers for an electron state includes the spin s, the quantum number of the superlattice band i, the wave vector along the axis of the superlattice q, and the two-dimensional in-plane wave vector $\vec{k}_{\parallel}$. Thus, from Eqs. (3.26) and (3.94), we obtain

$$\varrho(E) = \sum_{s,i,q,\vec{k}_{\parallel}} \delta\left(E - \epsilon_i - s_i + 2|t_i|\cos qd - \frac{\hbar^2 k_{\parallel}^2}{2m^*}\right). \tag{3.97}$$

The sums over s and $\vec{k}_{\parallel}$ can be calculated in the manner discussed in Sections 3.2–3.4; the result is

$$\varrho(E) = \frac{L_x L_y m^*}{\pi\hbar^2} \sum_{i,q} \Theta(E - \epsilon_i - s_i + 2|t_i|\cos qd). \tag{3.98}$$

Hence, when the total energy is less than the bottom of the ith miniband, $E < \epsilon_i + s_i - 2|t_i|$, the contribution to the density of states from the ith miniband is zero. For $E > \epsilon_i + s_i + 2|t_i|$ the contribution to $\varrho(E)$ from the ith miniband is equal to the total number of states of the miniband, $\varrho_0(E)$, i.e., the number of wells in the superlattice N multiplied by the density of states in one well, as depicted in Fig. 3.21. For intermediate situations, one has to integrate the Θ function over q in the region $-q_E < q < q_E$, where

$$q_E = \frac{1}{d}\arccos\left(\frac{\epsilon_i + s_i - E}{2|t_i|}\right).$$

As a result, one obtains the contribution of the ith miniband to the density of states in the form

$$\Delta\varrho_i = \frac{L_x L_y m^*}{\pi\hbar^2} N \begin{cases} 0 & E < \epsilon_i + s_i - 2|t_i| \\ \frac{1}{\pi}\arccos\left(\frac{\epsilon_i + s_i - E}{2|t_i|}\right) & \epsilon_i + s_i - 2|t_i| < E < \epsilon_i + s_i + 2|t_i|. \\ 1 & E > \epsilon_i + s_i + 2|t_i| \end{cases} \tag{3.99}$$

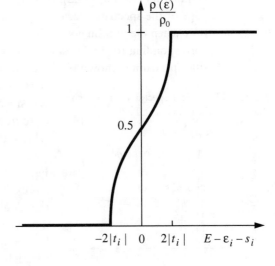

Figure 3.21. Density of states of a superlattice as a function of electron energy, $\varrho_0 = m^* N L_x L_y/\pi\hbar^2$. [After G. Bastard, *Wave Mechanics Applied to Semiconductor Heterostructures* (Halsted, New York, 1988).]

Figure 3.21 illustrates the energy dependence of $\varrho(E)$. One can see that the density of states in the superlattice exhibits a more complex form than the staircase of a single quantum well. The propagation of the electron across the barriers makes each step of the staircase smoother compared with that of the N noncoupled single quantum wells.

We have considered a very long superlattice, indeed, the previous discussion has been based on the assumption that $N \to \infty$. For most multiple-quantum-well systems approximately ten quantum wells are necessary for the system to behave as a superlattice. It must be stressed again that if the quantum wells and the barriers contain only a few atomic layers, the effective-mass approximation is not applicable for the description of the transverse electron motion in superlattices. New effects, such as folding of the Brillouin zone, have to be taken into account. These effects are studied in Chapter 4 for the example of a Si_n/Ge_m superlattice.

3.7 Excitons in Quantum Structures

Thus far in this chapter, we have studied the one-electron properties of quantum heterostructures and have neglected the interaction between carriers. However, such an interaction is important and leads to principally new properties of the materials and heterostructures. In this section we begin to study many-particle effects as embodied in the simplest two-particle case of excitons. These excitons are of great importance for most electric and optical properties of semiconductors and devices.

3.7.1 Excitons

The usual energy-band structure of bulk materials consists of conduction and valence bands separated by bandgap E_g. In dealing with optical spectra of semiconductors (see Fig. 3.22), this energy can be associated with the fundamental edge of the spectrum because phototransitions between the bands are possible only when the photon energy $\hbar\omega > \hbar\omega_g = E_g$; here ω_g is the threshold frequency corresponding to the fundamental absorption edge. The band-to-band absorption spectrum is shown by the dashed curve in Fig. 3.22 for the case in which the Coulomb interaction of the electron and the hole is disregarded. In fact, one usually observes a series of more or less narrow lines at frequencies $\omega < \omega_g$, which are especially pronounced at low temperatures. They are represented by the solid curve in Fig. 3.22. These lines are attributed to excitons. An exciton is a quasi-particle, which is formed by the Coulomb interaction between an electron and a hole. The concept of the holes was introduced in Subsection 2.6.2. This coupled pair can propagate through the crystal as a single particle. An exciton

Figure 3.22. Optical spectra of a semiconductor near the fundamental edge.

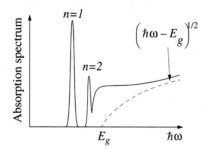

has also internal degrees of freedom, associated with relative distance between the electron and the hole comprising the exciton. These internal degrees of freedom are quantized, i.e., there are different quantum states of the exciton: a ground state with the lowest exciton energy and a number of the excited states. This quantization produces the previously mentioned series of peaks in the optical spectrum below the fundamental energy-band edge. The exciton energies depend on the energy spectra of the electrons and the holes, particularly on their effective masses, the dielectric constants of the materials, etc. This results in different exciton energies for different crystals. This picture is idealized, and it is applicable to such crystals as Si and Ge as well to III–V and II–VI compounds, etc. Furthermore, Fig. 3.22 also shows changes in the absorption spectrum above the fundamental edge at $\hbar\omega > E_g$. These changes are due to the Coulomb correlation of free (noncoupled) electrons and holes generated under the photon absorption.

An exciton is analogous to the bound state of an electron at a charged impurity, because of the same Coulomb character of the interaction. But, in contrast to the impurity case, all states of the electron–hole pair are excited states. Indeed, the ground state of a crystal corresponds to the valence band filled entirely by electrons, and accordingly there are no electrons in the conduction band and no holes in the valence band, as shown in Fig. 3.23. Any state of the crystal with a hole in the valence band and an electron in the conduction band is an excited state, i.e., the exciton corresponds to an excitation of the crystal. Therefore, as an excited state of the crystal, the exciton has a finite lifetime, which is in contrast to the impurity bound state considered in Subsection 3.7.2.

We start from the simplest bulk-material band structure, as depicted in Fig. 3.23, in which both the valence and the conduction bands are characterized by simple parabolic dispersion relations:

$$E_c(k) = E_g + \frac{\hbar^2 k_e^2}{2m_e}, \qquad E_v(k) = -\frac{\hbar^2 k_h^2}{2m_h}, \tag{3.100}$$

where m_e and m_h are the effective masses at the center of conduction and valence bands, respectively.

Figure 3.23. (a) Conduction (c) and valence (v) energy bands of a crystal, (b) dispersion dependences for electrons and holes and exciton levels, 1, 2, and 3. In the ground state of a crystal all energy states of the valence band are filled, while all states of the conduction band are empty.

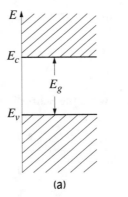
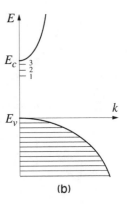

(a)   (b)

Let us take one electron from the completely filled valence band and put it in the conduction band. This leads to an excited state of the crystal. The process corresponds schematically to the following transition: $(v, k_v) \rightarrow (c, k_c)$, where k_v and k_c are the wave vectors of the electron in the valence and the conduction bands, respectively. According to the analysis given in Subsection 2.6.2, an empty place in the valence band may be considered to be a hole with positive mass m_h and wave vector $k_h = -k_v$. The energy of the excited state is thus

$$E_g + E_c(k_c) + |E_v(k_h)| \equiv E_g + \frac{\hbar^2 k_e^2}{2m_e} + \frac{\hbar^2 k_h^2}{2m_h}. \tag{3.101}$$

Equation (3.101) does not include the following Coulomb interaction energy between the electron and the hole:

$$V_{Cl} = -\frac{e^2}{\kappa |\vec{r}_e - \vec{r}_h|},$$

where κ is the dielectric permittivity of the crystal, $\vec{r}_e$ is the coordinate of the electron, and $\vec{r}_h$ is the coordinate of the hole.

The pair of particles, electron and hole, can be described by the two-particle Schrödinger equation:

$$(H_e + H_h + H_{Cl})\Psi(\vec{r}_e, \vec{r}_h) \equiv \left(\frac{\hat{p}_e^2}{2m_e} + \frac{\hat{p}_h^2}{2m_h} + V_{Cl} \right) \Psi(\vec{r}_e, \vec{r}_h)$$

$$= (E - E_g)\Psi(\vec{r}_e, \vec{r}_h), \tag{3.102}$$

where $H_{e,h} = \hat{p}_{e,h}^2/(2m_{e,h})$ are the Hamiltonians of the free electron (e) and the hole (h), respectively. The momentum operators are $\hat{p}_{e,h} = [-i\hbar(\partial/\partial\vec{r}_{e,h})]$. It is easy to see that the total Hamiltonian of the free electron and the hole has the energy given by Eq. (3.101) as an eigenvalue.

Let us introduce the relative coordinate $\vec{r}$,

$$\vec{r} = \vec{r}_e - \vec{r}_h, \tag{3.103}$$

and the coordinate $\vec{R}$ of the center of masses of the electron–hole pair,

$$\vec{R} = \frac{m_e \vec{r}_e + m_h \vec{r}_h}{m_e + m_h}. \tag{3.104}$$

Then we can rewrite the Schrödinger equation (3.102) in the form

$$\left[\frac{\hat{P}^2}{2(m_e + m_h)} + \frac{\hat{p}^2}{2m_r} - \frac{e^2}{\kappa r} \right] \Psi(\vec{r}, \vec{R}) = (E - E_g)\Psi(\vec{r}, \vec{R}), \tag{3.105}$$

where we have taken,

$$m_r = \frac{m_e m_h}{m_e + m_h}, \qquad \hat{\vec{P}} = -i\hbar \frac{\partial}{\partial \vec{R}}, \qquad \hat{\vec{p}} = -i\hbar \frac{\partial}{\partial \vec{r}}. \qquad (3.106)$$

Here, m_r is the reduced mass of the electron–hole pair, $\hat{\vec{P}}$ is the momentum operator of the pair as a whole, and $\hat{\vec{p}}$ is the internal momentum operator of the electron–hole pair.

The Schrödinger operator is the sum of the terms that are functions of either $\vec{r}$ or $\vec{R}$ only. This is why the internal and center-of-masses variables are separable and the wave function is a product:

$$\Psi = \frac{1}{\sqrt{V}} \exp(i\vec{k}\vec{R}) \, \psi(\vec{r}). \qquad (3.107)$$

The total exciton energy E is a sum:

$$E = E_g + \frac{\hbar^2 K^2}{2(m_e + m_h)^2} + E_{ex}, \qquad (3.108)$$

where the second term is the kinetic energy of the exciton. The important feature of the system under consideration is that the exciton with the effective mass $(m_e + m_h) = M^{ex}$ can propagate in the crystal as a plane wave with a wave vector $\vec{K}$. The exciton internal energy E_{ex} and the wave function ψ are related by

$$\left(-\frac{\hbar^2 \nabla_r^2}{2m_r} - \frac{e^2}{\kappa r} \right) \psi(r) = E_{ex} \psi(r), \qquad (3.109)$$

where ∇_r is an operator in the relative coordinate $\vec{r}$. Equation (3.109) is similar to the well-known equation for the hydrogen atom. The only difference is that the reduced mass of the electron–nucleus system is replaced by the reduced mass of the electron–hole pair m_r. Therefore the solution to the H problem may be adopted, and it follows immediately that the exciton wave function of the ground state has the form

$$\psi(r) = \frac{1}{\sqrt{\pi \left(a_0^{ex}\right)^3}} \exp\left(-r/a_0^{ex}\right),$$

where the exciton radius and the ground state energy are

$$a_0^{ex} = \frac{\hbar^2 \kappa}{m_r e^2}, \qquad E_{ex} \equiv -Ry^{ex}, \qquad (3.110)$$

where

$$Ry^{ex} = \frac{m_r e^4}{2\kappa^2 \hbar^2}.$$

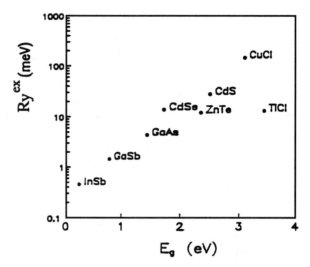

Figure 3.24. Experimental values for the exciton coupling energy E_0 versus bandgap E_g.

To exploit the analogy with the hydrogen atom further, the Rydberg $\mathrm{Ry} = me^4/2\hbar^2$, which determines ionization energy of the hydrogen atom, is replaced by an effective Rydberg constant Ry^{ex} appropriate to the solid-state case of the impurity or the exciton. From Eqs. (3.110) it follows that the exciton radius a_0^{ex} is proportional to the dielectric constant κ and inversely proportional to the reduced effective mass. These results predict a large value of a_0^{ex} that is generally greater than 100 Å for III–V compounds.

To give an idea of the values of the exciton energies, in Fig. 3.24 the energies of lowest (ground) exciton states $|E_{ex}|$ are presented versus the bandgap energy for different bulk semiconductors. The electron effective mass increases with increasing bandgap; this explains the increase in the exciton ground state energy with increasing bandgap.

3.7.2   Excitons in Quantum Wells

The results of Subsection 3.7.1 apply to the case of a bulk crystal. Let us apply the same approach to the study of excitons in double heterostructures. To be specific, we consider a structure that consists of a layer of material B embedded between two semi-infinite materials A. For excitons in bulk crystals it was found that the properties of the exciton depend strongly on the valence and the conduction bands. In the case of a heterostructure, the excitonic properties depend on the parameters in both materials A and B. There are at least two types of the energy-band diagrams in heterostructures.

In type-I heterostructures the conduction-band discontinuity $V_{b,e}$, or the band offset, and that of the valence band $V_{b,h}$ is such that electrons and holes are confined in the same layer as shown in Fig. 3.25(a); here the example is for AlGaAs/GaAs quantum structures.

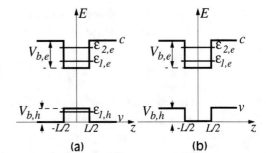

Figure 3.25. Two types of band diagrams of double heterostructures: (a) type-I, (b) type-II.

In type-II heterostructures each layer of the structure confines only one type of carrier, as illustrated in Fig. 3.25(b). An example is given by the InAs/GaSb system: The electrons are confined in the InAs layer, while the holes are confined in the GaSb layers.

An exciton in a type-I heterostructure has many features in common with the ordinary bulklike exciton. However, type-II heterostructures present a qualitatively different physical picture in which there are a spatially separated electron and hole of the electron–hole pair, the so-called interface exciton.

First, we consider type-I structures.

Type-I Structures

The Schrödinger equation should include the kinetic energies of both the electron and the hole, the potentials $V_e(r_e)$ and $V_h(r_h)$ that describe the wells for both particles, where

$$V_e(z) = \begin{cases} 0 & \text{for } |z| \leq L/2 \\ V_{b,e} & \text{for } |z| \geq L/2 \end{cases}, \tag{3.111}$$

$$V_h(z) = \begin{cases} 0 & \text{for } |z| \leq L/2 \\ V_{b,h} & \text{for } |z| \geq L/2 \end{cases}, \tag{3.112}$$

and the energy of interaction between the electron and the hole. Figure 3.25(a) illustrates the band structure for the system under consideration. Thus the Hamiltonian is

$$\mathcal{H} = E_g + \frac{\vec{p}_{\parallel e}^2}{2m_e} + \frac{\vec{p}_{\parallel h}^2}{2m_h} + V_e(\vec{r}_e) + V_h(\vec{r}_h) - \frac{e^2}{\kappa |\vec{r}_h - \vec{r}_e|}. \tag{3.113}$$

If we fix some electron and hole states inside the well, say $\chi_{n,e}(z_e)$ and $\chi_{l,h}(z_h)$, and neglect the electron–hole interaction, we can express the wave function of the pair in the form

$$\Psi(\vec{r}_e, \vec{r}_h) = \frac{1}{S} \exp[i(\vec{k}_{\parallel e}\vec{\rho}_e + \vec{k}_{\parallel h}\vec{\rho}_h)]\chi_{n,e}(z_e)\chi_{l,h}(z_h),$$

where $\vec{k}_{\parallel e}$ and $\vec{k}_{\parallel h}$ are the two-dimensional wave vectors of each of the particles, and $\vec{\rho}_e$ and $\vec{\rho}_h$ are two-dimensional coordinates of the electron and the hole, respectively; S is the area of the quantum-well layer. The energy corresponding to this wave function is

$$E = E_g + \epsilon_{n,e} + \epsilon_{l,h} + \frac{\hbar^2 k_{\parallel e}^2}{2m_e} + \frac{\hbar^2 k_{\parallel h}^2}{2m_h}.$$

It is clear that the last two equations describe decoupled and uncorrelated motion of the electron and the hole.

Since the quantum-well potentials break the translational symmetry in the z direction, instead of the transformations of Eqs. (3.103) and (3.104), new coordinates in the x, y plane can be introduced:

$$\vec{R}_\parallel = \frac{m_e \vec{\rho}_{e\parallel} + m_h \vec{\rho}_{h\parallel}}{m_e + m_h}, \qquad \vec{\rho} = \vec{\rho}_{e\parallel} - \vec{\rho}_{h\parallel}. \tag{3.114}$$

Both vectors characterize in-plane propagation. The z coordinates are still not transformed. As a result the Hamiltonian of Eq. (3.113) takes the form

$$\mathcal{H} = \frac{\hat{\vec{P}}_\parallel^2}{2M^{\mathrm{ex}}} + \frac{\hat{\vec{p}}_\parallel^2}{2m_r} - \frac{e^2}{\kappa\sqrt{\rho^2 + (z_e - z_h)^2}} + \frac{\hat{p}_{z,e}^2}{2m_e} + \frac{\hat{p}_{z,h}^2}{2m_h} + V_e(z_e) + V_h(z_h),$$

$$\tag{3.115}$$

where $\hat{\vec{P}}_\parallel$, $\hat{\vec{p}}_\parallel$, $\hat{p}_{z,e}$, and $\hat{p}_{z,h}$ are the momentum operators defined as in Eqs. (3.106). One can again apply the previously used approach of Eq. (3.107) by introducing partial factorization of the wave function $\Psi(\vec{r}_e, \vec{r}_h)$:

$$\Psi(\vec{r}_e, \vec{r}_h) = \frac{1}{\sqrt{S}} \exp(i\vec{k}_\parallel \vec{R}_\parallel) \psi(z_e, z_h, \vec{\rho}), \tag{3.116}$$

where $\vec{k}_\parallel$ characterizes the in-plane center-of-mass wave vector of the pair. The function $\psi(z_e, z_h, \vec{\rho})$ can be simplified if we consider some particular subbands of the electron and the hole and neglect the contributions of all other subbands. For an electron from the nth subband and a hole of the lth subband, we can approximate the wave function by

$$\psi(z_e, z_h, \vec{\rho}) = \chi_{n,e}(z_e) \chi_{l,h}(z_h) \phi(\vec{\rho}), \tag{3.117}$$

where $\phi(\vec{\rho})$ describes the relative motion of the electron and the hole.

The physical meaning of this approximation is that only the in-plane propagation of the particles is correlated as a result of the Coulomb interaction, while the transverse propagation is independent of the electron and hole coupling. Such an approximation is valid if the coupling energies of the particles are small compared with the energy of separation between subbands for both types of particles. In other words, the width of the well is supposed to be smaller than the effective Bohr radius. Note that each combination of subbands (n, e) and (l, h) gives the set of the coupled ground and excited states of the pair. Frequently the states of most physical interest are those formed from the lowest subbands of the electron and the hole.

Substituting the function ψ into the Schrödinger equation and integrating over coordinates z_e and z_h, we find that the effective potential energy has the form

$$V_{\text{eff}}^{(\text{ex})}(\rho) = -\frac{e^2}{\kappa} \iint dz_e \, dz_h \frac{|\chi_{1,e}(z_e)|^2 |\chi_{1,h}(z_h)|^2}{\sqrt{\rho^2 + (z_e - z_h)^2}}.$$

Let us apply the variational method discussed in Subsections 2.4.3 and 2.5.2. For the case in question, the functional that should be minimized is

$$\mathcal{E}_{\text{ex}}\{\phi\} = \iint dx \, dy \left[\frac{\hbar^2}{2m_r} (\nabla_2 \phi)^2 + V_{\text{eff}}^{(\text{ex})}(\rho) \phi^2 \right],$$

where, as before, ∇_2 is a two-dimensional operator. One can choose a simple trial function of the form

$$\phi(\rho) = \frac{1}{\lambda} \sqrt{\frac{2}{\pi}} e^{-\rho/\lambda}, \tag{3.118}$$

which allows one to find the energy of the exciton in the quantum well and its radius λ. In the approximation of Eqs. (3.117) and (3.118) the internal degrees of freedom of the electron and the hole are separated from their transverse propagation; moreover, the transverse propagation of the electron and the hole are completely separate from each other and uncorrelated. A more sophisticated approximation could include some correlation of the transverse degrees of freedom of the electron and hole, such as

$$\phi(\rho, z) = C \exp\left[-\frac{\sqrt{\rho^2 + (z_e - z_h)^2}}{\lambda} \right]. \tag{3.119}$$

In fact, for both trial functions the results can be obtained only by numerical calculations. In Fig. 3.26 the exciton binding energies E_{ex} are plotted for the two previously considered approximate cases as functions of the quantum-well layer thickness L. The binding energy increases in thin layers and in the extreme

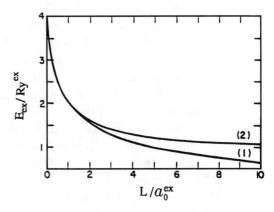

Figure 3.26. The exciton coupling energy for the type-I heterostructure calculated for two approximations of the trial function: curve 1 corresponds to Eq. (3.118), curve 2 corresponds to Eq. (3.119). [After G. Bastard, E. E. Mendez, L. L. Chang, and L. Esaki, "Exciton binding energy in quantum wells," Phys. Rev. B **26**, pp. 1974–1979 (1982).]

case, $L \to 0$, it reaches the value that is four times that in the bulk materials. The second approximation of Eq. (3.119) predicts a greater binding or, equivalently, coupling energy, for large thicknesses. Because both trial functions contain only one variational parameter λ, this result clearly indicates that there is a correlation between the transverse and the in-plane propagation of the pair. Note that only the second trial function yields the correct limit in the case $L \gg a_0^{ex}$, where a_0^{ex} is defined by Eqs. (3.110). In this limit the exciton takes on the bulklike behavior.

Thus for a type-I semiconductor heterostructure there are exciton states originating from a combination of the electron and the hole subbands. If energy separations between subbands are substantially larger than the energy of the exciton, only in-plane propagation of the electron and the hole exhibits strong correlations. This physical situation is typical for III–V compounds. In this case the transverse propagation characteristics are similar to those of a decoupled electron and hole, but there is some correlation between the transverse and the in-plane components of the propagation. An important feature of the heterostructure under consideration is that for the lowest subbands, the exciton (coupling) energies exceed those of excitons in bulk materials. Because of these larger exciton energies the excitons are present up to room temperature, in contrast to the case of bulk materials in which they are washed out at room temperature.

Type-II Structures

The diagram of electron and hole bands for this case is shown in Fig. 3.25(b). One can see that the embedded layer B confines only electrons and serves as a barrier for holes. In the previous case the coupling of the electron and the hole affects mainly only the in-plane propagation of the pair. Now the Coulomb interaction has to modify the hole transverse propagation dramatically in order to couple the pair. This is the main difference in the excitonic behavior in type-I and type-II heterostructures.

To take into account this fact, let us represent the wave function ψ of Eq. (3.116) in the form

$$\psi(\rho, z_e, z_h) = A\chi_{n,e}(z_e) \sum_{\vec{k}_h} C_{\vec{k}_h} \chi_{\vec{k}_h}(z_h) e^{-\rho/\lambda}, \tag{3.120}$$

where A is the normalization constant. The second multiplier describes transverse motion of the electron. The third one is the hole wave function, which should be localized because of an attraction by the localized electron, although it is constructed from the unbound valence-band wave functions of holes. The coefficients $C_{\vec{k}_h}$ are to be determined from the Schrödinger equation. The last multiplier describes the in-plane relative motion of the electron and the hole. We point out that this superposition of nonlocalized wave functions $\chi_{\vec{k}_h}(z_h)$ should result in confined states for holes. This is a complete analog to composing a localized wave packet from plane waves, which we studied in Chapter 2. The results of calculations are shown in Fig. 3.27 for the GaSb/InAs/GaSb double heterostructure. One can see that the hole is indeed confined near the electron-well layer, although the electron and the hole wave functions are

Figure 3.27. Band diagram of GaSb/InAs/GaSb double heterostructure and wave functions of the electron and the hole comprising the exciton in this type-II heterostructure. [After G. Bastard, E. E. Mendez, L. L. Chang, and L. Esaki, "Exciton binding energy in quantum wells," Phys. Rev. B **26**, pp. 1974–1979 (1982).]

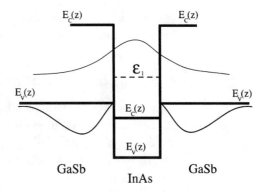

separated in space. Obviously this separation leads to a decrease in the exciton coupling energy E_{ex}^0. The bulk exciton in the InAs crystal is characterized by $Ry^{ex} = 1.3$ meV and $a_0^{ex} = 370$ Å. The coupling energy of the interface exciton is $\sim$0.6–0.8 of that of the bulk value, which is much smaller than the coupling in the type-I structure. Thus we can conclude that the excitonic effects in type-II quantum wells are suppressed substantially relative to those in type-I structures.

Before we close this section devoted to the exciton effects, we should make a remark concerning the exciton picture in the case of a complex valence band, as is typical for III–V compounds. The real valence band contains two types of energy dispersion $E_v(k_v)$: one for heavy holes and the other for light holes. Both of these particles can be excited in such crystals. The electron from the conduction band can couple with both types of holes, causing the formation of two types of excitons. In such quantum structures, heavy- and light-hole states are mixed and as a consequence the exciton physics becomes even more complicated. All these specific features are critical in establishing the optical spectra of quantum structures.

Another necessary remark is that the Coulomb attraction not only creates the bound states but also affects the free-electron and hole propagation. The interaction changes the wave functions of free states of electrons and holes in the case of small distances between them, and the particles become correlated. This effect of mutual correlation is important in photoexcitation of the electron–hole pairs since it leads to a significant modification of the absorption spectrum near the bandgap. This effect will be analyzed in Chapter 10, which is devoted to the optical properties of quantum semiconductor heterostructures.

3.8   Coulomb Bound States and Defects in Quantum Structures

In previous sections we have studied electron states in semiconductor heterostructures, taking into account only the interaction of electrons with the periodic potential of a crystal as well as with other electrons and holes. We have neglected the interaction of electrons with impurities and defects. However, these interactions play an important role in electrical and optical phenomena in semiconductor materials and devices. The processes of growing, doping, and processing heterostructures frequently create specific defects that do not occur in bulk materials, for example, different kinds of interface defects, etc.

Our goal in this section is to give a brief review of the electron states induced by impurities and other defects in heterostructures.

3.8.1 Coulomb Bound States of Impurities

Let us recall first the Coulomb bound states in bulk materials. For many actual cases, the interaction of an electron and an impurity having an electric charge e can be described by the Coulomb potential: $V(r) = -e^2/\kappa r$ in a manner similar to that discussed in Section 3.7 for an exciton. For this interaction potential the Schrödinger equation is

$$\left(-\frac{\hbar^2}{2m^*}\nabla^2 - \frac{e^2}{\kappa r} \right)\psi = E\psi, \tag{3.121}$$

where r is the distance between the impurity and the electron. It is assumed that the nondegenerate and isotropic electron band is characterized by effective mass m^*. Equation (3.121) is similar to Eq. (3.109) for the relative motion of the electron and the hole comprising an exciton. These two equations differ only through factors involving the effective masses. Equation (3.121) also has the form of the Schrödinger equation for the hydrogen atom and its solutions apply to the problem at hand. The ground state of an electron on an impurity has spherical symmetry and is described by the wave function

$$\psi_0(r) = \frac{1}{\sqrt{\pi a_0^3}} e^{-r/a_0}, \qquad a_0 = \frac{\kappa\hbar^2}{m^* e^2}. \tag{3.122}$$

The ground-state energy $E_1^{(3D)}$ is

$$E_1^{(3D)} = -\frac{1}{2}\frac{m^* e^4}{\kappa^2 \hbar^2} \equiv -\mathrm{Ry}^*. \tag{3.123}$$

Note that Ry^* is the effective Rydberg constant for the electron bound on the charge impurity. Equations (3.122) and (3.123) define the ground state of an electron bound on an impurity. As for the hydrogen atom, there is a set of excited bound states. Their energies are

$$E_n^{(3D)} = E_1^{(3D)}\frac{1}{n^2}, \quad n = 1, 2, \ldots. \tag{3.124}$$

At this point, it is important to emphasize that the energy $E_n^{(3D)}$ is negative for bound states. We have not changed the notation in this section; as in previous sections, n indicates the number of impurity levels. However, we add here an index $3D$ to avoid confusing this case with that of a quantum well in which the levels are also labeled by index n. In the case of bulk GaAs, with $\kappa = 12.8$ and $m^* = 0.067$ m, one obtains $a_0 \approx 100$ Å and $\mathrm{Ry}^* = 5.56$ meV.

For the case of heterostructures, the electron bound states are dependent on the impurity location. Impurities can be placed inside a quantum well, at the interface, or in the barrier layer; each different location results in different electron-energy levels, wave functions, etc. In addition, the characteristics of these bound electrons are also functions of the quantum-well parameters: the well depth V_b and its width L. To solve the problem of a hydrogenlike impurity in a quantum well, it is necessary to include the potential $V(z)$ that describes the quantum well in the Schrödinger equation (3.121):

$$\left[-\frac{\hbar^2}{2m^*}\left(\frac{\partial^2}{\partial x^2} + \frac{\partial^2}{\partial y^2} + \frac{\partial^2}{\partial z^2} \right) + V(z) - \frac{e^2}{\kappa}\frac{1}{\sqrt{\rho^2 + (z - z_i)^2}} \right]\psi = E\psi. \quad (3.125)$$

Here the interaction potential is written in cylindrical coordinates: The z axis is directed perpendicular to the plane of the well, and $\rho = \sqrt{x^2 + y^2}$ is the distance from this axis. The potential $V(z)$ is given by Eq. (3.1) and $\rho = 0$ and z_i determine the position of the impurity. Let

$$\Psi_{n,\vec{k}_\parallel} = \exp(i\vec{k}_\parallel\vec{\rho})\chi_n(z) \quad (3.126)$$

be the set of all eigenfunctions of the single-well problem, where n is the quantum-well level number. One can look for the exact solution of Eq. (3.125) as a superposition of these eigenfunctions with all possible $\vec{k}_\parallel$. In general, this procedure is complicated and can be done only numerically. For the sake of simplicity it is assumed that only the contribution from lowest ($n = 1$) quantized level of the well is important. Then it is possible to approximate the wave function for the electron on the impurity by

$$\psi(\rho, z) = \phi(\rho)\chi_1(z). \quad (3.127)$$

$\phi(\rho)$ satisfies the two-dimensional equation

$$\left[\frac{\hbar^2}{2m^*}\nabla_2 + V_{\text{eff}}(\rho) \right]\phi(\rho) = (E - \epsilon_1)\phi(\rho), \quad (3.128)$$

where ∇_2 is two-dimensional operator, ϵ_1 is the lowest electron energy in the quantum well, and the potential $V_{\text{eff}}(\rho)$ is

$$V_{\text{eff}}(\rho) = -\frac{e^2}{\kappa}\int_{-\infty}^{+\infty} \frac{dz\,|\chi_1(z)|^2}{\sqrt{\rho^2 + (z - z_i)^2}}. \quad (3.129)$$

It is evident that $V_{\text{eff}}(\rho)$ depends parametrically on z_i. One of the universal methods for the solution of the Schrödinger eigenvalue problem is the variational method discussed in Chapter 2. According to this method, we introduce a functional, which has the following form for the particular case of Eq. (3.129):

$$\mathcal{E}_i = \int\int \left[\frac{\hbar^2}{2m^*}(\nabla_2\phi)^2 + V_{\text{eff}}(\rho)|\phi|^2 \right] dx\,dy. \quad (3.130)$$

One can use the trial function of the form

$$\phi(\rho) = \frac{1}{\lambda}\sqrt{\frac{2}{\pi}}e^{-\rho/\lambda}, \tag{3.131}$$

where the only parameter λ represents radius of the state bound on the impurity in the well. After substitution of Eq. (3.131) into Eq. (3.130) for $\mathcal{E}_i$ we obtain

$$\mathcal{E}_i(\lambda \mid z_i) = \frac{\hbar^2}{2m^*\lambda^2} - \frac{2e^2}{\kappa\lambda}\int_0^\infty dx\, xe^{-x}\int_{-\infty}^{+\infty}dz\frac{\chi_1^2(z)}{\sqrt{x^2 + \frac{4}{\lambda^2}(z - z_i)^2}}. \tag{3.132}$$

Further progress in obtaining the solution is possible only through the application of numerical techniques. For the total energy of the electron bound to the impurity, it is possible to show that

$$E = \epsilon_1 + \min_\lambda \mathcal{E}(\lambda, z_i). \tag{3.133}$$

Analytically, it is possible to evaluate an extreme case in which the well is infinitely deep and much narrower than the radius of the bound state:

$$\lambda \gg L. \tag{3.134}$$

In this case for the electron confined in the well, the following inequality is valid: $2|z - z_i| \ll \lambda$. Therefore it is a good approximation to neglect the second term under the square root in Eq. (3.132). Then the integration in Eq. (3.132) can be performed analytically and yields

$$\mathcal{E} = \frac{\hbar^2}{2m^*\lambda^2} - \frac{2e^2}{\kappa\lambda}.$$

Now, the minimum condition

$$\frac{d\mathcal{E}}{d\lambda} = -\frac{\hbar^2}{m^*\lambda^3} + \frac{2e^2}{\kappa\lambda^2} = 0$$

gives the solution:

$$\lambda_{min} = \frac{\kappa\hbar^2}{2m^*e^2} \equiv \frac{1}{2}a_0.$$

The ground-state energy corresponding to λ_{min} is

$$\mathcal{E}_{min} \equiv E_{i1}^{(2D)} = -4\text{Ry}^*. \tag{3.135}$$

Thus in narrow and deep quantum wells the radius of localization of the electron around the charged impurity is half of the radius of the bound state in bulk crystals. The superscript $2D$ indicates that this is the energy of the electron ground state bound to the impurity in a two-dimensional quantum-well structure. In accordance with inequality (3.134), the distribution of the electron density on the impurity is shaped like a disk in the plane of the well. The ground-state energy of a two-dimensional impurity center is a factor of 4 greater than that of the bulk case.

The extreme case of a very thin $(L \to 0)$ and infinitely deep well is amenable to the foregoing approximate solution. The radius of the lowest orbit and an overestimate of the energy are obtained. If the height of the barrier is finite, the electron state is strongly dependent on the barrier height, the impurity position, and the well layer thickness.

In analogy to the case of an impurity in a bulk semiconductor, an impurity in a quantum well has a set of excited bound states for electrons. In the extreme case of a thin well given by inequality (3.134), these energies can be represented in a form similar to that of Eq. (3.124):

$$E_{in}^{(2D)} = E_i^{(3D)} \frac{1}{(n - 1/2)^2}, \quad n = 1, 2, 3, \ldots. \tag{3.136}$$

Here n refers to the bound state of an electron on the impurity associated with the lowest subband of a quantum well. In general, wave functions of the excited bound states do not have cylindrical symmetry, and a new quantum number m, the magnetic quantum number, is needed additionally to characterize the energy levels and the wave functions. The latter can be written in the form

$$\psi_{n,m}(\rho, \theta, z) = \phi(\rho)_{n,m} \, \chi_1(z) e^{im\theta},$$

where θ is an azimuthal angle, n is the orbital quantum number, and m is the integer azimuthal quantum number; compare with Eq. (3.127). The spectrum of energy levels is more complex. In Fig. 3.28, numerical results for the ionization energies of several lowest bound states are presented as functions of the quantum-well thickness and for two impurity positions. In the limits $L \gg a_0$ and $L \ll a_0$ these results tend to coincide with those of Eqs. (3.124) and (3.136), respectively. The ground state, $n = 1$, is not degenerate. The excited states are degenerate in both limiting cases, but for finite well widths they split into states with different azimuthal quantum numbers m. For example, the second excited state, $n = 2$, splits into $2s$ $(m = 0)$ and $2p$ $(m = 1)$ states. The third state splits into $3s$ $(m = 0)$, $3p$ $(m = 1)$, and $3d$ $(m = 2)$ states. As seen from Fig. 3.28 this splitting is more pronounced for asymmetric positions of the impurity.

These excited states determine the processes underlying electron capture by impurities and their ionization. These generation–recombination processes lead to additional fluctuations in the free-electron concentration and result in an excess current noise.

3.8.2 Interfacial Defects

Because of its extreme importance in the physics of semiconductors and solid-state electronics, the problem of electron bound states on impurities and defects is broad and could be the subject of a special book. In this textbook, attention is focused on one specific kind of defects – interfacial defects – that are of special

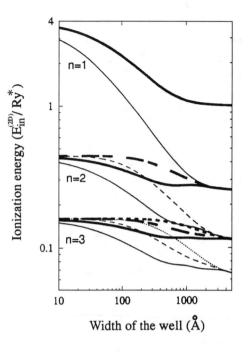

Figure 3.28. Ionization energies of excited states of the electron on the Coulomb impurity for the AlGaAs/GaAs quantum well versus the well width. Solid, long-dashed, and short-dashed curves corresponds to states with $m = 0$, 1, and 2, respectively. The thick curves are for an impurity situated at the center of the well, while thin curves are for an impurity at the interface.

interest in the field of heterostructure devices since they are present even in the most perfect heterojunctions.

At an interface between two materials, say A and B, there is usually a transition region that can be very narrow but nevertheless finite. In this region, material A can penetrate into material B and vice versa. In fact, for binary, ternary, and more complex materials, the adjacent layers may mix inhomogeneously near the heterojunction interface. This penetration is a rather random process, and it degrades the uniformity of the heterostructure along the layers and is responsible for interface imperfections and defects. For example, in the most perfect GaAs/AlGaAs structures, the interface fluctuation is of the order of one monolayer, 2.83 Å, in depth and a few hundred angstroms in lateral dimension. These so-called islands have compositions that are different from those of the basic layers. These islands can be characterized by local fluctuations in the Al content. Evidently the dimensions and the number of these defects depend on the growth conditions and further treatment of the structure. A universal quantitative theory of these defects cannot be formulated. Therefore it is necessary to describe them by simple models that describe the major effects of these imperfections. In these simple models, an interface defect can be characterized by a z-directed depth b and one or two sizes a and c in the x, y plane.

We will analyze the bound states associated with such a defect in a quantum well. Let the nominal interface between two materials A and B be situated at $z = 0$. We assume that material B is a barrier and layer A is a well. The barrier height between A and B is denoted by V_b. The defect, which attracts an electron, corresponds to the penetration of material A into B. The simplest models for the interface defects are

1. a shoebox defect with potential

$$V_{def}(x, y, z) = -V_b\, \theta(-z)\theta(b+z)\theta\left(\frac{a}{2} - |x|\right)\theta\left(\frac{c}{2} - |y|\right);$$

2. a pillbox defect of cylindrical symmetry,

$$V_{def}(\rho, z) = -V_b\, \theta(-z)\theta(b-z)\theta(a-\rho), \qquad \rho^2 = x^2 + y^2;$$

3. a semi-Gaussian defect of cylindrical symmetry,

$$V_{def}(\rho, z) = -V_b\, \theta(-z) \exp\left(-\frac{\rho^2}{2a^2}\right) \exp\left(-\frac{z^2}{2b^2}\right).$$

For the sake of definiteness we restrict our study to defects that attract electrons. The Schrödinger equation should now include the quantum-well potential $V(z)$ and the defect potential $V_{def}(x, y, z)$:

$$\left[-\frac{\hbar^2}{2m^*}\nabla^2 + V(z) + V_{def}(x, y, z)\right]\psi = E\psi. \tag{3.137}$$

For the case of an unperturbed quantum well we already know the eigenfunctions $\chi_n(z)e^{i\vec{k}\vec{\rho}_\parallel}$ and the energies $E_{n,k_\parallel} = \epsilon_n + \hbar^2 k_\parallel^2/(2m^*)$ of the corresponding quantized levels. The solution of Eq. (3.137) can be represented in the form

$$\psi(\vec{r}) = \sum_n \chi_n(\vec{r})\, f_n(x, y).$$

From the approach discussed previously for the charged-impurity case in which each subband generates its own set of bound states and in which these states are affected little by other subbands, it follows that the two-dimensional equation for the bound states attached to the nth subband is

$$\left[-\frac{\hbar^2}{2m^*}\nabla_2^2 + V_{eff}^{(n)}(x, y)\right] f_n(x, y) = (E - \epsilon_n)\, f_n(x, y), \tag{3.138}$$

where the effective defect potential for the nth subband is

$$V_{eff}^{(n)}(x, y) = \int_{-\infty}^{+\infty} dz\, V_{def}(x, y, z)|\chi_n(z)|^2.$$

The variational method may be used to find functions f_n and energies $\epsilon = E - \epsilon_n$. We restrict our calculations to the ground state of the electron on the defect and assume that this state is attached to the lowest quantum subband with $n = 1$. The shape of $f_n(x, y)$ depends on the type of defect. Let us consider the pillbox case and choose a probe function as

$$f_1(\rho) = \frac{1}{\lambda}\sqrt{\frac{2}{\pi}}e^{-\rho/\lambda}.$$

We have already used such a trial function for exciton and impurity problems; see Eqs. (3.118) and (3.131). The variational functional of Eq. (3.138) is

$$\mathcal{E}(\lambda) = \frac{\hbar^2}{2m^*\lambda^2} - V_b\left[1 - \left(\frac{2a}{\lambda} + 1\right)\exp\left(\frac{-2a}{\lambda}\right)\right]\int_{-b}^{0} dz\, |\chi_1(z)|^2,$$

where the function $\chi_1(z)$ was obtained in Section 3.2. We find that $\mathcal{E}(\lambda)$ can take on a minimum at

$$\lambda = \lambda_{\min} = 2a/\ln\left[\frac{4m^*a^2}{\hbar^2}V_bP_b\left(1 - e^{-2b\kappa_b}\right)\right],$$

where

$$P_b = \int_{-b}^{0} dz\, |\chi_1(z)|^2$$

is the probability of finding the electron in the barrier (see Fig. 3.15) and $\kappa_b = \sqrt{2m^*V_b/\hbar^2}$ serves as the parameter that characterizes the decay of the wave function in the barrier region. The radius of the bound state λ should be positive and $\ln[4m^*a^2V_bP_b(1 - e^{-2b\kappa_b})/\hbar^2]$ should also be positive. This requires that the argument of the logarithmic function must be greater than unity and the lateral dimension of the pillbox defect must exceed some critical value:

$$a > a_{\mathrm{cr}} = \frac{\hbar}{2\sqrt{m^*}}\frac{1}{\sqrt{V_bP_b(1 - e^{-2b\kappa_b})}}. \tag{3.139}$$

The energy of the bound state must be lower than the subband energy ϵ_n:

$$E_{1D} = \epsilon_n + \mathcal{E}(\lambda_{\min}), \quad \mathcal{E}(\lambda_{\min}) < 0.$$

This inequality also restricts the dimensions of the interface defects, which can generate additional bound states in the quantum well. For typical values of the effective mass $m^* = 0.067\,m$ (electrons) and $m^* = 0.4\,m$ (holes) and $V_bP_b = 3$ meV, it follows from Eq. (3.139) that the critical defect size is $a_{\mathrm{cr}} = 140$ Å for electrons and $a_{\mathrm{cr}} = 60$ Å for holes.

Another kind of interfacial defect – the trenchlike defect – is shown in Fig. 3.29. This defect is characterized by two geometrical parameters: its lateral dimension $2W$ and the depth of the trench $L' - L$. The results of calculation of the energies of electron states located near this defect are presented as functions of W/L for two depths of the trench. Note that for case 1 ($L'/L = 1.314$) there are two bound states.

From these calculations, we see that bound states on interface defects are important if the lateral size of the defect is of the order of or larger than the quantum-well thickness. Experimental studies show statistically that dominant defects have even larger sizes.

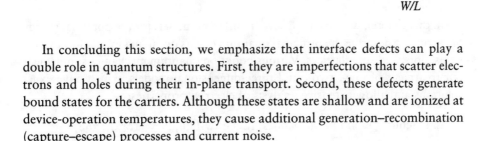

Figure 3.29. Bound states for the trenchlike defects with $V_b = \infty$. Curves 1 and 2 correspond to $L'_{1,2}/L = 1.314$ and 1.105, respectively. The dashed lines mark the positions of bottoms of the lowest two-dimensional subbands in quantum wells with widths $L'_{1,2}$. [After P. Voisin, *et al.*, "Luminescence from InAs-GaSb superlattices," Solid-State Commun. **39**, pp. 79–82 (1981).]

In concluding this section, we emphasize that interface defects can play a double role in quantum structures. First, they are imperfections that scatter electrons and holes during their in-plane transport. Second, these defects generate bound states for the carriers. Although these states are shallow and are ionized at device-operation temperatures, they cause additional generation–recombination (capture–escape) processes and current noise.

3.9 Closing Remarks

In this chapter we have studied electrons in different quantum structures: quantum wells, double quantum wells, quantum wires, quantum dots, and superlattices. To display the principal features of electrons in such structures we have simplified the problem by restricting ourselves to modeling heterostructures by steplike potentials and assuming a single effective mass for electrons throughout these structures. For these idealized systems, we have found drastic changes in electron spectra. With appropriate dimensional confinement, 1, 2, or even 3 electron degrees of freedom become quantized. Instead of the continuous-energy bands inherent in bulk materials, we have found a series of two- and one-dimensional energy subbands for quantum wells and quantum wires, respectively, and a series of discrete levels for quantum dots. For superlattices we have found a splitting of the bulk energy band into a series of one-dimensional minibands. At temperatures lower than or comparable with the subband distances these minibands lead to pronounced effects. The physics of low-dimensional electrons has essentially taken on the status of an entirely new field of physics because of the advent of nanodimensional heterostructures. In superlattices the electron motion becomes highly anisotropic and as a result the effective mass along the superlattice axis is enhanced substantially.

We have also studied energy spectra of excitons, impurities, and interfacial defect states in idealized quantum heterostructures. We have found that lowering the dimensionality of the electrons causes an increase in binding energies of excitons and Coulomb impurities. Furthermore, each of the low-dimensional subbands generates its own series of exciton and impurity levels. For Coulomb impurities, the energy spectrum depends strongly on the impurity position in a nanostructure.

By simplifying the description of quantum heterostructures, we have revealed the principal features of dimensionally confined heterostructures. In fact, in real heterostructures and devices there are a number of other, specific properties important for device applications as well as for understanding the physics of such nanostructures. In Chapter 4 we study these properties as well as particular parameters for different materials and heterostructures. Important effects related to the doping of heterostructures and to the population of low-dimensional states is studied in detail.

In this chapter, we have used several different models for the potential energy of carriers to describe the principal quantum effects occurring in quantum wells, wires, dots, and superlattices. Various books on quantum mechanics supplement our treatment of one-, two-, and three-dimensional models, which have exact solutions and allow extensive analysis of electron properties:

L. I. Schiff, *Quantum Mechanics* (McGraw-Hill, New York, 1968).
R. P. Feynman, P. R. Leighton, and M. Sands, *The Feynman Lectures on Physics* (Addison-Wesley, Reading, Mass., 1964), Vol. 3.

In our analysis, we have used the effective-mass approximation. Discussions of quantum heterostructures with more complex energy-band structures, as well as detailed analyses of particular examples of the heterostructures based on III–V compounds, can be found in the following references:

G. Bastard, *Wave Mechanics Applied to Semiconductor Heterostructures* (Halsted, New York, 1988).
C. Weisbuch and B. Vinter, *Quantum Semiconductor Structures* (Academic, New York, 1991).

Numerous results on electron quantization in SiO_2/Si structures are presented in

T. Ando, A. B. Fowler, and F. Stern, "Electronic properties of two-dimensional systems," Rev. Mod. Phys. **54**, 437–672 (1982).

PROBLEMS

1. In the limit of thick barriers, show that the equation that defines the energy levels of a double-well system defined by Eq. (3.89) reduces to the equation that defines the energy levels of a single quantum well.

2. Consider a 100-Å-wide GaAs quantum well with very thick $Al_{0.3}Ga_{0.7}As$ barriers. For this heterostructure, the conduction-band edge of the GaAs well is ~225 meV lower in energy than that of the barrier regions. On the other hand, Eq. (3.15) predicts an infinite number of bound states with $\epsilon_1 = 55$ meV, $\epsilon_2 = 220$ meV, and $\epsilon_3 = 495$ meV, etc. Thus ϵ_3 for the infinitely deep well exceeds the conduction-band offset of the $Al_{0.3}Ga_{0.7}As/GaAs/Al_{0.3}Ga_{0.7}As$ system. Based on these results, discuss the limits of applicability of Eq. (3.15) for realistic heterostructure devices.

3. Consider a spherical quantum dot of GaAs with a surrounding $Al_{0.3}Ga_{0.7}As$ barrier. What is the minimum radius for such a quantum dot if there is a bound state inside this spherical well?

4. For the zero-dimensional system defined by the potential of Eq. (3.61), it is possible to use the technique of separation of variables to obtain straightforwardly the wave functions of Eq. (3.62) and the eigenenergies of Eq. (3.63). For a more realistic case in which the potential discontinuity is finite, the wave functions are not equal to zero at the boundaries of the quantum box. Consider the formulation of the boundary conditions at the corners of the quantum box. Is it still possible to use the technique of separation of variables in the case in which the potential discontinuity is finite at the quantum-box boundaries?

Properties of Particular Quantum Structures

4.1 Introduction

In this chapter we study a variety of specific heterostructures that have promising applications in nanoelectronics. Of particular interest are heterostructures fabricated with semiconductor materials widely used in the electronic industry. Among the major such semiconductor material families are

> Group IV semiconductors and their compounds: Si is the basic material for the current microelectronics industry; Ge and the alloys Si_xGe_{1-x} have been used for special problems in electronics for many years, and currently it is found that their combination with Si opens new possibilities for Si-based technology, since it promises improved high-frequency devices and optoelectronic applications.
>
> III–V compounds: AlGaAs, InGaAs, etc., are used for high-speed microelectronics and optoelectronics and for different quantum devices. Currently new compounds based on GaN, AlN, and BN are evolving; they are promising for high-temperature electronics and as sources of short-wavelength light.
>
> II–VI compounds: These materials were used traditionally to fabricate model structures for studying different physical effects and for special device applications. In recent years, new promising sources of visible light have been designed on the basis of ZnSe/ZnSeS, ZnCdSe/ZnSe, and other such quantum heterostructures.
>
> IV–VI compounds: PbS, PbTe, GeTe, etc., are used for infrared photodetectors and lasers. Quantum IV–VI heterostructures have been successfully fabricated and studied.

Since the energy-band structure of materials underlies transport and optical properties, we study the energy spectra of some of these materials with emphasis on the parameters for their bandgaps, valence and conduction bands, effective masses, and electron affinities. Alloys of these materials are considered as well.

Next we consider the simplest single and double heterojunctions, introduce the concept of lattice-matched and lattice-mismatched materials and show that the quality of interfaces is strongly dependent on these characteristics. We study both matched and pseudomorphic heterostructures.

We pay special attention to different methods of the creation and control of carriers in quantum heterostructures. We show that a charge-transfer process can provide superb electron kinetic characteristics for heterostructures compared with those of homostructures. We discuss metal-dielectric-semiconductor structures, selective doping of single and double heterostructures, and super-lattices as important examples of structures exhibiting such charge-transfer processes.

4.2 Energy Spectra of Some Semiconductor Materials

In Chapter 3 we studied quantum-confinement effects for free carriers in quantum wells, wires, dots, and superlattices by using a single electron-energy band with a single effective mass. For real crystals the energy spectra have much more complex structures. For some physical effects and applications, these complications are not important; for others they lead to some corrections but do not change the processes qualitatively. However, there are also effects and related device applications for which specific features of the electron-energy spectra are of principal importance. As an example, we refer to the Gunn effect, i.e., negative differential resistance in high electric fields. Different types of microwave generators are based on the Gunn effect. At the same time, fundamental properties and effects such as multivalley electron-energy spectra and intervalley electron transitions underlie the Gunn effect. For quantum heterostructures that exploit electron-energy quantization in a straightforward manner, it is important to consider particular features of the energy-band structure.

4.2.1 Symmetry of Crystals and General Properties of Electron Spectra

As discussed in Chapters 2 and 3, electron-energy spectra are determined by several branches of energy with dispersion functions $E_i(\vec{k})$, where $\vec{k}$ is the electron wave vector of the first Brillouin zone. Usually these dependences are complex and can be obtained only numerically in the context of approximate methods.

Fortunately the Brillouin zone possesses a symmetry that directly reflects the symmetry of the unit cell of the crystal in coordinate space. If a crystal is mapped into itself because of transformations in the forms of certain rotations around the crystalline axes and of mirror reflections, one can speak about the point symmetry of directions in the crystal. In the Brillouin zone, this symmetry generates several points with high symmetry with respect to the transformations of the zone in $\vec{k}$ space. The extrema of the energy dispersion $E_i(\vec{k})$ always coincide with these high-symmetry points. In part, this fact allows one to simplify and solve the problem of obtaining electron spectra. Near extrema, the energy spectra can be approximated by expansions of $E_i(\vec{k})$ in series with respect to deviations from the symmetry points. Such an expansion can be characterized by several constants that define the reciprocal effective-mass tensor $(1/m^*)_{ij}$. The explicit form of this tensor has been derived in Subsection 2.6.3.

Table 4.1.
Symmetry Points in Group IV and III–V Compounds

Symmetry point	Position of extremum in the $\vec{k}$ space	Degeneracy
Γ	0	1
L	$\pm\frac{\pi}{a}[111], \pm\frac{\pi}{a}[\bar{1}11], \pm\frac{\pi}{a}[1\bar{1}1], \pm\frac{\pi}{a}[11\bar{1}]$	4
Δ	$\pm\gamma\frac{2\pi}{a}[100], \pm\gamma\frac{2\pi}{a}[010], \pm\gamma\frac{2\pi}{a}[001], \|\gamma\| < 1$	6
X	$\pm\frac{2\pi}{a}[100], \pm\frac{2\pi}{a}[010], \pm\frac{2\pi}{a}[001]$	3
K	$\frac{3\pi}{2a}[110]$ and 11 equivalent points	12
W	$\frac{\pi}{a}[210]$ and 5 equivalent points	6

Thus, because of the crystal symmetry, the problem of finding an electron-energy spectrum $E(\vec{k})$ is reduced to the following steps:

1. determination of the high-symmetry $\vec{k}$ points of the Brillouin zone,
2. calculation of energy positions of extrema,
3. analysis of effective masses or other parameters of an expansion of $E_i(\vec{k})$ within extrema.

The structure and the symmetry of the Brillouin zone for cubic crystals of group IV semiconductors and III–V compounds are similar. In Table 4.1 the symmetry points are presented for these semiconductor materials. Evidently, as a result of crystal symmetry, several points have the same symmetry; indeed, they are mapped into themselves under proper symmetry transformations. Such a degeneracy of the symmetry points is indicated in Table 4.1.

In particular, the Γ, L, and Δ points are of central importance. They give the positions of the lowest minima of the electron energy in III–V compounds, Ge and Si. In addition, the X points are important for III–V compounds because the upper minima in these compounds are located at the L and the X points. Figure 4.1 shows the surfaces of constant energy $E(\vec{k})$ for wave vectors near the symmetry points: Γ [Fig. 4.1(a)], L [Fig. 4.1(b)], and Δ [Fig. 4.1(c)]. Since electrons tend to be near the energy minima, one can think of electrons as being located inside marked regions of $\vec{k}$ space. Frequently these regions in $\vec{k}$ space are referred to as energy valleys or simply valleys. Materials with several valleys are also called many-valley semiconductors. For III–V compounds there is only one such valley around the point $\vec{k} = 0$; however, in the case of Si there are six valleys, in accordance with the degeneracy of the Δ points. It is worth emphasizing that in processes occurring far from equilibrium, other symmetry points can also play a considerable role.

The band structures are presented in Fig. 4.2 for Ge, Si, and GaAs. The two symmetric directions of the wave vectors [111] and [100] are shown. In each case, the energy $E = 0$ corresponds to the top of the highest valence band, which is located at the Γ point for each of these examples. Actually, the valence

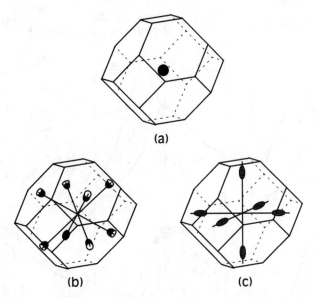

Figure 4.1. Surfaces of constant energy in the Brillouin zone: (a) GaAs, a sphere is situated at the center of the Brillouin zone; (b) Ge, eight half-ellipsoids are on [111]-type crystalline axes at the boundaries of the first Brillouin zone corresponding to four minima of the conduction band; (c) Si, six ellipsoids are along [100]-type crystalline axes at approximately three quarters of the distance from the center to the boundaries of the Brillouin zone.

bands have a complex structure even near the points $\vec{k} = 0$. For all group IV semiconductors and III–V compounds this structure is similar and includes two types of holes: light holes and heavy holes.

The highest valence bands and the lowest conduction bands are separated by bandgaps E_g. In contrast to the case of valence bands, conduction bands have different structures for the groups of materials considered in this section. The main difference is that for Si and Ge the lowest minima are in Δ and L points, respectively, while for most of III–V compounds there is only one lowest minimum, which is at the Γ point. The difference is not simply quantitative; indeed, it is qualitative and leads to a series of important consequences in the behavior of electrons.

We can see that in order to make a transition of an electron from the valence band to the conduction band in Si and Ge, we need not only to add some energy to excite an electron, but also to change its momentum by values comparable with the scale of the Brillouin zone. Such a semiconductor is a so-called indirect-bandgap semiconductor. On the other hand, in GaAs one can transfer an electron from the valence band to the conduction band directly without an appreciable change of its momentum. This type of crystal is a direct-bandgap semiconductor.

As discussed above, within the Γ, Δ, and L points the electron dispersion curves $E(\vec{k})$ can be expanded in series with respect to deviations from the minima $\vec{k}_\alpha$:

$$E(k) = E(k_\alpha) + \left(\frac{1}{m_\alpha^*}\right)_{ij} \hbar^2 (k_i - k_{i,\alpha})(k_j - k_{j,\alpha}). \tag{4.1}$$

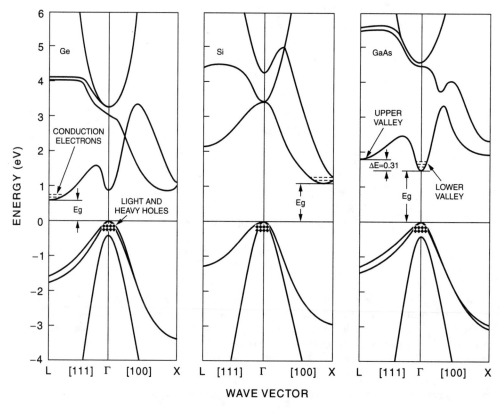

Figure 4.2. Energy-band structures for Ge, Si, and GaAs. Dispersion curves are given for two symmetric directions, [100] and [111]. The indirect characters of the bandgaps are clearly seen for Ge and Si, while for GaAs the direct bandgap occurs. [After S. M. Sze, *Physics of Semiconductor Devices*, 2nd ed. (Wiley, New York, 1981).]

For the Γ point, we have the simplest case of an isotropic effective mass:

$$\left(\frac{1}{m_\alpha^*}\right)_{ij} = \frac{1}{m^*}\delta_{ij}. \tag{4.2}$$

For Δ, X, and L points, i.e., for Δ, X, and L valleys, one can conclude from symmetry considerations that the reciprocal effective-mass tensor has only two independent components corresponding to longitudinal m_l and transverse m_t masses. For example, in the case of Si for a coordinate system with the z axis along the [100] crystal direction and the two other axes perpendicular to the first, one can obtain

$$\left(\frac{1}{m^*}\right)_{ij} = \begin{pmatrix} m_t^{-1} & 0 & 0 \\ 0 & m_t^{-1} & 0 \\ 0 & 0 & m_l^{-1} \end{pmatrix}. \tag{4.3}$$

For indirect group IV semiconductors, the energy parameters are presented in Table 4.2. The degeneracy marked in the table indicates the existence of six and

Table 4.2.
Energy-Band Parameters for Si and Ge

Group IV	Si	Ge
Type of bandgap	Indirect	Indirect
Lowest minima	Δ points	L points
Degeneracy	6	4
E_g (eV)	1.12	0.664
m_l/m	0.98	1.64
m_t/m	0.19	0.082
m_{hh}/m	0.50	0.44
m_{lh}/m	0.16	0.28
Δ_{so} (eV)	0.044	0.29

four equivalent energy valleys for Si and Ge, respectively. According to Fig. 4.2, the L points are situated at the edges of the Brillouin zone, i.e., only half of each energy valley lies inside the first Brillouin zone. This reduces the effective number of the valleys to four, as shown in Fig. 4.1. In Table 4.2 the bandgap E_g and the electron longitudinal m_l and transverse m_t effective masses are presented. In addition, for the case of holes, the effective masses of the light hole m_{lh} and the heavy hole m_{hh} bands as well as the distance to the split-off band Δ_{so} are given. Note that the next-highest minima in Si are situated at the L points and in Ge they are at the Γ and the Δ points.

For III–V compounds different situations occur for different materials: Some of these compounds are direct-bandgap crystals and others are indirect materials. Thus, for the conduction bands of these materials, the conduction-band edge can be found either at the Γ point or at the L or X-points. In Table 4.3 energy-band parameters are shown for two families of III–V compounds; the direct or the indirect nature of crystals is indicated.

For group IV semiconductors and for III–V compounds, the tops of the valence bands at $\vec{k} = 0$ have high degeneracies because these bands originate from the bonding of p orbitals of the atoms comprising the crystals. Thus, if one neglects the spin-orbit interaction of electrons, one obtains three degenerate valence bands, each of which is also doubly degenerate as a result of electron spin. In fact, the interaction between the spin of an electron and its motion – the so-called spin-orbit interaction – causes a splitting of these sixfold degenerate states. At $\vec{k} = 0$ they are split into (1) a quadruplet of states with degeneracy equal to 4, which are referred to as the Γ_8 valence bands, and (2) a doublet of states with degeneracy equal to 2, which are known as the Γ_7 valence bands. This splitting of valence bands Δ_{so} at $\vec{k} = 0$ into the Γ_8 and the Γ_7 bands can be seen in Fig. 4.2. Frequently one refers to the Γ_7 band as the split-off valence band. The energy parameter Δ_{so} of this splitting is presented in Tables 4.2 and 4.3. At finite $\vec{k}$, the spin-orbit interaction leads to further splitting of the Γ_8 bands into two branches: the heavy- and the light-hole bands. The degeneracy at $\vec{k} = 0$

Table 4.3.
Energy-Band Parameters for III–V Compounds

The As Family	InAs	GaAs	AlAs
Type of gap	Direct	Direct	Indirect
Lowest minima	Γ point	Γ point	X points
E_g (eV)	0.354	1.42	2.95
m_α^*/m	0.025	0.067	0.124
m_{hh}/m	0.41	0.50	0.50
m_{lh}/m	0.26	0.07	0.26
Δ_{so} (eV)	0.38	0.34	0.28
The Sb Family	**InSb**	**GaSb**	**AlSb**
Type of gap	Direct	Direct	Indirect
Lowest minima	Γ point	Γ point	X points
E_g (eV)	0.18	0.7	2.22
m_α^*/m	0.013	0.04	
m_{hh}/m	0.47	0.49	0.76
m_{lh}/m	0.015	0.076	0.15
Δ_{so} (eV)	0.81	0.34	0.29

makes the dispersion dependences $E_{h,l}(\vec{k})$ for the heavy- and the light-hole bands different from the simple form of Eq. (4.1):

$$E_{h,l}(\vec{k}) = A\vec{k}^2 \pm \sqrt{B^2\vec{k}^2 + C^2\left(k_x^2 k_y^2 + k_x^2 k_z^2 + k_y^2 k_z^2\right)}; \qquad (4.4)$$

here $A > 0$, B, and C are constants. An important feature of the dependences of Eq. (4.4) is that they have cubic symmetry. However, the spherical approximation of Eq. (4.4),

$$E_h(\vec{k}) = (A + \bar{B})\vec{k}^2 = \frac{\hbar^2 \vec{k}^2}{2m_{hh}}, \qquad E_l(\vec{k}) = (A - \bar{B})\vec{k}^2 = \frac{\hbar^2 \vec{k}^2}{2m_{lh}}, \qquad (4.5)$$

is frequently applicable. Parameters characterizing these approximations are the heavy-hole m_{hh} and the light-hole m_{lh} masses and $\bar{B} = \sqrt{B^2 + C^2/5}$. These masses are presented for Si, Ge, and some III–V compounds in Tables 4.2 and 4.3.

4.2.2 Band Structures of Semiconductor Alloys

As emphasized frequently in previous discussions, the energy structure of a particular semiconductor determines its optical and, in part, its electrical properties. For naturally existing semiconductor crystals like monoatomic Ge and Si, binary GaAs, etc., their fixed and unalterable energy-band structures restrict their applications. One of the powerful tools for varying the band structure is based on alloying two or more semiconductor materials. Some alloys exhibit well-ordered crystal structures. Although an alloy always has a disorder of the

Table 4.4.
Bandgaps for III–V Alloys

Alloy	E_g (eV)
$Al_xGa_{1-x}As$	$1.42 + 1.247x$
$Al_xIn_{1-x}As$	$0.360 + 2.012x + 0.698x^2$
$Ga_xIn_{1-x}As$	$0.36 + 1.064x$
$Ga_xIn_{1-x}Sb$	$0.172 + 0.139x + 0.415x^2$
$Al_xGa_{1-x}Sb$	$0.726 + 1.129x + 0.368x^2$
$Al_xIn_{1-x}Sb$	$0.172 + 1.621x + 0.430x^2$

constitutive atoms, contemporary technology facilitates partial control of this disorder and produces high-quality crystals. The properties of such materials can be interpreted in terms of nearly ideal periodic crystals.

Consider an alloy consisting of two components, A with a fraction x and B with a fraction $(1 - x)$. If A and B possess similar crystalline lattices, one can expect that the alloy A_xB_{1-x} has the same crystalline structure. Then the symmetry analysis can be extended to these types of alloys. For Si/Ge alloys and III–V compounds, this leads us to the previously discussed symmetry properties of the energy bands. Since the band structures are similar, one can characterize different parameters of the alloy as function of the fraction x. For example, the bandgap of an alloy can be represented as $E_g^{alloy} = E_g(x)$. Such approximate dependences are given in Table 4.4 for III–V compounds. They correspond to the bandgaps at Γ points.

As the composition of an alloy varies, the internal structure of energy bands changes significantly. For example, in the case of $Al_xGa_{1-x}As$ alloys, the lowest energy minimum of the Γ conduction band of GaAs is replaced by the six X minima of AlAs as the value of x is increased. The evolution of all three energy minima is shown in Fig. 4.3. One can see that near the composition $x \approx 0.4$, the alloy transforms from a direct- to an indirect-bandgap material. The x dependences of effective masses for different electron-energy minima as well as for heavy and light holes are presented in Table 4.5 for $Al_xGa_{1-x}As$. The

Table 4.5.
Effective Masses for the Alloy $Al_xGa_{1-x}As$

Type of minimum	Effective mass m_α^*/m
Γ point	$0.067 + 0.083x$
X minima	$0.32 - 0.06x$
L minima	$0.11 + 0.03x$
Heavy hole	$0.62 + 0.14x$
Light hole	$0.087 + 0.063x$

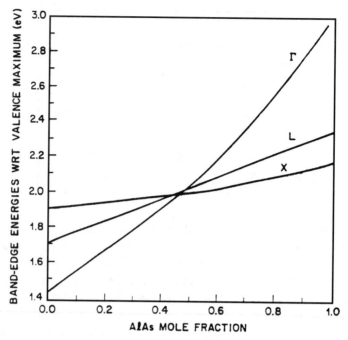

Figure 4.3. Positions of major energy minima of the conduction band with respect to (WRT) the valence band as a function of the fraction of AlAs in alloys $Al_xGa_{1-x}As$. The lowest minimum determines the bandgap of the alloy. [After S. M. Sze, *High-Speed Semiconductor Devices* (Wiley, New York, 1990).]

dependence of the bandgap of $Si_{1-x}Ge_x$ alloys, as a function of the Ge fraction, is shown in Fig. 4.4. $E_g^{alloy}(x)$ is a nonlinear function of x, as is seen for the unstrained alloys in Fig. 4.4; see the curve corresponding to unstrained material – the contribution of strain is discussed in Subsection 4.3.3. Conduction-band minima remain at Δ points until a Ge composition of approximately 85% is reached. At this point, the alloy conduction band becomes Ge-like with minima at L points and a faster decrease occurs in $E_g^{alloy}(x)$.

Clearly the ability to fabricate a variety of superb-quality materials provides an excellent tool for modifying the electron-energy structure.

4.2.3 Electron Affinities: Energy-Band Discontinuities of Heterostructures

In previous sections, by considering the energy bands of different semiconductors, we have analyzed energy structures in terms of relative positions of these bands. For example, in Fig. 4.2 all energies are presented with respect to the tops of valence bands. If we deal with a single homostructure, this approach is correct for most physical effects and many applications. In this case, absolute values of the energies are not important and only relative positions of the bands need to be taken into account. However, if two different materials are brought together the absolute values of energies become critically important.

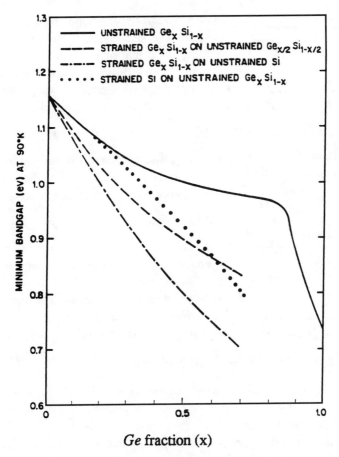

Figure 4.4. The bandgaps for Ge_xSi_{1-x} alloys. Unstrained and different kinds of strained alloys are presented. The strong nonlinear dependence of the bandgap for the unstrained alloy is caused by an interplay of Δ-type and L-type minima of the material. Two kinds of strained alloys correspond to different substrates, which impose a different strain on the alloy. The bandgap for strained Si materials grown on unstrained Ge_xSi_{1-x} alloys is shown for comparison. [After J. C. Bean, "The growth of novel silicon materials," Phys. Today 39(10), pp. 36–42 (1986).]

There is a simple way to compare energy bands of different materials. Let us introduce the vacuum level of the electron energy that coincides with the energy of an electron outside a material. It is obvious that the vacuum level is the same for any material. We can characterize the absolute energy position of the bottom of the conduction band with respect to this level, as shown in Fig. 4.5. The energy distance between the conduction-band bottom and the vacuum level χ is called the electron affinity. In other words, the electron affinity is the energy required for removing an electron from the conduction-band bottom to the outside of a material, i.e., to the vacuum level. The electron affinities allow us to compare energy bands of different materials with respect to each other. Another useful characteristic presented in Fig. 4.5 is the so-called work function. It is defined as

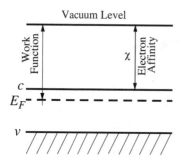

Figure 4.5. The diagram introducing the electron affinity and the work function in a crystal. The electron affinity may be used to find absolute values of energies of the valence and the conduction bands. The work function is useful for finding relative positions of bands for two adjacent doped semiconductors as well as for metal-to-semiconductor contacts.

the energy required for removing an electron from the Fermi level of a crystal to the vacuum level.

With this definition of electron affinity, one can calculate the discontinuity in the conduction band at an abrupt heterojunction of two materials, A and B:

$$\Delta E_c = E_{c,B} - E_{c,A} = \chi_A - \chi_B, \qquad (4.6)$$

where $\chi_{A,B}$ are the electron affinities of materials A and B. Similarly, one can calculate the discontinuity of the valence band for the same heterojunction:

$$\Delta E_v = E_{v,B} - E_{v,A} = \chi_B - \chi_A + \Delta E_g, \qquad (4.7)$$

where ΔE_g is bandgap discontinuity for the heterojunction. Thus, if this simple approach – the electron-affinity rule – is applicable to a pair of semiconductor materials, one can calculate band offsets for an ideal heterojunction. Furthermore, if three materials, say A, B, and C, obey this rule, the following transitivity property is valid:

$$\Delta E_v(A/B) + \Delta E_v(B/C) + \Delta E_v(C/A) = 0,$$

where $\Delta E_v(A/B)$ is the valence-band discontinuity at the A/B interface. Hence it is possible to calculate the band offset for one of three junctions if the parameters for two of them are known.

Unfortunately, this rule fails for many semiconductor pairs. One reason for this failure is the dissimilar character of chemical bonds in adjacent materials. The formation of new chemical bonds at such a heterojunction results in a charge transfer across this junction and the consequent reconstruction of energy bands, which leads to the breakdown of the rule. In real heterojunctions, band offsets can depend on the quality of interface, conditions of growth, etc.

Changes in discontinuity between the Γ minimum of the conduction band of GaAs and Γ, X, and L conduction-band minima of $Al_xGa_{1-x}As$ as functions of the fraction of AlAs are presented in Fig. 4.6. The valence-band offsets for these junctions can be found from these data and the bandgaps shown in Fig. 4.7.

Hence we conclude that if the energy-band structure of two semiconductors is known, energy discontinuities can be accounted for, at least for ideal heterojunctions.

4.3 Lattice-Matched and Pseudomorphic Heterostructures

In this section we consider some principal problems that arise in the fabrication of heterostructures. In general, one can grow any layer on almost any other material. In practice, however, the interfacial quality of such artificially grown structures can vary enormously. Even when one fabricates a structure from two materials of the same group or from compounds of the same family, the artificially

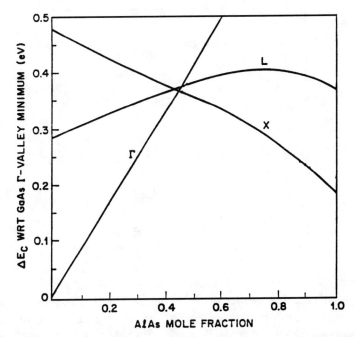

Figure 4.6. Discontinuity between the Γ minimum of the conduction band of GaAs and Γ, X, and L conduction-band minima of $Al_xGa_{1-x}As$ as functions of the AlAs mole fraction. The data facilitate the determination of conduction-band offsets at a heterojunction $Al_xGa_{1-x}As/Al_yGa_{1-y}As$ for all minima. [After S. M. Sze, *High-Speed Semiconductor Devices* (Wiley, New York, 1990).]

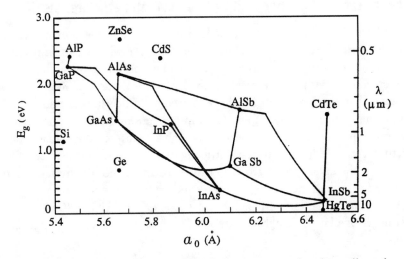

Figure 4.7. Room-temperature bandgaps E_g and absorption-wavelength cutoff λ as functions of lattice constant a_0 for III–V and II–VI compounds and group IV materials and their alloys. Each of the solid curves connecting two marked materials shows dependence $E_g(a_0)$ for a proper alloy. The right axis indicates the maximal wavelength of light that can be emitted or absorbed by a material because of interband phototransitions. [After S. M. Sze, *High-Speed Semiconductor Devices* (Wiley, New York, 1990).]

Table 4.6.
Lattice Constants for Cubic
Semiconductor Materials ($T = 300$ K)

Semiconductor	Lattice constant (Å)
SiC	3.0806
C	3.5668
Si	5.4309
GaP	5.4495
GaAs	5.6419
Ge	5.6461
AlAs	5.6611
InP	5.8687
InAs	6.0584

grown materials of the heterostructure may be different from the corresponding bulk materials. The quality of the materials near heterointerfaces depends strongly on the ratio of lattice constants for the two materials. In Table 4.6 lattice constants for several group IV semiconductors and III–V compound semiconductors are presented; all presented materials are cubic crystals.

Lattice constants for some other materials can be found from Fig. 4.7. Depending on the structural similarity and the lattice constants of the constitutive materials, there are two principally different classes of heterointerfaces.

4.3.1 Lattice-Matched and Lattice-Mismatched Materials

For lattice-matched structures, the lattice constants of the constitutive materials are nearly matched, i.e., the lattice constants are within a small fraction of a percent of each other. There is no problem, in general, in growing high-quality heterostructures with such lattice-matched pairs of materials. By high quality we mean that the interfacial structure is free of lattice imperfections such as dislocations, interface defects, etc. Such imperfections result in poor electrical and optical properties and may lead to fast and widespread degradation of the structure. Figure 4.8(a) illustrates a lattice-matched layer on a substrate. One can expect that the layer can be grown on the substrate if both materials are from the same family and the binding energies and crystal structures are similar.

According to the data of Fig. 4.7 and Table 4.6, the AlGaAs/GaAs system is an example of a lattice-matched material. The system has a very small mismatch

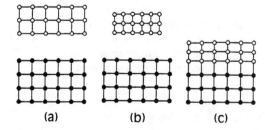

Figure 4.8. Schematic illustration of (a) lattice-matched, (b), (c) lattice-mismatched materials. For case (c) the resulting heterostructure is strained (pseudomorphic) if the upper layer adapts the substrate lattice.

(a)   (b)   (c)

of the lattice constants of only ~0.1% over the entire range of possible Al-to-Ga ratios in the AlGaAs. As a result, such heterostructures can be grown free of a mechanical strain and significant imperfections; hence these structures provide a practical way of tailoring band structures.

In addition to these tailored electronic parameters, elastic and other lattice properties can be different in layers comprising such a lattice-matched heterostructure. As discussed in Chapter 5, such a mismatch in elastic constants and dielectric permittivities leads to a modification of acoustic- and optical-phonon spectra in the heterostructures.

The case of lattice-mismatched structures is characterized by a finite lattice mismatch. Figure 4.8(b) depicts this case. If one tries to match these lattices, strain in the plane of growth and a distortion along the growth axis arise. Thus one obtains a strained layer with a lattice deformation.

Consider an elastic deformation of a lattice. It can be characterized by the vector of relative displacement $\vec{u}$. The displacement defines how any lattice point $\vec{r}$ moves to a new position, $\vec{r}' = \vec{r} + \vec{u}$, as a result of the deformation. Different points of the crystal can be deformed differently; thus the displacement depends on coordinates: $\vec{u} = \vec{u}(\vec{r})$. In fact, only relative displacements are important; they are determined by the strain tensor:

$$u_{ij} = \frac{1}{2}\left(\frac{\partial u_i}{\partial x_j} + \frac{\partial u_j}{\partial x_i}\right).$$

In Chapter 5 we will study acoustic vibrations of the lattice in terms of the displacement vector and the strain tensor. Here we consider only diagonal components of u_{ij}. They determine a change in the crystal volume, from V to V', produced by the strain:

$$\delta = \frac{V' - V}{V} = u_{xx} + u_{yy} + u_{zz}.$$

The elastic energy per unit volume of a crystal may also be expressed in terms of the strain tensor. For cubic crystals this energy is given by

$$U = \frac{1}{2}c_{11}\left(u_{xx}^2 + u_{yy}^2 + u_{zz}^2\right) + c_{44}\left(u_{xy}^2 + u_{xz}^2 + u_{yz}^2\right)$$
$$+ c_{12}(u_{xx}u_{yy} + u_{xx}u_{zz} + u_{yy}u_{zz}), \tag{4.8}$$

where c_{11}, c_{12}, and c_{44} are elastic constants or elastic moduli of the crystal. Derivatives of the elastic energy with respect to strain-tensor components give the so-called stress tensor:

$$\sigma_{ij} = \frac{\partial U}{\partial u_{ij}}. \tag{4.9}$$

Boundary conditions at a surface or at an interface may be formulated in terms of the stress tensor:

$$\sigma_{ij} N_j = f_i, \tag{4.10}$$

where $\vec{N}$ is a vector perpendicular to the surface and $\vec{f}$ is an external force applied to the surface.

Equations (4.8)–(4.10) are sufficient for calculations of the strain of a layer B grown on a mismatched substrate A. Let the lattice constants of these two materials be a_A and a_B, respectively. In this discussion, both materials are assumed to be cubic crystals and the direction of growth is along the [001] direction. If layer B adopts the lattice periodicity of substrate A, the in-plane strain of the layer is

$$u_{xx} = u_{yy} = u_\parallel = \frac{a_A}{a_B} - 1. \tag{4.11}$$

There should be no stress in the direction of growth; thus, from Eq. (4.10) it follows that $\sigma_{zz} = 0$. Calculating σ_{zz} from Eqs. (4.8), (4.9), $\sigma_{zz} = c_{11} u_{zz} + c_{12}(u_{xx} + u_{yy})$, we find the strain in the direction perpendicular to the layer:

$$u_{zz} = -\frac{2c_{12}}{c_{11}} u_\parallel. \tag{4.12}$$

Thus the strain can be found through the lattice-constant mismatch.

The strain results in two kinds of effects: first, the strain can generate different imperfections and defects; second, the strain in the layer leads to a change in the symmetry of the crystal lattice, for example, from cubic to tetragonal or to rhombohedral, etc. Of course, the latter effect can modify the energy-band structure of the layer.

To understand the nature of the formation of imperfections in a layered structure, let us consider the characteristic energies of the structure. First, a layer grown on a substrate with a mismatched lattice should possess extra elastic energy E_{el} caused by the strain. This energy is a function of the thickness of the layer d and increases with increasing d. In the simplest case of uniform strain, the elastic energy can be calculated through its density U: $E_{el} = UdS$, where S is the area of the layer. On the other hand, the generation of misfit dislocations or misfit defects requires some energy. Let us denote this energy by E_{im}. If the extra elastic energy exceeds the energy associated with the imperfection, i.e., if $E_{el}(d) > E_{im}$, the system will relax to a new state with lower energy and imperfections will be generated, that is, the extra strain energy is the main physical reason for the instability and degradation of heterostructures fabricated from materials with a large mismatch in lattice constants.

Since the value of E_{im} remains finite even in thin layers, for certain thicknesses we may get $E_{el}(d) < E_{im}$. Thus there is not sufficient strain energy and imperfections will not be generated. Such strained heterostructures can enjoy high quality. Hence, in some approximations, for each pair of materials there is a critical thickness of the layers d_{cr}; if $d < d_{cr}$ the lattice mismatch is accommodated by the layer strain without the generation of defects. The corresponding layered systems are called pseudomorphic heterostructures. In general, a pseudomorphic layer of material possesses some characteristics similar to those of the substrate and may possibly have the same lattice structure as the substrate material. In our case, a crystalline semiconductor layer grown on another semiconductor takes on the in-plane lattice periodicity of the substrate semiconductor. Figure 4.8(c) illustrates the case in which the deposited layer adopts the lattice periodicity of a substrate material.

Examples of such systems are the $Ga_{1-x}Al_xAs/Ga_{1-x}In_xAs$ and $GaAs/Ga_{1-x}In_xAs$ structures. In fact, these heterostructures are used to improve the characteristics of the heterojunction field-effect transistor (FET). In spite of significant mismatches of lattice constants, these structures are virtually free of interface defects because of the small thicknesses of the pseudomorphic layers required for fabricating functioning heterojunction FETs.

It is sometimes possible to grow defect-free systems with layer thicknesses exceeding the critical thickness. However, such systems are metastable and may lead to device degradation as a result of the generation of misfit dislocations and defects driven by temperature effects or other external perturbations. Central to the stability of pseudomorphic structures is the question of whether or not the strain energy leads to damage of the materials when the structures are subjected to various forms of external stress and processing. The experience accumulated in this field shows that in the case of small strain energy the heterostructures are stable. For example, in the case of the GaP/GaAsP layered system, the strain energy is $\sim 10^{-3}$ eV/atom. Since this quantity is rather small compared with the energy required for removing the atom from its lattice site in material, this system can be stable for sufficiently thin layers.

The above-discussed strain states are shown as functions of x for Ge_xSi_{1-x} layers grown on Si substrates in Fig. 4.9. The phase diagram – critical thickness of the layer versus the Ge fraction – consists of three regions: strained layers with defects at large thicknesses, nonequilibrium (metastable) strained layers without defects at intermediate thicknesses, and equilibrium and stable layers without defects at small thicknesses. According to these results, a stable Ge layer on Si (the largest misfit) cannot be grown with a thickness greater than 10 Å or so; however, this structure has the largest misfit possible for a Ge–Si structure.

4.3.2 Lattice-Matched Heterostructures

Let us return to Fig. 4.7 and discuss lattice-matched heterostructures in more detail. From this figure, we can determine the lattice constants of different compounds. First, we can see that the GaAs/AlAs system exhibits a really unique situation in that the lattice constants have almost identical values. To achieve a lattice match for other cases, it is possible to combine either a binary compound and a ternary compound or ternary–ternary compounds having appropriate ratios of atomic species within each layer. For example, in the case of GaInAs/InP structures, lattice matching is achieved exactly only if the ratio of Ga to In is 47 to 53 in the GaInAs layer; for other ratios, the GaInAs layer is not lattice matched with the InP. Another example, the wide-bandgap $Ga_{0.51}In_{0.49}P$ material is compatible with the narrow-bandgap GaAs material.

The data presented in Figs. 4.3, 4.6, and 4.7 facilitate the design of III–V heterostructures to achieve energy-band tailoring or engineering, that is, these data help us find the materials and possible alloys needed to identify structures with high-quality quantum interfaces and the necessary band structure. For example,

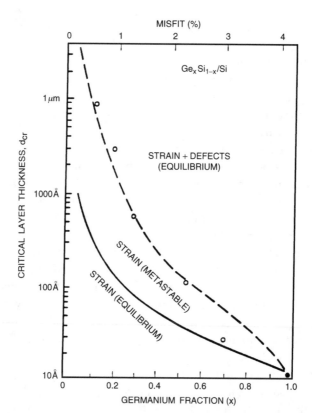

Figure 4.9. The phase diagram – the critical layer thickness versus the Ge fraction – for $Ge_x Si_{1-x}$ layers grown upon Si substrates. The upper region is favorable to form a phase with misfit defects in the strained layer. The middle region is a defect-free phase; it can be fabricated by a low-temperature technique and corresponds to metastable layers. The lower region is stable and a defect-free phase. [After R. People and J. C. Bean, "Erratum: Calculation of critical layer thickness versus lattice mismatch for $Ge_x Si_{1-x}$/Si strained-layer heterostructures [Appl. Phys. Lett. 47, 322 (1985)]," Appl. Phys. Lett. 49, pp. 229–229 (1986).]

by using these data, we can select a lattice-matched pair of materials for heterostructures with the bandgap necessary for particular optical applications, electronic bipolar devices, or unipolar devices with the necessary depths of quantum wells, heights of barriers, and even effective masses for electrons. A similar approach can be used to design devices based on other IV–IV and II–VI compounds.

Figure 4.3 also contains interesting information on higher minima of the $Al_x Ga_{1-x}As$ compounds and their dependences on composition. It is possible to trace the evolution of the minimum positions in AlGaAs/GaAs heterostructures as functions of AlAs content. The interplay of the three lowest minima leads to additional complications in the energy structure of a quantum well. Figure 4.10 illustrates the coordinate-dependent positions of the lowest minima of the AlAs/GaAs quantum well. One can see that, for these heterostructures, the GaAs layer provides a well for the Γ electrons. The L valley is weakly dependent of the composition, and the electrons from this valley are not really confined in the GaAs layer. On the other hand, if the electron transfers to the X valley it

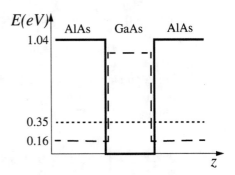

Figure 4.10. Spatial profiles of Γ (solid line), X (dashed line), and L (dotted line) minima in an AlAs/GaAs/AlAs heterostructure. Such a heterostructure creates a well for electrons from the Γ minimum. Electrons from X and L minima are not confined in the GaAs layer.

becomes free, because for electrons in the X valley of the GaAs layer, serves as a barrier instead of serving as a quantum well. Consequently, intervalley transitions that underlie a number of devices, such as Gunn diodes in bulk materials, lead to new effects in novel quantum-effect devices. As in the previous example, the intervalley Γ–X transition causes the escape of electrons from the quantum well. It is important to note that there is still a barrier of ∼0.16 eV between the bottom of the Γ valley in GaAs and the bottom of the X valley in AlAs.

In conclusion, the broad range of possibilities for controlling bandgaps, quantum-well depths, or barrier heights for both electrons and holes, energy-minima locations for the electron spectra, as well as electron and hole effective masses, provides the basis for energy-band engineering. Through such energy-band engineering, it is possible to design and fabricate high-quality heterostructures with designated optical and electrical properties. If one cannot achieve the desired properties by using lattice-matched compositions, it is possible to use strained pseudomorphic structures.

4.3.3 Strained Pseudomorphic Structures

As pointed out in Subsection 4.3.1 the strain in pseudomorphic systems causes important changes in their bandgap structure, including the shift of energy states and the breaking of energy-band degeneracies. Such effects provide additional flexibility in varying the energy levels in quantum structures and even in controlling the effective masses.

Let us consider briefly the modification of the energy structure in a strained crystal. If the energy band is not degenerate, the main result of the strain u_{ij} is to shift the energy minima; here we use the index α to label the energy bands. This shift can be written as a linear function of the strain tensor:

$$\delta E^\alpha = \mathcal{D}^\alpha{}_{ij} u_{ij}, \tag{4.13}$$

where $\mathcal{D}^\alpha{}_{ij}$ is the tensor of the deformation potential. The properties of the tensor $\mathcal{D}^\alpha{}_{ij}$ depend on the crystal symmetry and the type of the minimum. For example, for the Γ minimum, the tensor reduces to a scalar

$$\mathcal{D}^\Gamma{}_{ij} = \mathcal{D}\, \delta_{ij}. \tag{4.14}$$

The energy shift of the nondegenerate Γ minimum in the conduction band is

$$\delta E^\Gamma = \mathcal{D}(u_{xx} + u_{yy} + u_{zz}). \tag{4.15}$$

For Δ and X minima, the tensor reduces to two constants, $\mathcal{D}_l$ and $\mathcal{D}_t$. For example, in the case of the $X_{[100]}$ minimum (the same is true for Δ minima in Si), one obtains

$$\mathcal{D}^X_{ij} = \begin{pmatrix} \mathcal{D}_l & 0 & 0 \\ 0 & \mathcal{D}_t & 0 \\ 0 & 0 & \mathcal{D}_t \end{pmatrix}. \tag{4.16}$$

The energy shifts of this type of minimum are

$$\delta E^X_{[100]} = \delta E^X_{[\bar{1}00]} = \mathcal{D}_t(u_{xx} + u_{yy} + u_{zz}) + (\mathcal{D}_l - \mathcal{D}_t)u_{xx}, \tag{4.17}$$

$$\delta E^X_{[010]} = \delta E^X_{[0\bar{1}0]} = \mathcal{D}_t(u_{xx} + u_{yy} + u_{zz}) + (\mathcal{D}_l - \mathcal{D}_t)u_{yy}, \tag{4.18}$$

$$\delta E^X_{[001]} = \delta E^X_{[00\bar{1}]} = \mathcal{D}_t(u_{xx} + u_{yy} + u_{zz}) + (\mathcal{D}_l - \mathcal{D}_t)u_{zz}. \tag{4.19}$$

From Eqs. (4.17)–(4.19) and the conditions for isotropic strain (hydrostatic compression),

$$u_{xx} = u_{yy} = u_{zz}, \tag{4.20}$$

it follows that all Δ as well as X minima shift identically and no splitting occurs. In crystals subjected to anisotropic strain, the symmetry of energy minima is broken and they are split. For example, if a biaxial strain occurs in the x, y plane, the six initially equivalent Δ minima of a Si-like structure split into two minima of the [001] axis, called $\Delta_\perp$, and four minima of [100] and [010] axes, called $\Delta_\parallel$. A sign of the combination $(\mathcal{D}_l - \mathcal{D}_t)$ determines what kinds of minima go up or down. For Si and Ge crystals these combinations are positive and equal 9.2 and 15.9 eV, respectively. Thus, for a biaxial compression in the x, y plane, the strain is

$$u_{xx} = u_{yy} < 0, \qquad u_{zz} > 0, \tag{4.21}$$

and from Eqs. (4.17)–(4.19) we obtain an upward shift of the $\Delta_\perp$ minima and a decrease in energy of the $\Delta_\parallel$ minima. For a biaxial tension,

$$u_{xx} = u_{yy} > 0, \qquad u_{zz} < 0, \tag{4.22}$$

we find the opposite result: a decrease in energy of the $\Delta_\perp$ minima and an upward shift of the $\Delta_\parallel$ minima. Since the parameters of the deformation-potential tensor are of the order of magnitude of 10 eV, relatively small strains of 10^{-3}–10^{-2} produce energy-minima shifts of 10–100 meV.

The Γ valence bands that are degenerate at $\vec{k} = 0$ exhibit more complex behavior under a strain including a splitting of bands, a strain-induced anisotropy, and a variation of the effective masses. Under a biaxial strain in the x, y plane

$(u_{xx} = u_{yy} \neq u_{zz})$ the simplified dispersion relations of Eq. (4.5) for the heavy and the light holes transform to

$$E_{h,l} = a(u_{xx} + u_{yy} + u_{zz}) + A\vec{k}^2$$

$$\pm \sqrt{\bar{B}^2\vec{k}^4 + b\bar{B}(u_{xx} - u_{zz})(k_x^2 + k_y^2 - 2k_z^2) + b^2(u_{xx} - u_{zz})^2}. \quad (4.23)$$

Here a and b are two constants describing changes in energy spectra of the Γ valence bands. From Eq. (4.23) one can see that for a hydrostatic strain of Eq. (4.20), the valence bands do not split, masses do not change, and only a common shift of both the heavy and the light valence bands occurs. For biaxial strain there is splitting between the bands, even at $\vec{k} = 0$:

$$E_h(0) - E_l(0) = -2b(u_{xx} - u_{zz}).$$

From Eq. (4.23) it follows that the energy dispersions $E_h(\vec{k})$ and $E_l(\vec{k})$ depend on the strain; in particular, they become anisotropic with large changes in curvatures, i.e., in effective masses. Typically the constants of the deformation potential, a and b, are ~1–5 eV, respectively, that is, the splitting of the heavy- and the light-hole bands can reach 100 meV at strains of $\sim10^{-2}$.

At this point, we apply these results to pseudomorphic heterostructures. Using Eqs. (4.11) and (4.12), we can estimate the biaxial strain of the top layer. Then the energy spectra of electrons and holes can be calculated from Eqs. (4.17)–(4.19) and (4.23), respectively. In Fig. 4.11 the hole-energy bands are shown for an unstrained GaAs layer as well as for layers with biaxial tension and compression. The strain induces band splitting of up to 50 meV; considerable anisotropy of $E_{h,l}(\vec{k})$ and changes in curvatures of dispersion dependences are clearly evident.

Useful examples of the modification of energy bands in these heterostructures include the strain-induced reduction of the bandgap in the following binary and ternary compounds: GaP/GaAsP, GaSb/AlSb, GaAs/InAlAs, and Si/SiGe. Pseudomorphic systems of III–V compounds are promising for the design of efficient photodetectors, modulation-doped field-effect transistors (MODFETs), and injection lasers. A recent impressive example of the application of strained structures was the fabrication of the first blue–green laser diodes based on ZnSe/ZnSeS and ZnCdSe/ZnSe heterostructures. These systems are type-I pseudomorphic heterostructures and can provide quantum wells for electrons and holes. They have wider bandgaps than III–V compounds.

As indicated in Fig. 4.7, there is a small lattice mismatch of 0.28% between ZnSe and GaAs. Because of this and owing to virtually perfect GaAs technology, II–VI quantum structures are grown principally on GaAs substrates. Quantum structures with unique optical properties in the blue–green wavelength region have been produced, and qualitatively new laser systems have been developed as a result of a combination of technological achievements in the fabrication of III–V and II–VI compounds. These structures are so perfect that even at room temperature exciton emission is observed. By contrast, such excitonic emission has never been observed in bulk II–VI crystals.

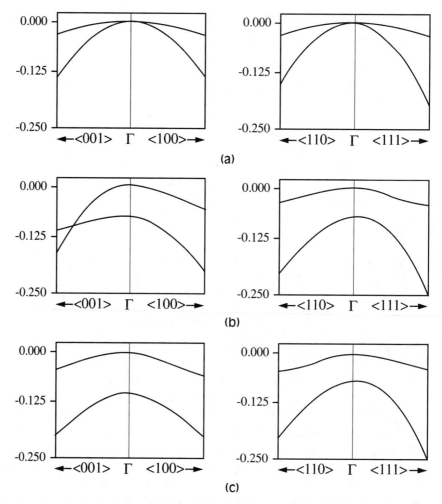

Figure 4.11. Energy dispersions of valence bands for (a) unstrained GaAs, (b) that under a biaxial tension, (c) biaxial compression. Results are presented for four directions of the wave vector, [100], [$\bar{1}$11], [110], and [111]. [After J. Singh, *Physics of Semiconductors and Their Heterostructures* (McGraw-Hill, New York, 1993).]

4.3.4 Si/Ge Strained Heterostructures

Si/Ge systems are interesting and important because they have opened new horizons for Si technology and Si-based applications, including optoelectronics. Let us consider these structures in more detail because they differ from the III–V quantum structures described in Subsection 4.3.3. The data of Table 4.6 show that heterostructures based on Si and Ge materials should always be pseudomorphic.

First, the stability and the quality of these Si/Ge pseudomorphic heterostructures are strongly dependent on the thicknesses of the strained layers, as discussed in Subsection 4.3.2. In fabricating Si/Ge structures one grows different numbers of Si and Ge monolayers. Thus layer thicknesses can be characterized by the numbers of these monolayers. Let n and m be numbers of monolayers of Si and Ge, respectively. This system is known as the Si_n/Ge_m superlattice. The second important factor determining the quality of these structures is the material of the substrate on which the superlattices are grown. We have discussed the case of Si/Ge layers grown upon Si substrates; see Fig. 4.9. For the fabrication of Si/Ge superlattices, the substrates of choice are frequently either Si_xGe_{1-x} alloys or GaAs materials.

Let us consider Si_xGe_{1-x} as a substrate. The elastic energy of a strained system depends on the alloy composition of the substrate. Figure 4.12 illustrates this dependence for different numbers of monolayers for the symmetric case $n = m$. In accordance with the previous discussion, the elastic energy increases with increasing thicknesses of strained layers for a given substrate material. Because of this, one uses superlattices with a few monolayers: $2 \le (n, m) \le 5$. Figure 4.12 shows also a nontrivial strain-energy dependence on the alloy composition of the substrate; the minimal strain energy is expected for $x = 0.4$–0.6.

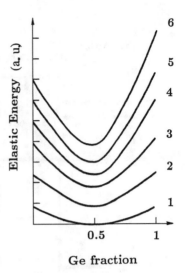

Figure 4.12. The elastic energy of strained Si_n/Ge_n superlattices with different numbers of monolayers n as a function of the Ge fraction in a substrate. The numbers at the curves indicate n.

Another important characteristic of pseudomorphic Si/Ge structures is the distribution of the elastic energy over the monolayers of the superlattice. It was shown that the most homogeneous distribution over layers occurs for the Si/Ge alloy with $x \approx 0.5$. From this point of view, the $Si_{0.5}Ge_{0.5}$ substrates are preferable. However, these results depend strongly on the orientation of the substrate material. Most often, the direction of growth on the substrate is chosen to be the [001] direction.

Although the technology underlying the growth and the fabrication of Si/Ge structures is still in a developing stage, from our short analysis one can see that there is qualitative and even quantitative knowledge concerning the behavior responsible for the stability and perfection of these structures.

Now let us discuss the energy-band structure of Si/Ge superlattices. For a Si crystal, the electron affinity and the bandgap at $T = 300$ K are 4.05 and 1.12 eV, respectively. For a Ge crystal, these parameters are 4.0 and 0.66 eV, respectively. Thus, for a heterojunction intermediate between pure Ge and pure Si, a comparison of the electron affinities indicates that the conduction-band edge of Si lies below the conduction band of Ge. The Ge valence-band edge lies ~0.5 eV higher than the Si valence-band edge. The strain imposed by the substrate will affect this offset further. As a result, the bandgap of Si/Ge layered systems tends to be type II. The conduction band is formed from [001] Δ-like states for a superlattice grown in the [001] direction. Figure 4.13 illustrates this point. For example, when a superlattice is grown on a Ge substrate, Ge layers are unstrained while Si layers are strained, as in the case of Fig. 4.13(a). According to Table 4.6

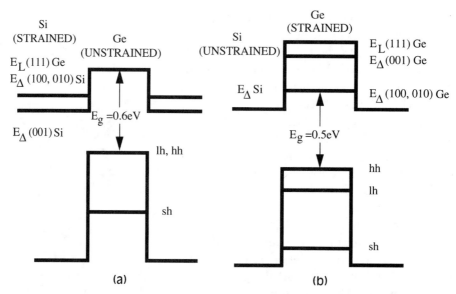

Figure 4.13. Energy structures of conduction and valence bands for Si/Ge superlattices grown on (a) a Ge substrate, (b) a Si substrate. The [001] orientation of growth is assumed. Only three layers of a superlattice are shown. The heterostructures tend to be the type-II superlattice. [After J. Zi, *et al.*, "Theoretical study of structures and growth of strained Si/Ge superlattices," Appl. Phys. Lett. 57, pp. 165–167 (1990).]

and Eqs. (4.11) and (4.12), the strain in the Si layer corresponds to biaxial tension ($u_\parallel > 0$). Equations (4.17)–(4.19) tell us that the $\Delta_\perp$ minima shift down. Thus the conduction-band edge will be formed from the strain-split $\Delta_\perp$ conduction minima in Si and L minima of unstrained Ge, which are only slightly higher in energy. Let us consider another case in which a superlattice is fabricated on an unstrained Si substrate, as in the case of Fig. 4.13(b). The strain in the Ge layers corresponds to biaxial compression ($u_\parallel < 0$). In the conduction band of the Ge layers, the $\Delta_\parallel$ minima go down and become the lowest ones, that is, they are below the usual [111] minima. As a result, the conduction band of the superlattice is formed from $\Delta_\parallel$-like valleys of Ge and Δ minima of unstrained Si; the conduction-band edge in Si is still lower in energy. Thus, for both the cases considered, the Si band edge should dominate in the conduction band and the Ge band edge dominates in the valence band, corresponding to a type-II superlattice. Figure 4.13 also depicts the character of the splitting of the valence band, showing qualitatively the positions of the heavy-hole (hh) valence band, the light-hole (lh) valence band, and the split-off hole band (sh).

More exact dependences of the energy structure of a Si/Ge superlattice can be calculated as functions of the strain. Let us introduce a planar strain $u_\parallel^{Si}$ in the Si layers. Since the strain is controlled by the substrate, $u_\parallel^{Si} = 0$ corresponds to the Si substrate and $u_\parallel^{Si} = 4.17\%$ corresponds to that of Ge. Figure 4.14 illustrates band edges in individual strained layers of Si and Ge as well as the minibands forming in the Si/Ge superlattice as functions of $u_\parallel^{Si}$. In the conduction band, under such a strain, the Δ-type minima of both materials, Si and Ge, split into $\Delta_\perp^{Si,Ge}$ and $\Delta_\parallel^{Si,Ge}$ minima. For Si layers, the two $\Delta_\perp$ minima, lying on the axes perpendicular to the layer, move down with increasing $u_\parallel^{Si}$, while the four $\Delta_\parallel$ minima move up. In the Ge layer, the splitting is the largest at $u_\parallel^{Si} = 0$. In the valence bands, the positions of heavy-hole bands are almost constant, but the light-hole band increases in energy for the Si layer; that of the Ge layer follows an opposite tendency.

In the Si/Ge superlattice, both types of layers contribute to the energy spectrum and a complex structure of minibands occurs. Each minimum results in a miniband. Thus in the conduction band there are minibands related to $\Delta_\perp$ minima; two of them – the narrow $\Delta_\perp^1$ and the wide $\Delta_\perp^2$ – are presented in Fig. 4.14. Other types of minibands originate from $\Delta_\parallel$ minima; one such wide miniband $\Delta_\parallel^1$ is shown in Fig. 4.14. The dependences of the $\Delta_\perp$- and the $\Delta_\parallel$-like minibands on $u_\parallel^{Si}$ are different; at small $u_\parallel^{Si}$, the $\Delta_\parallel$-like miniband is the lowest one, but for $u_\parallel^{Si} \geq 1\%$, the $\Delta_\perp$-like miniband becomes the lowest. In the valence band, there is a wide miniband originating from the light-hole band lh^1 and a narrower heavy-hole miniband hh^1. Figure 4.14 shows the evolution of these minibands with $u_\parallel^{Si}$.

These results reveal the shifts of the energy levels of strained layers and superlattices. But they do not tell us about dispersion dependences, the direct or indirect character of the band structures, etc. To determine them it is necessary to perform full-band-structure calculations. However, a qualitative understanding of the energy dispersion in minibands can be obtained by exploitation of the Kronig–Penney approximation and the so-called energy-band folding procedure.

Let us turn to the simplest Kronig–Penney model, which was used in Section 3.6. Consider the case of low barriers in a superlattice. Let d be the period of the superlattice, that is, d is the total width of one quantum well and one barrier. The barriers modify the dispersion curve primarily around the points $k = \pi n/d$, $n = \pm 1, \pm 2, \ldots$, which are related to the backscattering of electrons. Bandgaps appear around these points, and, instead of a continuous curve, a set of segments of the curve that have nearly the same shape is obtained. This procedure is equivalent to having a set of zones in wave-vector space. Fortunately, as a result of the periodicity of the system, we can reduce the $E(k)$ dependence to that of

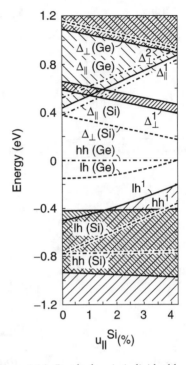

Figure 4.14. Band edges in individual layers (dashed and dashed–dotted lines) and super-lattice minibands (hatched regions) calculated for a Si_4Ge_4 superlattice as functions of the biaxial strain $u_\parallel^{Si}$ in Si layers. The interval of variation in $u_\parallel^{Si}$ corresponds to the Si_xGe_{1-x} substrate with x from 1 to 0. [After R. Zachai, *et al.*, "Photoluminescence in short-period Si/Ge strained-layer superlattices," *Phys. Rev. Lett.* **64**, pp. 1055–1058 (1990).]

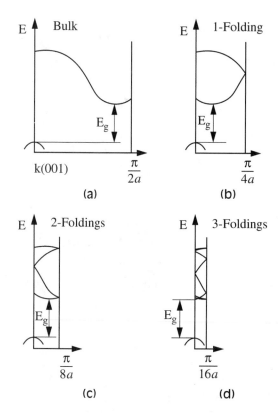

Figure 4.15. Effect of multiple-band folding for Si/Ge superlattices with a Si-like conduction band. For comparison, the bulk case is shown (a). Different numbers (n, m) of monolayers Si and Ge correspond to different foldings.

the first zone, which, as we already know, is the first Brillouin zone. The model discussed in this paragraph is known as a folded-zone representation, and the technique is referred to as the folding procedure.

The same easy-to-interpret procedure may be applied to the case of superlattices. Figure 4.15 illustrates the effect of multiple-band folding for several superlattices with Si-like conduction-band structures. The arrows on the figures indicate the position of the minimum of the lowest resulting conduction miniband. One can see that this position is shifted to the center of the Brillouin zone, $\vec{k} = 0$, as the number of layers per superlattice period increases. The top of the valence minibands is always situated at $\vec{k} = 0$; thus each successive folding makes the energy-band structure look more like a direct-bandgap structure. Although this model illustrates only a trend toward the formation of the band structure of Si/Ge superlattices, we can conclude that a [001]-orientation Si/Ge superlattice can be similar to direct-bandgap materials only if the band edge originates from Δ minima along the growth direction, i.e., from $\Delta_\perp$ minima.

Let us turn to the energy-band edges presented in Fig. 4.14. The $\Delta_\perp$ miniband becomes the lowest at $u_\parallel^{Si} \geq 1\%$. This means that quasi-direct-bandgap Si/Ge superlattices can be fabricated on either a Ge or a SiGe substrate. However, Si/Ge superlattices matched to Si cannot be direct-bandgap superlattices because a strain exceeding 1% is needed for the direct-bandgap structure; this magnitude of lattice mismatch is not possible for a Si substrate.

From these results, one can see that it is possible to combine and satisfy different requirements so that stable superlattices with desired physical properties – like a quasi-direct bandgap – may be realized. Accordingly, such an approach provides the basis for fabricating new heterostructures by use of Si technology for various applications, including optoelectronics.

4.4 Single-Heterojunction Devices: Selective Doping

4.4.1 Introduction

The results of the previous discussion allow us to begin our consideration of particular properties of heterostructure devices. The simplest of these are single-heterojunction devices. To this class, one may assign both metal-oxide-semiconductor (MOS) and semiconductor–semiconductor structures.

A question of importance, which was not discussed previously, is: how can one introduce carriers into heterostructures? By analyzing the band structure of semiconductor materials, we showed that the bandgap separates valence bands occupied by electrons from the lowest conduction band. Thus, at least for low temperatures, the conduction band is almost empty and a semiconductor should possess a high resistance. As the temperature increases some number of electrons is thermally excited from the valence bands to the conduction band. This creates the so-called intrinsic carriers – electrons and holes – that can carry an electric current.

The intrinsic concentrations of electrons n_i and holes p_i are given by Eqs. (2.119) and (2.122) under the condition

$$n_i = p_i. \tag{4.24}$$

Equation (4.24) implies that the hole concentration equals the concentration of electrons that have been removed from the valence band and placed in the conduction band. We designate the bottom of the conduction and the top of the valence bands by E_c and E_v, respectively, and use the simple dispersion dependences for electrons and holes given by Eqs. (2.136) and (2.137). The summation over $\vec{k}$ in Eqs. (2.119) and (2.122) can be replaced by an integration over the energy with the density of states represented by Eq. (3.31). The results of the calculation are

$$n = 2\left(\frac{m_e k_B T}{2\pi\hbar^2}\right)^{3/2} \exp\left(\frac{E_F - E_c}{k_B T}\right) = N_c \exp\left(\frac{E_F - E_c}{k_B T}\right), \tag{4.25}$$

$$p = 2\left(\frac{m_h k_B T}{2\pi\hbar^2}\right)^{3/2} \exp\left(\frac{E_v - E_F}{k_B T}\right) = N_v \exp\left(\frac{E_v - E_F}{k_B T}\right). \tag{4.26}$$

Here we have introduced the effective density of states for electrons N_c and holes N_v. In general, Eqs. (4.25) and (4.26) determine the carrier concentrations for any semiconductor if the Fermi level is known. For an intrinsic semiconductor the Fermi level E_{Fi} can be found by the substitution of Eqs. (4.25), (4.26) into

Eq. (4.24); the result is

$$E_{Fi} = \frac{1}{2} E_g + \frac{3}{4} k_B T \ln(m_h/m_e), \tag{4.27}$$

where $E_g = E_c - E_v$ is the bandgap. The Fermi level of an intrinsic material lies close to the middle of the bandgap. Finally, for the intrinsic concentrations we get

$$n_i = p_i = 2 \left(\frac{k_B T}{2\pi \hbar^2} \right)^{3/2} (m_e m_h)^{3/4} \exp\left(-\frac{E_g}{2k_B T} \right). \tag{4.28}$$

From Eq. (4.28) one can see that the concentrations n_i and p_i and therefore the current and other electrical and optical properties associated with intrinsic carriers are strongly dependent on the temperature. In most cases, they cannot be controlled by an external electric field, etc. Thus they are not useful for device operation.

In semiconductor devices, controllable methods of introducing carriers are used rather than relying on intrinsic carriers. One such method is the doping of materials by donor or acceptor impurities, which can generate free electrons (the so-called n-type material) or free holes (p-type material). Thus defects and impurities not only cause the deterioration of heterostructures but can also be important elements of real heterostructure devices.

Now we study specially doped heterostructures and devices based on them. We start with the concept of selective doping of heterostructures. Selective (or modulation) doping has found various applications and brings about major advantages for high-performance devices. As examples of such devices, we have to mention only several technologically important, selectively doped devices of microelectronics and optoelectronics: quantum-well photodetectors, quantum-cascade lasers, and transistors that use selective doping of the heterostructure to achieve better performance. These heterostructure FETs (HFETs) are referred to as high-electron-mobility transistors (HEMTs), two-dimensional electron-gas FETs (TEGFETs), selectively doped heterostructure transistors (SDHTs), and MODFETs.

The fundamental basis of electronic devices is electrical conduction of electrons and holes that behave as free carriers in various modes of device operation. Although these carriers may be supplied by donors or acceptors, these impurities cause a high rate of carrier scattering, which leads to low mobilities, increased resistivities, and other undesirable effects. When only bulk or bulklike materials were used in devices, this conflict between the necessity to control the concentration of carriers and the possibility of achieving high mobility was a major, unsolved problem.

The idea of spatially separating free carriers from their parent-charged impurities could not be realized in bulk materials because of the large electrostatic forces arising in the case of this separation. The first step in this direction was made when MOS structures began to be exploited in the so-called inversion-layer regime. Strictly speaking, a MOS structure with an inversion layer is not a selectively doped structure. But it illustrates the idea and is still the most applicable regime in Si electronics in which its principal embodiment takes the form of M–SiO$_2$–Si structures (M represents a metal). Let us therefore first study the major aspects of MOS structures.

4.4.2 Metal-Oxide-Semiconductor Structures

To explain the properties of a MOS structure, let us consider a metal and a semiconductor separated by an insulator. The energy diagram for a MOS system is presented in Fig. 4.16. In Subsection 4.2.3 we introduced the concepts of vacuum level, electron affinity, and work function. Let us choose some energy as the vacuum level. The electron affinity allows us to compare relative positions of energy bands of two crystals when they are brought together but before a charge transfer has occurred. If the charge transfer through the junction does occur, both materials become electrically charged – one positively, the other negatively. An electrostatic potential arises and manifests itself as band bending near the junction. Since the work function is the energy required for removing an electron from the Fermi level of a crystal to the vacuum level, the work functions of crystals determine the band bending; indeed, the charge transfer occurs until the Fermi levels of the two materials become aligned.

Let us introduce the work function for a metal $e\Phi_m$ and for a semiconductor $e\Phi_s$. We assume that

$$\Phi_m < \Phi_s;$$

we will explain this choice discussing Eq. (4.31). We also introduce the Fermi level in the metal E_{Fm} and in the semiconductor E_{Fs}, which is assumed to be of p type.

Figure 4.16. Energy diagrams of (a) electrically isolated metal and semiconductor, (b) equilibrium MOS structure after a charge transfer. The metal and the semiconductor are separated by a thin SiO$_2$ layer.

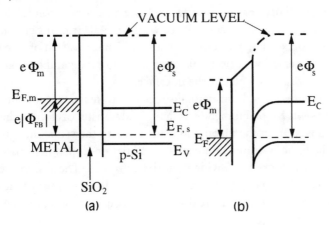

We can characterize the semiconductor by the electron and the hole concentrations, n and p, respectively. According to Eqs. (4.25)–(4.28) these concentrations in the bulklike region of the semiconductor can be written as

$$n = n_i \exp[(E_{Fs} - E_{Fi})/k_B T], \tag{4.29}$$

$$p = p_i \exp[-(E_{Fs} - E_{Fi})/k_B T], \tag{4.30}$$

where n_i and $p_i = n_i$ are given by Eq. (4.28) and E_{Fi} is determined by Eq. (4.27). For a p-type semiconductor, $p > n$, and from Eqs. (4.29) and (4.30) we obtain $E_{Fs} < E_{Fi}$ in the bulklike region of the p-type semiconductor.

Now let us allow electron exchange between the metal and the semiconductor. Under thermal-equilibrium conditions some electrons are transferred from the metal to the semiconductor until a constant Fermi level is established throughout the system. The metal is charged positively, and the semiconductor surface is charged negatively. The vacuum level should be continuous, as indicated in Fig. 4.16(b). We can see from this diagram that the edges of both conduction and valence bands E_c and E_v, respectively, bend down at the surface by $e|\Phi_{FB}|$, where

$$\Phi_{FB} = \Phi_m - \Phi_s. \tag{4.31}$$

The latter quantity can be interpreted as a voltage on the metal required for reaching the flat-band condition in the MOS structure. If the value of $e|\Phi_{FB}|$ is greater than the difference $E_i - E_{Fs}$, it follows from Eqs. (4.29) and (4.30) that the electron concentration exceeds the concentration of holes at the interface:

$$n > p.$$

Thus we have the situation in which the electron concentration near the surface may be high and the electrons are now present in the lightly doped p-type material. In such a material the scattering by impurities practically does not exist.

Usually the electrons accumulate inside layers 100–1000 Å thick. This phenomenon is called an inversion, and the layer with high carrier concentration near the interface is called the inversion layer. The presence of the inversion implies that we have changed the p-type conductivity of the acceptor-doped material into an n-type conductivity near the interface. Hence an electron-conduction channel is formed at the interface. In accordance with Fig. 4.16(b), this channel is separated from the bulk semiconductor with a small concentration of free holes by a nonconducting depletion region; this occurs because the band bending creates a potential barrier for the holes at the interface.

MOS structures provide an example of charge transfer. In this context, charge transfer represents the phenomenon of the spatial transfer of electric charge: electrons are removed from the metal and transferred to the semiconductor. Charge transfer is the basic phenomenon used in modulation-doped heterostructures.

MOS structures are used widely in Si electronics. The difference of work functions between Si and many metals – for example, Al – satisfies the condition $\Phi_{FB} < 0$, which is necessary to realize an electron inversion layer. A dielectric

(SiO$_2$) between the metal and the semiconductor is required for electrically iso-lating these two conductive materials. Note that another type of junction, the junction between a heavily doped n^+ polysilicon and a slightly doped p Si, is also characterized by $\Phi_{FB} < 0$, that is, this junction also causes the inversion layer. It is important technologically that contemporary Si technology can be used to fabricate MOS systems with SiO$_2$ insulators characterized by a low den-sity of interface traps, especially at the [100] surface. In many cases, achieving a low density of interface states is sufficient to yield a high-quality interface. This combination of materials – SiO$_2$/Si – is stable and practically does not degrade.

The MOS structure just discussed was the first system in which carrier-quantization effects were studied. At low temperatures the carriers are quan-tized in a triangularlike quantum well formed by the high SiO$_2$ barrier and the band bending of Si. This system attracted much attention in the 1970s. Re-cent achievements in Si technology have provided new means of fabricating high-quality MOS systems with many unique properties. Specifically, two-, one-, and even zero-dimensional quantum structures are now fabricated from M/SiO$_2$/Si systems. However, quantum behavior and relatively high mobility are observed at only low temperatures, of the order of 77 K and below. At room tem-perature, the typical mobility in inversion layers of M/SiO$_2$/Si systems is low and has a value of $\sim$750 cm^2/V s for a bulk acceptor concentration of $\sim$10^{15} cm^{-3}. In bulk Si, the electron mobility is $\sim$1500 cm^2/V s, i.e., at room temperature the mobility in the inversion layer is even less than that in the bulk.

4.4.3 Single Modulation-Doped Heterojunction

First, let us analyze qualitatively a single semiconductor heterojunction. We consider a junction of two semiconductors with work functions and bandgaps such that they result in a discontinuity of the conduction band, as indicated by Fig. 4.17(a). Then we assume that the barrier semiconductor material is n doped while the narrow-bandgap material on the right-hand side of the structure is undoped. In real situations, the latter material is, usually, lightly doped by acceptors, i.e., it is of p type.

The doped region of the system fixes the position of the Fermi level E_F. At equilibrium, the energy levels below E_F are occupied by electrons. Accordingly,

Figure 4.17. The charge-transfer effect in a selec-tively doped single heterostructure. The steplike pro-file of the bands is unstable. Charge transfer results in ionization of a layer of the doped material, band bending, lining up the Fermi level across the struc-ture, and formation of an electron channel at the interface.

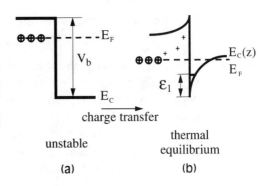

(a) (b)

the scheme sketched in Fig. 4.17(a) is unstable. The electrons will move toward the undoped crystal until an electrostatic field, brought about by the electric-charge redistribution, bends the band edges so that the Fermi level becomes constant across the materials. Instead of having an energy step, as in Fig. 4.17(a), one obtains the situation shown in Fig. 4.17(b), that is, the band edges are bent, there is ionization of impurities in some region of the doped part of the system, and there are free electrons inside the potential well. This well is formed by both the bandgap discontinuity on one side and the electrostatic potential on the other side.

Three important conclusions can be derived immediately from this consideration:

1. Although both the initial materials were insulators, at least at low temperatures, now at the interface, near the junction, one obtains an electron channel and the electron concentration in it is finite even at temperature $T = 0$ K.
2. Charge carriers are separated spatially from their parent impurities of the barrier side of the structure. Therefore charged impurities, which usually lead to large scattering rates and low mobilities, are only the sources of carriers. Moreover, electron scattering by impurities is suppressed as a result of the spatial separation.
3. The bending of the energy band creates a confining potential for carriers in one direction, say along the z axis. So there is a quantization of electrons in the z direction and the establishment of two-dimensional electron gas is quite possible, at least for low temperatures.

To complete the discussion of this simple model, let us note again that in the real situation the side of the heterostructure with the lower conduction-band edge is usually doped lightly with acceptors. In this case, the right barrier of the well is higher than half of the bandgap. Of course, the residual acceptors reduce the electron-channel mobility but their effect is weak.

Here we have presented the case in which the Fermi level is defined by modulation doping and cannot be controlled. These are so-called ungated heterostructures. For gated heterostructures the physical picture is slightly different. This case is studied in Subsection 4.4.4.

4.4.4 Basic Equations and Quantitative Results for a Single Heterostructure

Let us consider the problem of a modulation-doped semiconductor heterostructure more rigorously. To write the basic equations describing a modulation-doped heterostructure, it is necessary to take into account the many-body character of the problem. In fact, we begin by considering a steplike discontinuity of the energy band and find that the potential well is formed by a charge transfer in space. Thus the shape of the potential is determined by the charges of many electrons and ionized impurities. On the other hand, this many-body potential determines the motion of each electron and the total number of ionized impurities. Thus we face the self-consistent problem: The potential is defined by the concentration of electrons and the ionized impurities and it, in turn, affects their redistribution. The simplest approach to this self-consistent problem is to treat the electron quantization within the framework of the Hartree scheme, in which the potential is described by a self-consistent electrostatic field, as discussed in Section 2.5. The Poisson equation for the electrostatic potential Φ is

$$\nabla^2\Phi(\vec{r}) = \frac{4\pi e}{\kappa}\left[\sum_{(\text{acc})}\delta(\vec{r} - \vec{R}_A) - \sum_{(\text{don})}\delta(\vec{r} - \vec{R}_D) + \sum_\nu |\psi_\nu(\vec{r})|^2\right]. \quad (4.32)$$

Here e is the elementary electrical charge, κ is the dielectric constant, and $\vec{R}_A$ and $\vec{R}_D$ are the positions of the acceptors and the donors, respectively. The first two sums are calculated over all charged acceptor and donor atoms. The last term represents the electron contribution to the electric charge. The index ν runs over all occupied electron states.

Because $|\psi_\nu(\vec{r})|^2$ represents the probability of finding the electron of the state labeled by ν at the point $\vec{r}$, the sum over all occupied states ν gives exactly the total number of electrons at this point. For the self-consistent problem, it is also necessary to solve the Schrödinger equation for the electron wave functions $\psi_\nu(\vec{r})$:

$$\left[-\frac{\hbar^2}{2m^*}\left(\frac{\partial^2}{\partial x^2} + \frac{\partial^2}{\partial y^2} - \frac{\partial^2}{\partial z^2}\right) + V_b(z) - e\Phi(\vec{r})\right]\psi_\nu(\vec{r}) = E_\nu\psi_\nu(\vec{r}). \quad (4.33)$$

Here the total potential consists of two contributions: the built-in potential of the heterostructure $V_b(z)$ and the self-consistent potential $-e\Phi(\vec{r})$. In our case, $V_b(z)$ corresponds to the energy-band discontinuity at the junction [see Fig. 4.17(a)] that depends on only the z coordinate. According to Fig. 4.17(a) we can set

$$V_b(z) = -V_b\Theta(z),$$

where $\Theta(z)$ is the Heaviside step function.

The self-consistent electrostatic potential $-e\Phi(\vec{r})$ requires some additional explanation. According to the Poisson equation (4.32), the electrostatic potential is determined by the electrons and the charged impurities. These impurities are randomly distributed over the space; thus, in general, the potential $\Phi(\vec{r})$ is a random function of coordinates. We can divide the contribution of charged acceptors and donors to the potential into two parts, regular and random ones. To accomplish this, we introduce concentrations of the ionized acceptors $N_A^-(\vec{r})$ and donors $N_D^+(\vec{r})$, summing up the impurities in the small, but macroscopic (i.e., containing many impurities) volume ΔV around the point $\vec{r}$:

$$N_D^+(\vec{r})\Delta V = \int_{\Delta V} d^3R_D\delta(\vec{r} - \vec{R}_D)$$

$$N_A^-(\vec{r})\Delta V = \int_{\Delta V} d^3R_A\delta(\vec{r} - \vec{R}_A).$$

These quantities are primarily functions of the z coordinate but there are small contributions that depend on $\vec{r}$. The main contributions, $N_D(z)$ and $N_A(z)$, can be considered as results of averaging over the x, y coordinates. The deviations of the concentrations from their average values, i.e.,

$$\delta N_D(\vec{r}) \equiv N_D^+(\vec{r}) - N_D(z), \qquad \delta N_A(\vec{r}) \equiv N_A^-(\vec{r}) - N_A(z),$$

are, in fact, random and they create a fluctuating part of the electrostatic field $\Phi(\vec{r})$. The average concentrations $N_D(z)$ and $N_A(z)$ make contributions to the regular z-dependent potential $\phi(z)$. This potential confines the electrons. The deviations from the average concentrations $\delta N_A(\vec{r})$ and $\delta N_D(\vec{r})$ generate a fluctuating field, $\tilde{\phi}(\vec{r})$, i.e., the total field is $\Phi(\vec{r}) = \phi(z) + \tilde{\phi}(\vec{r})$. The fluctuating field breaks the translational symmetry in the plane of the junction, and it should be accounted for as a source of electron scattering. Hence, if we assume that $|\delta N_D|$ and $|\delta N_A|$ are much less that $N_{A,D}$, the contribution $\tilde{\phi}(\vec{r})$ can be neglected in the energy calculations; in this case $\tilde{\phi}(\vec{r})$ can also be taken into account as perturbations that cause scattering of the electrons. This type of electron scattering is important, and its influence on the electron mobility will be studied in Chapter 6. After dropping the fluctuating part of the potential, we obtain a potential $\phi(z)$ that is independent of the x and the y coordinates. Thus $\psi_\nu(\vec{r})$ can be factorized as discussed in Subsection 2.4.2:

$$\psi_\nu(\vec{r}) = \frac{1}{\sqrt{S}} e^{i(k_x x + k_y y)} \chi_i(z), \quad \nu \equiv \{i, k_x, k_y\}, \tag{4.34}$$

where S is the area of the junction. Now we immediately obtain the one-dimensional Schrödinger problem for the direction perpendicular to the plane of the junction:

$$\left[-\frac{\hbar^2}{2m^*} \frac{d^2}{dz^2} + V_b(z) - e\phi(z) \right] \chi_i(z) = \epsilon_i \chi_i(z). \tag{4.35}$$

According to Section 3.2, the total electron energy is

$$E(i, k_x, k_y) = \epsilon_i + \frac{\hbar^2}{2m^*} (k_x^2 + k_y^2).$$

It is easy to simplify the Poisson equation (4.32) by averaging both parts of the equation over all possible positions of donors and acceptors and find the equation for the potential $\phi(z)$:

$$\frac{d^2\phi(z)}{dz^2} = \frac{4\pi e}{\kappa} \left[\sum_\nu |\psi_\nu(z)|^2 - N_D(z) + N_A(z) \right]. \tag{4.36}$$

In the context of the above discussion, we consider $\phi(z)$ as the averaged, non-fluctuating part of the electrostatic potential.

To calculate the contribution of electrons, let us introduce the energy-dependent electron distribution function $\mathcal{F}(E_\nu)$. Because of the above assumed separation of the parallel and the transverse wave functions for the electrons, we can simplify considerably the electron contribution to the space charge that appears on the right-hand side of the Poisson equation (4.36). Using the particular form of the wave functions of Eq. (4.34), we can rewrite this contribution as

$$\sum_\nu |\psi_\nu|^2 \, \mathcal{F}(E_\nu) = \frac{1}{S} \sum_{s,i,k_x,k_y} |\chi_i(z)|^2 \, \mathcal{F}\left[\epsilon_i + \frac{\hbar^2}{2m}(k_x^2 + k_y^2) \right]$$

$$= \sum_i |\chi_i(z)|^2 n_i. \tag{4.37}$$

In Eq. (4.37) the summation is to be taken over all possible quantum numbers. The set of these quantum numbers consists of spin s, the wave vector of in-plane motion $\vec{k}$, and the quantum number i. We also introduce the total number of electrons in level i, n_i.

Under equilibrium conditions, the number of electrons occupying the state $\nu \equiv \{s, \vec{k}, i\}$ is given by the Fermi distribution of Chapter 2:

$$\mathcal{F}(s, \vec{k}, i) = \frac{1}{1 + \exp\left[-\frac{E_F - E(\vec{k},i)}{k_B T} \right]},$$

where T is the crystal temperature and $E(\vec{k}, i)$ is the total electron energy. Let us introduce dimensions of the junction in the x, y plane: L_x and L_y. Obviously, $S = L_x \times L_y$. Then, as shown in Chapter 2, the wave vector runs over the set $2\pi j/L_x, 2\pi l/L_y, j, l = 1, 2, \ldots$. Taking into account the spin degeneracy, we obtain

$$n_i \equiv \frac{1}{S} \sum_{s,k_x k_y} \mathcal{F}(s, \vec{k}, i) = \frac{1}{2\pi^2} \int dk_x \, dk_y \frac{1}{1 + \exp\left[-\frac{E_F - E(\vec{k},i)}{k_B T} \right]}$$

$$= \frac{m^* k_B T}{\pi \hbar^2} \ln\left(1 + \exp\frac{E_F - \epsilon_i}{k_B T} \right). \tag{4.38}$$

Equation (4.38) gives the surface density of electrons, which is also known as the area or sheet density. This quantity is a function of temperature and takes the simplest form for $T \to 0$:

$$n_i(T = 0) = \frac{m^*}{\pi \hbar^2}(E_F - \epsilon_i)\Theta(E_F - \epsilon_i). \tag{4.39}$$

Equation (4.39) indicates that the level i is occupied if the Fermi level exceeds the corresponding energy of quantization of transverse electron propagation.

Now one can formulate the boundary conditions for Eqs. (4.35) and (4.36). For localized electron states, one should set

$$\chi_i(z) \to 0 \quad \text{for } z \to \pm\infty. \tag{4.40}$$

For the electrostatic potential, in the case of an ungated heterostructure, we obtain

$$d\phi/dz \to 0 \quad \text{for } z \to \pm\infty. \tag{4.41}$$

Integrating Eq. (4.36) over an infinite range of z and using Eq. (4.41), one can find a neutrality equation in the form

$$\sum_i n_i + \int_{-\infty}^{+\infty} dz\,[N_A(z) - N_D(z)] = 0. \tag{4.42}$$

This result implies that despite the charge transfer the entire system remains electrically neutral.

The final set of self-consistent equations describing a selectively doped heterostructure, which is sketched in Fig. 4.17, can be written as

$$\left[-\frac{\hbar^2}{2m^*}\frac{d^2}{dz^2} + V_b(z) - e\phi(z) \right]\chi_i(z) = \epsilon_i(z), \tag{4.43}$$

$$\frac{d^2\phi}{dz^2} = \frac{4\pi e}{\kappa}\left[\sum_i |\chi_i(z)|^2 n_i - N_D(z) + N_A(z) \right], \tag{4.44}$$

$$n_i = \frac{m^* k_B T}{\pi \hbar^2}\ln\left[1 + \exp\left(\frac{E_F - \epsilon_i}{k_B T} \right) \right]. \tag{4.45}$$

In thermodynamic equilibrium, the Fermi level E_F of an ungated system is controlled by the temperature and the concentration of impurities in the bulk of the system. The ion concentrations N_D and N_A can be expressed through concentrations of dopant donor and acceptor atoms, $\mathcal{N}_d$ and $\mathcal{N}_A$, respectively:

$$N_D = \mathcal{N}_D \frac{1}{1 + 2\exp\left[\frac{E_F + e\phi(z) - E_D}{k_B T} \right]}, \tag{4.46}$$

$$N_A = \mathcal{N}_A \frac{1}{1 + g\exp\left[\frac{E_A - e\phi(z) - E_F}{k_B T} \right]}, \tag{4.47}$$

where E_D and E_A are the energies of donor and acceptor levels. The donors are supposed to be monovalent atoms, which gives the factor of 2 in the denominator of Eq. (4.46). In Equation (4.47) the factor g is the degeneracy of the valence band at the Γ point. For III–V compounds and Si we can set $g = 4$. Expressions (4.40) and (4.41) define the boundary conditions for Eqs. (4.43)–(4.45).

Equations (4.43)–(4.47) define self-consistently the profile of the quantum-well potential energy, the electron wave functions and energies, and the electron concentration at the interface.

4.4.5 Simple Analytical Estimates for a Selectively Doped Single Heterostructure

In the general case, the system of nonlinear coupled equations describing modulation-doped heterojunctions, which we have derived in Subsection 4.4.4, can be solved only numerically. But one can simplify the analysis considerably for the case of low temperatures $(T \to 0)$ and for steplike dependences of the dopant atoms $\mathcal{N}_D(z)$ and $\mathcal{N}_A(z)$:

$$\mathcal{N}_D(z) = \begin{cases} \mathcal{N}_D & \text{for } z < 0 \\ 0 & \text{for } z > 0 \end{cases}, \tag{4.48}$$

$$\mathcal{N}_A(z) = \begin{cases} \mathcal{N}_A & \text{for } z > 0 \\ 0 & \text{for } z < 0 \end{cases}. \tag{4.49}$$

At $T = 0$, the electron concentration on the energy level i is given by Eq. (4.39) and depends on only the Fermi level. The Fermi level coincides with the energy of the acceptors in the p-doped right-hand side of the structure shown in Fig. 4.17 and with the energy of donors in the n-type material on the left-hand side. Thus, for the case of Fig. 4.17, one gets

$$E_F = E_v(z \to \infty) + E_A = E_c(z \to -\infty) - E_D, \tag{4.50}$$

where $E_v(z)$ and $E_c(z)$ are the z-dependent top of the valence band and the bottom of the conduction band, respectively, and E_A and E_D are the ionization energies of acceptors and donors, respectively. To calculate the concentration n_i in each subband i, we need to know the position of the Fermi energy with respect to the bottom of subbands, which is $(E_F - \epsilon_i)$.

At low temperatures even a small decrease of the electrostatic potential ϕ in the region $z < 0$, that is, an increase of the electron-potential energy $-e\phi$, leads to the total ionization of donors in this region:

$$N_D = \mathcal{N}_D.$$

Applying this result to the case of Fig. 4.17, one finds that a depletion layer $(l_d < z < 0)$ is formed near the interface. In this depletion region electrons are removed from the layer. The layer contains uniform positive charge $+eN_D$, and its width l_d has to be calculated. The electrostatic potential in this layer can be found easily from the Poisson equation (4.44) with the constant right-hand side, $-4\pi e N_D/\kappa$:

$$\phi(z) = -\frac{2\pi e}{\kappa} N_D(z + l_d)^2. \tag{4.51}$$

Thus, at the junction $z = 0$, the top of the energy barrier is

$$-e\phi_D(z = 0) = \frac{2\pi e^2}{\kappa} N_D l_d^2$$

with respect to the band edge at $z \to -\infty$. Similarly, the bottom of the narrow-bandgap semiconductor is shifted up at $z \to +0$; note that we disregard charged acceptors. The total number of charged impurities per unit area of the interface is $l_d N_D \equiv N_s$. This charge creates an electric field near $z = 0$ given by

$$F = \frac{4\pi e}{\kappa} N_D l_d = \frac{4\pi e}{\kappa} N_s.$$

In the simplest self-consistent approach, one can approximate the quantum well by an infinite triangular potential, as in Fig. 3.6. The discontinuity at $z = 0$ is considered to be infinitely high, and the slope of the potential is determined by the electric field at $z = 0$. Such a potential profile has been studied in Section 3.2. For this case the wave function is given by the Airy function (3.18) and the lowest energy level is higher than the band edge $E_c(z \to +0)$ by

$$\epsilon_1 = -\left(\frac{e^2 \hbar^2}{2m^*}\right)^{1/3} \left(\frac{e}{\kappa} N_s\right)^{2/3} p_1, \tag{4.52}$$

where $p_1 \approx -2.35$ is the first root of the Airy function and we have used the equation for the electric field F. From the neutrality condition of Eq. (4.42) one can establish a relationship between the total number of ionized impurities per unit area N_s and the surface electron density n_s: $n_s = N_s$. Hence, according to Eq. (4.52), we can write $\epsilon_1 = \gamma n_s^{2/3}$ with $\gamma = |p_1|[(e^4 \hbar^2)/(2m^* \kappa^2)]^{1/3}$. On the other hand, γ can be considered as a phenomenological quantity to be extracted from experimental data. Let us restrict ourselves to the case in which only the lowest quantized energy level is occupied by electrons. Then it is easy to find the position of the Fermi energy E_F with respect to the bottom of the first subband. To do so, we can estimate the number of electrons in the level ϵ_1 by using Eq. (4.39). As a result, we find the following relationship between E_F and n_s [use Fig. 4.17(b) for reference]:

$$E_F = E_c(z \to -\infty) - V_b + \frac{2\pi e^2}{\kappa} \frac{n_s^2}{N_D} + \epsilon_1(n_s) + \frac{\pi \hbar^2}{m^*} n_s, \tag{4.53}$$

where V_b is the discontinuity of the conduction band. Here $\epsilon_1(n_s)$ is defined by Eq. (4.52) with $n_s = N_s$. At low temperatures, the Fermi level of the heterostructure is pinned by the donors at their energy position, as given by Eq. (4.50). Finally, we can exclude the Fermi energy:

$$V_b - E_D = \frac{2\pi e^2}{\kappa} \frac{n_s^2}{N_D} + \epsilon_1(n_s) + \frac{\pi \hbar^2}{m^*} n_s. \tag{4.54}$$

Equation (4.54) is simply the algebraic equation: The left-hand side is a positive constant equal to the offset between the conduction-band energy of the p-doped material and the donor energy of the n-doped material and the right-hand side is zero at $n_s = 0$ and is an increasing function of n_s. [Note, however, that Eq. (4.52) is valid only if charge accumulation in the narrow-bandgap material is due to the confined electrons; strictly speaking, at very low n_s it is not appropriate.] It is easy to see that Eq. (4.54) always has a single solution. It self-consistently defines the surface concentration of the electrons n_s on the interface. For a particular junction the solution is a function of only one parameter, namely the concentration of donors N_D. Solving Eq. (4.54), we can determine all parameters of the heterostructure including the thickness of the depletion layer $l_d = n_s/N_D$, the electric field F in the space region adjacent to the junction, etc.

The approach just discussed does not take into account several physical peculiarities of the heterojunction. First, there is always some degree of the compensation due to the presence of both donors and acceptors in the same material layers, which slightly affects the charge transfer. Second, the narrow-bandgap semiconductor is usually p doped for the case of formation of an n channel and n doped for the case in which a p channel is formed. For the n channel, the junction gives rise to a depletion of these residual acceptors in some region of the narrow-bandgap material. This depletion also contributes to the electrostatic potential. However, if the concentration of acceptors is small or comparable with the donor concentration in the region $z < 0$, this contribution does not affect the region of the electron accumulation: It is more important that it provides a high barrier on the narrow-bandgap semiconductor side of the structure ($z \to \infty$). Third, the doped region is usually separated from the interface by some thin, undoped layer referred to as the spacer layer. As a result, the electric voltage drops in the spacer region although it does not correspond to any transferred charge. The voltage drop on the spacer can be estimated as $V_{sp} = 4\pi e n_s d_{sp}/\kappa$, where d_{sp} is the thickness of the spacer. Consequently the absolute value of eV_{sp} should be added to the right-hand side of Eq. (4.54).

In conclusion, the simple model of selectively doped single heterojunctions described in this section can be used to make rough estimates quickly of the parameters of these structures as well as to study the major trends qualitatively. The advantage of this model is the reduction of the system of differential Eqs. (4.43) and (4.44) and the algebraic Eq. (4.45) to the single algebraic result of Eq. (4.54).

4.4.6 Numerical Analysis of a Modulation-Doped Single Heterojunction

A more accurate analysis of the problem of the electron channels forming at the junction can be made through the numerical solution of the system of Eqs. (4.43)–(4.45). This system of equations was solved first for the SiO_2/Si structure. The Schrödinger equation was solved by the variational method. The trial wave function was chosen in the form

$$\chi_1(z) = 0 \quad \text{for } z \le 0,$$

$$\chi_1(z) = \sqrt{\frac{b^3}{2}} z e^{-\frac{bz}{2}} \quad \text{for } z \ge 0, \tag{4.55}$$

to calculate the wave function and the energy of the lowest level. The approximation that the wave function vanishes in the barrier ($z < 0$) is reasonably accurate for the SiO_2/Si system, because the insulator barrier is high, with a height of

≈ 2 eV. For a junction based on III–V compounds, the barriers are much lower and are typically in the range 0.2–0.4 eV. To take into account the penetration of electrons into the barrier, we can change the trial function slightly by assuming the form

$$\chi_1(z) = Ae^{k_b z} \quad \text{for } z \leq 0,$$
$$\chi_1(z) = B(z + z_0)e^{-(bz)/2} \quad \text{for } z \geq 0. \tag{4.56}$$

Here we introduce an additional variational parameter k_b. The normalization rule for χ_1 and the boundary conditions requiring the continuity of both χ_1 and $d\chi_1/dz$ at $z = 0$ give three equations linking constants A, B, and z_0. In fact, the value of k_b can be defined as

$$k_b \approx \sqrt{2m_b V_b/\hbar^2}, \tag{4.57}$$

where V_b is the conduction-band offset at the heterojunction and m_b is the electron effective mass in the barrier material. In this case, b is the only variational parameter in the wave function (4.56).

Before we discuss numerical results, let us briefly turn to the picture of electron energy at the heterojunction in order to present both the conduction- and the valence-band profiles for the specific heterojunction of importance. The left-hand side of Fig. 4.18 shows the band edges of separated AlGaAs and GaAs; the distance to the vacuum level is shown conditionally. It is assumed that AlGaAs

Figure 4.18. Schematic of the energy-band diagram of a selectively doped AlGaAs/GaAs heterostructure before (left) and after (right) charge transfers have occurred. Relative positions of the valence and the conduction bands are given for both materials. The electron affinities are shown conditionally (left). The Fermi level in the $Ga_{1-x}Al_xAs$ material is supposed to be pinned on the donor level. The narrow-bandgap GaAs is slightly p doped. [After C. Weisbuch and B. Vinter, *Quantum Semiconductor Structures* (Academic, San Diego, 1991).]

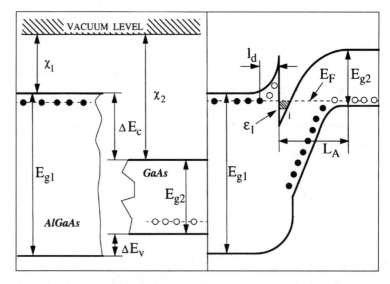

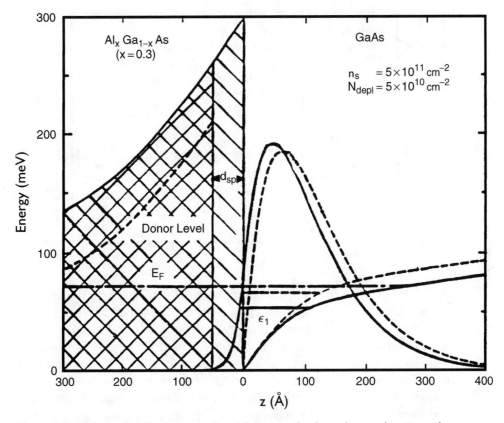

Figure 4.19. Calculated self-consistent potentials, energy levels, and wave functions of an $Al_{0.3}Ga_{0.7}As/GaAs$ selectively doped heterostructure. The junction is situated at $z = 0$. The spacer thickness is 50 Å, and the donor binding energy of AlGaAs is chosen as $E_D = 50$ meV. Solid curves correspond to the trial function of Eqs. (4.56), dashed curves to that of Eqs. (4.55). [After T. Ando, "Self-consistent results for a $GaAs/Al_xGa_{1-x}As$ heterojunction. I. Subband structure and light-scattering spectra," J. Phys. Soc. Jpn. **51**, pp. 3893–3899 (1982).]

is heavily n doped and GaAs is doped lightly by acceptors. This doping brings about the pinning of the Fermi level at the donor level on the AlGaAs side of the structure, as was discussed in Subsection 4.4.5. On the right-hand side of the figure, one can see the energy structure of the junction of AlGaAs/GaAs with two depletion regions: one positively charged on the AlGaAs side and another negatively charged of the width L_A on the GaAs side. Since the AlGaAs/GaAs heterojunction is type I (according to the classification of Chapter 2), for such a doping the quantum well occurs only for electrons.

In Fig. 4.19 the results of the numerical solution of Eqs. (4.43)–(4.45) are presented for a selectively doped $Al_{0.3}Ga_{0.7}As/GaAs$ heterojunction. Heavily n-doped $Al_{0.3}Ga_{0.7}As$ and lightly p-doped GaAs materials are separated by a thin undoped $Al_{0.3}Ga_{0.7}As$ spacer layer with thickness d_{sp}. The self-consistent potentials, energy levels, and wave functions are shown for two different cases in which the trial wave functions are defined by Eqs. (4.55) and (4.56), respectively. One can see that this more accurate treatment of Eqs. (4.56) yields results

Table 4.7.
Results of Calculations of Electron Parameters for Two
Wave-Function Approximations: Case A Corresponds
to Eqs. (4.56), Case B to Eqs. (4.55)

Case	n_s (10^{11} cm^{-2})	2	4	8
A	ϵ_1 (meV)	32.00	45.6	66.3
	z_0 (Å)	13	13	13
	$\langle z \rangle$ (Å)	100	82.6	65.6
	P_b (%)	0.7	1.11	1.95
B	ϵ_1 (meV)	38.6	56.6	86.1
	$\langle z \rangle$ Å	116	99	82
	P_b (%)	0	0	0

appreciably different from those obtained with the simple trial function defined by Eqs. (4.55). First, it is seen clearly that introducing self-consistency leads to profound changes of the potential at $z > 0$. This is due to electron screening that was not considered correctly in the simplest model. The wave functions for all approximations are close to each other, especially in the region of classically allowed motion. However, they are drastically different in the barriers. These differences can be important for the calculation of electron scattering by impurities, tunneling effects, capture by impurities, etc. The surface electron concentration n_s in the channel at the junction is equal to the number of ionized donors per unit area, $N_s = 5 \times 10^{11}$ cm^{-2}. The surface concentration of ionized acceptors, $N_{depl} = 5 \times 10^{10}$ cm^{-2}, is much smaller.

Other results obtained for different concentrations n_s of the electrons on the interface are summarized in Table 4.7. In this table the lowest energy level ϵ_1, the spatial scale of the electron confinement $\langle z \rangle$, and the probability of finding electrons in the barrier P_b are presented for different electron concentrations. The value of $\langle z \rangle$ was calculated as the quantum-mechanical average of the z coordinate of an electron (see Subsection 2.4.1); P_b is defined by Eq. (3.82). One can see that the relative position of the first energy level ϵ_1 increases as the concentration increases. The variational parameter z_0 in the approximation of Eqs. (4.56) is almost constant; hence expression (4.57) provides a good approximation. The value of $\langle z \rangle$ decreases with increasing n_s and the confining electrostatic potential. Thus we can see that the width of the electron channel is in the range 60–100 Å. The probability of finding electrons in the barrier of the wide-bandgap materials is small, but increases with increasing electron confinement. For the wave function of Eqs. (4.55), P_b is equal to zero. We can see that the latter model gives the energy ϵ_1 with an accuracy in the range 15%–25%. The accuracy decreases as the concentration n_s increases. Note that, as is evident from Fig. 4.19, the approximation of Eqs. (4.55) leads to a large discrepancy in the self-consistent potential at large z.

The examples given in our discussions illustrate the following major features of a selectively doped heterojunction:

1. formation of electron-conduction channels with concentrations in the range $n_s = 10^{11}$–10^{12} cm^{-2} at any temperature, including $T = 0$
2. spatial separation of the electrons from their parent donors with a very low probability of electron penetration into the barrier – less than or ~1%; spatial isolation of the electrons from the p-doped narrow-bandgap material
3. formation of a potential well for electrons with the potential profile self-consistently dependent on the electron concentration
4. quantization of electrons inside the potential well with the resultant two-dimensional character of the electron spectrum; the electrons are confined in the two-dimensional channel of a width less than 100 Å.

4.4.7 Control of Charge Transfer

In the previous subsections we have considered a heterojunction formed by two semi-infinite semiconductors with fixed concentrations of donors and acceptors (see Fig. 4.18). This results in a conducting channel at the interface with a constant and uncontrollable electron concentration. The control of the conductivity, or more exactly, the control of the resistance – or its inverse quantity conductance of the structure – is necessary to realize useful devices.

Let us consider the possibility of changing the conductance of a heterojunction by controlling the electron concentration. For that purpose, we study the so-called gated heterojunction that is presented schematically in Fig. 4.20(b). For comparison, Fig. 4.20(a) shows the ungated heterojunction considered above. The only difference between them is the metal contact placed on top of the n^+ layer of the AlGaAs barrier material in the gated structure.

This metal-semiconductor system is called a MES structure. For GaAs-like materials, MES structures are the most important for device applications, because these materials do not have a stable natural oxide as is the case of SiO$_2$ on Si. Therefore most of the electronic devices based on GaAs use MES structures. One also refers to these structures as Schottky gate structures.

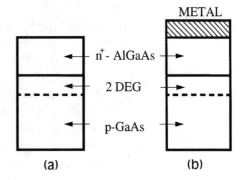

Figure 4.20. Schematics of (a) ungated, (b) gated heterostructures with two-dimensional electron gas.

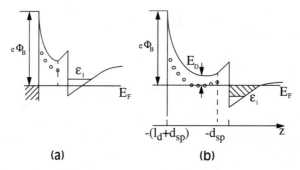

Figure 4.21. Conduction-band diagram for M/AlGaAs/GaAs heterostructures. The built-in Schottky voltage controls the depletion region under the metallic gate. It results in (a) normally off, (b) normally on devices. [After C. Weisbuch and B. Vinter, *Quantum Semiconductor Structures* (Academic, San Diego, 1991), p. 46.]

In contrast to the M/SiO$_2$/Si system considered above, in which the electron-conduction channel is formed in the absence of an external voltage, under a metallic gate in GaAs-like materials there are extended depletion regions that occur because of a high built-in voltage of $\sim$0.8 V. Such a depletion region is known as a Schottky depletion region.

The conduction-band energy diagram for a M/AlGaAs/GaAs heterostructure is presented in Fig. 4.21 for two thicknesses of the AlGaAs layer. The n-doped region is separated from the junction by an undoped spacer. For the thin AlGaAs layer the structure is shown in Fig. 4.21(a); the thicker layer corresponds to Fig. 4.21(b). For both cases there is an extended depletion region that affects the electron channel formed at the AlGaAs/GaAs interface and provides two possibilities of controlling the structure.

1. The normally off structure corresponds to Fig. 4.21(a). The depletion region extends through both a thin AlGaAs layer and the junction. The bottom of the quantum well shifts up. The Fermi level lies under the lowest energy subband. Thus there are no electrons inside the channel and the conductivity along the heterostructure is almost zero. The donors in the AlGaAs-doped region are ionized, and the electrons have left the semiconductor part of the structure that is charged positively. To turn on the conductivity of the device, it is necessary to apply a positive voltage to the metal gate and decrease the built-in Schottky voltage. Normally off structures can be fabricated by use of a thin AlGaAs barrier.

2. The normally on structure is illustrated in Fig. 4.21(b). In this case the built-in voltage drops across a thick AlGaAs layer so that the Fermi level lies above the lowest subband and electrons populate the channel without an external voltage bias. This channel has a finite conductivity under normal conditions. This case can be realized for sufficiently thick AlGaAs layers. Thus in normally on devices one can control conductance of the channel by applying a negative voltage to the metal. A large voltage leads to a depopulation of the channel and can switch off the device.

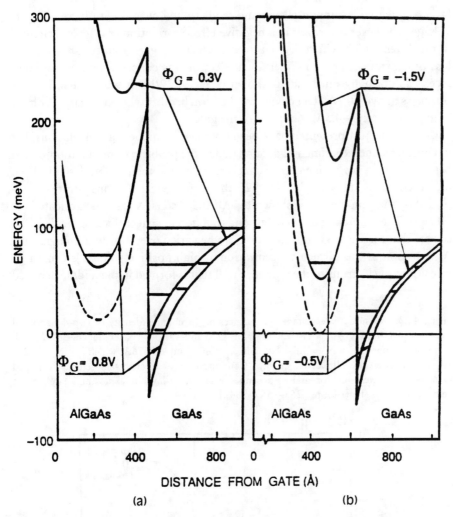

Figure 4.22. Calculated self-consistent potentials for conduction electrons in two M/AlGaAs/GaAs heterostructures. (a) corresponds to the normally-off device; (b) corresponds to the normally on device at room temperature. The Fermi level is at $E = 0$. Horizontal lines indicate the bottom energy of the lowest four subbands. Dashed curves indicate the donor levels. [After B. Vinter, "Subband in back-gated heterojunctions," Surf. Sci. **142**, pp. 452–455 (1984).]

The calculated electron-potential energy and quantized levels for both normally off and normally on devices are shown in Fig. 4.22 for different gate voltages Φ_G. The Fermi level is taken to be at the zero of energy. The normally off device is calculated for an AlGaAs layer thickness of ~ 400 Å; see Fig. 4.22(a). The quantum well formed on the interface contains up to four quantized levels. A positive voltage lowers the bottom of the conduction band of GaAs at the interface. The bottom touches the Fermi level at $\Phi_G = +0.3$ V and the device is turned on at a threshold voltage of approximately $+0.8$ V when the first quantized level touches the Fermi level. It is clearly seen that in the barrier AlGaAs layer a potential minimum occurs and tends to be lowered with increased gate voltage. The latter can result in a negative effect: formation of a second channel in this layer, which will collect electrons, screen the gate voltage, and result in loss of control of the two-dimensional electron-gas concentration at the interface. The normally on device is shown in Fig. 4.22(b). It has a AlGaAs layer thickness above 600 Å. The device can be switched off when a negative voltage of approximately $-0.5\,\mathrm{V}$ is applied to the gate.

The results just presented have been obtained by numerical calculations. Let us consider some experimental data related to the problem of modulation-doped heterostructures for which carrier concentrations and their mobilities have been measured simultaneously. In Fig. 4.23 the surface electron concentration controlled by the gate voltage is shown for Al/AlGaAs/GaAs systems fabricated for HEMTs. The different curves correspond to different spacer thicknesses d_{sp}. One can see that the electron concentration can be varied over 1 order of magnitude. A saturation of the surface concentration at high positive voltage is caused by transitions of electrons to the potential well that is formed in the middle depleted

Figure 4.23. Measured gate-voltage dependences of the channel density of two-dimensional electrons in $Ga_{0.3}Al_{0.7}As/GaAs$ structures for 12 K and various spacer thicknesses d_{sp}. All samples are doped with $N_D = 4.6 \times 10^{17}\,\mathrm{cm}^{-3}$, except sample R-76A, which has $N_D = 9.2 \times 10^{17}\,\mathrm{cm}^{-3}$. [After K. Hirakawa, *et al.*, "Concentration of electrons in selectively doped GaAlAs/GaAs heterojunction and its dependence on spacer-layer thickness and gate electric field," Appl. Phys. Lett. 45, pp. 253–255 (1984).]

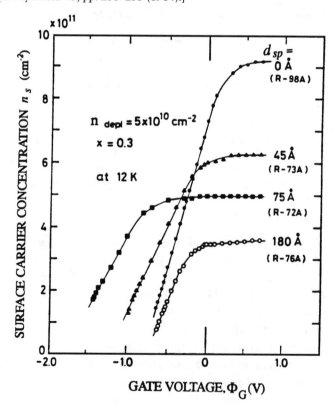

barrier region, as discussed above. Figure 4.23 shows that considerable changes in the properties of the structures occur for different spacer thicknesses. Among the various structures presented in Fig. 4.23, the particular structure with $d_{sp} = 0$ is most likely to be useful for fabricating a normally off device. Indeed, by applying a positive voltage to the gate, one can increase the electron concentration in the channel. Structures with a thick spacer are well suited for normally on devices; in these structures positive voltage does not change the concentration while a negative voltage reduces it sharply.

Note that a spacer is an important element of modulation-doped structures because it partially prevents the scattering of channel electrons by the heavily doped side of the heterostructure and increases the electron mobility. On the other hand, there is one negative effect of a spacer. A spacer leads to the weakening of the electrostatic potential that confines electrons near the interface. Hence a thick spacer causes a decrease in the electron concentration. This trade-off between the mobility and the carrier concentration requires an optimization of the structural design for each particular device application. As a result of such optimization, it is found that spacers with thicknesses in the 25–100-Å range are the most applicable for high-speed devices.

Here we have considered systems with a single heterojunction. These systems can be fabricated by a relatively simple technology, and they have numerous applications. But they suffer from greatly limited carrier concentration in the conduction channel. As can be seen from Fig. 4.23, the typical surface concentrations are less than 10^{12} cm^{-2} for single-heterojunction devices. Higher concentrations of carriers in the conductive channel can be obtained in double-junction systems. These systems are studied in Section 4.5.

4.5 Modulation-Doped Quantum Structures

In Section 4.4 we have studied the modulation doping of a single heterojunction. Modern semiconductor technology facilitates the fabrication of much more complex semiconductor devices, which combine special types of modulation doping with several heterojunctions. There are a number of such systems for various purposes and applications. In this section, we consider a few such systems that illustrate the wide possibilities for varying structural parameters. We study modulation-doped quantum wells, $n–i–p$ systems, and δ-doped structures.

4.5.1 Modulation-Doped Quantum Wells

The principle of modulation doping, or selective doping, can be applied to heterostructures with single or multiple quantum wells. The advantages of these types of selectively doped systems are a combination of electron quantization with a wide range of kinetic parameters, large electron concentrations, etc. From the technological point of view, the growth of a selectively doped quantum well is a much more complicated procedure, because it requires that an undoped narrow-bandgap material – for example, GaAs – be embedded between heavily

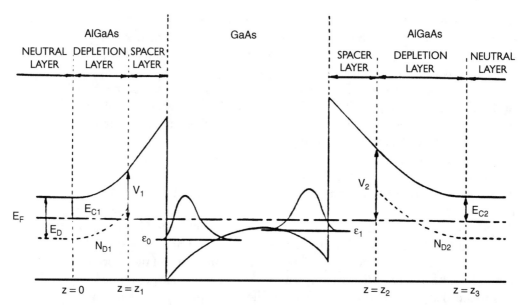

Figure 4.24. The energy-band diagram for a heterostructure with two AlGaAs/GaAs junctions. The resultant quantum well is asymmetric because of an asymmetric doping. The wave functions are shown in the limit of a thick well in which, in fact, two electron channels appear. [After C. Weisbuch and B. Vinter, *Quantum Semiconductor Structures* (Academic, San Diego, 1991), p. 49.]

doped wide-bandgap layers – AlGaAs, for the example at hand. Thus one has to fabricate double heterojunctions, and one of them must have an inverted interface with respect to the other. We leave aside any discussion of the challenging technological complications related to the growth of such structures. In Fig. 4.24, the structure and the potential profile of a modulation-doped quantum well are presented. It is assumed that the left part, $z < z_1$, and the right part, $z > z_2$, of the structure are doped by donors. Between the doped domains and the quantum well there are two spacer layers. The system, as depicted, is asymmetrical. From Fig. 4.24, one can see the formation of two depletion regions on the left-hand and the right-hand sides of the quantum well and the spatial modulation of the quantum-well bottom by the electrostatic potential. The latter effect is due to a charge transfer from impurities of the depletion region into the well. It is interesting to compare the properties of selectively doped quantum wells with those of the simple rectangular wells studied in Chapter 3. The initial rectangular profile changes as the electron concentration increases. The additional electrostatic potential inside the quantum well affects the energy levels.

From simple physical considerations, it is possible to understand the qualitative behavior of the energies and the wave functions of the electrons. The nonmonotonous contribution of the electrostatic potential lifts both the energy position of the bottom of the well and the positions of the energy levels relative to the bottom of the well. The separation between the first and the higher levels decreases. When the electrostatic potential inside the well exceeds the lowest level, two spatial regions of electron wave-function localization arise, as shown in Fig. 4.24; they are separated by the electrostatic barrier and are coupled only by tunneling or thermionic processes.

To describe the behavior of the energy levels and the corresponding wave functions, one can use a simple approach based on perturbation theory. We can start from the electron states obtained for the idealized rectangular quantum well in Section 3.2. We consider the electrostatic potential induced by modulation doping as a perturbation. This approach can be successful, because the energy spacing of a rectangular quantum well varies with the well thickness L as L^{-2} while the contribution of the electrostatic field is a linear function of L and the surface electron concentration n_s. Thus, at least when L is sufficiently small, the electrostatic potential $\phi(z)$ can be treated as a perturbation.

For the case in question, one can represent the well potential provided by the band offset as

$$V(z) = -V_b(z)\, \theta(-z + L/2)\, \theta(z + L/2), \tag{4.58}$$

where perfect symmetry of the heterostructure is assumed. We assume the same symmetry for the doping. Now we write the Schrödinger equation in the form

$$[\mathcal{H}_0 - e\phi(z)]\, \chi_i(z) = \epsilon_i \chi_i(z), \tag{4.59}$$

where $\mathcal{H}_0$ is the unperturbed Hamiltonian and the notations for the energy, electrostatic potential, and wave functions are the same as those introduced in Section 4.4. The electrostatic potential is determined by the Poisson equation (4.44) with $N_A = 0$ and the following z-dependent donor distribution:

$$N_D(z) = \begin{cases} N_D & \text{for } |z| > L/2 \\ 0 & \text{for } |z| < L/2 \end{cases}. \tag{4.60}$$

Let the quantum states of the Hamiltonian $\mathcal{H}_0$ be described by the energy levels ϵ_i^0 and the wave functions $\chi_i^{(0)}(z)$. These states are mixed by the perturbation $-e\phi$. The coupling of state j with state i is determined by the matrix element $\langle \chi_i^{(0)} | -e\phi | \chi_j^{(0)} \rangle$ and is inversely proportional to the energy difference $|\epsilon_i - \epsilon_j|$, as discussed in Subsection 2.4.4. It is evident that the contributions of delocalized unbound states to the shift of the lowest bound levels are small because energy separations are large. Thus we need to take into consideration the mixing among only the lowest bound levels. Since perfect symmetry of the system is assumed, the potential $\phi(z)$ is an even function of z. This leads us immediately to the conclusion that only levels initially having the same parity can be coupled by the perturbation. Thus the first state mixes with the third, the fifth, etc. The second state mixes with the fourth, the sixth, etc.

Let us restrict ourselves to considering only the first four levels. Then we can represent the unknown wave functions as

$$\chi_i(z) = \sum_j c_{ij} \chi_j^{(0)}(z), \quad i, j = 1, 2, 3, 4. \tag{4.61}$$

Multiplying the Schrödinger equation (4.59) by $\chi_j^{(0)}(z)$ and integrating over z, we obtain the equations for c_{ij} :

$$c_{11}\left(\epsilon_1^0 + \delta\epsilon_1 - \epsilon\right) + c_{13} V_{13} = 0,$$

$$c_{22}\left(\epsilon_2^0 + \delta\epsilon_2 - \epsilon\right) + c_{24} V_{24} = 0,$$

$$c_{33}\left(\epsilon_3^0 + \delta\epsilon_3 - \epsilon\right) + c_{31} V_{31} = 0,$$

$$c_{44}\left(\epsilon_4^0 + \delta\epsilon_4 - \epsilon\right) + c_{42} V_{42} = 0,$$

where we have introduced the notations for corrections to energies and matrix elements [see Eq. (2.64)]:

$$\delta\epsilon_i = \langle \chi_i^{(0)} | -e\phi | \chi_i^{(0)} \rangle, \qquad V_{ij} = V_{ji} = \langle \chi_i^{(0)} | -e\phi | \chi_j^{(0)} \rangle.$$

Because the set of initial functions $\{\chi_i^{(0)}\}$ consists of real functions, it is straightforward to find relationships between nondiagonal coefficients c_{ij}: $c_{12} = c_{21}$, $c_{13} = c_{31}$. As has already been emphasized, only states with the same parity are mixed. Thus solutions for $\epsilon_{1,3}$ are

$$\epsilon_{1,3} = \frac{\epsilon_1^0 + \epsilon_3^0 + \delta\epsilon_1 + \delta\epsilon_3}{2}$$

$$\pm \sqrt{\left(\frac{\epsilon_1^0 + \epsilon_3^0 + \delta\epsilon_1 + \delta\epsilon_3}{2}\right)^2 + V_{13}^2 - \left(\epsilon_3^0 + \delta\epsilon_3\right)\left(\epsilon_1^0 + \delta\epsilon_1\right)}. \qquad (4.62)$$

Expressions for $\epsilon_{2,4}$ can be written in a similar fashion. The wave functions of 1 and 3 states have the form

$$\chi_{1,3}(z) = \sqrt{\frac{\left(\epsilon_1^0 + \delta\epsilon_1 - \epsilon_{1,3}\right)^2}{\left(\epsilon_1^0 + \delta\epsilon_1 - \epsilon_{1,3}\right)^2 + V_{13}^2}} \left[\frac{V_{13}}{\epsilon_1^0 + \delta\epsilon_1 - \epsilon_{1,3}}\chi_1^{(0)}(z) - \chi_3^{(0)}(z)\right].$$

$$(4.63)$$

A similar form can be written for $\chi_{2,4}(z)$. Thus we have found the explicit formulas for the energies and the wave functions in the well that is distorted by a redistribution of the electric charge across the structure. The electrostatic potential $\phi(z)$ should be calculated numerically. Consistent with the first approximation, we should not correct the electric charge of the well, which we calculate by using the initial wave functions $\chi_i^{(0)}(z)$.

Using this approach, one can design an iterative procedure for calculating the energies and the wave functions of the structures more accurately. The results of an iterative, self-consistent numerical procedure are shown in Fig. 4.25. In this figure the first four energy levels are plotted as functions of the surface

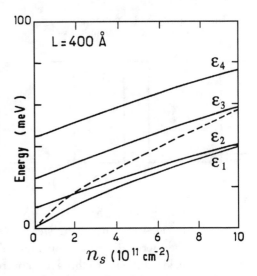

Figure 4.25. The energies of the four lowest sub-bands as functions of the surface concentration in the $Al_{0.4}Ga_{0.6}As/GaAs$ quantum well of 400-Å width. The dashed curve shows the Fermi level. [After F. Stern and J. N. Schulman, "Calculated effects of interface grading in GaAs-Ga$_{1-x}$Al$_x$As quantum wells," Superlatt. Microstr. **1**, pp. 303–305 (1985).]

concentration with respect to the position of the lowest level ϵ_1 at zero concentration. The Fermi level corresponding to this concentration is shown as well. At concentrations lower than $2 \times 10^{11} \, cm^{-2}$ only the lowest energy level is occupied. At higher concentrations, electrons populate the second level. The third and the fourth levels are empty. Figure 4.25 represents the case of a relatively wide quantum well, in which the lowest energy levels at high surface concentrations are below the maximum of the electrostatic potential and two regions of localization of the electron wave function occur, as illustrated by Fig. 4.24. When these levels are well below the maximum, they almost coincide and we get a double-quantum-well situation.

In the above analysis we have considered one particular case of modulation-doped heterostructures, namely, a single quantum well free of impurities. In that case electrons in the well are supplied by donors in the bulklike barrier regions. This situation is typical in high-electron-mobility devices.

Other applications may require notably different structure designs. For example, infrared photodetectors based on light absorption by the confined electrons may be designed as follows. The device consists of a multiple-quantum-well structure, with narrow single-subband quantum wells and thick barriers. The total number of quantum wells for this type of device is $\sim$50–70. Since perpendicular transport of the photocarriers is involved, the barrier regions should be flat and possess high mobility to minimize the time of flight of the electrons through the structure. These requirements can be combined if the well materials are doped and the barriers are undoped.

Another interesting example of selectively doped heterostructures is illustrated by a design that enables the control of mobility in the case of parallel transport. For this purpose, it is possible to fabricate a double-quantum-well structure with high mobility in one of the wells and poor mobility in the second well. The low-mobility conduction channel is heavily doped, while the

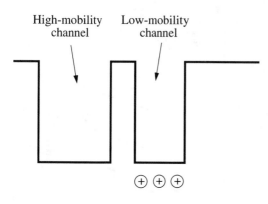

High-mobility channel Low-mobility channel

Figure 4.26. Schematic of an asymmetric double-quantum-well structure with doped low-mobility and undoped high-mobility channels.

high-mobility channel is free of impurities. Such an asymmetric structure is shown in Fig. 4.26. Charge transfer between the quantum wells leads to the distortion of the energy diagram, which initially consists of flat wells and barrier. An electric field transverse to the layers controls the electron wave functions and their redistribution between the wells. One can change the wave function and therefore efficiently switch the system between states with high or low conductance. The characteristic time of the switching is determined by the tunneling of electrons across the barrier separating these two wells. This tunneling time can be made to be very short (see more detailed discussion in Section 9.7). Such modulation-doped double-well structures are the basis for a new type of high-speed transistors – velocity-modulation transistors.

4.5.2 *n–i–p* Structures

In the previous subsections we have considered selectively doped systems with one, two, or more heterojunctions. The combination of heterojunctions and electrostatic fields or two interfaces inverted with respect to each other has resulted in a potential well. Thus the doped layer has been a source of carriers. But there is another possibility for using modulation doping as the basis for quantum structures: *n–i–p* structures or *n–i–p–i* superlattices. To explain these types of quantum systems, let us consider the modulation-doped semiconductor in which an insulator layer follows a donor-doped layer, the next layer is an acceptor-doped layer, etc., as illustrated in Fig. 4.27. The situation presented in Fig. 4.27, with electrons bound to their native donors, is unstable: the electrons must transfer from the donors to the acceptors. As a result, the electrostatic potential and the modulation of both the conduction-band edge and the valence-band edge occurs. In principle, the alternating potential wells for electrons and holes can lead to the quantization of the carriers.

Because we have analyzed similar systems previously, we can easily write the main equations for this case. The physical variables are the electrostatic potential $\phi(z)$ and the wave functions for electrons and holes $\chi_{e,h}(z)$, where z is the coordinate perpendicular to the layers. It is convenient to separate the two major contributions to $\phi(z)$: that from ionized impurities ϕ_{im} and that associated

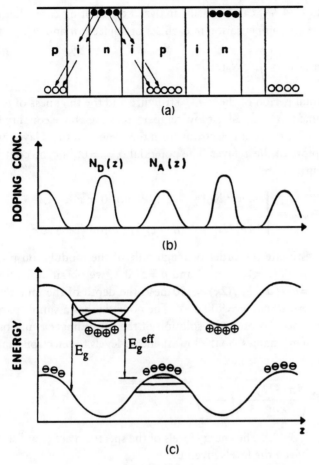

Figure 4.27. Schematic for the formation of an $n-i-p-i$ structure: (a) the sequence of the layers and the channels of the charge transfer; (b) the net space charge coinciding with the doping profile; (c) the resultant bandgap variation, carrier confinement, and the effective bandgap. [After C. Weisbuch and B. Vinter, *Quantum Semiconductor Structures* (Academic, San Diego, 1991), p. 52.]

with the free carriers ϕ_{carr}. Then we obtain

$$\frac{d^2 \phi_{\text{im}}}{dz^2} = -\frac{4\pi e}{\kappa}[N_D(z) - N_A(z)],$$

$$\frac{d^2 \phi_{\text{carr}}}{dz^2} = \frac{4\pi e}{\kappa}\left[\sum_i n_{ei}|\chi_{ei}(z)|^2 - \sum_j n_{hj}|\chi_{hj}(z)|^2\right],$$

$$\left[-\frac{\hbar^2}{2m_{e,h}^*}\frac{d^2}{dz^2} + V_{e,h}(z)\right]\chi_{ei,hj} = \epsilon_{ei,hj}\,\chi_{ei,hj},$$

$$V_{e,h}(z) = \mp e\left[\phi_{\text{im}}(z) + \phi_{\text{carr}}(z)\right],\tag{4.64}$$

where the $-$ sign corresponds to electrons and the $+$ sign to holes. For simplicity we consider a system without spacers. The impurities are assumed to be ionized.

The set of Eqs. (4.64) can be easily analyzed for limiting cases. Let us consider the case of exact compensation in each set of adjacent n and p layers. Hence,

$$\int_0^{d/2} N_D(z)\,dz = \int_{d/2}^d N_A(z)\,dz,$$

where d is total period of the n–i–p structure and the thickness of each n and p layer is assumed to be equal. At low temperatures the above equality leads to the absence of free carriers; all electrons from the donors of the n layer are captured by the acceptors of the p layer. The potential $\phi_{\mathrm{carr}} = 0$, but the contributions of ionized impurities are

$$V_{e,h} \equiv \mp e\phi_{\mathrm{im}} = \begin{cases} -2\pi e^2 N_D\left(\frac{d^2}{4} - z^2\right)/\kappa & \text{for } 0 < z < \frac{d}{2} \\ 2\pi e^2 N_A\left[\frac{d^2}{4} - (d-z)^2\right]/\kappa & \text{for } \frac{d}{2} < z < d \end{cases}.$$

It is easy to estimate the order of magnitude of the modulation potential. For GaAs with $N_D = N_A = 10^{18}$ cm^{-3} and $d = 500$ Å, we obtain a maximum of the potential $V_m = (\pi e^2 d^2 N_D)/(2\kappa) = 450$ meV; the depth of the modulation of the band edges is $\pi e^2 d^2 N_D/\kappa = 900$ meV. The potential has a simple parabolic profile. For a parabolic profile, the solution of the Schrödinger equation is known from quantum mechanics. So the lowest energy levels for electrons and holes can be immediately expressed as

$$\epsilon_{i,e,h} = 2\hbar\sqrt{\frac{\pi e^2 N_{D,A}}{\kappa m_{e,h}^*}}\left(i - \frac{1}{2}\right), \tag{4.65}$$

where $i = 1, 2, 3, \ldots$. The energy levels of the spectrum are equidistant with the separation between the levels given by

$$\Delta\epsilon = 2\hbar\sqrt{\frac{\pi e^2 N_{D,A}}{\kappa m_{e,h}^*}}.$$

For electrons, this separation is equal to 40.2 meV.

For optical applications, the width of the bandgap of the structure is important. Instead of an initial, unperturbed bandgap E_g, we obtain

$$E_g^{\mathrm{eff}} = E_g - 2V_m + \epsilon_{1,e} + \epsilon_{1,h} < E_g. \tag{4.66}$$

The corresponding energy separation is shown in Fig. 4.27(c).

If the compensation is not absolute, some excess carriers will accumulate in the wells; this case is illustrated in Fig. 4.27(c). One obtains conductive layers with a specific type of electrical conductivity: n type if $N_D > N_A$ and p type if $N_D < N_A$. From Eq. (4.66), which actually defines the optical bandgap, one can see that if

$$2V_m > E_g + \epsilon_{1,e} + \epsilon_{1,h},$$

the effective bandgap is negative, i.e., even for full compensation there are finite concentrations of electrons and holes in the structure at any temperature. Thus, in this case, instead of an insulator/semiconductor system one obtains a semimetal structure, which preserves its conductivity (both p and n types in corresponding layers) even at $T = 0 \, \mathrm{K}$. The necessary condition for the formation of such a semimetal is

$$\frac{\pi e^2 N_D d^2}{\kappa} > E_g.$$

Hence it requires a large layer thickness or a large period of the structure. For GaAs material with $N_D = 10^{18} \, \mathrm{cm}^{-3}$ one can obtain a critical period of this artificial semimetal superlattice of $d_c = 7000 \, \text{Å}$. Thus it is possible to create an insulator for the case of exact compensation, any desirable type of the electrical conductivity, and even a semimetal system by means of selective doping of the same semiconductor materials. These conclusions are valid for equilibrium conditions.

The class of nonequilibrium effects is much wider. For example, if one excites electrons and holes in the system, they will occupy appropriate wells, neutralize the ionized impurities, and strongly affect the optical bandgap. Thus one can expect strongly nonlinear optical phenomena that can be controlled by the external illumination of the structure. Because of the spatial separation of electrons and holes, both radiative and nonradiative channels of their recombination are suppressed and the system exhibits large characteristic transport times and lengths. These effects bring about a number of advantages in photodetector applications of p–i–n structures.

Because the impurities and the free carriers are not separated in space, the mobilities of the modulation-doped n–i–p superlattices are lower than that of conventional superlattices considered in Section 3.6. Some further development of these structures can be done, and additional heterojunctions can be added to the insulator region. These structures become more complicated, as indicated in Fig. 4.28, but they are characterized by much higher mobilities of the carriers. It must be pointed out that the technology even permits the fabrication of individual contacts to each of the conductive layers of the system under consideration.

Figure 4.28. Heterojunction doped superlattice. Quantum wells are fabricated in insulator regions of the n–i–p–i superlattice shown in Fig. 4.27. The carrier transport occurs in the high-purity undoped potential wells, separated from doped layers by spacers. [After K. Ploog and G. H. Döhler, "Compositional and doping superlattices in III–V semiconductors," Adv. in Phys. **32**, pp. 285–359 (1983).]

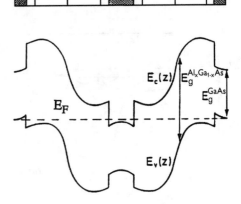

4.5.3　Delta Doping

In addition to the previously discussed selectively doped systems with abrupt interfaces and rather thick alternating donor/acceptor layers, the confinement of impurities within an atomic plane yields principally new electron properties and provides a number of novel applications. This doping within an atomic layer is called δ doping. The δ-function-like doping profile can be obtained by molecular-beam epitaxy, which allows one to insert dopant atoms into an atomic monolayer of the host crystal. By means of δ doping a high level of doping can be achieved.

The δ doping can be characterized by a sheet concentration of dopant atoms $\mathcal{N}_{Ds}$. If the sheet concentration is large, so that the wave functions of the individual dopants overlap, i.e., $\mathcal{N}_{Ds}a_0^2 > 1$, the carriers move in the potential of many ionized dopants. Here, a_0 is the Bohr radius for the donor as defined by Eq. (3.122). One can consider that the ionized δ layer of donors or acceptors provides a continuous sheet of the electric charge that creates an electrostatic, V-shaped potential for the carriers. The latter are localized around this sheet because of Coulomb attraction. The actual potential for carriers should be calculated self-consistently, taking into consideration screening effects, spreading wave functions, etc. If we set $N_{Ds}(z) = \mathcal{N}_{Ds}\delta(z - z_i)$, $N_{As} = 0$ in Eqs. (4.43)–(4.45) we obtain the set of equations necessary to describe a physical picture of δ doping with donors for the case in which the position of the sheet is at $z = z_i$. The boundary conditions for the electrostatic field and the wave functions are given by expressions (4.40) and (4.41). In Fig. 4.29 a schematic illustration of δ doping is presented for the example of a GaAs host material and Si dopant atoms. The formation of a V-shaped potential well for electrons as well as the

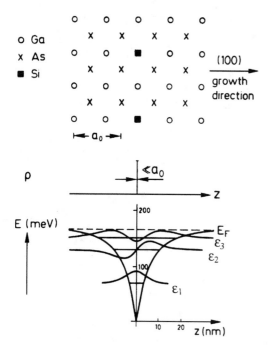

Figure 4.29. Impurity locations for δ doping (top), the charge of ionized donors ρ is located in one monolayer and treated as the δ function (middle), and the potential well due to impurity potential (bottom). Circles, crosses, and black boxes mark the positions of Ga, As, and dopant Si, respectively. The directions of the growth for the structures are indicated. Results of the calculation of the three lowest states and the wave functions versus the coordinate are shown. [After K. Ploog, "Delta-(δ)-doping in MBE-grown GaAs: concept and device application," J. Cryst. Growth **81**, pp. 304–313 (1987).]

electron wave functions must be modeled by self-consistent calculations. One can discern the appearance of two-dimensional subbands with transverse quantum numbers $n = 1$, 2, and 3. The spatial scale of localization of the electrons ranges from 100 to 200 Å, which is comparable with confinement in quantum-well heterostructures. In the case under consideration, the confinement increases with increasing surface-doping concentration $\mathcal{N}_{Ds}$. For these types of quantum structures, very high surface concentrations of the carriers – up to 10^{13} cm^{-2} – can be reached. The mobilities of carriers are not high because the carriers are confined in the doped region; mobilities are of the order of $(1–10) \times 10^3$ cm^2/V s in the range of temperatures from 80 to 300 K. Nevertheless, high conductance can be realized for parallel transport. It is amazing that the contacts to the δ-doped layer can be fabricated for measurements and applications of these structures.

A combination of several δ-doped layers or a single δ-doped layer with a heterostructure provides the means to realize new electron properties with these artificial systems.

For example, the sequence of δ-doped donor and acceptor layers gives a system of a sawtooth-doped superlattice of type II, which is the extreme analog to the alternating donor/acceptor layers presented in Fig. 4.27.

The δ doping is frequently used to populate a two-dimensional channel at the interface. In Table 4.8 the concentrations and the mobilities of the electrons on the AlGaAs/GaAs interface are presented for various structure designs and δ doping by Si.

Case 1 is presented for comparison and corresponds to the single δ layer in a bulklike GaAs. This case is characterized by the highest-area electron

Table 4.8.
Material Design, Concentrations, and Mobilities of Selectively δ-doped n-Al$_{0.3}$Ga$_{0.7}$As/GaAs Heterostructures

	Density of δ-doping (cm^{-2})	Configuration and spacer	300 K		77 K		4 K	
			n	μ	n	μ	n	μ
1	3×10^{12}	Bulklike GaAs	3.1	2.5	3	3	3.5	2.4
2	3×10^{12}	60-Å spacer	0.87	6.6	0.86	4.4	0.8	5.5
3	3×10^{12}	δ layer embedded in 20-Å GaAs quantum well with 60-Å spacer	7.2	9.3	0.59	150	0.57	380
4	4×10^{12}	δ layer embedded in 20-Å GaAs quantum well with 100-Å AlAs spacer	0.26	7.2	0.2	200	0.2	1100

The concentration n_S is in units of 10^{12} cm^{-2} and the mobility μ is in units of 10^3 cm^2/V s.

concentrations, but small mobilities. Case 2 corresponds to the design with a δ layer separated by a spacer from the AlGaAs/GaAs interface. The electron concentrations are smaller, but mobilities increase. Examples 3 and 4 are given for δ layers embedded in narrow quantum wells separated by different spacers from the interface. For the latter cases the concentrations decrease slightly, but the mobilities increase considerably. One can see that the structures of examples 2, 3, and 4 of selectively δ-doped heterostructures allow one to achieve high two-dimensional concentrations of carriers and the highest mobilities – above 10^6 cm^2/V s in the structure of example 4.

4.6 Closing Remarks

In this chapter we have studied the electronic properties of a number of particular quantum structures. Starting with energy spectra of semiconductor materials, we have used the crystal symmetries and concluded that for the most interesting situations we need to know the energy spectra near the high-symmetry points in $\vec{k}$ space. Several parameters are necessary to describe electrons in these cases: positions of the energy minima, the bandgap, and the effective masses. The actual hole-energy spectra were found to be at the Γ point; these spectra can be described by a few parameters. We have discussed these parameters for particular semiconductors of the group IV and III–V compounds.

We have analyzed semiconductor alloys and learned that for many practically interesting cases the alloy-energy spectra can be described similarly to those of pure crystals. These alloys may be engineered to produce considerable variations in the electron parameters.

Next, we explained the formation of discontinuities in both valence and conduction bands, in which two materials are brought together to form a heterojunction. This effect provides the practical basis for modifying electron-energy spectra, and it is particularly useful for a spatial modulation of the potential profiles experienced by electrons and holes as well as for the creation of various artificial heterostructures with energy barriers, quantum wells, wires, boxes, superlattices, etc.

When the state of the art in the fabrication of heterostructures was analyzed, it was established that high-quality heterostructures can be produced by use of materials with similar crystal properties, for example, materials from the same family. The ratio of lattice constants of both materials is a critical parameter for such heterostructures. If these lattice constants almost coincide, as in the case of lattice-matched materials, one can produce a heterostructure without the strain that is due to lattice mismatch or misfit imperfections. An example of such a lattice-matched system is the AlGaAs/GaAs heterostructure. For lattice-mismatched materials we have found that only thin layers can accommodate the lattice mismatch and retain a near-perfect crystalline structure. The resultant structures are strained layers that are pseudomorphic. The strain in pseudomorphic heterostructures leads to a set of new phenomena. In particular, it affects the energy spectra in the strained layers. An example is the Si/Ge heterostructure. Thus we have established that selected techniques for fabricating alloys and heterostructures facilitate energy-band engineering and the control of electron parameters.

We have considered methods of supplying carriers in the quantum heterostructures. We presented the idea of selective doping and the charge-transfer effect. It has been shown that by such a selective-doping method, a two-dimensional electron gas can be created at a heterointerface. The advantages of such a gas include the possibility of obtaining high electron concentrations (up to several units of 10^{12} cm^{-2}), the spatial separation of the electrons (holes) and their parent donors (acceptors), and considerable increase in the mobility of the carriers. We have also studied the filling of quantum wells by carriers and the possibility of controlling the electron concentration in gated heterostructures by external bias. The technique of δ doping, an extreme example of the selective-doping method, has been analyzed and found to be both flexible and capable of yielding the highest carrier concentrations at superb mobilities. We have formulated the basic equation that self-consistently describes charge-transfer effects and the quantization of electron-energy spectra. Several examples of particular heterostructures were analyzed in details.

More information on the crystal symmetry, electron-energy spectra, and the general theory of strain effects can be found in the following books:

C. Kittel, *Quantum Theory of Solids* (Wiley, New York, London, 1963).

I. Ipatova and V. Mitin, *Introduction to Solid-State Electronics* (Addison-Wesley, New York, 1996).

G. E. Picus and G. E. Bir, *Symmetry and Strain Induced Effects in Semiconductors* (Wiley, New York, 1974).

Detailed treatments of semiconductor alloys and heterojunctions as well as reviews of methods of their fabrication and doping are presented in the following references:

R. H. Hendel, *et al.*, "Molecular-beam epitaxy and the technology of selectively-doped heterostructure transistors," in *Gallium Arsenide Technology*, D. K. Ferry, ed. (Howard W. Sams, Indianapolis, IN, 1985).

K. Ploog, "Delta-(δ)-doping in MBE-grown GaAs: concept and device application," J. Cryst. Growth **81**, pp. 304–313 (1987).

S. M. Sze, *High-Speed Semiconductor Devices* (Wiley, New York, 1990).

J. Singh, *Physics of Semiconductors and Their Heterostructures* (McGraw-Hill, New York, 1993).

The following references are devoted to calculations and discussions of electron-energy spectra and wave functions in various quantum structures:

T. Ando, A. B. Fowler, and F. Stern, "Electronic properties of two-dimensional systems," Rev. Mod. Phys. **54**, pp. 437–672 (1982).

G. Bastard, *Wave Mechanics Applied to Semiconductor Heterostructures* (Halsted, New York, 1988).

C. Weisbuch and B. Vinter, *Quantum Semiconductor Structures* (Academic, San Diego, 1991).

PROBLEMS

1. From the values of lattice constants given in Table 4.6, explain why it is feasible to grow stable Si/Ge, $Al_xGa_{1-x}As/GaAs$, and $In_xAl_{1-x}As/In_yGa_{1-y}As$ heterostructures; explain why it is difficult to grow stable heterostructures of GaP/SiC and InP/SiC.

2. Consider an AlAs/GaAs superlattice with AlAs layers sufficiently thick that quantum confinement does not appreciably shift the X-conduction-band edges in the AlAs layers. For GaAs quantum wells with thicknesses less than ~35 Å, the lowest confined level in the GaAs Γ band has an energy greater than the lowest confined level of the AlAs X band. It is known that such a situation results in the mixing of Γ and X bands in superlattices. Assuming that the valence bands are bulklike, draw a schematic of the valence band, the Γ conduction band, and the X conduction band for an AlAs/GaAs superlattice with 30-Å-thick GaAs and 70-Å-thick AlAs regions. Assuming strong Γ–X coupling, indicate on the drawing where the electrons will reside and where the holes will reside.

3. Consider a quantum wire that is open on the end at $z = 0$ and terminated with a Schottky barrier at the end $z = a$. Suppose that the Schottky barrier creates a depletion region of thickness δ so that the effective length of the quantum wire is $a - \delta$. Suppose electrons with de Broglie wavelength λ are injected into the wire at $z = 0$ and suppose that these electrons propagate ballistically except for perfect reflections at $a - \delta$. Discuss how the Schottky barrier can be tuned to cause the constructive or destructive interference of de Broglie waves at $z = 0$.

Lattice Vibrations
in Quantum Structures

5.1 Introduction

In Chapters 3 and 4 we have considered electrons in quantum structures and have derived significant and important modifications of electron wave functions, energy spectra, and densities of states compared with the corresponding characteristics in bulk materials. In Chapter 4, we have also studied some peculiarities of the lattices of materials comprising a heterostructure. We have found that in the case of lattice-mismatched materials, the layer grown upon the substrate is strained. The lattice of such a layer adapts the periodicity of the substrate. The strain is of great importance, but it characterizes only static properties of the lattice. In this chapter, we turn our attention to lattice dynamics in quantum structures. Usually lattice dynamics is described in terms of a special kind of quasi-particle: phonons. We introduce phonons in bulk materials and heterostructures and show that phonons are also affected significantly by heterojunctions.

The description of lattice vibrations (phonons) is essential for understanding various electronic and optical phenomena in solids. Frequently phonons control electron characteristics, for example, mobility, intervalley scattering, processes of electron capture in quantum wells and wires, different kinds of tunneling processes, interband phototransitions in indirect semiconductors, etc. On the other hand, there are many applications based straightforwardly on the dynamics of phonons: acoustoelectronics, optoacoustics, thermoelectricity, etc. Finally, phonons play a dominant role in thermal budget and heat removal in electronic and optoelectronic devices. Knowledge of the lattice dynamics of heterostructures facilitates the selection of parameters and conditions to suppress or enhance electron–phonon interactions, to improve electrical and optical characteristics, to optimize the thermal budget of devices, etc. In principle, with these goals, special phonon-engineered structures can be designed and fabricated.

In the first stage of our consideration, we intentionally avoid introducing a definition of phonon because we want to start with the classical description of lattice vibrations. Next we recall the introduction of phonons in bulk crystals, and only then do we investigate the modification of lattice vibrations and phonon spectra in quantum structures. Our goal is to provide a simple picture of phonon-confinement in heterostructures. In many cases such an intuitive understanding of this picture is instructive and advantageous.

5.2 Vibrations of Atomic Linear Chains

5.2.1 Monoatomic Chain

First we construct the simplest model of lattice vibrations, which is based on a monoatomic linear infinitely long chain governed by the laws of classical mechanics. This simple model yields general conclusions and reveals the basic trends applicable to real crystal lattices. Let the equilibrium distance between atoms be the lattice constant a. The equilibrium position of the nth atom is na, and the displacement of this atom from this position is denoted by u_n. Figure 5.1 depicts such a linear chain of identical atoms of mass M. We assume that these atoms are connected by massless springs; these springs represent interatomic forces. If the displacements of atoms from their equilibrium positions are not too large, the restoring forces on the chain obey Hooks's law:

$$F = -\beta u, \tag{5.1}$$

where u is a change of the spring length, β is the quasi-elastic-force coefficient, and F is the force exerted by the spring. Now we can apply Eq. (5.1) for the total force F_n acting on the nth atom coupled with its two nearest neighbors by two springs as

$$F_n = -\beta(u_n - u_{n+1}) - \beta(u_n - u_{n-1}). \tag{5.2}$$

Hence the Newton equation of motion for the nth atom is

$$M\frac{d^2 u_n}{dt^2} + \beta(2u_n - u_{n-1} - u_{n+1}) = 0. \tag{5.3}$$

At first glance, this infinite set of linear differential equations, Eq. (5.3), appears difficult to solve. However, physical considerations simplify our task greatly. Indeed, this linear chain is similar to a string. The equation for a string can be found from Eq. (5.3) if we make the replacement $(2u_n - u_{n-1} - u_{n+1}) \rightarrow -a^2(\partial^2 u/\partial z^2)$:

$$\rho\frac{\partial^2 u}{\partial t^2} - \lambda\frac{\partial^2 u}{\partial z^2} = 0, \tag{5.4}$$

where $\rho = M/a$ is the linear density of the string and $\lambda = \beta a$ is the elastic modulus. The solution of Eq. (5.4) for an infinite string is a propagating wave:

$$u = A e^{i(qz - \omega t)}, \tag{5.5}$$

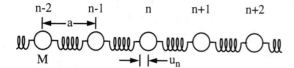

Figure 5.1. The monoatomic linear chain: masses are connected by springs. [After D. K. Ferry, *Semiconductors* (Macmillan, New York, 1991).]

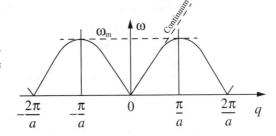

Figure 5.2. The dispersion relation for a monoatomic linear chain. [After J. M. Ziman, *Electrons and Phonons* (Oxford U. Press, London, 1960).]

where the amplitude A is generally a complex quantity, ω is the frequency, and q is the wave vector of the elastic wave. The relationship between ω and q is the so-called dispersion relation: $\omega = sq$, where $s = \sqrt{\lambda/\rho}$ is the velocity of the wave.

Let us use the Fourier transform for Eq. (5.3):

$$u_n = A e^{i(qan - \omega t)}. \tag{5.6}$$

Here the parameter q is still not defined. Substituting the u_n of Eq. (5.6) into Eq. (5.3), we get

$$M\omega^2 = \beta(2 - e^{iqa} - e^{-iqa}) = 2\beta(1 - \cos qa) = 4\beta \sin^2 \frac{qa}{2},$$

which can be rewritten as

$$\omega = \sqrt{\frac{4\beta}{M}} \left| \sin \frac{qa}{2} \right|. \tag{5.7}$$

In Eq. (5.7) we select the positive sign of the vibration frequency ω. Thus Eq. (5.6) is a solution of Eq. (5.3) for any integer n if the frequency ω is related to the wave vector q by the dispersion relationship of Eq. (5.7). Note that ω and q are independent of n. The solution, which is found, is a collective vibration of the chain. Figure 5.2 depicts this dispersion relation. Unlike in the case of a continuous string, ω is not proportional to q. However, for small arguments, $\sin x \approx x$ and for the long-wave (small-q) limit we have almost linear dispersion. In general, the waves (modes) with a linear dispersion $\omega(q) = sq$ are called acoustic modes, where s is the sound velocity in the crystal. From the asymptotic behavior at $q \to 0$, the dispersion relationship of Eq. (5.7) is that corresponding to acoustic modes. Other type of modes, namely optical modes, can be obtained only in a more complex diatomic chain; such modes are considered in Subsection 5.2.2.

For the solution of Eq. (5.6), both q and ω must be real to represent a steady state. For real quantities, because of the periodicity of the solution of Eq. (5.6) and the dispersion relationship of Eq. (5.7), q and $q' = q + 2n'\pi$ are identical if n' is an integer, since they correspond to the same frequency and to the same solution. The point is that the same displacements of atoms can be described by several different waves but only modes with the largest possible wavelengths correspond to the physical vibrations of the masses. Modes with wavelength $\lambda = 2\pi/q < 2a$ are therefore meaningless and do not exist physically. Hence all physically meaningful modes have wave vectors within the interval

$$-\frac{\pi}{a} \leq q \leq \frac{\pi}{a}, \tag{5.8}$$

which is the first Brillouin zone of the one-dimensional chain. If the wave vector changes within the Brillouin zone, the frequency of the vibrations covers a band with a finite width $\Delta\omega \equiv \omega_m = \sqrt{4\beta/M}$ independently of the lattice constant a, that is, unlike the acoustic like behavior at $q \to 0$, at the boundaries of the Brillouin zone, the vibration frequency has maxima and a quadratic-type dependence in the vicinity of the boundaries.

It is instructive to mention that the physical situation just discussed is analogous to that of electron waves in the crystalline potential (see Section 2.6) and the superlattice periodic potential (see Section 3.6). The point is that although the physical phenomena of electrons in a periodic potential and atomic vibrations are completely different, the behavior of both phenomena is constrained by a periodic structure; accordingly, the mathematical approaches and the forms of solutions are similar for both cases.

Keeping in mind our intention of applying the theory of lattice vibrations to nanometer-scale quantum structures, we anticipate that the allowed values of q depend on the geometry of the structure, its dimensional confinement characteristics, and the boundary conditions. For macroscopic crystals (bulk semiconductors), the boundary conditions should not affect the lattice dynamics in the bulk far from the surfaces (interfaces). In macroscopic systems we can impose cyclic conditions as the boundary conditions; we have already used this approach for electrons in Chapters 3 and 4. We can imagine the infinite chain as a large circle containing N atoms, where $N \gg 1$, and require that $u_{N+n} = u_n$. From Eq. (5.6) we see that the above condition leads to

$$e^{iqNa} = 1, \qquad q = \frac{2\pi l}{Na}, \tag{5.9}$$

where $-N/2 < l < N/2$ is an integer. Hence the wave number q becomes discrete and there are N possible values of it, corresponding to N different standing waves, since there are N degrees of freedom – one for each atom. When dealing with macroscopic systems of huge numbers N, we find that q is practically continuous, that is, q is quasi-continuous. The dispersion relation of Fig. 5.2 is valid for the case of a macroscopic system.

However, when we deal with microscopic systems, particular boundary conditions and dimensions of the system are of crucial importance. Strictly speaking, we can impose no boundary conditions on the solution of Eq. (5.3), since we have assumed that the chain is infinite. If we considered a one-dimensional finite chain, we would have to solve a finite number of equations similar to Eq. (5.3) with strict boundary conditions. The result would be a finite number of standing waves that depend on the boundary conditions. For example, in the case of an

Figure 5.3. The diatomic linear chain with alternating masses. [After D. K. Ferry, *Semiconductors* (Macmillan, New York, 1991).]

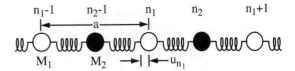

N-atom chain with hard boundary conditions ($u_1 = u_N = 0$), there would be $N - 2$ different vibrational modes. For free boundaries, there would be N vibrational modes in contrast to the previous case.

5.2.2 Diatomic Chain

In Subsection 5.2.1 we assumed that all atoms are identical. This model cannot reveal certain important features of lattice vibrations in complex crystals. Figure 5.3 shows the diatomic chain that serves as a model for diatomic crystals. Sites n_1 are occupied by atoms with mass M_1 and sites n_2 by atoms with mass M_2. In general, the distance between atoms associated with sites n_1 and n_2 may be different from the distance between atoms associated with sites n_1 and $n_2 - 1$. Therefore we define the lattice constant a as the period of this chain, i.e., a is the minimum distance between identical atoms. The coefficients of the quasi-elastic force β may also be different because the distances between atoms and atoms are different. Let us denote the coefficient of the quasi-elastic force between the atoms n_1 and n_2 by β_1 and that between the atoms n_1 and $n_2 - 1$ by β_2. Now within the approximation of small displacements we can write the equations of motion in a form similar to that of Eq. (5.3):

$$M_1 \frac{d^2 u_{n_1}}{dt^2} = -\beta_1 (u_{n_1} - u_{n_2}) - \beta_2 (u_{n_1} - u_{n_2-1}),$$

$$M_2 \frac{d^2 u_{n_2}}{dt^2} = -\beta_1 (u_{n_2} - u_{n_1}) - \beta_2 (u_{n_2} - u_{n_1+1}). \tag{5.10}$$

Again, we look for wavelike solutions. Since we are dealing with an quasi-infinite connected system of atoms, the wave numbers and the frequencies of moving masses must be the same but their amplitudes and phases may differ. We assume that

$$u_{n_1} = A_1 e^{i(qan_1 - \omega t)}, \qquad u_{n_2} = A_2 e^{i(qan_2 - \omega t)}. \tag{5.11}$$

The substitution of Eqs. (5.11) into Eqs. (5.10) leads to the relations

$$\left(\omega^2 - \frac{\beta_1 + \beta_2}{M_1} \right) A_1 + \left(\frac{\beta_1 + \beta_2 e^{-iaq}}{M_1} \right) A_2 = 0,$$

$$\left(\frac{\beta_1 + \beta_2 e^{iaq}}{M_2} \right) A_1 + \left(\omega^2 - \frac{\beta_1 + \beta_2}{M_2} \right) A_2 = 0. \tag{5.12}$$

This set of homogeneous linear equations has nontrivial solutions if and only if the determinant is equal to zero. This requirement leads to a biquadratic equation with respect to ω,

$$\left(\omega^2 - \frac{\beta_1 + \beta_2}{M_1}\right)\left(\omega^2 - \frac{\beta_1 + \beta_2}{M_2}\right) - \left(\frac{\beta_1 + \beta_2 e^{iaq}}{M_2}\right)\left(\frac{\beta_1 + \beta_2 e^{-iaq}}{M_1}\right) = 0,$$

with solutions

$$\omega_1 = \frac{\omega_0}{\sqrt{2}}\sqrt{\left(1 - \sqrt{1 - \gamma^2 \sin^2 \frac{aq}{2}}\right)},$$

$$\omega_2 = \frac{\omega_0}{\sqrt{2}}\sqrt{\left(1 + \sqrt{1 - \gamma^2 \sin^2 \frac{aq}{2}}\right)}, \qquad (5.13)$$

where

$$\omega_0 = \sqrt{\frac{(\beta_1 + \beta_2)(M_1 + M_2)}{M_1 M_2}},$$

$$\gamma^2 = 16\frac{\beta_1\beta_2}{(\beta_1 + \beta_2)^2}\frac{M_1 M_2}{(M_1 + M_2)^2}.$$

The quantity γ has a maximum equal to 1 when $\beta_1 = \beta_2$ and $M_1 = M_2$. Hence the second terms in the square roots of Eqs. (5.13) are not larger than 1 and the frequencies ω_1 and ω_2 are real. It is obvious from Eqs. (5.13) that for $\gamma = 1$ the first equation can be transformed to a form similar to that of Eq. (5.7). Note that for $\gamma = 1$, the parameter a in Eqs. (5.13) represents twice the interatomic distance. This is obvious when Figs. 5.1 and 5.3 are compared. Thus, in order to compare Eqs. (5.13) for this degenerate ($\gamma = 1$) case with Eqs. (5.7), we have to replace a in Eqs. (5.13) with $2a$. In analogy with the case of a monoatomic chain, this mode is referred to as an acoustic mode, $\omega_1 = \omega_a$. As $q \to 0$ the frequency of this mode tends to zero. The other mode is referred to as an optical-mode with $\omega_2 = \omega_{op}$, and its frequency tends to a constant as $q \to 0$. Again, the periodicity and the evenness of the solutions of Eqs. (5.11) and the frequencies of Eqs. (5.13) suggest that we can restrict our consideration to the q interval $(0, \pi/a)$, while the Brillouin zone will be the same as that of relation (5.8): $-\pi/a \le q \le \pi/a$.

Let us first determine the extreme values of ω_a and ω_{op} at $q = 0$ and $q = \pi/a$:

$$\omega_a(0) = 0, \qquad \omega_a(\pi/a) = \frac{\omega_0}{\sqrt{2}}\sqrt{1 - \sqrt{1 - \gamma^2}},$$

$$\omega_{op}(0) = \omega_0, \qquad \omega_{op}(\pi/a) = \frac{\omega_0}{\sqrt{2}}\sqrt{1 + \sqrt{1 - \gamma^2}}. \qquad (5.14)$$

Hence $\omega_{op}(0) = \omega_0 > \omega_{op}(\pi/a) \ge \omega_a(\pi/a) > \omega_a(0) = 0$. In the long-wavelength limit, $\lambda = 2\pi/q \gg a$ or $aq \ll 1$, and Eqs. (5.13) can be rewritten as

$$\omega_a \approx \frac{1}{4}\omega_0\gamma aq = sq, \qquad \omega_{op} \approx \omega_0\left(1 - \frac{\gamma^2 a^2}{32}q^2\right). \qquad (5.15)$$

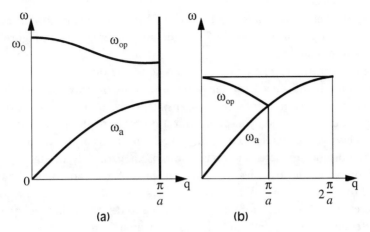

Figure 5.4. The dispersion relations for the (a) nondegenerate ($\gamma \neq 1$), (b) degenerate ($\gamma = 1$) cases.

Here $s = \omega_0 \gamma a / 4$ is the sound velocity for the diatomic chain. Taking derivatives of ω_1 and ω_2, as given by Eqs. (5.13), with respect to q one can show that both branches have extrema at edges of the Brillouin zone:

$$\left(\frac{d\omega_a}{dq}\right)_{\pm\pi/a} = \left(\frac{d\omega_{op}}{dq}\right)_{\pm\pi/a} = 0$$

if $\gamma^2 \neq 1$. Now taking into consideration the latter equality and Eqs. (5.14) and approximations (5.15), it is clear that dispersion curves have the general functional behavior depicted in Fig. 5.4. A noteworthy feature of the frequency spectrum is the appearance of a forbidden gap between $\omega_{a,\max} = \omega_a(\pi/a)$ and $\omega_{op,\min} = \omega_{op}(\pi/a)$. The forbidden region corresponds to frequencies at which vibrations cannot propagate unattenuated through the linear chain.

The gap decreases when γ increases. It vanishes in the degenerate case of $\gamma^2 = 1$. In this case, Eqs. (5.13) can be rewritten as

$$\omega_a = 2\sqrt{\frac{\beta}{M}} \left|\sin\frac{aq}{4}\right|, \qquad \omega_{op} = 2\sqrt{\frac{\beta}{M}} \left|\cos\frac{aq}{4}\right|. \tag{5.16}$$

Hence,

$$\omega_a(\pi/a) = \omega_{op}(\pi/a) = \frac{\omega_0}{\sqrt{2}}.$$

As pointed out above, this degenerate case corresponds to the case of identical atoms considered in Subsection 5.2.1. Therefore, for the degenerate case in this new notation, the lattice-constant is $a/2$, as is evident from Fig. 5.3. Consequently, the Brillouin zone is now defined by $-2\pi/a \leq q \leq 2\pi/a$. A natural question arises as to how we could have missed the optical branch when considering the monoatomic chain, if we get this branch now for the same case ($\gamma^2 = 1$). In reality, we have not missed any vibrations. The reason is that ω_a

replicates the behavior of ω given by Eq. (5.7) in the interval $(0, \pi/a)$ and ω_{op} replicates the behavior of ω, Eq. (5.7), in the interval $(\pi/a, 2\pi/a)$. Figure 5.4(b) illustrates this behavior.

By analyzing the ratio u_{n_1}/u_{n_2}, one can conclude that (1) if $M_1 \neq M_2$, the acoustic wave is influenced primarily by vibrations of heavier atoms and the optical wave by vibrations of lighter atoms; (2) if $M_1 = M_2$, the acoustic wave for $q \to 0$ is characterized by in-phase vibrations of both atoms in the cell, while the optical wave is characterized by antiphase vibrations of the atoms.

An important characteristic of phonons is their density of states. For the sake of simplicity, let us consider again the case of a monoatomic chain. The results, however, will have quite a general meaning. According to Eqs. (5.9), q is a discrete variable. Therefore it is worthwhile to determine how many modes with different values of q fall into the frequency interval from ω to $\omega + d\omega$. From Eqs. (5.7) and (5.9) it follows that

$$d\omega = a\sqrt{\frac{\beta}{M}} \left| \cos \frac{qa}{2} \right| dq, \qquad dq = \frac{2\pi}{aN} dl. \tag{5.17}$$

Hence, taking into account both branches $q < 0$ and $q > 0$, we find that the number of modes $d\nu$ in the interval $d\omega$ is

$$d\nu = 2dl = \frac{N}{\pi}\sqrt{\frac{M}{\beta}} \frac{d\omega}{\left| \cos \frac{aq}{2} \right|}. \tag{5.18}$$

Now, expressing $\cos[(aq)/2]$ in terms of ω by using Eq. (5.7), we get

$$\cos \frac{aq}{2} = \sqrt{1 - \sin^2 \frac{aq}{2}} = \sqrt{1 - \frac{\omega^2 M}{4\beta}},$$

and the density of states of vibrational modes per unit length is

$$\varrho_{ph}(\omega) = \frac{1}{aN}\frac{d\nu}{d\omega} = \frac{2}{\pi a}\frac{1}{\sqrt{\omega_m^2 - \omega^2}}, \qquad \omega_m = 2\sqrt{\frac{\beta}{M}}. \tag{5.19}$$

In the limit $\omega \to 0$, we find that $\varrho_{ph} = \text{const.}$; in addition, $\varrho \to \infty$ as $\omega \to \omega_m$. In a similar manner, we can derive the density of states for the modes of a diatomic chain.

Note that the density of states of lattice vibrations is a measurable quantity. The density of states affects the infrared absorption of light, electron interactions with lattice vibrations, thermal properties, etc.; hence the density of states may be determined through measurements of these quantities.

5.3 Normal Coordinates. Three-Dimensional Case

The equations of lattice motion for atomic chains are linear; therefore a superposition of solutions is also a solution. Hence the most general displacement of the nth atom can be written as a linear combination of the solutions that we obtained in Section 5.2:

$$u_n = \frac{1}{\sqrt{N}} \sum_q Q_q e^{iqna}, \tag{5.20}$$

where the summation extends over the entire Brillouin zone. The Fourier coefficients (amplitudes of different modes) Q_q are, in general, time dependent. There are just N such coefficients. Instead of real displacements u_n, $n = 1, 2, \ldots, N$, the coefficients Q_q can be treated as new time-dependent variables (coordinates). Substituting Eq. (5.20) into Eq. (5.3), we get the equation of motion for the coordinate Q_q in the form of that of a harmonic oscillator:

$$\frac{d^2 Q_q}{dt^2} + \omega^2(q) Q_q = 0. \tag{5.21}$$

Equation (5.21) has an appealing form since it includes only one kind of coordinate Q_q and does not include any other coordinate $Q_{q'}$. Such coordinates Q_q are called normal coordinates of the problem. It is worth emphasizing that the normal coordinates have nothing to do with real spatial coordinates of atoms. They are only the Fourier coefficients of the expansion of a real displacement in the Fourier series of Eq. (5.20). These coefficients have complex values. The convenience of using the normal coordinates lies in the simplest form of their dynamic equations as given by Eq. (5.21).

Now, let us discuss briefly the nature of three-dimensional lattice vibrations. In the simplest case of a cubic monoatomic lattice, each atom has spring connections to its six nearest-neighbor atoms; there are two such atoms along each coordinate. In such a lattice, the atomic motion becomes three dimensional. Along with longitudinal waves, which are analogous to those of the one-dimensional chain, now several transverse waves occur: the longitudinal wave propagates in the direction of atom displacements associated with this wave, while the transverse wave propagates perpendicularly to atom displacements. The velocity of the longitudinal mode is always greater than that of transversal modes. Strictly speaking, for short waves these longitudinal and transverse modes are coupled and cannot be separated easily. Depending on the mutual phases between neighboring atoms, acoustic and optical modes can be distinguished. The general rule is that the acoustic-mode frequency tends to zero when $q \to 0$, while the optical-phonon frequency tends to a nonzero value. If the number of atoms per primitive crystal cell is $\mathcal{N}$, there are $3\mathcal{N}$ different kinds of vibrational modes. Among them only three are acoustic modes; one is longitudinal and the other two are transversal. The rest of the $3\mathcal{N} - 3$ modes are optical; $\mathcal{N} - 1$ are longitudinal and $2\mathcal{N} - 2$ are transversal.

Formulating a microscopic description of three-dimensional lattice dynamics is a complex problem. However, formally one can generalize the canonical form of Eq. (5.21) for the three-dimensional case. Formal extension to three dimensions of the canonical form of Eq. (5.21) is simple enough. Let us consider a crystal with N unit cells containing $\mathcal{N}$ atoms each, with masses M_j, $j = 1, 2, \ldots,$ $\mathcal{N}$. Again we assume small displacements so that forces obey Hooke's law. Following the analogy with the one-dimensional chain the displacements can be written as

$$\vec{u}_{j,n} = \frac{1}{\sqrt{N}} \sum_{\vec{q},s} Q_{\vec{q},s} e^{i\vec{q}\vec{a}_{n,j}} \, \vec{b}_s(\vec{q}). \tag{5.22}$$

Here $\vec{q}$ is the wave vector of the mode, $\vec{b}_s(\vec{q})$ is a unit vector that defines the polarization of the wave branch characterized by the index s, and $\vec{a}_{n,j}$ is the vector that defines the position of the jth atom in the nth primitive cell. The summation extends over all directions of q and all wave branches s. In the same way as for the one-dimensional lattice, the equations of motion in these normal coordinates $Q_{\vec{q},s}$ can be reduced to the canonical form of the harmonic oscillator:

$$\frac{d^2 Q_{\vec{q},s}}{dt^2} + \omega^2(\vec{q}, s) Q_{\vec{q},s} = 0, \qquad Q_{\vec{q},s} = Q^*_{-\vec{q},s}. \tag{5.23}$$

If the crystal has only one atom per unit cell, there are three possible directions of polarization of the wave for each value of the wave vector $\vec{q}$. As we have already mentioned, for an isotropic case and in the long-wavelength limit they appear as one longitudinal and two transverse modes. If the crystal contains more than one atom per unit cell, the index s runs from 1 to $3\mathcal{N}$, where $\mathcal{N}$ is the number of atoms per unit cell. $\mathcal{N}$ is equal to 2 for many of the important III–V, II–VI, and IV–VI compounds, as well as for Si, Ge. When $\mathcal{N} = 2$ there are three acoustic modes [one longitudinal acoustic (LA) and two transverse acoustic (TA)] and three optical modes [one longitudinal optical (LO) and two transverse optical (TO)].

The Brillouin zone of relation (5.8) can be generalized from the case of a linear chain to that for a three-dimensional crystal. Since the Brillouin zone reflects only periodicity and symmetry properties of a crystal, it is analogous to the Brillouin zone for electron wave vectors $\vec{k}$. In Chapter 4 we studied the Brillouin zone including high-symmetry points Γ, X, and L. In Fig. 5.5 and Table 5.1 the general view of the Brillouin zone and the symmetry points are presented for phonons. These results can also be used for analysis of lattice vibrations in cubic crystals.

Figure 5.5 shows dispersion curves $\omega_{LA}(\vec{q})$, $\omega_{TA}(\vec{q})$, $\omega_{LO}(\vec{q})$, and $\omega_{TO}(\vec{q})$ for lattice vibrations of GaAs. Two different crystalline directions (111) and (100) are shown. As illustrated in Fig. 5.5, the actual dependence of the frequency on the wave vector is more complicated than that obtained from linear-chain analysis. The TA and the TO modes are doubly degenerate. Each branch depends not only on absolute value, but also on the direction of wave vector $\vec{q}$.

The phonon frequencies are collected in Table 5.1 for cubic crystals for which electron parameters were given in Tables 4.2 and 4.3. In Table 5.1 the frequencies are presented for three high-symmetry points, Γ, X, and L. Since the acoustic modes have zero frequency exactly at the Γ point, only LO and TO modes are given for this point. The X points correspond to the boundaries of the first Brillouin zone.

Table 5.1.
Phonon Frequencies for High-Symmetry Points (in Terahertz)

Symmetry points	$\Gamma(000)$		$X(100)$				$L(111)$			
Type of mode	LO	TO	LO	TO	LA	TA	LO	TO	LA	TA
Si	15.6	15.6	12.3	13.9	12.3	4.5	12.6	14.7	11.3	3.42
Ge	9.03	9.03	6.9	8.25	6.9	2.46	7.41	8.4	6.45	1.95
InAs	7.3	6.57	4.9	6.4	4.4	3.3	5.8	6.5	4.4	2.2
GaAs	8.76	8.07	7.22	7.56	6.8	2.36	7.15	7.84	6.26	1.86
AlAs	12.1	10.8		3.1	3.1					2.43
InSb	6.0	5.37	3.9	5.3	3.6	1.1	4.8	5.1	3.0	1.0
GaSb	7.3	6.9								
AlSb	5.91	5.55			8.9	2.4	7.0	9.2	6.4	1.86

The results shown in Table 5.1 indicate that phonon frequencies cover a wide range from 0 to tens of terahertz. These frequencies are almost isotropic at small $\vec{q}$: $\omega_{LA,TA}(\vec{q}) = s_{l,t}|q|$ and $\omega_{LO,TO} \approx \omega_{LO,TO}(0) - \text{const.} \times q^2$. At high $\vec{q}$ these frequencies become strongly dependent on wave-vector direction. Another conclusion of practical importance is that the phonon frequencies of the two materials comprising the heterostructures are considerably different; for example, LO phonon frequencies at the Γ point for AlAs and GaAs are 12.1 and

Figure 5.5. The phonon frequency dependency versus reduced wave vector, $qa/2\pi$, for GaAs. [After J. L. T. Waugh and G. Dolling, "Crystal dynamics of gallium arsenide," Phys. Rev. **132**, pp. 2410–2412 (1963).]

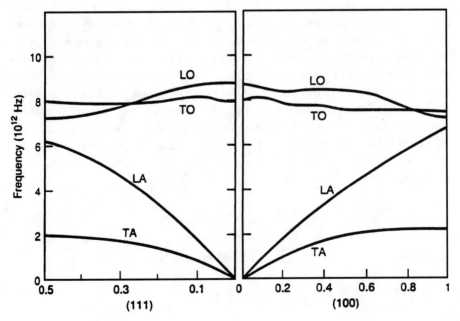

8.76 THz, respectively. This means that the LO vibrations of AlAs material are not resonant with those of GaAs material and therefore cannot propagate through the AlAs/GaAs junction. As we discuss in Section 5.5 new types of the lattice vibrations – interface and confined phonons – occur in the heterostructures.

In the three-dimensional case, the density of states of lattice vibrations can be written as

$$\varrho_{\text{ph}}(\omega) = \frac{1}{8\pi^3} \sum_s \int \frac{d\vec{\sigma}}{|\nabla_{\vec{q}}\,\omega|}. \tag{5.24}$$

Here the summation extends over all modes and the integral is taken over the surface of constant ω in q space. In an isotropic medium the isoenergetic surface is spherical and Eq. (5.24) can be simplified greatly. For LA and TA modes one gets, respectively,

$$\varrho_l(\omega) = \frac{\omega^2}{2\pi^2 s_l^3}, \qquad \varrho_t(\omega) = \frac{\omega^2}{\pi^2 s_t^3},$$

where s_l is the velocity of the longitudinal mode and s_t is the velocity of the transverse mode. Thus the density of states for a three-dimensional crystal in the limit of long-wavelength acoustic modes is proportional to ω^2. Note that ϱ_t has an additional factor of 2, since there are two independent transverse modes with different polarizations.

5.4 Phonons

So far we have considered a completely classical picture of lattice vibrations. For many purposes the classical picture is entirely adequate. However, the interaction of an electron with lattice vibrations can be described only in quantum-mechanical language since an electron itself is a quantum-mechanical object. Therefore it is extremely advantageous to describe lattice vibrations in terms of the same quantum-mechanical language. The transition to quantum mechanics is simple enough within the framework of harmonic oscillators, in which the displacements of atoms from their equilibrium positions are small so that the force between atoms is a linear function of the displacements. We can use the description of lattice vibrations in terms of normal coordinates $Q_{\vec{q},s}$.

Let us introduce the generalized momentum $\tilde{P}_{\vec{q},s}$, which is the conjugate to the normal coordinate, $\tilde{Q}_{\vec{q},s} = \sqrt{M}Q_{\vec{q},s}$,

$$\tilde{P}_{\vec{q},s} = \frac{d\tilde{Q}_{\vec{q},s}}{dt}.$$

Now the classical Hamiltonian for a harmonic oscillator is

$$H(\tilde{Q}, \tilde{P}) = \frac{1}{2} \sum_{\vec{q},s} \left[\tilde{P}_{\vec{q},s}^2 + \omega^2(\vec{q}, s)\tilde{Q}_{\vec{q},s}^2\right]. \tag{5.25}$$

The transition to quantum mechanics is achieved when the momenta $\tilde{P}_{\vec{q},s}$ are replaced by operators:

$$\tilde{P}_{\vec{q},s} \rightarrow \frac{\hbar}{i} \frac{\partial}{\partial \tilde{Q}_{\vec{q},s}}.$$

Then the quantum-mechanical Hamiltonian operator in the $\tilde{Q}$ representation takes the following form:

$$H = \frac{1}{2} \sum_{\vec{q},s} \left[-\hbar^2 \frac{\partial^2}{\partial \tilde{Q}_{\vec{q},s}^2} + \omega^2(\vec{q}, s) \tilde{Q}_{\vec{q},s}^2 \right]. \tag{5.26}$$

Each term in Eq. (5.26) has the form of the Hamiltonian of a harmonic oscillator with an effective mass equal to 1. Thus the lattice vibrations can be described as a set of harmonic oscillators. Therefore, to obtain a quantum-mechanical description of the lattice vibrations, we have to solve the Schrödinger equation for each oscillator. Let us for simplicity omit the indices $\vec{q}, s$ that characterize the different modes. The Schrödinger equation for a harmonic oscillator can be written as

$$-\frac{\hbar^2}{2} \frac{\partial^2 \psi}{\partial \tilde{Q}^2} + \frac{1}{2} \omega^2 \tilde{Q}^2 \psi = \epsilon \psi, \tag{5.27}$$

which has the following solutions for eigenenergies and eigenfunctions:

$$\epsilon = \epsilon_N = \left(N + \frac{1}{2} \right) \hbar\omega,$$

$$\psi = \psi_N(\tilde{Q}) = \left(\frac{\omega}{\pi\hbar} \right)^{1/4} \frac{1}{\sqrt{2^N N!}} e^{-\omega\tilde{Q}^2/2\hbar} H_N(\xi),$$

$$\xi = \left(\frac{\omega}{\hbar} \right)^{1/2} \tilde{Q}, \tag{5.28}$$

where N is a positive integer or zero, and $H_N(\xi)$ is the Hermite polynomial of the Nth order. Returning to the description of separate modes, one can write the energy of the mode denoted by $\{\vec{q}, s\}$:

$$\epsilon_N(\vec{q}, s) = \left(N_{\vec{q},s} + \frac{1}{2} \right) \hbar\omega(\vec{q}, s). \tag{5.29}$$

The quantities $N_{\vec{q},s}$, in Eq. (5.29) are called occupation numbers; they specify the degree of excitation of each mode $\{\vec{q}, s\}$. We see that the energy of lattice vibrations can change by only an integer multiple of $\hbar\omega$ $(\vec{q}, s)$, i.e., the lattice-vibrational energy is quantized. The quantum of lattice vibration has been given the name phonon. The phonon is a quasi-particle representing one basic unit of excitation of lattice vibration and is characterized by energy $\hbar\omega$ $(\vec{q}, s)$ and quasi-momentum $\hbar\vec{q}$. Then $N_{\vec{q},s}$ is called the phonon occupation number. Because phonons corresponding to the same vibrational mode are identical quantum mechanically, the vibrational state of the crystal is fully described by the phonon occupation numbers $N_{\vec{q},s}$.

In contrast to the electrons, the crystal phonons obey the Bose–Einstein statistics; see Subsection 2.5.1. In Bose–Einstein statistics, any energy-level may be populated without restriction on the number of particles. The equilibrium distribution function, or average number of phonons with the energy $\hbar\omega(\vec{q}, s)$, is the Planck distribution:

$$N_{\vec{q},s} = \frac{1}{\exp\left(\frac{\hbar\omega(\vec{q},s)}{k_B T}\right) - 1}. \tag{5.30}$$

Particles obeying the Bose–Einstein distribution are known as bosons.

For our future applications it is necessary to introduce operators:

$$c_{\vec{q},s} = \left(\frac{\omega}{2\hbar}\right)^{1/2} \tilde{Q}_{\vec{q},s} + \left(\frac{\hbar}{2\omega}\right)^{1/2} \frac{\partial}{\partial \tilde{Q}_{\vec{q},s}}, \tag{5.31}$$

$$c_{\vec{q},s}^+ = \left(\frac{\omega}{2\hbar}\right)^{1/2} \tilde{Q}_{\vec{q},s} - \left(\frac{\hbar}{2\omega}\right)^{1/2} \frac{\partial}{\partial \tilde{Q}_{\vec{q},s}}. \tag{5.32}$$

It is possible to show that

$$c_{\vec{q},s}\psi_N = \sqrt{N}\psi_{N-1}, \qquad c_{\vec{q},s}^+\psi_N = \sqrt{N+1}\psi_{N+1}, \tag{5.33}$$

where $N = N_{\vec{q},s}$ and matrix elements calculated on wave function ψ_N^*, ψ_N are

$$\langle N_{\vec{q},s}' | c_{\vec{q},s} | N_{\vec{q},s}\rangle = \begin{cases} \sqrt{N_{\vec{q},s}} & \text{if } N_{\vec{q},s}' = N_{\vec{q},s} - 1 \\ 0 & \text{otherwise} \end{cases}; \tag{5.34}$$

$$\langle N_{\vec{q},s}' | c_{\vec{q},s}^+ | N_{\vec{q},s}\rangle = \begin{cases} \sqrt{N_{\vec{q},s}+1} & \text{if } N_{\vec{q},s}' = N_{\vec{q},s} + 1 \\ 0 & \text{otherwise} \end{cases}. \tag{5.35}$$

Hence the application of operator $c_{\vec{q},s}$ or $c_{\vec{q},s}^+$ to the system of phonons decreases or increases, respectively, the phonon occupation number by 1. For this reason, the operators $c_{\vec{q},s}$ and $c_{\vec{q},s}^+$ are called annihilation and creation operators, respectively. From quantum mechanics we know that the conjugate momentum and coordinate obey the following commutation rule:

$$[\tilde{Q}_{\vec{q},s}, \tilde{P}_{\vec{q}',s'}] = \left[\tilde{Q}_{\vec{q},s}, \frac{\hbar}{i}\frac{\partial}{\partial \tilde{Q}_{\vec{q}',s'}}\right] = i\hbar\delta_{ss'}\delta_{\vec{q}\vec{q}'}. \tag{5.36}$$

Hence for the creation and the annihilation operators we can write

$$[c_{\vec{q},s}, c_{\vec{q}',s'}^+] = \delta_{ss'}\delta_{\vec{q}\vec{q}'}. \tag{5.37}$$

In quantum mechanics quasi-particles with creation and annihilation operators satisfying these commutation rules are called bosons, which is in accordance with our previous conclusion.

The Hamiltonian of Eq. (5.26) can be expressed compactly through the creation and the annihilation operators:

$$H = \frac{1}{2} \sum_{\vec{q},s} \hbar\omega(\vec{q}, s) \left[c_{\vec{q},s} c_{\vec{q},s}^{+} + c_{\vec{q},s}^{+} c_{\vec{q},s} \right]. \tag{5.38}$$

We will see that the interaction of an electron with lattice vibrations can be described in terms of phonon-emission or absorption by an electron. In concluding this brief introduction on bulk phonons, we emphasize that phonons are important not only in explaining the interaction of an electron with a lattice, but also in providing the means for the adequate descriptions of acoustic, acoustoelectronic, acousto-optic, thermoelectric, and thermal properties of solids.

5.5 Acoustic Vibrations in Quantum Structures

In previous sections we have considered bulk materials, so that boundary conditions on the equations describing atomic motion have not been important. For bulk materials, the assumption of cyclic boundary conditions is usually sufficient and it is unnecessary to consider the actual boundary conditions. A similar approach is common for electrons in bulk crystals. However, we have already seen that in quantum structures the electron spectrum is determined by the boundary conditions imposed on the electron wave function. In the same way, boundary conditions modify the phonon modes and their spectra in quantum structures. The exact microscopic treatment of lattice vibrations in quantum structures requires numerical methods that often do not provide transparent results. Therefore, in the following discussion we use the so-called long-wavelength limit and elastic-continuum approaches. The main criterion of such an approach is that length scales characterizing the elastic waves should significantly exceed the lattice constants of materials. It must be stressed that long-wavelength phonons are primarily responsible for interactions with electrons.

5.5.1 Acoustic Modes in the Long-Wavelength Limit

In the long-wavelength limit, we can apply the theory of elasticity for the description of lattice vibrations. Small elastic vibrations of a solid body can be described by a vector of relative displacement $\vec{u} = \vec{u}(\vec{r}, t)$, which was introduced in Subsection 4.3.1. The equations of motion of an elastic-continuum follow from Newton's second law and have the form

$$\rho \frac{\partial^2 u_i}{\partial t^2} = \frac{\partial \sigma_{ij}}{\partial x_j}, \tag{5.39}$$

where ρ is the density of a semiconductor and σ_{ij} is the stress tensor. The sums in the above equations are assumed to be taken over repeated italic subscripts, that is, the Einstein summation convention is adopted. We have introduced this tensor in Subsection 4.3.1 through the derivatives of the elastic-energy density $U(u_{ij})$ with respect to the strain tensor u_{ij}. For a cubic crystal the elastic-energy is given by Eq. (4.8). For an isotropic elastic-continuum, the elastic-energy can be written as

$$U = \frac{1}{2}\lambda\,\delta_{ij}\,\delta_{kl}\,u_{ij}\,u_{kl} + \frac{1}{2}\mu\,u_{ij}{}^2. \tag{5.40}$$

Then the stress tensor is

$$\sigma_{ij} = \lambda u_{kk}\,\delta_{ij} + 2\mu u_{ij}, \tag{5.41}$$

where λ and μ are elastic moduli, or Lamè constants, and δ_{ij} is the Kronecker delta. The stress tensor is related to the force $\vec{f}$ acting on unit surface through Eq. (4.10).

For an isotropic medium, Eq. (5.39) can be rewritten in vector form as

$$\frac{\partial^2 \vec{u}}{\partial t^2} = s_t^2 \nabla^2 \vec{u} + \left(s_l^2 - s_t^2\right)\mathrm{grad}\,\mathrm{div}\,\vec{u}, \tag{5.42}$$

where $s_l = \sqrt{(\lambda + 2\mu)/\rho}$ and $s_t = \sqrt{\mu/\rho}$ are the velocities of the LA and the TA waves in bulk semiconductors. The vector equation (5.42) may be reformulated in terms of scalar equations; specifically, from elasticity theory it is known that $\mathrm{div}\,\vec{u}$ is the relative change of volume $\Delta V/V$ and $\mathrm{rot}\,(\vec{u}/2)$ represents an angle of rotation of the cell. Equation (5.42) may be transformed into two wave equations for $\mathrm{div}\,\vec{u}$ and $\mathrm{rot}\,\vec{u}$. This technique proves to be convenient for many applications. Indeed, applying a divergence operator to both sides of Eq. (5.42), one gets

$$\frac{\partial^2}{\partial t^2}\,\mathrm{div}\,\vec{u} = s_l^2 \nabla^2\,\mathrm{div}\,\vec{u}, \tag{5.43}$$

and applying a rotor operator, one gets

$$\frac{\partial^2}{\partial t^2}\,\mathrm{rot}\,\vec{u} = s_t \nabla^2\,\mathrm{rot}\,\vec{u}. \tag{5.44}$$

Equations (5.43) and (5.44) are wavelike and they have wavelike solutions with amplitudes A_d and $\vec{A}_r$:

$$\mathrm{div}\,\vec{u} = A_d \exp[i(\vec{q}\vec{r} - \omega_l t)], \quad \omega_l = s_l|\vec{q}|; \tag{5.45}$$

$$\mathrm{rot}\,\vec{u} = \vec{A}_r \exp[i(\vec{q}\vec{r} - \omega_t t)], \quad \omega_t = s_t|\vec{q}|. \tag{5.46}$$

Equation (5.45) yields longitudinal waves and Eq. (5.46) yields torsional waves that can be expressed as a sum of two transverse modes.

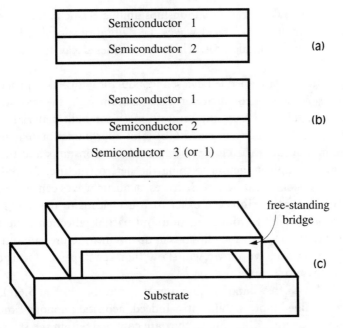

Figure 5.6. Schematic of the three possible structures with different phonon-confinement properties.

For bulk crystals, the boundary conditions do not have much effect on the lattice vibrations although, formally, the solutions of Eqs. (5.45) and (5.46) can be rewritten as standing waves. In quantum structures, boundary conditions reflect the composition and the dimensions of the quantum structure. Indeed, there are many different types of quantum structures and, in general, it is necessary to impose different boundary conditions for each structure. For an interface between materials 1 and 2, general boundary conditions include the continuity of the displacement vector,

$$\vec{u}_1 = \vec{u}_2, \tag{5.47}$$

and the equality of traction forces (forces per unit area) across the interface. According to Eq. (5.41) the latter conditions are

$$\sigma_{iz,1} = \sigma_{iz,2}, \quad i = x, y, z, \tag{5.48}$$

where we assume that the interface is perpendicular to the z axis.

Typical heterostructures of interest are (1) two semi-infinite semiconductors with a heterointerface between them [Fig. 5.6(a)], (2) a layered semiconductor structure [Fig. 5.6(b)], and (3) a free-standing semiconductor structure [Fig. 5.6(c)]. The acoustic-phonon states and spectra depend on the types of the semiconductor structures.

For a single heterointerface, Fig. 5.6(a), there is a class of elastic waves propagating in the bulk of both materials, partly transmitting and reflecting from the interface. For certain relationships among elastic parameters of materials, there are interface (surface) waves, i.e., vibrations localized near the interface and propagating along it. Their magnitudes decay far away from the interface.

For quantum wells embedded in a semiconductor with elastic properties similar to those of the semiconductor in the quantum well, the acoustic-phonon modes may be taken to have the same form as for bulk material, i.e., plane waves; this case is illustrated in Fig. 5.6(b). This approximation may be justified if we take into account the closeness of the elastic constants and the material densities of the semiconductors forming the heterostructure. Accordingly, for many III–V semiconductor heterostructures, acoustic waves can easily penetrate through the interfaces. However, exact treatment leads to the modification of acoustic waves resulting from refraction and partial reflection from the interfaces. For this reason, acoustic phonons in quantum structures are often treated as bulklike modes. However, we will show that new types of localized modes can arise in these structures.

If we deal with free-standing quantum structures, as in Fig. 5.6(c), acoustic modes are modified more significantly. Indeed, acoustic phonons cannot propagate into the vacuum; accordingly, they are confined within the structure. The case of free-standing structures provides interfacial conditions that lead to the extreme confinement limit for acoustic phonons. This case is considered in detail in Subsection 5.5.2.

5.5.2 Acoustic-Mode Localization in Heterostructures with Quantum Wells and Wires

The Simplest Model of Mode Localization

First, let us perform a semiquantitative analysis that demonstrates that acoustic phonons in principle can be localized in a layered semiconductor structure like that presented in Fig. 5.6(b). For the sake of simplicity we consider isotropic materials that have zero-displacement moduli, $\mu_1 = \mu_2 = 0$. In such a medium, there is no rotation ($\mathrm{rot}\,\vec{u} = 0$) and we should analyze only the equation for $\mathrm{div}\,\vec{u}$, i.e., only the longitudinal vibrations are considered.

Introducing a new function, $\xi = \lambda\,\mathrm{div}\,\vec{u}$, for the medium with the modulus λ, we can rewrite Eq. (5.43) as

$$\frac{1}{s_l^2}\frac{\partial^2 \xi}{\partial t^2} = \nabla^2 \xi. \tag{5.49}$$

This single wave equation describes our simplified case. The general boundary conditions at the interfaces between two continuum media, Eq. (5.48), require that the stress-tensor components σ_{zz} be continuous. According to Eq. (5.41), this is identical to the continuity of the variable ξ. By integrating Eq. (5.49) over z in a very narrow region around the interfaces, we find that the derivative $\partial\xi/\partial z$ is also

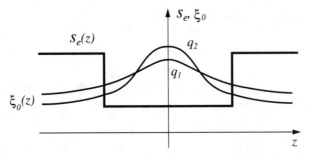

Figure 5.7. An embedded layer with a smaller sound velocity s_l localizing acoustic modes. The profile of the sound velocity $s_l(z)$ and the spreading of the localized wave $\xi_0(z)$ for two wave vectors $q_1 < q_2$ are shown.

continuous at the interfaces. Thus we get the continuity of ξ and its derivative as the boundary conditions for Eq. (5.49). On the other hand, $s_l = \sqrt{\lambda/\rho(z)}$ and $\lambda = \lambda(z)$ are, in general, discontinuous at the interface between two different elastic media.

Consider the solutions of Eq. (5.49) in the form of plane waves propagating along the layers. We can assume that the wave propagation is along the x axis without loss of generality:

$$\xi = \xi_0(z)e^{i(qx-\omega t)},$$

where q is the wave vector. The function $\xi_0(z)$ determines the spreading of the wave across the structure. Substituting this equation into Eq. (5.49), we can find for $\xi_0(z)$

$$\left[-\frac{\hbar^2}{2m}\frac{d^2}{dz^2} - \frac{\hbar^2\omega^2}{2ms_l^2(z)} \right]\xi_0 = -\frac{\hbar^2}{2m}q^2\xi_0. \tag{5.50}$$

Here we intentionally multiplied the left-hand and the right-hand sides by $\hbar^2/2m$ to transform the equation to exactly the same structure as that of the Schrödinger equation, where $-(\hbar^2/2m)(d^2/dz^2)$ and $-\hbar^2\omega^2/2ms_l^2(z)$ serve as analogs of an operator of kinetic and potential energy, respectively, while $-\hbar^2q^2/2m$ can formally be interpreted as an energy and $\xi_0(z)$ as a wave function. This situation is analogous to the one-dimensional potential-well problem in the context of the Schrödinger equation: layer 2 creates a well for acoustic vibrations if in middle layer 2 [see Fig. 5.6(b)] the sound velocity s_l is less than that in the side layers,

$$s_{l,2} < s_{l,1}, s_{l,3}.$$

Consequently, there is a solution $\xi_0(z)$ localized in middle layer 2, as shown in Fig. 5.7. According to the analysis given in Section 3.2, at least one of these localized solutions always occurs. For these localized acoustic waves the frequencies

ω and the wave vectors $\vec{q}$ satisfy the condition

$$\frac{\omega}{s_{l,2}} > q > \frac{\omega}{s_{l,1}}. \tag{5.51}$$

The localization is strong if the the well is deep, i.e., if the discontinuity associated with the effective well depth,

$$V_0 = \omega^2 \left(\frac{1}{s_{l,2}^2} - \frac{1}{s_{l,1}^2} \right), \tag{5.52}$$

is sufficiently large. Unlike the potential well for electrons, the elastic discontinuity of Eq. (5.52) depends on the frequency of the wave ω. This implies that high-frequency, short-wavelength modes tend to be localized more strongly than low-frequency, long-wavelength modes. Such an increase in the localization is shown schematically in Fig. 5.7 for two values of the wave vector q. The condition of strong localization inside layer 2 is

$$q \geq \frac{1}{L} \frac{s_{l,1}}{\sqrt{s_{l,1}^2 - s_{l,2}^2}}, \tag{5.53}$$

where L is the width of the layer. Another result of the frequency dependence of $V_0(\omega)$ is the occurrence of new localized modes when the frequency increases. Hence the lowest localized mode $\omega_0(q_{c,0})$ exists in the range $q > q_{c,0}$. The next mode arises at some critical frequency $\omega_1(q_{c,1})$ and wave vector $q_{c,1}$; it exists at $q > q_{c,1}$. Then, at greater ω and q, a third mode arises. In the model of an elastic-continuum the number of localized modes increases infinitely when $q \to \infty$. In conclusion, this simple model shows that an embedded layer with smaller sound velocity always leads to a localization of the acoustic vibrations near the layer.

Analysis of Localized Modes near Quantum Wells

A more rigorous analysis should include both LA and TA modes and should be based on Eqs. (5.43) and (5.44) with proper boundary conditions. Let us suppose that layers 1 and 3 are identical. We also assume wave propagation along the x axis:

$$\vec{u} = \vec{u}(z)e^{iqx - i\omega t}. \tag{5.54}$$

Then Eqs. (5.43) and (5.44) take forms analogous to that of Eq. (5.50):

$$\frac{d^2 \operatorname{div} \vec{u}}{dz^2} + \left(\frac{\omega^2}{s_l^2} - q^2 \right) \operatorname{div} \vec{u} = 0, \tag{5.55}$$

$$\frac{d^2 \operatorname{rot} \vec{u}}{dz^2} + \left(\frac{\omega^2}{s_t^2} - q^2 \right) \operatorname{rot} \vec{u} = 0. \tag{5.56}$$

Boundary conditions are given by Eqs. (5.47) and (5.48). Equation (5.48) can be represented in terms of components of the displacement vector:

$$\mu_1 \left(\frac{\partial u_{x1}}{\partial z} + \frac{\partial u_{z1}}{\partial x} \right) = \mu_2 \left(\frac{\partial u_{x2}}{\partial z} + \frac{\partial u_{z2}}{\partial x} \right),$$

$$\mu_1 \left(\frac{\partial u_{y1}}{\partial z} + \frac{\partial u_{z1}}{\partial y} \right) = \mu_2 \left(\frac{\partial u_{y2}}{\partial z} + \frac{\partial u_{z2}}{\partial y} \right),$$

$$\lambda_1 \operatorname{div} \vec{u}_1 + 2\mu_1 \frac{\partial u_{z1}}{\partial z} = \lambda_2 \operatorname{div} \vec{u}_2 + 2\mu_2 \frac{\partial u_{z2}}{\partial z}, \tag{5.57}$$

at $z = \pm(L/2)$; indexes 1 and 2 refer to medium 1 and 2.

Two different classes of the waves are possible for the structure presented in Fig. 5.6(b). These waves have different polarizations of the displacement vector $\vec{u}$. The shear-horizontal (SH) waves are polarized along the layers, i.e., for the displacement vector we can write

$$\vec{u} = [0, u_y(z), 0]. \tag{5.58}$$

The shear-vertical (SV) waves have two components parallel and perpendicular to the propagation direction:

$$\vec{u} = [u_x(z), 0, u_z(z)]. \tag{5.59}$$

The SV class can be also divided into symmetrical and asymmetrical waves with respect to the plane $z = 0$. For a symmetrical wave, we have an even u_x component and an odd u_z component of the displacement vector. For asymmetrical waves the u_x component is odd, while the u_z component is even. Symmetric SV waves are frequently referred to as dilatational waves. Asymmetric SV waves are called flexural waves. The first class, SH, corresponds to purely transverse waves, while the second, SV, is composed of a superposition of longitudinal and transverse displacements.

Let us introduce four different sound velocities for the materials of layers 1 and 2: $s_{l,1}$, $s_{t,1}$, $s_{l,2}$, and $s_{t,2}$. Different relationships among these four quantities define possible waves for layered structures. Since $s_t < s_l$, the following six distinct configurations are important:

A: $s_{t,2} < s_{l,2} < s_{t,1} < s_{l,1}$;

B: $s_{t,2} < s_{t,1} < s_{l,2} < s_{l,1}$;

C: $s_{t,2} < s_{t,1} < s_{l,1} < s_{l,2}$;

D: $s_{t,1} < s_{l,1} < s_{t,2} < s_{l,2}$;

E: $s_{t,1} < s_{t,2} < s_{l,1} < s_{l,2}$;

F: $s_{t,1} < s_{t,2} < s_{l,2} < s_{l,1}$.

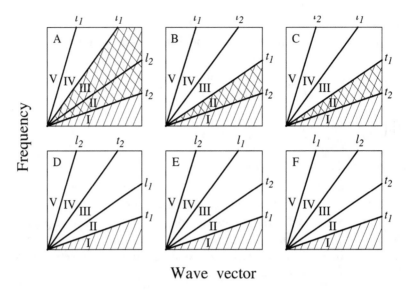

Figure 5.8. The ω–q diagrams for analysis of acoustic waves in the layered structure presented in Fig. 5.7. Labels on the curves correspond to longitudinal, l_1 and l_2, and transverse, t_1 and t_2, waves in materials 1 and 2. The shaded areas correspond to regions where modes localized within the central layer can exist.

Each of these configurations can be analyzed in terms of the ω–q diagram. Such diagrams are presented in Fig. 5.8. Four dispersion curves for longitudinal and transverse modes in two materials divide the ω–q plane into five regions, from I to V. Solutions localized near layer 2 should decay exponentially far away from the layer. From Eqs. (5.55) and (5.56), it follows that the latter is possible if

$$q^2 > \frac{\omega^2}{s_{t,1}}, \qquad q^2 > \frac{\omega^2}{s_{l,1}}.$$

Consequently, all the dispersion curves for localized modes should lie below the dispersion curve of the transverse modes of medium 1, i.e., in regions I, II, and III for configuration A, in regions I and II for configurations B and C, and in region I for configurations D, E, and F in Fig. 5.8.

The SH solutions are transverse modes and involve only Eq. (5.56), i.e., again, we get an equation similar to Eq. (5.50), which can be analyzed like the one-dimensional Schrödinger equation. Solutions of this equation with real ω and q exist only if

$$\frac{\omega}{s_{t,1}} < q < \frac{\omega}{s_{t,2}}. \tag{5.60}$$

Thus the SH modes can exist for configuration A in regions II and III and for configurations B and C in region II. For configurations D, E, and F no SH modes occur.

The SV modes involve both Eqs. (5.55) and (5.56). An analysis of the SV modes shows that these solutions exist with real ω and q for configuration A in regions I, II, and III and for configurations B and C they can exist in regions I and II. For configurations D, E, and F, these types of modes can exist only in region I.

Using particular parameters for the sound velocities of bulk materials that comprise heterostructures, one can see that, for many cases, the quantum-well layer confining the electrons (holes) also causes a localization of acoustic vibrations. Such examples are AlAs/GaAs/AlAs and GaP/GaAs/GaP (quantum wells for both electrons and holes), AlGaSb/InSb/AlGaSb (quantum wells for electrons), etc. Consider the $Al_{0.25}Ga_{0.75}As/GaAs$ double heterostructure that creates quantum wells for electrons and holes. The necessary elastic parameters are $s_l = 4.707 \times 10^5$ cm/s and $s_t = 3.329 \times 10^5$ cm/s for GaAs and $s_l = 4.901 \times 10^5$ cm/s and $s_t = 3.457 \times 10^5$ cm/s for $Al_{0.25}Ga_{0.75}As$, that is, heterostructures based on these materials correspond to configuration B. The SH modes exist in region II of Fig. 5.8(b). Calculations show that the SV modes exist only in the same region. In Fig. 5.9, SH types of modes are presented for this heterostructure for the case in which the width of the GaAs confining layer is

Figure 5.9. The dispersion relations for localized SH acoustic modes near the 200-Å quantum-well layer of the structure $Al_{0.25}Ga_{0.75}As/GaAs/Al_{0.25}Ga_{0.75}As$. All curves lie in a narrow region. Up to three modes of each type can be seen. [After L. Wendler and V. G. Grigorian, "Acoustic interface waves in sandwich structures," Surf. Science **206**, pp. 203–224 (1988).]

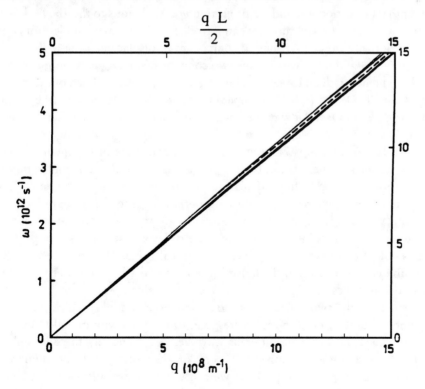

200 Å. The lowest mode and several excited modes can be seen at high values of q for both types of modes. Although the ω–q region for the existence of the localized modes is very narrow (see Fig. 5.9), they can be of importance. For high frequencies, these modes are localized within the quantum-well layer and contribute considerably to effects related to electron–phonon interactions.

Acoustic Modes near Quantum Wires

The previous analysis of the localization of acoustic modes can be extended to the case of a quantum-wire embedded in another material, if the quantum-wire and the ambient material have different elastic parameters. Let us discuss the results of such an analysis for a cylindrical quantum wire of radius R. Since rotational symmetry is assumed, one can use cylindrical coordinates (r, θ, z) with the z axis along the wire instead of Cartesian coordinates (x, y, z): Instead of Eq. (5.54) for a wave in a layered structure, one can assume a displacement vector of the form

$$\vec{u}_{\alpha,n}(r, \theta, z) = \vec{w}_{\alpha,n} e^{in\theta} e^{iqz - \omega t}. \tag{5.61}$$

Here n is an azimuthal number and the index α labels different solutions. Equations (5.55) and (5.56) should be rewritten in cylindrical coordinates with basic unit vectors $\vec{e}_r, \vec{e}_\theta$, and $\vec{e}_z$, and the displacement vector should be represented in terms of three basic vectors: a displacement along the radius w_r, an azimuthal projection w_θ, and a z projection w_z. Analogous to the above classification for a layered structure, we can introduce three types of modes: torsional (T) modes with only an azimuthal component of displacement, $\vec{w} = (0, w_\theta, 0)$; dilatational (D) modes, which have two components, $\vec{w} = (w_r, 0, w_z)$; and flexural (F) modes with all three components, $\vec{w} = (w_r, w_\theta, w_z)$. Each of these three types, $\alpha = T, D$, and F, has a number of modes, which can be labeled by a pair of integers P and n and a wave vector $\vec{q}$. Thus the total set of quantum numbers is $\{\alpha, P, n, \vec{q}\}$.

Possible relationships between the elastic parameters of the quantum wire and the ambient materials can be classified according to the A–F configurations given above in Fig. 5.8. In particular, for a GaAs wire embedded in an AlGaAs ambient material, configuration B is appropriate. A general feature of localized acoustic waves near a quantum-wire is that localization is not possible for small wave vectors. This is in contrast to the case of layered structures. Localization arises only if the wave vector q exceeds a critical value. The critical values $q_{\alpha,c}$ are different for torsional, dilatational, and flexural vibrations. Let us discuss dilatational waves. In Fig. 5.10 amplitudes of both components w_r and w_z are presented for the two lowest dilatational modes P_1 and P_2, which have zero azimuthal number. Both modes correspond to the same vibrational frequency, $\omega/2\pi = 0.6$ THz, but have different wave vectors. Generally there is a phase shift ϕ between these components. This makes the motion of the medium more complex. Actually, from Eq. (5.61) we can write the real displacement of the

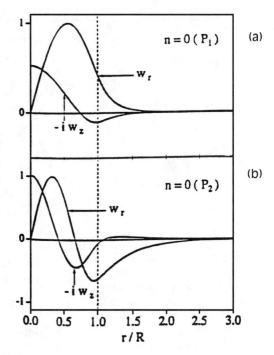

Figure 5.10. Dilatational waves localized near the cylindrical quantum wire of 100-Å radius. Two components of the displacement, w_r and w_z, are represented. Case (a) corresponds to the lowest mode P_1, case (b) corresponds to first excited mode P_2. The dotted vertical line marks the cylindrical interface position. [After N. Nishiguchi, "Confined and interface acoustic phonons in a quantum wire," Phys. Rev. B 50, pp. 10970–10980 (1994).]

medium (real part of complex vector $\vec{u}$) as

$$\vec{e}_r w_r(r) \cos(qz - \omega t) + \vec{e}_z w_z \cos(qz - \omega t + \phi).$$

This result implies that at any particular point (r, z), the element of the medium executes motion along an elliptic trajectory, excluding points on the wire axis, where $w_r(r) = 0$. Figure 5.11 depicts the pattern of the displacement field in the wire and the ambient medium for the lowest dilatational mode. Lines correspond to the displacements with respect to the undeformed state. The shaded region

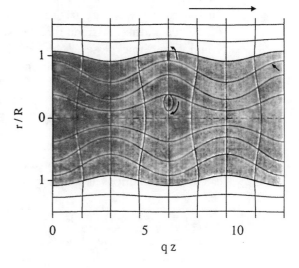

Figure 5.11. The field pattern of the displacement for the dilatational wave P_1 in the quantum wire. The wire is denoted by the shaded region. Two elliptic trajectories of elements of the wire are shown. [After N. Nishiguchi, "Confined and interface acoustic phonons in a quantum wire," Phys. Rev. B 50, pp. 10970–10980 (1994).]

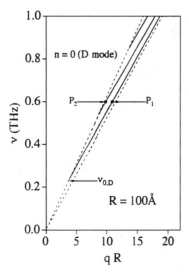

Figure 5.12. The three lowest dispersion dependences for dilatational waves, P_1, P_2, P_3, in the quantum wire. The frequencies $\nu = \omega/2\pi$ are given in terahertz. The critical frequency of arising the first mode, P_1, is marked by $\nu_{0,D}$. The results presented in Figs. 5.10 and 5.11 correspond to the frequency and the wave vectors marked for modes P_1 and P_2 in this figure. [After N. Nishiguchi, "Confined and interface acoustic phonons in a quantum wire," Phys. Rev. B **50**, pp. 10970–10980 (1994).]

denotes the wire. Both radial and axial displacements are clearly seen. Near the wire axis ($r \to 0$) only axial deformation exists. The lowest mode is almost localized in the wire: outside the wire ($r > R$) the displacement vanishes quickly. According to Fig. 5.10, the second mode is less localized. The previously mentioned elliptic motion is indicated in Fig. 5.11 by arrows for two elements of the wire.

The spectrum of dilatational waves, $\omega/2\pi$ versus qR, in the wire is presented in Fig. 5.12. Localized modes are situated between the two dispersion lines for transversal phonons in GaAs and AlAs, in accordance with the criterion of Eq. (5.60). All modes arise at finite wave vectors q. It is seen that the lowest mode has the wave length $\lambda = 2\pi/q \sim R$, that is, localized modes are relatively short-wavelength and high-frequency vibrations; the continuum approach should still be valid. Similar results can be proved for all three types of vibrations in the system under consideration.

Thus we can conclude that in heterostructures with quantum wires there are localized acoustic vibrations that tend to appear with relatively short wavelengths and high frequencies. Other features are similar to those of layered heterostructures.

Elastic Properties of Superlattices

Acoustic waves in superlattices can be investigated with Eqs. (5.43) and (5.44) and the boundary conditions of Eqs. (5.47) and (5.48). For simplicity, let us focus on longitudinal waves propagating in the z direction, perpendicular to the layers of the superlattice. Assume that the superlattice consists of alternative layers 1 and 2 with thicknesses d_1 and d_2, respectively. In such a case, the problem is reduced to a one wave equation for the displacement $u_z(z)$ with boundary conditions

$$u_{z,1} = u_{z,2}, \qquad \Lambda_1 \frac{\partial u_{z,1}}{\partial z} = \Lambda_2 \frac{\partial u_{z,2}}{\partial z}$$

at interfaces, where we have introduced $\Lambda_{1,2} = \lambda_{1,2} + 2\mu_{1,2}$. In each layer we can look for solutions as standing waves:

$$u_{z,1} = (A e^{iq_1 z} + B e^{-iq_1 z}) e^{-i\omega t},$$
$$u_{z,2} = (C e^{iq_2 z} + D e^{-iq_2 z}) e^{-i\omega t}.$$

Here, $q_{1,2} = \omega/s_{1,2}$ and $s_{1,2} = \sqrt{(\Lambda_{1,2})/(\rho_{1,2})}$ are the longitudinal sound velocities in materials 1 and 2. In accordance with the Bloch theorem formulated in
Section 2.6 for electrons and valid for any periodical systems, solutions for an
infinite superlattice as a whole should have the form

$$u_z(z, t) = e^{iqz - i\omega t} u(z),$$

where $u(z)$ is a periodic function, i.e., $u(z) = u(z + d)$, and q is the wave vector of
the vibrations in the superlattice. The range of q is appropriate for the Brillouin
zone. As a result, we get the following dispersion relation between the frequency
ω and the wave vector q for the superlattice:

$$\cos qd = \cos \frac{\omega d_1}{s_1} \cos \frac{\omega d_2}{s_2} - \frac{1}{2} \left(\frac{\rho_2 s_2}{\rho_1 s_1} + \frac{\rho_1 s_1}{\rho_2 s_2} \right) \sin \frac{\omega d_1}{s_1} \sin \frac{\omega d_2}{s_2}, \qquad (5.62)$$

where $d = d_1 + d_2$ is the period of the superlattice. Since the behavior of the wave
in the superlattice should be determined by the difference in elastic properties,
let us introduce a new parameter that characterizes the modulation of these
properties throughout the system:

$$\eta = \frac{\rho_2 s_2 - \rho_1 s_1}{\sqrt{\rho_1 \rho_2 s_1 s_2}}. \qquad (5.63)$$

With such a parameter, Eq. (5.62) can be transformed to the more instructive
form,

$$\cos qd = \cos \left[\omega \left(\frac{d_1}{s_1} + \frac{d_2}{s_2} \right) \right] - \frac{\eta^2}{2} \sin \frac{\omega d_1}{s_1} \sin \frac{\omega d_2}{s_2}. \qquad (5.64)$$

The parameter η has nonzero but small values for conventional superlattices of
III–V and II–IV compounds. Thus one can use perturbation theory to analyze
Eq. (5.64). If we omit the term proportional to η^2 in the lowest approximation,
we get

$$\cos qd = \cos \left[\omega \left(\frac{d_1}{s_1} + \frac{d_2}{s_2} \right) \right]. \qquad (5.65)$$

Solutions of Eq. (5.65) are

$$q = \pm \frac{\omega}{d} \left(\frac{d_1}{s_1} + \frac{d_2}{s_2} \right) + \frac{2\pi n}{d}. \qquad (5.66)$$

This result corresponds to a folding procedure similar to the one discussed in
Subsection 4.3.4 for electrons. In the case under consideration, this procedure is
applied to an average elastic dispersion curve:

$$\omega^{(0)}(q) = dq \frac{s_1 s_2}{s_1 d_2 + s_2 d_1}.$$

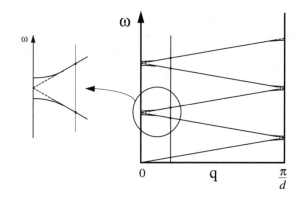

Figure 5.13. Schematic diagram of a folded dispersion $\omega(q)$ for acoustic waves in a super-lattice. The inset shows the splitting to two branches at the center of the Brillouin zone.

In Fig. 5.13 a sketch of this folding is shown. The first Brillouin zone is chosen as $0 < q < \pi/d$.

In such an approximation, a double degeneracy appears at the center and the edge of the Brillouin zone for frequencies

$$\omega_n = \pi n \frac{s_1 s_2}{s_1 d_2 + s_2 d_1},$$

where n is an even (odd) integer for the zone center (edge). To exclude the double degeneracy appearing in this approximation, one can take into account omitted terms as a perturbation. This produces a splitting of center and edge frequencies defined by $\omega = \omega_n \pm \Delta\Omega$, where for $\Delta\Omega$ we find

$$\Delta\Omega \approx \eta \frac{s_1 s_2}{s_1 d_2 + s_2 d_1} \sin\left(\pi n \frac{d_1 s_2 - d_2 s_1}{d_1 s_2 + d_2 s_1}\right). \tag{5.67}$$

One can see that $\Delta\Omega$ depends on the number of the branch n. This splitting is shown in Fig. 5.13. It is proportional to the small parameter η, while corrections to the dispersion inside the Brillouin zone are of the order of η^2.

One of the remarkable features of $\omega(q)$ for superlattices is the forbidden gaps at $q = 0, \pi/d$. These gaps mean that there are frequency intervals where the elastic waves cannot propagate through the superlattice. For example, consider a finite-dimension superlattice embedded into a material. An elastic wave with a frequency corresponding to that of one of the allowed bands passes through the superlattice; the transmission coefficient should be finite, although it may be less than 1 because there is a reflection. But for frequencies corresponding to the gaps, waves should be strongly reflected. In Fig. 5.14(a), calculations of $\omega(q)$ are represented for the AlAs/GaAs superlattice with $d_1 = d_2$. Both type of vibrations, longitudinal L and transverse T, are shown. The transmission coefficients are given for both types of vibrations in Figs. 5.14(b) and 5.14(c). The superlattice is supposed to have 15 periods. From these results, the appearance of the frequency gaps is clearly seen. These gaps depend on the number of the branch. Note that the transmission coefficient is small for the gaps, but it is not exactly zero. This is due to the finite dimension of the superlattice. It is instructive to recall the analogy

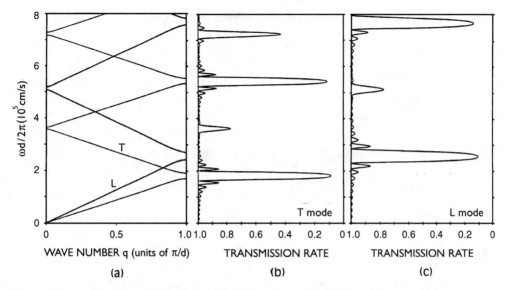

Figure 5.14. (a) Dispersion of LA (L) and TA (T) waves in a AlAs/GaAs infinite superlattice; transmission coefficients for T and L waves through a finite superlattice (b), (c) [After S. Tamura, *et al.*, "Acoustic phonon propagation in superlattices," Phys. Rev. B 38, pp. 1427–1449 (1988).]

with the propagation of the light through layered systems like the Fabry–Perot etalon. It displays well-known resonant transmission and reflection, depending on the frequency of the light.

Acoustic Modes in Free-Standing Thin Slabs

Now let us consider the localization (confinement) of acoustic vibrations in a free-standing semiconductor slab such as the one depicted in Fig. 5.6(c). This is the case of extreme confinement and we analyze it in more detail. The boundary conditions on the free surface of the slab imply that the components of the stress tensor normal to the surface vanish. If we specify the coordinate system in such a way that the axis z is perpendicular to the semiconductor slab and the surfaces of the slab have coordinates $z = \pm(L/2)$, where L is now the width of the slab, the boundary conditions take the form $\sigma_{xz} = \sigma_{yz} = \sigma_{zz} = 0$ at $z = \pm(L/2)$; accordingly, in terms of the components of the displacement vector, it follows that

$$\sigma_{xz} = \mu \left(\frac{\partial u_x}{\partial z} + \frac{\partial u_z}{\partial x} \right) = 0,$$

$$\sigma_{yz} = \mu \left(\frac{\partial u_y}{\partial z} + \frac{\partial u_z}{\partial y} \right) = 0,$$

$$\sigma_{zz} = \lambda \operatorname{div} \vec{u} + 2\mu \frac{\partial u_z}{\partial z} = 0, \tag{5.68}$$

at $z = \pm(L/2)$. We look for solutions in the form of Eq. (5.54) and should obtain the set of equations for eigenmodes $\vec{u}_n(\vec{q}, z)$ and eigenfrequencies $\omega_n(q)$.

The system of these equations can be represented in the operator form,

$$\mathcal{D}\,\vec{u}_n(\vec{q}_\parallel, z) = -\omega_n^2 \vec{u}_n(\vec{q}_\parallel, z), \tag{5.69}$$

where $\mathcal{D}$ is the matrix differential operator:

$$\mathcal{D} = \begin{bmatrix} s_t^2 \frac{d^2}{dz^2} - s_l^2 q_x^2 & 0 & (s_l^2 - s_t^2)iq_x \frac{d}{dz} \\ 0 & s_t^2 \frac{d^2}{dz^2} - s_t^2 q_x^2 & 0 \\ (s_l^2 - s_t^2)iq_x \frac{d}{dz} & 0 & s_l^2 \frac{d^2}{dz^2} - s_t^2 q_x^2 \end{bmatrix}. \tag{5.70}$$

The boundary conditions of Eqs. (5.68) become

$$\frac{du_x}{dz} = -iq_x u_z,$$

$$\frac{du_y}{dz} = 0,$$

$$\frac{du_z}{dz} = -iq_x \frac{s_l^2 - 2s_t^2}{s_l^2} u_x \tag{5.71}$$

at $z = \pm(L/2)$. It can be proved straightforwardly that operator $\mathcal{D}$ of the eigenvalue problem of Eqs. (5.69)–(5.71) is Hermitian, so the eigenfunctions $\vec{u}_n(\vec{q}_\parallel, z)$ corresponding to nondegenerate eigenfrequencies ω_n are orthogonal.

A major feature of the confined modes is their quantization in the z direction. Roughly speaking, the z components of the confined-mode wave vectors q_z take on only some discrete set of values at each particular in-plane wave vector q. As discussed above, there are different types of confined acoustic modes: SH waves and SV symmetrical (dilatational) and asymmetrical (flexural) waves.

The SH waves have only one nonzero component $u_y(z)$, for which one can find the following solutions:

$$u_y = \begin{cases} \cos(q_{z,n}\, z) & \text{for even } n \\ \sin(q_{z,n}\, z) & \text{for odd } n \end{cases}, \tag{5.72}$$

where $q_{z,n} = \pi n / L, n = 1, 2, 3, \ldots$. The dispersion relation for SH waves is

$$\omega_n = s_t \sqrt{q_{z,n}^2 + q^2}\,.$$

These modes are similar to the transverse modes in bulk semiconductors, and their quantization is based on a simple rule that states that an integer number of half-wavelengths fits in a semiconductor slab of width L.

The dilatational waves have two nonzero components:

$$u_x = iq\left[(q^2 - q_t^2)\sin\frac{q_t L}{2}\cos q_l z + 2q_l q_t \sin\frac{q_l L}{2}\cos q_t z\right], \tag{5.73}$$

$$u_z = q_l\left[-(q^2 - q_t^2)\sin\frac{q_t L}{2}\sin q_l z + 2q^2 \sin\frac{q_l L}{2}\sin q_t z\right]. \tag{5.74}$$

The parameters q_l and q_t are determined from the system of two algebraic equations,

$$\frac{\tan(q_t L/2)}{\tan(q_l L/2)} = -\frac{4q^2 q_l q_t}{\left(q^2 - q_t^2\right)^2}, \tag{5.75}$$

$$s_l^2 \left(q^2 + q_l^2\right) = s_t^2 \left(q^2 + q_t^2\right). \tag{5.76}$$

Equations (5.75) and (5.76) have many solutions for q_l and q_t at each particular q, and we label them by an additional index n: $q_{l,n}$ and $q_{t,n}$. These solutions are either real or pure imaginary, depending on q and n. One can use term, branches of solutions, to denote the functions $q_{l,n}(q)$ and $q_{t,n}(q)$, graphs of which are continuous singly connected curves. The frequencies of the dilatational waves are given by

$$\omega_n = s_l \sqrt{q^2 + q_{l,n}^2} = s_t \sqrt{q^2 + q_{t,n}^2} . \tag{5.77}$$

It is necessary to use a numerical approach to solve Eqs. (5.75) and (5.76) and find ω_n. However, it is useful to make use of an analytical analysis initially in order to identify different branches and understand their general behavior.

If $q = 0$, but $\omega_n \neq 0$, the roots of Eqs. (5.75) and (5.76) may be obtained from the conditions $\tan(q_t L/2) = 0$ and $s_l q_l = s_t q_t$. The appropriate solutions have the following form:

$$q_{t,n} = \frac{2\pi n}{L}, \qquad q_{l,n} = \frac{s_t}{s_l} 2\pi n L,$$

$$\omega_n = s_t \, q_{t,n} = s_l \, q_{l,n}, \quad n = 1, 2, 3, \ldots . \tag{5.78}$$

In this result, each integer n identifies a branch of solutions $\omega_n(q)$. The second set of branches may be obtained from the conditions $\tan(q_l L/2) = \infty$ and $s_l q_l = s_t q_t$. In this case, the solutions are

$$q_{l,n} = \frac{\pi + 2\pi n}{L}, \qquad q_{t,n} = \frac{s_l}{s_t} \frac{\pi + 2\pi n}{L},$$

$$\omega_n = s_t \, q_{t,n} = s_l \, q_{l,n}, \quad n = 0, 1, 2, \ldots . \tag{5.79}$$

For the case $q \to 0$ and $\omega \to 0$, it follows from Eq. (5.77) that both q_l and q_t should also go to zero. So one may use the series expansion to obtain an approximate solution for Eqs. (5.75) and (5.76). The result is

$$q_l = i \frac{\left(s_l^2 - 2 s_t^2\right)}{s_l^2} q, \qquad q_t = \frac{\sqrt{3 s_l^2 - 4 s_t^2}}{s_l} q, \qquad \omega = \frac{2 s_t}{s_l} \sqrt{s_l^2 - s_t^2} \, q. \tag{5.80}$$

Symmetric or dilational waves   Antisymmetric or flexural waves

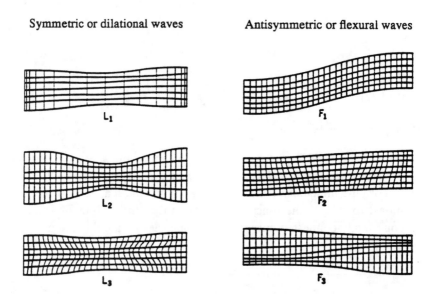

Figure 5.15. Wave-velocity field distributions for the four lowest-order SV modes in a free-standing thin slab. The left-hand part represents dilatational modes and the right-hand part flexural waves. [After B. A. Auld, *Acoustic Fields and Waves in Solids*, 2nd ed. (Krieger, Malabar, 1989), Vol. II.]

Thus we get a linear dispersion law for the lowest dilatational mode for small q and the velocity of this mode is smaller then s_l but larger than s_t. A schematic illustration of dilatational waves is presented on the left-hand side of Fig. 5.15. The dispersion law for dilatational phonons, calculated for a 100-Å-thick free-standing GaAs slab, is plotted in Fig. 5.16. Note that this slab is a quantum well for electrons; that is why it is possible to refer to a thin slab as a free-standing quantum well.

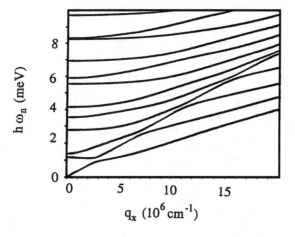

Figure 5.16. The dispersion relation for dilatational modes in a 100-Å-thick GaAs free-standing slab.

The remaining types of waves in slabs are flexural or antisymmetric waves. Flexural waves have two nonzero components:

$$u_x = iq \left[(q^2 - q_t^2) \cos \frac{q_t L}{2} \sin q_l z + 2q_l q_t \cos \frac{q_l L}{2} \sin q_t z \right], \tag{5.81}$$

$$u_z = q_l \left[(q^2 - q_t^2) \cos \frac{q_t L}{2} \cos q_l z - 2q^2 \cos \frac{q_l L}{2} \cos q_t z \right]. \tag{5.82}$$

Here q_l and q_t are determined from the solution of the transcendental equation,

$$\frac{\tan(q_l L/2)}{\tan(q_t L/2)} = -\frac{4q^2 q_l q_t}{(q^2 - q_t^2)^2}, \tag{5.83}$$

and Eq. (5.76). Solutions of these equations can be analyzed as was done above for dilatational waves. Consider the case $q \to 0$ and $\omega \to 0$. From Eq. (5.77) it follows that both q_l and q_t should also go to zero. So we may use a series expansion to obtain an approximate solution for Eqs. (5.83) and (5.76):

$$q_l = iq - i \frac{L^2 s_t^2 (s_l^2 - s_t^2)}{6 s_l^4} q^3,$$

$$q_t = iq - i \frac{L^2 (s_l^2 - s_t^2)}{6 s_l^2} q^3,$$

$$\omega = \frac{L s_t}{\sqrt{3} s_l} \sqrt{s_l^2 - s_t^2} q^2. \tag{5.84}$$

In this case, both q_l and q_t are pure imaginary numbers; thus the acoustic vibrations have essentially a surface-bound character and their amplitudes decrease exponentially from the surface to the interior of the slab; see Eqs. (5.81) and (5.82).

It is worth emphasizing the different behavior of the lowest dilatational and flexural modes determined by Eqs. (5.80) and (5.84). The dilatational mode is of an acoustic type with a finite phase velocity $s_{ph} = \omega/q$ and group velocity $s_{gr} = d\omega/dq$. [These two velocities were introduced in Section 2.4 in order to characterize the propagation of wave packets; see Eqs. (2.32) and (2.33).] The flexural mode has a quadratic dependence $\omega(q)$, which means that its phase and group velocities quench at small q and depend on the layer width:

$$s_{ph} = (Lq) \frac{s_t}{\sqrt{3} s_l} \sqrt{s_l^2 - s_t^2}, \qquad s_{gr} = 2s_{ph}.$$

The right-hand side of Fig. 5.15 qualitatively illustrates field distributions for several flexural modes. It is seen that only the lowest flexural mode corresponds to a bending of the free-standing quantum structure as a whole. This feature is reflected in the quadratic dependence of frequency on the magnitude of the

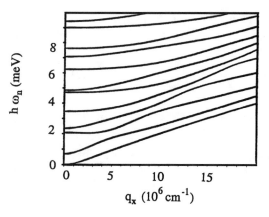

Figure 5.17. The dispersion relation for flexural modes in a 100-Å-thick GaAs free-standing slab.

in-plane wave vector. The set of branches for flexural modes in a 100-Å-thick GaAs slab (quantum well) is presented in Fig. 5.17.

Both Figs. 5.16 and 5.17 demonstrate the complex character of spectra of elastic vibrations in free-standing quantum structures. It is remarkable that there are modes (for example, second dilatational and third flexural) with negative dispersion, when the frequency decreases with the wave vector. One can see that the energy spacing between different branches is ~ 1 meV. This is clearly seen also in Fig. 5.18, where the densities of states of all three types of acoustic vibrations, shear, dilatational, and flexural modes, are presented as functions of the phonon energy $\hbar\omega$. The density of states is considerably different from that of the bulk acoustic phonons. The fine structure in the densities of states of Fig. 5.18 is related to the onset of higher modes as the frequency increases and clearly indicates the quantized nature of acoustic phonons in a free-standing quantum structure. The peaks in the densities of states in Figs. 5.18(b) and 5.18(c) correspond to contributions of the second dilatational and the third flexural modes in the regions with zero derivatives of ω with respect to q; see Eq. (5.19).

In conclusion, we have to say that elastic long-wavelength vibrations are considerably changed in quantum heterostructures. Their spatial dependences differ from those of plane waves. In particular, the vibrations can be localized or suppressed in regions of electron confinement. The frequency dependences and densities of states are modified and depend on the type of the structure, its geometry, dimensions, etc. Superlattices not only modify the frequency dependence, but also demonstrate resonant features in transmission and reflection of elastic waves.

5.6 Short-Wavelength and Optical Vibrations in Quantum Structures

5.6.1 Qualitative Analysis of Short-Wavelength Vibrations

Previous consideration of acoustic vibrations is restricted essentially to the long-wavelength limit. Short-wavelength vibrations can be calculated only

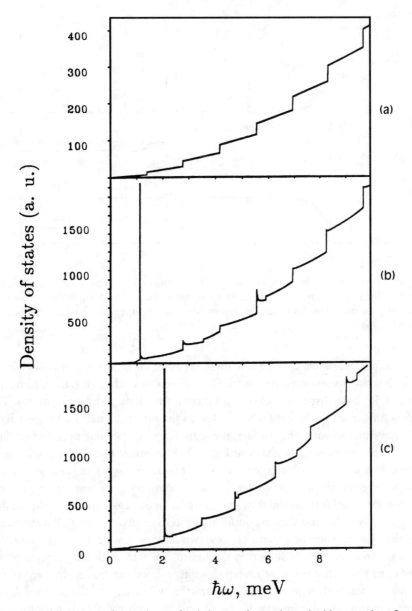

Figure 5.18. The density of states for confined phonons for (a) shear, (b) dilatational, (c) flexural modes in a 100-Å-thick GaAs free-standing slab.

numerically. But they can be interesting in the cases of narrow layers, short-period superlattices, etc. Assuming a long-wavelength character of vibrations along heterointerfaces $(\vec{q}_{\parallel} \rightarrow 0)$, we can qualitatively analyze their transverse structure by using the following method. Let interfaces be perpendicular to the z axis, which coincides with one of the crystalline axes, say (001). Materials 1 and 2, which comprise the heterostructure, are assumed to be of cubic symmetry. For such a case, longitudinal and transverse vibrations are not coupled by

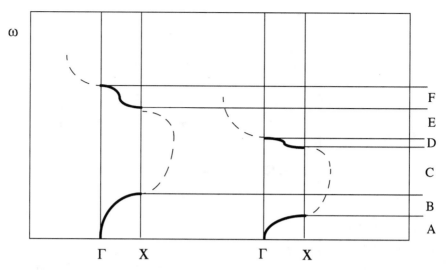

Figure 5.19. Sketch of phonon dispersion dependences for two bulk cubic materials, 1 (left) and 2 (right). The wave vector is assumed to be along the Γ–X direction. Frequencies for complex values of q are shown by dashed curves. Intervals A to F correspond to different relationships between $\omega_1(q)$ and $\omega_2(q)$.

the interface. Thus we can discuss these types of vibrations separately. We assume that the dispersion curves of bulk phonons in both materials are known. In Fig. 5.19, both dispersion curves are sketched for longitudinal vibrations. The dispersion curves for real wave vectors q_z in the intervals marked by Γ–X have clear physical meaning and, in fact, replicate the results of our previous studies, for example, the diatomic chain of Fig. 5.3. Remember that these results were obtained as solutions of algebraic Eqs. (5.12) at real q_z. The same equations always have solutions for complex or pure imaginary q_z. These solutions were omitted for an infinite crystal, but can be of use in our case. In Fig. 5.19, the dispersion curves $\omega(q_z)$ are shown conditionally as extended to complex wavelength intervals. Comparing the spectra of these two materials, we can get six frequency intervals, from A to F, as shown in the figure. We start with frequency interval A. For any particular frequency from this interval, we can find real wave vectors q_{z1} and q_{z2} for both materials. This means that vibrations of the interval A can propagate through the interface, i.e., extend through the whole structure. Of course, their amplitudes can be found only by rigorous calculations. In intervals B and F solutions with real wave vector q_{z1} exist only in material 1, while in material 2 these solutions have complex values of q_{z2}. Waves with frequencies from this interval can propagate through material 1, but decay in material 2. Thus, in the case of single interface, these waves are reflected from the interface. If a layer of material 1 is embedded in material 2, these waves are standing (confined) modes. A similar result is valid for frequency interval D: vibrations with the frequencies corresponding to propagating or confined modes in material 2. The last case corresponds to frequency intervals C and E, in which vibrations do not exist in both materials, but interface modes can occur. Figure 5.20

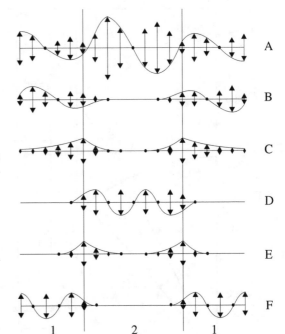

Figure 5.20. Sketch of lattice vibrations in a double-heterojunction structure for different frequency intervals, A–F, presented in Fig. 5.19.

schematically illustrates possible solutions for the case in which layer 2 is embedded in material 1.

This analysis is valid at any wave vectors, including those of the order of the inverse lattice-constant. One of the conclusions following from this analysis is that phonon confinement in a layer arises for vibrations with frequencies not existing in the ambient medium. Using the data of Table 5.1, we can easily find these cases for different heterostructures. For example, frequencies of LO phonons at the Γ points in GaAs and AlAs are 8.76 and 12.1 THz, respectively. Generally frequency bands of these phonons do not overlap. This means that a quantum-well GaAs layer in an AlGaAs/GaAs structure should confine LO phonons. Interface phonons are also possible. In Subsection 5.6.2 we study optical vibrations in III–IV compounds, taking into account that they are partly ionic crystals.

Generally, if some branch $\omega_\alpha(\vec{q})$, say an optical-phonon branch, is well isolated in the layer from other phonons in the ambient material, one can suppose that the amplitude of vibrations of such an isolated branch vanishes at interfaces, neglecting the slight penetration into the surrounding regions. Such a supposition may be associated with the so-called mechanical boundary conditions for confined phonons. As a result, one can get cosinelike (sinelike) dependences of these amplitudes on the z coordinate for even (odd) modes. These standing waves are equivalent to the bulklike solutions with wave vectors

$$q_{z,n} = \frac{\pi}{L}n, \quad n = 1, 2, \ldots,$$ (5.85)

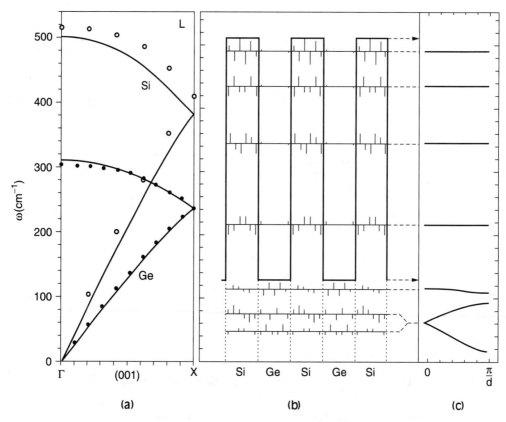

Figure 5.21. Longitudinal vibrations in the Si_6/Ge_6 superlattice. (a) Si and Ge vibrations in bulk materials, (b) amplitudes of the vibrations for different frequency intervals, (c) dispersion dependences. [After E. Molinari and A. Fasolino, "Phonons of ideal (001) superlattice: application to Si/Ge," in *Spectroscopy of Semiconductor Microstructures*, G. Fasol, A. Fasolino, and P. Lugli, eds. (Plenum, New York, 1989), pp. 195–205.]

for a layer bounded by $-L/2 < z < L/2$. Thus frequencies of these confined modes can be found from the dispersion of the bulk material:

$$\omega_{\alpha,n} = \omega_\alpha(q_{z,n}). \tag{5.86}$$

Now let us consider a short-period Si/Ge superlattice. We restrict ourselves to longitudinal vibrations. In Fig. 5.21(a), dispersion relations are shown for both Si and Ge along the (001) crystalline direction. From this figure and Table 5.1, we can see that all vibrations of Ge are in the frequency interval from 0 to 9.03 THz, while those of Si cover the interval from 0 to 15.6 THz. Thus there is an interval from 9.03 to 15.6 THz where vibrations are possible only in Si, that is, they can exist only in Si layers of a Si/Ge superlattice. In Fig. 5.21(b), high-frequency edges of the above intervals are presented for each of the layers of the Si_6/Ge_6 superlattice; each layer contains six monolayers. The dispersion curves for this superlattice are given in Fig. 5.21(c), but the lowest acoustic-type branch is not shown. Calculated amplitudes of vibrations in the superlattice are shown as functions of the position of the atomic layers along the (001) directions [Fig. 5.21(b)]. One can see that in accordance with the above analysis for the frequencies below 9.03 THz (310 cm^{-1}) vibrations propagate through the superlattice, although their amplitude is different in Si and Ge layers. Beyond this edge, there are only modes confined in Si layers. No interface modes appear for such a superlattice. The dispersion curves and the mode structure depend significantly on the superlattice period.

5.6.2 Optical Vibrations in Polar Crystals

Optical Phonons in Bulklike Materials

Now we focus our discussion on optical modes in III–V quantum structures. These compounds are ionic crystals, and the formation of a correct phonon picture in them requires a consideration of the long-range electrostatic interaction between ions. We start by considering bulk phonons in such crystals. For this purpose we use an isotropic-continuum approach, which was advantageous in our previous consideration of acoustic vibrations. This continuum approach is valid for optical phonons in the long-wavelength limit since such waves and all parameters vary little from one lattice site to the next. The approach assumes that one can allot a small, but macroscopic (i.e., containing many primitive cells) element around any point $\vec{r}$, so that inside this small element, all parameters are almost constant. In addition, we characterize lattice vibrations in terms of functions of the macroscopic coordinate $\vec{r}$. We consider an ionic crystal, each cell of which includes two oppositely charged ions with effective charges $\pm e^*$. Let the relative displacement of these ions be $\vec{u}$; $\vec{u}$ is considered to be a function of $\vec{r}$. Vibrations of the ions generate an electric field, which should be taken into account. The ions of a particular primitive cell are subjected to the influence of the local electric-field $\vec{F}$. Thus we can write Newton's equations for the relative motion of the ions in the form

$$\mathcal{M}\frac{d^2\vec{u}}{dt^2} = -\mathcal{M}\omega_0^2\vec{u} + e^*\vec{F}, \tag{5.87}$$

where $\mathcal{M}$ is the reduced mass of the ion pair M_1 and M_2, $1/\mathcal{M} = 1/M_1 + 1/M_2$, and ω_0 is the frequency associated with the short-range part of the interaction between atoms. The continuum approach uses the so-called macroscopic polarization vector $\vec{P}(\vec{r})$, which corresponds to the dipole moment of the crystal per unit volume. The polarization vector consists of two contributions: one comes from dipoles of ion pairs, and the second is due to polarization of the cores of the ions:

$$\vec{P} = \frac{N}{V}e^*\vec{u} + \frac{N}{V}\alpha\vec{F} = \frac{e^*}{V_0}\vec{u} + \frac{\alpha}{V_0}\vec{F}, \tag{5.88}$$

where N and V are the total number of primitive cells and the volume of the crystal, respectively; V_0 is the volume of the primitive cell and α is the electronic polarizability per primitive cell. The value α/V_0 can be expressed through the so-called high-frequency or optical permittivity: $\alpha/V_0 = (\kappa_\infty - 1)/4\pi$. The local electric field $\vec{F}$ is a superposition of all electric fields created by the ions in the crystal. We need one more equation for $\vec{F}$. In our long-wavelength approach this can be the following electrostatic equation:

$$\text{div}\,\vec{D} = \text{div}(\vec{F} + 4\pi\,\vec{P}) = 0. \tag{5.89}$$

Now it is convenient to represent each of the vectors $\vec{u}$, $\vec{P}$, and $\vec{F}$ as a sum of longitudinal components labeled by $\parallel$ and transverse components labeled by $\perp$. Since $\text{rot}\,\vec{F} = 0$ for the electrostatic field, for an infinite crystal the vector $\vec{F}$ has only a longitudinal component. From Eq. (5.89), it follows that

$$\vec{F} + 4\pi\,\vec{P}_\parallel = 0. \tag{5.90}$$

Equations (5.90) and (5.88) give the relationships between longitudinal vectors:

$$\vec{P}_\parallel = \frac{e^*}{\kappa_\infty V_0}\vec{u}_\parallel, \qquad \vec{F} = -\frac{4\pi e^*}{\kappa_\infty V_0}\vec{u}_\parallel. \tag{5.91}$$

The physical meaning of the first equation of Eqs. (5.91) is that the dipole of the primitive cell, $e^*\vec{u}$, is screened by the polarization of the cores of other ions in the crystal. Looking for solutions in the form

$$\vec{u}_\parallel, \vec{P}_\parallel, \vec{F} \propto e^{i\vec{q}\vec{r} - i\omega t}$$

and substituting Eqs. (5.91) into Eq. (5.87), we get the frequency of longitudinal vibrations:

$$\omega_{LO}^2 = \omega_0^2 + \frac{4\pi e^{*2}}{\kappa_\infty \mathcal{M} V_0}. \tag{5.92}$$

For the transverse component of the displacement $\vec{u}_\perp$, the electric field $\vec{F}$ does not enter Newton's equation; hence the frequency of transverse vibrations just equals ω_0:

$$\omega_{TO} = \omega_0. \tag{5.93}$$

We can eliminate the parameters of the crystal, $\mathcal{M}$, e^*, and V_0, from Eq. (5.92) by using the following method. Let us apply Eq. (5.87) to relate the external static electric field $\vec{F}_{st}$ to other quantities. We find that the displacement and the polarization induced by this field are

$$\vec{u}_{st} = -\frac{e^*}{\mathcal{M}\omega_{TO}{}^2}\vec{F}_{st}, \qquad \vec{P}_{st} = \left(-\frac{e^{*2}}{\mathcal{M}\omega_{TO}{}^2 V_0} + \frac{\kappa_\infty}{4\pi}\right)\vec{F}_{st}.$$

Combining these results with the electrostatic equation

$$\vec{D}_{st} = \vec{F}_{st} + 4\pi\,\vec{P}_{st} = \kappa_0\vec{F}_{st},$$

where κ_0 is the static dielectric permittivity, we find

$$\frac{e^{*2}}{MV_0} = \frac{\kappa_0 - \kappa_\infty}{4\pi} \omega_{TO}^2.$$

Now we can get the so-called Lyddane–Sachs–Teller relation between frequencies of longitudinal and transverse vibrations:

$$\frac{\kappa_0}{\kappa_\infty} = \frac{\omega_{LO}^2}{\omega_{TO}^2}. \tag{5.94}$$

Because the static permittivity is greater than the optical permittivity, the longitudinal-wave frequency is always higher than the transverse-wave frequency. It is important to emphasize that the splitting between frequencies of longitudinal and transverse vibrations in an ionic crystal is due exclusively to the long-range electrostatic interaction between charged ions.

Another important result can be obtained from Eqs. (5.87), (5.88), and (5.90). Let us express the polarization $\vec{P}$ in terms of the electric field $\vec{F}$ at arbitrary frequencies:

$$\vec{P} = \frac{\omega_0^2(1 - \kappa_0) + \omega^2(\kappa_\infty - 1)}{4\pi(\omega^2 - \omega_0^2)} \vec{F}. \tag{5.95}$$

Using the definition of the electrical displacement $\vec{D}$, we get

$$\vec{D} = \vec{F} + 4\pi\vec{P} = \kappa(\omega)\vec{F} = \kappa_\infty \frac{\omega^2 - \omega_{LO}^2}{\omega^2 - \omega_{TO}^2} \vec{F}. \tag{5.96}$$

Now we find the frequency-dependent permittivity of the crystal:

$$\kappa(\omega) = \kappa_\infty \frac{\omega^2 - \omega_{LO}^2}{\omega^2 - \omega_{TO}^2}. \tag{5.97}$$

It is easy to see that in both limits, $\omega \to 0$ and $\omega \to \infty$, Eq. (5.97) gives correct results for κ_0 and κ_∞, respectively. In Fig. 5.22, a sketch of $\kappa(\omega)$ is presented. Three important features of the permittivity can be discerned from this figure and Eq. (5.97). First, there is an interval $(\omega_{TO}, \omega_{LO})$ within which the permittivity is negative. Second, the permittivity is infinitely large at $\omega = \omega_{TO}$. Third, the permittivity is zero at $\omega = \omega_{LO}$. Two latter conclusions are important and can be interpreted as

$$\kappa(\omega) = 0 \text{ gives the frequency of longitudinal phonons,} \tag{5.98}$$

$$\frac{1}{\kappa(\omega)} = 0 \text{ gives the frequency of transverse phonons.} \tag{5.99}$$

Equations (5.98) and (5.99) are used in subsequent analyses to find proper optical vibrations in polar-quantum structures.

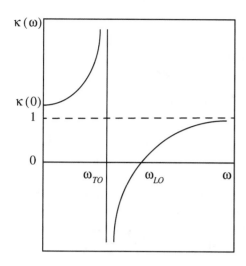

Figure 5.22. Sketch of the frequency dependence of the permittivity of a polar crystal.

It is worth emphasizing that the dielectric permittivity $\kappa(\omega)$ determines all infrared optical properties of crystals associated with lattice vibrations. In particular, in the frequency interval with a negative κ, the electromagnetic wave cannot propagate through the crystal; in this case, the light should be reflected totally.

Confinement of Longitudinal-Optical Phonons in Polar Heterostructures

Now let us turn to quantum structures and first make some qualitative physical speculations. In Subsection 5.6.1, in which we neglected the long-range electrostatic interaction, which is valid only for nonpolar crystals like Si, Ge, etc., we found that mechanical boundary conditions give us the wave vectors of Eq. (5.85) and frequencies of Eq. (5.86), which are appropriate for confined phonons. For polar crystals it is necessary to introduce additional variables corresponding to the polarization $\vec{P}(\vec{r})$ and the electric $\vec{F}$. Thus one should use additional boundary conditions. They can be chosen as electrostatic boundary conditions: continuity of the electric-field component parallel to the layer(s) $\vec{F}_{\parallel}$ and continuity of the electrical-displacement component perpendicular to the layer(s) D_z. Since confusion arises frequently when two types of boundary conditions – mechanical and electrostatic – are used in the continuum model, let us analyze this problem in detail.

Consider two materials, 1 and 2, with LO and TO phonon frequencies $\omega_{LO,1}$, $\omega_{TO,1}$ and $\omega_{LO,2}$, $\omega_{TO,2}$, respectively. These frequencies are supposed to be different for both materials. Dielectric permittivities $\kappa_1(\omega)$ and $\kappa_2(\omega)$ are introduced in the form of Eq. (5.97). We concentrate on LO vibrations, for which the electrostatic field is completely characterized by a scalar potential without a solenoidal contribution. In such a case, for each of the media we can write

$$\vec{F} = -\nabla\phi. \tag{5.100}$$

The field of Eq. (5.100) has to satisfy the equation $\mathrm{div}(\kappa\,\vec{F}) = 0$, that is

$$\kappa(\omega)\nabla^2\phi = 0. \tag{5.101}$$

In Eqs. (5.100) and (5.101) we omit the indices 1 and 2. Now we can see that Eq. (5.101) is automatically satisfied for $\kappa(\omega) = 0$. According to Eqs. (5.97) and (5.98), this condition gives the frequencies of longitudinal modes having nonzero electric fields in each of the materials. For example, the condition $\kappa_2(\omega) = 0$ corresponds to $\omega = \omega_{\mathrm{LO},2}$ and a nonvanishing vector $\vec{F}$ in layer 2. Another solution of Eq. (5.101) corresponds to $\kappa(\omega) \neq 0$, but $\nabla^2\phi = 0$. As we see below, this leads to interface modes.

Let us consider a confinement of polar-optical vibrations. As used above, we assume that waves propagates along the x direction. Then, for longitudinal vibrations, we can classify solutions for the electrostatic potential into two families with different symmetry:

$$\phi_s = \phi_{0,s}e^{iq_x x}\cos q_z z, \qquad \phi_a = \phi_{0,a}e^{iq_x x}\sin q_z z, \tag{5.102}$$

where $\phi_{0,s}$ and $\phi_{0,a}$ are some amplitudes of the waves. To obtain localized modes in layer 2 (see Fig. 5.6) we should assume that the electrostatic field in the surrounding area is zero, $\vec{F}_1 = 0$, and impose proper (electrostatic) boundary conditions at the interfaces:

$$F_{x,1} = F_{x,2} = 0, \qquad \kappa_1(\omega)F_{z,1} = \kappa_2(\omega)F_{z,2} \quad \text{at } z = \pm L/2, \tag{5.103}$$

where L is the width of the layer. With Eqs. (5.100) and (5.102), the boundary conditions at $z = \pm L/2$ give us

$$iq_x\phi_{0,s}e^{iq_x x}\cos\frac{q_z L}{2} = 0, \qquad iq_x\phi_{0,a}e^{iq_x x}\sin\frac{q_z L}{2} = 0, \tag{5.104}$$

$$q_z\kappa_2(\omega)\phi_{0,s}e^{iq_x x}\sin\frac{q_z L}{2} = 0, \qquad q_z\kappa_2(\omega)\phi_{0,a}e^{iq_x x}\cos\frac{q_z L}{2} = 0. \tag{5.105}$$

Equations (5.105) are satisfied at $\omega = \omega_{\mathrm{LO},2}$, as discussed above, while the pair of Eqs. (5.104) gives the conditions of the quantization of transverse wave vectors, which coincide with Eq. (5.85). For the symmetric and the asymmetric solutions they are, respectively,

$$q_{z,s} = \frac{\pi(2n+1)}{L}, \qquad q_{z,a} = \frac{2\pi n}{L}, \quad n = 1, 2, \ldots. \tag{5.106}$$

For purposes of comparison, we highlight the fact that Eq. (5.85) was obtained from mechanical boundary conditions, which imply that the ion displacement $\vec{u}$ is zero at interfaces.

Let us check this requirement for the results just obtained. As we have shown above, $u_x \propto F_x$ and $u_z \propto F_z$; see for example, Eqs. (5.91). Hence the condition

$$u_x\left(z = \pm\frac{L}{2}\right) = 0 \tag{5.107}$$

is consistent with Eqs. (5.104). The explicit forms of the conditions

$$u_z\left(z = \pm\frac{L}{2}\right) = 0 \tag{5.108}$$

in terms of $\phi_{s,a}$ are

$$q_z\phi_{0,s}e^{iq_x x}\sin\frac{q_z L}{2} = 0, \qquad q_z\phi_{0,a}e^{iq_x x}\cos\frac{q_z L}{2} = 0, \tag{5.109}$$

which, in contrast to Eqs. (5.105), do not contain the multiplier $\kappa_2(\omega)$ and are not automatically zero at $\omega = \omega_{LO,2}$. Solutions of Eqs. (5.109) for the two types of waves are

$$q_{z,s} = \frac{2\pi n}{L}, \qquad q_{z,a} = \frac{\pi(2n+1)}{L}, \qquad n = 1, 2, \ldots. \tag{5.110}$$

A comparison of Eqs. (5.106) and (5.110) shows that the boundary conditions of electrostatic equations (5.103) and mechanical equations (5.107) and (5.108) cannot be satisfied simultaneously, at least for $q_x \neq 0$ or $q_z \neq 0$.

However, if $q_x = 0$, we get $u_x = 0$, so that the electrostatic boundary condition of Eqs. (5.105) is fulfilled for $\omega = \omega_{LO,2}$, while Eqs. (5.110) coincide with the result obtained with the mechanical boundary conditions of Eq. (5.85). Remember that Eq. (5.85) was derived just for the case of $q_x = 0$. For $q_x \neq 0$ the conditions of Eqs. (5.110) are applicable if

$$q_z \gg q_x. \tag{5.111}$$

Indeed, for this case we get $|F_z| \gg |F_x|$ and Eqs. (5.105) are more stringent than Eqs. (5.104). The field component F_x is small, and Eqs. (5.104) can be disregarded in the lowest approximation.

Summarizing these results, one can state that in the limit of inequality (5.111) only the second electrostatic condition of Eqs. (5.105) should be kept for the lowest approximation. This coincides with the mechanical conditions imposed on the displacement component u_z and generally leads to a series of confined longitudinal modes. A similar analysis can be done for TO modes. In this case, the rule of the quantization is determined by the mechanical boundary conditions of Eqs. (5.110).

In Fig. 5.23, the results of the microscopic calculations are represented for the three lowest LO modes of a 20-Å layer of GaAs embedded in AlAs. These results are obtained in the limit of inequality (5.111). Atomic displacements u_z are shown for each monolayer; thus one can trace the displacements of the Al, Ga, and As ions. It is clearly seen that the modes are entirely confined in the GaAs layer and the ions in the AlAs layers are almost motionless, that is, these calculations support the previously derived results of the dielectric-continuum model.

$$q = (0, 0, q_z \to 0)$$

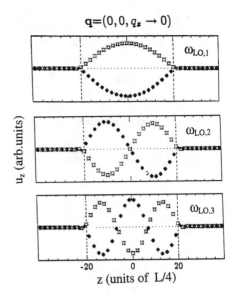

Figure 5.23. The amplitudes of atomic displacements of the several lowest confined optical-phonon modes in a GaAs slab embedded in the AlAs material. The wave vector is supposed to be $\vec{q} = (0, 0, q_x \to 0)$. The dotted vertical lines denote interfaces. Diamonds, stars, and asterisks indicate the positions of the Ga, Al, and As planes, respectively. [After E. Molinari, *et al.*, "Electron-phonon interaction in two-dimensional systems: a microscopic approach," Semicond. Sci. Technol. 7, pp. B67–B72 (1992).]

Interface Optical Phonons in Polar Materials

We conclude our discussion of optical vibrations in polar heterostructures by applying the electrostatic-continuum model to characterize interface phonons. For simplicity, we start by considering a single-junction heterostructure. As mentioned previously, interface phonons can be obtained from Eq. (5.101) if $\kappa_{1,2}(\omega) \neq 0$. Thus, for the electrostatic potential, we get the Laplace equation $\nabla^2 \phi = 0$. Solutions of this equation for a heterojunction with its origin at $z = 0$ are given by

$$\phi_{1,2}(x, z) = \phi_0 e^{i q_x x} e^{\pm q_z z} \tag{5.112}$$

for both sides 1 and 2 of the structure. Here the ($\pm$) signs have to be chosen so that the wave amplitude decays far away from the junction. Such solutions correspond to vibrations localized near the interface. Even though the electrostatic potential of the form of Eq. (5.112) is symmetric for both sides of the heterostructure, the corresponding amplitudes of electric fields and polarizations are considerably different, because of the different values of the dielectric permittivities $\kappa_{1,2}(\omega)$. Equation (5.112) provides the continuity of the electrostatic potential ϕ through the junction and also fulfills the first electrostatic condition of Eqs. (5.103). The second boundary condition of Eqs. (5.103) immediately gives us the dispersion relation for the interface phonons of a single heterostructure:

$$\kappa_1(\omega) + \kappa_2(\omega) = 0. \tag{5.113}$$

Using Eq. (5.97), we can represent this last result in the explicit form

$$\kappa_{\infty,1} \frac{\omega^2 - \omega_{LO,1}^2}{\omega^2 - \omega_{TO,1}^2} + \kappa_{\infty,2} \frac{\omega^2 - \omega_{LO,2}^2}{\omega^2 - \omega_{TO,2}^2} = 0. \tag{5.114}$$

Equation (5.114) is a quadratic equation in ω^2, which always has two positive solutions, that is, there are two frequencies $\omega_{IF}^{\pm}$ for interface modes. Using Fig. 5.22 as an example of the frequency dependence of $\kappa(\omega)$, we can see that, depending on the four parameters $\omega_{LO,1,2}$ and $\omega_{TO,1,2}$, the following hierarchy of the frequencies occurs:

$$\omega_{TO,1} < \omega_{IF}^- < \omega_{TO,2} < \omega_{LO,1} < \omega_{IF}^+ < \omega_{LO,2}, \qquad (5.115)$$

if $\omega_{TO,1} < \omega_{TO,2}, \omega_{LO,1} < \omega_{LO,2}$, and

$$\omega_{TO,2} < \omega_{IF}^- < \omega_{TO,1} < \omega_{LO,1} < \omega_{IF}^+ < \omega_{LO,2}, \qquad (5.116)$$

if $\omega_{TO,2} < \omega_{TO,1}, \omega_{LO,1} < \omega_{LO,2}$. Here we have assumed a particular order of frequencies in the bulk materials. For AlAs/GaAs heterostructures, $\omega_{TO,GaAs} < \omega_{LO,GaAs} < \omega_{TO,AlAs} < \omega_{LO,AlAs}$. For this case, one of the interface modes should be an AlAs-like mode, while the another should be a GaAs-like mode. This means, for example, that the AlAs-like interface mode has the frequency somewhere between $\omega_{TO,AlAs}$ and $\omega_{LO,AlAs}$. However, this mode is a joint mode of both materials.

Microscopic calculations of interface modes for an AlAs/GaAs/AlAs double heterostructure are presented in Fig. 5.24. Results are given for three longitudinal wave vectors, and they support the above conclusion made for a single-heterojunction. In accordance with Eq. (5.112), the localization of the interface mode increases with q_x. The mode in Fig. 5.24 is the AlAs-like mode: Vibrations have frequencies nearly equal to those of AlAs.

Figure 5.24. The amplitudes of atomic displacements of the AlAs-like interface modes in the AlAs/GaAs/AlAs double-heterojunction structure. The notation is the same as that in Fig. 5.23. [After E. Molinari, *et al.*, "Electron–phonon interaction in two-dimensional systems: a microscopic approach," Semicond. Sci. Technol. 7, pp. B 67–72 (1992).]

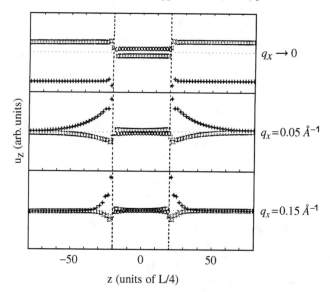

In conclusion, we have to say that an AlAs/GaAs/AlAs double-barrier structure has three distinct types of optical modes. At high frequencies, there are AlAs-like modes, which decay exponentially as they enter the GaAs region. At frequencies close to the frequency of the GaAs bulk phonon, there is a series of modes (with different q_z) that are almost entirely confined in the GaAs well layer. In addition, there are one or two interface modes located preferentially either in the AlAs or in the GaAs. As shown, these modes create electrostatic fields, and, because of this, they are strongly coupled with electrons confined in the GaAs quantum-well layer.

5.7 Closing Remarks

In this chapter, we have studied lattice vibrations in semiconductors as well as in semiconductor heterostructures. Starting with the simplest examples, one and two atoms per primitive cell of a linear chain, we have established that small-amplitude vibrations are described by linear equations. They correspond to a quadratic expansion of the crystal energy with respect to atomic (ionic) displacements. Such a quadratic expansion is the principal reason for the introduction of independent oscillators that can be treated as new quasi-particles – phonons.

The phonons are characterized by their wave vectors $\vec{q}$ and frequencies $\omega(\vec{q})$ and can be associated with different branches of the vibrations: LA and TA phonons and LO and TO phonons. The number of these branches equals the total number of atoms (ions) per primitive cell of the crystal. The universal character of the phonon concept allows one to quantize lattice vibrations easily: if one can find the frequencies $\omega(\vec{q})$ for a branch, the phonon energy is equal to $\hbar\omega(\vec{q})$; the wave solution corresponding to this energy is the phonon wave function. Phonons obey Bose–Einstein statistics, and their distribution function under equilibrium conditions is given by the Planck distribution.

For each type of crystal phonon, frequencies can cover a wide interval from 0 to tens of terahertz. We have seen that the properties of acoustic and optical phonons are quite different in the physically important region of small wave vectors. In particular, acoustic vibrations have a linear dependence on the dispersion $\omega(\vec{q}) \propto q$, while the optical frequencies can be treated frequently as being approximately constant independently of q.

Long-wavelength acoustic vibrations can be described by well-known equations for elastic waves in crystals. These equations provide powerful tools for investigating acoustic phonons. We have applied these equations to the study of acoustic phonons in quantum structures like quantum wells, wires, and superlattices. We have found considerable modifications in both dispersion dependences and wave solutions. In particular, localized modes occur near quantum wells and wires. The degree of their localization increases with frequency, so, despite relatively small variations in the elastic properties of III–V compounds, at high frequencies these modes are well localized within the quantum structures. For an acoustic phonon, we use the term localized modes, in contrast to the terms confined electrons and confined optical phonons, since the latter are really confined inside the layer, wire region, etc., while the localized modes carry a considerable portion of their energy outside these regions.

The modification of long-wavelength vibrations in superlattices is even more pronounced. Indeed, these modifications include the folding of the $\omega(\vec{q})$ curves as well as the opening of frequency gaps where the waves are forbidden totally. These gaps define stop bands and passbands that are manifested through the resonant character of wave propagation; indeed, these gaps result in resonant reflection and transmission like that of the Fabry–Perot etalon for light waves.

For short-wavelength vibrations, we applied a qualitative analysis, which led to the classification of the possible vibrations: waves propagating through whole structure, confined modes, and interface modes. For example, in a Si/Ge short-period superlattice we have found that the high-frequency optical vibrations of Si layers are confined, while the low-frequency optical vibrations of Ge layers and all acoustic phonons can propagate through the superlattice, etc.

In polar (ionic) materials like III–V compounds, long-range electrostatic interaction between ions from different crystal cells leads to a fundamental result: finite splitting between longitudinal ω_{LO} and transverse ω_{TO} phonon frequencies at $\vec{q} \to 0$. In particular, the dielectric permittivity, which has a frequency dispersion in the infrared region due mainly to the phonon contribution, is negative in the interval $(\omega_{TO}, \omega_{LO})$. The negative permittivity forbids any propagation of electromagnetic waves in this frequency interval. The splitting also complicates the investigation of phonons in polar heterostructures. We have given a comparative analysis of two models – electrostatic and mechanical – and reconciled the results following from these models for confined and interface modes in heterostructures. We found that, because of the difference in the phonon frequencies of AlGaAs and GaAs, the GaAs layers confine optical phonons. In addition, interface optical phonons occur at an AlGaAs/GaAs interface.

The modification of lattice vibrations in heterostructures is of importance for electronic and optoelectronic devices. First, the modification affects the electron–phonon interaction. The modification of the electron–optical-phonon interaction in III–V heterostructures is especially consequential, since both electrons and phonons are confined in the same narrow spatial layer – the quantum well. In general, by engineering of the structures, one can suppress, enhance, or optimize the electron–phonon interaction, and consequently it is possible to control related properties of the electrons.

Although the problem of confined and localized phonons arose initially to describe correctly the electron–phonon interaction in quantum structures, these phonons can be of interest by themselves for device applications. Actually, contemporary technology permits the fabrication of heterostructures so perfect that large free paths of phonons can be expected for them, at least at low temperatures. This, along with the fact that quantum wells and wires can serve as waveguides for localized (confined) modes, can be used in devices for special purposes.

The reader can learn more details on the general theory of the lattice vibrations in the following textbooks:

M. Born and K. Huang, *Dynamical Theory of Crystal Lattices* (Oxford U. Press, London, 1954).

C. Kittel, *Quantum Theory of Solids* (Wiley, New York, London, 1963).

G. P. Srivastava, *The Physics of Phonons* (Hilger, Bristol, Philadelphia, and New York, 1990).

I. Ipatova and V. Mitin, *Introduction to Solid-State Electronics* (Addison-Wesley, New York, 1996).

Additional information on phonons in heterostructures and data on specific materials can be found in current literature:

B. Jusserend and M. Cardona, "Raman spectroscopy of vibrations in super-lattices," in *Light Scattering in Solids*, M. Cardona and G. Guntherodt, eds. (Springer-Verlag, Berlin, 1989), pp. 49–146.

S. Tamura, D. C. Hurley, and J. P. Wolfe, "Acoustic phonon propagation in superlattices," Phys. Rev. B 38, 1427–1449 (1988).

M. A. Stroscio, "Interaction between longitudinal optical phonon modes of a rectangular quantum wire and charge carriers of a one-dimensional electron gas," Phys. Rev. B 40, 6428–6431 (1989).

N. Nishiguchi, "Confined and interface acoustic phonons in a quantum wire," Phys. Rev. B 50, 10970–10980 (1994).

M. A. Stroscio and K. W. Kim, "Piezoelectric scattering of carriers from confined acoustic modes in cylindrical quantum wires," Phys. Rev. B 48, 1936–1938 (1993).

L. Wendler and V. G. Grigorian, "Acoustic interface waves in sandwich structures," Surf. Sci. 206, 203–224 (1988).

B. K. Ridley, "Hot electrons in low dimensional structures," Rep. Prog. Phys. 54, 169–256 (1991).

E. Molinari and A. Fasolino, "Phonons of ideal (001) superlattices: application to Si/Ge," in *Spectroscopy of Semiconductor Microstructures*, by G. Fasol, A. Fasolino, and P. Lugli, eds. (Plenum, New York, 1989), pp. 195–205.

E. Molinari, C. Bungaro, M. Gulia, P. Lugli, and H. Rucker, "Electron–phonon interactions in two-dimensional systems: a microscopic approach," Semicond. Sci. Technol. 7, B67–B72 (1992).

PROBLEMS

1. The derivation of confined optical modes in a quantum-well structure with boundaries at $z = \pm L/2$ is given by Eqs. (5.101)–(5.109). Show that a similar

derivation is possible for the confined optical modes in quantum wire structures with boundaries at $y = \pm L_y/2$ and $z = \pm L_z/2$; assume that the quantum wire is unbounded in the x direction. How are the functional forms of the confined optical modes of such a quantum wire similar to those of a quantum well?

2. Consider a one-dimensional continuum, LA mode propagating along a single axis, say the z axis. Now confine this mode to the interval $(-L/2, L/2)$ by applying free and clamped boundary conditions to the mode displacement $u(z)$. Compare the unnormalized, confined modes for three cases: (a) free boundary conditions at both ends of the quantum wire, $\partial u(z)/\partial z|_{z=\pm L/2} = 0$; (b) clamped boundary conditions at both ends of the quantum wire, $u(\pm L/2) = 0$; and (c) free boundary at $z = L/2$ and clamped boundary at $z = -L/2$. (These modes are similar to the standing modes of organ pipes; at open boundaries of an organ pipe the modes behave as modes subjected to the free boundary condition of this example.)

3. Suppose a pseudomorphic layer has an in-plane spring constant that is 1% larger than the in-plane spring constant for the unstrained layer material. Assume that the layer is a cubic crystal with a [001] axis perpendicular to the interface. In the lowest approximation, by how much do the in-plane phonon frequencies of the strained and unstrained layers differ?

4. Figure 5.23 provides sketches of the amplitudes of atomic displacements of the several lowest confined optical-phonon modes in a GaAs slab embedded in AlAs. Suppose that the same structure is grown with only two or three new layers of AlAs in the center of the GaAs well. Assuming the GaAs well is much wider than the thin central layer of AlAs, sketch the approximate mode amplitudes for the two highest-energy confined optical modes in this modified structure.

5. Symmetric and antisymmetric interface optical-phonon modes in a GaAs quantum well surrounded by AlAs have nonzero and zero values of the potential in the center of the quantum well, which extends from $z = -L/2$ to $z = +L/2$. Suppose a new M/GaAs/AlAs structure is fabricated with the M/GaAs interface at $z = 0$ and the GaAs/AlAs interface at $z = +L/2$. At $z = 0$, the phonon potentials must satisfy the usual boundary conditions for a metal (M) surface. Assuming that the Thomas–Fermi screening distance in the metal is negligible, sketch the interface optical-phonon potential envelope that remains for an antisymmetric optical-phonon mode. What is the fate of the symmetric mode in this M/GaAs/AlAs structure?

Electron Scattering
in Quantum Structures

6.1 Introduction

In Chapters 3 and 4, we have studied the electron states and equilibrium properties of the electron gas in quantum structures. We have found that these properties are determined for each particular type of structure by the Coulomb interaction, the electron statistics, etc. The effects of electron–phonon interactions, fluctuations of doping concentrations, and scattering by defects normally do not change the equilibrium properties.

On the contrary, nonequilibrium processes, such as electric conductivity, electron diffusion, optical response, etc., involve primary changes in the electron motion under external perturbation and electron scattering. In this chapter, we concentrate on those features of electron scattering that are brought about by electron and phonon confinement in quantum heterostructures.

Away from equilibrium, the kinetics of electrons consists of intervals of propagation interrupted by scattering events. During the intervals between scattering events, acceleration may occur as a result of external fields. As a result of scattering, electrons dissipate their momentum and energy, can change the subband and the spin numbers, etc. To characterize such a process, one can introduce the mean duration of the propagation intervals and a set of other times describing the relaxation of the average electron momentum, energy, etc. These times depend on the scattering mechanisms and their intensity. It is important that scattering establish the upper limits on the electron mobility and the drift velocity and thus define the speed of operation of semiconductor devices. On the other hand, because scattering is a random process it also establishes lower limits on electron noise and thus defines the threshold of sensitivity of semiconductor devices. Therefore a key problem of solid-state electronics is understanding various scattering mechanisms present in semiconductors and semiconductor structures. There are a number of such mechanisms. They include electron interactions with bulklike and confined phonons, crystal defects and imperfections, and neutral and ionized impurities. In quantum structures these scattering processes can be significantly different from those in bulk materials.

Within the framework of quantum mechanics, scattering processes are described in terms of probabilities of transitions $W_{i,f}$ between an initial state i and a final state f per unit time. In Section 2.4, the general formula known as Fermi's golden rule has been derived for $W_{i,f}$. If we consider the semi-classical electron transport, quantum numbers characterizing initial and final states include initial and final wave vectors $\vec{k}$ and $\vec{k}'$. The probability of scattering between quantum states can be written as $W(n, \vec{k}; m, \vec{k}')$, where n and m are, in general, sets of other discrete quantum numbers. For a three-dimensional system, n coincides with the spin number s and $\vec{k}$ is a three-dimensional vector. For a two-dimensional case, $n \equiv \{i, s\}$, where i is the subband number and $\vec{k}$ is a two-dimensional vector, etc. As shown in Section 2.8, these probabilities define the collision integral in the Boltzmann transport equation(s). If probabilities $W(n, \vec{k}; m, \vec{k}')$ are known, the semi-classical electron transport can be calculated with the Boltzmann equation(s) or another equivalent computational approach – the Monte Carlo method. The latter method is based on modeling the electron trajectories. In the simplest variant of this method, one calculates an electron trajectory in an external field, introducing scattering mechanisms according to the probabilities $W(n, \vec{k}; m, \vec{k}')$. As a result of many scattering events, an electron distribution function is formed and any average physical quantity can be calculated. One can prove that in the limit of an infinite number of scattering events, the Monte Carlo method is identical to the Boltzmann equation approach.

Below, we analyze values $W(n, \vec{k}; m, \vec{k}')$ for specific scattering mechanisms. Before this analysis, it is instructive to obtain simple estimates for parameters characterizing the electron kinetics without specifying the scattering mechanisms.

Probe-particle approach: Let an electron gas be perturbed so that the electron distribution function $\mathcal{F}(n, \vec{k}, t)$ differs slightly from its equilibrium value $\mathcal{F}_F(n, \vec{k})$:

$$\mathcal{F}(n, \vec{k}, t) = \mathcal{F}_F(n, \vec{k}) + \Delta\mathcal{F}(n, \vec{k}, t).$$

This means that all parameters also deviate from their equilibrium values. If $A(n, \vec{k})$ is a physical quantity, its average value is $\bar{A} = \sum_{n,\vec{k}} A(n, \vec{k})\mathcal{F}(n, \vec{k}, t)$ and the deviation from the equilibrium value is

$$\Delta A = \sum_{n,\vec{k}} A(n, \vec{k})\Delta\mathcal{F}(n, \vec{k}, t). \tag{6.1}$$

One can calculate the rate of relaxation of this quantity:

$$\frac{d\Delta A}{dt} = \sum_{n,\vec{k}} A(n, \vec{k})\frac{\partial}{\partial t}\Delta\mathcal{F}(n, \vec{k}, t). \tag{6.2}$$

The temporary evolution of $\Delta\mathcal{F}(n, \vec{k}, t)$ is to be found from the simplest Boltzmann equation:

$$\frac{\partial}{\partial t}\Delta\mathcal{F}(n, \vec{k}, t) = \sum_{m,\vec{k}'}[W(m, \vec{k}'; n, \vec{k})\Delta\mathcal{F}(m, \vec{k}', t) - W(n, \vec{k}; m, \vec{k}')\Delta\mathcal{F}(n, \vec{k}, t)],$$

$$\tag{6.3}$$

where we assume that the electron gas is uniform and nondegenerate and external forces are absent. Substituting Eq. (6.3) into Eq. (6.2) we get

$$\frac{\mathrm{d}\Delta A}{\mathrm{d}t} = \sum_{n,\vec{k}} \Delta \mathcal{F}(n, \vec{k}, t) R_A(n, \vec{k}) \tag{6.4}$$

with

$$R_A(n, \vec{k}) \equiv -\sum_{m,\vec{k}'} [A(n, \vec{k}) - A(m, \vec{k}')] W(n, \vec{k}; m, \vec{k}'). \tag{6.5}$$

The form of Eq. (6.4) tells us that $R_A(n_0, \vec{k}_0)$ has the meaning of the rate of change in ΔA for the case in which the distribution function is disturbed near the electron state $\{n_0, \vec{k}_0\}$:

$$\Delta \mathcal{F}(n, \vec{k}, t = 0) = \delta_{n,n_0} \delta_{\vec{k}, \vec{k}_0} \Delta \mathcal{F}(n_0, \vec{k}_0). \tag{6.6}$$

Such a distribution corresponds to the so-called probe-particle approach. In this approach one assumes that at the initial moment all disturbed particles are concentrated in one electron state $\{n_0, \vec{k}_0\}$. Then, because of scattering, the system relaxes to an equilibrium state. In the probe-particle approach, electron-relaxation characteristics are especially simple. Substituting Eq. (6.6) into Eq. (6.3), we find the evolution of the population of the initial state $\{n_0, \vec{k}_0\}$ in the form

$$\frac{\partial}{\partial t} \Delta \mathcal{F}(n_0, \vec{k}_0, t) = -\frac{\Delta \mathcal{F}(n_0, \vec{k}_0)}{\tau_s(n_0, \vec{k}_0)},$$

where we introduce $\tau_s(n_0, \vec{k}_0)$ by the relationship

$$\frac{1}{\tau_s(n_0, \vec{k}_0)} = \sum_{m \neq n_0, \vec{k}' \neq \vec{k}} W(n_0, \vec{k}_0; m, \vec{k}'). \tag{6.7}$$

Thus $\tau_s(n_0, \vec{k}_0)$ determines the relaxation of the population of state $\{n_0, \vec{k}_0\}$. Another meaning of this quantity is obviously the average time between scattering events. The inverse value $1/\tau_s$ frequently is called as the total scattering rate.

The momentum-relaxation time $\tau_p(n_0, \vec{k}_0)$, can be introduced by means of the relaxation rate of the momentum of the electron gas with the initial distribution of Eq. (6.6). The momentum-relaxation rate is calculated by the substitution of $A \equiv \hbar\vec{k}$ into Eq. (6.2). The result is

$$\frac{\hbar\vec{k}_0}{\tau_p(n_0, \vec{k}_0)} = \hbar \sum_{n,\vec{k}} (\vec{k}_0 - \vec{k}) \, W(n_0, \vec{k}_0; n, \vec{k}).$$

If the probability of scattering is a function of $\vec{k}_0 - \vec{k}$, the momentum-relaxation time can be found in the form

$$\frac{1}{\tau_p(n_0, \vec{k}_0)} = \sum_{n, \vec{k}} \frac{k_0 - k \cos\theta}{k_0} W(n_0, \vec{k}_0; n, \vec{k}), \tag{6.8}$$

where θ is the angle between the wave vectors $\vec{k}$ and $\vec{k}_0$.

We also have to introduce the energy-relaxation rate, as it determines the efficiency of energy losses during scattering and is an important characteristic of hot electrons, which will be addressed in Chapter 7. Setting $A(\vec{k}) = E_n(\vec{k})$ in Eq. (6.5), we can derive the energy-relaxation time within the probe-particle approach as

$$\frac{1}{\tau_E(n_0, \vec{k}_0)} = \sum_{n, \vec{k}} \frac{E_{n_0}(\vec{k}_0) - E_n(\vec{k})}{E_{n_0}(\vec{k}_0)} W(n_0, \vec{k}_0; n, \vec{k}), \tag{6.9}$$

where $E_{n_0}(\vec{k}_0)$ and $E_n(\vec{k})$ are the electron energies before and after scattering, respectively.

Sometimes we need to calculate the ensemble average of any of the previously considered times. We can define these ensemble averages as

$$\frac{1}{\tau_s} = \sum_{n_0, n, \vec{k}_0, \vec{k}} W(n_0, \vec{k}_0; n, \vec{k}) \mathcal{F}_F(n_0, \vec{k}_0), \tag{6.10}$$

$$\frac{1}{\tau_p} = \sum_{n_0, n, \vec{k}_0, \vec{k}} \frac{k_0 - k \cos\theta}{k_0} W(n_0, \vec{k}_0; n, \vec{k}) \, \mathcal{F}_F(n_0, \vec{k}_0), \tag{6.11}$$

$$\frac{1}{\tau_E} = \sum_{n_0, n, \vec{k}_0, \vec{k}} \frac{E_{n_0}(\vec{k}_0) - E_n(\vec{k})}{E_{n_0}(\vec{k}_0)} W(n_0, \vec{k}_0; n, \vec{k}) \mathcal{F}_F(n_0, \vec{k}_0). \tag{6.12}$$

At low temperatures, when electrons become degenerate, in Eqs. (6.10)–(6.12) we should make the replacement

$$\mathcal{F}_F(n_0, \vec{k}_0) \to \mathcal{F}_F(n_0, \vec{k}_0)[1 - \mathcal{F}_F(n, \vec{k})]. \tag{6.13}$$

This means that for a degenerate electron gas only electron scattering around the Fermi energy takes place. Therefore there is no need to perform ensemble averaging in this extreme case.

Equations (6.10)–(6.12) give us simple estimates of the characteristic times for the relaxation of population, momentum, and energy of a particular state.

The rest of this chapter is organized as follows. In Section 6.2 we study general features of elastic scattering in a low-dimensional electron gas. Then the two main reasons for such scattering – remote ionized impurities and interface roughness – are examined in Sections 6.3, 6.4 and 6.5. In order to introduce different lattice scattering mechanisms, in Section 6.6 we interpret this scattering as transitions between different quantum states of the coupled electron–phonon system; we also present two classes of electron–phonon coupling: short- and long-range interaction mechanisms. In Sections 6.7 and 6.8 we analyze mechanisms for acoustic- and optical-phonon modes in bulklike crystals. In Section 6.9 we examine electron-acoustic scattering in quantum wells, wires, and free-standing structures. Finally, Section 6.10 is devoted to the study of optical-phonon scattering in heterostructures.

6.2 Elastic Scattering in Two-Dimensional Electron Systems

We start with a study of the scattering two-dimensional electrons by various defects and imperfections. In quantum heterostructures, such scattering processes have several new peculiarities. The first is related to the fact that in quantum structures selective doping is widely used, as discussed in Chapter 4. Such doping leads to a spatial separation of low-dimensional carriers and their parent-ionized impurities, i.e., the impurity distribution is highly nonuniform and the carriers are scattered by the remote impurities. The second peculiarity is due to the specific screening by two-dimensional carriers. Besides, new elastic-scattering mechanisms due to interface roughness and alloy fluctuations can affect the electron kinetics.

For an elastic process, the scattering probability can be written in the form of Fermi's golden rule:

$$W(m, \vec{k}', n, \vec{k}) = \frac{2\pi}{\hbar} |\langle m, \vec{k}' \,|\, \mathcal{H}_{\text{def}} \,|\, \vec{k}, n \rangle|^2 \delta[E_m(\vec{k}') - E_n(\vec{k})], \qquad (6.14)$$

where $E_n(\vec{k})$ is the energy of an electron from the nth subband with a two-dimensional wave vector $\vec{k}$, $\mathcal{H}_{\text{def}}$ is the Hamiltonian of interaction between an electron and defects, and $\langle m, \vec{k}' \,|\, \mathcal{H}_{\text{def}} \,|\, \vec{k}, n \rangle$ is the matrix element for electron scattering by defects. One can write $\mathcal{H}_{\text{def}}$ as a sum over all defects:

$$\mathcal{H}_{\text{def}} = \sum_j V_{\text{def}}(\vec{r} - \vec{R}_j), \qquad (6.15)$$

where $\vec{R}_j = (\vec{\rho}_j, Z_j)$ are the positions of the defects. Consider a layered structure with a two-dimensional electron gas. According to Chapter 3, the wave function $\psi_n(\vec{k}, \vec{R})$ should be taken in the form

$$\psi_{n,\vec{k}}(\vec{r}) = \frac{1}{\sqrt{S}} \chi_n(z) e^{i\vec{k}\vec{\rho}}, \qquad (6.16)$$

where S is the area of the two-dimensional electron gas and $\vec{\rho}$ is the in-plane vector. The matrix element of Eq. (6.15) is

$$\langle m, \vec{k}' \,|\, \mathcal{H}_{\text{def}} \,|\, \vec{k}, n \rangle = \sum_j e^{i(\vec{k}-\vec{k}')\vec{R}_j} \langle m, \vec{k}' \,|\, V_{\text{def}}(\vec{\rho}, z - Z_j) \,|\, \vec{k}, n \rangle, \qquad (6.17)$$

where the latter matrix element is calculated for an individual defect and is

$$\langle m, \vec{k}' \mid V_{\text{def}}(\vec{\rho}, z - Z_j) \mid \vec{k}, n \rangle = \frac{1}{S} \int_V d^3 r \, \chi_m^*(z) e^{-i\vec{k}'\vec{\rho}} V_{\text{def}}(\vec{\rho}, z - Z_j) \, \chi_n(z) e^{i\vec{k}\vec{\rho}},$$

(6.18)

where $\int_V d^3 r (\cdots)$ is an integral over the entire system. This result explicitly depends on the Z coordinate of the defects.

According to Eq. (6.14), the intensity of scattering is determined by the square of the absolute value of the matrix element of Eq. (6.17):

$$|\langle m, \vec{k}' \mid \mathcal{H}_{\text{def}} \mid \vec{k}, n \rangle|^2 = \sum_j \sum_{j'} e^{i(\vec{k}-\vec{k}')(\vec{R}_j - \vec{R}_{j'})} \langle m, \vec{k}' \mid V_{\text{def}}(\vec{\rho}, z - Z_j) \mid \vec{k}, n \rangle$$
$$\times \langle n, \vec{k} \mid V_{\text{def}}(\vec{\rho}, z - Z_{j'}) \mid \vec{k}', m \rangle.$$

(6.19)

In preparation for further calculations, we analyze the spatial scales of different factors on the right-hand side of Eq. (6.19). The exponential factor depends on the change in the momentum during scattering, $|\vec{k} - \vec{k}'|$. For scattering with large momentum changes, we can estimate $|\vec{k} - \vec{k}'| \sim k_F$, where k_F is the Fermi wave vector. Thus we can assume that the characteristic scale of the exponential factor is of the order of $2\pi/k_F$. The scale of the matrix elements is determined by the scale of the potential $V_{\text{def}}(\vec{r})$. The potential of charged impurities is the long-range Coulomb potential, and we can set its scale in terms of the order of the distance to the two-dimensional gas, Z. Let us suppose that the doped region of the structure is separated from the two-dimensional electron gas by a spacer of width Z_0. Thus we get $Z > Z_0$. We compare these two scales, $2\pi/k_F$ and Z_0, with the average distance between impurities, $N_{\text{im}}^{-1/3}$ (N_{im} is the average impurity concentration).

In the case of a high doping density with

$$\frac{2\pi N_{\text{im}}^{1/3}}{k_F} \gg 1, \qquad N_{\text{im}}^{1/3} Z_0 \gg 1,$$

a great number of impurities with different $|\vec{R}_j - \vec{R}_{j'}|$ contribute to Eq. (6.19). We can characterize the spatial impurity distribution by the concentration

$$N_{\text{im}}(\vec{R}) = N_{\text{im}}(Z) + \Delta N_{\text{im}}(\rho, Z),$$

(6.20)

where $N_{\text{im}}(Z)$ is the average concentration and $\Delta N_{\text{im}}(\rho, Z)$ is its fluctuating part, as introduced in Chapter 4. Note that the average concentration $N_{\text{im}}(Z)$ contributes to the regular confining potential and that the spatial scale of the fluctuations ΔN_{im} is much larger than the average distance between impurities $N_{\text{im}}^{-1/3}$. In Eq. (6.19), the summation over impurity positions $\vec{R}_j$ can be replaced by an integration of the impurity concentration over volume:

$$\sum_j (\cdots) \rightarrow \int_V d^3 R \, N_{\text{im}}(\vec{R})(\cdots).$$

(6.21)

Thus the right-hand side of Eq. (6.19) is transformed to

$$\int_V \int_{V'} d^2\rho_j dZ_j d^2\rho'_j dZ'_j N_{im}(\vec{\rho}_j, Z_j) N_{im}(\vec{\rho}'_j, Z'_j) \exp[i(\vec{k} - \vec{k}')(\vec{\rho}_j - \vec{\rho}'_j)]$$
$$\times \langle m, \vec{k}' \mid V_{def}(\vec{\rho}, z - Z_j) \mid \vec{k}, n \rangle \langle n, \vec{k} \mid V_{def}(\vec{\rho}, z - Z'_j) \mid \vec{k}', m \rangle. \tag{6.22}$$

Substituting Eq. (6.20) into expression (6.22), we can easily conclude that for $\vec{k} \neq \vec{k}'$ the terms proportional to the regular part of $N_{im}(\vec{R})$ vanish. As a result, we find that in this limit only the fluctuating part ΔN_{im} contributes to the square of the matrix element:

$$\int_V \int_{V'} d^2\rho_j dZ_j d^2\rho'_j dZ'_j \Delta N_{im}(\vec{\rho}_j, Z_j) \Delta N_{im}(\vec{\rho}'_j, Z'_j) \exp[i(\vec{k} - \vec{k}')(\vec{\rho}_j - \vec{\rho}'_j)]$$
$$\times \langle m, \vec{k}' \mid V_{def}(\vec{\rho}, z - Z_j) \mid \vec{k}, n \rangle \langle n, \vec{k} \mid V_{def}(\vec{\rho}, z - Z'_j) \mid \vec{k}', m \rangle. \tag{6.23}$$

A further reduction of expression (6.23) can be achieved for particular properties of $\Delta N_{im}(\vec{R})$. One of the methods for the description of random fluctuations is to introduce the so-called correlator of the fluctuations. The correlator describes spatial correlations of the fluctuations. In the case under consideration, the in-plane impurity distribution is uniform, on average, as seen from Eq. (6.20). One can introduce the correlator as follows:

$$\mathcal{K}_N(\vec{\rho}_j, Z_j, Z_{j'}) \equiv \frac{1}{S} \int_S d^2\rho \Delta N_{im}(\vec{\rho}, Z_{j'}) \Delta N_{im}(\vec{\rho}_j + \vec{\rho}, Z_j), \tag{6.24}$$

where $\int_S d^2\rho(\cdots)$ is an integral over the plane, that is, the correlator is a product of fluctuations at arbitrary points separated by the vector ρ_j and averaged over the plane. Now, if we use a two-dimensional Fourier transformation of the correlator,

$$\mathcal{K}(\vec{q}, Z_j, Z_{j'}) = \int_S d^2\rho \mathcal{K}(\vec{\rho}, Z_j, Z_{j'}) e^{i\vec{q}\vec{\rho}}, \tag{6.25}$$

we finally transform expression (6.23) into the form

$$S \int\int dZ_j dZ'_j \mathcal{K}(\vec{k} - \vec{k}', Z_j, Z_{j'})$$
$$\times \langle m, \vec{k}' \mid V_{def}(\vec{\rho}, z - Z_j) \mid \vec{k}, n \rangle \langle n, \vec{k} \mid V_{def}(\vec{\rho}, z - Z'_j) \mid \vec{k}', m \rangle. \tag{6.26}$$

The general conclusion, which can be made from Eqs. (6.24) and (6.25) and expressions (6.23) and (6.26), is that for selectively doped heterostructures at high impurity concentrations, the probability of electron scattering is proportional to the square of the impurity fluctuations.

Another limiting case corresponds to a small doping concentration:

$$\frac{2\pi N_{im}^{1/3}}{k_F} < 1$$

for so-called diluted impurities. In this limit, each impurity contributes separately to the scattering processes. Their contribution to the confining potential can be neglected. In this case we can estimate Eq. (6.19), omitting highly oscillating terms with $j \neq j'$. The result is

$$|\langle m, \vec{k}' | \mathcal{H}_{\text{def}} | \vec{k}, n \rangle|^2 = \sum_j |\langle m, \vec{k}' | V_{\text{def}}(\vec{\rho}, z - Z_j) | \vec{k}, n \rangle|^2. \tag{6.27}$$

Equation (6.27) can be significantly simplified for the typical condition $N_{\text{im}}^{1/3} Z_0 \gg 1$. Then we can substitute

$$\sum_j (\cdots) \rightarrow \int_V d^2 \rho_j \, dZ \, N_{\text{im}}(\vec{\rho}_j, Z)(\cdots) = S \int dZ \, \mathcal{N}(Z)(\cdots),$$

where $\mathcal{N}(Z) \equiv \frac{1}{S} \int_S d^2 \rho_j \, N_{\text{im}}(\vec{\rho}_j, Z)$. Finally, Eq. (6.27) takes the form

$$|\langle m, \vec{k}' | \mathcal{H}_{\text{def}} | \vec{k}, n \rangle|^2 = S \int dZ \, \mathcal{N}(Z) |\langle m, \vec{k}' | V_{\text{def}}(\vec{\rho}, z - Z) | \vec{k}, n \rangle|^2. \tag{6.28}$$

It is convenient to rewrite this result in another form. The matrix element is proportional to the total impurity concentration per unit area of the two-dimensional electron gas, $N_{\text{def}}^{\text{2D}}$:

$$|\langle m, \vec{k}' | \mathcal{H}_{\text{def}} | \vec{k}, n \rangle|^2 = N_{\text{def}}^{\text{2D}} (|\langle m, \vec{k}' | V_{\text{def}} | \vec{k}, n \rangle|^2)_{\text{av}}, \tag{6.29}$$

where we introduce

$$(|\langle m, \vec{k}' | V_{\text{def}} | \vec{k}, n \rangle|^2)_{\text{av}}$$
$$\equiv \frac{S}{N_{\text{def}}^{\text{2D}}} \int dZ \, \mathcal{N}(Z) |\langle m, \vec{k}' | V_{\text{def}}(\vec{\rho}, z - Z) | \vec{k}, n \rangle|^2. \tag{6.30}$$

Thus we conclude that, for diluted impurities, the probability of scattering is proportional to the total impurity concentration.

It is worth pointing out that the results for both limiting cases allow us to calculate electron scattering by remote and nonuniformly distributed defects.

To specify $V_{\text{def}}(\vec{\rho})$ for a charged impurity, we need to analyze the screening effect in a two-dimensional electron gas.

6.3 Screening of a Two-Dimensional Electron Gas

We begin with the remark that the screening of an electric field by free carriers plays an important role in electron scattering and transport as well as in other phenomena in semiconductors. It is impossible to consider the interaction of an electron with other electrons, holes, ionized impurities, or electrostatic potentials of other origins without taking into account electron-screening effects. In heterostructures with low-dimensional electrons, screening effects differ strongly from those of the three-dimensional system. We consider a simple model of screening by two-dimensional electrons.

We assume that an external potential $\phi^{\text{ext}}(\vec{\rho}, z)$ is small and varies slowly in the quantum-well layer. This allows us to examine the extreme case in which the thickness of the quantum well reaches zero. Thus we consider the electrons to be confined inside an infinitely thin layer at $z = 0$. The actual electron concentration is denoted by $n(\vec{\rho})$. The potential induces a redistribution of the electrons $\delta n(\vec{\rho})$. To calculate this redistribution we use a semiclassical approximation and introduce a $\vec{\rho}$ dependence on the electron energy:

$$E_n(k, \vec{r}) = \epsilon_n + \frac{\hbar^2 k^2}{2m} - e\phi^{\text{tot}}(\vec{r}, 0), \tag{6.31}$$

where ϵ_n is the energy level in the quantum well, $\vec{k}$ is the two-dimensional wave vector describing the in-plane propagation of the electron, and $\phi^{\text{tot}}(\vec{\rho}, z)$ is the total potential. Then

$$\phi^{\text{tot}} = \phi^{\text{ext}} + \phi^{\text{ind}}, \tag{6.32}$$

which takes into account the redistribution of electrons and the consequent induction of changes in the electrostatic potential. Hence, it is clear that $\delta n(\vec{\rho})$ and $\phi^{\text{ind}}(\vec{\rho}, z)$, where ϕ^{ind} is the induced potential, should be found self-consistently.

Based on the Fermi distribution of Chapter 2, the two-dimensional electron concentration may be written as

$$n(\vec{\rho}) = \frac{2}{S} \sum_{n,k} \left(1 + \exp\left\{ \frac{1}{k_B T} \left[\epsilon_n + \frac{\hbar^2 k^2}{2m} - e\phi^{\text{tot}}(\vec{\rho}) - E_F \right] \right\} \right)^{-1}$$
$$\equiv n_e[E_F + e\phi^{\text{tot}}(\vec{\rho})].$$

Using the fact that the magnitude of the potential ϕ^{ext} is small compared with E_F, for the induced redistribution of the electrons we find

$$\delta n(\vec{\rho}) = e\phi^{\text{tot}} \frac{\partial n_e(E_F)}{\partial E_F}. \tag{6.33}$$

The redistribution predicted by Eq. (6.33) corresponds to a change of the electric charge by $-e\delta n(r)$. Taking into account that the electrons are confined within the thin layer at $z = 0$, we can write the space charge in three-dimensional form as $-e\delta n(\vec{\rho})\delta(z)$, where $\delta(z)$ is the Dirac delta function. It is clear that the induced potential should be determined from the three-dimensional Poisson equation:

$$\Delta\phi^{\text{ind}}(\vec{\rho}, z) = \frac{4\pi e}{\kappa} \delta n(\vec{\rho})\delta(z). \tag{6.34}$$

Note that all potentials, ϕ^{ext}, ϕ^{ind}, and ϕ^{tot}, are three-dimensional electrostatic potentials, i.e., they are extended in the space outside the plane $z = 0$ and are dependent on the three coordinates, $\vec{\rho}, z$.

Since Eq. (6.34) is linear, we can use a two-dimensional Fourier expansion for all variables:

$$\phi(\vec{\rho}, z) = \frac{1}{S}\sum_{\vec{q}}\phi(\vec{q}, z)e^{i\vec{q}\vec{\rho}}; \qquad \delta n(\vec{\rho}) = \frac{1}{S}\sum_{q}\delta n(\vec{q})e^{i\vec{q}\vec{\rho}}. \tag{6.35}$$

Then the Poisson equation (6.34) can be transformed to the form

$$\frac{d^2\phi^{\text{ind}}(\vec{q}, z)}{dz^2} - q^2\phi^{\text{ind}}(\vec{q}, z) = \frac{4\pi e}{\kappa}\delta n(\vec{q})\delta(z). \tag{6.36}$$

The solution of Eq. (6.36) can be found easily:

$$\phi^{\text{ind}}(\vec{q}, z) = -\frac{2\pi e}{\kappa q}\delta n(\vec{q})e^{-q|z|}, \tag{6.37}$$

where we have imposed boundary conditions that make the induced potential vanish far away from the plane of the quantum well with the electric charge.

To find the distribution of the potentials in the plane, let us use the general solution of the Poisson equation known from electromagnetic theory:

$$\phi^{\text{ind}}(\vec{\rho}, z) = -\frac{4\pi e}{\kappa}\int_S d^2\rho' dz' \frac{\delta n(\vec{\rho}')\delta(z')}{\sqrt{(\vec{\rho} - \vec{\rho}')^2 + (z - z')^2}}.$$

Therefore the potential in the plane may be written as

$$\phi^{\text{ind}}(\vec{\rho}, 0) = -\frac{4\pi e}{\kappa}\int_S d^2\rho' \frac{\delta n(\vec{\rho}')}{|\vec{\rho} - \vec{\rho}'|}.$$

The Fourier expansion of this expression and Eq. (6.33) allows us to obtain an explicit relationship between $\phi^{\text{ind}}(\vec{q}, z = 0)$ and $\phi^{\text{tot}}(\vec{q}, z = 0)$:

$$\phi^{\text{ind}}(\vec{q}) = -\frac{2\pi e^2}{\kappa q}\frac{\partial n_e(E_F)}{\partial E_F}\phi^{\text{tot}}(\vec{q}). \tag{6.38}$$

As a result, from Eqs. (6.32) and (6.38), we can readily obtain the relationship between external potential $\phi^{\text{ext}}(q)$ and total potential $\phi^{\text{tot}}(q)$:

$$\phi^{\text{tot}}(\vec{q}) = \frac{1}{1 + q_{\text{sc}}/q}\phi^{\text{ext}}(\vec{q}) = \frac{1}{\kappa_{\text{el}}(q)}\phi^{\text{ext}}(\vec{q}). \tag{6.39}$$

Here, we have introduced the characteristic wave vector

$$q_{\text{sc}} \equiv \frac{2\pi e^2}{\kappa}\frac{\partial n_e}{\partial E_F} \tag{6.40}$$

and the function

$$\kappa_{\text{el}}(q) \equiv 1 + \frac{q_{\text{sc}}}{q}. \tag{6.41}$$

Now we can write the induced potential as a function of z:

$$\phi^{\text{ind}}(\vec{q}, z) = -\frac{2\pi e^2}{\kappa} \frac{1}{q + q_{\text{sc}}} \frac{\partial n_e}{\partial E_F} \phi^{\text{ext}}(\vec{q}, z = 0) e^{-q|z|}. \tag{6.42}$$

We see that any Fourier component of the external field $\phi^{\text{ext}}(\vec{q}, z = 0)$ redistributes the two-dimensional electron gas, which, in turn, induces a three-dimensional electrostatic potential. The three-dimensional nature of the induced potential and its extension into the surrounding space can cause interactions between separated quantum wells and wires as well as other effects.

It is evident from Eq. (6.39) that the quantity $\kappa_{\text{el}}(q)$ is the dielectric permittivity of the two-dimensional electron gas. It depends on the wave vector $\vec{q}$. The quantity q_{sc} determines the range of wave vectors in which the electrons effectively screen the external fields: at $q \ll q_{\text{sc}}$, we find $\kappa_{\text{el}} \approx q_{\text{sc}}/q \gg 1$, i.e., $\kappa_{\text{el}}(q)$ increases as q decreases. This result implies that there is a suppression of the long-wavelength components of the field. In the low-temperature limit and for the case in which only the lowest subband is occupied, we find,

$$q_{\text{sc}} = \frac{2m^* e^2}{\kappa \hbar^2} = \frac{2}{a_0^*}, \tag{6.43}$$

where a_0^* is the bulk effective Bohr radius, as discussed in Chapter 3. Remember that it was shown that a_0^* is ~ 100 Å for AlGaAs/GaAs heterostructures, i.e., the screening length of a two-dimensional gas is of the order of a typical quantum-well width. In the limit $T \to 0$, q_{sc} is independent of the temperature and the electron concentration in the quantum well.

It is instructive to compare this result with the result known for the three-dimensional case, in which the screening is described by the dielectric constant:

$$\kappa_{\text{el}}^{\text{3D}}(q) = 1 + \left(\frac{q_D}{q}\right)^2,$$

where q_D^{-1} is the screening length. It can be written in a form analogous to that of Eq. (6.40):

$$q_D^2 = \frac{4\pi e^2}{\kappa} \frac{\partial n_{3D}}{\partial E_{F,3D}},$$

where the three-dimensional concentration of electrons n_{3D} should be used as a function of the Fermi level $E_{F,3D}$ (see Subsection 2.5.1). In general, this screening length depends on both the electron concentration and the temperature. In the long-wavelength limit, where $q \ll q_D$, we find that $\kappa_{\text{el}}^{\text{3D}}$ increases in proportion to q^{-2}. A comparison of two- and three-dimensional cases indicates that the screening is less effective for low-dimensional electron systems. The physical reason for this fact is evident. In the case of electron confinement, the electrons are affected by only the potential within the confining layer. In contrast, for the bulk case, the electrons are perturbed in any space region and their redistribution leads to the more effective weakening of the external potential. This trend of decreased screening becomes even more pronounced in a one-dimensional electron gas.

To conclude this brief discussion on the screening by two-dimensional electron systems, let us note that the above results were obtained through semiclassical considerations. A quantum approach is required for correcting the results in the short-wavelength region. For example, in the case of only one occupied subband and zero temperature, instead of Eq. (6.41), we can obtain

$$\kappa_{\text{el}}(q) = 1 + \frac{q_{\text{sc}}}{q}\alpha(q), \tag{6.44}$$

with

$$\alpha(q) = 1 - \theta(q - 2k_F)\sqrt{1 - \left(\frac{2k_F}{q}\right)^2},$$

where q_{sc} is also determined by Eq. (6.40) and k_F is the Fermi wave vector as defined by $\hbar^2 k_F^2/2m^* = E_F$. The additional factor α in Eq. (6.44) plays a role for $q > 2k_F$, where it is always less than 1. The physical origin of this value is the reduction of the polarizability of a two-dimensional electron gas in the short-wavelength region.

6.4 Scattering by Remote Ionized Impurities

As analyzed in Chapter 4, electron scattering by residual impurities is weak in quantum structures because, in practice, their concentration in the conduction channels of quantum wells and quantum wires is small: typically, it does not exceed 10^{15} cm^{-3}. In selectively doped structures, which we have studied in Chapter 4, the main scattering process is due to remote impurities. The theory of this process is quite complicated. The complications come from the spatially nonhomogeneous character of the systems, the screening effect, and other factors. Here we consider the key physical aspects essential for understanding the suppression of this type of scattering.

To evaluate the scattering probabilities and characteristic scattering times τ_s and τ_p, it is necessary to determine the scattering potential $\mathcal{H}_{\text{def}}$. When we studied the electron states in modulation-doped semiconductors in Chapter 4, we represented the electrostatic potential of the remote ionized impurities as a sum of two contributions. The first is the average potential that, together with band discontinuities, forms the quantum well. The second is a fluctuating part of the potential that breaks translation symmetry parallel to the heterojunctions of the well. It is evident that only the fluctuating part of the potential causes scattering of the electrons.

In this context, the scattering causes a change in the electron momentum. According to Eq. (6.8), only the matrix elements corresponding to $\vec{q} \neq 0$ contribute to the time τ_p. That means that terms related to the average potential, which do not change momentum, vanish automatically in the formula for τ_p. Let $\vec{\rho}_i$ and Z_i

be coordinates of the ith ionized donor; then the electrostatic potential generated by all impurities at the point $(\vec{\rho}, z)$ (the external electrostatic potential) is

$$\phi(\vec{\rho}, z) = -\frac{2\pi e}{\kappa S} \sum_{\vec{\rho}_i, Z_i} \sum_{\vec{q}} \frac{1}{q} e^{-q|z - Z_i|} e^{i\vec{q}(\vec{\rho} - \vec{\rho}_i)}.$$

Here we use a two-dimensional Fourier transformation of the Coulomb potential. Let us transform the expression for an external potential ϕ^{ext} to the form

$$\phi^{\text{ext}}(\vec{\rho}, z) = \frac{1}{S} \sum_{\vec{q}} \phi^{\text{ext}}(\vec{q}, z) e^{i\vec{q}\vec{\rho}}.$$

In this formula, $\phi^{\text{ext}}(\vec{q}, z)$ is the Fourier component of an external potential in some plane z. Consider the case in which there are transitions within only the lowest subband, $n = 1$. According to Eq. (6.14), one has to calculate a diagonal matrix element of the defect potential. In other words, we have to introduce the external potential averaged over the quantum well:

$$\phi^{\text{ext}}(\vec{\rho}) = \int dz\, \chi_1^2(z) \phi^{\text{ext}}(\vec{\rho}, z) = \frac{1}{S} \sum_{\vec{q}} \int dz\, \chi_1^2(z) \phi^{\text{ext}}(\vec{q}, z) e^{i\vec{q}\vec{\rho}}. \tag{6.45}$$

In Eq. (6.45) we may interpret the quantity

$$\int dz\, \chi_1^2(z) \phi^{\text{ext}}(\vec{q}, z) \tag{6.46}$$

as a Fourier component of the external potential affecting two-dimensional electrons.

Our previous analysis of the screening effect revealed that it is necessary to divide each of these Fourier components by the dielectric permittivity of the two-dimensional electron gas, $\kappa_{\text{el}}(q)$. Then we obtain the total potential:

$$-e\phi^{\text{tot}}(\vec{\rho}) = -\frac{1}{S} \sum_{\vec{q}} \frac{1}{\kappa_{\text{el}}(q)} \int dz\, \chi_1^2(z) \phi^{\text{ext}}(\vec{q}, z) e^{i\vec{q}\vec{\rho}}. \tag{6.47}$$

Calculation of the matrix element of this potential finally gives

$$\langle 1, \vec{k} \,|\, \mathcal{H}_{\text{def}} \,|\, 1, \vec{k} + \vec{q} \rangle = -\frac{2\pi e^2}{\kappa S} \frac{1}{q + q_{\text{sc}}} \sum_{i} e^{-i\vec{q}\vec{\rho}_i} F(q, Z_i), \tag{6.48}$$

where we have introduced the form factor $F(q, z)$ through the definition

$$F(q, Z) = \int dz'\, \chi_1^2(z') e^{-q|Z - z'|}. \tag{6.49}$$

For low temperatures, when impurity scattering is dominant, it is possible to identify mechanisms for suppressing remote impurity scattering even at this stage of the calculations. Actually, in this case only the electrons with the energy equal to the Fermi energy contribute to the transport phenomena. According to Eq. (6.48), $\vec{q}$ represents the change in the momentum during a scattering event. It is convenient to define the angle θ of elastic scattering of an electron with the Fermi energy as

$$q = 2k_F \sin \frac{\theta}{2}; \tag{6.50}$$

thus $F(q, z) = (k_F, \theta, z)$. Now we can see that the form factor of Eq. (6.49) decreases exponentially as the distance from impurities increases. Then, for selectively doped heterostructures, typically $k_F Z \gg 1$, which means that only small-angle scattering occurs. However, this type of scattering is much less effective for causing relaxation of the electron momentum.

To present the momentum-scattering time given by Eq. (6.11) in a simpler form, let us assume that the impurities are located in a thin doped layer at $z = Z_0$, that is, we assume δ doping. Furthermore, we assume that the impurities are dilute enough so that each ion scatters electrons independently. Thus we can use the result of Eq. (6.28). Let a random impurity distribution in this δ layer be characterized by the area impurity concentration N_D^{2D} so that $\mathcal{N}(Z) = N_D^{2D} \delta(Z - Z_0)$. The electrons are taken to be confined by a heterostructure potential in so thin a quantum well that we can replace $\chi_1^2(z)$ with $\delta(z)$ in Eq. (6.49). As used in Subsection 6.2.1, instead of summing over R_i, Z_i, we can integrate over impurity coordinates. As a result of simple algebra we find

$$\frac{1}{\tau_{p,\text{im}}} = \frac{\pi m^* N_D^{2D}}{\hbar^3} \left(\frac{2e^2}{\kappa}\right)^2 \int_0^\pi \frac{d\theta(1 - \cos\theta)}{(q_{\text{sc}} + 2k_F \sin\frac{\theta}{2})^2} \exp\left(-4k_F|Z_0|\sin\frac{\theta}{2}\right). \tag{6.51}$$

Based on this relaxation time, it is easy to estimate the electron mobility associated with scattering by remote impurities through application of the approximate relationship discussed in Chapter 2: $\mu_{\text{im}} = e\tau_{p,\text{im}}/m^*$. Because $k_F|Z_0| \gg 1$, we can replace $\sin(\theta/2)$ by $\theta/2$ and $1 - \cos\theta$ by $\theta^2/2$; then, under the condition that $q_{\text{sc}} \sim k_F$, it results in

$$\mu = 16|Z_0|^3 \frac{e}{\hbar} \sqrt{2\pi n},$$

where we assume that the density of ionized impurities N_D^{2D} equals the electron density n and $k_F = \sqrt{2\pi n}$. We see that the mobility increases as the third power of the distance to the doped layer. As a result of screening effects, the dependence of μ on N_D^{2D} is weaker than a linear dependence. For the purpose of illustrating a numerical estimate, let us choose $|Z_0| = 150$ Å and $n_s = 10^{12}$ cm^{-2}. Then we obtain $\mu_{\text{im}} \approx 2 \times 10^5$ cm^2/V s, which is a reasonable value for a low-temperature mobility. The results of more rigorous calculations of low-field mobilities will be discussed in Chapter 7.

Let us turn our attention once more to the form factor of Eq. (6.49). Under the integral, there is a product of two functions: One function decays exponentially far away from the impurity, while another, $\chi_1^2(z)$, decreases steeply outside the electron channel. Their product and the whole integral are sensitive to the shape of the wave function in the region outside the quantum well. By controlling this shape, we can change the mobility considerably. This method of controlling electron mobility can be used in so-called velocity-modulation transistors. We will analyze this type of transistor in Chapter 9.

6.5 Scattering by Interface Roughness

Another scattering mechanism important for quantum heterostructures is scattering by interface roughness. Interface scattering has a long history of investigation, primarily in metal microstructures such as metal films, thin wires, etc. The simplest example of this mechanism was introduced in the forties by Fuchs, who assumed that there is some probability P of scattering the electron momentum because of an interaction with the surface, so that the electron forgets its initial momentum completely. Hence P is the probability of diffusive electron scattering. Although the principles of electron scattering at the surfaces have been developed at a fundamental level on the basis of quantum mechanics, they are related for the most part to classical electron motion.

To explain electron scattering by interface roughness in a low-dimensional system, let us consider a quantum well with slightly variable positions of the walls, as in Fig. 6.1. We assume that this interface roughness occurs in isolated regions of the quantum-well layer. Thus the electron motion in the well can be thought of as almost a free motion of low-dimensional electrons, which seldom are scattered by the roughness region. Let $z_l(\vec{\rho} - \vec{\rho}_j) \equiv z_{l,j}(\vec{\rho})$ and $z_r(\vec{\rho} - \vec{\rho}_j) \equiv z_{r,j}(\vec{\rho})$ be the positions of the left and the right walls, respectively, for the roughness region with its center at $\vec{\rho}_j$. The wall positions depend on in-plane vector coordinates $\vec{\rho}$. Since any roughness region is isolated far away from this region, we get

$$z_{l,j} \to -\frac{L}{2}, \qquad z_{r,j} \to \frac{L}{2} \quad \text{for } |\vec{\rho} - \vec{\rho}_j| \to \infty. \tag{6.52}$$

Figure 6.1. Schematics of a two-dimensional electron channel with a variable width.

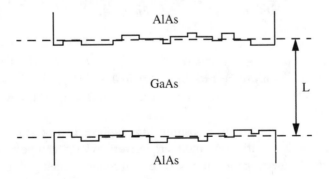

The well is supposed to be infinitely deep. Thus we should solve the general Schrödinger equation

$$-\frac{\hbar^2}{2m^*}\left(\vec{\nabla}_{\vec{\rho}}^2 + \frac{\partial^2}{\partial z^2}\right)\psi(\vec{\rho}, z) = E\psi(\vec{\rho}, z) \qquad (6.53)$$

for $z_{l,j}(\vec{\rho}) \leq z \leq z_{r,j}(\vec{\rho})$ with the boundary conditions

$$\psi[\vec{\rho}, z_{l,j}(\vec{\rho})] = \psi[\vec{\rho}, z_{r,j}(\vec{\rho})] = 0. \qquad (6.54)$$

As a result of these boundary conditions, this problem is different from the simplest case with ideal plane walls of an infinitely deep quantum well analyzed in Section 3.2.

The problem of electron scattering by such an isolated region with the interface roughness can be analyzed as follows. We introduce a new system of coordinates ζ, $\vec{\rho}$ by substituting ζ for z:

$$\zeta = \zeta(\vec{\rho}, z) = -\frac{L}{2} + L\frac{z - z_{l,j}(\vec{\rho})}{L + d_j(\vec{\rho})},$$

where $d_j(\vec{\rho}) + L = z_{r,j}(\vec{\rho}) - z_{l,j}(\vec{\rho})$ is a local width of the well. In these new coordinates the boundary conditions of Eq. (6.54) take the simple form:

$$\psi\left(\vec{\rho}, \zeta = -\frac{L}{2}\right) = \psi\left(\vec{\rho}, \zeta = \frac{L}{2}\right) = 0, \qquad (6.55)$$

while Eq. (6.53) can be transformed to

$$-\frac{\hbar^2}{2m^*}\left(\vec{\nabla}_{\vec{\rho}}^2 + \frac{\partial^2}{\partial\zeta^2}\right)\psi(\vec{\rho}, \zeta) + \hat{V}_{\text{def},j}\psi(\vec{\rho}, \zeta) = E\psi(\vec{\rho}, \zeta), \qquad (6.56)$$

where we have introduced the operator of an effective defect potential in the form

$$\hat{V}_{\text{def},j}(\vec{\rho}, \zeta) = \frac{\hbar^2}{2m^*}\left(2\frac{d_j}{L}\frac{\partial^2}{\partial\zeta^2}\right) + \frac{\hbar^2}{2m^*}\left[2\zeta\frac{(d_j)'_{\vec{\rho}}}{L}\frac{\partial^2}{\partial\vec{\rho}\partial\zeta} + \zeta\frac{(d_j)''_{\vec{\rho}^2}}{L}\frac{\partial}{\partial\zeta}\right]$$
$$+ \frac{\hbar^2}{2m^*}\left\{[(z_{l,j})'_{\vec{\rho}} + (z_{r,j})'_{\vec{\rho}}]\frac{\partial^2}{\partial\vec{\rho}\partial\zeta} + \frac{1}{2}[(z_{l,j})''_{\vec{\rho}^2} + (z_{r,j})''_{\vec{\rho}^2}]\frac{\partial}{\partial\zeta}\right\}. \qquad (6.57)$$

Here $(z_{l,j})'_{\vec{\rho}}$, $(z_{r,j})'_{\vec{\rho}}$, $(d_j)'_{\vec{\rho}}$ and $(z_{l,j})''_{\vec{\rho}^2}$, $(z_{r,j})''_{\vec{\rho}^2}$, $(d_j)''_{\vec{\rho}^2}$ are the first and the second derivatives with respect to the vector $\vec{\rho}$: $\partial/\partial\vec{\rho} \equiv \vec{\nabla}_{\vec{\rho}}$, $\partial^2/\partial\vec{\rho}^2 \equiv \vec{\nabla}_{\vec{\rho}}^2$. In Eq. (6.57) we restrict ourselves to terms linear with respect to deviations of the positions of the walls.

Since variations of the wall positions are supposed to be small, the terms proportional to these variations and their derivatives $\hat{V}_{\text{def},j}$ can be considered as perturbations. Neglecting these terms, we get an equation identical to Eq. (6.53).

Then, taking into account the boundary conditions of Eq. (6.55), we see that in such a zero-order approximation, the solutions $\psi_{n,\vec{k}}^{(0)}(\vec{\rho}, \zeta)$ are given by Eq. (6.16) with the replacement $z \to \zeta$, that is, the wave functions have a more complex form than that of a free electron propagating along the quantum-well layer. However, according to expressions (6.52), for $|\vec{\rho} - \vec{\rho}_j| \to \infty$ the wave functions almost coincide with Eq. (6.16) and consequently the electron states can be characterized by the wave vector and the subband number.

The operator $\hat{V}_{\mathrm{def},j}$ induces scattering between these states. We shall calculate matrix elements (6.17) with the operator of Eq. (6.57). Different terms of this operator are responsible for different types of scattering. The first term proportional to $\partial^2/\partial\zeta^2$ leads to only intrasubband transitions:

$$\langle n, \vec{k}' | \hat{V}_{\mathrm{def},j} | n, \vec{k} \rangle = -\epsilon_n \frac{2}{S} \int_S d\vec{\rho} \frac{d_j(\vec{\rho})}{L} e^{i(\vec{k}-\vec{k}')\vec{\rho}}, \tag{6.58}$$

while the second and the third terms lead to intersubband transitions:

$$\langle m, \vec{k}' | \hat{V}_{\mathrm{def},j} | n, \vec{k} \rangle = -(\epsilon_n - \epsilon_m) \frac{1}{S} \int_S d\vec{\rho} \frac{d_j(\vec{\rho})}{L} e^{i(\vec{k}-\vec{k}')\vec{\rho}} \int_{-L/2}^{L/2} d\zeta\, \zeta\, \chi_m(\zeta) \frac{d\chi_n(\zeta)}{d\zeta}$$

$$- (\epsilon_n - \epsilon_m) \frac{1}{S} \int_S d\vec{\rho} \frac{z_{l,j}(\vec{\rho}) + z_{r,j}(\vec{\rho})}{2} e^{i(\vec{k}-\vec{k}')\vec{\rho}}$$

$$\times \int_{-L/2}^{L/2} d_j \zeta\, \chi_m(\zeta) \frac{d\chi_n(\zeta)}{d\zeta}. \tag{6.59}$$

Integrals over ζ in Eq. (6.59) give additional selection rules: The first contribution is not zero for transitions between subbands with the same parity (even $\leftrightarrow$ even, odd $\leftrightarrow$ odd transitions), and the last contribution corresponds to transitions between subbands with different parity (even $\leftrightarrow$ odd transitions). These results can be interpreted as follows. The first contribution to Eq. (6.59) is proportional to variations of the well width d_j. Such variations do not change the initial symmetry of subbands and parity-changing transitions are forbidden. The last contribution is proportional to a flexural deformation of the quantum-well layer. (Actually, $z_{l,j} + z_{r,j}$ describes a shift of the center of the well.) This type of variation does change the symmetry, and transitions are possible between subbands with different parity.

To find the total probability of the scattering, one should account for all roughness regions similarly to those of Eq. (6.17) and the screening effect studied in Section 6.3:

$$\langle m, \vec{k}' | \hat{\mathcal{H}}_{\mathrm{def}} | n, \vec{k} \rangle = \sum_j e^{-i(\vec{k}-\vec{k}')\vec{\rho}_j} \frac{\langle m, \vec{k}' | \hat{V}_{\mathrm{def},j} | n, \vec{k} \rangle}{\kappa_{\mathrm{el}}(\vec{k} - \vec{k}')}, \tag{6.60}$$

where we divide the matrix element by the dielectric permittivity of the two-dimensional electron gas, $\kappa_{\mathrm{el}}(\vec{k} - \vec{k}')$, given by Eq. (6.41).

Since the roughness regions are assumed to be isolated, further calculations are similar to those for the case of scattering by dilute impurities studied in

Section 6.4. For the square of the absolute value of matrix element (6.60) we get

$$|\langle m, \vec{k}' | \hat{\mathcal{H}}_{\text{def}} | n, \vec{k}\rangle|^2 = \sum_j \frac{|\langle m, \vec{k}' | \hat{V}_{\text{def},j} | n, \vec{k}\rangle|^2}{\kappa_{\text{el}}^2}. \tag{6.61}$$

For the sake of simplicity, we present results for intrasubband scattering. Substituting Eq. (6.58) into Eq. (6.61), we find

$$|\langle n, \vec{k}' | \hat{\mathcal{H}}_{\text{def}} | n, \vec{k}\rangle|^2 = \epsilon_n^2 \frac{1}{S^2} \sum_j \frac{4}{\kappa_{\text{el}}^2} \int_S \int_S d\vec{\rho} \, d\vec{\rho}' \frac{d_j(\vec{\rho})}{L} \frac{d_j(\vec{\rho}')}{L} e^{i(\vec{k}-\vec{k}')(\vec{\rho}-\vec{\rho}')}. \tag{6.62}$$

Now we can define the correlator of fluctuations of the well width as

$$\mathcal{K}_d(\vec{\rho}) = \frac{1}{S} \sum_j \int_S d\vec{\rho}' \frac{d_j(\vec{\rho}')}{L} \frac{d_j(\vec{\rho}' + \vec{\rho})}{L}.$$

Obviously this result is averaged over all roughness-region spatial correlations in the well width. Frequently, for this correlation function, it is possible to assume the Gaussian form,

$$\mathcal{K}_d(\vec{\rho}) = \frac{D^2}{L^2} e^{-\rho^2/\Lambda^2},$$

where the parameter D characterizes the average fluctuation in the well width and Λ is the correlation length describing the spatial correlations of the roughness. Now we can rewrite Eq. (6.62) as

$$|\langle n, \vec{k}' | \hat{\mathcal{H}}_{\text{def}} | n, \vec{k}\rangle|^2 = \epsilon_n^2 \frac{4}{S} \frac{\mathcal{K}_d(\vec{k} - \vec{k}')}{\kappa_{\text{el}}^2(\vec{k} - \vec{k}')},$$

where we use the Fourier transform of the correlator $\mathcal{K}_d(\vec{k} - \vec{k}')$ as defined by Eq. (6.25): $\mathcal{K}_d(\vec{q}) = (D^2 \Lambda^2 \pi / L^2) e^{-(\Lambda q)^2/4}$.

Finally, according to Eq. (6.8) for electrons with the Fermi energy, the momentum-relaxation time takes the form

$$\frac{1}{\tau_{p,n}} = \frac{4\pi m^* D^2 \Lambda^2}{\hbar^3 L^2} \epsilon_n^2 \mathcal{I}\left(\Lambda k_F, \frac{k_F}{q_{\text{sc}}}\right), \tag{6.63}$$

where we define the function

$$\mathcal{I}\left(\Lambda k_F, \frac{k_F}{q_{\text{sc}}}\right) = \frac{1}{2\pi} \int_0^{2\pi} d\theta \frac{(1 - \cos\theta)}{[1 + q_{\text{sc}}/2k_F \, \sin(\theta/2)]^2} \exp\left(-\Lambda^2 k_F^2 \sin^2\theta/2\right).$$

The latter integral depends on a single parameter characterizing the roughness, Λk_F. It is easy to see that this type of scattering occurs if the correlation length is small, i.e., $\Lambda k_F < 1$. In this limit the scattering is almost isotropic and the integral is approximately equal to 2π. The momentum-relaxation time in the lowest subband is

$$\frac{1}{\tau_{p,1}} = \frac{2\pi^3 D^2 \Lambda^2 \epsilon_1}{L^4 \hbar} \, \mathcal{I}\left(\Lambda k_F, \frac{k_F}{q_{\rm sc}}\right), \tag{6.64}$$

where we have used Eq. (3.15) for ϵ_1 in an infinitely deep well. In the limit, $\Lambda k_F > 1$, the exponential factor in integral (6.63) suppresses the scattering.

The form of Eq. (6.64) is convenient for numerical estimates. Let us set $L = 125$ Å, $m^* = 0.067$ m, $D/L = 0.03$, and $\Lambda/L = 0.2$. If the electron concentration is 10^{12} cm^{-2}, the Fermi wave vector is $k_F = 2.5 \times 10^6$ cm^{-1} and $\Lambda k_F \approx 0.39$. Using Eq. (6.43), we estimate $q_{\rm sc}/k_F \approx 0.8$ and $\mathcal{I} = 1.86$. Finally we get $\tau_{p,1} = 1.08 \times 10^{-12}$ s. This scattering time leads to a mobility of $\mu = \tau_{p,1}/m^* \approx 1.8 \times 10^5$ cm^2/V s. This estimate shows that interface roughness scattering can dominate over other scattering mechanisms at low temperatures.

6.6 Electron–Phonon Interaction

In principle, we can eliminate electron scattering by crystal defects, impurities, and imperfections by using selective doping and making the crystal perfect. However, the scattering mechanism that is inherent in the nature of a crystal and cannot be eliminated is electron scattering by lattice vibrations – phonons. Different kinds of phonons were studied in Chapter 5.

Electron–phonon scattering is crucially important in systems of any dimensionality. However, in low-dimensional structures with perfect interfaces and impurities separated by thick spacers, electron–phonon scattering may become the only scattering mechanism. Moreover, as we have seen in Chapter 5, not only the properties of electrons, but also those of phonons, are modified substantially in nanoscale structures. Therefore, for adequate treatment of electron kinetics in these structures, we must take into account electron–phonon interaction with modified phonons.

6.6.1 Transitions due to Electron–Phonon Interactions

The physical reason for the electron–phonon interaction can be understood as follows. In Section 2.6, we defined the electron states as states in an ideal periodic crystalline potential $W(\vec{r})$. These states are characterized by discrete number(s) α and a wave vector $\vec{k}$. Lattice vibrations modify the crystalline potential and break the translation invariance of the system. This disturbance in the crystalline potential δW can be identified with the electron–phonon interaction $H_{\rm e-ph}$. If this interaction is relatively weak, it results in transitions between different states of the crystal. The states shall be chosen as states of noninteractive electron and phonon subsystems. Let electron and phonon wave functions be $\psi_{\alpha,\vec{k}}$ and Φ, respectively. Then the total wave function of the crystal is $\Phi\psi_{\alpha,\vec{k}}$. The electron–phonon interaction leads to transitions

$$\Phi\psi_{\alpha,\vec{k}} \to \Phi'\psi_{\alpha',\vec{k}'} .\qquad(6.65)$$

We emphasize that electron–phonon interactions cause changes in states of both electrons and phonons.

If the interaction is strong, it can lead to new effects like the formation of a new quasi-particle – the polaron – for a single electron and structural phase transitions in multielectron systems, etc. Below, we concentrate on electron–phonon scattering effects as the most important for electron transport and optical properties as well as for their application.

To calculate transition probabilities due to electron–phonon interactions, we can apply Fermi's golden rule:

$$W(\alpha, \vec{k}, \alpha', \vec{k}' \,|\, \vec{q}) = \frac{2\pi}{\hbar} |\langle \Phi'\psi_{\alpha',k'} \,|\, H_{e-ph} \,|\, \Phi\psi_{\alpha,k}\rangle|^2 \,\delta[E_\alpha(k) - E_{\alpha'}(k') \mp \hbar\omega(\vec{q})],$$

$$(6.66)$$

where $\hbar\omega(q)$ is the phonon energy in a given mode. If $\{\alpha, \vec{k}\}$ is an initial electron state and $\{\alpha', \vec{k}'\}$ is a final electron state of a transition, the upper sign in Eq. (6.66) corresponds to emission and the lower sign to absorption of the phonon. Thus initial Φ and final Φ' states of the phonon subsystem differ by the phonon numbers. The Dirac delta function provides the energy-conservation law in the crystal:

$$E_{\alpha'}(k') - E_\alpha(k) = \mp\hbar\omega(\vec{q}) .\qquad(6.67)$$

Equation (6.66) is written for the electron–phonon interaction in a uniform bulklike material. The results need to be generalized for heterostructures. For the latter systems, we used the effective-mass approximation. Thus we shall adapt Eq. (6.66) to this approach. As discussed in Section 2.6, the effective-mass approximation is valid for slowly varying external potentials. For the electron–phonon interaction, the scale of the potential disturbance is obviously characterized by the inverse phonon wave vector q^{-1}. This means that, when analyzing electron–phonon scattering between electron states found in the effective-mass approximation, we should restrict ourselves to the consideration of long-wavelength phonons: $qa_0 \ll 1$, where a_0 is the lattice constant.

Now we can generalize our analysis for studying the transitions in quantum structures. Suppose the transitions are to be within the same electron band α. According to Eq. (2.158), general expressions of electron wave functions of initial and final states can be presented as

$$\psi_\beta = F_\beta u_{\alpha,\vec{k}_0}, \qquad \psi_{\beta'} = F_{\beta'} u_{\alpha,\vec{k}_0},\qquad(6.68)$$

where $u_{\alpha,\vec{k}_0}$ is the Bloch function corresponding to an energy minimum at $\vec{k} = \vec{k}_0$ in the band α, F_β, and $F_{\beta'}$ are envelope functions, and β and β' are sets of quantum numbers. For example, in the case of a quantum well, β contains a subband number and a two-dimensional wave vector, etc. In terms of these wave functions, we have to consider transitions

$$\Phi F_\beta u_{\alpha,\vec{k}_0} \rightarrow \Phi' F_{\beta'} u_{\alpha,\vec{k}_0}. \tag{6.69}$$

Following the conventional procedure, we obtain a generalization of expression (6.66) in which the transition probabilities have the form

$$W(\beta, \beta' \mid \vec{q}) = \frac{2\pi}{\hbar} |\langle \Phi' F_{\beta'} u_{\alpha,\vec{k}_0} \mid H_{\text{e-ph}} \mid \Phi F_\beta u_{\alpha,\vec{k}_0}\rangle|^2 \delta[E_\beta - E_{\beta'} \mp \hbar\omega(\vec{q})] \tag{6.70}$$

and the energy-conservation law has the form

$$E_{\beta'} - E_\beta = \mp\hbar\omega(\vec{q}). \tag{6.71}$$

6.6.2 Short-Range and Long-Range Electron–Phonon Interactions

Long-wavelength phonons, which are now the focus of our analysis, affect the electrons in two different ways. First, a short-range disturbance in the crystalline potential can be described as a local change in the electron potential by the so-called deformation potentials, which can exist for different vibrational modes. Second, a lattice strain can break the local electrical neutrality and produce an electrical polarization of the crystal. This leads to a long-range electric field, which in turn disturbs the electron motion. Such macroscopic fields induce piezoelectric electron–phonon interactions associated with acoustic modes and polar electron–phonon interactions associated with optical modes.

To describe local and long-range electron–phonon interactions one can present the disturbance of the crystalline potential $\delta W(\vec{r})$ as

$$H_{\text{e-ph}} = \delta W(\vec{r}) = \overline{\delta W}(\vec{r}) + \widetilde{\delta W}(\vec{r}), \tag{6.72}$$

where $\overline{\delta W}(\vec{r})$ is an average over a small crystal region with dimensions larger than a_0 (i.e., a long-range contribution) and $\widetilde{\delta W}(\vec{r})$ is a local potential. Throughout the region of averaging, $\overline{\delta W}$ is almost constant. The local contribution $\widetilde{\delta W}(\vec{r})$ is approximately a periodic potential. The latter means that $\widetilde{\delta W}$ leads to a local change in the dispersion $E(\vec{k})$. In general, this change depends on the coordinate $\vec{r}$.

Using the wave functions in the general form of Eqs. (6.68), we represent the electron matrix element as

$$\langle \psi_{\beta'} \mid \delta W \mid \psi_\beta \rangle = \langle \psi_{\beta'} \mid \overline{\delta W} \mid \psi_\beta \rangle + \langle \psi_{\beta'} \mid \widetilde{\delta W} \mid \psi_\beta \rangle,$$

with

$$\langle \psi_{\beta'} \mid \overline{\delta W} \mid \psi_\beta \rangle = V_0 \sum_n F_{\beta'}^*(\vec{a}_n) F_\beta(\vec{a}_n) \frac{1}{V_0} \int_{V_0} d^3r \, |u_{\alpha,\vec{k}_0}(\vec{r})|^2 \overline{\delta W}(\vec{r}), \tag{6.73}$$

$$\langle \psi_{\beta'} \mid \widetilde{\delta W} \mid \psi_\beta \rangle = V_0 \sum_n F_{\beta'}^*(\vec{a}_n) F_\beta(\vec{a}_n) \frac{1}{V_0} \int_{V_0} d^3r \, |u_{\alpha,\vec{k}_0}(\vec{r})|^2 \widetilde{\delta W}(\vec{r}). \tag{6.74}$$

Here n runs over all crystal cells and V_0 is the volume of the primitive cell. Since the long-range contribution is constant across the cell, in the first formula we can use

$$\frac{1}{V_0} \int_{V_0} d^3 r \, |u_{\alpha, \vec{k}_0}|^2 \overline{\delta W}(\vec{r}) \approx \overline{\delta W}(\vec{a}_n).$$

Replacing the summation by an integration in Eq. (6.73), we get the matrix element of the long-range contribution in the form

$$\langle \psi_{\beta'} | \overline{\delta W} | \psi_\beta \rangle = \int d^3 r \, F_{\beta'}^*(\vec{r}) \overline{\delta W}(\vec{r}) F_\beta(\vec{r}), \tag{6.75}$$

that is, the matrix element of this type of interaction may be calculated directly from the envelope functions. From Eq. (6.75) it follows that $\overline{\delta W}(\vec{r})$ is interpreted as a macroscopic field that is inducing the transitions. In the following discussion, we find explicit relationships between $\overline{\delta W}(\vec{r})$ and the atomic displacements for different phonon modes.

As for the integral in Eq. (6.74), for long-wavelength phonons, we can expect this integral to be a smooth function of the position of the cell a_n:

$$\frac{1}{V_0} \int_{V_0} d^3 r \, |u_{\alpha, \vec{k}_0}(\vec{r})|^2 \widetilde{\delta W}(\vec{r}) \equiv D(\vec{a}_n).$$

Here we have introduced the deformation potential $D(\vec{r})$. Then the matrix element of the local contribution is

$$\langle \psi_{\beta'} | \widetilde{\delta W} | \psi_\beta \rangle = \int d^3 r \, F_{\beta'}^*(\vec{r}) D(\vec{r}) F_\beta(\vec{r}).$$

Hence, if we use electron envelope functions, the Hamiltonian of the short-range electron–phonon interaction should be written as

$$H_{\text{e-ph}}(\text{short-range}) = D(\vec{r}). \tag{6.76}$$

In the subsections below we determine the deformation potential for different types of phonon modes.

6.6.3 The Interaction with Different Phonon Modes

As defined above, the Hamiltonian $H_{\text{e-ph}}$ depends on the displacements of the atoms (ions) comprising the crystal from their equilibrium positions. To describe these displacements, in Section 5.3 we introduced different vibrational modes: 3 acoustic and $3\mathcal{G} - 3$ optical modes; $\mathcal{G}$ is the number of atoms per primitive crystal cell. Let Q_s characterize the displacement(s) associated with a given mode s; the total number of modes is equal to $3\mathcal{G}\mathcal{N}$, where $\mathcal{N}$ is the number of primitive cells in the crystal. The Hamiltonian $H_{\text{e-ph}}$ is expanded in a series with respect to all the displacements Q_s. For most effects, we can restrict ourselves to the first linear terms in such an expansion. Each term proportional to Q_s – the linear strain – can be interpreted in terms of the electron interaction with this mode s. Thus we can think about electron interactions with LA, TA, LO, and TO modes:

$$H_{\text{e-ph}} = \sum_s H_{\text{e-ph},s}. \tag{6.77}$$

These physical considerations allow us to find the electron–phonon interaction with a given mode as follows. The crystalline potential $W(\vec{r})$ can be presented as the sum of the contributions from all atoms (ions):

$$W(\vec{r}) = \sum_{j,n} V_j(\vec{r} - \vec{a}_{n,j}),$$

where $V_j(\vec{r} - \vec{a}_{n,j})$ is the potential created by a jth atom in the nth primitive cell with the position $\vec{a}_{n,j}$. Let $\vec{u}_{j,n}(t)$ be a displacement of this atom that is dependent, generally, on time t. A disturbance in the crystalline potential is

$$H_{\text{e-ph}} \equiv \delta W(\vec{r}) = -\sum_{j,n} \frac{\partial V_j}{\partial \vec{r}} \vec{u}_{j,n}(t) = \sum_{j,n} \vec{\mathcal{V}}_{j,n}(\vec{r})\vec{u}_{j,n}(t). \tag{6.78}$$

Obviously, $\vec{\mathcal{V}}_{j,n}(\vec{r}) = \partial V_j(\vec{r} - \vec{a}_{n,j})/\partial \vec{r}$ has three components that determine changes in $W(\vec{r})$ at various displacements along the coordinate axes. The displacement $\vec{u}_{j,n}$ can be expressed in terms of the normal coordinates $Q_{\vec{q},s}$ introduced in Section 5.3:

$$\vec{u}_{j,n} = \frac{1}{\sqrt{\mathcal{N}}} \sum_{\vec{q},s} \vec{b}_s(\vec{q}) Q_{\vec{q},s} \exp(i\vec{q}\vec{a}_{n,j}).$$

Here, $\vec{b}_s(\vec{q})$ is a unit vector that defines the polarization of the wave branch characterized by the index s. Now we can rewrite Eq. (6.78) as

$$
\begin{aligned}
H_{\text{e-ph}} &= \frac{1}{\sqrt{\mathcal{N}}} \sum_s \sum_{\vec{q}} \left[\sum_{j,n} \vec{\mathcal{V}}_{j,n}(\vec{r}) e^{i\vec{q}(\vec{a}_{n,j} - \vec{r})} \right] \vec{b}_s(\vec{q}) Q_{\vec{q},s} \exp(i\vec{q}\vec{r}) \\
&\equiv \frac{1}{\sqrt{\mathcal{N}}} \sum_s \sum_{\vec{q}} \bar{w}_{\vec{q},s}(\vec{r}) \vec{b}_s(\vec{q}) Q_{\vec{q},s} \exp(i\vec{q}\vec{r}),
\end{aligned}
\tag{6.79}
$$

where $\bar{w}_{\vec{q},s}(\vec{r}) = \sum_{j,n} \vec{\mathcal{V}}_{j,n}(\vec{r}) \exp[i\vec{q}(\vec{a}_{n,j} - \vec{r})]$.

Hence we find $H_{\text{e-ph}}$ in the form of Eq. (6.77), i.e., as a linear function of the normal coordinates corresponding to different phonon modes.

Recall that, according to Section 5.4, in the normal coordinates, the lattice vibrations are an ensemble of independent modes (oscillators) numerated by quantum numbers $\{\vec{q}, s\}$. Each mode is characterized by a number of the phonons $N_{\vec{q},s}$ and a wave function $\Phi_{N_{\vec{q},s}}(Q_{\vec{q},s})$. Thus the total wave function of the phonon subsystem is

$$\Phi = \prod_{\vec{q},s} \Phi_{N_{\vec{q},s}}(Q_{\vec{q},s}). \tag{6.80}$$

To use this wave function, let us rewrite Eq. (6.79) in terms of the phonon creation and annihilation operators, $c_{\vec{q},s}^+$ and $c_{\vec{q},s}$, respectively. The operators were defined by Eqs. (5.31) and (5.32) through the normal coordinates so that

$$Q_{\vec{q},s} = \sqrt{\frac{\hbar V_0}{2 M_0 \omega_s(\vec{q})}}(c_{\vec{q},s} + c_{\vec{q},s}^+),\tag{6.81}$$

where M_0 is the mass of the primitive cell and $V = V_0 \times \mathcal{N}$ is the crystal volume. Now the general expression for the electron–phonon interaction with a given type of vibration s can be rewritten as

$$H_{\text{e-ph},s} = \sum_{\vec{q}} \Gamma(\vec{r} \mid \vec{q}, s) e^{i\vec{q}\vec{r}}[c_s(\vec{q}) + c_s^+(\vec{q})],\tag{6.82}$$

where

$$\Gamma(\vec{r} \mid \vec{q}, s) = \sqrt{\frac{\hbar V_0}{2 V M_0 \omega_s(\vec{q})}}\, \bar{w}_{\vec{q},s}(\vec{r})$$

can be called the electron–phonon coupling factor. In the sum of Eq. (6.82) each term proportional to $[c_s(\vec{q}) + c_s^+(\vec{q})]$ affects only one element in the product in Eq. (6.80):

$$\langle \Phi' | [c_s(\vec{q}) + c_s^+(\vec{q})] | \Phi \rangle = \langle \Phi'_{N_{\vec{q},s'}}(Q_{\vec{q},s}) | [c_s(\vec{q}) + c_s^+(\vec{q})] | \Phi_{N_{\vec{q},s}}(Q_{\vec{q},s}) \rangle.$$

The latter matrix element is given by Eqs. (5.34) and (5.35). Using these results for the transition defined by expression (6.65) in bulklike material, we get

$$|\langle \Phi' \psi_{\alpha',\vec{k}'} | H_{\text{e-ph},s} | \Phi \psi_{\alpha,\vec{k}} \rangle|^2 = \left(N_{\vec{q},s} + \frac{1}{2} \pm \frac{1}{2}\right)|\Gamma(\alpha, \vec{k}, \alpha', \vec{k}' \mid \vec{q}, s)|^2,$$

with

$$\Gamma(\alpha, \vec{k}, \alpha', \vec{k}' \mid \vec{q}, s) = \int \mathrm{d}^3 r\, \psi_{\alpha',\vec{k}'}^*(\vec{r}) \Gamma(\vec{r} \mid \vec{q}, s) e^{i\vec{q}\vec{r}} \psi_{\alpha,\vec{k}}(\vec{r});\tag{6.83}$$

here $+$ and $-$ correspond to the emission and the absorption, respectively, of the phonon in the mode $\{\vec{q}, s\}$. Finally, the probabilities of emission and absorption of the phonon $\{\vec{q}, s\}$ are

$$W^{(+)}(\alpha, \vec{k}, \alpha', \vec{k}' \mid s, \vec{q}) = \frac{2\pi}{\hbar}|\Gamma(\alpha, \vec{k}, \alpha', \vec{k}' \mid \vec{q}, s)|^2(N_{\vec{q},s} + 1)\delta[E_\alpha(k) \\ - E_{\alpha'}(k') - \hbar\omega_s(\vec{q})],\tag{6.84}$$

$$W^{(-)}(\alpha, \vec{k}, \alpha', \vec{k}' \mid s, \vec{q}) = \frac{2\pi}{\hbar}|\Gamma(\alpha, \vec{k}, \alpha', \vec{k}' \mid \vec{q}, s)|^2 N_{\vec{q},s}\delta[E_\alpha(k) \\ - E_{\alpha'}(k') + \hbar\omega_s(\vec{q})].\tag{6.85}$$

We can interpret $W^{(+)}$ as the sum of two processes: induced phonon emission (proportional to the phonon number $N_{\vec{q},s}$) and spontaneous emission. Absorption is always proportional to the phonon number (i.e., it is always an induced process). This result for electron interaction with phonons is similar to the result of light–matter interaction discussed in Chapter 10.

The summation of Eqs. (6.84) and (6.85) over all possible phonon wave vectors $\vec{q}$ gives us the total probability of electron scattering between the states $\{\alpha, \vec{k}\}$ and $\{\alpha', \vec{k}'\}$ that is due to the interaction with the phonons of the mode s:

$$W(\alpha, \vec{k}, \alpha', \vec{k}' \mid s) = \sum_{\vec{q}} \left[W^{(+)}(\alpha, \vec{k}, \alpha', \vec{k}' \mid s, \vec{q}) + W^{(-)}(\alpha, \vec{k}, \alpha', \vec{k}' \mid s, \vec{q}) \right].$$

(6.86)

These results apply to the case of uniform bulklike materials. For a quantum heterostructure, we have to make two principal changes. First, in heterostructures scattering occurs between electron states that are not plane waves. These states are described in terms of the wave functions of Eqs. (6.68), and the transition probabilities are given by Eq. (6.70). Second, if the modification of the phonon subsystem in a heterostructure is significant, we have to specify the coupling factor Γ. For long-wavelength phonons, it can be done in terms of either macroscopical displacements or macroscopic fields. In Section 6.9 we describe the main idea of such an approach, which is referred to as the method of envelope phonon wave functions. Thus for heterostructures we can rewrite the equations for $W^{(\pm)}$ as

$$W^{(+)}(\beta, \beta' \mid s, \vec{q}) = \frac{2\pi}{\hbar} |\Gamma(\beta, \beta' \mid \vec{q}, s)|^2 (N_{\vec{q},s} + 1) \delta[E_\beta - E_{\beta'} - \hbar\omega_s(\vec{q})],$$

(6.87)

$$W^{(-)}(\beta, \beta' \mid s, \vec{q}) = \frac{2\pi}{\hbar} |\Gamma(\beta, \beta' \mid \vec{q}, s)|^2 N_{\vec{q},s} \delta[E_\beta - E_{\beta'} + \hbar\omega_s(\vec{q})],$$

(6.88)

where matrix elements are calculated with the wave functions of Eqs. (6.68):

$$\Gamma(\beta, \beta' \mid \vec{q}, s) = \int d^3 r F_{\beta'}^*(\vec{r}) \Gamma(\vec{r} \mid \vec{q}, s) e^{i\vec{q}\vec{r}} F_\beta(\vec{r}) |u_{\alpha, \vec{k}_0}(\vec{r})|^2.$$

(6.89)

As defined by Eq. (6.72), the separation of δW into macroscopic and microscopic fields facilitates the representation of the coupling factor $\Gamma(\vec{r} \mid \vec{q}, s)$, as the sum of two contributions: $\Gamma(\vec{r} \mid \vec{q}, s) = \bar{\Gamma}(\vec{r} \mid \vec{q}, s) + \tilde{\Gamma}(\vec{r} \mid \vec{q}, s)$. Here, $\bar{\Gamma}(\vec{r} \mid \vec{q}, s)$ can be interpreted as the coupling with a mode $\{\vec{q}, s\}$ by means of the long-range interaction while $\tilde{\Gamma}(\vec{r} \mid \vec{q}, s)$ is that determined by the deformation potential. Thus we can calculate the transition probabilities of Eqs. (6.84) and (6.85) as well as those of Eqs. (6.87) and (6.88) for these two mechanisms.

Again, the total probability of electron–phonon scattering between states β and β' with phonons of the mode s can be calculated by summation over $\vec{q}$:

$$W(\beta, \beta' \mid s) = \sum_{\vec{q}} \left[W^{(+)}(\beta, \beta' \mid s, \vec{q}) + W^{(-)}(\beta, \beta' \mid s, \vec{q}) \right].$$

(6.90)

In conclusion, we have derived general results for electron–phonon interactions, which can be applied for bulklike materials as well as for confined electrons and phonons. In the first case, in Eqs. (6.84) and (6.85), the wave vectors $\vec{k}, \vec{k}'$, and $\vec{q}$ are three-dimensional vectors, α and α' indicate the electron bands, and s designates the type of bulk mode. If electrons are confined in a quantum well or quantum wire, in Eqs. (6.87) and (6.88), $\vec{k}$ and $\vec{k}'$ are two or one-dimensional vectors and β and β' include the quantum numbers of the low-dimensional sub-bands. If phonons are confined, s includes a quantum number of the confined branch and $\vec{q}$ is either a two- or a one-dimensional vector. We have shown that the interactions with different types of phonon modes can be accounted for independently. Moreover, for any given mode we have separated the interaction into short-range and long-range mechanisms.

There are different phonon modes and mechanisms; however, for a given material or heterostructure and for particular temperatures and other conditions, only one or a few types of electron–phonon interactions can be dominant.

In the following sections, we specify electron interactions with bulk and (or) confined acoustic and optical phonons in bulk materials, quantum wells, and wires.

6.7 Interaction with Acoustic Phonons

In this section, we analyze the coupling of electrons with acoustic phonons. We begin with a study of the short-range coupling by means of the deformation potential of Eq. (6.76). As discussed in Section 6.6, for long-wavelength phonons, this potential can be described in terms of changes in the electron-energy spectrum:

$$E(\vec{k}) = E_\alpha(\vec{k}) + \mathcal{D}_{ij}^\alpha(\vec{k})u_{ij},$$

where $E_\alpha(\vec{k})$ is the energy of the electron band α in the absence of a strain, u_{ij} is the strain tensor defined in Section 4.3, and $\mathcal{D}_{ij}^\alpha(\vec{k})$ is a tensor of the second rank that, in general, is dependent on the wave vector $\vec{k}$. In the long-wavelength phonon limit, we can neglect this dependence and find the deformation potential as

$$H_{\text{e-ph,A}}(\text{short-range}) = D^\alpha(\vec{r}) = \mathcal{D}_{ij}^\alpha u_{ij}; \tag{6.91}$$

see also Eq. (4.13). That is, for each of the electron bands, we reduce the description of the short-range interaction to a constant tensor $\mathcal{D}_{ij}^\alpha$. We can specify the properties of this tensor by taking into account the crystal symmetry, as we did in Section 4.2 for the electron spectra.

Consider, for example, a cubic crystal and carriers within a nondegenerate Γ point. From symmetry considerations, it is evident that in such a case the local disturbance of the crystalline potential should be proportional to the relative change of volume, $\Delta V / V = \operatorname{div} \vec{u}$:

$$D(\vec{r}) = \mathcal{D}\Delta V / V = \mathcal{D} \operatorname{div} \vec{u}, \tag{6.92}$$

where $\mathcal{D}$ is the only constant of the deformation potential and we have omitted the index α.

Table 6.1.
Scattering Processes by Means of the Deformation Potential

Symmetry Point	Phonons for Intravalley Processes	Phonons for Intervalley Processes
Γ	LA	LO + LA ($\Gamma \leftrightarrow L$)
X	LA + TA	LO (g scattering)
		LA + TO (f scattering)
L	LA + TA + LO + TO	LO + LA ($L \leftrightarrow L$)
		LO + LA ($L \leftrightarrow X$)

It is easy to estimate an order of magnitude for the value of $\mathcal{D}$. Actually, let us assume that the lattice strain is so large that the resultant change in the lattice constant $\delta a_0 \sim a_0$, i.e., $\mathrm{div}\,\vec{u} \sim 1$. For such a strain we should get a shift of the electron-energy spectrum of $\sim e^2/a_0$, which is an estimate for the Coulomb interaction between ions. Applying this consideration to Eq. (6.92), we can estimate that $\mathcal{D}$ is of the order of 1–10 eV.

In Section 5.5, we found that, for bulk materials, $\mathrm{div}\,\vec{u} \neq 0$ only for the LA mode. Thus the form of Eq. (6.92) tells us that for a nondegenerate Γ point, carriers interact with only longitudinal phonons. Since the relative displacement is wavelike, it follows that $\mathrm{div}\,\vec{u} \propto q$, that is, the potential is proportional to the phonon wave number q. Rewriting Eq. (6.92) in terms of the phonon creation and annihilation operators, as in Eq. (6.81), we get the electron-acoustic phonon coupling factor as

$$\widetilde{\Gamma}(\vec{r} \,|\, \vec{q}, \mathrm{LA}) \equiv \frac{1}{\sqrt{V}} B_{\mathrm{DLA}}(q) = i\sqrt{\frac{\hbar V_0}{2 V M_0 \omega(q)}} \mathcal{D}q, \tag{6.93}$$

where $\omega(q) = v_l q$ is the LA phonon dispersion and v_l is the longitudinal sound velocity.

For a cubic crystal with energy minima situated at X or L symmetry points (see Section 4.2), the deformation-potential tensor $\mathcal{D}_{ij}$ does not reduce to a constant. If a minimum α is on an axis along the unit vector $\vec{e}$, the tensor can be represented as

$$\mathcal{D}_{ij} = \mathcal{D}_l \delta_{ij} + \mathcal{D}_t e_i^\alpha e_j^\alpha.$$

Here, $\mathcal{D}_l$ and $\mathcal{D}_t$ are constants; see Eq. (4.16). In such a case, electrons interact with both LA and TA phonons. It is easy to find equations for $\widetilde{\Gamma}$ for each type of these phonons. Table 6.1 summarizes possible scattering processes due to electron–phonon deformation coupling.

Consider now the long-range interaction or the macroscopic potential ϕ induced by acoustic phonons. The corresponding electron–phonon interaction is

$$H_{\mathrm{e-ph}}(\text{long-range}) \equiv \overline{\delta W} \equiv -e\phi. \tag{6.94}$$

The macroscopic potential ϕ can be found from the Poisson equation:

$$\kappa \nabla^2 \phi = 4\pi \, \text{div} \, \vec{P}, \tag{6.95}$$

where $\vec{P}$ is the polarization of the crystal induced by the strain. For acoustic modes (A), if the strain can be represented as a smooth function of the position, the polarization $\vec{P}_j^{(A)}$ can be expressed through the strain tensor:

$$\vec{P}_j^{(A)} = \kappa \beta_{j,kl} u_{kl}. \tag{6.96}$$

Here the vector $\vec{P}_j^{(A)}$ and the tensor u_{kl} are related through the piezoelectric tensor $\beta_{j,kl}$ of rank three. Since the strain tensor u_{kl} is symmetric with respect to the permutation $k \leftrightarrow l$, the tensor $\beta_{j,kl}$ is also symmetric with respect to that permutation. The properties of the piezoelectric tensor depend on the symmetry of the crystal under consideration. In cubic crystals with a center of the inversion – Si, Ge, etc. – this tensor is identically zero, that is, the piezoelectric effect does not appear. In cubic III–V compounds like GaAs and InP, in which there are two different atoms in the primitive cell, a center of inversion is absent and the piezoelectric effect does exist. In the latter case, the crystal symmetry reduces the tensor to only one nonzero constant:

$$\beta_{i \neq j \neq k} = \beta_{1,23} = \beta_{2,13}, \text{ etc.} \tag{6.97}$$

Values of the piezoelectric constants $\beta_{1,23}$ for some semiconductors are given in Table 6.2.

Substituting Eq. (6.96) into Eq (6.95), we get the equation for the macroscopic potential ϕ. For example, in the case of a bulk crystal, the interaction with the phonon mode $\{s, \vec{q}\}$ is

$$-e\phi_{s,\vec{q}}^{(A)} = i \frac{4\pi e}{q^2} \left[\vec{q} \vec{P}_{s,\vec{q}}^{(A)} \right], \tag{6.98}$$

where the polarization $\vec{P}_{s,\vec{q}}^{(A)}$ is induced by this mode. The interaction can be rewritten as

Table 6.2. Piezoelectric Constants for Some III–V Compounds	
Compound	$\beta_{1,23}$ (C/m^2)
InSb	0.004
GaAs	0.012
InAs	0.003
GaSb	0.008

$$-e\phi_{s,\vec{q}}^{(A)} = -\frac{4\pi e}{q^2} \beta_{j,kl} q_j \frac{1}{2}(u_k q_l + u_l q_k)$$

$$= -\frac{4\pi e}{q^2} \beta_{j,kl} q_j q_k u_l. \tag{6.99}$$

Let $\vec{b}_{s\vec{q}}$ be the unit polarization vectors and $Q_{\vec{q},s}$ be the normal coordinates for a given mode:

$$\vec{u}_s = \frac{1}{\sqrt{N}} \sum_{\vec{q}} \vec{b}_{\vec{q},s} \left(Q_{\vec{q},s} e^{i\vec{q}\vec{r}} + Q_{\vec{q},s}^* e^{-i\vec{q}\vec{r}} \right). \tag{6.100}$$

Then we find

$$-e\phi^{(A)} = -\frac{4\pi e}{\sqrt{N}q^2} \sum_{\vec{q}} \beta_{j,kl}q_jq_kb_{l,\vec{q},s}\left(Q_{\vec{q},s}e^{i\vec{q}qvr} + Q^*_{\vec{q},s}e^{-i\vec{q}\vec{r}}\right).$$

Comparison of this equation with Eq. (6.82) gives us the electron–phonon coupling factor that is due to the piezoelectric effect for a given mode $\vec{q}, s$:

$$\bar{\Gamma}(\vec{r}\,|\,\vec{q}, \mathrm{PA}) = -\frac{4\pi e}{q^2} \sqrt{\frac{\hbar V_0}{2VM_0\omega_{\vec{q},s}(\vec{q})}}\,\beta_{j,kl}q_jq_kb_{l,\vec{q},s}. \qquad (6.101)$$

This coupling depends on the angle between the phonon wave vector $\vec{q}$ and the crystal axis. The strong angular dependence can be illustrated as follows. For III–V compounds, because of Eq. (6.97), for all phonons with $\vec{q}$ parallel to any axis of the [100] type, we find $\bar{\Gamma} = 0$, that is, the coupling factor is zero and interaction is absent. If $\vec{q}$ is parallel to directions of the [111] type, the coupling factor is zero for transverse phonons, while for longitudinal phonons it is finite.

Keeping in mind that the transition probabilities are proportional to the square matrix element of Γ, one can introduce the spherical average of the square coupling factor over all propagation directions. Then for the coupling factor with longitudinal and transverse modes we can set

$$\bar{\Gamma}(\vec{r}\,|\,\vec{q}, \mathrm{PLA}) \equiv \frac{1}{\sqrt{V}}B_{\mathrm{PLA}}(\vec{q}) = -4\pi e\sqrt{\frac{\hbar V_0}{2VM_0\omega_l(\vec{q})}}\,\beta_{1,23}\sqrt{\frac{12}{35}}, \qquad (6.102)$$

$$\bar{\Gamma}(\vec{r}\,|\,\vec{q}, \mathrm{PTA}) \equiv \frac{1}{\sqrt{V}}B_{\mathrm{PTA}}(\vec{q}) = -4\pi e\sqrt{\frac{\hbar V_0}{2VM_0\omega_t(\vec{q})}}\,\beta_{1,23}\sqrt{\frac{16}{35}}, \qquad (6.103)$$

respectively.

The coupling factors of Eqs. (6.93), (6.102), and (6.103) are not dependent on the electron position $\vec{r}$. Thus the calculation of matrix elements in Eqs. (6.84) and (6.85) and (6.87) and (6.88) reduce to the evaluation of the following integrals:

$$\int \mathrm{d}^3\vec{r}\,\psi^*_{\alpha'\vec{k}'}(\vec{r})e^{i\vec{r}\vec{q}}\psi_{\alpha\vec{k}}(\vec{r}), \qquad \int \mathrm{d}^3\vec{r}\,F_{\beta'}(\vec{r})e^{i\vec{r}\vec{q}}F_{\beta}(\vec{r})$$

for a bulk material and a heterostructure, respectively. These integrals give $\delta_{\vec{k}',\vec{k}+\vec{q}}$. As a result, in the bulk case we get contributions to the intraband $(\alpha' = \alpha)$ matrix element from both the short- and the long-range interactions in the form

$$\bar{\Gamma}(\alpha\vec{k}, \alpha\vec{k}'\,|\,\vec{q}, \mathrm{LA}) \equiv \frac{1}{\sqrt{V}}B_{\mathrm{DLA}}(q)\delta_{\vec{k}',\vec{k}+\vec{q}}, \qquad (6.104)$$

$$\bar{\Gamma}(\alpha\vec{k}, \alpha\vec{k}'\,|\,\vec{q}, \mathrm{LA}) \equiv \frac{1}{\sqrt{V}}B_{\mathrm{PLA}}(q)\delta_{\vec{k}',\vec{k}+\vec{q}}, \qquad (6.105)$$

$$\bar{\Gamma}(\alpha\vec{k}, \alpha\vec{k}'\,|\,\vec{q}, \mathrm{TA}) \equiv \frac{1}{\sqrt{V}}B_{\mathrm{PTA}}(q)\delta_{\vec{k}',\vec{k}+\vec{q}}, \qquad (6.106)$$

where we designate the interaction by means of the deformation potential with LA phonons by DLA and the piezoelectric interaction with longitudinal and transverse phonons by PLA and PTA, respectively.

Note first, that the short-range contribution of Eq. (6.104) is purely imaginary, while the long-range contribution is real. This leads to the conclusion that the square of the total matrix element for interaction with acoustic phonons is

$$|\Gamma|^2 = |\tilde{\Gamma} + \bar{\Gamma}|^2 = |\tilde{\Gamma}|^2 + |\bar{\Gamma}|^2,$$

that is, deformation-potential scattering and piezoelectric scattering do not interfere with each other.

Another important conclusion that follows from the Kronecker deltas of Eqs. (6.104)–(6.106) is the selection rule for the wave vectors. Since the electron–phonon interaction of Eq. (6.82) contains positive $\vec{q}$ and negative wave vectors $-\vec{q}$, we can present this rule as

$$\vec{k}' - \vec{k} = \pm\vec{q}. \tag{6.107}$$

Thus, for electron–phonon scattering involving long-wavelength phonons, we obtain the conservation law for the wave vectors of electrons and phonons in a bulklike material.

The energy- and the momentum-conservation laws given by Eqs. (6.67) and (6.107) determine the kinematics of electron–phonon scattering. Consider a parabolic electron dispersion $E_\alpha(\vec{k}) = (\hbar^2 k^2)/(2m^*)$. Let θ_q be the angle between $\vec{k}$ and $\vec{q}$, respectively. The conservation laws can be presented as

$$k'^2 = k^2 \pm 2kq\cos\theta_q + q^2, \qquad \frac{\hbar^2 k'^2}{2m^*} = \frac{\hbar^2 k^2}{2m^*} \pm \hbar\omega_s(\vec{q}). \tag{6.108}$$

The upper sign corresponds to phonon absorption, and the lower corresponds to phonon emission. Eliminating k', we get

$$\cos\theta_q = \mp\frac{q}{2k} + \frac{m^*\omega_s(\vec{q})}{\hbar kq}. \tag{6.109}$$

Since $|\cos\theta_q| < 1$, Eq. (6.109) implies significant restrictions on the phonon characteristics q and ω_s.

For acoustic modes, we set $\omega = sq$. As usual, $s = s_l$ and $s = s_t$ for longitudinal and transverse phonons, respectively. Then the analysis of Eq. (6.109) gives results that can be classified in terms of the initial electron energy:

1. $E(k) < \frac{1}{2}m^*s^2$. Phonon emission is forbidden; scattering with phonon absorption occurs with any angle θ_q. The phonon wave vector should satisfy the conditions $q_{min}^{abs} \le q \le q_{max}^{abs}$, with $q_{min}^{abs} = 2k(sm^*/\hbar k - 1)$ and $q_{max}^{abs} = 2k(sm^*/\hbar k + 1)$.

2. $\frac{1}{2}m^*s^2 < E(k) < m^*s^2$. Phonon emission occurs only for forward scattering, $0 \le \theta_q \le \cos^{-1}(sm^*/\hbar k)$, with wave vectors in the interval from 0 to

$q_{max}^{em} = 2k(1 - sm^*/\hbar k)$. Phonon absorption occurs for $\cos^{-1}(sm^*/\hbar k) \leq \theta_q \leq \pi$ in the wave-vector interval from 0 to q_{max}^{abs}.

3. $E(k) > m^* s^2$. Phonon emission and absorption occur for any angle θ in the wave-vector intervals $(0, q_{max}^{em})$ and $(0, q_{max}^{abs})$, respectively.

It is instructive to estimate the parameter $m^* s^2$ that appears in this analysis. Since typical sound velocities are $\sim 5 \times 10^5$ cm/s, setting $m^* = 0.1$ m we find that $m^* s^2 \approx 0.01$ meV, which corresponds to $T \approx 0.1$ K. Hence we can conclude that only the last of the three cases occurs at all practical temperatures. For this case, maximum transferred momenta are $q_{max}^{em} \approx q_{max}^{abs} \approx 2k$. It is easy to account for energy changes involved in the processes of emission and absorption of the phonons: $E' - E = \hbar\omega(q) \leq \hbar s q_{max} \approx 2sk$. The relative change in the energy is

$$(E' - E)/E \leq 4sm^*/\hbar k = 2\sqrt{2m^* s^2/E} .$$

Using the above estimate, one can see that relative changes of electron energy are, in fact, very small. For example, at nitrogen temperatures electrons have a thermal energy of ~ 10 meV; thus, for the case of scattering by acoustic phonons the relative changes in electron energy do not exceed 0.02. Such a scattering is almost elastic and may be referred to as quasi-elastic scattering.

Thus the electron–phonon interaction causes electron scattering with significant changes in the electron momentum, but the relative change in the electron energy is small. This quasi-elastic character of the scattering allows us to simplify the evaluation of scattering times for such an interaction. Consider the momentum-scattering time τ_p given by Eq. (6.8). For intraband processes in a bulk crystal we should consider initial and final wave vectors $\vec{k}$ and $\vec{k}'$ as three-dimensional vectors and omit the discrete quantum number n_0. All processes of phonon emission and phonon absorption contribute to this equation and we shall use $W(\vec{k}, \vec{k}' | s)$ as defined by Eq. (6.86). In general, all three mechanisms, DLA, PLA, and PTA, should be taken into account. Their contributions are additive and can be calculated similarly. For quasi-elastic scattering the phonon energy $\hbar\omega$ is small compared with the initial $E(\vec{k})$ and final $E(\vec{k}')$ electron energies. Hence in the δ functions of Eqs. (6.84) and (6.85) we can neglect $\hbar\omega$. Then the contribution to τ_p from any of the mechanisms takes the form

$$\frac{1}{\tau_p(\vec{k})} = \sum_{\vec{k}'} \frac{k - k' \cos\theta}{k} W(\vec{k}, \vec{k}' | s)$$

$$= \frac{2\pi}{\hbar V} \sum_{\vec{k}', \vec{q}} \frac{k - k' \cos\theta}{k} |B(\vec{q})|^2 (2N_{q,s} + 1) \delta_{\vec{k}', \vec{k}+\vec{q}} \, \delta[E(\vec{k}') - E(\vec{k})],$$

$$(6.110)$$

where θ is the angle between the vectors $\vec{k}$ and $\vec{k}'$, N_q is the phonon number of the phonon mode under consideration, and $B(\vec{q})$ is obtained from Eqs. (6.104)–(6.106).

From the conservation law of Eq. (6.107), we find

$$\cos\theta = \frac{k'^2 + k^2 - q^2}{2k'k}.$$

Then Eq. (6.110) can be rewritten as

$$\frac{1}{\tau_p} = \frac{2\pi}{\hbar}\frac{V}{(2\pi)^3}\int d^3q \frac{q^2 - (k'^2 - k^2)}{2k^2}\frac{1}{V}|B(q)|^2(2N_{q,s}+1)\delta[E(\vec{k}') - E(\vec{k})].$$

Here $k' = \sqrt{k^2 + q^2 + 2\vec{k}\vec{q}}$ as a result of the factor $\delta_{\vec{k}',\vec{k}+\vec{q}}$. To perform the integration, it is convenient to use the spherical coordinates, where θ_q is the angle between vectors $\vec{k}$ and $\vec{q}$:

$$\frac{1}{\tau_p} = \frac{1}{2\pi\hbar}\int_0^\infty dq\, q^2 \int_0^\pi d\theta_q \sin\theta_q \frac{q^2 - (k'^2 - k^2)}{2k^2}|B(q)|^2(2N_{q,s}+1)$$

$$\times \delta\left[\frac{\hbar^2}{2m^*}(q^2 + 2kq\cos\theta_q)\right]$$

$$= \frac{m^*}{4\pi\hbar^3 k^3}\int_0^{2k} dq\, q^3 |B(q)|^2(2N_{q,s}+1). \tag{6.111}$$

Here we have taken into account that the δ function determines θ_q in such a way that $\cos\theta_q = -q/2k$ if $q < 2k$; otherwise the integrand is zero.

Similarly, we can obtain the time of scattering defined by Eq. (6.7):

$$\frac{1}{\tau_s} = \frac{m^*}{2\pi\hbar^3 k}\int_0^{2k} dq\, q |B(q)|^2(2N_{q,s}+1). \tag{6.112}$$

Equations (6.111) and (6.112) are dependent on the phonon occupation numbers $N_{q,s}$. Usually the phonon subsystem is supposed to be in equilibrium and the phonon occupation numbers are given by the Planck distribution of Eq. (5.30):

$$N_{q,s} = \frac{1}{\exp(\hbar\omega_{q,s}/k_B T) - 1}.$$

It is worth noting that $N_{q,s}$ depends strongly on the phonon energy and temperature T. Let us use the previously discussed typical values of energies of the phonons participating in the scattering event: $\hbar\omega \leq 2\hbar sk = 2\sqrt{2m^* s^2 E(k)}$. Using the Planck distribution, we can find two energy regions: $E(\vec{k}) \ll (k_B T)^2/ms^2$ and $E(\vec{k}) \gg (k_B T)^2/ms^2$, which have significantly different numbers of phonons. In the first region, we get a large number of phonons: $N_q \approx k_B T/\hbar\omega \gg 1$, while in the second the phonon numbers are exponentially small. Thus, in accordance with Eq. (6.111), for the first region the scattering is due primarily to absorption and induced emission of phonons. In the second energy region the scattering is due to spontaneous-emission processes. Thermalized electrons with energy $E \sim 3k_B T/2$ are typically in the first energy region.

Table 6.3.
Acoustics-Scattering Times for Electrons from a Γ Minimum

Relaxation Times	DA	PA
$\dfrac{1}{\tau_p}$	$\dfrac{\sqrt{2}D^2 m^{*3/2} k_B T}{\pi\hbar^4 C_l} E^{1/2}$	$\dfrac{e^2 \beta_{1,23}^2 m^{*1/2} k_B T}{2^{2/3}\pi\hbar^2}\left(\dfrac{12}{35 C_l} + \dfrac{16}{35 C_t}\right)\dfrac{1}{E^{1/2}}$
$\dfrac{1}{\tau_s}$	$\dfrac{1}{\tau_p}$	$\dfrac{1}{\tau_p}\ln\dfrac{\sqrt{2m^* E}}{\hbar q_D}$

The scattering times given by Eqs. (6.111) and (6.112) are determined by the electron energy, crystal temperature, and the particular scattering mechanism. In Table 6.3, the times τ_p and τ_s are presented for the deformation potential and piezoelectric mechanisms in the actual limiting cases $E(\vec{k}) \ll (k_B T)^2/ms_{l,t}^2$ and the quasi-elastic scattering. For the piezoelectric mechanism, both longitudinal and transverse phonons are taken into account. In Table 6.3, we introduce the notations

$$M_0 s_{l,t}/V_0 = C_{l,t}, \tag{6.113}$$

where $C_{l,t}$ represent the mean elastic constants of the material. It is interesting to note that the deformation potential and the piezoelectric mechanisms cause different energy dependences for τ_p and τ_s. Note that, according to Eq. (6.112), for the piezoelectric mechanism, the time scattering τ_s diverges at $k \to 0$. This result of Table 6.3 is obtained for integration in Eq. (6.112) from q_D to $2k$, where q_D is the screening length for the three-dimensional electron gas introduced in Section 6.3.

6.8   Interaction with Optical Phonons

Now we consider electron–phonon coupling for optical phonons. For actual cases of III–V compounds, Si, SiGe, and other materials having two atoms per primitive cell, there are three optical modes – one is longitudinal, the other two are transverse. First, we analyze the short-range interaction with optical phonons.

In Section 5.6, we showed that an optical distortion of a crystal is described by a relative displacement of atoms in the cell $\vec{u}^{(O)}$ (here we add a superscript to $\vec{u}$ to indicate that we deal with the optical mode of vibrations). The short-range deformation potential of Eq. (6.76) can be presented as a linear function of $\vec{u}^{(O)}$:

$$H_{\text{e–ph,O}}(\text{short-range}) = D^{(\text{Opt})}(\vec{r}) = \vec{D}^{(O)}\vec{u}^{(O)}, \tag{6.114}$$

where the deformation potential $D^{(\text{Opt})}(\vec{r})$ is a scalar and the displacement $\vec{u}^{(O)}$ is a vector. Thus the coefficient in Eq. (6.114), $\vec{D}^{(O)}$, has to be a vector. It is instructive to compare with the case of the acoustic deformation potential of Eq. (6.91). $\vec{D}^{(O)}$ is in units of electron volts per centimeter.

The form of Eq. (6.114) immediately tells us that in a cubic crystal with a Γ extremum when no isolated vector exists, the only possibility is $\vec{D}^{(O)} = 0$, that is, the short-range interaction with optical phonons is absent. According to Chapter 4, this is valid for many III–V compounds, like GaAs, InSb, etc.

On the other hand, for a cubic crystal and electrons within energy extrema in X or L points, there naturally exists an isolated vector. For a given extremum, this vector can be defined in momentum space and directed from the origin to the position of a given extremum. For example, if an electron belongs to the [111] L extremum, the vector is along the [111] axis and Eq. (6.114) takes the form

$$D^{(\mathrm{Opt})}(\vec{r}) = D^{(O)}\left[u_x^{(O)} + u_y^{(O)} + u_z^{(O)}\right]. \tag{6.115}$$

This case requires only the constant parameter $D^{(O)}$.

The absolute value of $\vec{D}^{(O)}$ depends on the specific properties of the crystal in question. As was done above for acoustic phonons, one can easily obtain an order-of-magnitude estimate for $|\vec{D}^{(O)}|$. For this, let us imagine that the optical displacement $\vec{u}^{(O)}$ is approximately equal to the lattice constant a_0; such a displacement should produce a change in the electron energy of the order of e^2/a_0. From Eq. (6.114), we find that $|\vec{D}^{(O)}|$ is ~ 1–10 eV/Å.

Thus we can conclude that for indirect-bandgap semiconductors like Ge, SiGe, AlAs, etc., the short-range interaction with optical phonons in the form of Eq. (6.114) does occur. Note that the particular symmetry of the primitive cell of Si crystals – the diamond lattice with two identical atoms in the cell – is such that the optical deformation potential for X extrema is zero. Possible scattering processes due to the coupling between electrons and optical phonons by means of the deformation potential are pointed out in Table 6.1.

Next, let us represent the optical displacement in the form of Eq. (6.100),

$$\vec{u}_s^{(O)} = \frac{1}{\sqrt{N}} \sum_{\vec{q}} \vec{b}_{\vec{q},s}^{(O)} \left[Q_{\vec{q},s}^{(O)} e^{i\vec{q}\vec{r}} + Q_{\vec{q},s}^{(O)*} e^{-i\vec{q}\vec{r}} \right], \tag{6.116}$$

and introduce the optical-phonon annihilation and creation operators similarly to those of Eq. (6.81). The only difference is that for the optical-phonon case, the mass of the primitive cell M_0 has to be replaced by the reduced mass of the ions participating in the optical displacement. Then we obtain the electron–phonon interaction in the form of Eq. (6.82) with the short-range electron–phonon coupling with an optical mode s:

$$\tilde{\Gamma}(\vec{r}\,|\,\vec{q}, DO) = \sqrt{\frac{\hbar V_0}{2VM\omega_{s,\vec{q}}^{(O)}}}\, \vec{D}^{(O)} \vec{b}_{\vec{q},s}, \tag{6.117}$$

where $\omega_{s,\vec{q}}^{(O)}$ is the optical-phonon frequency and $\mathcal{M}$ is the reduced mass. Since the dispersion of $\omega_{\vec{q},s}^{(O)}$ is relatively weak, this coupling factor depends not on the phonon wave vector $\vec{q}$, but on the optical-mode polarization. For example, in the case of Eq. (6.117), the coupling factor can be rewritten as

$$\tilde{\Gamma}(\vec{r}\,|\,\vec{q},\,DO) = \sqrt{\frac{\hbar V_0}{2VM\omega_{s,\vec{q}}^{(O)}}}\,D^{(O)}\cos\tilde{\theta}_s,$$

where $\tilde{\theta}_s$ is the angle between the polarization of the mode s and the [111] crystalline direction. Keeping in mind that the scattering is determined by the square matrix element of $\tilde{\Gamma}$, we can average this square over all possible polarizations and types of optical modes and rewrite the coupling factor in terms of an effective constant D:

$$\tilde{\Gamma}(\vec{r}\,|\,\vec{q},\,DO) = \sqrt{\frac{\hbar V_0}{2VM\omega_{s,\vec{q}}^{(O)}}}\,D \equiv \frac{1}{\sqrt{V}}B_{DO}(\vec{q}). \qquad (6.118)$$

The long-range interaction with optical phonons can be described in terms of the macroscopic potential $\phi^{(O)}$ as in Eq. (6.94). The potential satisfies Eq. (6.95) with the polarization $P = P^{(O)}$ due to the optical vibrations. Note that this type of interaction is frequently referred to as the electron–polar-optical-phonon interaction. $P^{(O)}$ can be represented as a linear function of the optical displacement:

$$P_i^{(O)} = \gamma_{ij}u_j^{(O)}, \qquad (6.119)$$

where γ_{ij} is a tensor with constant components. Using the solution for $P^{(O)}$ in the form of Eq. (6.99) and representing $u^{(O)}$ through Eq. (6.116), we find the interaction:

$$H_{e\text{-ph},O}(\text{long-range}) = -e\sum_{\vec{q},s}\phi_{\vec{q},s}^{(O)}e^{i\vec{q}\vec{r}},$$

where

$$-e\phi_{\vec{q},s}^{(O)} = i\frac{4\pi e}{\sqrt{N}q^2}q_i\gamma_{ij}\big[b_{\vec{q},s}^{(O)}\big]_j Q_{\vec{q},s}.$$

Generally, for a cubic crystal, the tensor γ_{ij} reduces to the single scalar value γ, so that

$$-e\phi_{\vec{q},s}^{(O)} = i\frac{4\pi e}{\sqrt{N}q^2}\gamma q_i\big[b_{\vec{q},s}^{(O)}\big]_i Q_{\vec{q},s}^{(O)}.$$

The last result shows that only the LO vibrations $\vec{b}_{\vec{q},s}\,\|\,\vec{q}$ interact with electrons. Recalling the results of Section 5.6 and particularly Eq. (5.91), we find that $\gamma = e^*/(\kappa_\infty V_0)$, where e^* is an effective charge of the ions and κ_∞ is the high-frequency permittivity of the crystal. Then, with Eq. (6.81) and the definition of the coupling factor, the latter can be represented as

$$\bar{\Gamma}(\vec{r}\,|\,\vec{q},\,PLO) \equiv \frac{1}{\sqrt{V}}B_{PLO} = i\frac{4\pi e}{q}\gamma\sqrt{\frac{\hbar V_0}{2VM\omega_{q,s}^{(O)}}}. \qquad (6.120)$$

Thus the polar long-range interaction with optical phonons is inversely proportional to the wave vector.

Finally, the matrix elements produced by both the short- and the long-range interactions with the optical phonons are

$$\tilde{\Gamma}(\alpha\vec{k}, \alpha'\vec{k}' \,|\, \vec{q}, DO) = \frac{1}{\sqrt{V}} B_{DO}(q)\delta_{\vec{k}',\vec{k}+\vec{q}}, \tag{6.121}$$

$$\bar{\Gamma}(\alpha\vec{k}, \alpha'\vec{k}' \,|\, \vec{q}, PLO) = \frac{1}{\sqrt{V}} B_{PLO}(q)\delta_{\vec{k}',\vec{k}+\vec{q}}. \tag{6.122}$$

According to Eqs. (6.117) and (6.120), the short-range and the long–range mechanisms do not interfere with each other:

$$\left|\tilde{\Gamma}^{(O)} + \bar{\Gamma}^{(O)}\right|^2 = \left|\tilde{\Gamma}^{(O)}\right|^2 + \left|\bar{\Gamma}^{(O)}\right|^2 .$$

Now we can analyze the carrier-scattering times due to optical phonons. For several limiting cases, it is possible to obtain explicit forms for these scattering rates.

If the electron energy E is so large that $E \gg \hbar\omega_0$, we can use the quasi-elastic approximation, which results in Eqs. (6.111) and (6.112).

For the deformation-potential mechanism $B(q)$ does not depend on q, and simple calculations give

$$\frac{1}{\tau_s^{(DO)}} = \frac{1}{\tau_p^{(DO)}} = \frac{1}{\tau^{(DO)}} \left(\frac{E}{\hbar\omega_0}\right)^{1/2} [2N^{(O)} + 1],$$

where we have introduced the number of optical phonons $N^{(O)}$ and the parameter

$$\frac{1}{\tau^{(DO)}} = \frac{m^*\sqrt{m^*}}{2^{1/2}\pi\hbar^2 \mathcal{M}/V_0\sqrt{\hbar\omega_0}} D^2.$$

For the polar-optical-phonon mechanism we find that the momentum-scattering time has the form

$$\frac{1}{\tau_p^{(PLO)}} = \frac{1}{\tau^{(PLO)}} \left(\frac{\hbar\omega_0}{E}\right)^{1/2} [2N^{(O)} + 1],$$

where we have introduced the parameter

$$\frac{1}{\tau^{(PLO)}} = \frac{(e^*)^2\sqrt{m^*\omega_0}}{2^{1/2}\hbar^{3/2}} \left(\frac{1}{\kappa_\infty} - \frac{1}{\kappa_0}\right).$$

The total scattering time of Eq. (6.112) diverges at small q as $\ln q$. Thus we can write the approximate expression

$$\frac{1}{\tau_s^{(PLO)}} = \frac{1}{\tau_p^{(PLO)}} \ln\left(\frac{4E}{\hbar\omega_0}\right).$$

The energy-relaxation time defined by Eq. (6.9) can be evaluated straightfor-wardly if we set $|E' - E| = \hbar\omega_0$:

$$\frac{1}{\tau_E^{(DO)}} = \frac{1}{\tau_p^{(DO)}} \frac{\hbar\omega_0}{E}, \qquad \frac{1}{\tau_E^{(PLO)}} = \frac{1}{\tau_p^{(PLO)}} \frac{\hbar\omega_0}{E}.$$

Then, if the electron energy E is of the order of the phonon energy $\hbar\omega_0$, the quasi-elastic approximation is no longer valid. One can calculate the total scat-tering time τ_s. It consists of contributions coming from optical-phonon emission $\tau_s^{(+)}$ and optical-phonon absorption $\tau_s^{(-)}$:

$$\frac{1}{\tau_s} = \frac{1}{\tau_s^{(+)}} + \frac{1}{\tau_s^{(-)}}.$$

The results for $\tau_s^{(\pm)}$ are

$$\frac{1}{\tau_s^{(DO,\pm)}} = \frac{1}{\tau^{(DO)}}\left[N^{(O)} + \frac{1}{2} \pm \frac{1}{2}\right]\left(\frac{E \mp \hbar\omega_0}{\hbar\omega_0}\right)^{1/2},$$

$$\frac{1}{\tau_s^{(PLO,\pm)}} = \frac{1}{\tau^{(PLO)}}\left[N^{(O)} + \frac{1}{2} \pm \frac{1}{2}\right]\ln\left|\frac{\sqrt{E} + \sqrt{E \mp \hbar\omega_0}}{\sqrt{E} - \sqrt{E \mp \hbar\omega_0}}\right|.$$

One must remember that $\tau_s^{(+)}$ is nonzero only if $E > \hbar\omega_0$. Near the threshold of phonon emission, $E - \hbar\omega_0 \ll \hbar\omega_0$, both scattering mechanisms lead to the same energy dependence:

$$\frac{1}{\tau_s} \propto \sqrt{\frac{E - \hbar\omega_0}{\hbar\omega_0}}.$$

In a strict sense, we cannot introduce the momentum-scattering time for inelastic processes. But the total scattering time τ_s can be considered to be a good qualitative estimate for τ_p, since phonon emission and absorption result in large changes in the electron momentum. The energy-relaxation time can be written by means of $\tau_s^{(\pm)}$:

$$\frac{1}{\tau_E} = \frac{\hbar\omega_0}{E}\left[\frac{1}{\tau_s^{(+)}} - \frac{1}{\tau_s^{(-)}}\right].$$

For different III–V compounds, the polar in-teraction of optical phonons provides the dominant contribution to the scattering. Numerical values for $\tau^{(PLO)}$ at room temperature are presented in Table 6.4 for several polar materials.

It is interesting to compare the contributions of different phonon-scattering mechanisms and their

Table 6.4.
Scattering Time Parameters for Polar-Optical Phonons

Compound	$\tau^{(PLO)}$ (ps)
InSb	0.7
GaAs	0.14
InAs	0.2
GaSb	0.4

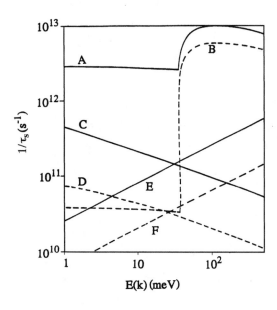

Figure 6.2. The scattering rates due to different mechanisms of electron–phonon interaction in a GaAs crystal at two temperatures. Polar optical: A, 300 K; B, 77 K. Piezoelectric: C, the carrier concentration equals 10^{15} cm^{-3}, 300 K; D, 10^{15} cm^{-3}, 77 K. Nonpolar acoustic: E, 300 K; F, 77 K. [After B. K. Ridley, "Hot electrons in low-dimensional structures," Rep. Prog. Phys. 54, pp.169–256 (1991).]

roles at different temperatures. In Fig. 6.2, such a comparison is presented for a GaAs crystal. The scattering rates $1/\tau_s$ for all types of mechanisms – the acoustic deformation potential, piezoelectric scattering, and scattering by polar-optical phonons – are given as functions of the electron energy at two crystal temperatures, 77 and 300 K. The steplike increase in the scattering rate for optical phonons is due to the threshold character of phonon emission; indeed, such emission events are possible only when the electron energy $E > \hbar\omega_0$, which is $\sim$36 meV for GaAs. From these data, one can see that at room temperature the optical phonons completely dominate in the scattering. At nitrogen temperature, the piezoelectric-acoustic mechanism is comparable with that of optical phonons for $E < \hbar\omega_0$. In general, acoustic-phonon scattering becomes the main mechanism at low temperatures. In bulk GaAs materials, the deformation-potential mechanism does not compete with the piezoelectric mechanism. In low-dimensional systems, this situation changes considerably.

We concentrated previously on the case of the simplest structure of the electron- (hole-) energy band. However, it is essential to take into account that many crystals, including Si, Ge, and their alloys, have several electron-energy minima that play a dominant role in their electron properties; in particular, phonon-induced transitions between valleys (intervalley processes) should be taken into account. For example, in Si, two kinds of intervalley processes occur. The first is the so-called g process, which corresponds to transitions between two opposite valleys (like [100] $\leftrightarrow$ [$\bar{1}$00] transitions). The second is the so-called f process between nonopposite valleys ([100] $\leftrightarrow$ [010], [001]). These processes involve phonons with large wave vectors (short-wavelength or non-zone-center phonons). For example, the g process is possible with the participation of phonons with wave vectors along the [100]-type direction and of a magnitude of $\sim$0.3 of the Brillouin zone edge value. Particular selection rules for these transitions show that the g process is controlled by LO phonons, while the f process is associated with LA and TO phonons. These processes are indicated in Table 6.1, in which other possible scattering processes that occur by means of the deformation potential are collected as well. In the case of a complex valence-band structure associated with light- and heavy-hole bands, there are additional scattering processes between the valence bands. In such interband transitions, both acoustic and optical phonons participate.

In conclusion, we have studied the nature of the electron–phonon interaction and obtained general formulas describing this interaction for electrons of any dimensionality. Then, we have analyzed particular properties of electron–phonon scattering in bulklike semiconductors. We have found that it is possible to introduce an interaction between electrons and each of the phonon modes. For long-wavelength phonons – Brillouin zone-center phonons – we have separated short-range and long-range mechanisms of interactions. The first is described in terms of deformation potentials for both acoustic and optical phonons. This mechanism is dominant for elemental semiconductors like Si, Ge, etc. The long-range interactions (piezoelectric-acoustic-phonon and polar-optical-phonon couplings) are characteristic for ionic materials like III–V compounds, although short-range interaction with acoustic phonons can be important for them as well.

6.9 Scattering of Electrons by Acoustic Phonons in Quantum Wells and Wires

Now we are prepared to study electron–phonon scattering of dimensionally confined structures. In the preceding discussion, we presented a crystal as two subsystems of electrons and phonons. Then the interaction between these subsystems was taken into account. We can apply the same approach to quantum heterostructures. First, we have to take into consideration all modifications for heterostructures in both subsystems. Second, the electron–phonon interaction, in general, should be changed as well. We can divide these changes into two types. One type of change comes from modifications in the wave functions of electrons and phonons in heterostructures; all matrix elements will be modified since they depend on these functions. Another type of change originates from the possible variations in parameters of the Hamiltonian H_{e-ph} for different layers or different parts of the structure, etc. We will neglect variations of these parameters and concentrate on the effects caused by the modifications of the electron and the phonon subsystems.

Let us formulate the electron–phonon interaction, taking into account these modifications. Electron wave functions enter the matrix elements for electron–phonon scattering in an explicit form and do not require special analysis. We simply use the envelope wave functions F_β and $F_{\beta'}$ with

$$\beta = \{n_0, \vec{k}_0\}, \qquad \beta' = \{n, \vec{k}\} \tag{6.123}$$

$$\beta = \{l_1, l_2, \vec{k}_0\}, \qquad \beta' = \{l_1', l_2', \vec{k}\} \tag{6.124}$$

for cases of a quantum well and a quantum wire, respectively. Here, n and n' and l_1, l_2, l'_1, and l'_2 are the subband quantum numbers, as introduced in Chapter 3, while $\vec{k}_0$ and $\vec{k}$ are either in-plane vectors or one-dimensional vectors.

Consider the role of the phonon subsystem. This can be done in the framework of the so-called envelope phonon wave-function method. The basic idea is the following. Electron–phonon interactions introduced in the preceding sections were defined essentially as linear functions of atomic displacements. Since only long-wavelength phonons are important for electron–phonon scattering, one can use a macroscopic description of the interaction in terms of slowly varying acoustic and optical displacements, that is, one should use Eqs. (6.91) and (6.114) for the deformation potential and Eqs. (6.94)–(6.96) and (6.119) for the electrostatic potential induced by the lattice vibrations. Let $\vec{U}(\vec{r})$ be a displacement associated with either acoustic ($\vec{U} = \vec{u}$) or optical [$\vec{U} = \vec{u}^{(O)}$] lattice vibrations. We assume that for a heterostructure all phonon modes related to these kinds of vibrations are known, that is, the equations of the lattice vibrations and electrostatic potential are solved with proper boundary conditions. In Chapter 5, this problem has already been analyzed for acoustic and optical phonons in heterostructures and free-standing structures. Let the solutions be denoted by $\vec{w}_\nu(\vec{r})$, where ν contains all phonon numbers and accounts for all modes. The solutions should be normalized and orthogonal. Then they can be considered as the phonon form factors or the phonon envelope functions. The arbitrary displacement $\vec{U}(\vec{r})$ can be represented as a series:

$$\vec{U}(\vec{r}) = \sum_\nu Q_\nu \vec{w}_\nu(\vec{r}). \tag{6.125}$$

The coefficients of such an expansion Q_ν are the so-called normal coordinates, introduced on the basis of the complete set of the functions $\vec{w}_\nu(\vec{r})$. Obviously the expansion of Eq. (6.125) is a generalization of Eqs. (6.100) and (6.116). The latter were used for uniform bulk materials in which the set of functions corresponds to the plane waves of different polarizations. The normal coordinates Q_ν introduced in Eq. (6.125) behave like those studied in Section 5.3. In particular, the creation and the annihilation operator for phonons of mode ν can be defined as in Eqs. (5.31) and (6.81):

$$Q_\nu = \sqrt{\frac{\hbar V_0}{2M\omega_\nu}}(c_\nu + c_\nu^+), \tag{6.126}$$

where ω_ν stands for the frequency of the mode ν and $M = M_0$ for acoustic vibrations and $M = \mathcal{M}$ for optical modes. Within such an approach, the energy of the phonon subsystem takes the canonical form:

$$H_{\mathrm{ph}} = \sum_\nu \hbar\omega_\nu\left(c_\nu^+ c_\nu + \frac{1}{2}\right);$$

that is, each mode contributes to the total energy as an independent oscillator with energy $\hbar\omega_\nu$.

Next, one has to calculate the Hamiltonian of the electron–phonon inter-action in terms of the operators c_ν and c_ν^+. Because the Hamiltonian is always supposed to be a linear function of the displacements, one can find a linear form similar to that of Eq. (6.82).

For example, applying this procedure to the case of the short-range electron–phonon interaction, for the different mechanisms one gets

$$H_{\text{e-ph}, A}(\text{short-range}) = \sum_\nu \Gamma(\vec{r} \mid \nu, A)[c_\nu^{(A)} + c_\nu^{(A)+}],$$

$$\Gamma(\vec{r} \mid \nu, A) = \sqrt{\frac{\hbar V_0}{2VM_0\,\omega_\nu^{(A)}}}\, \mathcal{D}[\vec{\nabla}\vec{w}_\nu^{(A)}(\vec{r})],$$

(6.127)

$$H_{\text{e-ph}, O}(\text{short-range}) = \sum_\nu \Gamma(\vec{r} \mid \nu, O)[c_\nu^{(O)} + c_\nu^{(O)+}],$$

$$\Gamma(\vec{r} \mid \nu, O) = \sqrt{\frac{\hbar V_0}{2VM\omega_\nu^{(O)}}}\, [\vec{\mathcal{D}}^{(O)}\vec{w}_\nu^{(O)}(\vec{r})],$$

(6.128)

where we designate the envelope wave functions and frequencies for acoustic and optical phonons as $\vec{w}_\nu^{(A)}$ and $\omega_\nu^{(A)}$ and $\vec{w}_\nu^{(O)}$ and $\omega_\nu^{(O)}$, respectively. $\Gamma(\vec{r} \mid \nu, A)$ and $\Gamma(\vec{r} \mid \nu, O)$ obviously represent the coupling between the electrons and the phonons of the modes ν.

For long-range interactions, the Hamiltonian cannot be found explicitly in the general case. Actually, according to Eqs. (6.96) and (6.119), it is easy to calculate the polarizations $\vec{P}^{(A)}$ and $\vec{P}^{(O)}$ in terms of c_ν and c_ν^+, but then one needs to solve the Poisson equation (6.95), which can be done only by specifying the functions $\vec{w}_\nu(\vec{r})$. In Section 6.10, such a calculation will be analyzed for polar-optical phonons.

Now we return to the case of acoustic phonons. The general analysis of acous-tic vibrations given in Section 5.5 has shown that, in layered heterostructures, acoustic waves consist of extended and confined modes. The extended modes propagate throughout the whole heterostructure in any direction, while the con-fined modes are typically localized near quantum-well layers and propagate along them. Frequently, as a result of small differences in the elastic parameters of III–V compounds, acoustic confinement is relatively weak and the extended modes are approximately plane waves. This allows one to neglect the confine-ment effect and consider the electron–phonon scattering in the simple model of bulklike acoustic phonons.[†] In such an approximation, we can directly use the results of Eqs. (6.87)–(6.89) with the coupling factors in the form of Eqs. (6.93) and (6.101). Note that in Eq. (6.89) the phonon wave vector $\vec{q}$ is a three-dimen-sional vector. According to Eqs. (6.93) and (6.101), the coupling factors for both short- and long-range interaction mechanisms are independent of electron

---

[†] The confinement of the acoustic mode can be of great importance for other phenomena like the propagation of acoustic waves, acoustoelectronic effects, amplification of the waves by electron drift, etc.

coordinates. Thus the evaluation of the matrix element of Eq. (6.89) for the wave functions F_β and $F_{\beta'}$ is reduced to the calculation of the overlap integral:

$$\mathcal{I} = \int d\vec{r}\, F_{\beta'}^* e^{i\vec{q}\vec{r}} F_\beta. \tag{6.129}$$

In bulk materials, when the wave functions are plane waves, this integral yields $\mathcal{I} = |\mathcal{I}|^2 = \delta_{\vec{k}\pm\vec{q},\vec{k}'}$.

6.9.1 Quantum Wells

In quantum structures, the overlap integrals are modified considerably and depend on the geometry of the structure. For quantum wells, we use wave functions of the form of Eq. (6.16). To get $\mathcal{I}$ in an explicit form, let us use the model of a rectangular quantum well with an infinitely deep potential. This case was analyzed as the limiting case in Section 3.2. The subband energies are given by Eq. (3.15), and the wave functions are

$$F_{n,\vec{k}}(\vec{R}) = \sqrt{\frac{2}{SL}}\cos\frac{2\pi nz}{L} e^{i\vec{k}\vec{\rho}}, \tag{6.130}$$

where L is the width of the well and $\vec{\rho}$ is the position of an electron in the x, y plane. Calculation of the overlap integral gives

$$|\mathcal{I}_{2D}|^2 = G_{n_0,n,q_z}\delta_{\vec{k}_0\mp\vec{q}_{||},\vec{k}}, \tag{6.131}$$

where $\vec{q}_{||}$ is the projection of the three-dimensional phonon wave vector on the x, y plane and the so-called form factor G_{n_0,n,q_z} is

$$G_{n_0,n,q_z} = \frac{(\pi^2 n_0 n)^2 (q_z L/2)^2 \, \sin^2[q_z L/2 + (\pi/2)(n_0 + n)]}{\left[(q_z L/2)^2 - (\pi^2/4)(n_0 - n)^2\right]^2 \left[(q_z L/2)^2 - (\pi^2/4)(n_0 + n)^2\right]^2}. \tag{6.132}$$

Thus, according to Eq. (6.131), we obtain the conservation law for only planar projections of the momenta:

$$\vec{k}_0 \pm \vec{q}_{||} = \vec{k}.$$

The form factor G replaces the Kronecker delta with respect to z components of the wave vectors. Equation (6.132) shows that, in principle, for any transverse component of the phonon wave vector $\vec{q}$, electron–phonon scattering is possible. This is one of the results that brings about new features for scattering in heterostructures.

The scattering probabilities of Eqs. (6.87)–(6.88) and the characteristic times given by Eqs. (6.7)–(6.9) are determined by several factors: the form factor G, the energy-conservation law, the in–plane momentum-conservation law, and the

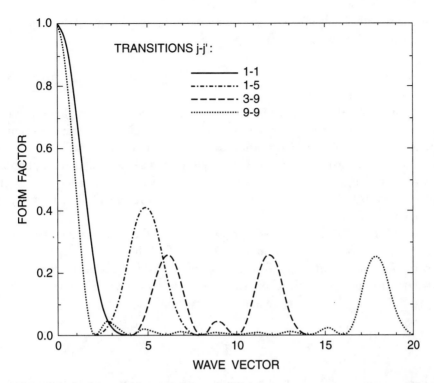

Figure 6.3. The form factors G_{n_0, n, q_z} as functions of the phonon wave-vector component $q_z L / \pi$. Results are shown for intrasubband $(1 \leftrightarrow 1, 9 \leftrightarrow 9)$ and intersubband $(1 \leftrightarrow 5, 3 \leftrightarrow 9)$ transitions.

phonon occupation numbers N_q. Because these probabilities are not amenable to analytical calculations, let us analyze these factors qualitatively.

Figure 6.3 demonstrates the form factors G as functions of the dimensionless phonon wave-vector component $q_z L / \pi$ for different subband numbers. One can see that the form factor for scattering within the lowest subband $(n_0 = n = 1)$ has a maximum only at $q_z = 0$ and the spreading of this function is given approximately by $\Delta q_z = 2\pi / L$. According to Eq. (6.132), for higher subband indices n_0, intrasubband scattering generally is characterized by the form factor G_{n_0, n, q_z} with two main maxima at $q_z = 0$ and $q_z = 2\pi n_0 / L$, as illustrated in Fig. (6.3) for $n_0 = n = 9$. Intersubband scattering events also have maxima at $q_z L = \pm \pi (n_0 \pm n)$, as shown for the $1 \leftrightarrow 5$ and $3 \leftrightarrow 9$ intersubband transitions. One can see that the form factors are continuous functions that are nonzero over a finite range of phonon wave vectors. Indeed, the electron can emit or absorb a phonon with any q_z from the range of nonzero form factor G.

It is instructive to recall the result known from general physics that the momentum-conservation law follows from the translation symmetry of space. When we deal with crystals with periodic lattice structures, translation invariance is responsible for electron wave vector conservation. In quantum heterostructures, we do not have translation invariance in the directions perpendicular to heterointerfaces; in particular, this is the case when electrons are located within the quantum-well layer. The transverse-momentum components of such electrons can be defined with an uncertainty given by

$$\Delta k_z \geq \frac{2\pi}{L}, \tag{6.133}$$

as discussed in Section 2.2. On the other hand, bulk acoustic phonons have well-defined momentum components in all directions. Hence the transverse component(s) of the phonon momentum cannot be conserved precisely during the interaction of an acoustic phonon with an electron. The dispersion in the phonon wave vector is approximately equal to the electron wave vector uncertainty given by relation (6.133). This agrees with the conclusion reached following analysis of the detailed structure of the form factor.

In Section 6.7, we used two conservation laws – the conservation of electron energy and three-dimensional momentum – to find the kinematics of electron-acoustic-phonon scattering. Now, the kinematics is determined by the energy-conservation and the in-plane momentum-conservation laws. This change makes a big difference. Actually, the energy-conservation equation is

$$\epsilon_n + \frac{\hbar^2 (\vec{k} \mp \vec{q}_\parallel)^2}{2m^*} - \epsilon_{n_0} - \frac{\hbar^2 \vec{k}^2}{2m^*} = \mp \hbar s_l \sqrt{q_\parallel^2 + q_z^2} \,,$$

where $-$ (+) corresponds to phonon emission (absorption). For intraband transitions ($n_0 = n$), we can rewrite the last equation in the form

$$\frac{q^2}{2k} \mp q_\parallel \cos\theta_q = \mp \left(\sqrt{\frac{m^* s_l^2 / 2}{\hbar^2 k^2 / 2m^*}} \right) \sqrt{q_\parallel^2 + q_z^2} \,,$$

where θ_q is the angle between $\vec{k}$ and $\vec{q}_\parallel$. According to the numerical estimates of Section 6.7, for typical electron energies the dimensionless factor in the right-hand side of the last equation is substantially smaller than unity. For $q_\parallel \ll k$ this immediately implies strong inequality between actual q_z and $q_\parallel$:

$$q_z \gg q_\parallel, \tag{6.134}$$

that is, two-dimensional electrons interact primarily with the phonons that have perpendicular components of the wave vectors that are much larger than the in-plane components. In other words, electrons effectively interact with phonons propagating primarily perpendicular to the interfaces.

The previously discussed features of low-dimensional electron–acoustic-phonon interaction result in an additional inelasticity in electron scattering in quantum structures. Indeed, the wave vector of interacting acoustic phonons can be estimated as $q \approx q_z \approx \Delta k_z \geq 2\pi/L$. The phonon energy associated with this wave vector is

$$\Delta E = \hbar s_l q \geq \hbar s_l \frac{2\pi}{L} \,. \tag{6.135}$$

One can see that the smaller the size of a quantum structure, the greater the transferred phonon wave vector and the acoustic-phonon energy associated with it. As discussed in Section 6.7 for bulk materials, the typical energy of an acoustic phonon interacting with an electron of kinetic energy E is $\sim 2\sqrt{2ms_l^2 E}$. The transferred energy of Eq. (6.135) can be represented similarly: $\Delta E \approx 2\sqrt{2ms_l^2 \epsilon_1}$, where ϵ_1 is the lowest subband energy. Since for quantum wells of widths less than $L = 100$ Å we estimated that ϵ_1 sufficiently exceeds the electron kinetic energy E, the uncertainty in question leads to an additional inelasticity of electron scattering by acoustic phonons compared with that of bulk materials. The inelasticity increases when the width L decreases. As a result, acoustic-phonon scattering turns out to be essentially inelastic and may become a channel of the energy relaxation of nonequilibrium electrons.

As the quantum-well width decreases, the uncertainty in the electron wave vector grows and more phonons are available for interaction with an electron. As a result, the scattering probabilities and rates increase inversely proportional to the well width. However, there is no divergence of the scattering rate as the width of the quantum well formally goes to zero. Indeed, the phonon occupation number $N_{\vec{q}}$ is responsible for the cutoff of high-energy phonons: their numbers decrease exponentially with increasing phonon energy. This factor becomes dominant when the width L becomes less than some critical value. This value is easy to estimate; to do this, we compare the energy of the phonons with the maximum wave vectors of $\sim 2\pi/L$ with the thermal phonon energy $k_B T$. This limits the further increase in the scattering rates for the widths $L \approx 2\pi\hbar s_l/k_B T$.

Summarizing this qualitative analysis of the key factors in Eqs. (6.87) and (6.88), we note that the electron-acoustic-phonon scattering in quantum wells involves phonons propagating perpendicularly to interfaces and having large wave vectors of the order of $2\pi/L$. The latter brings two important effects: (1) increasing transferred energy (inelasticity of scattering) and (2) increasing the rate scattering in proportion to L^{-1} in an extended interval of the well widths.

For the total scattering time τ_s, one can obtain analytical results by neglecting the inelasticity of electron-acoustic-phonon scattering, as was done previously for Eq. (6.110). The time of scattering of an electron from subband n_0 is defined by Eq. (6.7) and now can be specified as follows:

$$
\frac{1}{\tau_s(n_0, \vec{k}_0)} = \sum_{n,\vec{k}} \left[W^{(+)}(n_0, \vec{k}_0; n, \vec{k}) + W^{(-)}(n_0, \vec{k}_0; n, \vec{k}) \right]
$$

$$
= \frac{2\pi}{\hbar} \sum_{n,\vec{k},\vec{q}} \left| \Gamma(n_0, \vec{k}_0, n, \vec{k} \,|\, A) \right|^2 (2N_{\vec{q}} + 1)\delta[E_n(\vec{k}) - E_{n_0}(\vec{k}_0)].
$$

$$(6.136)$$

Here $N_{\vec{q}}$ is the acoustic-phonon number and $\Gamma(n_0, \vec{k}_0, n, \vec{k} \,|\, A)$ is defined by Eq. (6.89) and can be represented in the explicit form using Eqs. (6.93) and (6.101) for the deformation potential and the piezoelectric mechanisms, respectively. For the case of interaction by means of the deformation potential, we get

$$\frac{1}{\tau_s(n_0, \vec{k}_0)} = \frac{\mathcal{D}^2}{8\pi^2 C_l} \sum_n \int dq_z \, dq_{\|} q_{\|} \, d\theta_q q \, \sin\theta_q \, G_{n_0,n,q_z} (2N_q + 1)$$

$$\times \delta[E_n(\vec{k}) - E_{n_0}(\vec{k}_0)],$$

where C_l is defined by Eq. (6.113). For temperatures $k_B T \gg \hbar\omega_q$ [which, according to Eq. (6.135) is equivalent to $k_B T \gg 2\pi\hbar s_l/L$], we can approximate $(2N_{\bar{q}} + 1) \approx 2k_B T/\hbar s_l q$. Then integration over $q_{\|}$ and θ_q can be done easily:

$$\frac{1}{\tau(n_0, \vec{k}_0)} = \frac{\mathcal{D}^2 m^* k_B T}{\pi\hbar^3 C_l} \sum_n^* \int_{-\infty}^{+\infty} G_{n_0,n,q_z} dq_z,$$

where the $*$ means that the sum includes only those subbands corresponding to allowed transitions from the (n_0)th subband, i.e., $\epsilon_n \leq E_{n_0}(\vec{k}_0)$. In the model of infinitely deep wells, the last integral can be calculated as

$$\int_{-\infty}^{+\infty} dq_z G_{n_0,n,q_z} = \frac{\pi}{L} \left(2 + \delta_{n_0,n}\right).$$

Finally, we get the simple result

$$\frac{1}{\tau_s(n_0, \vec{k}_0)} = \frac{2\mathcal{D}^2 m^* k_B T}{\hbar^3 C_l L} \sum_n^* \left(1 + \frac{1}{2}\delta_{n_0,n}\right). \tag{6.137}$$

Thus the total scattering rate increases as the well narrows. This result does not depend on the electron kinetic energy explicitly (compare with the result of Table 6.3), but it is subject to a steplike increase when the electron energy reaches the bottom of the next subband. This behavior simply reflects the steplike dependence of the two-dimensional density of states.

For the same quasi-elastic approximation, the total scattering time for the piezoelectric interaction mechanism takes the form

$$\frac{1}{\tau_s(n_0, \vec{k}_0)} = \frac{e^2 m^* k_B T}{2\pi\hbar^3} \beta_{1,23}^2 \left(\frac{12}{35 C_l} + \frac{16}{35 C_t}\right) \sum_n^* \int_{-\infty}^{+\infty} dq_z \frac{G_{n_0,n,q_z}}{q_z\sqrt{q_z^2 + 4k_0^2}}, \tag{6.138}$$

where C_t and C_l are defined by Eq. (6.113). Further calculations of Eq. (6.138) are possible only numerically. But from the form of Eq. (6.138) we can draw the conclusions that the scattering rate for the piezoelectric mechanism decreases with the electron wave vector $\vec{k}_0$ and drops for transitions to higher subbands.

Numerical estimates made for AlGaAs/GaAs quantum wells show that acoustic-phonon scattering becomes important for temperatures below 10–20 K.

6.9.2 Quantum Wires

Let the propagation of an electron in a quantum wire be restricted in the y and the z directions: in other words, we take the wire to be along the x axis. The electron wave functions should be taken in the form

$$\psi_{l_1,l_2,k}(x, y, z) = \frac{1}{L_x}\chi_{l_1,l_2}(y, z)e^{ikx},$$

where $\chi_{l_1,l_2}(y, z)$ and $\{l_1, l_2\}$ are the subband wave functions and the subband numbers, respectively, and $k = k_x$ is the only component of the electron wave vector. Assuming the well to be of a rectangular cross section $L_y \times L_z$ and infinitely deep, we can use the wave functions of Eqs. (3.52) and (3.53). The overlap integral of Eq. (6.129) becomes

$$\mathcal{I}_{1D} = G_{l_1,l_1',q_y}G_{l_2,l_2',q_z}\delta_{k',k+q_x}, \tag{6.139}$$

where the form factors are given by Eq. (6.132) with the replacements $\{n_0, n, q_z\} \rightarrow \{l_1, l_1', q_y\}, \{l_2, l_2', q_z\}$. The Kronecker delta and consequently the wave-vector-conservation law refer only to wave vectors or their projections directed along the x axis. Because there is free propagation along only one axis, we can distinguish forward or backward scattering of an electron with a finite wave vector $\vec{k}$.

All qualitative conclusions made for quantum wells are valid for the wires. Furthermore, the effects of the increase in a transferred phonon wave vector $\sqrt{q_y^2 + q_z^2}$ and inelasticity are enhanced even more for quantum wires. Analogous to the previous case, acoustic-phonon scattering intensifies inversely proportional to the area of the wire cross section. The quasi-elastic approach used above leads to a divergence of the scattering rate near the bottom of each of the one-dimensional subbands, which is proportional to the one-dimensional density of states. More realistic evaluations can be performed numerically. In Fig. 6.4, the inverse total scattering times – the rates – are presented for two quantum wires of AlAs/GaAs with cross sections 250 Å $\times$ 150 Å and 40 Å $\times$ 40 Å. The results are calculated for the deformation-potential mechanism and intrasubband processes at $T = 30$ K. Forward and backward scatterings are shown; they almost coincide and differ only for electrons with high energies. Absorption and emission rates are plotted separately. The curves with maxima correspond to the emission rates. One can see that the scattering rates increase significantly for a thinner wire. Absolute values of the rates are large and in the range 10^{11}–10^{12} s^{-1}; as a result, this type of scattering becomes predominant at temperatures below 30–40 K.

6.9.3 Free-Standing Quantum Structures

As mentioned in Subsection 6.9.1 the confinement effect on electron-acoustic-phonon scattering is not important in typical III–V heterostructures. The

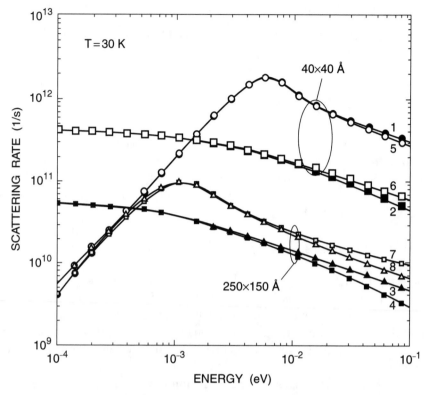

Figure 6.4. Acoustic-phonon-scattering rates for AlAs/GaAs quantum wires. The dimensions of rectangular cross sections are indicated. The temperature is taken to be 30 K. The rates of absorption (monotonous curves) and emission (nonmonotonous curves) are shown separately. Backward (1,2,3,4) and forward (5,6,7,8) scatterings of an electron with a finite wave vector are different, as is seen for large wave vectors (energies).

situation is completely different in free-standing quantum structures. In Subsection 5.5.2 we showed that for these structures, only confined phonons exist. The confinement leads to significant changes of acoustic-phonon spectra $\omega_m(\vec{q}_\parallel)$ and phonon form factors (phonon wave functions), which become

$$\vec{w}_\nu(\vec{r}) = \frac{1}{\sqrt{S}}\vec{u}_{m,\vec{q}_\parallel}(z)e^{i\vec{q}_\parallel\vec{\rho}}$$

with orthonormalized vectors $\vec{u}_{m,\vec{q}_\parallel}(z)$. On the basis of previously presented results, the modifications in both the spectra and the form factors should be taken into account in calculating electron–phonon scattering in free-standing quantum structures. In particular, for the deformation-potential mechanism that dominates at low temperatures, the coupling is

$$\Gamma(\vec{\rho}, z \mid m, \vec{q}_\parallel, A) = \mathcal{D}\sqrt{\frac{\hbar V_0}{2VM_0\omega_n(\vec{q}_\parallel)}}\left\{i\vec{q}_\parallel\vec{u}_{n,\vec{q}_\parallel}(z) + \frac{d}{dz}[u_{n,\vec{q}_\parallel}(z)]_z\right\}e^{i\vec{q}\vec{\rho}}.$$

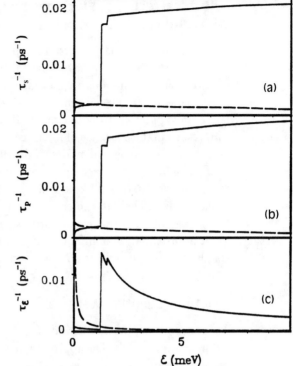

Figure 6.5. The scattering rates due to confined acoustic phonons in a GaAs freestanding quantum structure with a width of 100 Å. (a) The total scattering rate $1/\tau_s$, (b) the inverse momentum-relaxation time $1/\tau_p$, (c) the inverse energy-relaxation time $1/\tau_E$. The solid and the dashed curves correspond to acoustic-phonon emission and absorption, respectively. The results are obtained for nondegenerate electrons at $T = 4$ K.

Figure 6.5 presents the results of calculations of the total scattering time and the momentum- and the energy-scattering times. Calculations are made for a GaAs free-standing quantum well of 100-Å width at $T = 4$ K. Contributions of both processes, emission and absorption of confined phonons, are shown separately. The emission rates have steplike shapes and an additional structure due to the onsets of the confined modes; see Figs. 5.16 and 5.17. Keeping in mind that the intensity of scattering is an approximately linear function of the temperature, as in Eqs. (6.137) and (6.138), we can estimate the total scattering rate to be of the order of that for a quantum well with similar dimensions. The fact that all three times, τ_s, τ_p, and τ_E, are of the same order of magnitude demonstrates a large transferred momentum and a strong inelasticity of electron scattering by the confined phonons.

In conclusion, we have shown that the major features of electron scattering by acoustic phonons in quantum structures are a strong dependence on geometry scales and an increase in inelasticity. In some cases, the energy of an acoustic phonon involved in the scattering process may be of the order of the optical-phonon energy and exceed the mean energy of electrons. Therefore electron scattering by acoustic phonons in quantum structures becomes an important channel of energy relaxation. For electrons with energies lower than the optical-phonon energy, the electron-acoustic-phonon interaction may become the dominant scattering mechanism. This interaction defines the low-field electron mobility in perfect heterostructures at low temperatures.

6.10 Scattering of Electrons by Optical Phonons in Quantum Wells and Wires

In this section, we study electron scattering by optical phonons in heterostructures. In Section 5.6, we have learned that, in contrast to the case of acoustic phonons, optical phonons are modified significantly in III–V heterostructures. In particular, if the frequency bands of optical phonons in two materials do not overlap, optical phonons from one material cannot penetrate into another material. This results in the existence of interface optical modes localized near heterointerfaces and confined optical modes within quantum wells and wires. Thus electrons confined in a quantum well or a quantum wire interact with confined modes, interface modes, and bulklike optical phonons of the materials surrounding the wells or wires. The latter interaction can be negligible if an electron is confined so strongly that its penetration into its surroundings is small. This corresponds to the approximation of infinitely high barriers surrounding wells and wires. For simplicity this approximation is assumed in Subsections 6.10.1 and 6.10.2 .

6.10.1 Quantum Wells

First, consider the interaction of an electron with longitudinal confined phonons in quantum wells. The approximate solutions for longitudinal confined optical phonons have been analyzed in Section 5.6. Such modes are characterized by in-plane wave vectors $\vec{q}$ and discrete (transverse) numbers m. It is worth emphasizing that the confined phonons are two-dimensional phonons. Their frequencies, in general, depend on $\vec{q}$ and m, but because the dispersion of optical phonons is small in the long-wavelength limit, we can attribute the same frequency ω_{LO} to all confined modes. In the spirit of the general analysis of Section 6.9, we should base our calculation on the complete set of orthogonal and normalized solutions for longitudinal confined modes $\vec{w}_{\vec{q},m}(\rho, z)$. We use the solutions of Eqs. (5.102) that satisfy the boundary conditions of Eqs. (5.106) and have the quantized q_z projections given by Eqs. (5.105). We get the following projections of the vectors $\vec{w}_{\vec{q},m}(\rho, z)$ to the x, y plane and the z axis:

$$(\vec{w}_{\vec{q},m})_{\parallel} = \sqrt{\frac{2}{SL}} \frac{i\vec{q}\, H_m(z)}{\sqrt{q^2 + \left(\frac{\pi m}{L}\right)^2}}\, e^{i\vec{q}\vec{\rho}}, \qquad (\vec{w}_{\vec{q},m})_z = \sqrt{\frac{2}{SL}} \frac{\frac{\pi m}{L}\frac{dH_m(z)}{dz}}{\sqrt{q^2 + \left(\frac{\pi m}{L}\right)^2}}\, e^{i\vec{q}\vec{\rho}}.$$

Here the function $H_m(z)$ corresponds to the symmetric and antisymmetric solutions in terms of Eqs. (5.102):

$$H_m(z) = \begin{cases} \cos(\pi mz/L) & \text{odd } m \\ \sin(\pi mz/L) & \text{even } m \end{cases}.$$

Having the explicit form of $\vec{w}_{\vec{q},m}$, we can introduce the optical displacement associated with a particular solution $\vec{u}^{(O)} = Q_{\vec{q},m}\vec{w}_{\vec{q},m}$, as well as the polarization $P = \gamma\vec{u}^{(O)}$; see Eq. (6.119). Then we calculate the electrostatic potential $\phi_{\vec{q},m}$ from

Poisson's equation (6.95) and the contribution to $H_{e-ph,LO}$:

$$-e\phi_{\vec{q},m}(\vec{\rho}, z) = -4\pi e\gamma\sqrt{\frac{2}{SL}}\frac{Q_{\vec{q},m}H_m(z)}{\sqrt{q^2 + (\frac{\pi m}{L})^2}}\,e^{i\vec{q}\vec{\rho}}.$$

From this procedure, we get the coupling of electrons and confined LO phonons:

$$\Gamma(\vec{r}\,|\,q, m, \text{LO}) = -4\pi e\gamma\sqrt{\frac{\hbar V_0}{SLM\omega_{\text{LO}}}}\frac{H_m(z)}{\sqrt{q^2 + (\frac{\pi m}{L})^2}}\,e^{\vec{q}\vec{\rho}}. \tag{6.140}$$

Next we calculate the matrix elements of this coupling by using the wave functions of the confined electrons as given by Eq. (6.130):

$$\Gamma(n_0, \vec{k}_0, n, \vec{k}\,|\,\vec{q}, m, \text{LO}) = -4\pi e\gamma\sqrt{\frac{\hbar V_0}{SLM\omega_{\text{LO}}}}\sqrt{\frac{G_{n_0,n,m}}{q^2 + (\frac{\pi m}{L})^2}}\,\delta_{\vec{k},\vec{k}+\vec{q}}.$$

Here, $G_{n_0,n,m}$ is the square of the overlap integral that can be calculated analytically:

$$\begin{aligned}
G_{n_0,n,m} &= \left[\frac{2}{L}\int_{-L/2}^{L/2}dz\cos\frac{\pi n_0 z}{L}\cos\frac{\pi n}{L}H_m(z)\right]^2 \\
&= \frac{32(n_0 nm)^2\left[1 - (-1)^{n_0+n+m}\right]}{\pi^2\left[m^2 - (n_0 + n)^2\right]^2\left[m^2 - (n_0 - n)^2\right]^2}.
\end{aligned}$$

Finally, for the scattering rate we get

$$\begin{aligned}
\frac{1}{\tau_s(k_0, n_0)} &= \frac{2e^2\omega_{\text{LO}}}{L}\left(\frac{1}{\kappa_\infty} - \frac{1}{\kappa_0}\right)\left(N_{\text{LO}} + \frac{1}{2} \pm \frac{1}{2}\right) \\
&\quad\times \sum_{n,m}G_{n_0,n,m}\int d^2q\frac{\delta\left[E_n(\vec{k}_0 + \vec{q}) - E_{n_0}(\vec{k}_0) \mp \hbar\omega_{\text{LO}}\right]}{q^2 + (\frac{\pi m}{L})^2}, \tag{6.141}
\end{aligned}$$

where we exclude γ by means of the static and high-frequency permittivities (see Section 5.6). Further calculations can be made only numerically. But we can draw qualitative conclusions from these formulas. First, we get the in-plane wave-vector-conservation law: $\vec{k} = \vec{k}_0 + \vec{q}$. Then, from the form of the overlap integral $G_{n_0,n,m}$, some selection rules follow. Intraband transitions ($n_0 = n$) occur if one of the symmetric modes (m is odd) is involved. Scattering between the electron states with the same symmetry [n_0 is odd (even) and $n(\neq n_0)$ is odd (even)] is possible with a phonon mode of the same symmetry [m is odd (even)]. From the coupling factor, we can see that, in contrast to the case of acoustic phonons, scattering by confined optical phonons decreases when the width of the well decreases. This

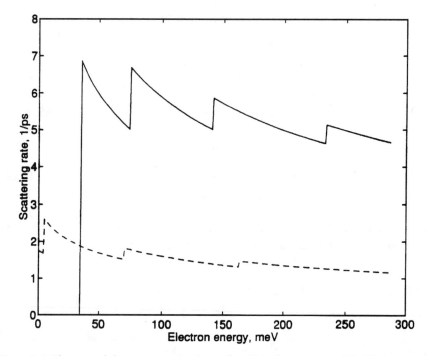

Figure 6.6. The rates of electron scattering by confined LO phonons in an AlAs/GaAs quantum well at $T = 30$ K for $L = 200$ Å within and from the first electron subband. The emission and the absorption rates are shown by the solid and the dashed curves, respectively.

result originates directly from the electric character of electron–polar-phonon interaction.

As pointed out, interaction with the LO phonons is the primary phonon mechanism responsible for scattering at temperatures above several tens of degrees Kelvin. In Fig. 6.6, the scattering rate is shown for a AlAs/GaAs quantum well at $T = 30$ K. The figure provides plots of the LO phonon-scattering rates for transitions within or from the first subband for a well width of 200 Å. Up to ten confined modes are accounted in Eq. (6.141). Contributions of emission and absorption are presented separately. The overall energy dependence of the scattering rate is reminiscent of that in bulk materials, although the former exhibits the sawtooth fine structure due to intersubband transitions.

Consider briefly the scattering by the interface modes. In Section 5.6, we have found that at a single heterojunction, two interface modes appear. Each is characterized by the in-plane wave vector $\vec{q}$ and exponentially decays far from the heterojunction, as given by Eq. (5.112). The two have different frequencies $\omega_{\text{IF},1,2}(\vec{q})$ and amplitudes. For the case of a quantum well, there are two heterojunctions, and each of them generates interface modes. Combinations of the modes coming from two junctions can be represented as symmetric (s) and antisymmetric (a) functions of the z coordinate. Applying the previously described procedure to introduce normal coordinates corresponding to the interface modes, we can get the coupling between these modes and the electrons in the forms

$$\Gamma(\vec{r}\,|\,\vec{q},s,\text{IF}) = \sqrt{\frac{2\pi\hbar^2\omega}{S}}\left[\beta_1^{-1}(\omega)\tanh\left(\frac{qL}{2}\right) + \beta_2^{-1}(\omega)\right]^{-1/2}\frac{1}{\sqrt{2q}}\,f_s(z)e^{i\vec{q}\vec{\rho}},$$

$$\Gamma(\vec{r}\,|\,\vec{q},a,\text{IF}) = \sqrt{\frac{2\pi\hbar^2\omega}{S}}\left[\beta_1^{-1}(\omega)\coth\left(\frac{qL}{2}\right) + \beta_2^{-1}(\omega)\right]^{-1/2}\frac{1}{\sqrt{2q}}\,f_a(z)e^{i\vec{q}\vec{\rho}},$$

for the symmetric and the antisymmetric solutions, respectively. For the two different types of the interface modes, one should use an appropriate frequency from the pair $\omega = \omega_{\text{IF},1,2}$. The index $i = 1, 2$ refers to the parameters of the well and the surrounding material, respectively. Here, we also define

$$\beta_i = \left(\frac{1}{\kappa_{\infty i}} - \frac{1}{\kappa_{0i}}\right)\frac{\omega_{\text{LO}i}^2}{\omega^2}\left(\frac{\omega^2 - \omega_{\text{TO}i}^2}{\omega_{\text{TO}i}^2 - \omega_{\text{LO}i}^2}\right)^2,$$

$$f_s(z) = \begin{cases} e^{q(L/2-|z|)} & |z| > L/2 \\ \dfrac{\cosh(qz)}{\cosh(qL/2)} & |z| \le L/2 \end{cases},$$

$$f_a(z) = \begin{cases} -e^{q(L_z/2-|z|)} & |z| > L/2 \\ \dfrac{\sinh(qz)}{\sinh(qL_z/2)} & |z| \le L/2 \end{cases}.$$

Note that because the interface modes penetrate both materials, their parameters determine the coupling with electrons.

Using these coupling constants, one can calculate the scattering by the interface phonons. Qualitative analysis can be done similarly to what has been done previously for the confined optical phonons. Importantly, the conservation of in-plane wave vectors is still valid. In contrast to preceding case, the coupling with the interface phonons depends significantly on the in-plane phonon wave vector. Then, another difference is that the interface phonon scattering increases when the well width diminishes. This follows from the simple fact that there is more overlap between wave functions of electrons and phonons in narrow wells.

Figure 6.7 illustrates the energy dependence of the scattering rate due to the emission of an interface phonons in a quantum-well structure. To stress the importance of these phonons, at least in narrow wells, we present the results for a AlAs/GaAs well with a 30-Å width at a temperature of 30 K. Because of the presence of two different interface phonon modes, GaAs-like and AlAs-like, with different frequencies and because some intersubband transitions are forbidden, the fine structure of the interface phonon-scattering rate differs slightly from that of a confined phonon mode. Also, the scattering rate for the latter is presented. The rate of acoustic-phonon emission is shown to facilitate comparison of all phonon mechanisms. In the case of such a narrow well, interface phonon scattering is dominant. In the case presented in Fig. 6.6, scattering by confined optical phonons dominates. Thus, for quantum wells of different widths, different kinds of polar optical modes play the most significant role.

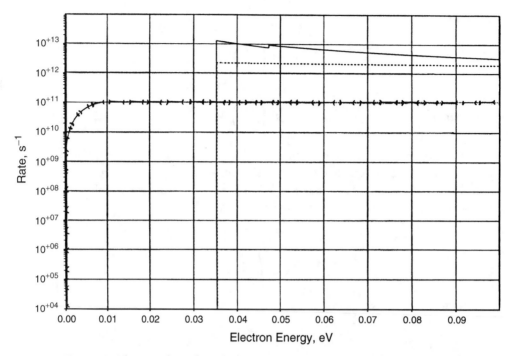

Figure 6.7. The rate of interface phonon emission in a GaAs/AlAs quantum well (solid curve) at $T = 30\,\mathrm{K}$ for a well width of 30 Å. The rates of emission of confined optical phonons (dotted line), and bulklike acoustic phonons (dotted–dashed curve) are shown for comparison.

6.10.2 Quantum Wires

Let us restrict ourselves to brief remarks on polar-optical scattering in quantum wires. As for the previous case, this scattering is due primarily to confined and interface (one-dimensional) phonons existing in quantum wires. Analysis shows that there is the one-dimensional wave-vector-conservation law and the phonons assist both intrasubband and intersubband electron transitions. The relative contribution to the scattering rate of confined and interface phonons depends on the wire cross section: the confined phonons dominate in wide wires, while the interface phonons play a dominant role in narrow wires. Figure 6.8 presents plots of the rates of electron scattering by the confined phonons from the first subband to all available states. Numerical calculations are done for an AlAs/GaAs quantum wire with a cross section of 150 Å × 250 Å. The unique features of this energy dependence are the multiple divergent peaks, each of which is related to a particular intersubband transition. The origin of these peaks is evidently the divergent electron density of states at the bottom of each subband. It must be stressed here that this divergence is integrable, i.e., the average scattering intensity in the equilibrium electron system is finite. Another important feature associated with the integrable properties of the rate is that electrons accelerated by an electric field fly through the energy corresponding to such an infinitely high scattering region with a probability of less than unity of being scattered. This is

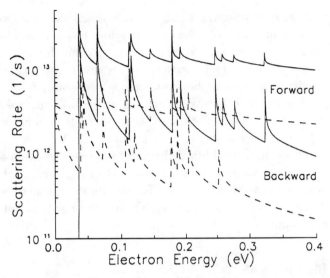

Figure 6.8. The electron-scattering rates by confined optical phonons in an AlAs/GaAs quantum wire as functions of the electron energy. The wire cross section is 150 Å × 250 Å. The temperature is assumed to be 30 K. The solid curves represent emission rates, while the dashed curves represent absorption rates. Forward and backward scattering are shown separately. The complex nonmonotonous structure is due to a number of interband transitions.

reminiscent of electron tunneling through an infinitely high and narrow potential barrier. Figure 6.9 demonstrates the same rates of electron scattering by the interface phonon modes. The same divergent peaks are present but their number is smaller than in the case of the confined phonons. The point is that some intersubband transitions due to the interface optical phonons are forbidden because of

Figure 6.9. The electron-scattering rates by interface optical phonons in a quantum wire with the parameters referred to in Fig. 6.8 as functions of the electron energy.

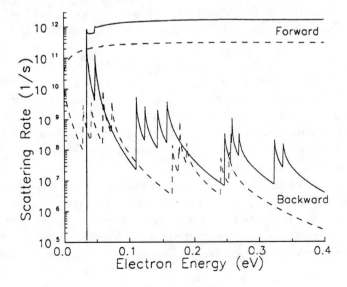

the symmetry of the electron and the phonon wave functions. As we mentioned in Section 6.9, for a quantum wire one can distinguish forward- and backward-scattering processes. In both Figs. 6.8 and 6.9, backward- and forward-scattering rates are plotted separately. The forward-scattering rate is considerably higher. This is because the polar optical-phonon-scattering probability is a function of the phonon wave number $\vec{q}$. Forward-scattering events involve smaller q than backward-scattering events do and thus are more probable. In other words, we can conclude that scattering in the wires is extremely anisotropic.

In conclusion, we have established that electron-optical-phonon scattering is strongly modified in quantum wells and wires. Only modified, confined, and interface optical phonons scatter electrons. The modifications lead to more complex energy dependences of scattering rates, which have to be taken into consideration in electron kinetics.

6.11 Closing Remarks

Electron scattering is a factor of vital importance for any electronic or optoelectronic devices. Scattering determines the electric resistivity, current-voltage characteristic, the speed of operation of the device, laser efficiency, sensitivity of detectors, etc. In quantum structures, scattering processes are modified significantly compared with those in bulklike materials. In this chapter, we have studied the most common processes of electron scattering in bulk materials and heterostructures. Each of such processes was defined first for the bulk case and then necessary modifications were made in order to get a proper description of the scattering in heterostructures.

We evaluated electron scattering in terms of probabilities of transitions between different electron states and also defined the different characteristic times: total scattering time (the inverse scattering rate) and momentum- and energy-scattering times. The results obtained for scattering probabilities and scattering rates are to be used in the Boltzmann transport equation and other kinetic equations describing the electric and the optical properties in electronic and optical devices.

We began with an analysis of elastic scattering by impurities, defects, and interface roughness. For this case, the main differences in scattering low-dimensional electrons and bulk electrons are the changes in properties of confined electrons, the different characters of screening effects, considerable spatial nonuniformity of the impurity distributions, etc. In particular, we have found that the spatial separation of electrons and charged impurities provides a significant suppression of the scattering rates.

We have studied the electron–phonon interaction as one of the important channels of electron relaxation. We have established that, for typical situations, electrons interact with long-wavelength (center-zone) phonons. For such phonons, it is possible to separate short-range and long-range interaction mechanisms. Then we found that the interactions with different phonon modes can be introduced independently. For acoustic-phonon scattering in heterostructures, we neglected the confinement effect and formulated the results for scattering by means of the deformation potential and piezoelectric mechanisms. We found that this type of scattering is intensified significantly in quantum wells and wires. For example, if for a pure GaAs material, the deformation potential does not play a role at any temperature, in AlGaAs/GaAs quantum wells and especially in quantum wires this mechanism is dominant up to 30 K. For the case of free-standing quantum structures, we took into consideration the confined acoustic phonons, which are important for low-temperature scattering.

For the case of the electron–optical-phonon interaction, we introduced interaction by means of the deformation potential in atomic semiconductors and polar-optical-phonon scattering in ionic semiconductors. Scattering by optical phonons dominates at temperatures above several tens of degrees Kelvin in materials such as III–V compounds. In heterostructures, the scattering by bulklike optical phonons is almost negligible; instead, the confined and the interface optical phonons contribute dominantly to the scattering rates. We analyzed scattering by these phonons and found that it is different from that in bulk materials. In particular, the confined and the interface phonons bring about quite complex energy dependences of scattering times and a strong dependence on geometric factors of heterostructures.

In this chapter, to illustrate the general results, we have frequently used as examples GaAs materials and free-standing structures and AlGaAs/GaAs quantum wells and wires. As for the semiconductors of group IV, the data for bulk Si and Ge can be found in, E. M. Conwell, *High Speed Transport* (Academic, New York, 1967). More details on scattering in bulk crystals may be found in the following books:

V. E. Gantmakher and Y. B. Levinson, *Carrier Scattering in Metals and Semiconductors* (North-Holland, Amsterdam, 1987).

K. Hess, *Advanced Theory of Semiconductor Devices* (Prentice-Hall, Englewood Cliffs, NJ, 1988).

B. K. Ridley, *Quantum Processes in Semiconductors*, 3rd ed. (Clarendon, Oxford, 1993).

Detailed discussions on the electron–phonon interaction and scattering in quantum wells are presented in the following publications:

B. K. Ridley, "Hot electrons in low-dimensional structures," Rep. Prog. Phys. **54**, 169–256 (1991).

B. K. Ridley, "Electron–phonon interaction in 2D-systems," in *Hot Carriers in Semiconductor Nanostructures (Physics and Applications)*, J. Shah, ed. (Academic, Boston, 1992), pp. 17–51.

B. K. Ridley, *Electrons and Phonons in Semiconductor Multilayers* (Cambridge U. Press, Cambridge, 1997).

One can find the analysis of scattering by acoustic phonons in quantum wires and free-standing structures in the following papers:

N. Bannov, V. V. Mitin, and M. Stroscio, "Confined acoustic phonons in free-standing quantum wells and their interaction with electrons," Phys. Status Solidi (b) **183**, pp.131–142 (1994).

B. K. Ridley and N. A. Zakhleniuk, "Hot electrons under quantization conditions, I–III," J. Phys. Condensed Matter **8**, pp. 8525–8581 (1996).

A general review on confined phonons that contains 170 references is

M. A. Stroscio, G. J. Iafrate, H. O. Everitt, K. W. Kim, Y. Sirenko, S. Yu, M. A. Littlejohn, and M. Dutta, "Confined and interface optical and acoustic phonons in quantum wells, superlattices, and quantum wires," in *Properties of III–V Quantum Wells and Superlattices*, Pallab Bhattacharya, ed. EMIS Data Review Series (INSPEC, Institution of Electrical Engineers, London, 1996), pp. 194–208.

PROBLEMS

1. Derive the selection rules for electron intersubband scattering by confined LO phonons in a quantum well.

2. Based on the Hamiltonian for electron-confined LO phonons in a rectangular quantum wire and Fermi's golden rule, derive the corresponding scattering rate from the state k_x in subband $\{j, l\}$ to any state in the subband $\{j', l'\}$.

3. Consider the case in which electrons are injected into a polar semiconductor with an energy close to the LO phonon energy. In such a case, the electron distribution forms a spherical shell in $\vec{k}$ space. Now consider the situation in which an external electric field is applied. Electrons with positive longitudinal components of the wave vector – with the wave vector against the electric field – will gain energy and will be capable of emitting or absorbing a LO phonon. On the other hand, those electrons with negative longitudinal components of the wave vector will be shifted down in energy and will not be able to undergo LO phonon emission. In the absence of rapid randomization of the three-dimensional electron momentum due to electron–electron, acoustic-phonon, and ionized impurity scattering, what would be the net sign of the electron velocity? This effect has not yet been observed in bulk semiconductors. Would this effect be easier to observe in a quasi-one-dimensional quantum wire?

4. Calculate the scattering rate due to the screened Coulomb potential of an impurity of charge Ze:

$$U(\vec{r}) = \frac{Ze^2}{\kappa} \frac{e^{-\lambda r}}{r} \; ;$$

assume that the initial and the final electron wave functions are characterized by the nominal plane-wave solutions in a bulk crystal.

5. The electron-polar-optical-phonon-scattering rate may be derived to be of the form

$$W(k) = \left(\frac{2e^2\omega_0}{\hbar v \kappa_{eff}} \right) \left[n(\omega_0) \ln \left(\frac{q_{max}}{q_{min}} \right) + [n(\omega_0) + 1] \ln \left(\frac{q_{max}}{q_{min}} \right) \right],$$

where

$$\frac{1}{\kappa_{eff}} = \frac{1}{\kappa_\infty} - \frac{1}{\kappa_s}, \qquad q_{max} = k \left[1 + \sqrt{1 \pm \frac{\hbar\omega_0}{E(k)}} \right],$$

$$q_{min} = k \left[\mp 1 \pm \sqrt{1 \pm \frac{\hbar\omega_0}{E(k)}} \right],$$

with the upper and the lower signs corresponding to phonon emission and absorption, respectively, and $E(k) > \hbar\omega_0$ for phonon emission. As discussed in the text, LO phonons produce a space-charge modulation at the phonon frequency ω_0, which interacts with the scattered electron; the electric displacement associated with this displacement is proportional to the displacement amplitude of the anion and the cation. Indeed, Newtons's law for this system may be written as $e^* \kappa_{eff} \vec{D} = m^* \omega^{2(O)} \vec{u}$, where e^* is the effective charge associated with the anion–cation interaction. For carrier densities greater than $\sim 10^{17}\,\mathrm{cm}^{-3}$, there is another longitudinal space-charge mode that causes significant scattering of electrons. This second mode is known as a plasmon and its frequency ω_p is determined by the resonant frequency of oscillation of the electron gas with respect to the fixed ions. Indeed, from $\nabla \vec{D} = 4\pi n$, it follows that for a density of carriers n and a dielectric constant κ that $\vec{D} = 4\pi n e \vec{u} / \kappa$. Thus

$$nm^* \frac{d^2\vec{u}}{dt^2} = -ne\vec{D} = -\frac{4\pi n^2 e^2 \vec{u}}{\kappa}, \qquad \omega_p = \left(\frac{4\pi n e^2}{\kappa m^*} \right)^{1/2}.$$

Hence,

$$e\kappa \vec{D} = m^* \omega_p^2 \vec{u}.$$

Furthermore, the Hamiltonian for an electron interacting with a longitudinal polarization field is proportional to the volume integral of the electric displacement of the charge time of the field associated with the polarization of the longitudinal mode. From these similarities, would you expect the electron-plasmon-scattering rate to have the same form as the electron LO phonon-scattering rate?

6. Show that for the case of dilute impurities located in a thin doped layer that the mobility associated with scattering by remote impurities is given approximately by

$$\mu = 16 |Z_0|^3 \frac{e}{\hbar} \sqrt{2\pi n},$$

if it is assumed that the density of ionized impurities N_D^{2D} equals the electron density n and $k_F = \sqrt{2\pi n}$. Z_0 denotes the location of the impurity layer.

7. As stated in Section 6.8,

$$\frac{1}{\tau_s^{(\text{PLO},\pm)}} = \frac{1}{\tau^{(\text{PLO})}} \left[N^{(O)} + \frac{1}{2} \pm \frac{1}{2} \right] \ln \left| \frac{\sqrt{E} + \sqrt{E \mp \hbar\omega_0}}{\sqrt{E} - \sqrt{E \mp \hbar\omega_0}} \right| .$$

The notation in this expression is exactly as in Section 6.8. Provide a complete derivation of this result that gives all the intermediate steps necessary to obtain this result.

8. Evaluate $\sqrt{\hbar/(2m^*\omega_{\text{LO}})}$ for 36-meV optical phonons and for the GaAs effective mass. What is the physical significance of this quantity?

7

Parallel Transport in Quantum Structures

7.1 Introduction

In previous chapters, we have studied electron-scattering rates and the electron and phonon spectra of various semiconductor structures. In doing so, we have established a foundation for studying various nonequilibrium electronic phenomena in quantum structures. Among the large number of nonequilibrium electron phenomena – which determine the principles and regimes of device operation as well as the limitations of their applications – we will focus our attention on electron transport in electric fields and optical effects. Whole classes of processes – such as thermoelectricity and acoustoelectronic phenomena, which also have broad specific applications – are not within the scope of this text.

The variety of physically realizable regimes of electron transport in quantum structures is much broader than that for bulk materials. Let us classify the different transport regimes of quantum structures.

In Chapter 2, we have defined both classical and quantum electron-transport regimes. Now we take another look at electron transport in quantum structures from the standpoint of device geometry.

For layered structures, such as structures with interfaces, quantum wells, and superlattices, as well as for wire-containing structures, we can distinguish two types of electron transport. The first type is related to a motion of electrons along the layers or the wire axis. This case is called parallel or horizontal transport, as shown schematically in Fig. 7.1(a). The second type of the transport is related to the electron transfer in the direction(s) perpendicular to the layers or to the wire axis, as depicted in Fig. 7.1(b). One refers to this type of electron motion as perpendicular or vertical transport. For quantum dots and boxes, electron transport in electric fields is always perpendicular to some interface.

Depending on physical circumstances, each of these two types of electron transport can be of either classical or quantum character, as introduced in Chapter 2. Classical parallel transport is, in some aspects, similar to transport in bulk materials. An example of classical electron transport along the layers of a multiple-quantum-well structure is shown in Fig. 7.2(a). However, even classical parallel transport can differ from that in bulk materials because of a difference in electron and phonon spectra, magnitudes and energy dependences of scattering rates, etc. Thus, for classical parallel transport, such characteristics as low-field mobility, diffusivity, etc., can be drastically different from those of bulk materials.

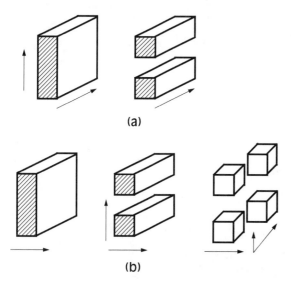

(a)

Figure 7.1. (a) Parallel (horizontal), (b) perpendicular (vertical) electron transport in quantum structures.

(b)

Another example of different transport properties is given by the hot-electron problem. Hot electrons in quantum structures manifest many phenomena such as intersubband transitions and real-space transfer (escape from wells, wires, and dots), etc., which are not present in bulk systems.

The quantum transfer of electrons through a barrier or through several barriers by means of tunneling is typical for perpendicular transport in quantum-well systems, as illustrated in Fig. 7.2(b). Moreover, such tunneling processes lead to new transport phenomena and entirely new applications like resonant-tunneling diodes, low-threshold lasers, etc. Classical perpendicular transport in quantum wells is associated with the excitation of the electrons to upper subbands and further thermionic emission of the electrons over the barrier, as in Fig. 7.2(b). Quantum and classical types of perpendicular transport are also characteristic for the cases of quantum wires and boxes.

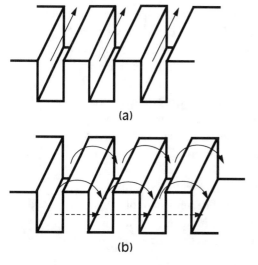

(a)

(b)

Figure 7.2. (a) Parallel classical electron motion along the layers of a multiple quantum well, (b) perpendicular thermionic (classical) transport over barriers (solid curves) and quantum tunneling (dashed curves).

Superlattices exhibit another important example of perpendicular transport. In Subsection 8.3.2 we show that, for superlattices, one can introduce a specific time scale: $\tau_{sl} = \pi \hbar/(eFd)$, where d is the period of the superlattice and F is the external electric field. At a low electric field, when τ_{sl} is much greater than the scattering time τ_s, vertical transport is almost conventionally classical. However, for a strong electric field, when τ_{sl} becomes smaller than τ_s, the quantum character of electron transport manifests itself and leads to new effects such as the Wannier-Stark energy ladder and Bloch oscillations.

To complete this classification of electron-transport regimes in the heterostructures, we should mention that the separation of parallel- and perpendicular-transport regimes are somewhat conditional. In some cases, parallel transport gives rise to perpendicular electron transfer. A typical example of this is real-space transfer of hot electrons in which quantum-well electrons heated in a parallel electric field leave the well. These electrons may contribute to perpendicular transport. In other cases, vertical transport, redistributing electrons among layers, affects the parallel component of electron current.

In this chapter we study classical parallel electron transport in quantum structures. First, we consider low electric fields.

7.2 Linear Electron Transport

Linear transport occurs in relatively low electric fields in which the current-voltage characteristic is linear or, in other words, in which Ohm's law is valid. In the linear regime, electrons are in quasi-equilibrium: the mean electron energy does not differ appreciably from the thermal-equilibrium energy. The quasi-equilibrium means that the electron distribution function introduced in Sections 2.8 and 2.9 differs slightly from the equilibrium Fermi (or Boltzmann) distribution. According to the discussion in Section 2.9, such distribution functions can be presented as in Eqs. (2.198) with the symmetric part identical to the Fermi (Boltzmann) function and the asymmetric part satisfying inequality (2.200). According to the analysis of Section 2.9, in a low-field limit, electrons are characterized by a mobility and diffusion coefficient. It is worth noting that the time τ_p in Eq. (2.209) includes all scattering mechanisms. It is possible to characterize each mechanism by some particular scattering time τ_i, and the total scattering rate $1/\tau_p$ can be represented as the sum of scattering rates due to various mechanisms:

$$\frac{1}{\tau_p} = \sum_i \frac{1}{\tau_i}. \tag{7.1}$$

According to Eqs. (2.209) and (7.1), each scattering mechanism contributes to the mobility. Figure 7.3 shows the contributions of impurity- and lattice-scattering mechanisms for Si; each of these mechanisms leads to a different temperature dependence of the mobility, as shown in the inset. In Fig. 7.4, the low-field mobilities $\mu_{n,p}$, and diffusion coefficients $D_{n,p}$ of both electrons and

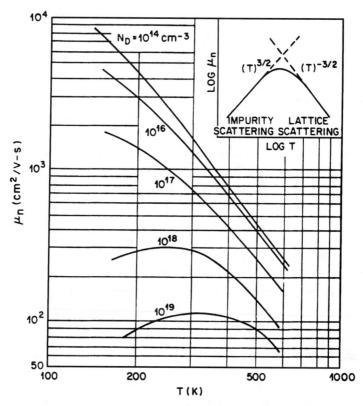

Figure 7.3. The electron mobility in Si versus the lattice temperature T for different donor concentrations. The inset illustrates the transition from impurity scattering to lattice scattering with increasing temperature. The intensities of impurity and lattice scatterings are proportional to $T^{3/2}$ and $T^{-3/2}$, respectively. [After S. Sze, *Semiconductor Devices. Physics and Technology* (Wiley, New York, 1985).]

holes are presented as functions of impurity concentrations for the cases of Si and GaAs bulk materials at room temperature. One can see that the electron mobility of GaAs always exceeds that of Si.

In Fig. 7.4, the diffusion coefficient is calculated by means of the Einstein relation of Eq. (2.212). In general, under nonequilibrium conditions the Einstein relation is no longer valid and one should use Eq. (2.211). In Chapter 2, we have also introduced the system of Boltzmann transport equations for the set of distribution functions $\mathcal{F}_n(\vec{k})$ corresponding to the number of electrons with a two-dimensional wave vector $\vec{k}$ in the nth subband. Now we can use these equations to investigate the mobility in quantum structures.

In the low-field limit, which we are now interested in, we can simplify Eq. (2.197) to a form suitable for mobility calculations. For this purpose, we introduce the following form of $\mathcal{F}_n$:

$$\mathcal{F}_n(\vec{k}) = \mathcal{F}_n^0(\vec{k}) + \frac{e\hbar}{m^*} \vec{k}\vec{F} \frac{\partial \mathcal{F}_n^0(k)}{\partial E_n} \tau_n[E_n(\vec{k})], \qquad (7.2)$$

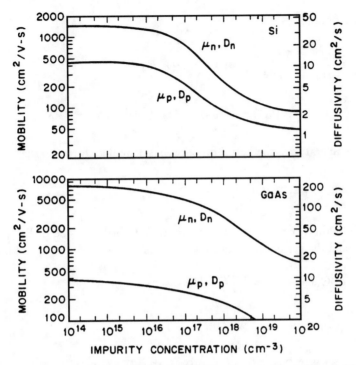

Figure 7.4. Comparison of electron and hole mobilities (μ_n, μ_p, left scale) and diffusion coefficients (D_n, D_p, right scale) in Si (upper part) and GaAs (lower part) at room temperature. [After S. Sze, *Semiconductor Devices. Physics and Technology* (Wiley, New York, 1985).]

where $\mathcal{F}_n^0(\vec{k})$ is the Fermi function for electrons in nth subband, τ_n is a function of the electron energy of this subband, and $E_n(\vec{k}) \equiv \epsilon_n + \hbar^2 \vec{k}^2/2m^*$. The function τ_n is independent of the electric field, and it can be interpreted as a scattering time for the electrons of the nth subband. In general, it includes both intrasubband and intersubband scattering processes. Substituting Eq. (7.2) into Eq. (2.197) and taking into account Eq. (6.14) and the equality

$$\vec{\nabla}_{\vec{k}} \mathcal{F}_n^0(\vec{k}) = \frac{\hbar^2 \vec{k}}{m^*} \frac{\partial \mathcal{F}_n^0(\vec{k})}{\partial E_n},$$

we find the equations for a set of unknown functions $\tau_n[E_n(\vec{k})]$:

$$-\hbar(\vec{k}\vec{F})\frac{\partial \mathcal{F}_n^0(\vec{k})}{\partial E_n} = 2\pi \sum_{m,\vec{k}'} |\langle m, \vec{k}' | \mathcal{H}_{\text{def}} | n, \vec{k} \rangle|^2$$

$$\times \left[\vec{k}'\vec{F} \frac{\partial \mathcal{F}_m^0}{\partial E_m} \tau_m(E_m) - \vec{k}\vec{F} \frac{\partial \mathcal{F}_n^0}{\partial E_n} \tau_n(E_n) \right] \delta[E_m(\vec{k}') - E_n(\vec{k})].$$

Choosing a coordinate system with the x axis oriented along the applied electric field $\vec{F}$, multiplying both sides of the last equation by k_x, and calculating the sum over $\vec{k}$, we obtain

$$-\sum_{\vec{k}} k_x^2 \frac{\partial \mathcal{F}_n^0}{\partial E_n} = \frac{2\pi}{\hbar} \sum_{m,\vec{k},\vec{k}'} |\langle m, \vec{k}' | \mathcal{H}_{\text{def}} | n, \vec{k} \rangle|^2$$

$$\times (k_x k_x' - k_x^2 \delta_{nm}) \frac{\partial \mathcal{F}_m^0}{\partial E_m} \tau_m(E_m) \delta[E_m(\vec{k}') - E_n(\vec{k})]. \tag{7.3}$$

For low temperatures, the Fermi function has a steplike form, so that

$$\frac{\partial \mathcal{F}_n^0}{\partial E_n} \cong -\delta[E_F - E_n(\vec{k})],$$

where E_F is the Fermi energy. Calculating the right-hand side of Eq. (7.3), we can represent the system of equations for τ_n as

$$(E_F - \epsilon_n)\Theta(E_F - \epsilon_n) = \sum_m K_{m,n}(E_F)\tau_m(E_F), \tag{7.4}$$

where for coefficients $K_{n,m}$, we obtain

$$K_{nm} = \frac{2\pi^2 \hbar^3}{m^{*2} S} \sum_{\vec{k}',\vec{k}} |\langle m, \vec{k}' | \mathcal{H}_{\text{def}} | n, \vec{k} \rangle|^2 [\vec{k}\vec{k}_z' - \vec{k}k_z \delta_{n,m} \delta_{\vec{k},\vec{k}'}]$$

$$\times \frac{\partial \mathcal{F}_m^0}{\partial E_m} \delta[E_m(\vec{k}') - E_n(\vec{k})]. \tag{7.5}$$

If only the lowest subband is filled by electrons, as in the so-called quantum limit, we need to find one function, τ_1. This can be done in an explicit form, setting $\mathcal{F}_{n>1} = 0$:

$$\tau_1 = \frac{E_F - \epsilon_1}{K_{11}}, \qquad \epsilon_1 \le E_F < \epsilon_2. \tag{7.6}$$

For elastic scattering one obtains from Eq. (7.6)

$$\frac{1}{\tau_1(E_F)} = \frac{m^* S}{2\pi \hbar^3} \int_0^{2\pi} d\theta (1 - \cos\theta)(|\langle 1, k | \mathcal{H}_{\text{def}} | 1, k + q \rangle|^2). \tag{7.7}$$

In Eq. (7.7) $q = 2k|\sin\theta/2|$, $k = \sqrt{2mE_F}/\hbar$, and one should use the matrix elements analyzed in Sections 6.2 and 6.5 for impurity and roughness scattering, respectively. Together with those results, Eq. (7.7) shows that, at low temperatures, the contribution to the inverse relaxation time comes from elastic collisions taking place on the Fermi circle.

When the second subband begins to fill, the value of τ_1 changes abruptly:

$$\tau_1 = \frac{E_F - \epsilon_1}{K_{11} - K_{21}K_{12}/K_{22}}, \qquad \epsilon_2 \le E_F \le \epsilon_3, \tag{7.8}$$

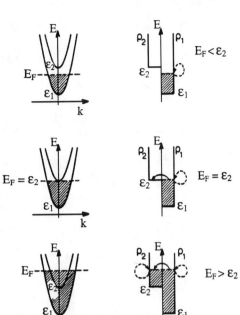

Figure 7.5. Schematic of intersubband transitions in a quantum well at $T = 0$ K for three different positions of the Fermi level with respect to the bottom of the second subband. The energy $\epsilon_{1,2}$ versus the wave vector and the density of states $\rho_{1,2}$ versus energy are shown. The intraband and the interband transitions are indicated by dashed and solid lines, respectively

where K_{11} is also modified; more specifically, K_{11} increases. This finite change in τ_1 occurs as a result of opening an additional channel of electron scattering because of intrasubband transitions: $(1, k) \rightarrow [2, k'(\approx 0)]$. Because the density of states of a two-dimensional subband is constant, this additional scattering channel gives a finite contribution to τ even at $E_F = \epsilon_2(0)$.

Figure 7.5 provides schematic representations for both intrasubband and intersubband scattering. The electron kinetic energy versus the in-plane wave vector and the density of states versus the energy are presented for three cases: (1) one subband is occupied and only intrasubband processes are possible, (2) near a threshold when intrasubband scattering is accompanied by a new scattering channel, and (3) two subbands are filled and there are contributions of both types of processes in each subband.

Because we are now interested in estimating the electron mobility, let us calculate the density of the electric current, which is expressed through the distribution function of Eq. (7.2):

$$\vec{j} = \frac{2e^2\hbar^2}{S} \sum_{n,k} \vec{k}\frac{\vec{k}\vec{F}}{m^{*2}} \tau_n(E_F)\delta\left[E_F - E_n(k)\right].$$

If we define the surface electron concentration n_n associated with the nth subband as $n_n = (m^*/\pi\hbar^2)(E_F - \epsilon_n)$ for $E_F > \epsilon_n$, we can rewrite the formula for the current density in the form

$$\vec{j} = \sum_n \frac{n_n e^2 \tau_n(E_F)}{m^*} \vec{F},$$

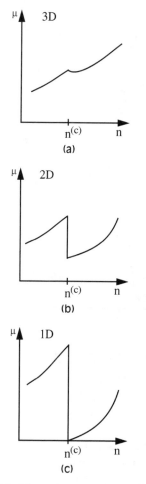

Figure 7.6. Schematic comparison of electron mobilities versus concentration in (a) bulk, (b) two-dimensional, (c) quasi-one-dimensional semiconductors. The onset of the upper subband (band) as indicated on the abscissa corresponds to the crossing by the Fermi level of the bottom of the upper subband (band) (see Fig. 7.5).

i.e., we express the mobilities of electrons from different subbands in terms of τ_n. This formula means that the contributions to the conductivity of the subbands populated by electrons can be interpreted as contributions of independent conductive channels that are characterized by their own mobilities.

Using the previously derived results for threshold changes in the scattering time τ_1, which are associated with the opening of intersubband channels, we can find an unusual behavior of the mobility as a function of the electron concentration; this behavior is illustrated in Fig. 7.6. The opening of each new scattering channel gives rise to a threshold drop in the mobility of the two-dimensional gas. For comparison, the mobility in a bulk material is shown for a similar situation in which the Fermi level becomes equal to additional minima of the electron spectra. In the case of one-dimensional electrons, in which the density of states diverges at the bottom of the second subband, one can expect the mobility to drop to zero. However, in this latter case, the τ approximation is no longer valid and a more accurate treatment is needed.

A decrease of the mobility due to the onset of additional channels of intersubband scattering has been observed experimentally. A gated AlGaAs/ GaAs heterointerface, similar to that analyzed in Section 4.4, was used to create two-dimensional electrons on the interface and to control their concentrations by the gate voltage. The results of measurement of the surface electron concentration in two subbands, n_1 and n_2, are presented in Fig. 7.7(a) as functions of the gate voltage. The concentration in the first subband increases with the voltage; then the Fermi level reaches the bottom of the second subband, which becomes populated. At this moment, the mobility drops in accordance with our analysis. It is interesting to note that the mobility of electrons in the second subband is lower than that in the first.

These results clearly demonstrate the importance of both intrasubband and intersubband scattering channels, regardless of the specific mechanisms of electron scattering.

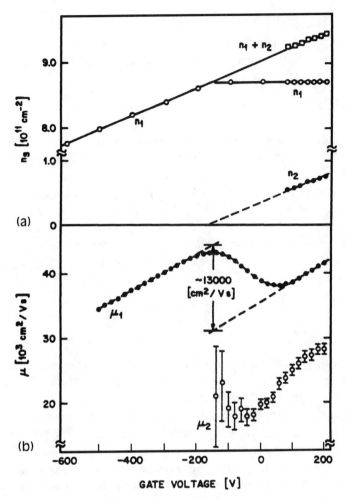

Figure 7.7. Experimental observation of the decreasing mobility effect that is due to the opening of an intersubband channel for electrons on a gated GaAs/AlGaAs heterointerface. Results for the electron concentration and mobility are given as functions of the gate voltage. [After C. Weisbuch and B. Vinter, *Quantum Semiconductor Structures* (Academic, San Diego, 1991).]

7.3 High-Field Electron Transport

Many heterostructure devices operate in high electric fields to achieve the desirable high-speed or high-frequency performance of these devices. However, being accelerated in high electric fields, the electron mean energy and the average momentum acquired are far greater than those associated with thermal equilibrium. This leads to a strong nonequilibrium state of the electron gas. Strong nonequilibrium electrons not only move fast, but also exhibit a number of specific effects that find various practical applications. We consider this heating of an electron gas in the following subsections.

7.3.1 Hot, Warm, and Cold Electrons

We begin our analysis with qualitative considerations. We describe the changes in the electron energy in an electric field through the relationship

$$\frac{\mathrm{d}E}{\mathrm{d}t} = e|\vec{F}\vec{v}| - \frac{E - E_{\mathrm{eq}}}{\tau_E}. \tag{7.9}$$

The first term on the right-hand side corresponds to the power gained by the electron from the electric field; the second term represents the rate of the electron-energy losses. The losses are proportional to the deviation of electron energy from its equilibrium value E_{eq} and τ_E is the energy-relaxation time given by Eq. (6.7). The electron-energy relaxation is due solely to phonon emission and absorption, because scattering by impurities is an elastic process.

The electron energy increases in the electric field; thus the rate of electron-energy loss due to the phonon emission must also increase for the overall energy balance to be maintained. In the stationary case, Eq. (7.9) yields the electron energy

$$E = E_{\mathrm{eq}} + e|\vec{F}\vec{v}|\tau_E. \tag{7.10}$$

Here the velocity $\vec{v}$ also depends on the electric field. Thus the mean electron energy increases with increasing electric field and exceeds the equilibrium value E_{eq}. It is convenient (and conventional in semiconductor electronics) to consider the electron effective temperature T_e, instead of the mean electron energy. The relationship between the temperature and the mean energy can be found in thermal equilibrium and is $E = nk_B T_e/2$, where the factor n is the dimensionality of the structure. Obviously, under thermal equilibrium the electron temperature T_e coincides with the lattice temperature T. Under nonequilibrium conditions these two temperatures may differ. Therefore the effective electron temperature expressed through the mean electron energy serves as a gauge of the nonequilibrium state. If T_e exceeds T only slightly and the electron transport still obeys Ohm's law, we have warm electrons. The case with $T_e \gg T$ corresponds to situations in which the electrons are far from equilibrium; such a situation is frequently referred to as the hot-electron problem.

The complete description of a nonequilibrium-electron gas can be realized only on the basis of the distribution function discussed in Chapter 2. In high electric fields, the electron distribution function differs strongly from that of the equilibrium state. To find the distribution one must solve the Boltzmann transport equation written in Section 2.8 in its general form, without linearization. All mechanisms of scattering should be included in the collision term, as represented by the right-hand side of the Boltzmann equation (2.191).

Here we merely give a simple analysis of the possible electron distributions for the bulk case. Qualitatively, one can distinguish between two extreme cases. The first case corresponds to quasi-elastic scattering, in which the electron momentum-relaxation rate is much greater than the rate of energy relaxation.

As a result, an almost spherically symmetric distribution function occurs, as illustrated in Fig. 7.8(a). This electron–electron scattering dictates the form of distribution function: it should be similar to the Maxwell distribution with electron temperature T_e. For the frequently encountered case of high electron concentration, the exchange of the momentum among the electrons dominates over other mechanisms of the momentum and the energy relaxations. (The criteria of applicability of the electron temperature approximation were studied in Section 2.9.) In practice, electron concentrations should be more than or approximately $10^{14}\,\mathrm{cm}^{-3}$. In this case it is convenient to represent the distribution function as a sum of two parts, as in Eqs. (2.198). For the symmetric part of the function we obtain

$$\mathcal{F}_0 = C\exp(-E/k_B T_e).$$

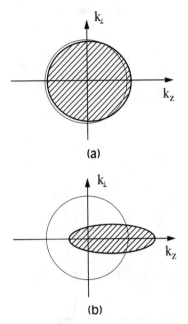

Figure 7.8. Schematic of (a) quasi-isotropic, (b) sharply anisotropic (streaminglike) electron distributions in momentum space for the electric field oriented along the z axis.

The asymmetric part of the distribution function is relatively small, but it determines the mobility and the diffusivity of hot electrons. Note that the mobility and the diffusivity are now functions of the applied electric field in contrast to the previously considered quasi-equilibrium situation.

Figures 7.9 and 7.10 illustrate hot-electron phenomena within the electron-temperature approximation. The electron-temperature and the electric-current dependences on the electric field for bulk samples are calculated for electron scattering by optical and acoustic phonons in a bulk sample. The parameters are chosen for n-Si at $T = 77\,\mathrm{K}$. One can see that the electron temperature can exceed the lattice temperature up to 1 order of magnitude. The current-voltage characteristics clearly exhibit nonlinearities due to the hot-electron effects.

Another type of the distribution function occurs when both an electron–electron interaction and quasi-elastic scattering are relatively weak compared with the inelastic processes. In this case, the distribution function and all the electron characteristics are strongly anisotropic. As an example, we can refer to the low-temperature case. An electron starts from the low-energy region and gains energy – it accelerates – up to the energy of the optical phonon almost without collisions. Then this electron emits an optical phonon and scatters back to the low-energy region. This process repeats itself again and again. As a result, quasi-periodic-electron motion in energy space with alternating acceleration periods and scatterings, the so-called streaming-transport regime, occurs. The streaming regime is characterized by an extremely anisotropic (needlelike) electron distribution, as illustrated in Fig. 7.8(b). We discuss electron streaming in more detail later in Subsection 7.3.2.

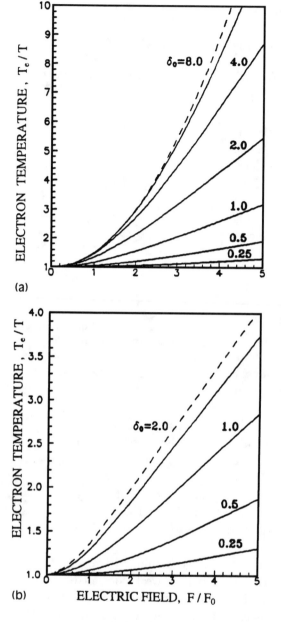

(a)

(b)

ELECTRON TEMPERATURE, T_e/T

ELECTRIC FIELD, F/F_0

Figure 7.9. Relative electron temperature T_e/T versus the dimensionless applied electric field F/F_0 for two mechanisms of electron scattering and $T = 77$ K. (a) Scattering by nonpolar optical phonons, (b) scattering by acoustic phonons. The effective mass and parameters of electron–phonon interaction are chosen as for n-Si. The dashed curves present results for bulk samples. The solid curves are given for different sample thicknesses d in units L_E with $\delta_0 = d/L_E$ and $F_0 = k_B T/eL_E$. [After V. A. Kochelap, *et al.*, "Limitation and suppression of hot-electron fluctuations in submicrometer semiconductor structures," Phys. Rev. B 48, pp. 2304–2311 (1993).]

Naturally, besides these extreme cases of the electron distributions for high electric fields, there is a variety of intermediate regimes of nonequilibrium electron transport. In general, all phenomena caused by electron heating are known as hot-electron phenomena. Hot-electron effects in bulk materials have been an active subject of the study starting in the 1960s. Now most of the hot-electron phenomena in bulk material have been investigated in detail. We briefly overview them and then study how they are modified and what new effects appear in quantum structures.

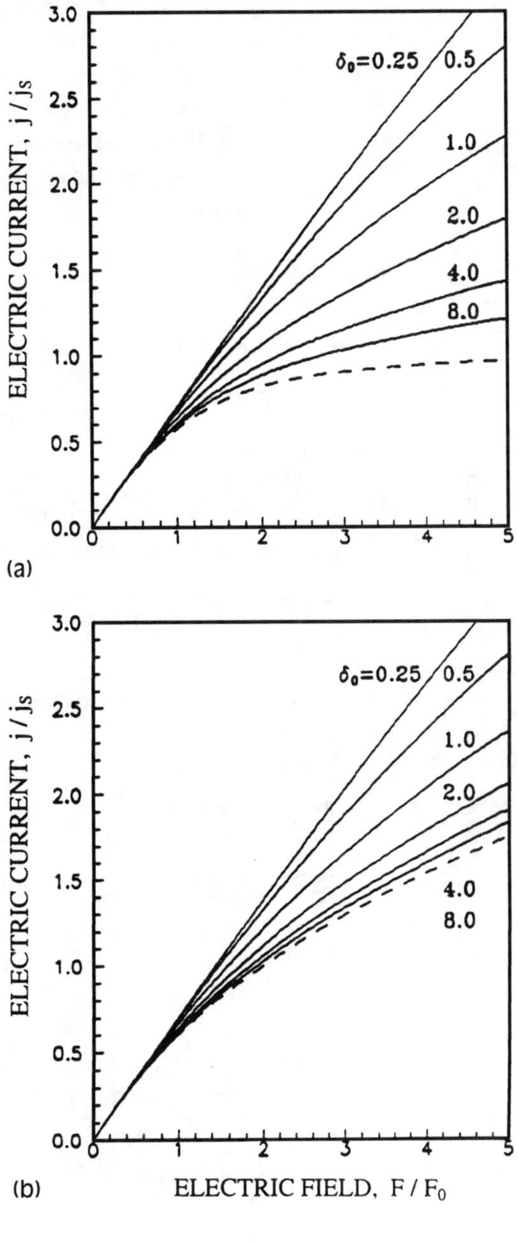

Figure 7.10. Relative electric current j/j_S versus the dimensionless electric field for two mechanisms of scattering as in Fig. 7.9. The dashed curves present results for bulk samples; the others are for different sample thicknesses $d = \delta_0 L_E$. [After V. A. Kochelap, *et al.*, "Limitation and suppression of hot-electron fluctuations in submicrometer semiconductor structures," Phys. Rev. B **48**, pp. 2304–2311 (1993).]

7.3.2 Velocity Saturation

One of the most common hot-electron effects in semiconductors is drift-velocity saturation. An example of this effect is given by the previously mentioned electron-streaming regime. For the sake of simplicity, let us assume that the scattering of electrons at energies below the optical-phonon energy is negligible (as is the case for very low temperatures ($kT \ll \hbar\omega_0$) in perfect lightly doped semiconductors), while the rate of optical-phonon emission above the optical-phonon energy is very high compared to any other scattering rate. This means that the electron reaches the optical-phonon energy without scattering, and after the immediate emission of an optical phonon it transits almost exactly to the band bottom. This process repeats itself over and over, with the result that the electron motion is periodic in both momentum and energy spaces. The solution of Eq. (2.142) for the absolute value of electron velocity $v = p/m^*$ and external force $\vec{f} = -e\vec{F}$ in the passive energy region (below the optical-phonon energy) yields to its linear increase,

$$v(t) = \frac{e}{m^*}Ft, \tag{7.11}$$

from $v = 0$ at $t = 0$ to $v = v_0$,

$$v_0 = \sqrt{\frac{2\hbar\omega_0}{m^*}}, \tag{7.12}$$

when the electron reaches the optical-phonon energy $\hbar\omega_0$. From Eqs. (7.11) and (7.12), we obtain the period of the streaming motion as

$$T = \frac{m^* v_0}{eF} = \frac{\sqrt{2m^*\hbar\omega_0}}{eF}. \tag{7.13}$$

To find an average drift velocity, we average Eq. (7.11) over one streaming period. The result is

$$\langle v \rangle = \frac{1}{T}\int_0^T v(t)\,dt = \frac{v_0}{2}. \tag{7.14}$$

Now applying our simple model of parabolic electron dispersion, where

$$E(t) = \frac{m^* v^2(t)}{2}, \tag{7.15}$$

and taking into account the fact that the electron energy does not exceed the optical phonon energy, we can obtain the mean electron energy by averaging Eq. (7.15) over one streaming period; it is found to be

$$\langle E \rangle = \frac{1}{T}\int_0^T E(t)\,dt = \frac{\hbar\omega_0}{3}. \tag{7.16}$$

Thus both the drift velocity and the mean energy cease to depend on the electric field in the streaming regime, i.e., they saturate. It is also known that the higher the electric field, the better the streaming conditions. Hence, even if some scattering in the passive energy region exists, an increase in the electric field will lead to the saturation of the electron's velocity and energy because of streaming. Strictly speaking, streaming can be realized if certain conditions are met: (1) The temperature must be low enough, generally $k_B T \ll \hbar\omega_0$, so that phonon-absorption processes are frozen out, (2) the phonon-emission rate must exceed all other scattering rates near the phonon-emission threshold, and (3) the electric field should be strong enough to accelerate electrons up to the phonon-emission threshold without scattering, but be weak enough to avoid deep electron penetration beyond the emission threshold and thus ensure scattering by phonon emission down to the conduction-band bottom.

However, streaming alone cannot explain the velocity saturation effects, especially at higher temperatures. It has been demonstrated theoretically that electron transport in a parabolic conduction band is not stable if the only scattering mechanism is polar-optical-phonon scattering. On average, the electrons will gain more energy from even the weakest electric field than they lose by polar-optical-phonon scattering. Therefore the electron energy will increase indefinitely. In a nonparabolic band such an unstable situation occurs only above some threshold electric field. Hence the nonparabolicity of the conduction band and the onset of other scattering mechanisms play important roles in electron transport. Usually the streaming velocity given by Eq. (7.14) is only a benchmark. At high temperatures, when the scattering in the passive region is effective, the actual velocity is lower. In contrast, when the electric field is so strong that electrons penetrate far into the active energy region beyond the optical-phonon energy, the actual velocity may exceed the streaming velocity. Each particular situation requires special treatment by means of an accurate solution of the Boltzmann equation. Qualitatively, velocity saturation occurs as a result of the following reasons. As the electric field increases, so does the mean electron energy. As a rule, the scattering rates increase (the mobility decreases) as the electron energy increases; see Chapter 6. Thus we come to the conclusion that an increase in the electric field leads to a decrease in the mobility. It happens that in many common semiconductors the mobility decreases exactly as $1/F$, so that the drift velocity, $v_d = \mu F$, is independent of the electric field.

Usually, saturation effects are easier to measure experimentally or obtain numerically than to derive analytically. For this reason, empirical expressions are usually used to describe the velocity-saturation effect. There is a simple approximation for the velocity saturation:

$$v = \frac{v_s}{[1 + (F_s/F)^\gamma]^{1/\gamma}}. \tag{7.17}$$

Here, v_s is the saturated velocity, F_s is a constant characterizing the electric field necessary to reach the saturation effect, and γ is some constant parameter. For pure Si, $v_s \simeq 10^7$ cm/s; this value is approximately the same for electrons and holes, while the characteristic fields are 20 and 10 kV/cm, respectively. Typical curves measured for drift velocities as functions of the applied electric field are presented in Fig. 7.11 for n-Si. The results are obtained for temperatures T of 77, 193, and 298 K. The linear low-field ($v \propto F$), sublinear ($v \propto \sqrt{F}$), and saturation ($v \approx$ const.) regimes are indicated separately. One can see that increasing the temperature suppresses the drift velocity, including saturated values. For this material, the saturation velocity falls in the range from 0.7 to 1×10^7 cm/s. The saturation is reached for fields above several kilovolts per centimeter.

Table 7.1.
Values of the Saturation Velocities for
Semiconductor Materials at Room Temperature

Material	Saturation velocity $(10^7$ cm/s)
Si	1
SiC	2
SiO$_2$	1.9
AlAs	0.65
GaP	1.1
GaAs[†]	2
InP[†]	2.5
InAs[†]	4.4
InSb	6.5

[†]Material exhibits negative differential conductivity. These
values correspond to the maximum drift velocities.

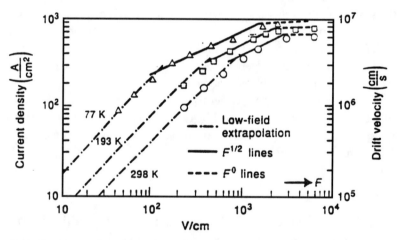

Figure 7.11. Experimental results on the drift electron velocity and current density as functions of the electric field in *n*-Si at different temperatures.

It is instructive to compare the saturated velocities for different materials. Such a comparison is given in Table 7.1. One can see that some III–V compounds, including GaAs, InP, and InSb, have saturation velocities that are several times larger than those of Si, GaP, and AlAs. One of the reasons for this is smaller electron effective masses for Γ minima in direct-energy-bandgap compounds.

7.3.3 Transient Overshoot Effect

Another hot-electron effect important for high-speed device operation is the so-called velocity overshoot. To explain this phenomenon, let us mention that our previous conclusions have been reached for the steady-state case in which the electron distribution is stationary. In other words, the previous analyses have referred to electron properties averaged over times much greater than the characteristic times of the system such as the mean-free-flight time, momentum- and energy-relaxation times, intervalley scattering time, streaming period, etc. Now we consider processes that occur in the electron system immediately following its deviation from the equilibrium state. In this particular case, we focus on the response to a step in the electric field.

In general, the momentum-relaxation time τ_p of Eq. (6.8) is shorter than the energy-relaxation time τ_E of Eq. (6.9). Therefore the velocity response to the electric-field step is faster than the energy response for the case described by Eq. (7.9). If the scattering rate increases with increasing electron energy, as is typical, then the electron velocity may exceed the stationary velocity during a time interval of the order of τ_E. In other words, the transient velocity is not just a function of the electric field but also of the electron energy. Indeed, the velocity adjusts itself quickly to the quite slowly changing energy, and it follows that energy until the energy reaches the steady state. Initially, when the electron energy has still not reached the stationary value, the electron velocity corresponding to the transient energy is higher than the velocity corresponding to the stationary energy.

This overshoot effect is particularly pronounced if there is an onset of new mechanisms for electron scattering at higher energies ($E \gg \hbar\omega_0$) that lead to additional momentum relaxation. A typical example is the onset of intervalley electron scattering in III–V and II–VI group semiconductors. Figure 7.12 shows schematically the plot of electron energy versus wave vector for the [100] and the [111] directions. The curvatures associated with $E(\vec{k})$ for the side valleys are much lower than those of the central valley, in accordance with the ratio of the effective masses. The transitions are possible when heated electrons in the Γ valley reach the energy of the bottoms of the side valleys. This phenomenon can be interpreted in terms of electron transfer in momentum space. As a result of the smaller times for electron acceleration in the central Γ valley to energies above the bottom of the X or L valleys relative to times for electron transfer from the lowest Γ valley to the upper X or L valleys, the electrons can reach very high velocities in the Γ valley before they are transferred into the valleys that have larger effective masses. The overshoot effect takes place in many semiconductor materials, not only in those with heavy-mass upper valleys, where it is most pronounced.

The overshoot effect is shown in Fig. 7.13 for Si and GaAs. The drift velocity is presented as a function of time. The electric field is assumed to be switched on at time $t = 0$. The results are given for different electric fields. The overshoot effect is pronounced in high electric fields. The decrease of the drift velocity is explained by the transition to the heavy-mass valleys. The maximum of the

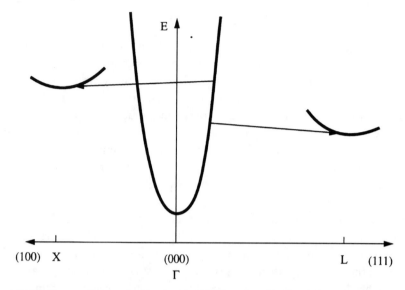

Figure 7.12. Many-valley conduction band of III–V semiconductors. The arrows indicate hot-electron transitions from the Γ to the X and L valleys due to phonon emission or absorption.

transient velocity can exceed the stationary saturated velocity by as much as 2–4 times.

From a physical explanation of the overshoot effect, we can understand how it is possible to utilize this effect. Let us imagine that cold electrons enter an active region of a semiconductor device through a contact. If there is a high electric field in the active region, the electrons will be accelerated. At some distance from the injecting contact the electrons will reach the maximum overshoot velocity, and after that their velocity will gradually decrease to the stationary value. If the active region of the device is short and comparable with the distance over which the overshoot effect takes place, electron transit through this active region will occur at a velocity higher than the stationary velocity, and the overall transit time will be shorter. Consequently the device will be able to operate at higher speeds and frequencies.

7.3.4 Gunn Effect

The transient overshoot effect just discussed is similar to the stationary overshoot effect; this effect exhibits a nonmonotonous dependence on the stationary velocity-field characteristic. However, if the transient overshoot effect takes place in many semiconductors and over a wide range of electric fields, the nonmonotonous velocity-field dependence occurs in materials with lighter effective mass in the lowest valley and with heavier masses in the upper valleys. Otherwise, as we already know, the velocity simply saturates as we increase the electric field. In Fig. 7.12, transitions of hot electrons from a central Γ valley to the upper heavy-mass L and X valleys are shown schematically on the plot of electron energy versus wave vector for the [100] and the [111] directions. As mentioned in

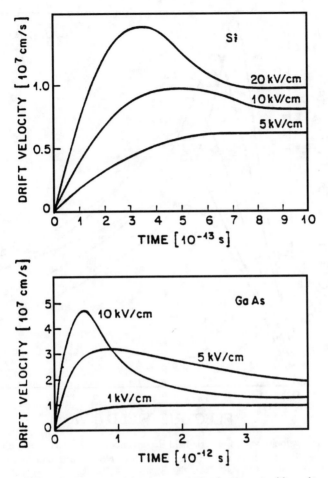

Figure 7.13. The transient response of the electron drift velocity to steplike pulses of the electric field at room temperature. The magnitudes of the field are indicated for each plot. The upper panel is for Si, while the lower is for GaAs. Different time scales of the panels correspond to different scattering characteristics in these materials.

Subsection 7.3.3 the electron transfer to the upper valleys leads to an increase of the electron effective mass. This results in a current-voltage characteristic with a portion that has a negative slope. This effect is referred to as negative differential resistance.

The electron drift velocity in GaAs and in other III–V semiconductors is a typical example of the N-type nonmonotonous dependence on the electric field: The velocity reaches a maximum, then decreases as the field increases further, as depicted in Fig. 7.14. In this figure, the electron-drift velocities versus voltage are presented for $In_yGa_{1-y}As$ alloys. All curves exhibit negative differential resistance for electric fields above 1–3 kV/cm. One can see that in a semiconductor with a narrower bandgap than GaAs and a lighter Γ-valley effective mass, the maximum velocities are larger and the negative differential resistance is much more pronounced.

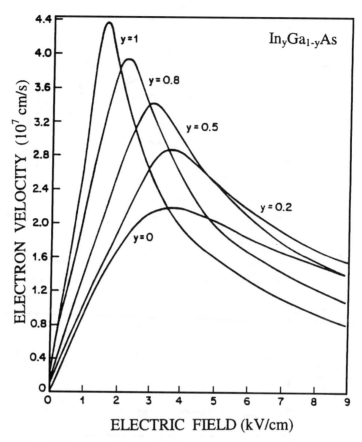

Figure 7.14. The velocity versus electric-field dependences for n-$In_yGa_{1-y}As$ alloys at room temperature for different content In. [After S. M. Sze, *High-Speed Semiconductor Devices* (Wiley, New York, 1990).]

From electric-circuit theory, it is well known that systems with negative differential resistance can be unstable. Indeed, if voltage fluctuation – say an increase – occurs while the device is biased so that the work point is on the negative slope, this voltage increase will lead to a reduction in the current and further increase in the voltage. If the fluctuations arise in some point of the device active region they will lead to the formation of high-electric-field domains at the end of the active region, and these domains will transit through the active region. Thus a device with N-type current-voltage characteristics incorporated in an electric circuit with controlled voltage will lead to the current oscillations. These oscillations can be made periodic if one controls the place where domains are formed. Such controlled spatial formation of the electric domains is usually achieved by the fabrication of a notch in the doping profile near the injecting contact. This phenomenon of current oscillations in semiconductors is known as the Gunn effect. The Gunn effect is the basis for a whole class of devices for microwave generation and amplification. The effect has been observed in many

III–V materials. The Gunn effect is also possible in Si. In this case, Si has to be strained to produce a crystal deformation that breaks the degeneracy of the valleys. The deformation along the [100] direction results in splitting the degeneracy of three pairs of valleys into one and two pairs. The upper valleys should have larger effective masses in the current direction. The transitions of hot electrons into these valleys cause a decrease in the current and result in negative differential resistance.

7.3.5 Nonequilibrium Phonons

In principle, for high electric fields, hot-electron relaxation through the emission of phonons should drive the phonon subsystem out of equilibrium. Indeed, electrons, as a rule, emit phonons that do not obey the Planck distribution of Eq. (5.30); see Chapter 5. These nonequilibrium phonons eventually decay into thermal phonons as a result of anharmonic interactions. As an example, optical phonons in GaAs decay into a pair of short-wavelength acoustic phonons with a lifetime of roughly 10 ps. If the generation rate of nonequilibrium phonons is slower than their decay rate, the approximation of phonon equilibrium is justified. However, if the electron concentrations are high and their excess energy is high as well, the generation of nonequilibrium phonons becomes effective and cannot be compensated for by phonon decay. As a result, the nonequilibrium-phonon populations grow rapidly and exceed the thermal-equilibrium populations by orders of magnitude. This situation is frequently referred to as the nonequilibrium-phonon or hot-phonon problem.

In turn, nonequilibrium phonons affect all major electron-transport and optical characteristics by dramatically modifying the scattering rates. Nonequilibrium phonons are responsible for the slow relaxation of photoexcited hot electrons and phonon drag effects in transport.

To make rough estimates of phonon population growth in high electric fields, one can write the crystal-energy-balance equation as

$$\frac{dN}{dt} = \frac{jF}{\hbar\omega} - \frac{N - N_0}{\tau_{ph}}, \tag{7.18}$$

where N is the total phonon population in a unit volume of the crystal, N_0 is the equilibrium-phonon population, j and F are the current density and the electric field, respectively, and τ_{ph} is the phonon-decay or thermalization time. The stationary solution of Eq. (7.18) in steady electric fields is

$$N = N_0 + \frac{jF\tau_{ph}}{\hbar\omega} = N_0 + \frac{env_d F\tau_{ph}}{\hbar\omega}, \tag{7.19}$$

where n and v_d are the electron concentration and the drift velocity, respectively.

For III–V semiconductors, the major electron-scattering mechanism is scattering by longitudinal-optical phonons. Therefore the predominant growth is in the optical-phonon population. The thermalization time τ_{ph} accounts for the decay of nonequilibrium phonons into the thermal equilibrium of background acoustic phonons; its numerical value generally ranges from 1 to 15 ps. Therefore, under normal device operation conditions and for typical currents and voltages, nonequilibrium phonons manifest themselves profoundly. The nonequilibrium-phonon problem can be solved rather accurately by the ensemble Monte Carlo method.

7.3.6 Hot-Electron Size Effect

The previously discussed hot-electron phenomena are characteristic of both bulk materials and heterostructures. There is an important intermediate case of classical transport in active layers of submicrometer thicknesses. Indeed, according to the general analysis of possible transport regimes given in Section 2.3, if characteristic diffusion lengths become comparable with the layer thickness L, one can expect the appearance of classical size effects. In terms of Table 2.1, these effects correspond to transverse size effects. For hot electrons, the energy-relaxation time τ_E corresponds to the diffusion energy-relaxation length $L_E = \sqrt{D\tau_E}$, where D is the diffusion coefficient. The meaning of this length is the following. A hot electron diffuses in space for distances of $\sim L_E$ before losing its excess energy. If the layer thickness is $\sim L_E$ or less, the energy relaxation on layer boundaries should affect the electron distribution over the energy. For example, if there is an additional mechanism of energy relaxation on the boundaries, the heating of electrons can be suppressed.

As an example, we can consider a Si sample with a modest donor doping, $n_D \approx 10^{13}$ cm^{-3}. For such a case, at nitrogen temperature or below, L_E exceeds 1 μm. Thus, for actual submicrometer active layers, the size effect can be an important factor in hot-electron phenomena. In Figs. 7.9 and 7.10, the results of calculations of the electron temperature and current for such layers are presented. It is realistic to suppose that there is additional energy dissipation at both of the boundaries. This gives rise to a nonuniform electron-temperature distribution across the layers. The maximum electron temperatures are shown in Fig. 7.9 as functions of the electric field for different layer thicknesses $\delta_0 = d/L_E$. The units of the electric field, $F_0 = k_B T/e L_E$, correspond to the characteristic heating electric field. The results are presented for electron scattering by nonpolar optical phonons and by acoustic phonons, as discussed in Subsection 7.3.1. The uniform electron temperature for bulk samples is shown for comparison. One can see that electron heating is suppressed considerably for narrow layers. For layers with $d \approx L_E$, the maximum electron temperature decreases as many as 3–10 times. If $\delta_0 \ll 1$, the heating is small and electrons can be considered to be almost in equilibrium.

The suppression of electron heating improves the current-voltage characteristic, as seen in Fig. 7.10. It becomes ohmiclike with larger currents. Since electron concentrations are supposed to be constant, this last result implies an increase in the electron-drift velocity, that is, a cancellation of the velocity-saturation effect.

7.4 Hot Electrons in Quantum Structures

To consider the hot-electron problem for the case of heterostructures, one should take into account the set of new features peculiar to heterostructures; these include

1. changes in the scattering rates due to the quantization of electrons and phonons
2. additional mechanisms of scattering
3. intersubband transitions
4. electron escape from the confining potential wells

The scope of this book does not allow us to consider hot electrons in heterostructures in much detail, so we concentrate on several of the most important effects.

7.4.1 Nonlinear Transport in Two-Dimensional Electron Gases

As in bulk systems, in quantum heterostructures the application of an electric field heats the electrons and changes the electron mobility. This results in a nonlinear dependence of the electron-drift velocity on the electric field.

Consider the typical case of a single modulation-doped n-AlGaAs/GaAs heterojunction. According to Chapter 4, the electron quantization and two-dimensional electron gas occur as a result of the charge transfer in such a system. The drift velocity of such electrons on the interface is presented in Fig. 7.15 for both room temperature and nitrogen temperature. The drift velocity of bulk electrons in GaAs is shown for comparison. First, one can see that in the modulation-doped heterostructure, the velocity of electrons is greater than that of the bulk material. The velocity increases substantially with decreasing temperature. Then a nonlinearity in the velocity plot appears at relatively low electric fields (hundreds of volts per centimeter at $T = 300\,\text{K}$); in the bulk material the transport is almost linear since, according to Fig. 7.14, for GaAs nonlinearities appear at electric fields of thousands of volts per centimeter. At low temperatures, the nonlinearity occurs for electric fields below 100 V/cm. The contribution of electrons from the lowest subband to the overall electron velocity can be extracted, as in Fig. 7.15(b). As is evident, the intersubband transitions also become important for fields of $\sim$100 V/cm. Electron transitions into higher subbands lead to a decrease in the velocity, because scattering by remote impurities increases as a result of the fact that the electron wave function of those subbands penetrates deeply under the barrier into the doped region.

With sophisticated experimental methods, direct measurements of the electron temperature are possible. In Fig. 7.16, the results of precise measurements of the electron temperature for two-dimensional electrons are presented as a function of the electric input power per electron. The experiment was performed at helium temperature for AlGaAs/GaAs heterojunctions with different areal electron concentrations in the range of 3–$8 \times 10^{11}\,\text{cm}^{-2}$. These concentrations and

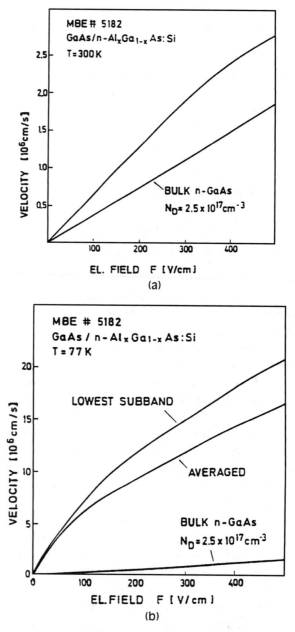

Figure 7.15. The electron-drift velocity as a function of the electric field in modulation-doped GaAs/AlGaAs heterostructures. The velocity of the bulk GaAs is presented for comparison. (a) corresponds to $T = 300\,\mathrm{K}$, (b) is for $T = 77\,\mathrm{K}$. In case (b) the velocity of electrons in the lowest subband and the overall velocity are shown separately. [After C. Weisbuch and B. Vinter, *Quantum Semiconductor Structures* (Academic, San Diego, 1991).]

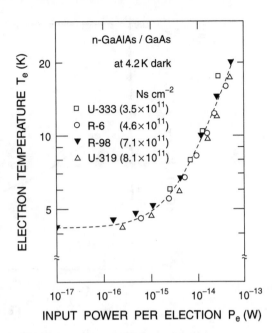

Figure 7.16. The measured electron temperature versus input power per electron in GaAs/AlGaAs heterostructures with different areal electron concentrations. [After C. Weisbuch and B. Vinter, *Quantum Semiconductor Structures* (Academic, San Diego, 1991).]

temperature are such that only the lowest electron two-dimensional subband is populated; this is true even for an electron temperature of $\sim$40 K. As can be seen, electron heating is not dependent on the concentration. The heating becomes pronounced beginning at an electric power of 10^{-15} W per electron, which corresponds to an electric field as low as 0.3 V/cm. The electron heating threshold is so low because of the very high electron mobility at the low temperatures. The results demonstrate that the electron temperature increases to more than 30 K. A further increase in the field will lead to intersubband transitions, the population of higher subbands, and the escape of electrons from the quantum well. The latter effect is known as the real-space transfer effect and is discussed below.

7.4.2 Nonlinear Electron Transport in Quantum Wires

In Chapter 6, we mentioned that acoustic-phonon scattering in low-dimensional structures and particularly in quantum wires is far more important for electron momentum and energy relaxation than in bulk materials. Indeed, the momentum conservation for the electron-acoustic-phonon system holds with an accuracy of only $2\pi\hbar/L$, where L is the effective thickness of the structure: $L = L_z$ for the quantum wells or $L^{-2} = L_y^{-2} + L_z^{-2}$ for quantum wires. For example, in a GaAs quantum wire with $L_y = L_z = 80$ Å and sound velocity $u = 5.2 \times 10^5$ cm/s, the phonon energy $2\pi\hbar u/L$ corresponding to the momentum uncertainty is equal to 3.8 meV. This energy corresponds to a thermal-equilibrium one-dimensional electron energy at $T = 88$ K. Thus, over a wide range of the system parameters, an electron can absorb or emit an acoustic phonon with energy comparable with its

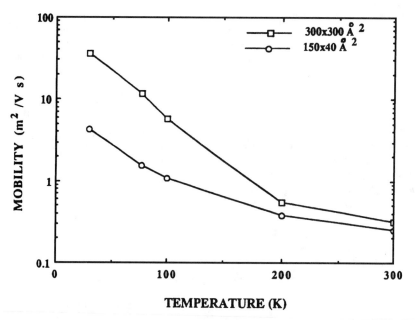

Figure 7.17. Low-field electron mobilities as functions of the temperature for two cross sections of the wires. Calculations were performed for the temperatures indicated by symbols.

own energy. As a result, at low and intermediate temperatures and at intermediate electric fields, the transport is controlled primarily by inelastic acoustic-phonon scattering.

Most of the results discussed below have been obtained by the Monte Carlo simulation of electron transport in GaAs/AlGaAs quantum wires. The probabilities of scattering by acoustic and optical phonons were taken to be those studied in Chapter 6. Impurity scattering was ignored. The low-field electron mobilities as functions of the temperature for two wire cross-sections are presented in Fig. 7.17. For $T = 30\,K$, low-field mobility values of 3.5×10^5 and $4 \times 10^4\,cm^2/Vs$ are obtained for the cases in which the quantum-wire cross sections are $300\,Å \times 300\,Å$ and $150\,Å \times 40\,Å$, respectively. Because of an increase in acoustic-phonon scattering as well as the confined and interface optical-phonon-absorption rates, the mobilities decrease dramatically for higher temperatures. Remarkably, for thinner wires, the electron mobility is suppressed. This can be understood in terms of the increasing intensity of acoustic-phonon scattering in thinner wires, as discussed in Chapter 6.

The electron-drift velocity as a function of the applied electric field is shown in Fig. 7.18 for three wires at $T = 30$ K with different cross sections. For comparison, the drift velocities estimated from low-field mobilities are also presented. At low fields all wires exhibit almost linear (ohmic) behavior of the drift velocities. Note that the velocity value decreases substantially for thinner wires. For thick quantum wires, a region with a weak superlinearity can be seen on the velocity-field dependence. This effect takes place for an electric field of the order of 1–10 V/cm. As the thickness of the quantum wire decreases, this superlinear

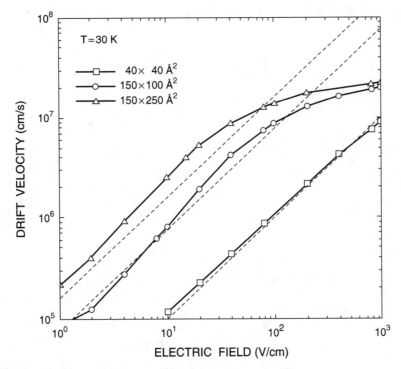

Figure 7.18. Electron-drift velocity as a function of the applied electric field for three different cross sections of the quantum wire. The dashed lines show the drift velocity estimated from the low-field mobility. Wire cross sections and lattice temperature are also indicated.

region moves up in electric fields and finally disappears for a quantum wire with a cross section of 40 Å × 40 Å. This superlinear dependence of the electron velocity is a purely one-dimensional effect that is caused by the reduction of the efficiency of acoustic-phonon scattering as the electron gas is heated in the quantum wire. The electron heating is shown in Fig. 7.19, in which the mean electron energy is plotted versus electric field for the same three quantum wires as in Fig. 7.18. In a wire, the energy of thermal-equilibrium electrons with 1 degree of freedom is equal to $k_B T/2$, which is 1.3 meV at $T = 30$ K. This value corresponds to the zero electric field in Fig. 7.19. When the field increases, the electron mean energy begins to grow after some critical field. These critical fields increase as the wire thickness decreases. Obviously this is again due to more intensive electron scattering by acoustic phonons in thinner wires. One can see that a one-decade increase in the field above the critical value gives rise to a mean energy five to eight times larger than the thermal-equilibrium value. Analysis of different scattering mechanisms shows that electrons with energies higher than several $k_B T$ easily escape from acoustic scattering and run away up to the optical-phonon-emission threshold. A further increase in the electric field leads to near saturation in the drift velocity, as seen from Fig. 7.18. The saturation occurs at the streaming value of v [Eq. (7.12)]. For GaAs, this value is approximately 2.2×10^7 cm/s. Thus the figure shows a transition from the superlinear to the streaming regime of electron transport.

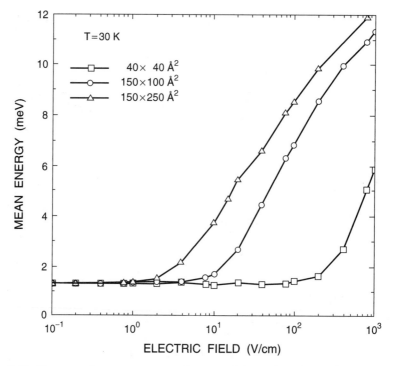

Figure 7.19. The mean electron energy as a function of the applied electric field for the same quantum wires as those in Fig. 7.18.

Note that in a quantum wire the ideal streaming – infinitely high electron-optical-phonon-scattering rate and the absence of any other scattering mechanism – has another simple interpretation. For such a case, electrons oscillate forth and back in one-dimensional $\vec{k}$ space indefinitely long, which implies that there are permanent oscillations of the instantaneous electron-drift velocity and its mean energy. This phenomenon should be manifested as a resonant behavior of the system at high frequencies. For actual conditions, both the penetration through the optical-phonon-scattering threshold and the acoustic-phonon scattering randomize the phase of the electron ensemble, and even though each electron under certain conditions continues to oscillate, the mean parameters approach stationary values. The streaming regime in quantum wires is reached at lower electric fields than in bulk materials. This is due to higher electron mobility in quantum wires than in bulk samples at low temperatures.

Another important parameter – the electron diffusion coefficient – is shown in Fig. 7.20 as a function of the electric field. The previously discussed superlinear region is reflected as a broad maximum in the diffusivity-field dependence. The maximum of the diffusivity curve is pronounced for thick quantum wires and almost disappears for a $40 \overset{\circ}{A} \times 40 \overset{\circ}{A}$ quantum wire. The transition from the superlinear region to the electron-streaming regime manifests itself as a decrease in the diffusivity.

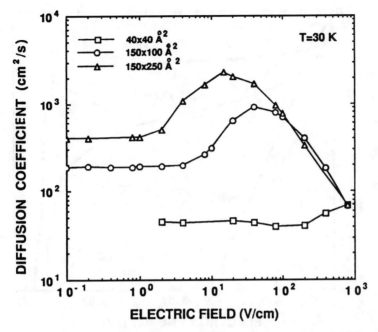

Figure 7.20. The electron diffusion coefficient as a function of the electric field for the same quantum wires as Figs. 7.18 and 7.19.

It is instructive to consider the electron distribution function under different electric fields. Figure 7.21 depicts this function for a quantum wire with a $250 \text{ Å} \times 150 \text{ Å}$ cross section. Three values of the field, $F = 0, 20,$ and 200 V/cm, are analyzed. The zero-field distribution function is just the Boltzmann distribution with the temperature taken to be that of the system. At finite fields, the distribution function in the intermediate energy region is flattened and extends up to the optical-phonon energy. There is no electron penetration beyond the optical-phonon energy for the fields presented, so that the distribution function is cut off at the phonon energy. Both types of existing optical phonons manifest themselves. At lower electric fields, the cutoff energy in the electron distribution coincides with the lowest interface optical-phonon energy, which is equal to 34.5 meV; this is slightly below the energy of the bulk GaAs longitudinal-optical phonons. At higher electric fields, some fraction of the electrons penetrate beyond the interface phonon energy. However, there is practically no electron penetration above the confined optical-phonon energy. The point is that electron scattering by the latter phonons in this thick quantum wire is much stronger than electron scattering by interface phonons, as is evident from the discussions in Chapter 6. The distribution function for electric fields displayed in Fig. 7.21 also has two distinguishable slopes, so that it may be characterized by two effective temperatures: very low temperature (lower than the equilibrium temperature) close to the subband bottom and high temperature (much higher than the equilibrium temperature) in the intermediate energy region. However, there are not enough

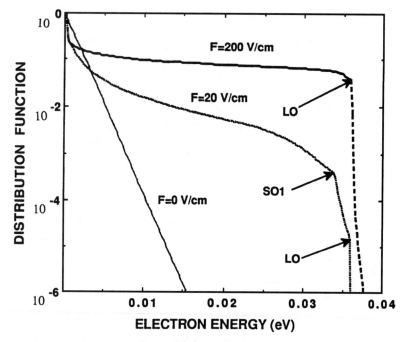

Figure 7.21. The electron distribution function versus electron energy for three different electric fields. The quantum-wire cross section is $150\,\text{Å} \times 250\,\text{Å}$ and $T = 30\,\text{K}$. Energy positions of the interface and confined optical phonons in the well are indicated.

electrons accumulated at the subband bottom to ensure a significant influence of the electron-cooling effect; hence, the mean energy always increases with the growing field in this particular quantum wire.

7.4.3 Real-Space Transfer of Hot Electrons

We mentioned previously that in quantum structures the heating of the electrons by an external electric field gives rise to electron transitions between the low-dimensional energy levels or, in other words, the electron redistribution over the low-dimensional subbands of quantum structures. But if the energy becomes sufficiently large, some electrons reach the top of the barriers and escape over these barriers into the surrounding material. This process is illustrated by Fig. 7.22(a).

There is some similarity between such processes of emission over barriers and intervalley transitions shown in Fig. 7.12. However, intervalley transitions occur in momentum space, i.e., electrons do not change their positions in real space during the transitions. In contrast, emission over the barrier leads to electron transfer in real space. Accordingly, it is referred to as the real-space transfer of electrons. This transfer occurs in the direction perpendicular to the heating electric current.

Originally, the effect was proposed as a method to obtain intentionally a negative differential conductivity and an N-type current-voltage characteristic. Let

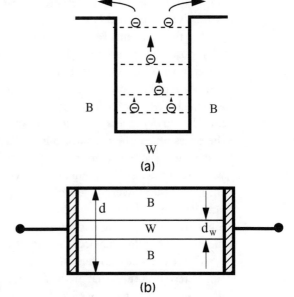

Figure 7.22. (a) Schematic of the level repopulation in a quantum well and the real-space transfer of electrons, (b) the layered structure for parallel transport with the real-space transfer. W and B denote the quantum well and barrier layers, respectively.

us consider the structure shown in Fig. 7.22(b) and assume that the mobilities of the quantum-well material and the barrier material are μ_W and μ_B, respectively. Since, in the general case, $\mu = e\tau/m^*$, the difference in the mobilities can be due to the different scattering times or (and) effective masses. If the concentrations of the electrons in the well and in the barriers are n_W and n_B and mobilities μ_W and μ_B, respectively, one can obtain for the total current density along the heterostructure

$$j = e\left(\mu_B n_B d_B + \mu_W n_W d_W\right) F/d, \qquad d_B = d - d_W, \qquad (7.20)$$

where d_W and d_B are the thicknesses of the well and the barrier layers, respectively. In the general case, the concentrations n_B and n_W are functions of the electric field. Although the electrons redistribute between the barrier and the well materials, their total concentration does not change: $n_B d_B + n_W d_W = \text{constant}$. When the electric field increases the electrons become hot, real-space-transfer processes take place, and the electrons are redistributed between the wells and the barriers. As a result, the concentration n_B increases at the expense of n_W. If $\mu_W \gg \mu_B$, the electron redistribution can bring about a sharp decrease of the effective mobility. The latter can lead to a negative differential resistance.

Although there is a close analogy between this effect and the Gunn effect, the case of real-space transfer has some advantages. Indeed, one can vary the characteristics of the layers artificially and over very wide ranges of μ_W/μ_B relative to unity; thus it is possible to control the current-voltage characteristic by shifting the threshold electric field of the negative resistance and changing the peak-to-valley ratio, which is defined as the ratio of the current maximum to the current minimum. In quantum structures based on many-valley semiconductors, both types of negative differential resistance can occur and, in such cases, they will be superimposed.

For the first time, real-space transfer and negative differential resistance have been observed experimentally in multilayer AlGaAs/GaAs heterostructures. Because of the electrical instability, microwave oscillations have been observed as well. When these oscillations occur the electrons have to cycle between the well and the barrier layers. Two processes are involved in this cycle: electron emission into the barrier layers and electron capture back into the wells. For hot electrons, the emission can be considered as a kind of thermionic current over the barrier and its time is much shorter than the time of the inverse process – capture into the well. Therefore the bottleneck for cyclic electron transfer out of and back into the well is the slower process of electron capture. At room temperature, the characteristic time of capture into the well is $\sim 10^{-11}$ s for AlGaAs/GaAs structures. Since this time determines the kinetics of the real-space transfer, the previously mentioned oscillations occur at high frequencies up to hundreds of gigahertz.

The real-space transfer effects are important because they occur in any quantum heterostructures that are biased electrically with high voltage. The first device proposed on the basis of this effect has been the negative-resistance field-effect transistor (NERFET). But possible applications of this effect can go far beyond the use of negative differential resistance. In Chapter 9 we will study devices based on real-space transfer in more detail.

7.4.4   Other Effects of High-Field Electron Transport in Quantum Structures

As we have already seen, many new effects appear in quantum structures and many conventional transport regimes and effects are modified. Because of the limited scope of this textbook, we are unable to review all these effects of lower dimensionality. Let us list just a few of them.

Nonequilibrium phonons in quantum heterostructures: Because of the optical-phonon confinement in quantum heterostructures, nonequilibrium optical phonons cannot escape from the active region. Even the intermediate product of optical-phonon decay, namely, short-wavelength nonthermal acoustic phonons, cannot penetrate through the interface because of strong reflection. As a result, the population of nonequilibrium phonons grows more rapidly than in bulk materials and can even lead to the quasi-thermal destruction of the crystal lattice. Obviously the effects of nonequilibrium phonons are much more pronounced in quantum heterostructures.

There is one aspect of electron interactions with nonequilibrium phonons that is associated exclusively with one-dimensional quantum wires. The wave vector of a phonon participating in scattering is defined by the energy- and the momentum-conservation equations and equals $q = \sqrt{k^2 + k'^2 - 2kk'\cos\theta}$, where k is the electron wave vector before scattering, θ is the angle between electron wave vectors before and after scattering, and $k' = \sqrt{k^2 \pm 2m^*\omega_0/\hbar}$ is the electron wave vector after absorption (sign +) or emission (sign −) of the phonon of frequency ω_0 (for optical phonons, for example). In one-dimensional structures, there are just two final states for scattered electrons: forward scattering with $\cos\theta = 1$ or backward scattering with $\cos\theta = -1$. Consequently, there are two possible phonon wave vectors available for emission (and two for absorption) by any single electron: $q_{min} = |k - k'|$ and $q_{max} = k + k'$. In contrast, in quantum wells (or bulk materials), because of the existence of 1 (or 2) additional degrees of freedom, $\cos\theta$ can take on any value in the range $(-1, +1)$, so that there is an entire range of q from q_{min} to q_{max} available for electron interactions. Therefore electrons in quantum wires, having appreciably different energies, generate nonequilibrium phonons in different q-space regions that do not overlap. In turn, these phonons can be reabsorbed only by the electrons that have generated them, unlike in bulk materials and in quantum wells, in which electrons can reabsorb phonons emitted by other electrons. Consequently, electrons that have different energies cannot redistribute their energy and momentum through the emission and the subsequent reabsorption of optical phonons.

Effects of different electron and phonon confinement: In layered structures different confinement of electrons and phonons can be realized. Indeed, very thin AlAs layers (of the order of 2–5 monolayers) separating GaAs layers can be nearly transparent to electrons that tunnel with near-unity efficiency but prevent free optical-phonon propagation as a consequence of the different optical-phonon frequencies in these materials. Another example of different confinement is electrostatically confined electrons and bulk phonons, as in the case of inversion layers, modulation doping, etc. By proper selection of structural materials and parameters, one can either suppress the phonon-scattering rates and thus enhance the mobility or increase the scattering rates and inelasticity and thus increase electron-energy relaxation. This opens new means of controlling the transport characteristics in quantum structures.

Heat dissipation from quantum structures: One of the consequences of high-field electron transport is the necessity for heat removal from active channels of devices. For quantum structures, two new aspects arise in connection with this problem. First, phonons confined in regions of quantum wells or wires do not participate in energy transfer from these regions. Thus heat removal is possible only after a decay of confined modes into other extended phonon modes, typically acoustic phonons. Second, the mean free path of these acoustic phonons can greatly exceed the characteristic spatial scales of the devices. Under such a condition, the heat removal is essentially a nonlocal process. The heat dissipation should be described in terms of kinetic equations for phonons rather than in terms of the thermal conductivity.

Mutual drag in parallel quantum structures: If two electron channels are separated by a barrier of finite width, electrons in these channels can interact either by means of an electrostatic potential or by an exchange of phonons. For the latter interaction, the phonons must be common for both electron channels. This is true for acoustic phonons and even for optical phonons if electron confinement in the channels does not coincide with the phonon confinement. Both mechanisms of electron interaction lead to drag effects whereby electron momentum from one channel can be transferred to electrons in another channel, thus inducing electron drift in the second channel even in the absence of an external field there. This effect opens other than electrical means of controlling electron transport in quantum structures, which may be important for future device applications.

7.5 Closing Remarks

In this chapter, we have introduced two types of electron transport in quantum structures: parallel (horizontal) transport and perpendicular (vertical) transport. Next, we analyzed parallel transport, which corresponds primarily to classical electron motion along layers or along a wire axis.

In a weak electric field, when the electron subsystem is near equilibrium, the transport is linear (the ohmic regime) and can be characterized by two field-independent parameters – the mobility and the diffusion coefficient. These parameters are related by the universal Einstein relation studied in Chapter 2. We have shown that high values of low-field mobilities are achieved in quantum structures because of the spatial separation of carriers and their parent impurities. Another feature of the linear transport in quantum structures is intersubband transitions, which result in a specific mobility versus field dependence.

An increase in the electric field drives the electron subsystem far from equilibrium. The transport becomes nonlinear. The electric field heats electrons and increases their mean energy. Generally, in high electric fields the electron distribution function can differ dramatically from that of the equilibrium case. For example, a highly anisotropic distribution occurs in the so-called streaming regime. Under high electric fields the current-voltage characteristic becomes nonlinear, and mobilities and diffusion coefficients are dependent on the field magnitude. For device applications, one of the most crucial high-field effects is velocity saturation in high electric fields, which limits the electron-drift velocity. For various materials, the saturated velocities differ because of different effective masses, electron–phonon couplings, etc. In direct-bandgap materials GaAs, InAs, and InSb, the saturated velocities are several times larger than those of AlAs, Si, etc. Velocity saturation limits the high-speed operation of devices. We have shown that in submicrometer active layers there can exist a hot-electron size effect – a dependence of the electron heating on layer thickness and scattering on the boundaries. This size effect can suppress the heating and improve the drift velocity.

To overcome limitations associated with the saturation-velocity effect, one can use the transient overshoot effect. Because of overshoot, the average velocity of electrons can exceed the saturation values as many as 2 to 4 times. Overshoot can be used in devices with short intercontact distances.

In this chapter, we have found that in quantum structures hot-electron effects are even more pronounced than in bulklike materials. In particular, these effects occur for smaller electric fields. The peculiarities of hot electrons in quantum structures are associated with changes in intensities of scattering mechanisms and intersubband transitions studied in this and previous chapters. A principally new phenomenon – real-space transfer – occurs as the result of the escape of hot electrons from quantum wells or wires. This kind of hot-electron effect has a number of interesting applications and will be discussed in more details in Chapter 9.

Additional information on the low-field mobility of electrons and holes in quantum wells and wires can be found in the following publications:

T. Ando, A. B. Fowler, and F. Stern, "Electronic properties of two-dimensional systems," Rev. Mod. Phys. **54**, 437–672 (1982).

G. Bastard, *Wave Mechanics Applied to Semiconductor Heterostructures* (Halsted, New York, 1988).

C. Weisbuch and B. Vinter, *Quantum Semiconductor Structures* (Academic, San Diego, 1991).

Supplementary treatments on hot-electron problems in bulk materials and quantum wells are found in the following reviews and books:

E. M. Conwell, *High-Speed Transport* (Academic, New York, 1967).

B. K. Ridley, "Hot electrons in low-dimensional structures," Rep. Prog. Phys. **54**, 169–256 (1991).

B. K. Ridley, "Electron-phonon interaction in 2D systems," in *Hot Carriers in Semiconductor Nanostructures (Physics and Applications)*, J. Shah, ed. (Academic, Boston, 1992), pp. 17–51.

J. Shah, ed., *Hot Carriers in Semiconductor Nanostructures (Physics and Applications)* (Academic, Boston, 1992).

W. Pötz and P. Kocevar, "Cooling of highly photo-excited electron-hole plasma in polar semiconductors and semiconductor quantum wells: a balance-equation approach," in *Hot Carriers in Semiconductor Nanostructures (Physics and Applications)* (Academic, Boston, 1992), pp. 87–120.

B. K. Ridley, *Quantum Processes in Semiconductors*, 3rd ed. (Clarendon, Oxford, 1993).

B. K. Ridley, *Electrons and Phonons in Semiconductor Multilayers* (Cambridge U. Press, Cambridge, 1997).

Nonlinear electron transport in quantum wires is treated analytically in the following paper:

B. K. Ridley and N. A. Zakhleniuk, "Hot electrons under quantization conditions, Parts I–III," J. Phys. Condens. Matter **8**, 8525–8581 (1996).

The scope of this book does not include an analysis of the role of nonequilibrium phonons in heat-removal processes. The following papers provide an introduction to this problem:

M. I. Flik, B. I. Choi, and K. E. Goodson, "Heat transfer regimes in microstructures," J. Heat Transfer **114**, 666–674 (1992).

R. Mickevicius, V. Mitin, V. Kochelap, M. Stroscio, and G. Iafrate, "Radiation of acoustic phonons from quantum wires," J. Appl. Phys. **77**, 5095–5097 (1995).

V. V. Mitin, G. Paulavicius, N. A. Bannov, and M. A. Stroscio, "Radiation patterns of acoustic phonons emitted by hot electrons in a quantum well," J. Appl. Phys. **79**, 8955–8963 (1996).

A quite complete discussion of different aspects of real-space-transfer effects is presented in the following review:

Z. S. Gribnikov, K. Hess, and G. A. Kosinovsky, "Nonlocal and nonlinear transport in semiconductors: real-space transfer effects," J. Appl. Phys. **77**, 1337–1373 (1994).

Key references on piezoelectric interactions during parallel transport are

P. J. Price, "Two-dimensional electron transport in semiconductor layers. I. Phonon scattering," Ann. Phys. **133**, pp. 217–239 (1981).

M. A. Stroscio and K. W. Kim, "Generalized piezoelectric scattering rate for electrons in a two-dimensional electron gas," Solid-State Electron. **37**, 181–182 (1994).

PROBLEMS

1. In modeling carrier transport in semiconductor devices, Eq. (7.17) is frequently approximated by taking $\gamma = 1$ or by assuming a two-segment piecewise linear relationship between v and F: $v = v_s(F/F_s)$ for $F < F_s$ and $v = v_s$ for $F \geq F_s$. Assume that the most accurate form results when $\gamma = 2$. When $F = 3F_s$, what are the values of v for $\gamma = 2$, for $\gamma = 1$, and for the two-segment piecewise linear relationship? When $F = F_s$, how do these three values compare?

2. Within the approximation of Fermi's golden rule, the probability W_{12} of the scattering of electrons from the state with energy ϵ_1 to the state characterized by energy ϵ_2, which due to piezoelectric interaction with the acoustic phonons in a zincblende crystal, may be written as

$$W_{12} = \frac{2\pi}{\hbar} \frac{S}{(2\pi)^3} 2\delta(\epsilon_1 - \epsilon_2),$$

where

$$S = \frac{k_B T (eh_{14})^2}{2} \frac{\pi}{q_\|} \left(\frac{B_l}{\rho s_l^2} + \frac{2B_t}{\rho s_t^2} \right),$$

with

$$B_l = \frac{q_\|}{\pi} \int_{-\infty}^{+\infty} \frac{A_l}{q_\|^2 + q_z^2} \, dq_z, \qquad B_t = \frac{q_\|}{\pi} \int_{-\infty}^{+\infty} \frac{A_t}{q_\|^2 + q_z^2} \, dq_z,$$

$$A_l = 36 \frac{q_z^2 q_x^2 q_y^2}{(q_z^2 + q_\|^2)^3}, \qquad 2A_t + A_l = \frac{4}{(q_z^2 + q_\|^2)} (q_z^2 q_x^2 + q_x^2 q_y^2 + q_z^2 q_y^2).$$

In these results W_{12} equals the rate of transitions from initial state 1 to final state 2 per unit volume of $\vec{k}$ space, q_z is the acoustic-phonon wave vector normal to a two-dimensional electron gas in the (100) plane, q_x and q_y are the phonon wave vectors in the (100) plane, $q_{\parallel}^2 = q_x^2 + q_y^2$, $h_{14} = \kappa\beta_{1,23}$ is the piezoelectric constant that has a value of roughly $0.157\,\text{C/m}^2$ for GaAs, ρ is the mass density, and s_l (s_t) is the longitudinal (transverse) velocity of sound in the lattice. In this problem, a linear relation between the phonon frequency and the wave vector has been assumed. On performing the indicated integrations over q_z, we find

$$\frac{\pi}{q_{\parallel}} B_l = \frac{q_z}{4}\pi \frac{q_x^2 q_y^2}{\left(q_x^2 + q_y^2\right)^{5/2}},$$

$$\frac{\pi}{q_{\parallel}} B_t = \frac{\pi}{2}\frac{1}{\left(q_x^2 + q_x^2\right)^{1/2}} - \frac{3\pi}{4}\frac{q_x^2 q_y^2}{\left(q_x^2 + q_y^2\right)^{5/2}}.$$

Thus

$$S = \frac{k_B T (e h_{14})^2}{2}\left\{\left[\frac{q_z}{4}\pi\frac{q_x^2 q_y^2}{\left(q_x^2 + q_y^2\right)^{5/2}}\right]\frac{1}{\rho s_l^2}\right.$$

$$\left. + \left[\frac{\pi}{2}\frac{1}{\left(q_x^2 + q_y^2\right)^{1/2}} - \frac{3\pi}{4}\frac{q_x^2 q_y^2}{\left(q_x^2 + q_y^2\right)^{5/2}}\right]\frac{1}{\rho s_t^2}\right\}.$$

In the limit where the average is taken over azimuthal directions, $q_x^2 q_y^2 \rightarrow q_{\parallel}^4/8$ and $q_x^2 = q_y^2 \rightarrow q_{\parallel}^2/2$.

Show that the piezoelectric scattering rate for electrons in a two-dimensional electron gas interacting with an isotropic acoustic phonon in a (100) plane of a zincblende crystal may be written as

$$W_{12} = \frac{2\pi}{\hbar}\frac{S}{(2\pi)^2}2\delta(\epsilon_1 - \epsilon_2),$$

where

$$S = \frac{k_B T (e h_{14})^2}{2}\frac{\pi}{q_{\parallel}}\left(\frac{9}{32}\frac{1}{\rho s_l^2} + \frac{13}{32}\frac{1}{\rho s_t^2}\right).$$

3. The Einstein relation between the diffusivity D and the mobility μ is given by Eq. (2.212) for the case of equilibrium, nondegenerate electrons. For the case of electrons, the sum of the drift and the diffusion currents is

$$j(x) = -e\mu n(x) F(x) - eD\frac{dn(x)}{dx}.$$

Assuming equilibrium conditions for the current as well as for the carrier distribution, show that $D = \mu k_B T/e$ for nondegenerate electrons. (This result also holds for holes.)

4. The power radiated by an electron emitting LO phonons in a polar material may be approximated by $P = (\hbar\omega_{LO}/\tau)e^{-\hbar\omega_{LO}/k_B T}$, where $\tau = (2\alpha\omega_{LO})^{-1}$ for a three-dimensional system ($\tau = 0.13\,\text{ps}$ in GaAs) and $\tau = (\pi\alpha\omega_{LO})^{-1}$ for a two-dimensional system ($\tau = 0.08\,\text{ps}$ in GaAs), and α is the Fröhlich coupling

constant. Consider the case in which an electron is injected into the uppermost energy region of a quantum well. This electron thermalizes to the ground state of the quantum well by emitting a sequence of LO phonons; however, these phonons are trapped in the quantum well and there is a high probability of LO phonon absorption by the electron since the phonon emission R_{em} and phonon absorption R_{ab} rates are related by $R_{\text{ab}} = (e^{\hbar\omega_{\text{LO}}/k_B T}) R_{\text{em}}$. This absorption process greatly reduces the net power radiated by the electrons in such a situation. In fact, another time scale must be considered in estimating the power radiated by an electron in such a quantum well; indeed, the LO phonons decay into two LA phonons with a lifetime of several picoseconds in many polar semiconductors. For GaAs at 77 K, this lifetime against the decay of a LO phonon to two LA phonons is $\tau_{\text{ph}} \approx 7 \, \text{ps}$. Electron-scattering probabilities with such LA phonons are at least 2 orders of magnitude weaker than those for electron LO phonon scattering. Present a plausibility argument that for the capture of an electron in a GaAs quantum well, $P \approx (\hbar\omega_{\text{LO}}/\tau_{\text{ph}})e^{-\hbar\omega_{\text{LO}}/k_B T}$, that is, $\tau \to \tau_{\text{ph}}$.

Perpendicular Transport in Quantum Structures

8.1 Introduction

In Chapter 7 we studied classical parallel electron transport in quantum structures. According to Section 2.3, quantum parallel transport can manifest itself in only very short electron channels that are shorter than the dephasing length of the electrons. The dephasing length decreases dramatically with increasing lattice temperature. Therefore parallel quantum transport is observable at very low temperatures. On the other hand, there are quantum heterostructures that exhibit quantum perpendicular transport up to room temperature. One such example is resonant electron transmission through a double-barrier heterostructure. Another example is given by perpendicular transport in superlattices.

In this chapter we first study resonant electron tunneling through double-barrier structures. Two principal mechanisms of resonant tunneling are considered: coherent and sequential tunneling. We develop a simple theoretical model describing these types of tunneling. Then we show that resonant tunneling results in a negative differential conductivity.

The current-voltage characteristics of different heterostructures with superlattice elements are analyzed for the collisionless (ballistic) and collisional transport regimes. For both cases, negative differential resistance can occur. We discuss possible ways to control the electric current in such structures. Then we analyze Bloch oscillations and Wannier–Stark ladders of energy levels for superlattices. We also consider a specific type of heterostructures – a ballistic-injection device – that uses electron injection over barriers. The latter structures exhibit classical perpendicular transport of electrons. Finally, we study single-electron transfer and Coulomb blockade effects, which are important for devices with small cross sections and low currents.

8.2 Double-Barrier Resonant-Tunneling Structures

We start by considering a double-barrier heterostructure as an example of resonant-tunneling diodes. Figure 8.1 shows a sequence of layers in such a structure. The top and the bottom parts of the structure are doped regions, while the barriers and the well layers are undoped. Figure 8.1(a) shows a specific structure in which a quantum-well layer of GaAs is embedded between two $Al_xGa_{1-x}As$

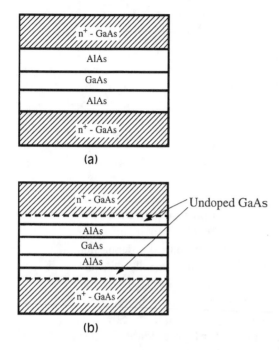

Figure 8.1. Layer designs for double-barrier resonant-tunneling structures. (a) Alternating layers: n-doped GaAs (substrate and one of the contacts) – undoped AlGaAs (barrier) – GaAs (quantum well) – undoped AlGaAs (barrier) – n-doped GaAs (top contact). (b) The same as (a) except for additional undoped spacer layers of GaAs between the contacts and the barriers.

barrier layers. The top and the bottom regions of GaAs serve as contacts. A slightly different design is shown in Fig. 8.1(b), in which two additional spacer layers separate the doped regions and the double-barrier part of the structure. The purpose of these spacer layers is to prevent scattering of tunneling electrons by impurities in the contact regions. The thicknesses of the well, barriers, and spacers may be varied substantially. Inside the quantum well, several quantized levels can exist. In fact, these levels are quasi-bound states, because there is a small but finite probability of electron tunneling out of the well. The tunneling is responsible for a finite lifetime of the electrons on those levels and leads to some broadening of the quantum-well states.

Thus a resonant-tunneling diode can be thought as a system with two contacts containing three-dimensional electrons and a quantum well with two-dimensional electrons (if any). These three subsystems are weakly coupled through tunneling. Energy-band diagrams of the structure are presented in Fig. 8.2 for three different voltage biases. Figure 8.2(a) corresponds to the equilibrium case, in which no voltage is applied. In the well, there is at least one quasi-bound level; the single level depicted in Fig. 8.2 has an energy ϵ_1. Actually, ϵ_1 is the bottom of lower two-dimensional subband. The parameters of the diode are chosen to be such that in the nonbiased state of the diode, the quasi-bound level ϵ_1 lies above the Fermi energy in the contacts, as in Fig. 8.2(a). By applying a voltage bias to the contacts, one can produce a downward shift of the level in the well. For electrons with arbitrary energies the probability of tunneling through the double-barrier structure is small. The structure is designed to prevent a thermal transfer of electrons over the barriers. Therefore the only situation favorable for electron transmission through the structure is when the quasi-bound level lies below the

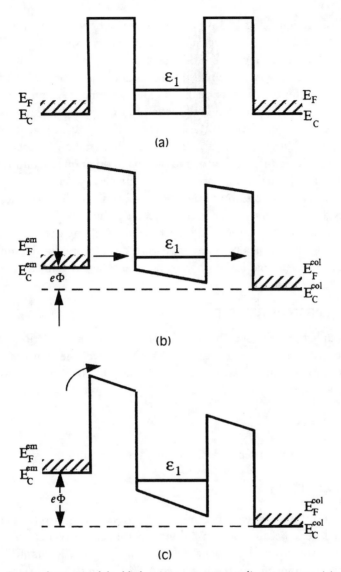

Figure 8.2. Energy diagrams of double-barrier resonant-tunneling structures: (a) equilibrium conditions; voltage-biased scheme; (b) in-resonance; (c) off-resonance. Note that the left and the right parts of the structure here correspond to the top and the bottom of the structure in Fig. 8.1.

Fermi energy but above the conduction-band bottoms of the contacts. In this case, those electrons from the emitter (left contact) whose kinetic energy of the perpendicular motion $E_\perp = \hbar^2 k_z^2 / 2m^*$ coincides with ϵ_1 are transmitted through the structure with finite probability. This is the so-called resonant-tunneling process, which has the important attribute of exhibiting negative differential resistance.

The following qualitative model of resonant tunneling explains the appearance of negative differential resistance. Before this level reaches the resonance position, the current through the diode is exponentially small because it is controlled by nonresonant tunneling and processes for transport over barriers. When the resonance is reached, as depicted in Fig. 8.2(b), the transmission coefficient and the electric current through the diode increase sharply. Further increase in the current with increasing voltage bias continues until the resonant level passes the bottom of the emitter conduction band. There are no electrons, as shown in Fig. 8.2(c), to tunnel resonantly and the current decreases in spite of increased voltage bias. Consequently the current-voltage characteristic of the structure contains a segment that exhibits negative differential resistance. Further increase of the current can be induced either by the shifting of other quasi-bound states (not depicted in Fig. 8.2) so that they are resonant with electron energies in the emitter or by the substantial evolution of the potential profile so that both nonresonant transports through and over the barrier become dominant.

The resonant transmission of the electrons through the double-barrier diode can be provided by physically different processes. Conceptually, the simplest way is direct quantum-mechanical tunneling, i.e., the coherent-tunneling processes.

8.2.1 Coherent Tunneling

In this case, the electron is characterized by a single wave function throughout the whole quantum structure. An electron that enters the structure from the emitter is described by the wave function $\psi^{em}(\vec{r}, z)$, which has the form of an incident plane wave,

$$\psi^{em}(\vec{r}, z) = A^{em} e^{i(\vec{k}\vec{r} + k_z^{em} z)},$$

and an outgoing wave after passing through the structure,

$$\psi^{em}(\vec{r}, z) = B^{em} e^{i(\vec{k}\vec{r} + k_z^{col} z)};$$

here, $\vec{k}$ and $\vec{r}$ are two-dimensional in-plane vectors and z is directed perpendicularly to the layers. Because of translational invariance the vector $\vec{k}$ is conserved; thus it is the same for both incident and outgoing waves. The quantities k_z^{em} and k_z^{col} are the z components of the wave vector in the emitter and in the collector, respectively. The relation between these two quantities can be obtained from the energy-conservation law:

$$E_{\perp}^{em} - E_{\perp}^{col} \equiv \frac{\hbar^2 (k_z^{em})^2}{2m^*} - \frac{\hbar^2 (k_z^{col})^2}{2m^*} = e\Phi,$$

where Φ is the voltage applied to the structure. We can find the coefficients A^{em} and B^{em} by solving the Schrödinger equation with the potential corresponding to the double-barrier structure under voltage bias, as we did in Subsection 3.2.2. The ratio of the coefficients B^{em} and A^{em} defines the transmission coefficient:

$$T(E_{\perp}) = \frac{|B^{em}(E_{\perp})|^2}{|A^{em}(E_{\perp})|^2}.$$

Instead of solving the Schrödinger equation, here we give a simplified formula for $T(E_\perp)$ that allows us to calculate the current in an explicit form.

As it follows from the qualitative analysis presented in the previous subsection, $T(E_\perp)$ has to have sharp peaks in the vicinity of resonant energies ϵ_n. For a structure with symmetric barriers we can approximate these peaks with

$$T(E_\perp) = \frac{1}{1 + \left(\frac{E_\perp - \epsilon_n}{\Delta E}\right)^2}, \tag{8.1}$$

where ΔE is the full width of the half-maximum of the transmission peak. In Eq. (8.1) we assume that the probability of tunneling through the structure is unity when the electron energy $E_\perp$ exactly coincides with a quasi-bound state ϵ_n. Besides the transmission coefficient, it is convenient to introduce the probability of tunneling of the electrons out of the well per second, $\Gamma = \Delta E / \hbar$, and the lifetime of electrons of the quasi-bound level $\tau = \hbar / 2\Delta E$. The tunneling probability per unit time can be expressed as the product of the attempt rate $v_z/2L_w$ and the probability of tunneling through a single barrier with one attempt T:

$$\Gamma = T \frac{v_z}{2L_w},$$

where L_w is the thickness of the well and v_z is the transverse velocity of the electron in the well, which can be estimated from the relationship $m^* v_z^2 / 2 = \epsilon_n$.

For the purpose of making illustrative numerical estimates, we choose the following set of parameters: the energy of the level ϵ_1 is 50 meV, the height and the thickness of the barriers are $V_b = 300$ meV and $L_b = 40$ Å, respectively, and the well width is $L_w = 100$ Å. For this structure, we obtain $v_z = 5 \times 10^7$ cm/s and $\Gamma = 10^{11}$ s^{-1}. Hence the lifetime of an electron in the well equals $\tau = 5 \times 10^{-12}$ s. Both τ and Γ depend critically on the height and the thickness of the barriers.

For an asymmetric double-barrier structure, we define two different transmission coefficients for the left-hand T_l and the right-hand T_r barriers, so that the total transmission can be approximated by

$$T(E_\perp) = \frac{4T_l T_r}{(T_l + T_r)^2} \frac{1}{1 + \left(\frac{E_\perp - \epsilon_n}{\Delta E}\right)^2}, \tag{8.2}$$

where $\Delta E \equiv \hbar\Gamma$ and $\Gamma = (T_l + T_r)v_z/2L_w$. For asymmetric barriers, the maximum transmission at $E_\perp = \epsilon_n$ is less than unity. It is necessary to stress that the above results are obtained under the assumption that the electrons transit through the system without phase-changing scattering.

Formulas (8.1) and (8.2) describe the transmission of an electron with fixed energy $E_\perp$. To calculate the electric current, we should take into account all electrons that tunnel from the emitter and the collector. We suppose that in the contact regions – heavily doped regions – the thermal equilibrium for electrons is established in a very short time compared to T. Therefore, to a reasonable degree of approximation, we can assume the Fermi distribution for electrons entering the double-barrier part of the structure. As shown in Fig. 8.2(b), the

Fermi energies of the emitter E_F^{em} and collector E_F^{col} are not equal if a voltage bias Φ is applied (analogous to the approach used in Section 2.7):

$$E_F^{em} - E_F^{col} = e\Phi.$$

We can now calculate the electron areal density (also referred before as the sheet or surface density) of the electrons having some fixed energy $E_\perp$ in the contacts. With the results of Eq. (2.181) it follows that

$$n^{em}(E_\perp) = \frac{k_B T m^*}{\pi \hbar^2} \ln\left(1 + \exp \frac{E_F^{em} - E_\perp}{k_B T}\right), \tag{8.3}$$

$$n^{col}(E_\perp) = \frac{k_B T m^*}{\pi \hbar^2} \ln\left(1 + \exp \frac{E_F^{col} - E_\perp}{k_B T}\right). \tag{8.4}$$

Then the contribution to the electric-current density of the emitter electrons is

$$j^+ = e \sum_{E_\perp > E_c^{em}} \frac{\hbar k_z^{em}}{m^*} T(E_\perp) n^{em}(E_\perp) = \frac{e}{2\pi\hbar} \int_{E_c^{em}}^\infty dE_\perp T(E_\perp) n^{em}(E_\perp), \tag{8.5}$$

where the lower limit of integration starts at the conduction-band minimum in the emitter E_c^{em}. The contribution of the collector electrons has a similar form:

$$j^- = e \sum_{E_\perp > E_c^{col}} \frac{\hbar k_z^{col}}{m^*} T(E_\perp) n^{col}(E_\perp) = \frac{e}{2\pi\hbar} \int_{E_c^{col}}^\infty dE_\perp T(E_\perp) n^{col}(E_\perp). \tag{8.6}$$

The net current is

$$j = j^+ - j^-. \tag{8.7}$$

For small biases we may expand Eqs. (8.3) and (8.4) with respect to the voltage bias; it then follows that

$$
\begin{aligned}
n^{em}&(E_\perp) - n^{col}(E_\perp) \\
&= \frac{k_B T m^*}{\pi \hbar^2} \left[\ln\left(1 + \exp \frac{E_F + e\Phi - E_\perp}{k_B T}\right) - \ln\left(1 + \exp \frac{E_F - E_\perp}{k_B T}\right)\right] \\
&\approx \frac{k_B T m^*}{\pi \hbar^2} \frac{e\Phi}{k_B T} \frac{e^{(E_F - E_\perp)/k_B T}}{1 + e^{(E_F - E_\perp)/k_B T}}.
\end{aligned}
$$

The current density j can be calculated easily for small voltage biases:

$$j = \frac{e^2 m^*}{2\pi\hbar^3} \Phi \int_{E_c}^\infty dE_\perp T(E_\perp) \frac{1}{1 + e^{(E_F - E_\perp)/k_B T}}, \tag{8.8}$$

where $E_c = E_c^{em} = E_c^{col}$ under zero voltage.

If the voltage bias is not small and the current from the emitter to the collector dominates, it is possible to drop the second term in Eq. (8.7), so that

$$j \approx j^+, \tag{8.9}$$

and the dependence on Φ originates from the variable position of the level ϵ_1 relative to the bottom of the conduction band in the emitter contact E_c^{em}.

Now let us consider a finite bias Φ in the case in which the temperature is zero. The restriction to zero temperature makes it possible to derive an explicit form for the current. Then, instead of Eq. (8.3), we obtain

$$n^{\text{em}}(E_\perp) = \frac{m^*}{\pi \hbar^2} (E_F - E_\perp). \tag{8.10}$$

The electron transverse energy is limited by $E_c^{\text{em}} < E_\perp < E_F^{\text{em}}$; otherwise the sheet density of the tunneling electrons is zero. We get the expression for the current:

$$j = \frac{em^*}{2\pi^2 \hbar^3} \int_{E_F^{\text{em}} - e\Phi}^{E_F^{\text{em}}} dE_\perp \left(E_F^{\text{em}} - E_\perp \right) T(E_\perp), \quad e\Phi < E_F^{\text{em}}. \tag{8.11}$$

This formula takes into account the fact that because all states in the collector with energy $E_\perp < E_F^{\text{em}} - e\Phi$ are occupied, tunneling is possible only for those electrons with energies that satisfy the following conditions: $E_F^{\text{em}} > E_\perp > E_F^{\text{em}} - e\Phi$. For $e\Phi > E_F^{\text{em}}$ all emitter electrons can tunnel through the barriers and we have

$$j = \frac{em^*}{2\pi^2 \hbar^3} \int_{E_c^{\text{em}}}^{E_F^{\text{em}}} dE_\perp \left(E_F^{\text{em}} - E_\perp \right) T(E_\perp), \quad e\Phi > E_F^{\text{em}}. \tag{8.12}$$

We suppose that the Lorentzians in Eqs. (8.1) and (8.2) are narrow compared with E_F^{em}. Then the integration over $E_\perp$ in Eqs. (8.11) and (8.12) is straightforward and yields

$$j = \frac{em^* v_z}{2\pi \hbar^2 L_w} \left[E_F^{\text{em}} - \epsilon_n(\Phi) \right] \frac{T_l T_r}{T_l + T_r}, \tag{8.13}$$

where we use Eq. (8.2) as for asymmetric barriers. Equation (8.13) represents the tunneling current through the bound state with energy $\epsilon_n(\Phi)$, which depends on the voltage bias as illustrated previously by the example of Fig. 8.2(b). A peak value of the current is reached when the bound-state energy is resonant with the bottom of the conduction band in the emitter E_c^{em}:

$$j_p = \frac{em^* v_z}{2\pi \hbar^2 L_w} E_F \frac{T_l T_r}{T_l + T_r}. \tag{8.14}$$

Then the current drops rapidly to the value determined by off-resonant tunneling processes. This value can be estimated if we assume that the off-resonant transmission coefficients are constant:

$$j_v = \frac{em^*}{2\pi^2\hbar^3} E_F^2 T_l T_r. \tag{8.15}$$

The results obtained for the coherent mechanism of tunneling through a double-barrier structure support the qualitative discussion given in the beginning of this section.

8.2.2 Sequential Tunneling

Another process responsible for the resonant-tunneling effect is the so-called sequential tunneling. In the sequential-tunneling scheme, electron transmission through the structure is regarded – somewhat artificially – as two successive transitions: first from the emitter to the quantum well and then from the well to the collector. It is important to stress the main difference between the previously studied coherent mechanism and the sequential mechanism of resonant tunneling. The first mechanism excludes any electron collisions during the transition from the emitter to the collector. The second case applies even when there is electron scattering inside the quantum well.

To introduce the sequential-tunneling mechanism, let us recall that we can characterize the electron state in the contacts by a set of two parameters: $\{\vec{k}, E_\perp\}$. In the quantum well, electrons are also described by two parameters $\{\vec{k}, \epsilon_n\}$. Note that the in-plane wave vector $\vec{k}$ is conserved during the tunneling. Let us introduce some occupation numbers for the contacts $\mathcal{F}^{em}(\vec{k}, E_\perp^{em})$ and $\mathcal{F}^{col}(\vec{k}, E_\perp^{col})$ and for the quantum well $\mathcal{F}^w(\vec{k}, \epsilon_n)$. According to the simplified physical picture of sequential tunneling, we can write straightforwardly an equation for these occupancies:

$$\frac{d\mathcal{F}^{em}(\vec{k}, E_\perp^{em})}{dt} = -\sum_{\epsilon_n} W_{E_\perp^{em},\epsilon_n}^{em,w}(\vec{k})\left[\mathcal{F}^{em}(\vec{k}, E_\perp^{em}) - \mathcal{F}^w(\vec{k}, \epsilon_n)\right], \tag{8.16}$$

$$\frac{d\mathcal{F}^w(\vec{k}, \epsilon_n)}{dt} = -\sum_{E_\perp^{em}} W_{\epsilon_n,E_\perp^{em}}^{w,em}(\vec{k})\left[\mathcal{F}^w(\vec{k}, \epsilon_n) - \mathcal{F}^{em}(\vec{k}, E_\perp^{em})\right]$$
$$- \sum_{E_\perp^{col}} W_{\epsilon_n,E_\perp^{col}}^{w,col}(\vec{k})\left[\mathcal{F}^w(\vec{k}, \epsilon_n) - \mathcal{F}^{col}(\vec{k}, E_\perp^{col})\right]. \tag{8.17}$$

Here we introduce the rates of transitions between the contacts and the well:

$$W_{E_\perp^{em},\epsilon_n}^{em,w}(\vec{k}) = W_{\epsilon_n,E_\perp^{em}}^{w,em}(\vec{k}),$$

$$W_{\epsilon_n,E_\perp^{col}}^{w,col}(\vec{k}) = W_{E_\perp^{col},\epsilon_n}^{col,w}(\vec{k}),$$

where the lower indices designate the transverse energies of the initial and the final states of the transitions. Knowing these occupancies, we can calculate any characteristic of the structure. For example, the sheet density of the electrons inside the quantum well is

$$n^{\text{tot}} = \frac{1}{S} \sum_{\vec{k}, \epsilon_n} \mathcal{F}^w(\vec{k}, \epsilon_n).$$

Here S is the area of the diode cross section. As discussed previously, tunneling through the structure is possible if the appropriate states in the collector are empty, i.e., $\mathcal{F}^{\text{col}} = 0$. Accordingly, we may express $\mathcal{F}^w$ through $\mathcal{F}^{\text{em}}$ from Eq. (8.17) in the stationary regime:

$$\mathcal{F}^w(\vec{k}, \epsilon_n) = \frac{\sum_{E_\perp^{\text{em}}} W_{\epsilon_n, E_\perp^{\text{em}}}^{w, \text{em}}(\vec{k}) \, \mathcal{F}^{\text{em}}(\vec{k}, E_\perp^{\text{em}})}{\sum_{E_\perp^{\text{em}}} W_{\epsilon_n, E_\perp^{\text{em}}}^{w, \text{em}}(\vec{k}) + \sum_{E_\perp^{\text{col}}} W_{\epsilon_n, E_\perp^{\text{col}}}^{w, \text{col}}(\vec{k})}. \tag{8.18}$$

The first term in the denominator is the rate of tunneling from the quantum well to the emitter. The second term is the rate of electron tunneling from the well to the collector. Thus we can introduce the times of electron transfer to the emitter and to the collector, respectively:

$$\frac{1}{\tau_l(\vec{k}, \epsilon_n)} = \sum_{E_\perp^{\text{em}}} W_{\epsilon_n, E_\perp^{\text{em}}}^{w, \text{em}}(\vec{k}), \qquad \frac{1}{\tau_r(\vec{k}, \epsilon_n)} = \sum_{E_\perp^{\text{col}}} W_{\epsilon_n, E_\perp^{\text{col}}}^{w, \text{col}}(\vec{k}). \tag{8.19}$$

We can rewrite Eq. (8.18) as

$$\mathcal{F}^w(\vec{k}, \epsilon_n) = \frac{\tau_l \tau_r}{\tau_l + \tau_r} \sum_{E_\perp^{\text{em}}} W_{\epsilon_n, E_\perp^{\text{em}}}^{w, \text{em}}(\vec{k}) \mathcal{F}^{\text{em}}(\vec{k}, E_\perp^{\text{em}}). \tag{8.20}$$

Now we can calculate the total current density through the total number of electrons leaving the emitter:

$$j = -\frac{e}{S} \sum_{\vec{k}, E_\perp^{\text{em}}} \frac{\mathrm{d}\mathcal{F}^{\text{em}}(\vec{k}, E_\perp^{\text{em}})}{\mathrm{d}t}. \tag{8.21}$$

Substituting Eqs. (8.16) and (8.20) into Eq. (8.21) and taking into account equalities (8.19), we find

$$j = \frac{e}{S} \sum_{\vec{k}, \epsilon_n, E_\perp^{\text{em}}, E_\perp^{\text{col}}} \mathcal{F}^{\text{em}}(\vec{k}, E_\perp^{\text{em}}) \, W_{E_\perp^{\text{em}}, \epsilon_n}^{\text{em}, w}(\vec{k}) \, W_{\epsilon_n, E_\perp^{\text{col}}}^{w, \text{col}}(\vec{k}) \frac{\tau_l \tau_r}{\tau_l + \tau_r}. \tag{8.22}$$

If the quasi-bound level is very narrow ($\Delta \epsilon_n \ll \epsilon_n$), the conservation of electron energy and in-plane momentum implies that the emitter electron can tunnel into the well only if its kinetic energy $E_\perp^{\text{em}}$ is equal to $E_\perp^{\text{em}} \equiv \epsilon_n(\Phi)$. Figure 8.2(b) illustrates this case. When one subband ϵ_1 contributes to the current, the times of electron transfer τ_l and τ_r may be easily expressed in terms of transmission coefficients for both barriers as in Subsection 8.2.1:

$$\frac{1}{\tau_l} = \frac{v_z}{2L_w} T_l, \qquad \frac{1}{\tau_r} = \frac{v_z}{2L_w} T_r.$$

Then the summation in Eq. (8.22) results in exactly the same equation (8.13) for the current density j. This means that sequential tunneling regime leads to the same results as coherent tunneling. The sequential-tunneling mechanism does not require a collisionless regime of tunneling. In principle, we could take into consideration scattering inside the well. For that case it is necessary to add appropriate terms to Eq. (8.17). For example, we can add the scattering rate in the form

$$\left[\frac{\mathcal{F}^w(\vec{k}, \epsilon_n)}{dt}\right]_{sc} = -\sum_{\vec{k}',n'} W^{w,w}_{\vec{k},\epsilon_n,\vec{k}',\epsilon_{n'}}[\mathcal{F}^w(\vec{k}, \epsilon_n) - \mathcal{F}^w(\vec{k}', \epsilon_{n'})].$$

The scattering would modify the final expression for the current given by Eq. (8.22), but the resonant-tunneling effect would remain for some range of the parameters.

8.2.3 Comparison of Two Mechanisms of Resonant Tunneling

Although both coherent and sequential processes result in the same behavior of the double-barrier resonant structures, it is possible and instructive to separate and compare these processes. Let us introduce the width of the quasi-bound state Γ_r as the full width of the half-maximum of the transmission peak. Let the collision broadening of this state be Γ_{sc}. From the above discussion, we conclude that coherent tunneling dominates if $\Gamma_r > \Gamma_{sc}$, and the sequential processes dominate if $\Gamma_r < \Gamma_{sc}$. Table 8.1 illustrates the different regimes of tunneling for several particular $Al_{0.3}Ga_{0.7}As/GaAs$ double-barrier structures at zero bias. Here L_w and L_B are the thicknesses of the wells and the barriers, respectively. The ratio Γ_r/Γ_{sc} is calculated with the estimate $\Gamma_{sc} = \hbar/\tau$, where τ is the scattering time. From the results of Chapter 6, this time may be either calculated from first principles or deduced from experimental mobility measurements. For Table 8.1, the following values are assumed: $T = 300$ K, mobility in the well $\mu = 7000$ cm^2/V s, $\tau \simeq 3 \times 10^{-13}$ s; when $T = 200$ K, $\mu = 2 \times 10^4$ cm^2/V s, $\tau = 10^{-12}$ s; and for $T = 70$ K, $\mu \geq 10^5$ cm^2/V s, $\tau \geq 5 \times 10^{-12}$ s. If $\Gamma_r/\Gamma_{sc} > 1$, elastic collisions

Table 8.1.
Resonance Width and Collision Broadening of Level ϵ_1 of $Al_{0.3}Ga_{0.7}As/GaAs$ Double-Barrier Structures for Three Temperatures and Different Thicknesses of Barriers

L_w (Å)	L_B (Å)	Γ_r (meV)	Γ_r/Γ_{sc} at $T = 300$ K	Γ_r/Γ_{sc} at $T = 200$ K	Γ_r/Γ_{sc} at $T = 70$ K
50	70	1.3×10^{-2}	6×10^{-3}	1.9×10^{-2}	2.6×10^{-1}
50	50	1.5×10^{-1}	7.5×10^{-2}	8.3×10^{-1}	3.1
50	30	1.76	0.88	1.3	3.6
20	50	6.03	3.02	4.56	124

can be neglected and resonant tunneling is highly coherent. From Table 8.1, one can see that this type of tunneling is typical for the case of low temperatures and thin barriers. In the opposite limit, $\Gamma_r / \Gamma_{sc} \ll 1$, the tunneling processes are more likely sequential ones. It is important to stress that scattering also affects the magnitude of the transmission coefficient. The maximum of the transmission probability decreases by the factor $\Gamma_r / (\Gamma_r + \Gamma_{sc})$ when scattering takes place. This explains why resonance tunneling is washed out at high temperatures or in structures with defects and impurities.

8.2.4 Negative Differential Resistance under Resonant Tunneling

As we have already seen from qualitative discussions, resonant-tunneling structures manifest current-voltage curves with negative differential resistance. Typical current-voltage characteristics for a double-barrier structure of AlInAs/ GaInAs are presented in Fig. 8.3. The results are shown for two temperatures, $T = 80$ K and $T = 300$ K. The structure is symmetric (similar barriers, equal thicknesses of spacers, and doping of contacts); thus there is an almost anti-symmetric current-voltage characteristic. For the lower temperature, it is evident that there is a portion with almost zero current at finite voltage biases, which corresponds to the position of the resonant quasi-bound level above the Fermi level of the emitter and very small nonresonant and overbarrier currents compared to the resonant current. When the level is shifted below the bottom of the emitter conduction band ($\Phi \approx \pm 0.75$ V), the current drops again to almost zero values. At room temperature small overbarrier currents exist at any finite voltage bias and the current drop is also considerably less than in the former case. An important parameter for characterizing N-type negative differential resistance is

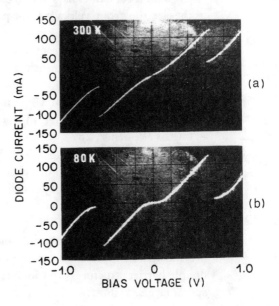

Figure 8.3. Measured current-voltage characteristics of a symmetric $Al_{0.48}In_{0.52}As/Ga_{0.47}In_{0.53}As$ double-barrier resonant-tunneling structure at 300 K (top) and 80 K (bottom). [After S. M. Sze, ed., *High-Speed Semiconductor Devices* (Wiley, New York, 1990).]

the ratio of the maximum and the minimum values of the current – the so-called peak-to-valley ratio. Achieving large peak-to-valley ratios greatly enhances the possibility of applications utilizing negative differential resistance. For the case presented in Fig. 8.3, the peak-to-valley ratio is 15:1 at 80 K. At room temperature, the ratio decreases to ~4:1 but still remains sufficient for applications. The peak-to-valley ratio depends not only on the physical nature of the negative differential resistance, but also on many technological and design factors. Although the development of the double-barrier system is still in progress, for the most optimized and perfect structures, ~20:1 and even higher values of the ratio may be achieved at room temperature.

Another important parameter of any system exhibiting a negative differential resistance is the characteristic time of the processes responsible for the negative differential resistance. This time determines the physical upper frequency limit at which the negative differential resistance disappears. For the previously considered resonant-tunneling device, the frequency limit is not easy to estimate, because this type of perpendicular transport has no classical analogy. There are many papers in which this problem is examined. It is clear that careful analysis of the frequency properties of the tunneling can be done only by a numerical self-consistent calculation involving the time-dependent Schrödinger equation, the kinetic equations describing processes in the contacts, and the Poisson equation. This complex problem has not yet been solved. However, when the barriers were modulated by small time-dependent voltages and the frequency response of the system was examined, it was shown that the characteristic time of the tunneling processes can be estimated by the following formula:

$$t_{tr} = m^* \int_0^d \frac{dz}{\sqrt{2m^*[V(z) - E]}} + \frac{2\hbar}{\Gamma} \equiv \frac{d}{v_g} + \frac{2\hbar}{\Gamma}, \tag{8.23}$$

where d is the total thickness of two barriers and the well and v_g can be interpreted as the electron group velocity. If the perpendicular energy of the incident electron $E_\perp$ is much greater than the resonance width Γ, it can be shown that the total transit time through the structure is approximately given by t_{tr}. The first term presents the semiclassical transit time across the structure and the second term is the so-called phase time.

For a typical example of a symmetric resonant-tunneling structure with 17-Å-thick AlAs barriers and a 45-Å-thick well, the quasi-bound level has an energy $\epsilon_1 \approx 0.13$ eV; hence, $2\hbar/\Gamma = 0.45 \times 10^{-12}$ s. For a drift velocity $v_d \geq 10^7$ cm/s, the first term gives only 0.8×10^{-13} s. Thus $\tau_{tr} \approx 0.5 \times 10^{-12}$ s. The cutoff frequency corresponding to this time exceeds 1 THz. In Section 9.7, we will discuss several specific resonant-tunneling devices that exhibit extremely high cutoff frequencies.

Recent progress in technology has made it possible to fabricate more complex heterostructures for resonant tunneling. For example, as we have mentioned previously, to suppress scattering in the double-barrier region, we can grow a double-barrier structures separated from the anode and the cathode by undoped spacer layers, as illustrated schematically in Fig. 8.1(b). To achieve multiple-peak current-voltage characteristics, they technologically integrate several double-barrier systems. There are two possibilities: (1) horizontal (parallel) integration and (2) perpendicular (in-series) integration. Both types of diode integration will be analyzed in Section 9.8.

In the previous discussions, we studied resonant-tunneling diodes by using layered heterostructures. A variety of different schemes based on the resonant-tunneling effect have been also proposed for the case of lateral electron motion. In this case, the electron gas in the contact regions is two dimensional and tunneling occurs through the barriers separated by these contacts. A one-dimensional electron channel – a quantum wire – is formed between these barriers. Thus the electron tunnels through the resonant state of this wire. The simplest structure proposed to realize such a lateral tunneling device is illustrated in Fig. 8.4. The sequence of horizontal layers in the central region of this device is completely similar to the conventional double-barrier resonant-tunneling structure shown in Fig. 8.1(b). But the GaAs undoped layers outside the double-barrier region are chosen to be quite wide (≥ 1000 Å) to prevent conventional (bulk) resonant tunneling. Instead, a tilted $Al_{0.75}Ga_{0.25}As$ region produces a two-dimensional electron gas on the $Al_{0.75}Ga_{0.25}As$/GaAs interface in the region of the wide GaAs layers; they are marked by 2-DEG(S) and 2-DEG(D), where (S) and (D) refer to source and drain, respectively. Upon the $Al_{0.75}Ga_{0.25}As$ layer, there is an additional gate with an associated potential Φ_G, which facilitates control of the

Figure 8.4. Schematic of a lateral double-barrier resonant-tunneling structure. [S. Luryi and F. Capasso, "Resonant tunneling of two-dimensional electrons through a quantum wire: A negative transconductance device," Appl. Phys. Lett. **47**, 1347–1349 (1985).]

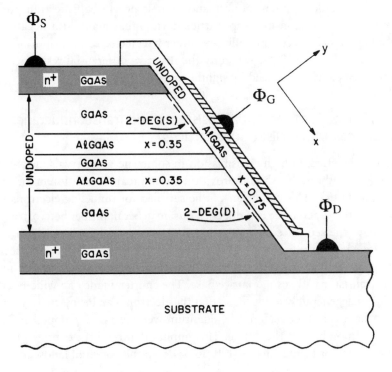

two-dimensional electron concentration on the interface, as discussed in Section 4.4. Through these means, one gets a pair of two-dimensional electron reservoirs provided by the contacts (the top and bottom n^+-GaAs layers) and separated by two barriers of $Al_{0.35}Ga_{0.65}As$. As seen from Fig. 8.4, a quantum wire is formed between the barriers on the $Al_{0.75}Ga_{0.25}As/GaAs$ interface. A voltage bias between the contacts designated by Φ_S and Φ_D gives rise to a resonant-tunneling current of two-dimensional electrons through the one-dimensional quasi-bound state and a current–voltage characteristic with negative differential resistance.

Finally, different resonant-tunneling structures exhibit negative differential resistance and are characterized by extremely short transit times of the carrier transport through the structures.

8.3 Superlattices and Ballistic-Injection Devices

As we know from Chapter 3, a superlattice is a periodic structure consisting of identical quantum wells separated by identical barriers. Electrons tunnel through these barriers from well to well. As a result, instead of the quantized energy levels of single quantum well, relatively narrow energy minibands are formed that correspond to collective electron states of the whole superlattice. These collective states have wave functions that extend along the superlattice axis. One often refers to these bands as minibands. The lowest minibands are usually separated by some energy gaps from one another. This energy spectrum leads to a resonant behavior of electron transmission through the superlattice at those energies where minibands are formed. This kind of transmission can be used to design a superlattice device with a negative differential resistance. To achieve such a goal, the superlattice should be quite perfect and exhibit nearly ballistic transport. In Subsection 8.3.1 we consider superlattice devices that manifest a negative differential resistance effect for conditions under which electron transport is nearly ballistic. In Subsection 8.3.4 we study the negative differential resistance for a transport regime characterized by significant scattering.

8.3.1 Negative Differential Resistance and Transconductance of Devices with Ballistic Superlattices

To use the effect of high transmission in some energy intervals and high reflection in the others, one should provide a flat miniband condition in the region of the superlattice. This condition can be satisfied for tunneling electrons if the region with the superlattice is p doped. Thus, in general, proper heterostructures should have the sequence of n–p–n-doped regions, which, as is well known, is typical for a bipolar transistor. Figure 8.5 depicts the energy diagram of a technologically interesting superlattice device. In fact, this structure is a heterojunction bipolar transistor with a superlattice base. The emitter, made of a wide-bandgap material, is degenerately n doped, so that the electrons can be injected by tunneling into the superlattice region. The quantum-well layers are p doped and comprise a common base of the device. This implies that both the emitter–base and the base–collector voltage biases shift up or down the potential landscape of the

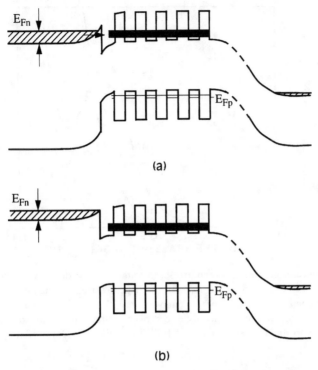

Figure 8.5. The energy-band diagram of a bipolar device with a superlattice embedded into the base p region. (a) The diagram under the emitter–base voltage lining up the Fermi energy in the emitter and the lowest miniband (an injection condition), (b) The diagram under the emitter–base voltage lowering the top of the first miniband below the emitter energy-band bottom (a reflection condition). [After F. Capasso, S. Sen, F. Beltram, *et al.*, "Quantum functional devices: resonant-tunneling transistors, circuits with reduced complexity and multi-valued logic," IEEE Trans. Electron Devices **36**, pp. 2065–2082 (1989).]

superlattice as a whole. Figure 8.5(a) shows an emitter–base voltage such that the electrons are injected into the miniband. For this case, the structure is transparent for electrons, and the emitter–collector current is large. If the emitter–base voltage exceeds the bias required for lining up the bottom of the emitter conduction band with the top of the miniband, as in Fig. 8.5(b), the electron transmission through the superlattice region is small and the collector current decreases. It is clear that if a ballistic regime occurs, the physical mechanisms governing the current are essentially similar to those of double-barrier structures. The collector current as a function of emitter–collector voltage has a negative differential resistance region. The peak-to-valley ratio may be controlled by the base potential. The third electrode – the base – provides an additional possibility of controlling the current.

Figure 8.6 shows the dependence of collector current on emitter–base voltage in the common-base configuration at two temperatures, 7 and 77 K, for an InP/GaInAs heterostructure. The 5000-Å-thick InP n^+ layer doped to an approximate density of 2×10^{18} cm^{-3} serves as an emitter of the device. The 20-Å-thick undoped InP layer separates the emitter from the superlattice. The latter is made

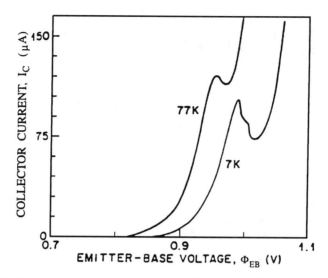

Figure 8.6. Collector current versus emitter–base voltage for the device with the superlattice embedded into the base region. Parameters of the structure are presented in the text. Results are obtained for $\Phi_{CB} \approx -2$ V and for two temperatures: 77 and 7 K. [After F. Capasso, S. Sen, F. Beltram, *et al.*, "Quantum functional devices: resonant-tunneling transistors, circuits with reduced complexity and multi-valued logic," IEEE Trans. Electron Devices **36**, pp. 2065–2082 (1989).]

of 70-Å-wide GaInAs wells with heavy p doping ($\approx 2 \times 10^{18}$ cm^{-3}) and 20-Å-thick InP barriers. There are 20 superlattice periods. The superlattice ends at an n type Ga$_{0.47}$In$_{0.53}$As collector, which has a thickness of 1.8 μm. To achieve electron injection through the base, the superlattice has relatively wide minibands. The ground state of the electron miniband extends from 36 to 75 meV. (The heavy-hole miniband is much narrower; it extends from 11.9 to 12 meV and plays no role in the transport.) At low temperatures, the thermionic emission – an overbarrier process – from the emitter is negligible. Carrier injection occurs as a result of the tunneling mechanism. The portion of the negative slope on the 7-K curve in Fig. 8.6 appears when the miniband top passes the bottom of the conduction band in the emitter. It is necessary to emphasize that current is a function of the emitter–base voltage, and the derivative of this current on voltage defines transconductance of the structure in contrast to conductance, which is the derivative of this current on emitter–collector voltage. This means that negative transconductance may be realized and controlled by the emitter-collector voltage. At higher temperatures, the negative transconductance effect is shifted to lower voltages and the magnitude of the effect decreases. The shift is caused by a decrease of the bandgap of the GaInAs. A negative differential resistance of the device ($\partial g_{EC}/\partial \Phi_{EC} < 0$) is observed along with the negative transconductance.

Another example of a superlattice device with negative differential resistance is a monopolar structure with the energy diagram depicted in Fig. 8.7. In this structure, a few-period superlattice is separated by undoped layers from the

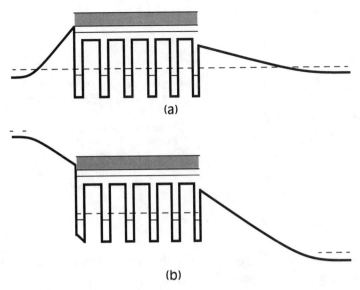

(a)

(b)

Figure 8.7. The energy-band diagram of a superlattice base monopolar transistor: (a) the device in equilibrium, (b) the device is biased so that electrons are injected from an emitter into a miniband. [After F. Capasso, S. Sen, F. Beltram, *et al.*, "Quantum functional devices: resonant-tunneling transistors, circuits with reduced complexity and multi-valued logic," IEEE Trans. Electron Devices **36**, pp. 2065–2082 (1989).]

emitter and the collector. The quantum wells are doped and comprise the common base. The barriers are relatively thick, so the tunneling current is suppressed. Because of the overbarrier electron reflection (see Chapter 3), some minibands are formed at energies exceeding the barrier height, i.e., the minibands appear in a classical continuum over the barriers. The electrons can be either injected into these minibands or reflected from minigaps. The first causes a high electric current. By applying a voltage bias between the emitter and the base Φ_{EB} one can shift the positions of the minibands and minigaps and control the current through the structure. As a result, this structure exhibits both negative differential resistance and transconductance.

Figure 8.8 exhibits the collector current in the monopolar device with the following layer design. The superlattice consists of 5.5 periods of 40-Å-thick n^+-GaAs and 200-Å-thick undoped $Al_{0.31}Ga_{0.69}As$ layers. An undoped and graded Al_xGa_{1-x} injector barrier is 5000-Å thick, and its height corresponds roughly to the bottom of one miniband. A similar graded undoped layer separates the superlattice from the collector. All layers – the emitter, the collector, and the quantum wells – are doped. The donor concentration is 2×10^{18} cm^{-3}. The width of the working miniband is 23 meV. When the voltage bias Φ_{EB} is applied, the triangular injector barrier is flattened and the injection current increases. When the energy of the injected electrons corresponds to the superlattice minigap, the injection efficiency decreases and so does the current. The total current as a function of Φ_{EB} exhibits the negative transconductance. This behavior is illustrated in Fig. 8.8 for several collector–base biases Φ_{CB}. For larger Φ_{CB}, the

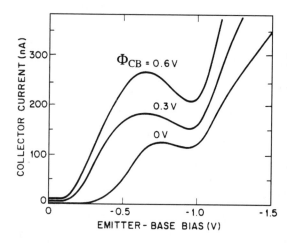

Figure 8.8. The collector current versus emitter–base voltage in the monopolar three-terminal device with the superlattice in the base as shown in Fig. 8.7. Results are obtained for $T = 30$ K at different collector–base biases. [After F. Capasso, S. Sen, F. Beltram, *et al.*, "Quantum functional devices: resonant-tunneling transistors, circuits with reduced complexity and multi-valued logic," IEEE Trans. Electron Devices 36, pp. 2065–2082 (1989).]

increase in the collector current is due to the lower injector barrier. The device just discussed has been proposed as a monopolar transistor.

Both types of the structures just studied, bipolar and monopolar, operate in a manner similar to that of double-barrier structures because of the near-ballistic nature of electron transport. Negative differential resistance and transconductance are achieved by the suppression of the electron injection into minibands. Both types of devices with a superlattice in a base region require much more advanced fabrication technology than double-barrier structures and operate at only low temperatures.

8.3.2 Bloch Oscillations

Another class of kinetic effects in superlattice structures originates from a small width of the allowed minibands. Let us consider the motion of an electron along the superlattice axis z. For an electron in some miniband n with energy $\epsilon_n(k)$, the semiclassical equations of motion in an electric field $F \equiv F_z$ are

$$\frac{dp_z}{dt} \equiv \hbar \frac{dk_z}{dt} = -eF, \tag{8.24}$$

$$v_z(k_z) = \frac{1}{\hbar} \frac{d\epsilon_n(k_z)}{dk_z}. \tag{8.25}$$

The solution of Eq. (8.24) for a constant electric field is

$$k_z(t) = k_z(0) - eFt/\hbar. \tag{8.26}$$

Therefore the wave vector k_z increases linearly with time. However, we know that for a superlattice, one can consider the wave vector k_z only inside the reduced Brillouin zone $\{-k_{Br}, k_{Br}\}$. (See Section 3.6 for a discussion of the basic properties of the superlattice Brillouin zone.) Once the electron reaches the boundary of this zone k_{Br} it appears at the point $-k_{Br}$ of the reduced Brillouin zone representation. The electron velocity $v_z(k_z)$ given by Eq. (8.25) in the interval $(-k_{Br}, 0)$ has one opposite sign to the velocity in the interval $(0, k_{Br})$. This is the reason why this phenomenon is closely reminiscent of a classical reflection from a rigid boundary and is referred to as Bragg reflection. Then the electron continues its motion according to Eqs. (8.24) and (8.25). Under these conditions, the electron motion is periodic, i.e., the velocity oscillates, as may be obtained from Eqs. (8.24), (8.25), and (3.94), and the energy is restricted by the miniband width. Because the width of the Brillouin zone is equal to $2\pi/d$, where d is the spatial period of the superlattice, one can find the time period T_B and the frequency f_B of the electron oscillations:

$$T_B = \frac{1}{f_B} = \frac{2\pi}{\omega_B} = \frac{2\pi}{d} \frac{\hbar}{eF}. \tag{8.27}$$

The electron motion in real space is also characterized by the same periodic oscillations. These oscillations are commonly referred to as Bloch oscillations and they are characterized by time $\tau_{sc} = \tau_1\hbar/(eFd)$ required for electron to go from $k = 0$ to $k = k_{Br} : T_B = 2\tau_{sc}$. Figure 8.9(a) illustrates the dispersion relation $\epsilon(k)$ of the lowest miniband of a superlattice in the absence of an electric field (compare with Fig. 3.20). Other energy minibands are shown schematically in Fig. 8.9(b). In an electric field, the semiclassical energy of the electron changes as

$$\epsilon^{(tot)}(z, k) = \epsilon(k) - eFz. \tag{8.28}$$

Figure 8.9. Schematic representation of energy spectra for a superlattice: (a) the reduced Brillouin zone, (b) several bands in the absence of an electric field, (c) the same as case (b), biased by an electric field.

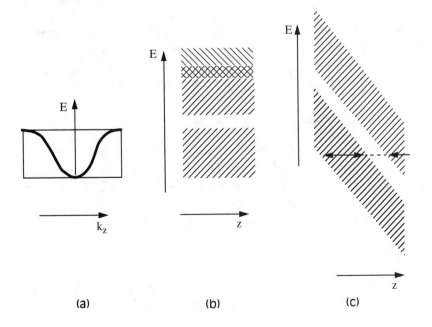

(a)   (b)   (c)

This spatial dependence is presented in Fig. 8.9(c). From this figure, one can see that an electron with fixed total energy $\epsilon^{(tot)}$ oscillates in the space region between two turning points corresponding to the bottom and the top of the tilted miniband. This implies that electrons are localized in some spatial region. If the electric field increases, the slope of the minibands increases and the degree of the electron localization increases as well. According to the semiclassical relationship of Eq. (8.28), the size of the localization region δz decreases as $1/F$: $\delta z = \Delta_{mb}/eF$, where Δ_{mb} is the miniband width.

This idealized picture assumes that during time intervals considerably exceeding the oscillation period, the electron is not scattered by crystal defects, phonons, etc. If one sets the spatial period of the superlattice d at 35 Å and the electric field at $F = 10$ kV/cm, it then follows from Eq. (8.27) that the oscillation period is $\sim 10^{-12}$ s. Therefore the scattering time τ_{sc} should be much longer:

$$\omega_B \tau_{sc} \gg 1. \tag{8.29}$$

The upper limit of the scattering time can be deduced from the mobility μ. For the parameters just assumed, one can obtain the requirement on the mobility of the structure with a superlattice: $\mu \gg 4 \times 10^3$ cm^2/V s. For such a complex artificial structure as a superlattice, it was difficult to achieve these values of the mobility until recently.

Since the left-hand side of inequality (8.29) is proportional to the electric field, it might be assumed that it is easy to satisfy the inequality by increasing the electric field. However, a restriction for the observation of Bloch oscillations in a very high field is imposed by the possibility of electron tunneling through the minigap. From Fig. 8.9(c), it is evident that an electron with some fixed energy can tunnel through the bandgap from one miniband to another. This process is known as Zener tunneling. The probability is small if the bandgap is large. Hence, the smaller the electric field, the smaller the probability. Consequently, in a chosen superlattice the interband-tunneling processes impose limits on the electric field. If the field exceeds this limit, Zener tunneling may destroy the Bloch oscillations. This is why Bloch oscillations are possible in a perfect superlattice with low scattering rates and large separations between the minibands.

8.3.3 Wannier–Stark Energy Ladder

We have seen that the electric field in a superlattice leads to electron localization; therefore, in a high electric field, the semiclassical considerations of Subsection 8.3.2 must be corrected from a quantum-mechanical point of view. In the so-called one-band approximation, the Schrödinger equation for the motion of the electrons along the superlattice axis takes the form

$$\left[\hat{\epsilon}\left(-i\frac{d}{dz} \right) - eFz \right] \chi(z) = \epsilon \chi(z), \tag{8.30}$$

where the first term is the kinetic-energy operator of this motion. In this operator, the wave vector k_z in the miniband energy dispersion $\hat{\epsilon}(k_z)$ is replaced by the operator $[-i(\mathrm{d}/\mathrm{d}z)]$. An example of the dispersion relation $\epsilon(k_z)$ for a superlattice is given by Eq. (3.94) and it has a more complicated form than the quadratic form $-\hbar^2 k_z^2/2m^*$ that we are accustomed to. The second term is the potential energy of the electrons in an electric field. An analysis of Eq. (8.30) shows that the electron-energy spectrum is no longer a continuum. Indeed, it splits into a series of levels with an equidistant energy separation proportional to the electric field. This splitting is known as Stark splitting. If the electron wave function $\chi(z)$ is a solution of Eq. (8.30) with some energy, say ϵ_0, then the function $\chi(z - nd)$ is also a solution of the same equation with energy

$$\epsilon = \epsilon_0 + neFd, \tag{8.31}$$

where n is an integer and d is the superlattice period. The set of these solutions constitutes the Wannier–Stark ladder of energy levels. The ladder is infinite, and the energy-level separation is $\Delta\epsilon = eFd$. The quantity $\Delta\epsilon$ may be expressed in terms of the Bloch oscillation frequency of Eq. (8.27):

$$\Delta\epsilon = \hbar\omega_B.$$

As we have discussed in Section 3.6, in the absence of an electric field, the electron wave function $\chi(z)$ is extended throughout the superlattice. If the field is nonzero, the wave function is localized in a region with a size that decreases as the field increases, as shown in Figs. 8.10(a) and 8.10(b). The electron is confined in one of the quantum wells for the high-field limit, as illustrated by Fig. 8.10(b). Hence the Wannier–Stark energy ladder corresponds to energy states, which can be approximately attributed to different quantum wells. A comparison of the previous analysis and that of this subsection leads us to the conclusion that the Bloch oscillations are the counterpart in the time domain of the stationary-state Wannier–Stark ladder in the energy domain.

Both of these superlattice effects have been beyond the possibility of experimental demonstration for a long time. But recently, as superlattice structures have become sufficiently better – constant period, small defect densities, high transverse mobility, etc. – the Wannier–Stark ladder has been observed. Experiments have been done mainly by use of optical methods. Measurements of photocurrent and light absorption, electroreflectance spectroscopy, and other methods have been applied to study superlattices of AlGaAs/GaAs with typical parameters as follows: a total of 20–40 periods per superlattice, a period of $\sim$40–55 Å, and a coherence length (the path of electron propagation without scattering events leading to the loss of their phase) that reaches $\sim$10–20 periods. Characteristic electric fields are $\sim$10–60 kV/cm. As a result of technological and research efforts, Wannier–Stark ladders have been observed even at room temperature. For example, in a 55-Å-period GaAs/AlGaAs superlattice it has been established that the coherence length of the electrons varies from 17 periods at 5 K to a minimum of 9 periods at room temperature. This is sufficient to observe the Wannier–Stark ladder and the extreme localization of the electrons in one of the quantum wells.

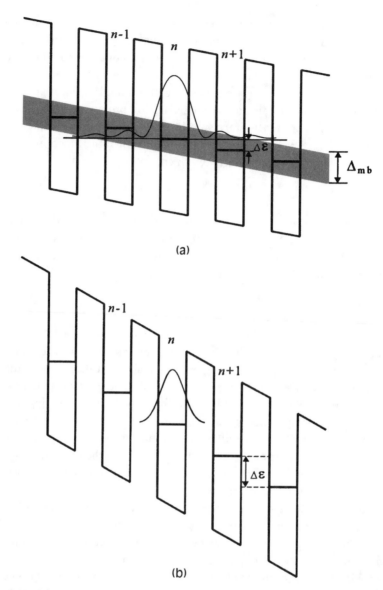

Figure 8.10. Schematic of electron wave functions in a biased superlattice: (a) electron wave function from the nth well extends over several periods of the superlattice in a relatively small electric field when $\Delta\epsilon < \Delta_{mb}$, (b) electron is localized in the nth well when $\Delta\epsilon > \Delta_{mb}$.

In our previous discussions, we have considered superlattices based on layered structures. Another approach, with considerable potential for becoming a reality, is being developed by use of lateral systems. For example, such a lateral Bloch oscillator can consist of a double periodic array of metal dot gates embedded in the metal-oxide-semiconductor field-effect transistor. These gates induce a two-dimensional superlattice potential in the inversion layer. The electron concentration and the minibands can be controlled by both the dot gates and the principal gate of the structure.

The substantial interest in Bloch oscillation effect is due not only to the fact that it represents a novel physical phenomenon, but also to its potential usefulness for applications. Indeed, the oscillations of charged particles are accompanied by the emission of electromagnetic waves. Bloch oscillations in superlattices should generate a high-frequency emission in the microwave region of the electromagnetic spectrum. From a quantum-mechanical point of view, this emission corresponds to transitions between the levels of the Wannier–Stark ladder. Optical measurements have shown that the distance between the levels of the ladder corresponds to oscillation frequency up to 1.5 THz. This frequency region is attractive for various applications and is still not covered by other semiconductor devices, because of the gap between conventional microwave emission sources like the Gunn diodes (hundred of gigahertz or fewer) and the infrared laser systems (tens of terahertz or more). However, the observation and study of the Bloch oscillations by means of the measurement of this microwave emission is still a great problem and requires development of a special type of detection technique. Microwave generation will be discussed further in Chapter 9 for specific examples of double-barrier resonant-tunneling diodes.

8.3.4 Negative Differential Resistance in Superlattices for Electron-Collision Regimes

The effects that we have just analyzed require sufficiently coherent (ballistic) electron transport. Another effect of importance, which suppresses the Bloch oscillations, takes place in the case of strong electron scattering. Let us consider semiclassical electron motion through a superlattice. For the electron velocity given by Eq. (8.25), one can write Newton's equation of motion of Eq. (8.24) in the form

$$dv_z = -\frac{eF}{\hbar^2} \frac{d^2\epsilon}{dk_z^2}\, dt. \tag{8.32}$$

In a collisional regime, we introduce the scattering time τ_{sc} and an additional term in Eq. (8.32) $-v_z dt/\tau_{sc}$. Then, in the stationary case, the average drift velocity determined by collisions is

$$v_d = \int_0^\infty \exp(-t/\tau_{sc}) \frac{eF}{\hbar^2} \frac{d^2\epsilon}{dk_z^2}\, dt. \tag{8.33}$$

Since k_z is changing with time [Eq. (8.24)], the quantity $d^2\epsilon/dk_z^2$ is also a function of time. Let the dispersion relation for the miniband be as defined by Eq. (3.94), $\epsilon = \epsilon_0 + \Delta_{mb} \cos k_z d/2$; then one can calculate the integral in Eq. (8.33). The result is

$$v_d = \frac{\Delta_{mb}d}{\hbar} \frac{eFd\tau_{sc}/\hbar}{1 + (eFd\tau_{sc}/\hbar)^2}. \tag{8.34}$$

According to this formula, at low fields, the drift velocity is $\mu_{eff}F$, where the effective mobility is given by the almost-conventional formula, $\mu_{eff} = e\tau_{sc}/M^*$, because for small energies, the effective mass of a superlattice is $M^* = 2\hbar^2/\Delta_{mb}d^2$. As the electric field is increased, the drift velocity v_d increases, reaches its maximum at $F_c = \hbar/ed\tau_{sc}$, and then decreases. Thus the velocity versus the electric-field relationship exhibits a region of negative differential resistance. The conditions for the observation of this negative differential resistance is much easier to realize than those for Bloch oscillations. For typical values of the superlattice period, $d < 100\,\text{Å}$, and when the scattering time $\tau_{sc} \leq 10^{-12}$ s, the threshold electric field F_c is of the order of 1 kV/cm. This kind of negative differential resistance has been observed at low temperatures.

8.3.5 Ballistic-Injection Devices

To complete our study of perpendicular electron transport, we consider devices exhibiting classical perpendicular motion. On the one hand, such motion occurs if characteristic sizes of a device L are greater than the electron wavelength λ. On the other hand, the most interesting devices are those with ballistic electrons. The latter means that device sizes should be shorter than the mean free path of electrons l_e. Thus devices with $\lambda \ll L \leq l_e$ are in this category. In Chapter 9, we will study a number of such devices. Here, we restrict ourselves to several general remarks.

The main idea is to achieve a higher velocity compared with those realized with common methods that use the electron drift in an electric field. One means of doing this is to inject electrons from an emitter over a barrier into an active region (the base). Then electrons should be collected by another contact (the collector). The velocity of such electrons in the base region is $v_B \cong \sqrt{2V_b/m^*}$. Here V_b is the height of the barrier. For a GaAs base with $m^* = 0.067$ m and a barrier height $V_b \approx 0.3$ eV, one obtains an injected electron velocity of up to 9×10^7 cm/s, which is considerably greater than the saturated and the overshoot velocities for parallel transport, as discussed in Chapter 7. Furthermore, the distribution function of the injected electrons is strongly anisotropic. Therefore the emitter is a source of a sharp, high-speed electron beam. Both factors are important for high-speed operation of the device.

Estimates of the time of flight of the injected electrons through a base with dimension $d_B \sim 1\ \mu$m gives a value of $\sim 10^{-12}$ s. This time is less than or of the order of the scattering time. Therefore it is expected that the electron transport from the emitter to the collector would be almost collisionless, i.e., ballistic. In practice, it is useful to control the voltage drop between the emitter and the base. This can be done if the base is doped. As a result, to avoid collisions, one should use an even thinner base. For example, a GaAs base with a doping level of $\sim 10^{18}$cm^{-3} or higher can be used for a ballistic device if the base is several hundred to a thousand angstroms thick.

From the above estimate of the velocity of injected electrons, it follows that a larger emitter barrier is preferable for achieving higher velocities. Actually, it is not desirable to increase the energy of the injected electrons above the energy of the next upper minima of the conduction band of the base material or above the impact ionization threshold. For example, in GaAs the energy should not exceed 0.3 eV since the *L*-valley minimum is 0.3 eV above that of the Γ valley, and for InSb it is not desirable to exceed the impact ionization threshold energy of 0.25 eV. Otherwise the threshold mechanism of scattering may destroy the ballistic regime.

The progress in technology makes it possible to extend the idea of ballistic injection to systems with perpendicular transport of low-dimensional electrons. In Section 9.6 we will study different aspects of these devices.

8.4 Single-Electron Transfer and Coulomb Blockade

In studying the perpendicular transport we assume that the number of electrons participating in this transport is so large that no role is played by the discrete nature of the electrons. For small devices operating with weak currents, this assumption is no longer valid.

In general, in bulklike materials and macroscopic devices the electron discreteness is not manifested in average characteristics such as the total electric current, etc. However, it is well known that this discreteness manifests itself by contributing to current noise (fluctuations); the so-called shot noise is entirely due to the electron and its fixed-charge discreteness. However, when dimensions of devices are scaled down, the role of the discreteness increases. In the case of ultrasmall devices, the discreteness of the electron charge gives rise to principally new effects in electron transport. This transport becomes correlated, i.e., a transfer of one electron depends on the transfer of others. A new class of devices referred to as single-electron devices is based on such processes.

The basic properties of these effects are extremely simple and can be described by the example of a customary system: an insulated tunnel junction between two conductive electrodes (heavily doped semiconductor regions, or metals, etc.), as shown in Fig 8.11 for a metal-insulator-metal system. Let this junction be characterized by the conductance G and the capacitance C. The conductance is supposed to be small enough for us to consider the system as a leaking capacitor. We can assume that the capacitance is roughly proportional to the cross section

Figure 8.11. The simplest system of single electronics: a tunnel junction with a small capacitance C. Single-electron transfer is shown schematically.

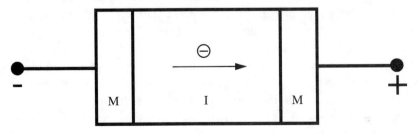

of this junction S. Thus a small S implies a small capacitance. Remember that the electrostatic energy E_{el} of a capacitor with a charge Q is $E_{el} = Q^2/2C$. Let a single electron pass through the junction; this event will change the initial electrostatic charge just by the elementary charge e. Hence the electrostatic energy of the junction changes by

$$\Delta E_{el} = \frac{(Q-e)^2}{2C} - \frac{Q^2}{2C} = \frac{e(e-2Q)}{2C}. \tag{8.35}$$

Then single-electron transfer leads to a voltage fluctuation across the junction equal to $\Delta\Phi_s = e/C$ and a corresponding current fluctuation $\Delta j = G\Delta\Phi_s$. Applying these results to a macroscopic junction, say with a cross section $S = 0.1$ mm $\times$ 0.1 mm, one can see that, even at helium temperature, $k_B T \gg \Delta E_{el}$, i.e., the role of single-electron transfer is negligible. However, if the junction is of nanoscale dimensions, say $S = 200 \text{Å} \times 200 \text{Å}$, the energy associated with one electron $E_s = e^2/2C$ is as large as 10 meV. Thus at low temperatures the energetics of a single-electron process becomes important.

The uncertainty relation (2.34) between the energy ΔE_s and a tunneling time τ_t allows us to estimate the extreme limit of τ_t, $\tau_t > 2\pi\hbar/\Delta E_{el}$, and the electric current associated with the tunneling of a single electron: $j_s \approx e/\tau_t \geq e^3/4\pi\hbar C$. If this current j_s exceeds the current fluctuation because of the voltage $\Delta\Phi_s$, i.e., $j_s \geq \Delta j = G\Delta\Phi_s$, the single–electron transfer will control the electric current through the junction. The latter leads to $e^2/h \gg G$. Recall that the value in the left-hand side of this inequality just coincides with the quantum of conductance, $G_0 \equiv e^2/h$, introduced by the Landauer formula, Eq. (2.186), in Section 2.7.

Using these results, one can conclude that if the capacitance and the conductance of a device are so small that

$$C \ll \frac{e^2}{k_B T}, \qquad G \ll \frac{e^2}{h}, \tag{8.36}$$

then electron transport is correlated and single-electron effects are important.

From Eq. (8.35), one can see that any electron transfer is prohibited for a small initial capacitor charge: $-e/2 < Q < e/2$. The physics of this so-called Coulomb blockade to tunneling is simple. If the conditions of inequalities (8.36) are met, the charging energy plays the dominant role in the system, the tunneling of an electron is energetically unfavorable, and at low temperatures the tunneling is not possible at all (it is blocked). This results in the specific current-voltage characteristic presented in Fig. 8.12. The main feature of such a characteristic is the total suppression of the current in some finite interval of external voltage biases, $-e/C < \Phi < +e/C$.

The manifestation of correlated transport is strongly dependent on the external circuit to which the single tunnel junction is attached. Let the external circuit impedance be $Z(\omega)$. If the impedance is as low as $|Z(\omega)| \ll G_0^{-1}$ for the most important frequencies, $\omega \sim 1/\tau_t \sim e^2/\hbar C$, charge fluctuations in the circuit are greater than the elementary charge e and all correlation effects are suppressed. If the external circuit impedance is in the intermediate interval $G_0^{-1} \ll |Z(\omega)| < G^{-1}$, the junction exhibits the Coulomb blockade in the bias range $-e/C$ to e/C, as shown in Fig. 8.12, but outside this range there is no correlation between tunneling events. Finally, if $|Z(\omega)| \gg G^{-1} \gg G_0^{-1}$, the external circuit can be considered as a source of a fixed dc current I. This current causes a recharging of the junction inside the range of the Coulomb blockade without electron tunneling; it corresponds to a linear change in the charge with time: $dQ/dt = I \approx$ constant. When the edge of the blockade range is reached, an electron tunnels through the junction. The system finds itself again in the blockade range – near the opposite edge – and the process repeats. Thus one gets temporal oscillations of the charging with a frequency that is determined by the current:

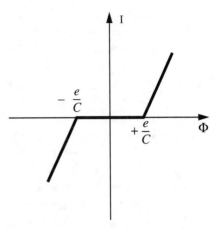

Figure 8.12. Current-voltage characteristic under the Coulomb blockade. At the voltage-bias range from $-e/C$ to e/C the current is suppressed by Coulomb correlations.

$$f_{\text{SET}} = (I/e).$$

This is the so-called single-electron tunneling (or Bloch) oscillations.

The previously discussed nonlinear current-voltage characteristics, the oscillations, and other single-electron effects give rise to a principally new approach to low-energy electronics. This field of single-electron devices is developing rapidly and has many potential applications. Different devices – the single-electron transistor, the turnstile, the single-electron pump, etc. – have been proposed and realized on the basis of these effects. Although these results were achieved at low temperatures, modern technology portends their extension to liquid nitrogen and even room temperatures.

8.5 Closing Remarks

In this chapter we have studied different effects associated with perpendicular electron transport in quantum structures. Both quantum and classical regimes occur in these structures and are used in devices.

An important example of perpendicular quantum transport is provided by resonant tunneling, which takes place in quantum structures with two or more barriers. There are two ways of considering this kind of the transport: coherent and sequential tunneling. The first is a single-quantum process associated with the transmission of electrons through the whole structure. The second involves two sequential–tunneling processes that can be separated in time as well as by intermediate scattering events. Double-barrier resonant-tunneling structures are the most promising in this class of devices. The main characteristics of the resonant-tunneling effect are negative differential resistance up to room temperature and a short tunneling time that provides the possibility of operating at very high frequencies up to the terahertz range.

Another example of perpendicular quantum transport is given by transport through superlattices. For this case, it is possible to have either coherent (ballistic) transport or transport in a collisional regime; both exhibit negative differential resistance. In Chapter 9, we will study applications of both double-barrier and superlattice structures in different devices of practical importance.

Perfect superlattice structures also manifest a new type of oscillations – Bloch oscillations – in electric fields. These oscillations are characterized by high frequencies and can be a source of microwave emission in the range from hundreds of gigahertz to a few terahertz. In Chapter 11 we will study optical manifestations of the Wannier–Stark ladder and Bloch oscillations in devices with superlattices.

We have found that classical perpendicular transport can be observed in ballistic-injection devices. For such a device, electrons are injected into a base region and pass through it, with the velocities largely exceeding the saturation values studied in Chapter 7.

Finally, we considered tunneling for devices with small cross-sectional areas, which results in correlated electron transport; the transfer of single electrons across such a device depends on the transfer of other electrons. The nonlinear current-voltage characteristics, single-electron oscillations, and other effects in circuits with such small cross-sectional devices give rise to the new field of low-energy electronics – single-electron electronics.

Additional information on resonant-tunneling effects and negative differential resistance in superlattices can be found in the following references:

S. Luryi, "Frequency limit of double-barrier resonant-tunneling oscillators," Appl. Phys. Lett. **47**, 490–492 (1985).

P. J. Price, "Simple theory of double-barrier tunneling," IEEE Trans. Electron Devices **36**, 2340–2343 (1989).

F. Capasso, S. Sen, F. Beltram, *et al.*, "Quantum functional devices: resonant-tunneling transistors, circuits with reduced complexity and multiple-valued logic," IEEE Trans. Electron Devices **36**, 2065–2082 (1989).

G. A. Toombs and F. W. Sheard, "The background to resonant-tunneling theory," in *Electronic Properties of Multilayers and Low-Dimensional Structures*, J. M. Chamberlain, *et al.*, eds. (Plenum, New York, 1990), pp. 257–282.

T. C. L. G. Sollner, E. R. Brown, C. D. Parker, *et al.*, "High-frequency applications of resonant-tunneling devices," in *Electronic Properties of Multilayers and Low-Dimensional Structures*, J. M. Chamberlain, *et al.*, eds. (Plenum, New York, 1990), pp. 283–296.

J. Shah, ed., *Hot Carriers in Semiconductor Nanostructures (Physics and Applications)* (Academic, Boston, 1992).

E. R. Brown, *High-Speed Resonant-Tunneling Diodes*, in "Heterostructures and quantum devices," N. G. Einspruch and W. R. Frensley, eds. (Academic, San Diego, 1994), pp. 306–350.

Discussions of electrostatic effects under tunneling, single-electron transfer and the Coulomb blockade are presented, for example, in the following references:

B. Ricco and M. Ya. Azbel, "Physics of resonant tunneling. One-dimensional case," Phys. Rev. B **29**, 1970–1981 (1984).

D. V. Averin and K.K. Likharev, "Coulomb blockade of the single-electron tunneling, and coherent oscillations in small tunnel junctions," J. Low Temp. Phys. **62**, 345–373 (1986).

K. K. Likharev, "Correlated discrete transfer of single electrons in ultrasmall tunnel junctions," IBM J. Res. Develop. **32**, pp. 144–158 (1988).

H. Koch and H. Lubbig, eds., *Single-Electron Tunneling and Mesoscopic Devices* (Springer-Verlag, Berlin, 1992).

PROBLEMS

1. Explain the physical origin of negative differential resistance observed in resonant–tunneling structures. For a double-barrier resonant-tunneling structure with three quasi-bound states, sketch the conduction-band edge at each extremum of the current-voltage characteristic. Also plot the current-voltage characteristic over the useful range of voltages for such a three-level double-barrier resonant-tunneling device.

2. Show that in the absence of inelastic scattering, the general expression for the total current of electrons tunneling through a double-barrier resonant-tunneling structure may be written as

$$ j = 2e \int d\vec{k} g(\vec{k}) v_{lz} [f(E) - f(E + e\Phi)] |T(E_\perp, \Phi)|^2 , $$

where v_{lz} is the electron velocity in the direction of tunneling, f is the Fermi–Dirac distribution function, Φ is the applied voltage, E is the electron kinetic energy, $E_\perp$ is the component of energy associated with the momentum in the direction of tunneling, $|T|^2$ is the tunneling probability, and $g(\vec{k})$ is the electron density of states.

3. For a superlattice with period d, the steady-state probability amplitude $\phi(x)$ obeys the relationship $|\phi(x)|^2 = |\phi(x+d)|^2$. Accordingly, $\phi(x) = u(x)e^{i\alpha x}$, where $u(x) = u(x + d)$ is a periodic envelope function and α is a phase. Discuss the analogy between the Bloch wave function for a bulk crystal with lattice constant a and that for a superlattice with period d (see Chapter 2).

4. Assuming that the resonance width Γ of the quasi-bound state in an AlAs-GaAs-AlAs quantum well is such that $2\hbar/\Gamma = 0.45 \times 10^{-12}$ s, what is the characteristic time for tunneling through the structure? Take the drift velocity to be 10^7 cm/s, and take the AlAs barriers and the GaAs quantum well to have thicknessess of 8 and 23 Å, respectively. What is the cutoff frequency corresponding to this characteristic tunneling time? Assuming that $2\hbar/\Gamma$ remains constant, is it possible to realize additional significant enhancement in the cutoff frequency by simply reducing the dimensions of the double-barrier quantum-well structure?

Electronic Devices Based on Quantum Heterostructures

9.1 Introduction

Previous chapters have prepared us to study microelectronic devices based on heterostructures. These devices use almost all possible types of electron transport: parallel and perpendicular transport, transport in classical and quantum regimes, and so on. Previously studied effects such as charge transfer in momentum space and real space, hot-electron effects, resonant tunneling, and others are exploited in new generations of devices. Different materials and their heterostructure combinations are used widely to test new principles and to design high-performance devices.

We start by studying heterostructure devices based on the various classical transport regimes. The operational principles of these devices are analogous to those of homostructure devices. In this context, we use a classification scheme based on an extension of that used for known classes of microelectronic devices. Then we continue with an analysis of electronic devices based on the resonant-tunneling effect and consider devices that combine resonant-tunneling and conventional operation principles. In addition, we consider three-terminal devices otherwise known as transistors. Depending on the principles of operation, transistors can be associated with one of two large classes: field-effect transistors (FETs) and potential-effect transistors.

Devices Controlled by the Field Effect

Devices of the first group are field-effect or voltage-controlled devices. A common feature of these devices is that a voltage is applied to a controlling electrode – a gate – that is capacitively coupled to the active region of the device. This electrode is spatially and electrically separated from carriers in the active region by an insulator or a depletion region. The electrode controls the resistance of the active region and consequently the current between two other terminals, the source and the drain.

Figure 9.1 illustrates the relationship between the various types of FETs as well as how these devices have evolved. This class of transistors is divided into three large groups of devices used in practice as well as a group of potentially new devices based on recent developments. The first group includes different variations of metal-oxide-semiconductor (MOS) FETs (MOSFETs). A

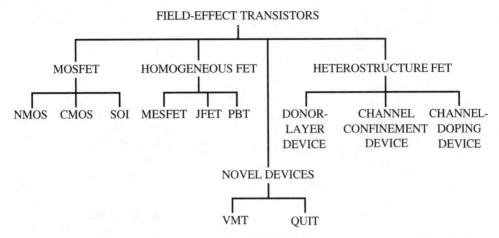

Figure 9.1. The family tree of FETs (see text for definitions of acronyms). Transistors based on Si technology, MOSFETs, transistors based on bulklike III–V compounds, MESFETs, and on heterostructures are set apart in three large groups as having much in common in their technologies, properties, and applications. New prospective developments in which field control is used are presented in a separate branch.

schematic of a MOSFET is presented in Fig. 9.2. The first MOSFET was fabricated in 1960 with a MOS structure, with the oxide layer being SiO_2; this combination of materials is still in use in these devices. Such a MOS structure was discussed briefly in Section 4.4. Over the years, the MOSFET has been improved significantly and a substantial scaling down of all dimensions of the device has been achieved. Figure 9.3 illustrates how device dimensions have been reduced as MOSFET technology has matured. Currently, devices with gate lengths of ~ 0.2 μm are fabricated for commercial purposes. Other device dimensions like the gate oxide thickness as well as the depths of junctions for the source and drain have been reduced as well. MOSFET technology has a lot of advantages; for example, one can produce different kinds of FETs on the same substrate; examples include n-channel MOS (NMOS)FETs and p-channel MOSFETs (PMOSFETs). When NMOSFETs and PMOSFETs are integrated on the same chip the resulting structure is known as a complementary MOS (CMOS) structure that has the great advantage of consuming less power; indeed, current flows through a CMOS circuit only during a switching operation. A MOSFET with a silicon-on-insulator (SOI) provides another example of the fabrication of devices on single-crystal Si; this is achieved through a nonepitaxial approach in which the oxide is formed by the implantation of oxygen. The SOI has several advantages for operating in high-speed regimes as well as for applications requiring the fabrication of stacked three-dimensional integrated circuits.

A second group of FETs in Fig. 9.1 may be classified as homogeneous FETs. This group includes the metal-semiconductor FET (MESFET), the junction FET (JFET), and other more sophisticated devices. MESFETs use a metal gate and homogeneous materials such as the n-type GaAs. As discussed in Section 4.4,

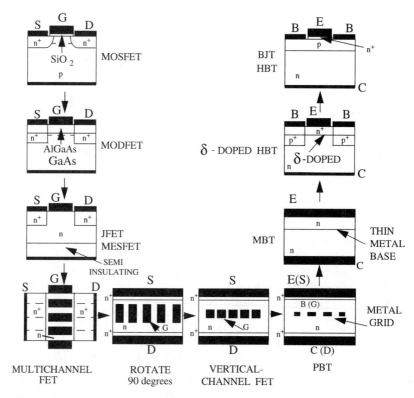

Figure 9.2. Schematics of transistors of practical importance. Devices are given in a sequence illustrating the transformation of their configurations from a MOSFET to a bipolar transistor. [After S. M. Sze, ed., *High-Speed Semiconductor Devices* (Wiley, New York, 1990).]

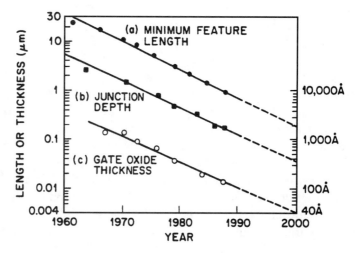

Figure 9.3. Scaling down MOSFET dimensions. A continuous decrease in all length scales of MOSFETs is presented as a function of time. The left axis is given in micrometers, the right in angstroms. The tendency is to reach the quantum limit of a hundred angstroms after 2005. [After S. M. Sze, ed., *High-Speed Semiconductor Devices* (Wiley, New York, 1990).]

the Schottky depletion region arises in the semiconductor under the metal in this system. This depletion region insulates the gate from the conducting channel and plays the same role as the SiO_2 layer in the MOSFET. The variant of FETs with a junction is quite similar to that of the MESFET, but the p–n junction is used as a gate. Since both MESFETs and JFETs are fabricated from high-mobility materials, like GaAs, they have various attractive properties for high-speed operation. In Fig. 9.1, besides the MESFET and JFET, the permeable-base transistor (PBT) subgroup is also presented. This latter subgroup of FETs has fine metal grids embedded into the semiconductor material between the source and the drain, as illustrated in Fig. 9.2. This type of FET is used for ultra-high-speed power applications.

A third group of FETs in Fig. 9.1 incorporates heterostructures and differs in the design of the doping and conducting channel. For example, the donor-layer heterostructure FET (HFET) subgroup has one or several doped layers that supply electrons to the conducting channel. If the channel is formed on a single heterointerface (see Section 4.4) the donor-layer device is called a modulation-doped FET (MODFET) or a high-electron-mobility transistor (HEMT). The case in which there is a specially designed electron confinement in a quantum well and modulation doping is usually classified as another subgroup of devices known as channel-confinement devices. If the conducting channel is the only doped channel, the device can be considered as a metal-insulator-semiconductor FET (MISFET). An example is a metal-undoped-AlGaAs–n-doped-GaAs structure.

One of the key parameters of contemporary devices is the maximum frequency of efficient operation – the cutoff frequency. At fixed device dimensions, the cutoff frequency depends on material characteristics and device design. As a basis for making comparisons, Fig. 9.4 illustrates representative parameters of all three of these groups of FETs. One can see that the HFET class exhibits superb

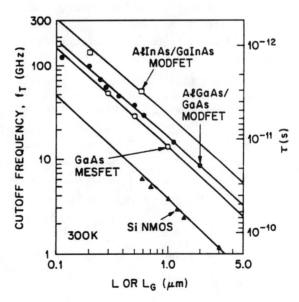

Figure 9.4. The cutoff frequency in gigahertz versus the device length L or the gate length L_G in micrometers for the three groups of FETs: MOSFETs, MESFETs and MODFETs. The latter are presented for AlGaAs/GaAs and AlInAs/GaInAs material combinations. The parameters are given at 300 K. [After S. M. Sze, ed., *High-Speed Semiconductor Devices* (Wiley, New York, 1990).]

high-frequency performance. Since the main subject of this book is the analysis of quantum heterostructures and their applications, we focus our attention on heterostructure-based FET devices.

We include two groups of novel devices based on parallel transport in heterostructures in the FET family. For all three groups – MOSFETs, homogeneous FETs, and HFETs – the voltage is applied to the gate controls, mainly the electron concentration in the active region. The velocity-modulation transistor (VMT) uses another idea: The voltage controls the potential profile(s) of the channel(s) with low-dimensional carriers, as well as the shape of their wave functions and, as a result, the mobility. We discuss these novel types of devices, which can be classified as voltage-controlled devices.

Another group of novel devices is that of the quantum-interference transistors (QUIT), which operate on the basis of interference of the electron waves. An external voltage can control the interference and the current through the device. This quantum device has no analogs with devices based on the classical physics. The QUIT is studied in Subsection 9.3.2.

Potential-Effect Transistors

The family of potential-effect transistors is shown in Fig. 9.5. In contrast to the case of FET devices, potential-effect transistors are current controlled. The controlling electrode is resistively coupled to the active region of the device, and the carriers are separated by an energy barrier.

The most important representative of this class is the bipolar transistor (BT). It was invented in 1947 and has undergone considerable and persistent transformations. The BT currently provides high speed in most circuit applications. BT operation is based on the principle of controlling the current by injecting minority carriers. For example, in an n^+-p-n device, the minority carrier injection from a forward-biased n^+-p junction (emitter base) into a reverse-biased $p-n$ junction (base collector) provides the controlling function of the collector current. The carriers travel from the emitter to the collector perpendicularly to junctions; thus the carrier transient time through the base determines the cutoff

Figure 9.5. The family tree of potential-effect transistors. The devices are set apart into two large groups, bipolar transistors and hot-electron transistors. [After S. M. Sze, ed., *High-Speed Semiconductor Devices* (Wiley, New York, 1990).]

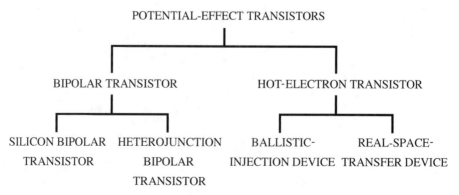

frequency of the device. Device scaling to reduce the transient time in a homojunction BT (JBT) cannot be achieved while simultaneously realizing all requirements for high-performance, high-doping levels in both the emitter and base, a small base region, etc. Using heterostructures helps to solve these problems and improve the device performance. Thus we can divide the BT class into two large and important groups: JBTs and heterojunction BTs (HBTs), as shown in Fig. 9.5.

Another group of potential-effect devices is represented by different types of hot-electron transistors, as illustrated in Fig. 9.5. In Chapter 7, we studied the hot-electron effect and found that the carriers can gain their energy in an electric field and have an energy exceeding a few $k_B T$, where T is the crystal temperature. Ballistic-injection devices use an emitter-barrier–base-barrier–collector structure. The hot-electrons are injected over or through an emitter barrier into a narrow-base region. Injected carriers have a large velocity and transit through the base almost ballistically; that is, they transit the base without significant scattering. The current through the device is controlled by changing the height of a collector barrier. This principle is used for a number of different configurations and combinations of materials; in particular, this subgroup includes the metal-base transistor (MBT) and the planar-doped barrier transistor.

The second representative of the hot-electron-transistor group is the subgroup of devices based on the real-space-transfer (RST) effect, studied previously in Section 7.4. In RST devices, carriers heated in a high-mobility quantum well can be transferred to neighboring layers with a low mobility, which can result in a negative differential resistance (NDR). In other terms, RST devices are electron-temperature devices: one modulates the electron temperature of the carriers, which are then transferred over the barrier. The subgroup of these devices includes the charge-transfer transistor and the negative-resistance FET (NERFET). In this chapter we study both subgroups of hot-electron transistors.

9.2 Field-Effect Transistors

We start with an analysis of FETs. Although each particular group of FETs has its own features and, in general, requires a special theory to describe it, all FETs exhibit similar dependencies of the current on the drain–source and gate-voltage biases. It is possible to define the set of parameters characterizing any device of this class. We introduce these parameters considering the regimes of operation of a conventional homostructure MESFET and then analyze different types of heterostructure FETs.

9.2.1 Principle of Field-Effect-Transistor Operation

Figure 9.6 presents a simple model for a FET. We assume that the device active region is made of an n-channel that can be designed, for example, by a homogeneous doping. The source and the drain are heavily doped n^+ regions that are assumed to serve as ohmic contacts. The gate upon the active layer forms a Schottky barrier and a depletion region under the gate. This is typical

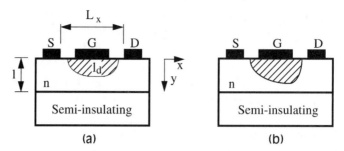

Figure 9.6. A schematic of the MESFET model. The conducting channel of width l is supposed to be grown on a semi-insulating substrate. The length of the channel is L_x. The depleted region under the gate, $l_d(x)$, is marked by a shadowed region. In case (a) there is no voltage bias between the source and the drain; in case (b) a positive voltage is applied to the drain.

for commonly used materials like GaAs, InP, etc. The active region is restricted by a nonconducting substrate. The length of the conducting channel (or simply channel) L_x is assumed to be substantially greater than its width l, so one can apply a quasi-one-dimensional treatment of the depletion region. This means that one can solve the one-dimensional Poisson equation at a fixed x coordinate, which can be considered as a parameter. For simplicity, we also disregard high-field effects in order to present the physics behind the operation of the FET.

If no voltage is applied to the contacts, the depletion region is almost uniform along the active region, as shown in Fig. 9.6(a). The characteristics of the depletion region are determined by the built-in Schottky voltage Φ_{FB}, as discussed in Section 4.4 [Eq. (4.31)]. If a negative voltage is applied to the gate and there is still no voltage between the source and the drain, the depletion region extends farther into the active region and decreases the width of the channel. At some voltage the channel is pinched off.

Let the gate voltage be fixed Φ_G and a small positive voltage Φ_D be applied to the drain. A current will flow through the channel in the ohmic linear regime. If the drain voltage is increased further, it will affect the distribution of the potential in the device: the depletion region width increases near the drain end of the channel, as shown in Fig. 9.6(b), because of the increasing potential difference between the gate and the drain. At a certain drain voltage, the channel begins to pinch off at the drain end. At this voltage the current saturates and remains nearly constant with a further increase in the drain voltage.

Let us introduce the system of coordinates for a device shown in Fig. 9.6; that is, the x axis is along the channel and the gate position corresponds to $y = 0$. The thickness of the n channel is taken to be l. With the above simplifications, the current through the device is

$$I_D = e\mu_n N_D \left[l - l_d(x) \right] L_z \frac{d\phi(x)}{dx}, \tag{9.1}$$

where L_z is the width of the gate along the z direction, N_D is the donor concentration, and $l_d(x)$ is the width of the depletion region. The electric field in the channel is calculated by means of the potential in the channel $\phi(x)$: $F_x = -d\phi/dx$. For a quasi-one-dimensional calculation, we can use the result of Eq. (4.51), obtained for a depletion region. It can be rewritten in the notation of this section as

$$\phi(x, y) = -\frac{2\pi e N_D}{\kappa} [y - l_d(x)]^2 + \frac{2\pi e N_D l_d{}^2(x)}{\kappa} + \Phi_G - \Phi_{BI}, \tag{9.2}$$

where we have taken into account that the built-in Schottky potential Φ_{FB} makes a negative contribution $-\Phi_{BI}$. From Eq. (9.2), we can easily obtain the potential in the channel: $\phi(x) \equiv \phi(x, y = l_d)$. Then we can find the width of the depletion region in terms of $\phi(x)$:

$$l_d(x) = \left\{ \frac{\kappa}{2\pi e N_D} [\phi(x) + \Phi_{BI} - \Phi_G] \right\}^{1/2}. \tag{9.3}$$

Let us introduce the characteristic potential

$$\Phi_P = \frac{2\pi e N_D l^2}{\kappa}. \tag{9.4}$$

Then Eq. (9.3) can be rewritten as

$$l_d(x) = l \left[\frac{\phi(x) + \Phi_{BI} - \Phi_G}{\Phi_P} \right]^{1/2}. \tag{9.5}$$

If no voltage is applied to the drain, the voltage in the channel is zero, $\phi(x) = 0$. Then Eq. (9.5) gives the depletion-layer width under equilibrium conditions (no current through the device). One can see that the depletion region extends over the whole channel when the condition $\Phi_{BI} - \Phi_G = \Phi_P$ is satisfied. Thus the value Φ_P can be interpreted as the internal pinch-off voltage. From these considerations we can conclude that the device is turned off if

$$\Phi_G < \Phi_{th} \equiv \Phi_{BI} - \Phi_P \tag{9.6}$$

and turned on if

$$\Phi_G > \Phi_{th}. \tag{9.7}$$

In accordance with Eq. (9.4), the pinch-off voltage depends on the doping concentration N_D and the channel width l. This allows us to fabricate two types of FETs: the normally off devices with

$$\Phi_{BI} > \Phi_P \tag{9.8}$$

and the normally on devices with

$$\Phi_{BI} < \Phi_P. \tag{9.9}$$

For example, at the same level of doping in the normally off device, a thinner channel is used. The normally off devices require an applied gate voltage to operate and are called enhancement-mode devices. The normally-on devices are called the depletion-mode devices.

For the GaAs MESFET, the value Φ_{BI} is approximately or higher then 0.8 V at room temperature, and for the GaAs JFET it has a value of ~ 1.4 V. If we assume a channel-doping concentration of $N_D = 10^{17}$ cm^{-3}, we can estimate that the device will be an enhancement-mode device for $l < 0.1$ μm (for a MESFET) and a depletion-mode device for the case in which $l > 0.1$ μm.

Substituting Eq. (9.5) into Eq. (9.1) we get a nonlinear differential equation of first order for $\phi(x)$:

$$\frac{I_D}{e\mu_n N_D l L_z} = \left\{ 1 - \left[\frac{\phi(x) + \Phi_{BI} - \Phi_G}{\Phi_P} \right]^{1/2} \right\} \frac{d\phi}{dx},$$

where the left-hand side is a constant, since the current does not change along the channel. We impose the boundary conditions

$$\phi(x = 0) = \Phi_S = 0, \qquad \phi(x = L_x) = \Phi_D.$$

This differential equation can be integrated in the implicit form; this procedure leads to $x = x(\phi)$. Here we are interested in the equation for the current, which can be found as a function of the voltage at the drain Φ_D:

$$I_D = \frac{e\mu_n N_D L_z l}{L_x} \left\{ \Phi_D - \frac{2\left[(\Phi_D + \Phi_{BI} - \Phi_G)^{3/2} - (\Phi_{BI} - \Phi_G)^{3/2} \right]}{3\Phi_P^{1/2}} \right\}, \quad (9.10)$$

where we assume the case of a normally on device. Let us introduce the conductance for a completely opened channel,

$$g_0 = \frac{e\mu_n N_D l L_z}{L_x}. \tag{9.11}$$

In the notation of Eq. (9.11), the drain current takes the form

$$I_D = g_0 \left\{ \Phi_D - \frac{2\left[(\Phi_D + \Phi_{BI} - \Phi_G)^{3/2} - (\Phi_{BI} - \Phi_G)^{3/2} \right]}{3\Phi_P^{1/2}} \right\}. \tag{9.12}$$

Since Eq. (9.12) is derived for the open channel, i.e., for $l_d(x) < l$, from Eq. (9.4) it follows that we can use it only if $\Phi_D < \Phi_P + \Phi_G - \Phi_{BI}$. At the drain voltage

$$\Phi_D = \Phi_{D_{sat}} \equiv \Phi_P + \Phi_G - \Phi_{BI},$$

despite the pinched-off channel, the current is finite:

$$I_{D_{sat}} = g_0 \left[\frac{\Phi_P}{3} - \Phi_{BI} + \Phi_G + \frac{2(\Phi_{BI} - \Phi_G)^{3/2}}{3\Phi_P^{1/2}} \right]. \tag{9.13}$$

Strictly speaking, the above model is not valid for $\Phi_D > \Phi_{D_{sat}}$, when the channel is pinched off. In Fig. 9.7, the domain of the applicability of the model – known as the linear region – is shown by the dashed curve in the I_D–Φ_D plane. Additional analysis is needed to extrapolate the model to higher drain voltages. The following analysis of results for the voltage near the pinch-off point suggests

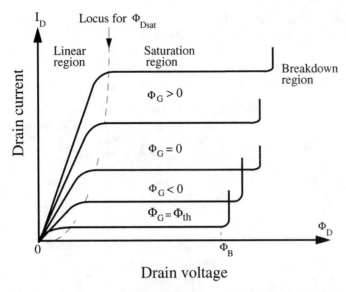

Figure 9.7. The drain-current versus drain-voltage characteristics for the MESFET at different gate voltages. The linear and saturation regions are indicated. The breakdown effect occurs at high voltages Φ_B and is presented conditionally. [After S. M. Sze, ed., *High-Speed Semiconductor Devices* (Wiley, New York, 1990).]

a way to obtain accurate interpolation results. Equation (9.13) tells us that the current is finite even for a channel of zero width, $l - l_d(x) \to 0$. Formally, in our model this is possible because the electric field tends to infinity, $-d\phi/dx \to \infty$. In fact, we should analyze the product $n(x)v_{dr}(x)$, where $n(x)$ is the electron concentration and $v_{dr} = \mu d\phi/dx$ is the carrier-drift velocity. In Chapter 7, we studied carrier properties at high electric fields and have found that, for Si, the drift velocity saturates, and for n-GaAs and n-InAs the velocity reaches a maximum and then saturates. Finite values of the drift velocities show that the pinch-off condition can be met only if the carrier concentration increases, that is, a charge accumulation, $n(x) > N_D$, should occur near the drain end. This charge accumulation leads to an almost constant current, even for the pinch-off regime. Thus we can extrapolate the value of Eq. (9.13) for the pinch-off regime assuming the current saturates and then remains constant:

$$I_D = I_{D_{sat}} \quad \text{at } \Phi_D > \Phi_{D_{sat}}. \tag{9.14}$$

In Fig. 9.7, the current-voltage characteristics corresponding to such a case are shown for different values of the gate voltage. Roughly speaking, there are two regimes of MESFET operation: the almost linear regime before the pinch off and the saturation regime when pinch off occurs. For a particular device, the gate voltage Φ_G controls intervals of the drain voltages Φ_D where each of the regimes occurs. The current-voltage characteristics are restricted by breakdown at high applied voltages to the drain. At negative gate voltages, the breakdown occurs at lower drain voltages since the channel is narrower and local electric fields are higher.

One of the main parameters of any FET is the so-called transconductance g_m, which characterizes the control of the drain current by the gate voltage:

$$g_m = \frac{dI_D}{d\Phi_G}\bigg|_{\Phi_D} \tag{9.15}$$

where the derivative is calculated at constant Φ_D. To analyze the transconductance, it is convenient to estimate it in two limiting cases. For a small drain voltage, $\Phi_D \ll \Phi_{BI} - \Phi_G$, we get from Eqs. (9.15) and (9.12),

$$g_m = g_0 \frac{(\Phi_D + \Phi_{BI} - \Phi_G)^{1/2} - (\Phi_{BI} - \Phi_G)^{1/2}}{\Phi_P^{1/2}} = g_0 \frac{\Phi_D}{2\Phi_P^{1/2}(\Phi_{BI} - \Phi_G)^{1/2}}, \tag{9.16}$$

i.e., the transconductance increases with Φ_D. In the saturation regime, as follows from Eqs. (9.15) and (9.13), it reaches its maximum value:

$$g_{m,\mathrm{sat}} = g_0 \left[1 - \left(\frac{\Phi_{BI} - \Phi_G}{\Phi_P}\right)^{1/2}\right]. \tag{9.17}$$

From Eqs. (9.16) and (9.17) and Eq. (9.11), one can see that for a high transconductance of a FET a semiconductor material with high mobility μ_n is needed and the channel length L_x should be as short as possible.

The model presented above gives the simplest description of a MESFET and allows us to understand its main characteristics such as the $I_D(\Phi_D, \Phi_G)$ dependences, the transconductance g_m, etc. A more accurate model should use two-dimensional calculations of the electrostatic problem, include a realistic $\mu_n(F)$ dependence, take into consideration the current leakage through the Schottky gate, and so on. In Fig. 9.8, such calculations are presented for a GaAs MESFET that has the gate parameters $L_x = 1\ \mu$m and $L_z = 20\ \mu$m. The GaAs active region is characterized by a mobility of 2500 cm^2/V s, a threshold voltage $\Phi_{\mathrm{th}} = 0.21$ V, and a doping level $N_D = 7.25 \times 10^{16}$ cm^{-3}. For comparison, experimental data for the device are shown as well. One can see that the device operates with a source–drain-voltage bias in the range from 0 to 2 V and a drain current from 0 to 1.2 mA. The controlling gate voltage Φ_{GS} is 0.4–0.8 V.

9.2.2 Amplification and Switching

FETs are active semiconductor devices that are capable of analog amplification and switching in digital circuits.

A FET operating as an amplifier converts the supply power – usually from a dc source – to power at the input-signal frequency. Let an input ac signal with a small amplitude $\Delta\Phi_G$ be applied to the gate of a FET (the input terminal). The device can produce an output signal ΔI_D with a high amplitude and power level at a load connected to the drain (the output terminal). The complex physical processes taking place in a FET operating as an amplifier can be represented by an equivalent circuit, depicted in Fig. 9.9. This equivalent circuit includes internal

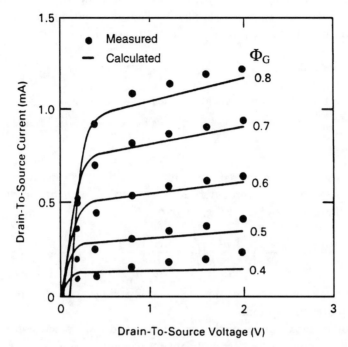

Figure 9.8. Measured and calculated current-voltage characteristics of the GaAs MESFET at different gate–source biases. [After C. H. Hyun, *et al.*, "Analysis of noise margin and speed of GaAs MESFET DCFL using μM-SPICE," IEEE Trans. Electron. Dev., **ED-33**, pp. 1421–1426 (1986).]

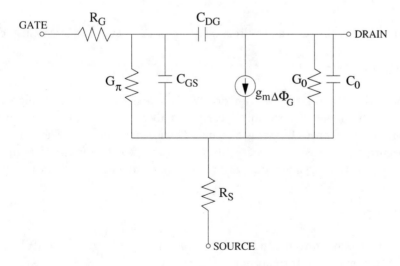

Figure 9.9. An equivalent-circuit model for a FET as an amplifier.

parasitic resistances of the gate R_G and the source R_S. The input conductance G_π describes the leakage current through the input diode (through the gate). In the case of FETs, this leakage is negligible. Thus the input small-signal current through the gate at the frequency ω is mainly the displacement current,

$$\Delta I_{\text{in}} = i\omega C_G \Delta \Phi_G, \tag{9.18}$$

where C_G is the gate-to-channel capacitance:

$$C_G = \frac{\kappa L_z L_x}{4\pi l}. \tag{9.19}$$

Note that there is a general relationship between the gate capacitance and the transconductance g_m. It can be derived if we consider a change in the electric charge at the gate Q and express g_m in terms of Q:

$$g_m \equiv \left.\frac{\partial I_D}{\partial \Phi_G}\right|_{\Phi_D} = \left.\frac{\partial I_D}{\partial Q}\right|_{\Phi_D} \left.\frac{\partial Q}{\partial \Phi_G}\right|_{\Phi_D} = \frac{C_G}{\Delta t} = \frac{C_G}{t_{\text{tr}}}, \tag{9.20}$$

where Δt can be interpreted as the average transit time t_{tr} for electrons to move from the source to the drain.

Two other capacitances of the circuit, C_{GS} and C_{DG}, describe the relation between the charge on the gate and gate–source and gate–drain biases, respectively. Between C_G, C_{GS}, and C_{DG}, there is the relation $C_{\text{GS}} = C_G - C_{\text{DG}}$. The drain–source conductance,

$$G_0 = \left.\frac{\partial I_D}{\partial \Phi_D}\right|_{\Phi_G}, \tag{9.21}$$

is introduced in the equivalent circuit. The output ac signal can be calculated as

$$\Delta I_D = g_m \Delta \Phi_G. \tag{9.22}$$

A comparison of Eq. (9.18) with Eq. (9.22) shows that at a relatively small frequency the input (displacement) current is always small with respect to the output (resistive) one. Thus we get an amplification of the current. The difference in the currents decreases with increasing ω. We can introduce the cutoff frequency $f_T = \omega_T/2\pi$ as the frequency at which both input and output currents are equal:

$$f_T = \frac{g_m}{2\pi C_G}. \tag{9.23}$$

Using the maximum transconductance in the saturation regime, Eq. (9.17), we get the maximum cutoff frequency $f_{T\text{max}}$ in our model:

$$f_{T\text{max}} = \frac{2e\mu_n N_D l^2}{\kappa L_x^2}. \tag{9.24}$$

Another useful form for f_T can be obtained from Eqs. (9.20) and (9.23) in terms of a model based on the saturation electron velocity v_s, when we can set $t_{tr} \approx L_x/v_s$:

$$f_T = \frac{v_s}{2\pi L_x}. \tag{9.25}$$

The results of Eqs. (9.24) and (9.25) support our qualitative analysis that led to the conclusion that higher cutoff frequencies can be reached for shorter conductive channels and higher carrier velocities.

The equivalent circuit presented in Fig. 9.9 also allows us to calculate the power gain U, which can be defined as the ratio of the output-signal power P_{out} to the input-signal power P_{in}. The maximum available gain in the power is achieved when the input and output are impedance matched. Neglecting G_π, we can show for the equivalent-circuit model of Fig. 9.9 that

$$U \equiv \frac{P_{out}}{P_{in}} \approx \frac{1}{4} \frac{(f_T/f)^2}{G_0(R_G + A^2 R_S) + g_m B R_G}, \tag{9.26}$$

where

$$A = \frac{C_{GS}}{C_{GS} + C_{DG}}, \qquad B = \frac{C_{DG}}{C_{GS} + C_{DG}}.$$

We can define the maximum frequency of oscillations as the frequency at which the power gain is equal to unity:

$$f_{max} \approx \frac{f_T}{2[G_0(R_G + R_S) + 2\pi f_T C_{DG} R_G]^{1/2}}. \tag{9.27}$$

We see that f_{max} depends directly on internal device capacitances and parasitic resistances. Note that f_{max} can be either above or below f_T.

We have established general parameters characterizing small-signal amplification by a MESFET. There are a number of other figures of merit that are important for particular circuits and applications.

We now consider circuit applications of FETs. In general, the main goal of a circuit for digital operations is to obtain the highest possible switching speed with the lowest overall static and dynamic dissipation of the power. MESFETs, mostly n-type GaAs FETs, are used widely for this goal as devices having high speeds. We illustrate FET circuit applications by considering only one important example – the simplest invertor. An invertor is used to provide a low output at a high-input signal and a high output at a low-input signal. Figure 9.10(a) depicts the circuit of an invertor based on a FET with a resistive load R_L. I_L and I_T are currents through the load and the transistor, respectively. The voltage Φ_{DD} supplies the power, and Φ_{in} and Φ_{out} are the input and the output voltages, respectively. Such an invertor is a key element of so-called direct-coupled FET logic (DCFL). In DCFL, the output of a logic stage – the drain of the switching FET – is connected directly to the gate of a FET of the following stage, etc. The

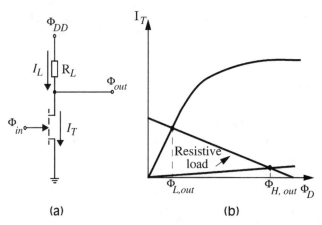

(a)   (b)

Figure 9.10. (a) A simple invertor composed of an enhancement-mode FET and a resistive load R_L; Φ_{in} and Φ_{out} are input and output signals, Φ_{DD} is the supply power voltage, I_L and I_T are currents through the load and the transistor. (b) The principle of operation of an invertor. The load-line and two current-voltage characteristics of the transistor are plotted. A low input Φ_{GL} corresponds to the transistor state with a low current and high-output voltage $\Phi_{H,out}$. A high input Φ_{GH} produces a higher current and a drop in the output voltage $\Phi_{L,out}$.

application of a FET in such a circuit is possible only if the transistor is initially turned off; see Eq. (9.6). For example, one can use the enhancement-mode device [see inequality (9.8)], which is in an off state when no gate voltage is applied. In Fig. 9.10(b), the I_T–Φ_D characteristic for a low gate voltage Φ_{GL} is represented by a direct line with a small slope. If a high-input voltage Φ_{GH} is applied to the gate, the channel is open, and the I_T–Φ_D characteristic corresponds to higher currents and exhibits a typical saturation behavior. From the Kirchhoff voltage law, we can find the load-line equation:

$$\Phi_{DD} = I_T R_L + \Phi_D,$$

which is plotted in Fig. 9.10(b) by the direct line with a negative slope. Here we neglect the output current of the next cascade. Intersections of the load line and the two I_D–Φ_D characteristics give the output signals: a high-output voltage $\Phi_{H,out}$ at a low-input gate voltage Φ_{GL} and a low-output voltage $\Phi_{L,out}$ at a high input Φ_{GH}, that is, the circuit inverts the signal.

This analysis is valid for a low-frequency input signal, when the act of switching can be considered as infinitely fast. In fact, the switching dynamic shows a time delay τ_d that depends on both the transistor and the circuit. The internal transistor time delay can be estimated with the previously discussed cutoff and maximum frequencies.

Another important parameter is the power dissipation, which consists of two contributions. The static power dissipation is

$$P_{st} = \Phi_{DD} I_{LL},$$

where I_{LL} is a low current in the initial turned-off state of the device. The dynamic

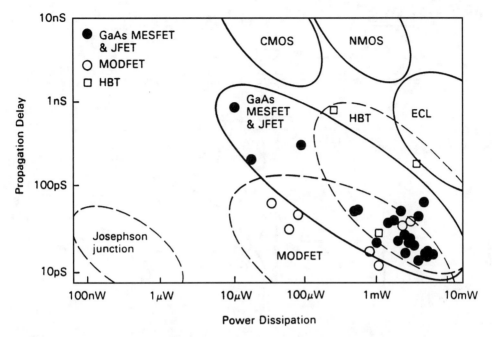

Figure 9.11. The delay time versus the power dissipation for different high-speed devices. [After H. Hasegawa, *et al.*, Extended Abstracts of the Sixteenth International Conference on Solid-State Devices and Materials, Kobe, Japan (1984), p. 413.]

power dissipation, which is associated with the discharge of the device and the load capacitances C_N, is

$$P_{\mathrm{dyn}} = f \frac{C_N \Phi_{\mathrm{DD}}^2}{2}.$$

Here it is assumed that the invertor is driven at frequency f. Since it is always true that $f\tau_d < 1$, the minimum dissipation energy required per logic gate can be estimated as

$$E_{\min} \approx \tau_d P_{\mathrm{dyn}} < \frac{C_N \Phi_{\mathrm{DD}}^2}{2}.$$

Figure 9.11 presents the time delay and the dynamic power dissipation for different high-speed systems. For MESFETs and JFETs made of GaAs, the time-delay–power-dissipation island covers a domain with τ_d ranging from 10 ps to 1 ns and P_{dyn} varying from 10 μW to 10 mW.

9.2.3 Heterostructure Field-Effect Transistors

We have studied the general operational principles of FETs; now we begin an analysis of HFETs. Contemporary semiconductor technology provides the means to vary widely the doping and heterojunction design of these devices. The principal material systems used for HFETs are as follows: n^+-AlGaAs/GaAs

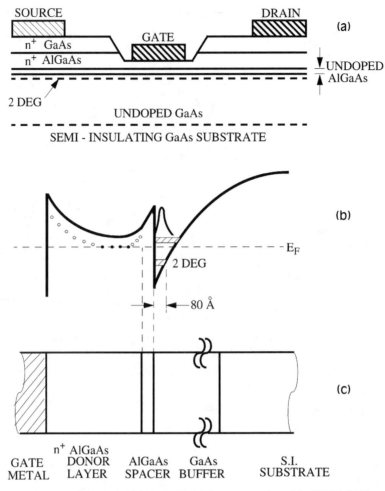

Figure 9.12. A recessed-gate n^+-AlGaAs FET: (a) The cross section of the HFET. (b) The energy-band diagram: The Fermi level and electron confined states are shown. In the donor layer there are two depletion regions: one is due to the Schottky built-in voltage; the second is due to the conduction-band offset at the junction. (c) The sequence of layers in the device.

(unstrained) heterostructures, strained-layer n^+-InGaAs channels on GaAs, and n^+-InAlAs/InGaAs heterojunctions grown on InP. Although the latter two structures are more complex, they offer better transport parameters in the conducting channel.

Figure 9.12(a) shows schematically one of the variants of the HFET – a recessed-gate n^+-AlGaAs/GaAs HFET. The device is fabricated on a semi-insulating GaAs substrate upon which an undoped buffer layer of GaAs is grown. A thin undoped AlGaAs spacer layer and an n-doped AlGaAs layer are grown on the buffer layer. Three electrodes, source, drain, and gate, are fabricated on the top of the structure. Under the electrodes, two heavily n^+-doped regions serve as the contacts to the two-dimensional electron gas formed under the heterojunction. Such a two-dimensional electron-gas channel was studied in Section

4.4 in detail. The gate length varies from 1.0 to 0.1 μm, or less, depending on the speed needed for applications. The dimensions in the vertical directions are those that were studied in Section 4.4 for both normally turned-on and normally turned-off devices. Figures 9.12(b) and 9.12(c) illustrate the energy-band coordinate dependence for the cross section of the structure under the gate. The major elements of the proper physical picture – the Schottky depletion layer under the gate, the depletion layer caused by the charge transfer to the potential well under the heterojunction, and the quantization of the electrons in the well – were analyzed in detail in Subsection 4.4.7. Figures 4.21 and 4.22 depict the energy-band diagrams for turned-off and turned-on devices as well as reconstructions of the diagrams under an external voltage applied to the gate. From that analysis it follows that the gate voltage strongly controls the electron sheet concentration on the interface n_s. The simple analytical approach developed in Subsection 4.4.5 for an ungated single heterojunction can be easily extended to this situation with the gate at the top of the heterostructure. The second Schottky depletion region shall be taken into consideration. For this purpose we use the following phenomenological approximation for n_s:

$$n_s = \frac{\kappa}{4\pi e(l_d + l_s + \Delta l)}(\Phi_G - \Phi_{\text{th}}), \tag{9.28}$$

where l_d and l_s are the widths of the donor and the spacer AlGaAs layers, respectively, and Δl is a correction factor. The threshold voltage Φ_{th} includes the built-in Schottky voltage Φ_{BI}, the conduction-band offset V_b, the shift up of the top of the heterojunction barrier $2\pi e^2 N_D l_d^2 / \kappa$ [see Eqs. (4.51) and (4.53)], and a correction, ΔE_{F0}:

$$\Phi_{\text{th}} = \Phi_{\text{BI}} - \left(V_b - \frac{2\pi e}{\kappa} N_D l_d^2 + \Delta E_{F0}\right)\bigg/ e. \tag{9.29}$$

The corrections Δl and ΔE_{F0} are introduced to take into account the quantization of electrons in the well and the concentration dependence of the Fermi level of the two-dimensional electrons. With such a threshold-voltage equation, any HFET can be described properly. Equation (9.29) defines the critical parameter Φ_{th}, which determines the depletion-mode $\Phi_{\text{th}} < 0$ or enhancement-mode $\Phi_{\text{th}} > 0$ device operation. Figure 9.13 presents the threshold voltage as a function of the doped layer thickness l_d at different doping concentrations. One can see that for concentrations such that $N_D \leq 5 \times 10^{17}$ cm^{-3}, devices remain in the enhancement mode for thicknesses $l_d < 0.04$ μm. For $N_D = 2 \times 10^{18}$ cm^{-3}, the devices are in the enhancement mode at $l_d < 0.02$ μm and in the depletion mode at larger thicknesses, etc. As discussed in Chapter 4, an important parameter for a HFET is the width of the spacer, which suppresses the scattering of the carriers by ionized impurities in the donor layer, but decreases the concentration of the two-dimensional carriers on the interface. Another factor that can lead to a decrease in the carrier concentration and gate control is that the donor layer in the HFET, say AlGaAs, is subjected to two different depletion mechanisms. One is due to the built-in Schottky barrier under the gate, and the second occurs

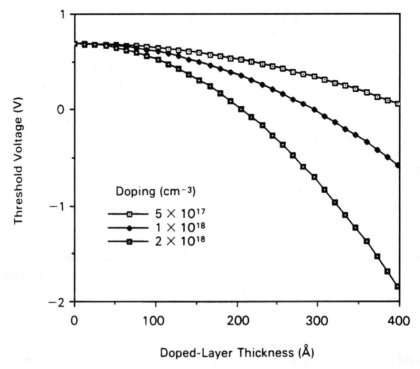

Figure 9.13. The threshold voltage of a AlGaAs/GaAs HFET versus the donor-layer thickness. Results are presented for different doping levels. [After P.-C. Chao, *et al.*, "DC and micro wave characteristics of sub-0.1-μm gate-length planar-doped pseudomorphic HEMT's," IEEE Trans. Electron Devices **36**, pp. 461–473 (1989).]

as a result of the energy-band offset at the junction. Figure 9.14 illustrates these two depletion mechanisms. One can see that an additional potential minimum is formed inside the donor layer. This minimum can lead to the accumulation of the carriers if it is not fully depleted by the built-in and the applied voltages. This second channel with the carriers in the donor layer, if formed, can screen the two-dimensional gas on the interface by controlling the gate voltage. Thus it is necessary to avoid the formation of the second channel in the donor level. In Fig. 9.14, the highest gate voltage corresponds to the case in which the second channel starts to open. The opening of the second conducting channel results in a complex dependence of the HFET transconductance on the gate voltage.

It can be shown that the relation of Eq. (9.20) between the transconductance and the gate capacitance is valid for the HFET, so the transit time t_{tr} determines the speed of the HFET. As we discussed in Chapters 4 and 5, the electron-drift velocities in two-dimensional channels can be larger than in bulk crystals. This leads to advantages in using HFETs over conventional homogeneous MESFETs, even at room temperature. Figure 9.4 provides a comparison of cutoff frequencies related to transit times for GaAs MESFETs and for AlGaAs/GaAs donor-layer MODFETs with different gate lengths. At the same gate length, HFETs demonstrate a lower transit time. At low temperature, HFETs should have much

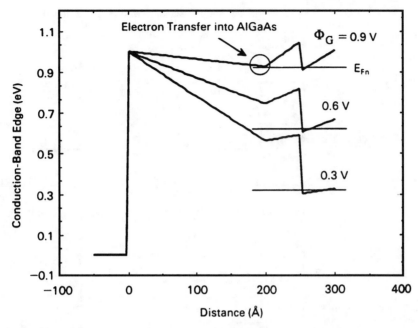

Figure 9.14. An illustration of the formation of a second electron channel due to the trade-off of two depletion regions. At a low gate voltage this channel is above the Fermi level and is empty; increasing the voltage leads to a line-up of the bottom of this channel with the Fermi level and opening it for electrons. [After P.-C. Chao, *et al.*, "DC and micro wave characteristics of sub-0.1-μm gate-length planar-doped pseudomorphic HEMT's," IEEE Trans. Electron Devices **36**, pp. 461–473 (1989).]

better characteristics. For example, as we saw, the two-dimensional gas mobility increases from approximately $8000 \, \text{cm}^2/\text{V s}$ at $T = 300 \, \text{K}$ to $\sim 100,000 \, \text{cm}^2/\text{V s}$ or more at nitrogen temperatures.

We have considered the properties of the single-heterojunction AlGaAs/GaAs FET. The possibilities offered by modern technology facilitate the fabrication of HFETs with improved parameters. Two principal methods are used to realize special device designs: selective doping and carrier confinement.

We have studied HFETs with a bulklike, doped, wide-bandgap AlGaAs donor layer. Another type of doping is δ doping. This method makes it possible to achieve the necessary carrier concentration in active two-dimensional channels. The confinement of carriers in a well-defined channel prevents the RST of carriers to the second conducting channel inside the donor layer, as discussed previously. Two types of heterostructures are used for this: a double heterostructure (a quantum well) and an inverted-structure HFET. The latter corresponds to the design with a wide-bandgap doped layer situated not between the gate and the conducting channel, but under the channel. Some of these combinations of selective doping and improved channel confinement are shown schematically in Fig. 9.15. Besides the previously mentioned single-quantum-well HFET, δ-doped HFET, and inverted-structure HFET, two additional HFETs are presented: an inverted-structure HFET with an additional barrier insulating the gate (HIGFET)

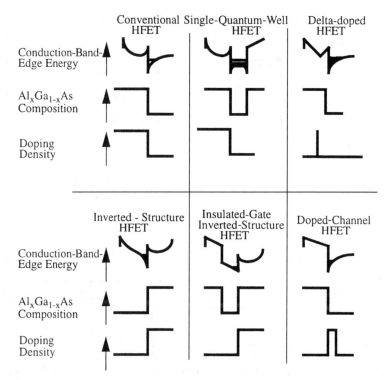

Figure 9.15. Schematic diagrams of the doping, composition, and the resultant energy-band structures for different designs of HFETs.

and a HFET with a doped channel. The wide variety of AlGaAs/GaAs HFET designs makes it possible to meet the requirements of different applications.

Now let us briefly consider the results obtained for HFETs based on materials other than AlGaAs/GaAs. Two of them, InP and $In_xGa_{1-x}As$, are the most studied because of their small effective electron masses, high electron mobilities, and high saturation velocities. In addition, the large energy distances from the lowest Γ minimum to higher valleys bring about advantages, since they prevent premature saturation in electron velocities that is due to intervalley electron transfer. For example, a mobility as high as $13,800 \, cm^2/V \, s$, a saturation velocity up to $2.95 \times 10^7 \, cm/s$, and a $\Gamma-L$ energy separation of $\sim 0.55 \, eV$ were observed for the $In_{0.53}Ga_{0.47}As$ alloy at room temperature. As pointed out in Subsection 4.3.2, this alloy is lattice matched to InP, which is a wide-bandgap material. HFETs with lattice-matched $InP/In_{0.53}Ga_{0.47}As$ heterostructures have demonstrated superb parameters compared with those of AlGaAs/GaAs devices.

In general, $In_xGa_{1-x}As/AlGaAs$ heterostructures are composed of lattice-mismatched material combinations when $x \neq 0.53$.

In general, device properties are improved with increasing In mole fractions in the conducting layer. According to our analysis in Section 4.3, these layers are strained except for the $In_{0.53}Ga_{0.47}As/InP$ system. Therefore, In-containing layers can be grown only with low thicknesses. In Fig. 9.16, the sequence of the layers and the energy-band diagram are presented for the $GaAs/In_{0.15}Ga_{0.85}As/Al_{0.15}$

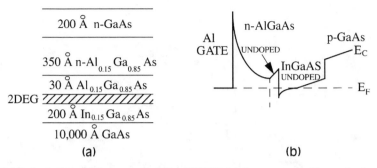

Figure 9.16. An InGaAs/AlGaAs strained-layer HFET: (a) the sequence of layers in the device, (b) the energy-band diagram for the device. [After H. Morçoç and H. Unlu, in *Microwave and Digital Applications, Semiconductors and Semimetals*, R. Dingle, ed. (Academic, New York, 1987), Vol. 24, Chap 2.]

$Ga_{0.85}As$ multilayer structure. The thick GaAs layer serves as a substrate upon which the 200-Å $In_{0.15}Ga_{0.85}As$ layer is grown. This layer is strained. Then the spacer and the donor layers of $Al_{0.15}Ga_{0.85}As$ follow. A narrow n-GaAs layer is placed before the Al gate. In the narrow-bandgap InGaAs layer, the two-dimensional electron gas is formed with concentrations from 2.4×10^{12} cm^{-2} to 4.5×10^{12} cm^{-2} and mobilities above 4000 $cm^2/V\,s$ at 300 K and 18,000 $cm^2/V\,s$ at 77 K. Using InP as a substrate and a combination of layers made of $In_xAl_{1-x}As$ and $In_yGa_{1-y}As$ with x and y at ~ 0.53, one can meet the requirement of having a high In mole fraction with a modest (less than 1%) lattice mismatch. For example, $In_{0.6}Ga_{0.4}As$ channels with n-type $In_{0.4}Al_{0.6}As$ donor layers that exhibit parameters desirable for high-speed application have been fabricated. Figure 9.4 shows that In-containing HFETs exhibit cutoff frequencies several times greater than those based on AlGaAs/GaAs heterostructures. The highest cutoff frequency reaches the value of 200 GHz for a 0.1-μm gate length. The current frequency limit for the InP/InGaAs/InAlAs HFET is 340 GHz. The fastest homostructure NMOS from the Si-based FETs exhibits a much lower cutoff frequency.

Si-based HFETs

Let us briefly review the different heterostructures used for FETs based on Si technology. Although many combinations of materials are investigated with the goal of fabricating high-quality Si-based heterostructures, the Si/SiGe system is the most studied and developed and has already found device applications. Properties of SiGe alloys and Si/SiGe heterostructures were studied in Chapter 4. These heterostructures can be used for the creation of two-dimensional electron and hole gases and improvement of device parameters.

In Si/SiGe systems, a low-dimensional n-type channel can be obtained if a strained Si layer is grown upon a strained SiGe layer. Figure 9.17(a) depicts the cross section of an n-channel Si/SiGe MODFET. The device is similar to the previously discussed AlGaAs/GaAs FETs with selective doping and improved channel confinement. The electron channel is formed in the undoped strained Si layer

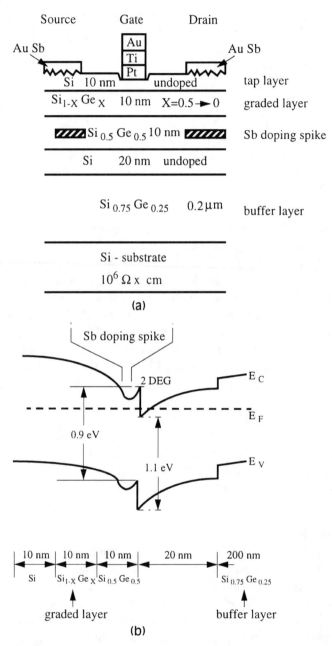

Figure 9.17. The Si/SiGe-based *n*-channel MODFET: (a) the cross-section of the device: the electron channel is formed in the strained Si layer; (b) the energy-band diagram of the device. [After H. Daembkes, *et al.*, "The *n*-channel SiGe/Si modulation-doped field-effect transistor," IEEE Trans. Electron Devices **ED-33**, pp. 633–638 (1986).]

situated between two SiGe layers. The lower $Si_{0.75}Ge_{0.25}$ buffer layer is grown upon Si and is strained and undoped. The upper $Si_{0.5}Ge_{0.5}$ layer is thin and δ doped by donors. Next, the $Si_{1-x}Ge_x$-layer is graded, with x varying from 0.5 to 0. The top thin Si layer is undoped. Such a structure is chosen to create the electron channel and avoid degradation of the characteristics of the strained layer. Figure 9.17(b) shows the energy-band diagram for this device. The electron channel on the $Si_{0.5}Ge_{0.5}$/Si heterointerface is marked. The $Si_{0.5}Ge_{0.5}$ donor layer is δ doped by Sb, which leads to the formation of a spike in the potential profile.

A p-channel MODFET requires only one strained layer of SiGe. This is illustrated by Fig. 9.18. The structure consists of a p^--Si substrate, a thin strained layer made of $Si_{0.8}Ge_{0.2}$, and the modulation-doped Si layer shown in the energy-band diagram of Fig. 9.18(b). An interesting feature of such a structure is that the two-dimensional hole gas in the strained SiGe layer is characterized by a high mobility of $8600\,cm^2$/V s at $T = 77\,K$ and a high-saturation hole velocity of 1.5×10^7 cm/s.

There are many other schemes for Si-based HFETs that combine the advantages of heterostructure bandgap engineering and selective-doping methods. Worldwide developments in these areas are now in progress.

9.3 Velocity-Modulation and Quantum-Interference Transistors

9.3.1 Velocity-Modulation Transistors

For all the previously analyzed FETs, the gate voltage controls the number of conducting electrons and the speed of operation is determined by the transport time of the electrons along the conductive channel. An alternative principle of device operation consists of modulating electron mobility, or possibly electron-drift velocity, without significant changes in the number of current carriers. To achieve such a modulation of the drift velocity one can use spatially nonuniform (selective) carrier scattering across the active channel. Then a redistribution of the carriers across the channel in response to a gate voltage would lead to control of the velocity. The time of such a redistribution across a narrow channel can be considerably smaller than that of longitudinal transport. A three-terminal device with a modulation of the drift velocity by an external voltage has been referred to as the velocity-modulation transistor (VMT).

Different particular schemes have been proposed for the VMT. To illustrate the main idea, let us consider the single heterojunction provided by two gates: the front gate is placed upon the wide-bandgap layer (say AlGaAs), while the back gate is situated behind the narrow-bandgap layer (say GaAs). The wide-bandgap layer is supposed to be doped, so as a result of the charge transfer, a potential quantum well and a conductive channel with two-dimensional electron gas are formed on the heterointerface. The distribution of electrons across the quantum well is determined by the square of their wave function, $|\psi(z)|^2$. Let us assume

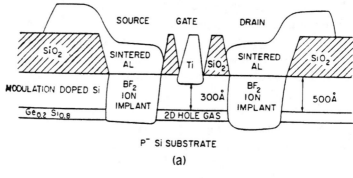

(a)

$$Eg(Si) = 1.1eV$$
$$Eg(Ge_{0.2}Si_{0.8}) = 0.95eV$$

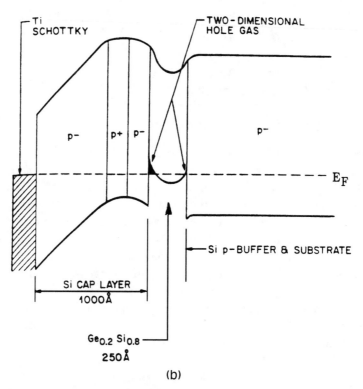

(b)

Figure 9.18. The Si/SiGe-based *p*-channel MODFET. (a) the cross-section of the device. Details of structures, doping, contacts, and insulators are shown. The $Si_{0.2}Ge_{0.8}$ layer is strained. (b) The band diagram of the device. [After T. P. Pearsall, *et al.*, "Enhancement- and depletion-mode *p*-channel Ge_xSi_{1-x} modulation-doped FET's," IEEE Electron Devices, Lett. EDL-7, pp. 308–310 (1986).]

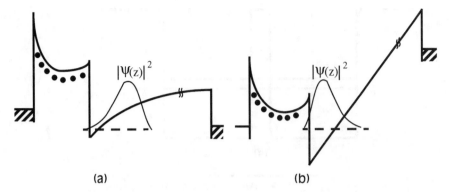

(a) (b)

Figure 9.19. The schematic of controlling the electron wave function by negative voltage applied to (a) front, (b) back gates.

that for such a selectively doped heterostructure, the main scattering mechanism is scattering by ionized impurities in the wide-bandgap layer. (This situation has been discussed in Section 4.5 and Chapters 6 and 7.) According to the analysis given in Section 4.5, the intensity of this scattering depends on the probability of finding carriers inside the depletion region. The extension of the wave function under the barrier is critically dependent on the potential profile of the well. A change in the voltage applied to the front and (or) back electrodes can significantly change the potential and redistribute the wave function across the well. Figure 9.19 illustrates this effect. Figure 9.19(a) corresponds to a negative voltage applied to the front gate. This results in a decrease in both the wave-function localization and the carrier penetration under the barrier. As a result, impurity scattering is suppressed and the carrier mobility is increased. Figure 9.19(b) corresponds to the situation in which a negative voltage is applied to the back gate. This leads to a shift of the maximum of the wave function close to the junction and an increase in the carrier penetration into the barrier region. The scattering increases and the mobility decreases.

The characteristic time of charge transfer between the well and the AlGaAs doped layer or contacts is much greater than the time associated with the redistribution of the wave function across the well. The latter time can be estimated to be as short as 10^{-13} s. Thus, at high-frequency operation, the velocity-modulation effect should be the dominant factor controlling electric signals.

Another type of heterostructure proposed for the VMT is shown in Fig. 9.20. The sequence of the layers, their width, and their doping are given in Fig. 9.20(b). The main new elements of the structure are a double GaAs quantum well presented in Fig. 9.20(a) and an asymmetric δ doping shown in Fig. 9.20(b). Electron states in the wells are coupled by means of tunneling, as analyzed in Section 3.5. For the particular double-quantum-well structure, the splitting $\Delta\epsilon$ of the levels due to such a coupling is approximately several millielectron volts. The levels are illustrated on the right-hand side of Fig. 9.20. If the broadening of the levels in each well is much less than the splitting $\Delta\epsilon$, one can introduce the mobilities in each of the wells, μ_1 and μ_2. Introducing the momentum-relaxation

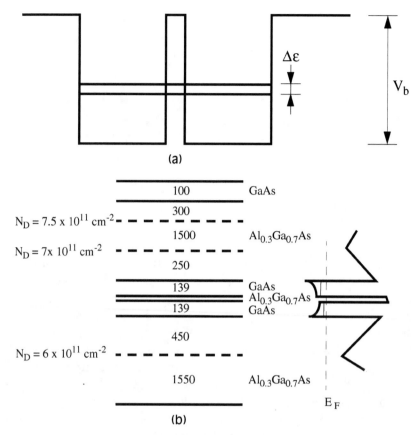

Figure 9.20. (a) Schematic potential profile of a double quantum well. $\Delta\epsilon$ is the energy splitting of the ground state of the system. (b) The sample structure and the equilibrium-band diagram (on the right-hand side). All widths are given in angstroms. The dashed lines show the δ doping; the rest of structure is undoped. The barrier between the wells is 40 Å wide. The gate electrode is not shown. [After A. Palevski, *et al.*, "Resistance resonance in coupled potential wells," Phys. Rev. Lett. **65**, pp. 1929–1932 (1990).]

times, $\tau_{1,2} = m^*\mu_{1,2}/e$, one can rewrite the criterion as

$$\frac{\hbar}{\Delta\epsilon} \ll \frac{m^*\mu_{1,2}}{e}.$$

Here we take into account that the doping breaks the initial symmetry of the wells and makes the mobilities in the wells different. The mobilities in the considered structure satisfy this criterion. At helium temperature, the ratio of the mobilities in the wells is $\mu_1/\mu_2 = 1.3$.

Only one gate is placed upon the structure. At some critical gate voltage (approximately $-0.8\,\text{eV}$) resonance occurs between the states in both wells. In a narrow range ($\sim 10\,\text{meV}$) around this critical voltage, electron states and wave functions in the wells can be controlled effectively by the gate voltage: The change in the voltage leads to a considerable redistribution of the wave function between the wells. That results in a modulation of the effective mobility of the structure between μ_1 and μ_2. The speed associated with this modulation process is limited by the lowest value among τ_1^{-1}, τ_2^{-1}, and $\Delta\epsilon/\hbar$. For the considered structure, these limitations are not important up to the terahertz frequency range.

9.3.2 Quantum-Interference Transistors

The remaining class of FETs presented in Fig. 9.1 is the so-called quantum-interference transistor (QUIT), which is based on quantum-ballistic electron transport. According to the classification given in Section 2.3, for such a transport regime, one has to provide for the coherence length l_ϕ to be greater than the characteristic device scale L:

$$l_\phi > L, \tag{9.30}$$

that is, this device is a mesoscopic device.

The principle of operation of a QUIT is the control of the interference pattern of conducting electrons by an external voltage. These patterns can be arranged as a result of interference of the waves traveling through two or more channels (arms).

A schematic of a two-channel quantum-interference device is shown in Fig. 9.21(a). Basically, it is an ordinary FET with a short channel and a barrier parallel to the current. The barrier is embedded in the middle region of the device. The barrier splits the main channel into two channels: 1 and 2. There are also two contacts to the main channel, the source and the drain. The gate is placed on the top of the device. Because of the gate, the symmetry of channels 1 and 2 is broken if a gate voltage is applied. The distance between contacts L satisfies inequality (9.30). The widths of the main channel and the split channels are small, so there is a quantization of transverse-electron states in the z direction. For simplicity, one can assume that only the lowest subband is populated in each of the device regions. Two-dimensional subbands are illustrated in Fig. 9.21(b); there the subbands $E(k_x\,k_y)$ and the Fermi energy E_F are plotted for three major device regions. In the middle region, the bottoms of the subbands ϵ_1 and ϵ_2 generally are different for channels 1 and 2. The electron-energy spectra are of the form

$$E = \epsilon_1 + \frac{\hbar^2}{2m^*}\left(k_{x,1}^2 + k_y^2\right) = \epsilon_2 + \frac{\hbar^2}{2m^*}\left(k_{x,2}^2 + k_y^2\right), \tag{9.31}$$

where E is the energy of an incident electron and k_x and k_y are the components of its wave vector in the plane of the device. The k_x components can be different in the channels.

Let an electron be injected from the source into the left region of the main channel. The electron's wave function is

$$\psi_L = \chi_L(z)e^{i(k_x x + k_y y)},$$

where $\chi_L(z)$ is the wave function of the lowest subband in the left region. We can introduce the amplitudes of the waves transmitted into channels 1 and 2:

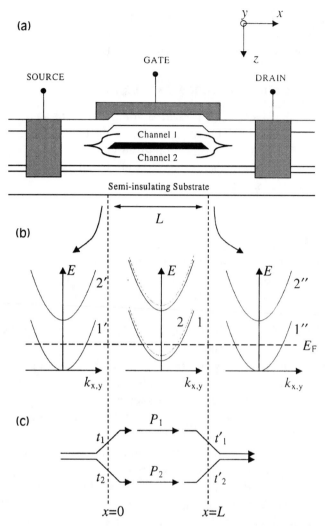

Figure 9.21. The schematic of the QUIT with two parallel channels. (a) The structure of the device. The barrier in the middle of the structure forms two electron channels. The solid lines show splitting (before the barrier) and subsequent converging of the electron ways. (b) Electron subbands in different regions of the device and the Fermi energy under equilibrium. In the middle of the device subbands in both channels are different (indicated by solid and dashed lines). (c) Illustration of the electron ways in the device; t_1, t_2 and t'_1, t'_2 indicate the amplitudes of split and interfering waves, respectively, and P_1 and P_2 indicate factors of different phase shifts in the channels.

t_1 and t_2. These amplitudes determine the electron wave functions inside the middle device region:

$$\psi_M = \begin{cases} t_1 \chi_{M,1} e^{i[k_{x,1}(x-x_L)+k_y y]} & \text{upper channel} \\ t_2 \chi_{M,2} e^{i[k_{x,2}(x-x_L)+k_y y]} & \text{lower channel} \end{cases},$$

where $\chi_{M,1}(z)$ and $\chi_{M,2}(z)$ are wave functions of the lowest subbands in channels 1 and 2, respectively, and x_L is the coordinate of the left end of the channels. Since potential profiles in both channels are different in general, the wave vectors are different: $k_{x,1} \neq k_{x,2}$. Then let t'_1 and t'_2 characterize the transmitted waves from the channels into the right device region; see Fig. 9.21(c). The wave function inside the right region can be written as

$$\psi_R = (t_1 P_1 t'_1 + t_2 P_2 t'_2) \chi_R(z) e^{i[k_x(x_L - x_R) + k_y y]}, \tag{9.32}$$

where $P_1 \equiv \exp(ik_{x,1}L)$, $P_2 \equiv \exp(ik_{x,2}L)$, and $\chi_R(z)$ is the wave function of the transverse motion in the right end of the channels. For the sake of simplicity, we neglect multiple reflections from the ends of the barriers.

The wave functions χ_L and χ_R are normalized to unity; thus the total transmission coefficient for the device is

$$T(E) = |t'_1 P_1 t_1 + t'_2 P_2 t_2|^2. \tag{9.33}$$

We can assume that both channels 1 and 2 are symmetrical about $z = 0$ in the absence of a gate voltage. Thus, for the lowest occupied subbands, we find

$$t_2 = t_1, \quad t'_2 = t'_1.$$

Then Eq. (9.33) gives us

$$|T(E)|^2 = 2|t_1 t'_1|^2 (1 + \cos\theta), \tag{9.34}$$

where

$$\theta = (k_{x,2} - k_{x,1}) L \tag{9.35}$$

is the relative phase shift in both channels. If we introduce the average electron velocity

$$v_x = \frac{\hbar (k_{x,1} + k_{x,2})}{2m^*},$$

we can represent the difference in the wave as

$$k_{x,2} - k_{x,1} = \frac{\epsilon_1 - \epsilon_2}{\hbar v_x}. \tag{9.36}$$

Now the phase shift takes the form

$$\theta = \frac{L}{v_x} \frac{\epsilon_1 - \epsilon_2}{\hbar}.$$

The origin of the phase shift is obvious: if $\epsilon_1 \neq \epsilon_2$, a difference in the kinetic energies for both channels gives rise to different phases of the waves coming into the right device region; of course, these different phases lead to quantum interference.

For symmetrical channels with respect to $z = 0$, in the absence of a gate voltage, we find $\epsilon_1 = \epsilon_2 = \epsilon_0$ and the phase shift equals zero. If we apply a gate voltage, the potential energy as a function of the transverse coordinate is modified:

$$V(z) = V_0(z) - e\Phi(z).$$

Here $\Phi(z)$ is the potential induced by the applied voltage. This leads to subband energies,

$$\epsilon_1 = \epsilon_0 - e\langle \chi_{M,1} \,|\, \Phi \,|\, \chi_{M,1} \rangle$$
$$\epsilon_2 = \epsilon_0 - e\langle \chi_{M,2} \,|\, \Phi \,|\, \chi_{M,2} \rangle.$$

Using Eqs. (9.36) and (9.35), we find the phase shift:

$$\theta = \frac{L}{v_x} \frac{e\Phi_{12}}{\hbar}, \tag{9.37}$$

where $\Phi_{12} \equiv \langle \chi_{M,2} \,|\, \Phi(z) \,|\, \chi_{M,2} \rangle - \langle \chi_{M,1} \,|\, \Phi(z) \,|\, \chi_{M,1} \rangle$. The value Φ_{12} represents the difference between the average potential in the channels. This difference determines the transmission through the device.

To calculate the electric current, one can use Eqs. (2.175) and (2.176), which were derived for quantum-ballistic transport. Equation (2.184) defines the device conductance G. At low temperatures, these results lead to the Landauer formula of Eq. (2.185), which, in the case under consideration, has the form

$$G = \frac{2e^2}{h} T(E) = \frac{4e^2}{h} |t_1 t_2|^2 (1 + \cos\theta). \tag{9.38}$$

The second term in the parentheses is due to electron interference. One can see that the interference controls the device conductance. If the phase shift of split waves $\theta = 0$ ($\Phi_{12} = 0$), the conductance reaches the maximum: $G_{\max} = 8e^2|t_1 t_2|^2/h$. If $\theta = \pi$, i.e.,

$$e\Phi_{12} = \frac{\hbar \pi v_x}{L}, \tag{9.39}$$

the interference is destructive and the conductance vanishes.

Let us introduce the characteristic transit time through the channels, $t_{tr} = L/v_x$. If we set $L = 2000$ Å, and $v_x = 2 \times 10^7$ cm/s, we get $t_{tr} = 1$ ps and for the destructive potential difference we get $\Phi_{12} \approx 2$ meV. Since the screening effects are not important for a typical channel width of ~ 100 Å, we can consider the former estimate of voltage to be the gate voltage necessary to destroy the device conductance. It is important that the shorter channels require the larger gate voltage. Obviously the transit time t_{tr} determines the cutoff frequency of the device, $\omega_{cf} = 1/2\pi t_{tr}$. Since the device channels are not doped and can be

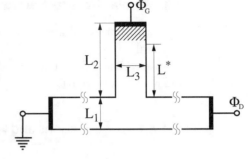

Figure 9.22. The schematic of a T-shaped QUIT. The source (grounded), drain, and gate electrodes are indicated.

made quite short, the devices portend operation up to the terahertz frequency region.

Our simple model of the QUIT allows us to compare this device with a conventional FET. If the latter is in a normally on state with the Fermi energy of electrons in the conducting channel equal to E_F, we can estimate the threshold voltage needed to deplete the FET channel as

$$e\Phi_{\mathrm{th,FET}} \approx E_F = \frac{\hbar^2 k_F^2}{2m^*}. \tag{9.40}$$

For quantum devices described by Eq. (9.39), we find

$$e\Phi_{\mathrm{th,QUIT}} \approx e\Phi_{12} = \frac{\pi \hbar v_x}{L} = \frac{\pi \hbar^2 k_F}{m^* L} = \frac{\hbar^2 k_F^2}{2m^*} \frac{2\pi}{k_F L} = e\Phi_{\mathrm{th,FET}} \frac{\lambda_F}{L}, \tag{9.41}$$

where λ_F is de Broglie wavelength. Hence $\lambda_F \ll L$; one can see that the quantum device can operate with a significantly smaller controlling gate voltage.

Another design of a structure for a QUIT is sketched in Fig. 9.22. The structure is T shaped and consists of a transverse arm (stub) and a channel connecting the source (grounded) and the drain. The transverse arm has finite dimensions and has a gate on the end of the arm; this end is commonly referred to as the top of the device. If the length L_2 and the width L_3 are less than the coherence length l_ϕ, the reflection of the electron wave from the arm produces de Broglie wave-interference patterns. If the width of the main channel L_1 also is small compared with l_ϕ, the pattern extends across the channel and determines the transmission coefficient of electrons through the device, i.e., the source–drain conductance. A voltage applied to the gate changes the penetration length L^* of the electron wave into the arm and, consequently, the interference pattern. The estimates show that the gate voltage can effectively control the conductance of such a T-shaped device for which the dimensions L_1, L_2, and L_3 are approximately several hundred angstroms.

In conclusion, the modulation-velocity effect and the quantum-interference effect provide new principles of operation for three-terminal devices. These devices are now in the early stage of study. However, these approaches portend the effective control by a small voltage as well as the development of high-speed transistors.

9.4 Bipolar Heterostructure Transistors

In terms of practical applications, BTs are the most important devices among potential-effect transistors; see Fig. 9.5. Our main goal is to discuss heterostructure-based BTs. We start with a review of the basic properties of p–n junctions and homostructure BTs.

9.4.1 p–n Junctions and Homostructure Bipolar Transistors

p-n Junctions

We assume that the reader is acquainted with the physics of p–n junctions, and here we summarize only the main facts necessary to understand BTs. At the end of this chapter, we provide references on this subject.

In Fig. 9.23, the schematic of a p–n junction is presented. An abrupt transition between the p- and the n-doped sides of the structure is assumed. The charge transfer across the junction lines up the various Fermi levels and results in a Fermi level E_F, depletes two regions in the p- and the n-doped sides, and creates a

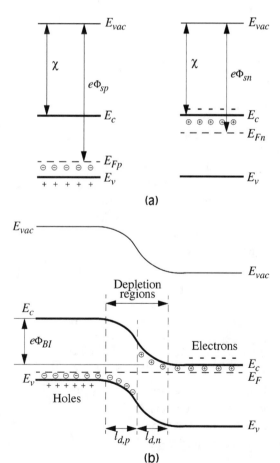

Figure 9.23. The formation of a p–n junction. (a) An energy schematic of a semiconductor with p and n doping (left- and right-hand sides, respectively). The electron affinities are the same, but the work functions $e\Phi_{sp}$ and $e\Phi_{sn}$ and Fermi levels E_{Fp} and E_{Fn} are different. (b) A p–n junction with abrupt coordinate dependence of doping concentration.

built-in potential Φ_{BI}, which prevents the penetration of electrons into the p-type side of the structure and holes into the n-type side of the structure.

The built-in voltage can be found by applying Eqs. (4.25) and (4.26), which establish relations between the concentrations of holes p and electrons n and the Fermi level E_F. Far away from the junction, these equations give us

$$E_F = E_v - k_B T \, \ln \frac{p}{N_v(T)}, \quad p \approx N_A, \; z \to -\infty,$$

$$E_F = E_c + k_B T \, \ln \frac{n}{N_e(T)}, \quad n \approx N_D, \; z \to \infty,$$

where E_v and E_c are the energies of the top of the valence and the bottom of the conduction bands, respectively; $N_v(T)$ and $N_c(T)$ are the effective densities of states for both bands, and N_A and N_D are acceptor and donor concentrations on the p and the n sides of the structure. Here we use the following estimates for the concentrations of electrons n and holes p:

$$p \equiv p_{p,0} \approx N_A, \qquad n \equiv n_{p,0} \approx \frac{n_i^2}{N_A} \ll p_p \text{ at } p \text{ side}, \tag{9.42}$$

$$n \equiv n_{n,0} \approx N_D, \qquad p \equiv p_{n,0} \approx \frac{n_i^2}{N_D} \ll n_n \text{ at } n \text{ side}, \tag{9.43}$$

where n_i is determined by Eq. (4.28).

Defining the built-in voltage so that $-e\Phi_{BI}$ equals the difference in the position of the conduction band at $z \to +\infty$ and $z \to -\infty$, we obtain

$$\Phi_{BI} = \frac{1}{e} \left[E_g + k_B T \, \ln \frac{N_D N_A}{N_e(T) N_p(T)} \right]. \tag{9.44}$$

According to our analysis of Section 4.4, the electrostatic potential in the p- and the n-depletion regions is a quadratic function of the z coordinate, and the built-in potential is related to the widths of the depletion regions $l_{d,n}$ and $l_{d,p}$ through

$$\Phi_{BI} = \frac{2\pi e}{\kappa} \left(N_A l_{d,p}^2 + N_D l_{d,n}^2 \right). \tag{9.45}$$

Using the quasi-neutrality condition

$$N_A l_{d,p} = N_D l_{d,n}$$

and Eq. (9.44), we find $l_{d,p}$ and $l_{d,n}$:

$$l_{d,p} = \left[\frac{\kappa}{2\pi e} \frac{N_D \Phi_{BI}}{N_A(N_A + N_D)} \right]^{1/2}, \tag{9.46}$$

$$l_{d,n} = \left[\frac{\kappa}{2\pi e} \frac{N_A \Phi_{BI}}{N_D(N_A + N_D)} \right]^{1/2}. \tag{9.47}$$

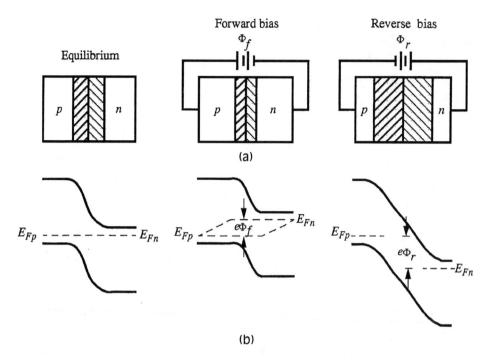

Figure 9.24. The scheme of a p–n junction at equilibrium, under forward Φ_f, and under reverse Φ_r, biases. (a) The biasing of the p–n junction, (b) energy bands and quasi-Fermi levels. For nonequilibrium cases instead the Fermi level, two quasi-Fermi levels for the holes and the electrons, E_{Fp} and E_{Fn}, are shown.

The electric field is the linear function of the z coordinate in each depletion region:

$$F(z) = \begin{cases} -4\pi e N_A \left(l_{d,p} + z\right)/\kappa & \text{if } -l_{d,p} < z < 0 \\ 4\pi e N_D \left(z - l_{d,n}\right)/\kappa & \text{if } 0 < z < l_{d,n} \end{cases},$$

that is, the field reaches the maximum, $|F_m| = 4\pi \, N_A l_{d,p}/\kappa$, at the junction where $z = 0$.

Let a voltage be applied to the p–n junction. The forward bias of the p–n junction corresponds to a negative potential $-\Phi$ applied to the n-type side. A positive potential leads to the reverse bias of the junction. Figure 9.24 illustrates equilibrium, forward-biased, and reverse-biased p–n junctions. Under the forward and the reverse biases, the potential differences between the p- and the n-type sides become $\Phi_{BI} \mp \Phi$. The small applied voltage is dropped mainly across the depletion regions because their resistance is much greater than that of neutral regions of the junction. The voltage breaks down the equilibrium; in particular, there is no longer a common Fermi level, and an electric current begins to flow through the structure. One can describe nonequilibrium carrier concentrations by using so-called quasi-Fermi levels, E_{Fn} and E_{Fp}, which are different for electrons and holes:

$$n = N_e(T) \exp\left(\frac{E_{Fn} - E_c}{k_B T}\right), \tag{9.48}$$

$$p = N_v(T) \exp\left(\frac{E_v - E_{Fp}}{k_B T}\right); \tag{9.49}$$

here the quasi-Fermi levels E_{Fn} and E_{Fp} in the depletion regions remain approximately the same as in quasi-neutral regions. The difference in the quasi-Fermi levels can be calculated by means of the applied voltage:

$$e\Phi = E_{Fn} - E_{Fp}. \tag{9.50}$$

Thus, from Eqs. (9.48) and (9.49), we find

$$pn = n_i^2 \exp\left(\frac{E_{Fn} - E_{Fp}}{k_B T}\right) = n_i^2 \exp\left(\frac{e\Phi}{k_B T}\right).$$

Let us define electrons as majority carriers and holes as minority carriers on the n-doped side of the structure and vice versa on the p-doped side. We can estimate the concentration of minority carriers at the boundaries of the depletion and the quasi-neutral regions:

$$p_n \approx \frac{n_i^2}{N_D} \exp\left(\frac{e\Phi}{k_B T}\right) = p_{n,0} \exp\left(\frac{e\Phi}{k_B T}\right) \quad \text{at } z = l_{d,n}, \tag{9.51}$$

$$n_p \approx \frac{n_i^2}{N_A} \exp\left(\frac{e\Phi}{k_B T}\right) = n_{p,0} \exp\left(\frac{e\Phi}{k_B T}\right) \quad \text{at } z = -l_{d,p}. \tag{9.52}$$

Approximations (9.51) and (9.52) indicate that the minority concentrations at the ends of the depletion regions increase exponentially with an increase in the forward bias ($\Phi > 0$) and decrease exponentially with an increase in a reverse bias ($\Phi < 0$).

Let us consider the quasi-neutral regions. As emphasized previously, in these regions the electric fields are small and can be neglected. For the minority carriers, we can use the drift-diffusion equations. This kind of equation was derived in Chapter 2 for the unipolar case, in which only one sort of carrier, say electrons, exists. For the bipolar case, we have two types of carriers, electrons and holes, and we need to generalize the drift-diffusion equations by introducing electron–hole recombination. For example, in the n-doped quasi-neutral region we can write the equation for the minority carriers, i.e., for holes, in the form

$$D_p \frac{\partial^2 p_n}{\partial z^2} - \frac{p_n - p_{n,0}}{\tau_p} = 0, \tag{9.53}$$

where D_p is the diffusion coefficient and τ_p is the electron–hole-recombination time. The second term in Eq. (9.53) represents the rate of recombination of nonequilibrium holes in the n region. Equation (9.53) should be solved for the quasi-neutral region $z > l_{d,n}$ with the boundary condition of approximation (9.51) at $z = l_{d,n}$. The result is

$$p_n - p_{n,0} = p_{n,0} \left[\exp\left(\frac{e\Phi}{k_B T}\right) - 1 \right] \exp\left(-\frac{z - l_{d,n}}{L_p}\right). \tag{9.54}$$

Here we introduce the so-called hole-recombination length for the n side of the structure, $L_p = \sqrt{D_p \tau_p}$.

According to the above assumptions, in the quasi-neutral regions the currents of minority carriers consist primarily of the diffusion component. Thus, on the n side, $z > l_{d,n}$, we get for the hole current

$$I_{p,n} = I_{p,D} = -e D_p \frac{\partial p_n}{\partial z} = \frac{e D_p p_{n,0}}{L_p} \left[\exp\left(\frac{e\Phi}{k_B T}\right) - 1 \right] \exp\left(-\frac{z - l_{d,n}}{L_p}\right). \tag{9.55}$$

In the same way, we can write the equation for the minority electrons on the p side of the structure:

$$D_n \frac{\partial^2 n_p}{\partial z^2} - \frac{n_p - n_{p,0}}{\tau_n} = 0. \tag{9.56}$$

Then, on the p-side, $z < -l_{d,p}$, we find the electron concentration,

$$n_p(z) - n_{p,0} = n_{p,0} \left[\exp\left(\frac{e\Phi}{k_B T}\right) - 1 \right] \exp\left(\frac{z + l_{d,p}}{L_n}\right), \tag{9.57}$$

and the electron current injected through the p–n junction:

$$I_{n,p} = I_{n,D} = e D_n \frac{\partial n_p}{\partial z} = \frac{e D_n n_{p,0}}{L_n} \left[\exp\left(\frac{e\Phi}{k_B T}\right) - 1 \right] \exp\left(\frac{z + l_{d,p}}{L_n}\right), \tag{9.58}$$

where L_n is the electron-recombination length on the p side defined by means of the electron recombination time τ_n: $L_n = \sqrt{D_n \tau_n}$. Note that L_n and L_p are different as a result of differences in the diffusion coefficients D_n and D_p and the recombination times τ_n and τ_p. Typically,

$$L_n, \ L_p \gg l_{d,n} + l_{d,p}. \tag{9.59}$$

This result implies that the recombination of holes and electrons in both depletion regions is negligible. This justifies the previous approximation of constant quasi-Fermi levels through the depletion regions. Now we can use Eqs. (9.55) and (9.58) to calculate the total current through the p–n junction as a sum of electron and hole currents injected through the depletion regions:

$$I = I_{n,p}\big|_{z=-l_{d,p}} + I_{p,n}\big|_{z=l_{d,n}} = I_s \left[\exp\left(\frac{e\Phi}{k_B T}\right) - 1 \right], \tag{9.60}$$

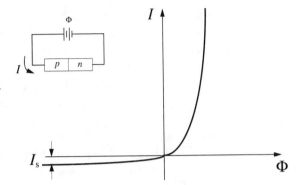

Figure 9.25. The current-voltage characteristic of the p–n junction.

where the saturation current through the p–n diode is

$$I_s = e\frac{D_p p_{n,0}}{L_p} + e\frac{D_n n_{p,0}}{L_n}. \tag{9.61}$$

Note that electron and hole currents in Eq. (9.60) are calculated at different space points. It is possible to do that only in the case of low recombination currents in the depletion regions; indeed, under these conditions the electron and the hole currents are constant across the depletion region.

The current-voltage characteristic of the p–n junction is presented in Fig. 9.25. The main features of this characteristic are a rectifying behavior and a strong nonlinearity. This nonlinearity is used in numerous applications of the p–n diode. Under reverse bias, the injection of minority carriers primarily determines the current through the junction. Under strong reverse voltages, the current saturates to quite a small value I_s, so that the p–n junction becomes essentially nonconducting. For a forward voltage, the junction becomes strongly conducting and exhibits an exponentially increasing current.

Homostructure Bipolar Transistor

A homostructure BT consists of two p–n junctions. Usually one of them – the emitter junction – is forward biased, while the other – the collector junction – is reverse biased. The energy-band diagram for a BT is shown schematically in Fig. 9.26. This diagram can be easily understood in terms of the charge-transfer

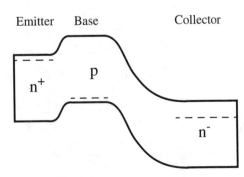

Figure 9.26. An energy diagram of a homostructure n–p–n BT along the direction of electron travel. Quasi-neutral emitter, base, and collector regions are indicated. The emitter junction is forward biased, and the collector junction is slightly reverse biased.

effect between the emitter and the base and the base and collector. The depicted device is of the n–p–n type, although almost all results are applicable to p–n–p devices. The voltage bias is supposed to be such that an electron travels from the left n region (the emitter) across the p region (the base) to the right n region (the collector). Applied forward bias to the emitter junction lowers the energy barrier for an electron flowing from the emitter to the base. Simultaneously, a similar flow of holes from the base to the emitter appears. Those electrons that overcome the barrier travel across the base. Recombination of the minority carriers (electrons) occurs in the base, but if the thickness of the base region is less than the diffusion length of the minority carriers, these electron losses are negligible. When the electrons reach the collector junction, they are swept into the collector because of the high electric field within the depletion regions of the latter junction and form the collector current.

We now consider the main parameters of BTs. Let $I_{E,n}$ and $I_{E,p}$ be the electron and the hole contributions to the emitter current, respectively. The collector current I_C is proportional to the electron contribution $I_{E,n}$:

$$I_C = B I_{E,n}, \tag{9.62}$$

where the factor B is the base-transport factor. It accounts for the fact that not all electrons can reach the collector ($B < 1$). We define the emitter efficiency γ_E as the electron fraction in the emitter current:

$$\gamma_E = \frac{I_{E,n}}{I_{E,n} + I_{E,p}}. \tag{9.63}$$

The ratio of the emitter current to the collector current is the so-called current-transfer ratio:

$$\frac{I_C}{I_E} = B\gamma_E \equiv \alpha. \tag{9.64}$$

For well-designed BTs, the parameters B, γ_E, and α are close to unity. The parameter of primary importance for a BT is the ratio of the collector current and the controlling base current. The base current I_B can be represented as the sum of the hole current injected into the emitter and the hole current that is due to the recombination of the holes and the electrons injected into the base. The latter equals $(1 - B)I_{E,n}$. The base–collector current is negligible since strong reverse bias between the base and the collector is assumed. Thus the base current may be written as

$$I_B = I_{E,p} + (1 - B)I_{E,n}. \tag{9.65}$$

We define the base–collector-current amplification in terms of the ratio

$$\frac{I_C}{I_B} = \frac{B I_{E,n}}{I_{E,p} + (1 - B)I_{E,n}} = \frac{B\gamma_E}{1 - B\gamma_E} = \frac{\alpha}{1 - \alpha}. \tag{9.66}$$

This ratio is the so-called collector-current-amplification factor or the current gain:

$$\beta = \frac{\alpha}{1 - \alpha}. \tag{9.67}$$

Thus, to reach a high current gain, one needs to design the device so that the current-transfer ratio α is close to unity.

An important parameter of the BT is the transconductance,

$$g_m = \frac{\partial I_C}{\partial \Phi_{BE}}. \tag{9.68}$$

The transconductance characterizes the control of the collector current by the voltage bias applied to the emitter junction, Φ_{BE}.

The current-voltage characteristics of BTs and the parameters γ_E, B, β, and α can be calculated in terms of the p–n junction theory considered previously. Let L_B and L_E be the base and the emitter widths, respectively, and let S be the area of the device cross section. The depletion regions for both junctions can be considered as infinitely narrow; thus the carrier transport has to be evaluated in only the three quasi-neutral regions of the emitter, base, and collector. Let us denote the collector–base voltage bias as Φ_{CB}. Both biases, Φ_{BE} and Φ_{CB}, are chosen to be positive for the forward-biased emitter junction and the reverse-biased collector junction. At the boundaries of the base regions the minority electron concentration is determined by the processes of charge transfer, i.e., injection, through the junctions. Applying the result of Eq. (9.57) for both junctions, we get in the base region

$$n_B(z = 0) = n_{B,0} \exp\left(\frac{e\Phi_{BE}}{k_B T}\right), \qquad n_B(z = L_B) = n_{B,0} \exp\left(-\frac{e\Phi_{CB}}{k_B T}\right),$$

$$\tag{9.69}$$

where the emitter junction is taken to be at $z = 0$. The distribution of the excess electrons across the base can be found as the solution of Eq. (9.56) with the boundary conditions of Eqs. (9.69):

$$n_B - n_{B,0} = \frac{n_{B,0}}{\sinh(L_B/L_{n,B})} \left\{ \sinh\left(\frac{L_B - z}{L_{n,B}}\right) \left[\exp\left(\frac{e\Phi_{BE}}{k_B T}\right) - 1 \right] \right.$$

$$\left. + \sinh\left(\frac{z}{L_{n,B}}\right) \left[\exp\left(\frac{-e\Phi_{CB}}{k_B T}\right) - 1 \right] \right\}, \tag{9.70}$$

where $\{n, B\}$ labels the electron parameters in the base region; for example, $L_{n,B} = \sqrt{D_{n,B}\tau_{n,B}}$ is the electron-recombination length in the base. Now we can calculate the electron contribution to the emitter current:

$$
\begin{aligned}
I_{E,n} &= e D_{n,B} S \frac{\partial n_B}{\partial z}\bigg|_{z=0} \\
&= -\frac{e D_{n,B} n_{B,0} S}{L_{n,B} \sinh(L_B/L_{n,B})} \\
&\quad \times \left\{ \cosh\left(\frac{L_B}{L_{n,B}}\right) \left[\exp\left(\frac{e\Phi_{BE}}{k_B T}\right) - 1 \right] - \left[\exp\left(\frac{-e\Phi_{CB}}{k_B T}\right) - 1 \right] \right\}.
\end{aligned}
\tag{9.71}
$$

In the emitter region, $z < 0$, we should find the minority hole concentration p_E. The results of Eqs. (9.53)–(9.55) can be rewritten in the notation of this section as

$$
p_E = p_{E,0} + p_{E,0} \left[\exp\left(\frac{e\Phi_{BE}}{k_B T}\right) - 1 \right] \exp\left(\frac{z}{L_{p,E}}\right),
\tag{9.72}
$$

where we assume that the emitter width L_E is much greater than the hole-recombination length in the emitter region $L_{p,E}$. The hole contribution to the emitter current is

$$
\begin{aligned}
I_{E,p} &= -e D_{p,E} S \frac{\partial p_E}{\partial z}\bigg|_{z=0} \\
&= -\frac{e D_{p,E} p_{E,0} S}{L_{p,E}} \left[\exp\left(\frac{e\Phi_{BE}}{k_B T}\right) - 1 \right].
\end{aligned}
\tag{9.73}
$$

Similarly, we can calculate the minority hole distribution in the thick ($L_C \gg L_{p,C}$) collector:

$$
p_C = p_{C,0} + p_{C,0} \left[\exp\left(-\frac{e\Phi_{CB}}{k_B T}\right) - 1 \right] \exp\left(-\frac{z - L_B}{L_{p,C}}\right),
\tag{9.74}
$$

where $L_{p,C}$ is the hole-recombination length in the collector region.

The collector current I_C in a homostructure transistor is limited by the rate at which electrons travel through the base as well as the injection rate of holes through the reverse-biased collector junction:

$$
I_C = I_{C,n} + I_{C,p},
\tag{9.75}
$$

with

$$
\begin{aligned}
I_{C,n} &= e D_{n,B} S \frac{\partial n_B}{\partial z}\bigg|_{z=L_B}, \\
I_{C,p} &= -e D_{p,C} S \frac{\partial p_C}{\partial z}\bigg|_{z=L_B}.
\end{aligned}
$$

Substituting Eqs. (9.70) and (9.74) into the two terms of Eq. (9.75), one can get for both contributions

$$
I_{C,n} = -\frac{e\, D_{n,B} n_{B,0} S}{L_{n,B}\, \sinh(L_B/L_{n,B})} \left[\exp\left(\frac{e\Phi_{BE}}{k_B T}\right) - 1 \right]
$$
$$
+ \frac{e\, D_{n,B} n_{B,0} S}{L_{n,B}}\, \coth\left(\frac{L_B}{L_{n,B}}\right) \left[\exp\left(-\frac{e\Phi_{CB}}{k_B T}\right) - 1 \right], \tag{9.76}
$$

$$
I_{C,p} = \frac{e\, D_{p,C} p_{C,0} S}{L_{p,C}} \left[\exp\left(-\frac{e\Phi_{CB}}{k_B T}\right) - 1 \right]. \tag{9.77}
$$

Thus we have now obtained all current components in the BT and can evaluate the basic parameters of the device and find the conditions that provide for high-performance device characteristics. In principle, the results obtained for the BT can be used for both forward and reverse biases of the emitter and the collector junctions. Depending on the biases of both junctions, one can classify the possible modes of operation. The active mode of operation corresponds to a forward bias of the emitter junction and a reverse bias of the collector junction. The cutoff mode corresponds to reverse biases of both junctions. The saturation mode of operation occurs when both junctions are forward biased.

Let us consider the active mode of operation. We assume that the collector junction is strongly reverse biased, $e\Phi_{CB} \gg k_B T$, so that contributions of the reverse current from the collector side are entirely negligible. First we begin with an analysis of the emitter efficiency of Eq. (9.63). A high emitter efficiency γ_E requires a suppression of the hole current through the emitter junction. Comparing Eqs. (9.71) and (9.73), we see that two conditions are favorable for high γ_E:

$$
p_{E,0} \ll n_{B,0}, \qquad L_B \ll L_{n,B}. \tag{9.78}
$$

According to approximations (9.42) and (9.43), the first condition can be met if the emitter is doped more heavily than the base:

$$
N_{D,E} \gg N_{A,B}.
$$

The second condition requires a smaller base width in comparison with the minority electron-recombination length in the base. In such a limit, we get

$$
\gamma_E \approx 1 - \frac{p_{E,0} D_{p,E} L_B}{n_{B,0} D_{n,B} L_{p,E}}. \tag{9.79}
$$

The base-transport factor B, defined by Eq. (9.62), can be evaluated easily from Eqs. (9.71)–(9.75). We find that the second condition of approximations (9.78) should be fulfilled for a high B:

$$
B = \frac{I_C}{I_{E,n}} \approx 1 - \frac{L_B^2}{2 L_{n,B}^2}. \tag{9.80}
$$

The current-transfer ratio of Eq. (9.64) can be calculated with approximation (9.79) and Eq. (9.80):

$$
\alpha \approx
\begin{cases}
1 - L_B^2/(2L_{n,B}^2) \\
\qquad\qquad\qquad\qquad \text{if } L_B \gg 2p_{E,0}D_{p,E}L_{n,B}^2/(n_{B,0}D_{n,B}L_{p,E}) \\
1 - p_{E,0}D_{p,E}L_B/(n_{B,0}D_{n,B}L_{p,E}) \\
\qquad\qquad\qquad\qquad \text{if } L_B \ll 2p_{E,0}D_{p,E}L_{n,B}^2/(n_{B,0}D_{n,B}L_{p,E})
\end{cases}.
$$

Within the same limits, the current gain of Eq. (9.67) gives

$$
\beta \approx
\begin{cases}
2L_{n,B}^2/L_B^2 & \text{if } L_B \gg 2p_{E,0}D_{p,E}L_{n,B}^2/(n_{B,0}D_{n,B}L_{p,E}) \\
n_{B,0}D_{n,B}L_{p,E}/(p_{E,0}D_{p,E}L_B) & \text{if } L_B \ll 2p_{E,0}D_{p,E}L_{n,B}^2/(n_{B,0}D_{n,B}L_{p,E})
\end{cases}.
$$

Hence, under the conditions of inequalities (9.78), the current gain is much greater than unity.

According to the definition of Eq. (9.68) and the results of Eqs. (9.76) and (9.77) for the collector current, the transconductance of a BT is

$$
g_m = \frac{e}{k_B T} I_C. \tag{9.81}
$$

When Eq. (9.81) is used, it is possible to represent the collector-current modulation ΔI_C in terms of the change in the base–emitter bias Φ_{BE}:

$$
\frac{\Delta I_C}{I_C} = \frac{e\Delta\Phi_{BE}}{k_B T},
$$

that is, the exponential dependence of the collector current on the base–emitter voltage causes a large value of the transconductance of the BT.

There are three possible biasing configurations for a BT in a circuit. In Fig. 9.27, these configurations are presented: Fig. 9.27(a), the common-base configuration, in which the input terminals are the emitter and the base, while the output terminals are the collector and the base; Fig. 9.27(b), the common-emitter configuration, in which the input terminals are the base and the emitter, while the collector and the base terminals are the output terminals; and Fig. 9.27(c), the common-collector configuration, with the base and the collector as the input terminals and the emitter and the collector as the output terminals.

The full current-voltage characteristic of a BT depends on the biasing configuration. In Fig. 9.28(a), the set of output current-voltage characteristics, I_C–Φ_{CB}, is shown for the common-base configuration. The characteristics depend parametrically on the emitter bias (current). All three operational modes are possible. The active mode occurs for $\Phi_{BE} \geq 0$. The characteristic at $\Phi_{BE} = 0$ separates the region of the cutoff operation mode in the I_C–Φ_{CB} plane. At forward collector–emitter biases, $\Phi_{CB} \leq 0$, the saturation mode of operation occurs. Similarly, in Fig. 9.28(b), the output current-voltage characteristics of the common-emitter configuration are shown for all modes of operation. The cutoff-mode region corresponds to $\Phi_{BE} \leq 0$, when the emitter junction is no longer forward biased.

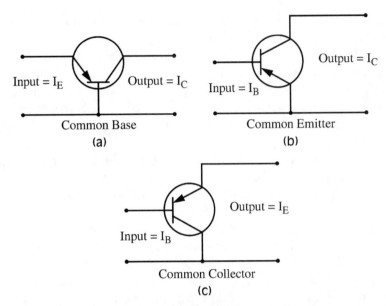

Figure 9.27. The three possible biasing configurations for a BT in a circuit: (a) common-base, (b) common-emitter, (c) common-collector configurations. The input and the output signals are indicated. In the transistor symbol all three terminals are marked differently.

The saturation mode occurs in the region of the I_C–Φ_{CE} plane restricted by the condition $\Phi_{BE} = \Phi_{CB}$, i.e., both junctions are forward biased.

Each of the configurations and modes of operation has its own set of applications. Consider the case of the common-emitter configuration; it can be used for the amplification of a small signal as well as for digital circuits. For example, a load line in the output circuit presented in Fig. 9.28(b) for the common-emitter configuration shows intersections of this line with the current-voltage

Figure 9.28. The collector-current-voltage characteristics for two biasing configurations for different input currents (the increasing number of the curve corresponds to the increase of the input current). (a) The common-base configuration: the current is a function of the collector-base voltage; (b) the common-emitter configuration: the current is a function of the collector-emitter voltage.

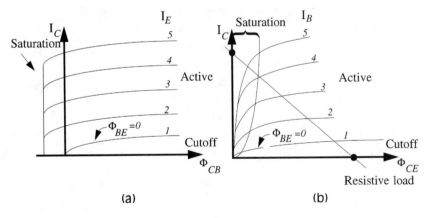

characteristic curves. If the input signal can vary the mode of operation from cutoff to saturation, we get an almost ideal switch. When the transistor is in the cutoff mode, no current passes through the load resistance; indeed, the voltage drops primarily across the transistor. If the transistor is switched to the saturation mode, its resistance is small, the output current is large, and the voltage drops principally in the load. Such an operation is possible in a quasi-stationary regime.

For high-speed operation, the physical processes in the transistor lead to a delay time and a frequency limitation. There are different reasons for signal delay: times for charging of capacitances of both junctions, the transit time for minority carriers through the base and the depletion regions, etc. The times of charging of the junction capacitances, τ_E and τ_C, can be defined as $\tau_E = R_E C_E$ and $\tau_C = R_C(C_C + C_S)$, where R_E and R_C are the resistances of the emitter and the collector regions, respectively, C_E and C_C are capacitances related to both junctions, respectively, and C_S is a collector-substrate parasitic capacitance. The base transit time $\tau_{\text{tr},B}$ is limited by electron diffusion through the base: $\tau_{\text{tr},B} \approx L_B^2/2D_{n,B}$. The collector depletion region l_d is the widest for a reverse-biased collector junction. Because of a high electric field in the depletion region, the transit time through this depletion region $\tau_{\text{tr},d}$ is limited by the saturation electron velocity v_s: $\tau_{\text{tr},d} = l_d/v_s$. The total delay time τ_{EC} can be represented by the sum

$$\tau_{EC} = \tau_E + \tau_C + \tau_{\text{tr},B} + \tau_{\text{tr},d}. \tag{9.82}$$

Similar to the case of FETs considered previously, the cutoff frequency is expressed through this time:

$$f_T = \frac{1}{2\pi \tau_{EC}}. \tag{9.83}$$

At high frequencies, the current gain is inversely proportional to frequency f_T.

Two other important figures of merit are the maximum power gain and the power-delay product. They were discussed for FETs in Subsection 9.2.2. The maximum available power gain in a BT is proportional to f^{-2}, as for FETs.

From our analysis, it follows that scaling down the basic dimensions of the device and variation of the doping in the BT can improve its performance. There are different limiting factors that were not discussed previously, and we now turn to them briefly as they relate to the performance of the BT. According to inequalities (9.78), the emitter region should be doped as high as possible. In fact, for high-doping levels in a semiconductor, there is the phenomenon of bandgap narrowing due to the Coulomb interaction of ionized impurities and free carriers. For example, for Si this bandgap narrowing is estimated to be ΔE_g (eV) $\approx 2 \times 10^{-11} \times N_{\text{imp}}^{1/2}$ (cm^{-3}). This bandgap narrowing leads to a great increase in the intrinsic carrier concentration n_i and, according to Eqs. (9.42) and (9.43), to the detrimental effect of increased hole injection into the emitter region. In turn, this injection of holes results in a decrease in the current gain.

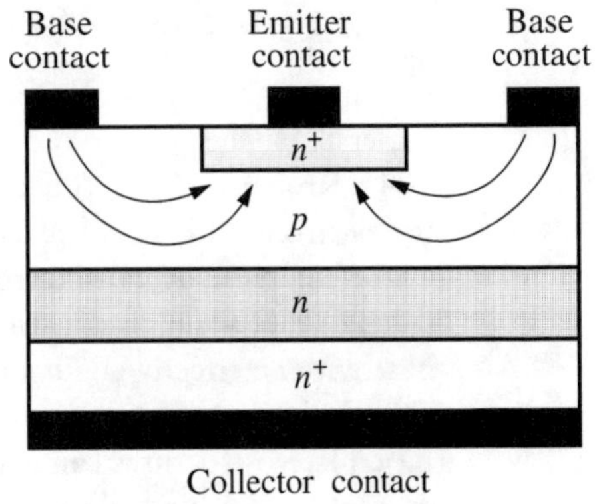

Base contact   Emitter contact   Base contact

n^+

p

n

n^+

Collector contact

Figure 9.29. The design schematic of a Si BT. The side contacts to the base lead to a lateral base current and a base-spreading resistance.

Another conclusion that follows from inequalities (9.78) is that the base width L_B, should be as small as possible. However, as we discussed previously, the base region is situated between two depletion regions. Thus an effective base width can be represented by

$$L_{B,\text{eff}} = L_B - l_{d,E} - l_{d,C},$$

where $l_{d,E}$ and $l_{d,C}$ are the widths of the emitter and the collector depletion regions in the base. A reverse collector–base bias increases $l_{d,C}$ and decreases the effective base width. At large biases, the two depletion regions can overlap and, as a result, $L_{B,\text{eff}}$ shrinks to zero. This effect is known as punch through. It leads to sharp increases in both the emitter and the collector currents and therefore to device breakdown. One of the methods for preventing the punch-through effect is the use of relatively small doping in the collector region, which causes an extension of the collector depletion region primarily toward the collector.

Another important limitation comes from the design of the BT that is shown schematically in Fig. 9.29, in which the emitter (n^+ region), base (p region), collector (n region), and contacts to them are presented. For such a design, the base current flows in the lateral direction and the so-called base-spreading resistance associated with such a current path becomes important. The voltage drops nonuniformly in the lateral direction along the base; this leads to a nonuniform voltage difference between the emitter and the base regions. As was discussed previously, the base current is an exponential function of Φ_{BE}. Thus the base-spreading resistance results in so-called emitter-current crowding at the edges of the emitter region and eliminates the conduction at the center of the emitter region. According to the previous discussion, the doping of the base region should be kept relatively low in order to prevent hole injection into the emitter. A low-doping level in the base region causes a large base-spreading resistance and degrades the parameters of the device. This and other constraints in scaling down the device can be overcome by combining heterojunctions and homojunctions in BTs.

9.4.2 Heterostructure Bipolar Transistors

The basic idea of a heterostructure BT (HBT) was to use a wide-bandgap material for the emitter and a narrow-bandgap material for the base (Shockley, 1949; Kroemer, 1957). A heterostructure can provide different potential barriers for electrons and holes because of the conduction- and the valence-band offsets at the interface. It can improve the control of the electron and hole injection through the p–n junction. Using heterostructures, one can also design a graded base with an effective built-in field for the minority electrons in the base region. This field drives the electrons toward the collector and decreases the transit time $\tau_{tr,B}$. Heterostructures can offer other possible improvements in the device. For example, a double heterostructure with a wide-bandgap collector operating in the saturation mode can eliminate the hole injection from the collector to the base under a forward bias of the collector junction.

Lattice-matched heterostructures, like AlGaAs/GaAs and InGaAs/InAlAs/InP, and pseudomorphic heterostructures, like Si/SiGe systems, are currently used in HBTs.

Wide-bandgap emitter: In Fig. 9.30(a), a schematic of the energy-band diagram is presented for an n–p–n HBT with a wide-bandgap emitter material and an abrupt heterojunction. This schematic is typical for a HBT with an abrupt emitter n-AlGaAs/p-GaAs heterojunction. The bandgap of the emitter is $E_{g,E}$, and in the base the bandgap is equal to $E_{g,B}$. The band offsets of conduction and valence bands are different and equal $V_{b,c}$ and $V_{b,v}$, respectively, where $\Delta E_g \equiv E_{g,E} - E_{g,B} = V_{b,c} + V_{b,v}$. The following analysis provides a way to generalize the results obtained previously for the homostructure transistor so that they can be applied to the heterostructure case. We assume that the depletion region at the p–n junction remains very narrow and minority carriers are injected through this region. According to Eqs. (9.71), (9.73), (9.76), and (9.77), one of the factors that determines the current is the set of equilibrium concentrations of the minority carriers, $n_{B,0}$, $p_{E,0}$, and $p_{C,0}$. Approximations (9.42) and (9.43) define these values by means of intrinsic concentrations in the materials; in the case in question, we deal with concentrations $n_{i,E}$, $n_{i,B}$, and $n_{i,C}$ in the emitter, base, and collector, respectively. According to Eq. (4.28), these characteristics are dependent on the bandgaps that, for the case of heterostructure devices, are assumed to be different for the emitter and the base. Thus we can rewrite Eq. (4.28) in the notation of this section as

$$n_{i,E}^2 = N_{c,E} N_{v,E} \exp\left(-\frac{E_{g,E}}{k_B T}\right), \qquad n_{i,B}^2 = N_{c,B} N_{v,B} \exp\left(-\frac{E_{g,B}}{k_B T}\right).$$

Here, $N_{c,E}$, $N_{c,B}$, $N_{v,E}$, and $N_{v,B}$ are the effective densities of states for electrons and holes in the emitter and the base regions. Using these equations, we can adapt all previous results for a HBT. For example, in the case of a narrow base, the current gain β can be represented by

$$\beta \approx \frac{N_{D,E}}{N_{A,B}} \frac{N_{c,B} N_{v,B}}{N_{c,E} N_{v,E}} \frac{D_{n,B} L_{p,E}}{D_{p,E} L_B} \exp\left(\frac{\Delta E_g}{k_B T}\right). \tag{9.84}$$

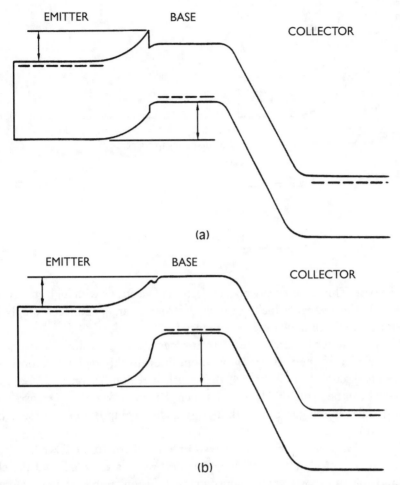

Figure 9.30. Two energy-band diagrams for HBTs with a wide-bandgap emitter. (a) HBT with abrupt emitter junction, (b) HBT with a graded-alloy composition at the emitter junction.

We see that the difference in the density of electron and hole states in wide- and narrow-bandgap materials may play a role, but the main factor comes from the difference in the bandgap ΔE_g. For example, as we already know from previous chapters, ΔE_g is typically greater than 250 meV, which is greater than 10 $k_B T$ at room temperatures. Thus the exponential factor provides an additional factor of $\sim 10^4$, and the current gain increases by an even larger factor. Having this additional factor for controlling the injection of minority carriers, one can choose an adequate high doping of the base region and considerably decrease the base-spreading resistance. At the same time, the doping of the emitter can be reduced to practical and realistic levels. In terms of approximation (9.84), this means that the exponential factor will compensate for an unfavorable $N_{D,E}/N_{A,B}$ ratio. As a result, the base resistance can be minimized even for an ultrathin base region.

Another interesting effect provided by the energy-band diagram of Fig. 9.30(a) is the appearance of a spike–notch region for electrons injected from the emitter

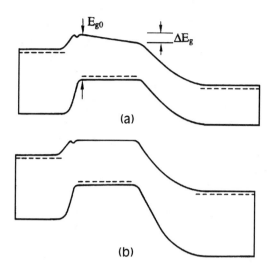

Figure 9.31. Energy-band diagrams for (a) HBT with a graded-base region, (b) a double HBT with a wide-bandgap collector.

into the base. This feature leads to an injection of electrons with high energy (hot electrons). The average velocity of such electrons is large, and, accordingly, the transit time through the base $\tau_{tr,B}$ is reduced.

The spike–notch feature of the electron-potential energy near the p–n junction can be avoided if one uses an alloy composition in this region. According to the results discussed in Chapter 4, the transient region between $Al_{x_0}Ga_{1-x_0}As$ and GaAs can be made of $Al_xGa_{1-x}As$ such that the value of x decreases from x_0 to 0. Such a configuration with a graded-alloy composition is presented in Fig. 9.30(b).

Graded base: In the previous analysis of a homostructure BT, we have used the fact that electric fields in the quasi-neutral regions are small and the electron and hole transports occur, mainly, in the diffusion regime. Heterostructures with a graded-alloy composition in the base can provide a gradual change in the energy bandgap across this region. It creates an effective built-in electric field for minority electrons, while an approximately flat valence band can be maintained. In Fig. 9.31(a), the energy-band diagram is presented for the case of a heterostructure transistor with a wide-bandgap emitter, a graded transient region between the emitter and the base, and a graded-base region. By using such a structure, one can decrease the transit time $\tau_{tr,B}$, in a controllable way; this results in a considerable improvement in the high-speed characteristics.

Wide-bandgap collector: BTs operating with a base–collector forward bias can be improved by use of a wide-bandgap collector design. For a double HBT, such a case is presented in Fig. 9.31(b). Both the emitter and the collector are made of wide-bandgap materials. This prevents the injection of holes from the base into the collector for a forward bias of the collector junction. Another advantage of a wide-bandgap collector is the suppression of the breakdown in the transistor. The combination of a wide-bandgap emitter and a wide-bandgap collector in the double HBT makes it possible to realize symmetric operations with the current, which leads to new circuit applications.

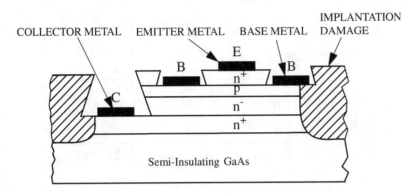

Figure 9.32. The schematic of the cross section of an AlGaAs/GaAs HBT.

AlGaAs/GaAs BTs: The excellent properties of AlGaAs/GaAs heterostructures and achievements of GaAs technology are used widely for the fabrication of HBTs based on this system. In Fig. 9.32, the cross section of such a device is presented. Semi-insulating GaAs is used as a substrate since it has a high resistivity. Upon the substrate, a relatively thick ($\approx 0.5\ \mu$m) and heavily doped (above $5 \times 10^{18}\ \text{cm}^{-3}$) GaAs layer is grown, which serves as a contact (subcollector) to the device. The collector is made by the n-GaAs layer with a width of $\sim 0.5\ \mu$m and a modest doping level of $\sim 3 \times 10^{16}\ \text{cm}^{-3}$. The base layer is very thin – typically 0.05 to 0.1 μm – and is made of GaAs or InGaAs. The use of a heterojunction facilitates doping this layer to extremely high levels in the range from 5×10^{18} to $10^{20}\ \text{cm}^{-3}$. The emitter is made of $\text{Al}_x\text{Ga}_{1-x}\text{As}$ with an Al fraction x, typically in the range from 0.2 to 0.3. The base–emitter junction is usually fabricated with a graded-alloy composition, as discussed previously. The emitter is n doped ($\sim 5 \times 10^{17}\ \text{cm}^{-3}$) and is 0.1–0.15 μm in width. A thin GaAs cap layer fabricated on the emitter layer has a high-doping level of $10^{19}\ \text{cm}^{-3}$. Metallic contacts are provided for all three terminals.

The ability to fabricate a very thin base opens the possibility of using a p-doped InGaAs strained layer as the base. The advantages of using InGaAs come from the large bandgap difference between the AlGaAs emitter and the InGaAs base as well as the higher electron mobility in the base. These HBTs show superb parameters; among them is a cutoff frequency that is greater than 100 GHz.

InGaAs/InAlAs and InGaAs/InP BTs: As we already know, the two alloys, $\text{In}_{0.53}\text{Ga}_{0.47}\text{As}$ and $\text{In}_{0.52}\text{Al}_{0.48}\text{As}$, are lattice matched with InP. The bandgaps for these three materials are 0.75, 1.5, and 1.35 eV, respectively. The conduction- and the valence-band offsets for these lattice-matched materials are shown in Fig. 9.33. The values of these parameters are such that it is possible to fabricate high-performance HBTs with a wide-bandgap emitter. In Fig. 9.34, the cross section of such a device is presented. The substrate is made of semi-insulating InP, the collector and base are fabricated of InGaAs, and the emitter is a heavily doped wide-bandgap InAlAs layer.

The base–emitter junction is usually graded, but an abrupt junction with a spike–notch barrier is also used. The injected electrons move across the base in

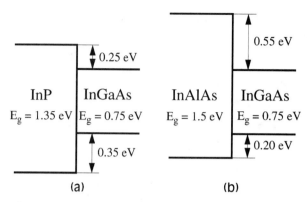

Figure 9.33. Conduction- and valence-band offsets for lattice-matched (a) InP/In$_{0.53}$Ga$_{0.47}$As, (b) In$_{0.52}$Al$_{0.48}$As/In$_{0.53}$Ga$_{0.47}$As heterojunctions.

the Γ minimum with the low effective mass ($m^*/m = 0.045$ in In$_{0.53}$Ga$_{0.47}$As). As we mentioned previously, this substantially reduces the base transit time. InGaAs HBTs demonstrate cutoff frequencies above 170 GHz, which are higher than those exhibited by GaAs-based devices.

The main applications of this group of BTs are high-speed digital and analog circuits as well as microwave power applications.

9.5 Si/SiGe Heterostructure Bipolar Transistors

The previous analyses have revealed some of the principal limitations in dimensional scaling of Si-based homostructure BTs to smaller sizes. Because of the extreme importance of this kind of BT, many efforts have been made to combine standard Si technology with heterostructure fabrication technology in order to overcome these limitations.

One such approach is to find a wide-bandgap material compatible with Si. Several candidate materials have properties desirable for such applications. Semi-insulating polycrystalline Si (SIPOS) has a bandgap of ~1.5 eV. High-current-gain transistors have been fabricated with SIPOS. A comparison of the current-gain parameters for SIPOS transistors and those of nominal Si BTs are presented

Figure 9.34. The device cross section and design of an InAlAs/InGaAs/InP HBT. The layers, substrate, contacts, and other details are indicated. [After S. M. Sze, ed., *High-Speed Semiconductor Devices* (Wiley, New York, 1990).]

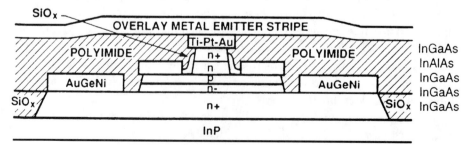

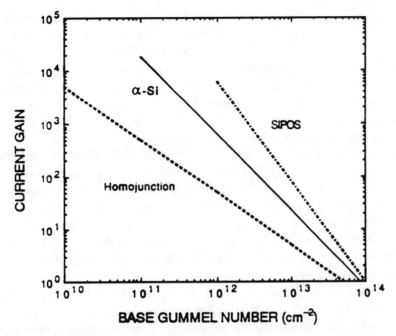

Figure 9.35. The current gain as a function of the base Gummel number for two types of Si-based BTs with wide-bandgap emitters. Results for devices with SIPOS emitters and amorphous Si emitters (α-Si) are presented. For comparison, the data for homostructure transistors are shown. [After S. M. Sze, ed., *High-Speed Semiconductor Devices* (Wiley, New York, 1990).]

in Fig. 9.35 as functions of the product of the base doping $N_{A,B}$ and the base width, L_B; this product is referred to as the base Gummel number: $G_B = N_{A,B}L_B$. The gain of SIPOS transistors is at least 1 order of magnitude higher than that of Si devices with the same Gummel number. But there are several disadvantages of this scheme; in particular, one important disadvantage is a high emitter resistance. The scheme is under further development. Another possibility is an amorphous Si emitter. In this case, the bandgap is even wider, $\sim$1.6 eV. Such transistors with a high current gain, as indicated in Fig. 9.35, have been fabricated; however, they still have a large series resistance. Microcrystalline Si, β-SiC, and GaP have also been used as high-bandgap emitters for Si-based BTs. All these developments are now in progress.

Another approach to improve BTs within the framework of Si technology is to use a narrow-bandgap material for the base region of the device. In Chapter 4 we found that the SiGe alloy has a reduced bandgap and therefore can be used for a narrow-bandgap base. The difficulty is that Si and SiGe alloys are lattice mismatched and SiGe layers can be grown only with relatively small thicknesses. At thicknesses less than the critical thickness, the SiGe alloy layer adopts the lattice periodicity of Si and can be grown quite perfectly without misfit defects, although it grows as a strained, pseudomorphic layer. This kind of Si-based heterostructure transistor has excellent characteristics and is analyzed in detail below.

Thus the main motivation for using a $Si_{1-x}Ge_x$ alloy as a partner for Si is that the bandgap may be reduced by increasing the Ge fraction. Figure 4.4 illustrates this effect. Since a $Si_{1-x}Ge_x$ alloy should be used as the base layer, i.e., it should be grown upon an unstrained Si collector, the bandgap dependence for this case is given by the curve corresponding to a strained layer of SiGe on unstrained Si. This curve shows that the bandgap gradually decreases with the Ge fraction and both effects – the alloy composition and the strain – contribute to decreasing the bandgap. Since serious limitations on the practically realizable Ge fraction come from associated decreases in the critical layer thicknesses, as illustrated by Fig. 4.9, only SiGe alloys with relatively small fractions of Ge can be used. It is necessary to take into account the effect of bandgap narrowing under high-doping concentrations, which was mentioned in Section 9.4. For example, if a strained $Si_{0.7}Ge_{0.3}$ layer is doped $\sim 7 \times 10^{18}$ cm^{-3}, the total bandgap discontinuity between this layer and lightly doped unstrained Si is ~ 0.27 eV, while the bandgap narrowing is ~ 0.05 eV. For a doping level of $\sim 10^{20}$ cm^{-3}, the latter effect is ~ 0.2 eV.

According to approximation (9.84), the current gain and, therefore, the collector current are both exponential functions of the factor $\Delta E_g / k_B T$. Direct measurements of the collector current exhibit this dependence for devices with base regions made of different SiGe alloys. In Fig. 9.36, the collector currents in the $Si/Si_{1-x}Ge_x$ heterostructure transistors are plotted as functions of the inverse temperature. They are normalized to the collector current of a Si BT to exclude the influence of other factors. The measurements are presented for devices with sufficiently thin $Si_{1-x}Ge_x$ layers and different Ge fractions. The

Figure 9.36. The normalized collector current versus the temperature for BTs with a SiGe base region. [After C. A. King, "Heterojunction bipolar transistors with $Si_{1-x}Ge_x$ alloys," in *Heterostructures and Quantum Devices*, N. G. Einspruch and W. R. Frensley, eds. (Academic, San Diego, 1994), pp. 157–187.]

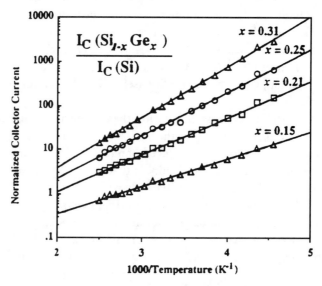

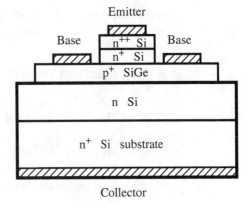

Figure 9.37. The schematic of the cross section of an n–p–n-Si/SiGe HBT. The p^+ SiGe base layer is strained.

results clearly indicate almost ideal exponential temperature dependence of the transistor characteristics. The increasing slopes of the curves correspond to the increasing bandgap offsets that results from an increased Ge fraction.

In Fig. 9.37, the cross section of a Si/SiGe double HBT is shown. The device is fabricated on an n^+-Si substrate (subcollector with doping of $\sim 2 \times 10^{19}$ cm^{-3}) contacted by an electrode. The collector layer is doped to approximately 2×10^{16} cm^{-3} and has a width of ~ 0.3 μm. The p^+ base is fabricated of a Si$_{1-x}$Ge$_x$ alloy, with x from 0.2 to 0.3, and has a width of typically 300–500 Å. The base doping is in the range of 2×10^{18} to 5×10^{19} cm^{-3}. Two-sided metallic electrodes provide the direct contact with the base region. The emitter is made of n^+ Si with doping of $\sim 5 \times 10^{17}$ cm^{-3}. At the top of the emitter, a heavily doped n^{++}-Si layer is placed to provide a good contact to the metallic electrode. The same structure has been used with a thicker base made of a Si/Si$_{0.5}$Ge$_{0.5}$ superlattice. Current gains from 10 to 20 and cutoff frequencies up to 30 GHz have been achieved in these devices. Note that a similar device of the p–n–p type has been fabricated based on these principles and the same range of cutoff frequencies was obtained.

Further improvement of this device may be obtained with two novel structures: a polysilicon wide-bandgap emitter may be used to increase the injection efficiency, and a graded bandgap in the base region may be used to introduce a built-in effective field to speed the minority electron motion across the base.

An extremely high cutoff frequency of 75 GHz at room temperature is measured for these devices with the Si/Ge graded base. At nitrogen temperature, a cutoff frequency as high as 94 GHz has been reported for this type of BT. Similar high-frequency results have been obtained for a p–n–p device with a compositionally graded n-SiGe base: $f_T = 55$ GHz. Modeling of these devices shows that optimizing the structural parameters makes it possible to reach very high current gains of up to 10,000. Calculations indicate an upper limit for the cutoff frequency in the range 100–130 GHz, which is approximately half of the high-frequency limit predicted for optimized InGaAs devices.

Thus both ideas – the wide-bandgap emitter and the graded-base region – are used in Si-based BTs. They facilitate significant improvements in device parameters, particularly in the high-speed operation. The further development of these transistors is in progress.

9.6 Hot-Electron Transistors

In classifying the electronic devices in Section 9.1, we have assigned hot-electron devices to the class of potential-effect transistors. In fact, hot-electron and potential-effect transistors are different in some aspects and they should be considered separately. In this section we study two groups of devices that rely primarily on the hot-electron effect: ballistic-injection devices and RST devices.

In Chapter 7, we studied the hot-electron effect that takes place in high electric fields when the electric power input of the electron system exceeds the rate of energy losses. This results in a nonequilibrium distribution function and field-dependent parameters of the electron gas. In general, hot-electron effects are present in any device that uses high electric fields. Furthermore, these effects also offer new principles of device operation. One such case – the Gunn diode – has been briefly studied in Subsection 7.3.4. Heterostructure technology opens a way to the use of nonequilibrium electrons. We start by considering ballistic-injection devices.

9.6.1 Ballistic-Injection Devices

A ballistic-injection device is primarily a unipolar device, that is, only one type of carrier, say electrons, is used. As for a BT, a ballistic device consists of the emitter, base, and collector. The role of the emitter is to inject electrons with high velocities into the base; the second electrode should collect these electrons. The input base voltage controls the electron injection and therefore the output emitter–collector current. If small changes in the input produce larger changes in the output, the device exhibits a current gain. Generally, in ballistic devices, the electrons are injected into the base with a high energy exceeding 0.1 eV. This should lead to a decrease in the time of flight through the base region. Another advantage of ballistic devices is related to their unipolar character; this means that it is possible to choose the fastest type of majority carriers (electrons) and avoid the participation of slower minority holes. Note that, in certain cases, both types of carriers can be used in ballistic-injection devices.

To develop a transistor that is faster than a BT or a FET, several schemes of ballistic-injection devices have been proposed. They differ in physical mechanisms of the electron injection as well as in the materials used in the devices. The first and perhaps simplest device is a metal-oxide-metal-oxide-metal heterostructure. In Fig. 9.38(a), this device is shown under a bias. Other similar device structures are the metal-oxide-metal-semiconductor structure and the semiconductor-metal-semiconductor (SMS) structure presented in Figs. 9.38(b) and 9.38(c), respectively. The principle of operation is the same for these three structures. Consider, for example, the case of a SMS structure. One can see a close analogy with the BT: A forward-biased semiconductor-metal junction serves as the emitter, a second metal-semiconductor junction serves as the collector, and a metal layer is the base. Both junctions are, in fact, Schottky diodes; one is forward biased and the other is reverse biased. Under such bias conditions, electrons are injected over the Schottky barrier with energies substantially exceeding the thermal energy in the base. If the base is narrow, the electrons fly across the base region without losing their energy. Their further destiny is controlled by the base–collector bias: a lowering in the collector barrier increases the fraction of electrons α coming into the collector electrode and consequently the collector current. According to the definition of Eq. (9.64), the fraction of electrons coming into the collector is the current-transfer ratio. If α is close to unity, the base–collector current gain can be large according to Eq. (9.66). Different materials were used for the structures shown in Fig. 9.38. In particular, the Si/CoSi$_2$/Si structure was investigated for the SMS scheme. An α factor as high as 0.6 was reported for this case. A common disadvantage of metal-base transistors is the high electron reflection from the metal. This reflection is mainly a quantum-mechanical effect in nature and therefore cannot be avoided. It occurs even for an ideal semiconductor-metal interface.

Semiconductor heterostructures may be used in other ways to realize ballistic-injection devices. Let us consider n-type devices. In semiconductor-based ballistic-injection devices, the emitter, base, and collector are doped regions separated by two barriers. The barriers can be formed by growing layers of materials with positive conduction-band offset, as shown in Fig. 9.39(a). This structure actually uses four heterojunctions. Another kind of barrier can be produced by planar acceptor doping in a homostructure, as illustrated in Fig. 9.39(b).

Consider typical parameters of structures for ballistic-injection devices. If the height of the emitter barrier is V_b, the velocity of electrons injected into the base can be estimated as

$$v_B \approx \sqrt{2V_b/m^*} = 5.9 \times 10^7 \sqrt{\tilde{V}_d/(m^*/m)} \ \text{cm/s},$$

where $\tilde{V}_d$ is the barrier energy in units of electron volts. For GaAs we can assume that $\tilde{V}_d \approx 0.3$ eV and $m^* = 0.067$ m. For the velocity we get $v_B = 1.3 \times 10^8$ cm/s. This value is appreciably larger than the characteristic electron velocity in devices such as FETs and BTs. Another important feature is that the injected electrons

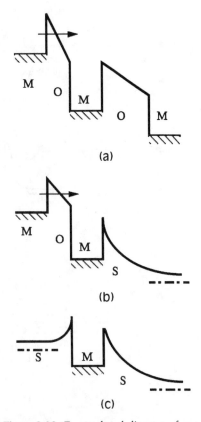

Figure 9.38. Energy-band diagrams for metal-base ballistic-injection transistors. Metal, oxide, and semiconductor layers are marked by M, O, and S, respectively. Semiconductor layers are doped. The diagrams are presented under operational bias conditions.

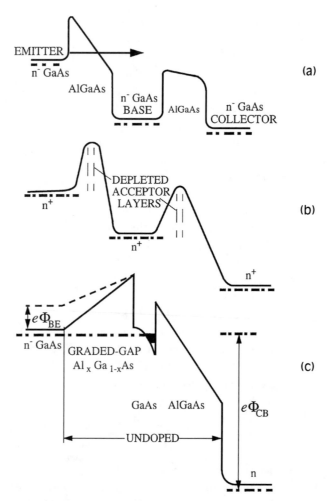

Figure 9.39. Energy-band diagrams for semiconductor ballistic transistors: (a) The AlGaAs/-GaAs structure with four heterojunctions and a doped base. Electrons are injected into the base due to tunneling. This kind of device is called the tunneling-hot-electron-transfer amplifier. (b) A homostructure electron device. Barriers are formed by a planar p doping. The base is n doped. (c) A device with an undoped base. [After S. M. Sze, ed., *High-Speed Semiconductor Devices* (Wiley, New York, 1990).]

exhibit velocity spreading in a very narrow velocity cone. Actually, the average value of the lateral component of electron momentum in the emitter is $p_{\parallel,E} \sim \sqrt{2m^* k_B T}$, where T is the device temperature. Because of the lateral translation symmetry this component does not change under electron injection through the barrier. Thus the characteristic angle for the velocity spreading can be evaluated as follows:

$$\theta \sim \frac{v_{\parallel,B}}{v_B} = \frac{p_{\parallel,B}}{m^* v_B} = \frac{p_{\parallel,E}}{m^* v_B} = \sqrt{\frac{k_B T}{V_b}}.$$

For example, in the case of helium temperature and $V_b \approx 0.3$ eV, this equation gives $\theta \approx 6°$.

Although the velocity of electrons coming into the base increases with the barrier height, the real energy-band structure restricts the optimal barrier height to a value below that of the energy of upper valleys in the conduction band. Actually, electron injection with higher energy would lead to intervalley transfer from the Γ valley with a small effective mass to the higher valleys with higher effective masses and lower velocities; that is, such a scattering process leads to degradation of the transistor performance parameters. For GaAs, the L valleys have energies of $\sim$0.3 eV, as indicated in Fig. 4.6. Thus the emitter-barrier height should not exceed this value. Another limitation can be related to the threshold for impact ionization.

Collisions in the base reduce the number of ballistic electrons. If the electron mean free path in the base is l_e and the base width is L_B, the fraction of ballistic electrons collected by the collector is

$$\alpha \approx \exp(-L_B/l_e).$$

Thus the base region should be quite narrow. In this case, the base region has to be heavily doped to reduce the lateral base resistance. This problem is similar to the base-spreading resistance effect discussed previously for BTs. Such ballistic devices are referred to as devices with a doped base.

A limitation of the base-doping technique comes about as a result of the fact that high doping gives rise to additional electron scattering and quenches the ballistic regime of the electron motion. In particular, for III–V compounds, if impurity concentrations exceed 10^{18} cm^{-3}, scattering by plasmons becomes very strong compared to scattering by phonons. As a result, for AlGaAs/GaAs devices, base regions with doping of $\sim 10^{18}$ cm^{-3} and widths of 300–800 Å are used. Optimization of the structure parameters facilitates realizing a ballistic device with a transfer ratio $\alpha = 0.9$ and a current gain $\beta \approx 9$ at helium temperature.

As we discussed previously, the energy separation between the Γ and the upper L and X valleys in InGaAs is larger than that of GaAs. This means that it is possible to use higher barriers in the transistor and to achieve better performance. An example is the GaAs/AlGaAs/InGaAs ballistic device. The emitter and the collector are made of GaAs and the barriers are of AlGaAs. The base region is made of a pseudomorphic In$_{0.12}$Ga$_{0.88}$As layer with a width of 200 Å. In such a base layer, the $\Gamma - L$ energy separation is $\sim$0.38 eV. For this device the following gains are reached: $\beta \approx 40$ at 4 K and $\beta \approx 27$ at 77 K.

Figure 9.39(c) presents another design for a ballistic device. The emitter barrier is made with a graded composition and the base region is undoped. The base is induced by the collector electric field, which leads to the formation of a two-dimensional electron gas at the undoped interface. This type of ballistic device is referred to as the induced-base transistor. The advantages of the induced-base transistor are the following. The two-dimensional electron gas in the base is characterized by a high mobility and large two-dimensional concentrations of up to 2×10^{12} cm^{-2}. Both effects cause a low base resistance; thus, the base can be chosen to be very narrow, typically 100 Å. This results in a high fraction of the ballistic electrons coming into the collector. For example, a well-designed AlGaAs/GaAs induced-base transistor results in $\alpha \approx 0.96$, even at room temperature. Similar induced-base transistors are fabricated with InGaAs/AlGaAs and p-doped Ge/SiGe heterostructures.

As we already pointed out, hot-electron transistors are characterized by a high operation speed. The electron transit time of the order of 0.1–0.5 ps corresponds to a cutoff frequency in a range from 2 to 10 THz.

9.6.2 Real-Space Transfer Devices

In Chapter 7, we discussed the effect of RST of hot electrons. Although the RST of a charge in an electric field occurs for a number of physical situations, currently the term RST is applied to the following case. Under equilibrium conditions, the carriers are assumed to be in a potential well (a quantum well, mostly) and to have a high mobility. If an electric field increases the carrier energy, some of the carriers transfer from the well to the ambient material, which has a different value of mobility. Thus the RST should result in a nontrivial spatial-carrier distribution and a nonlinear current-voltage characteristic. If the mobility of the ambient material is less than that of the well, the RST effect may lead to a negative differential conductivity. This phenomenon was the basis of the principal original idea: to create artificially controlled negative differential resistance (NDR), while for other cases – such as the Gunn effect – the negative differential conductivity is determined by the material properties. In fact, the idea of the RST effect is much more productive and has already led to the development of several different devices known as the RST transistors. Among them, the charge-injection transistor (CHINT), the negative-resistance FET (NERFET), the hot-electron memory, and the light-emitting RST transistor were experimentally demonstrated. They exhibited excellent characteristics and offered principally new operation regimes.

We now consider the devices presented in Fig. 9.40. The heterostructure of Fig. 9.40(a) is similar to a HFET structure: There are source and drain contacts, the n-AlGaAs/GaAs junction produces a two-dimensional electron gas at the interface, and the wide-bandgap electrode is, partly, used to control the lateral electric-field distribution. New elements in the device are an undoped and graded AlGaAs barrier under the GaAs layer and a conducting GaAs layer (substrate) contacted by the third electrode (bottom collector). The dimensions of all elements of the device are presented in Fig. 9.40(a). The energy-band diagram shows that the GaAs quantum well localizes the electrons, while the graded barrier creates the effective electric field that sweeps the electrons injected into this barrier region toward the substrate contact. Figure 9.40(b) depicts a similar three-terminal device. This is a simplified design of the device from Fig. 9.40(a).

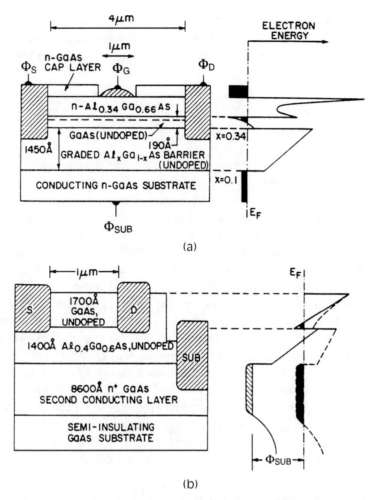

Figure 9.40. Cross-sections and energy-band diagrams of two CHINT/NERFET devices: (a) The AlGaAs/GaAs symmetric CHINT structure. The upper *n*-doped AlGaAs layer supplies electrons in the GaAs quantum well. (b) The undoped (asymmetric) CHINT structure. The electron channel is induced by a positive substrate voltage. [After S. M. Sze, ed., *High-Speed Semiconductor Devices* (Wiley, New York, 1990).]

The device is fabricated upon a semi-insulating GaAs substrate. This substrate is followed by the conducting GaAs layer (substrate layer) that has a side contact. Only one wide-bandgap undoped AlGaAs layer is grown upon the conducting layer and used as the barrier. The upper layer is undoped GaAs. Such a sequence of GaAs and AlGaAs layers is similar to the inverted heterostructures discussed in Section 9.2. A two-dimensional electron channel is induced at the interface when a positive bias exceeding a threshold value is applied to the substrate contact. The channel is supplied by the source and the drain electrodes.

Both devices of Fig. 9.40 operate similarly. A voltage is applied between the source and the drain to create an electric field that has to heat up the electrons.

The hot electrons are injected over the barrier. Then they move in the effective field toward the conducting substrate layer and are collected by that layer. The substrate current is limited by the injection rate and can be evaluated as a thermionic current

$$I_{\text{SUB}} = en_s \sqrt{\frac{k_B T_e}{2\pi m^*}} \frac{S}{L} \exp\left(-\frac{V_b}{k_B T_e}\right), \tag{9.85}$$

where n_s is the electron sheet concentration, L is the channel width, V_b is the energy height of the barrier separating the channel and the barrier region. T_e is the temperature of hot electrons in the channel; it depends on the source–drain voltage Φ_{SD}. Let the drain be grounded and the voltage Φ_{SUB} be applied to the substrate layer. The substrate current I_{SUB} is a complex function of both voltages, Φ_{SD} and Φ_{SUB}, because the electron concentration n_s, the barrier energy V_b, and the electron temperature T_e, depend on both these voltages. In Fig. 9.41, the substrate current versus source–drain voltage dependences are shown for a series of substrate voltages. The symmetry of the dependences with respect to a change in the drain–source polarity reflects the lateral symmetry of the device. One can see an extremely sharp enhancement of I_{SUB} with Φ_{SD}: the current varies from 4 to 6 orders of magnitude, when the voltage Φ_{SD} changes in the range of 0.1–0.8 V. This proves that hot electrons make a dominant contribution to the substrate current. Using Eq. (9.85) and some reasonable simplifications, one can determine the electron temperature in the two-dimensional channel. The electron temperature can be significantly greater than the lattice temperature, which equals 77 K. Thus this device operates like a vacuum diode with a controlled cathode temperature. Like any other three-terminal devices, a CHINT can be

Figure 9.41. The substrate current of a CHINT with a symmetric structure as a function of the source–drain voltage. Data are presented for different substrate voltages for $T = 77$ K. [After S. M. Sze, ed., *High-Speed Semiconductor Devices* (Wiley, New York, 1990).]

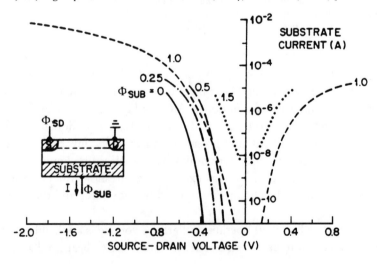

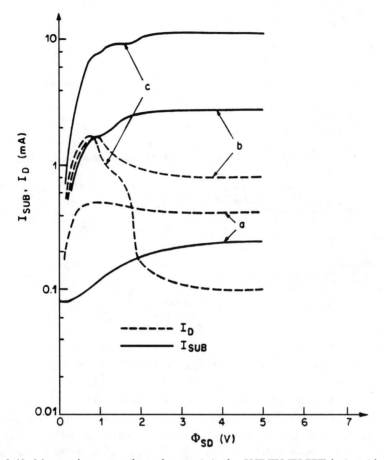

Figure 9.42. Measured current-voltage characteristics for CHINT/NERFET device with threshold voltage $\Phi_{th} = 5$ V. Curves marked by a, b, and c correspond to $\Phi_{SUB} - \Phi_{th} = 1, 2$, and 2.7 V. All substrate currents show a saturation, while drain currents have regions with a NDR. [After S. M. Sze, ed., *High-Speed Semiconductor Devices* (Wiley, New York, 1990).]

characterized by the transconductance

$$g_m = \left. \frac{\partial I_{SUB}}{\partial \Phi_{SD}} \right|_{\Phi_{SUB}}. \tag{9.86}$$

The sharp enhancement of I_{SUB} with Φ_{SD} results in a large transconductance.

The RST device presented in Fig. 9.40(b) has a threshold substrate voltage Φ_{th} that turns on the electron channel, the substrate current, and the RST effect. Generally this device does not exhibit the source–drain symmetry. In Fig. 9.42, both the substrate and the drain currents are shown as functions of the source–drain voltage for a particular device with $\Phi_{th} = 5$ V. These results are measured for three values of Φ_{SUB} at room temperature. The substrate and the drain currents behave differently. The substrate current saturates at large Φ_{SD}, while the drain current exhibits a drop and a region of NDR. These features can be explained as follows. The electron temperature increases as the source–drain voltage grows.

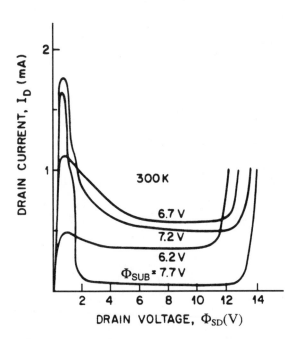

Figure 9.43. The NERFET source–drain current-voltage characteristics. Data were measured at room temperature. Values of substrate voltages are shown at the curves. [After S. M. Sze, ed., *High-Speed Semiconductor Devices* (Wiley, New York, 1990).]

This leads to an increased injection of electrons from the quantum well into the barrier layer. A portion of the injected electrons is collected by the substrate electrode. The appearance of a negative electric charge in the barrier region causes a change in the potential distribution such that there is a decrease in the electron concentration in the channel and a reduction in the conductivity of the channel. As a result, the drain current decreases with increasing source–drain voltage. The larger the Φ_{SUB}, the more pronounced these effects are, i.e., there is an enhanced saturation of I_{SUB} and an enhancement of the NDR of the source–drain current.

The large NDR in the electron-channel circuit controlled by the substrate electrode facilitates the use of this device as a NERFET. Figure 9.43 illustrates the source–drain current-voltage characteristics of a NERFET with a structure similar to that of Fig. 9.40(b) and with the following parameters: The 1-μm-thick, n^+-GaAs conducting substrate layer is grown upon semi-insulating GaAs, the 2000-Å barrier layer is made of undoped $Al_{0.45}Ga_{0.55}As$, and the 2000-Å channel layer is an undoped GaAs. The currents are measured at room temperature. The large range of the source–drain voltage makes it possible to realize a quasi-saturation regime in the source–drain current. A remarkable feature of this device in the saturation regime is the negative transconductance defined by Eq. (9.86). The peak-to-valley ratio for the device has been reported to be as high as 160 at room temperature. The device can be used as a microwave oscillator or as an amplifier for which efficient control of the oscillations can be achieved by use of a third substrate electrode. A NERFET can be used also in other numerous circuit applications.

Intrinsic limitations of the speed of CHINT/NERFET devices originate from the main physical processes involved: a finite time of electron heating in the channel and the charge-transfer time. For a GaAs electron channel, the first time can be estimated to be ~1 ps or even less. This time is required for increasing the electron temperature up to 1500 K. The charge-transfer time consists of two components: the time needed for electrons to overcome the barrier and the drift time to the substrate. The first time is longer and can be evaluated as 5–10 ps for a typical GaAs/AlGaAs junction. The electron-drift time in a strong electric field or a built-in effective electric field across a 2000-Å-thick region in the direction of the substrate is approximately several picoseconds. Modeling shows that the stationary state in this type of device is achieved at approximately 6–10 ps. Microwave operations with a current gain $\alpha > 1$ at frequencies up to 32 GHz have been achieved at room temperature. Circuit applications have been made in the gigahertz frequency range.

9.7 **Applications of the Resonant-Tunneling Effect**

The effect of resonant tunneling in heterostructures and the associated strongly nonlinear current-voltage characteristics and NDR are the bases for various novel electron devices. Superb device parameters and multifunctional applications are predicted for these devices in view of the small transport time of the order of 100 fs, the low intrinsic parasitic and specific capacitances, and the large current densities of 10^5 A/cm^2 or more.

In this section we consider high-frequency and circuit applications of the resonant-tunneling effect. The first group of applications includes oscillators and high-frequency multipliers. The second exploits the resonant-tunneling effect for improving transistors and developing new devices for multiple-valued logic and memory. Use of the effect in unipolar lasers is considered in Subsection 12.3.9.

9.7.1 Resonant-Tunneling Oscillators

The application of the resonant-tunneling effect in high-frequency oscillators is based on the existence of the NDR studied in Chapter 8. We have seen that a resonant-tunneling current with a well-developed NDR and a large peak-to-valley ratio can be achieved for various heterostructures over a wide temperature range.

To review the principle of using NDR for obtaining electric oscillations, we consider the simplest electric circuit containing a resistance R, a capacitance C, and an inductance L; see Fig. 9.44. Let us introduce the resistance R as the ratio

$$R = \frac{\Delta \Phi}{\Delta I},$$

where ΔI is a change in the current through the resistance when the voltage drop is changed by $\Delta \Phi$.
Thus, in fact, R is the differential resistance that can
have both positive and negative signs. If the alter-
nating current $\tilde{I}$ and voltage $\tilde{\Phi}$ applied to the circuit
are

$$\tilde{I} = I_0 e^{-i\omega t}, \qquad \tilde{\Phi} = \Phi_0 e^{-i\omega t},$$

Figure 9.44. A resonant electric circuit.

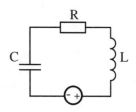

one has the following relation between the magnitudes I_0 and Φ_0:

$$I_0 = \frac{1}{Z(\omega)}\Phi_0, \qquad Z(\omega) = R - i\left(\omega L - \frac{1}{\omega C}\right), \tag{9.87}$$

where $Z(\omega)$ is the impedance of the circuit.

Even if $\Phi_0 = 0$, oscillations can exist in the circuit at frequencies for which $Z(\omega_c) = 0$:

$$\omega_c = -i\frac{R}{2L} \pm \sqrt{\frac{1}{LC} - \left(\frac{R}{2L}\right)^2}. \tag{9.88}$$

For $(L/C) > (R/2)^2$ and for a positive resistance R, there are oscillations with the frequency $\sqrt{1/(LC) - (R/2L)^2}$ and damping characterized by $\gamma = -\text{Im}\{\omega\} = R/2L$. Here, γ is positive and $\text{Im}\{\omega\}$ denotes the imaginary part of ω.

If the real part of the impedance is negative,

$$\text{Re}\{Z(\omega)\} = R < 0, \tag{9.89}$$

then γ is negative and the steady state of the circuit is unstable with respect to the generation of voltage oscillations.

The above results have been obtained for a resonant RCL electric circuit. In the high-frequency region, resonant oscillations can occur in an electromagnetic cavity. One such cavity, a waveguide oscillator, is shown in Fig. 9.45. The geometry of the cavity and its dimensions determine the resonant electromagnetic waves and their resonant frequencies. Let us place the resonant-tunneling diode (RTD) inside the cavity. A dc voltage can be applied to the diode by means of a pin and a whisker, as shown in Fig. 9.45. The whisker also couples the electromagnetic field of the cavity to the diode. This coupling can be optimized by a backshort that changes the longitudinal dimension of the cavity.

The principal approach for describing the amplification and generation of electromagnetic waves in the cavity should be based on Maxwell's equations. The density of the electric current flowing through the diode provides the interaction

Figure 9.45. The cross section of a standard waveguide resonator with a resonant-tunneling diode. The diode chip, whisker, dc bias pin, and backshort are shown.

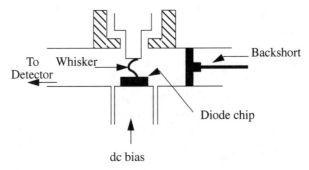

between the electromagnetic waves and the diode. Since this principal approach is complex, we use another approach that is both instructive and not as complex.

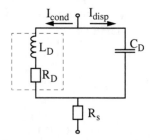

From the theory of electromagnetic fields, it is known that a system consisting of a cavity and an active element can be described by an equivalent circuit. To use this approach we have to find the impedance of the RTD. The total current flowing

Figure 9.46. The equivalent circuit of a RTD.

through the diode consists of two parts: a displacement current I_{disp} and a conduction current I_{cond}, as shown in Fig. 9.46. Let the area of the layers comprising the diode be S_D and the distance between the electrodes be d_D. Then the capacitance of the diode is $C_D = S_D \kappa / 4\pi d_D$. The displacement current can be evaluated as

$$I_{disp} = -\Phi(\omega) i\omega C_D.$$

To find the parameters of the ac conduction current, we consider the conduction current $I_{cond}(t)$ in response to a voltage $\Delta\Phi$ that is turned on suddenly. We assume that the electric charge tunneling through the diode is characterized by a single time constant τ, which is related to the finite lifetime of the quasi-bound state in the quantum well between the barriers. This time determines how the system approaches the steady state. Thus the conduction current as a function of the time is assumed to be

$$I_{cond}(t) = I_1 \Theta(-t) + [I_2 + (I_1 - I_2)e^{-t/\tau}]\Theta(t),$$

where $\Theta(t)$ is the Heaviside step function. Using the Fourier transform of this impulse response, we obtain the following equation for the impedance of the diode:

$$\frac{1}{Z_{cond}(\omega)} = \frac{1}{\Delta\Phi} \int_{-\infty}^{+\infty} dt \, e^{i\omega t} \frac{dI_{cond}}{dt}(t) = \frac{1}{R_D - i\omega L_D},$$

where $R_D = \Delta\Phi/(I_2 - I_1) = \Delta\Phi/\Delta I$ is the differential resistance of the diode; additionally, the value $L_D \equiv R_D \tau$ can be associated with an effective induction of the diode. Therefore we can write the conduction current as

$$I_{cond}(\omega) = \Phi(\omega)/Z_{cond}(\omega)$$

and conclude that the finite lifetime of the quasi-bound resonant state causes an additional induction – the quantum-well induction – in the equivalent circuit. The series resistance R_s presented in Fig. 9.46 is caused by a parasitic resistance of the contact regions and the lead wires. Note that the inductance L_D is negative in the region of NDR where oscillations can occur. Thus, in accordance with Eq. (9.88), it cannot be used to form a resonant circuit with capacitance C_D.

The impedance of the equivalent circuit presented in Fig. 9.46 can be calculated as

$$Z_D = \frac{R_D - i\left(\omega L_D - \omega C_D R_D^2 - \omega^3 C_D L_D^2\right)}{(1 - \omega^2 L_D C_D)^2 + \omega^2 C_D^2 R_D^2} + R_s. \tag{9.90}$$

The result of Eq. (9.90) demonstrates that Z_D depends strongly on the frequency. At low frequencies we get

$$Z_D \approx \text{Re}\{Z_D\} \approx R_D + R_s. \tag{9.91}$$

The impedance is negative for $R_D < -R_S$. At high frequencies, instead of approximation (9.91), we have

$$\text{Re}\{Z_D\} = \frac{R_D}{\omega^4 L_D^2 C_D^2} + R_s. \tag{9.92}$$

Equation (9.92) tells us that at high frequencies the first term on the right-hand side of Eq. (9.92) tends to zero; therefore for $\text{Re}\{Z_D\}$ it undoubtedly become positive. Thus there is always the highest frequency ω_{cf} at which the real part of the diode impedance is equal to zero, that is, $\text{Re}\{Z_D\} < 0$ for $\omega < \omega_{\text{cf}}$; both amplification and generation of electric oscillations are allowed.

The simplest equivalent circuit for a diode placed inside a cavity is presented in Fig. 9.47. The values C_c, L_c, and R_c are the capacitance, induction, and resistance, respectively, that characterize the cavity. Comparing the equivalent circuits of Figs. 9.44 and 9.47, one can see that, in accordance with Eq. (9.89), electric oscillations occur if

$$\text{Re}\{Z_D(\omega)\} + R_c < 0. \tag{9.93}$$

This inequality is strongly dependent on frequency ω. At low ω, from relations (9.91) and (9.93) we get the following condition for oscillations:

$$R_D + R_s + R_c < 0, \tag{9.94}$$

i.e., oscillations are possible if the diode is biased to a region with NDR and the absolute value of the differential resistance is sufficiently large. If we define the conductance $G = -R_D^{-1}$, we can rewrite inequality (9.94) as

$$G(R_s + R_c) < 1. \tag{9.95}$$

Combining Eq. (9.90) and inequality (9.93), one can find the cutoff (intrinsic) frequency of the diode oscillations:

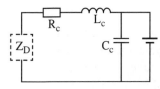

Figure 9.47. The equivalent circuit of the resonator containing the diode with impedance Z_D.

$$\omega_{cf}(RCL) = \left(\frac{1}{|L_D|C_D}\left\{\left(1 + \frac{C_D}{2|L_D|G^2}\right)\right.\right.$$

$$\left.\left. \times \sqrt{1 + \frac{1 - G(R_s + R_c)}{G(R_s + R_c)[1 + C_D/(2|L_D|G^2)]}} - 1\right\}\right)^{1/2}. \qquad (9.96)$$

In the limit of a short tunneling time, $\tau \ll |R_D|C_D$, Eq. (9.96) takes the form

$$\omega_{cf}(RC) = \frac{1}{C_D}\left(\frac{G}{R_s + R_c} - G^2\right)^{1/2}, \qquad (9.97)$$

i.e., if the effective induction of the diode is negligible, the cutoff frequency $\omega_{cf}(RC)$ is substantially influenced by the parasitic resistance of the diode and the equivalent-circuit resistance. Here we note that Eq. (9.96) takes into account the intrinsic resistance, capacitance, and induction of the RTD, while in Eq. (9.97), the induction is neglected.

In fact, the frequency of oscillations of a diode placed in a cavity depends also on the cavity parameters: the capacitance C_c, induction L_c, and resistance R_c. If the cavity has a resonant frequency of less than the limit given by Eq. (9.96), the RTD amplifies electromagnetic waves or generates microwave oscillations.

Small values of the transit time, the intrinsic parasitic resistance, and the capacitance facilitate operation in a high-frequency range up to hundreds of gigahertz. In Fig. 9.48, the power of the generated microwave emission is depicted as a function of frequency $f = \omega/(2\pi)$ for two double-barrier diodes. Diodes differ in their sets of parameters R_D, C_D, and L_D. The cutoff frequencies corresponding to Eqs. (9.96) and (9.97) are marked by f_{RCL} and f_{RC} and are presented for both diodes. These results demonstrate the importance of the quantum-well inductance: a finite lifetime of the quasi-bound level in the well decreases the cutoff frequency for both devices substantially. Lower cutoff frequencies are obtained from double-barrier heterostructures with thicker barriers, which result in a larger tunneling time τ. Figure 9.48 presents data on the measured power of microwave emissions. They can be compared with calculations based on the previously discussed equivalent-circuit approach. In particular, the experiments confirm that the cutoff frequencies are predicted accurately by Eq. (9.96). Figure 9.48 shows that oscillation frequencies up to 400 GHz are reached. For frequencies in the range above 100 GHz, the emitted power is of the order of 1–10 μW.

The lifetime of the quasi-bound level is crucially important for high-frequency operation of a RTD. It depends on heterostructure parameters and material combinations. In Fig. 9.49, the lifetimes of the quasi-bound states are depicted for five particular double-barrier structures. The structures are assumed to be made of GaAs/AlGaAs, InGaAs/AlAs, InGaAs/GaAs, and InAs/AlSb. The widths of the quantum wells are equal to 46 Å, and the barriers have thicknesses ranging from 10 to 60 Å. The heights of the barriers vary from 0.3 to 1.2 eV and are marked

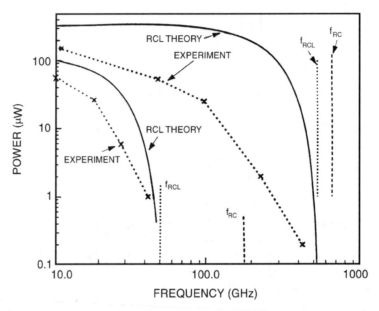

Figure 9.48. The microwave output power as a function of the frequency for two different double-barrier structures. Experimental data and calculations are presented. The theoretical limits are marked. The quantum-well inductance is included in f_{RCL}, but not in f_{RC}. [After T. C. L. G. Sollner, *et al.*, "High-frequency applications of resonant-tunneling devices," in *Electronic Properties of Multilayers and Low-Dimensional Structures*, J. M. Chamberlain, *et al.*, eds. (Plenum, New York, 1990), pp. 283–296.]

at the lines. This figure shows that the lifetime and therefore the cutoff frequency can be varied over a very wide range.

If the lifetime τ is sufficiently short, other effects can limit the speed and the oscillation frequency of the diode. Usually, in order to suppress scattering effects in the quantum well, spacer layers are grown. The transit time across such spacers can restrict the cutoff frequency if this time exceeds the tunneling time τ. Other parameters determining the oscillation frequency are differential conductance G and series resistance R_s. Figure 9.50 illustrates the effect of both of these parameters. In this figure the maximum oscillation frequency is calculated as a function of the conductance G. The curves are calculated without the transit-time effect (labeled by f_{RCL}) and taking the effect into account (labeled by f_{TT}). The series resistance is assumed to vary from $R_s = 4\,\Omega$ at dc current to $R_s = 6\,\Omega$ at 600 GHz. These parameters can currently be achieved for resonant-tunneling structures. This particular calculation shows that the transit-time effect decreases the maximal cutoff frequency from $f_{RCL} = 506$ to $f_{TT} = 457$ GHz. The two upper curves are calculated for structures with a small series resistance equal to $2\,\Omega$ at all frequencies. For this case, the oscillation frequencies are well above those of the previous case. The transit-time effect plays an even more important role and results in a decrease of the maximum cutoff frequency from $f_{RCL} = 1.1$ GHz to $f_{TT} = 835$ GHz. The calculations indicate a way of improving the structures so that operation frequencies increase: decrease both the series resistance and the transit time across the structure. It is expected that in the structures with smaller

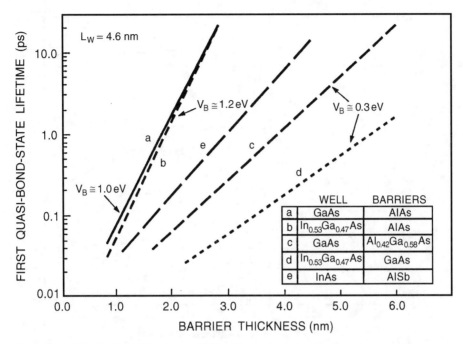

Figure 9.49. The lifetime of the first quasi-bound state in n-type double-barrier structures made from five of the material systems presented in the inset. In each structure the quantum-well width is fixed at 4.6 nm. The energy heights of the barriers are indicated at the curves. [After E. R. Brown, "High-speed resonant-tunneling diodes," in *Heterostructures and Quantum Devices*, N. G. Einspruch and W. R. Frensley, eds. (Academic, San Diego, 1994).]

transit distances, it will be possible to reach maximum oscillation frequencies above 1 THz.

The generated microwave power depends on the current through the device and on the peak-to-valley ratio ($\gamma_{pv} = j_p/j_v$). Three types of heterostructures currently used for RTD's – GaAs/AlAs, InGaAs/AlAs, and InAs/AlSb – are characterized by current densities exceeding 10^5 A/cm^2 and exhibit negative conductance at room temperature. In Fig. 9.51, the current density versus the applied voltage is depicted for these three cases. The largest γ_{pv} is reached for InGaAs/AlAs structures (above 10 at room temperature), but the highest currents are typically for InAs/AlSb, $j_m \approx (3.5\text{--}4) \times 10^5$ A/cm^2. For GaAs/AlAs structures, these parameters are relatively modest: $\gamma_{pv} \approx 1.4$ and $j_m \approx 1.5 \times 10^5$ A/cm^2. Figure 9.52 shows the dependence of the specific power on the frequency of microwaves generated by these three RTDs. The power decreases with the frequency. For example, for the GaAs/AlAs diode operating at the high frequency of 420 GHz, microwave power as small as 0.2 μW was measured. For the InAs/AlSb diode with the frequency $f = 712$ GHz, the power was $\sim 0.3\,\mu$W.

9.7.2 Resonant-Tunneling Diode as Frequency Multiplier

The strong nonlinearity of the current-voltage characteristics of a RTD can be used for frequency multiplication, frequency mixing, and other nonlinear

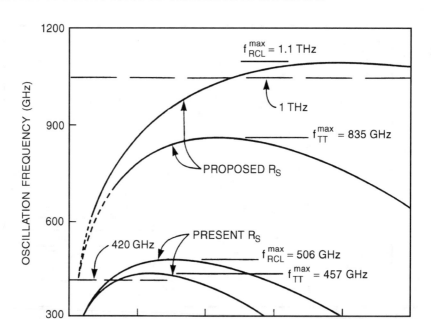

Figure 9.50. Calculations of maximum oscillation frequency as a function of differential conductance for two models: without (RCL) and with (TT) the phase taken into account. The maximum frequencies for both models are marked by f_{RCL}^{max} and f_{TT}^{max}. [After T. C. L. G. Sollner, *et al.*, "High-frequency applications of resonant-tunneling devices," in *Electronic Properties of Multilayers and Low-Dimensional Structures*, J. M. Chamberlain, *et al.*, eds. (Plenum, New York, 1990), pp. 283–296.]

high-frequency applications. Let us consider the operation of the diode as a frequency multiplier. Such a microwave application is based on an undulation of the diode current-voltage characteristics. Let us consider the diode with the current-voltage characteristics shown in Fig. 9.53(a) for the voltage range from −0.45 to +0.45 V. Assume that an ac input voltage signal with an amplitude of 0.45 V is applied to this diode. The wave front of the input signal – the voltage per period – is depicted in Fig. 9.53(b); see the left axis of this figure. A change in the voltage leads to a variation in the current through the diode. Since the current-voltage characteristics of Fig. 9.53(a) are highly nonlinear, during one period of the input signal the current versus the wave phase exhibits a more complex form with multiple peaks and minima. In the time domain this corresponds to a complex time dependence of the current. If we present this dependence as a sum over harmonics, a set of harmonics with relatively large amplitudes should appear. Because of the antisymmetric properties of the current-voltage characteristic, only odd harmonics can exist in this case. For this particular example, the output signal contains five subsequent odd harmonics (RTD quintupler) – the first, third, fifth, seventh, and ninth. The higher harmonics in the output signal occur as a result of frequency multiplication. Thus the diode can serve as the resistive multiplier.

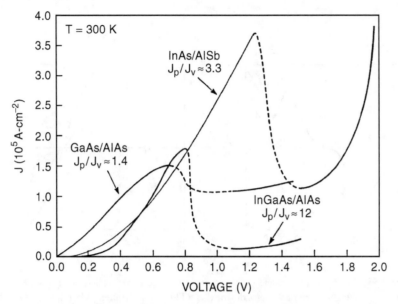

Figure 9.51. The room-temperature current density versus voltage curves for high-speed double-barrier RTDs made from three different material systems: GaAs/AlAs, InAs/AlSb, and InGaAs/AlAs. [After E. R. Brown, "Resonant tunneling in high-speed double barrier diodes" in *Hot Carriers in Semiconductor Nanostructures: Physics and Applications*, J. Shah, ed. (AT&T and Academic, Boston, 1992), pp. 469–498.]

Figure 9.54 demonstrates the spectrum of the multiplier output. The fundamental frequency here, referred to as the first harmonic, is equal to 4 GHz. Thus, for example, the fifth harmonic corresponds to oscillations with a frequency of 20 GHz. In Fig. 9.54 the measured data and calculations of the relative power are presented; there is good agreement between experiment and theory. For the optimized devices one can expect an up-conversion efficiency of $\sim 1\%$.

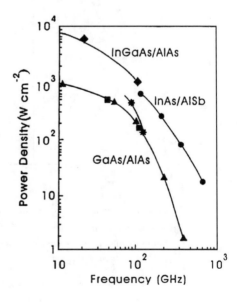

Figure 9.52. The generated microwave power per unit cross section of double-barrier resonant-tunneling transistors as a function of the frequency for the same three devices as in Fig. 9.51. [After E. R. Brown, "Resonant tunneling in high-speed double barrier diodes" in *Hot Carriers in Semiconductor Nanostructures: Physics and Applications*, J. Shah, ed. (AT&T and Academic, Boston, 1992), pp. 469–498.]

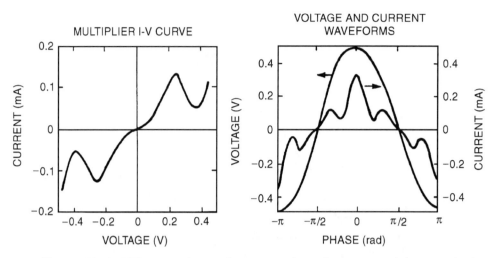

Figure 9.53. An RTD quintupler: (a) the current-voltage characteristic of the RTD, (b) the resulting voltage and current waveforms for this RTD as a multiplier. [After T. C. L. G. Sollner, *et al.*, "High-frequency applications of resonant-tunneling devices," in *Electronic Properties of Multilayers and Low-Dimensional Structures*, J. M. Chamberlain, *et al.*, eds. (Plenum, New York, 1990), pp. 283–296.]

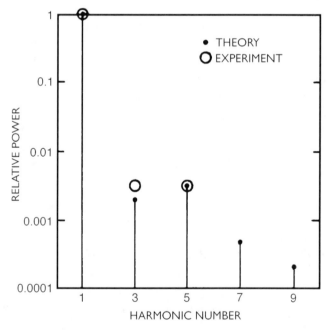

Figure 9.54. Theoretical and experimental spectrum of the multiplier output. For measurement the pump was at 4 GHz. [After T. C. L. G. Sollner, *et al.*, "High-frequency applications of resonant-tunneling devices," in *Electronic Properties of Multilayers and Low-Dimensional Structures*, J. M. Chamberlain, *et al.*, eds. (Plenum, New York, 1990), pp. 283–296.]

9.7.3 Resonant-Tunneling Transistors

We now consider promising applications of the resonant-tunneling effect in different transistor structures. One can incorporate a RTD into one of the junctions or transport regions of a BT, a FET, or a hot-electron device. These hybrid devices consist of a transistor and a RTD. They exhibit new properties and constitute a new class of circuits with greatly reduced complexity; indeed, fewer transistors per function compared with circuits containing the usual transistors.

As we discussed previously, transistor action can be obtained in different ways. In BTs the collector current of minority carriers injected into the base is controlled by the base current. In unipolar FETs the source–drain conductance is controlled by the gate voltage. An application of the resonant-tunneling effect in transistors can be understood by an extension of the RTD principle. Consider a BT in which the resonant-tunneling structure is introduced into the emitter junction. This transistor can be used in a common-emitter circuit; see Fig. 9.27(b). The collector current obviously increases with Φ_{EC}. At large Φ_{EC}, when almost all injected electrons are collected, the current saturates. Then the base – emitter bias Φ_{BE} controls the resonant-tunneling current. If a substantial emitter–collector voltage Φ_{EC} is applied, the collector current I_C is primarily determined by the carriers injected through the double-barrier structure, and the collector current I_C should exhibit the typical RTD dependence on Φ_{BE}, as presented in Fig. 9.55(a). Figure 9.55(b) shows schematically the expected I_C–Φ_{CE} characteristic. This characteristic is similar to that of conventional transistors. But a monotonous increase in the base–emitter voltage leads to a nonmonotonous emitter current and, as a consequence, to alternating regions of positive and negative transconductance. A sharp dependence of the resonant emitter current on

Figure 9.55. The principle of controlling the collector current by the base–emitter voltage in a resonant-tunneling BT. The common-emitter biasing configuration is assumed: (a) the collector current I_C versus base–emitter voltage Φ_{BE}, (b) collector currents versus the collector–emitter voltage Φ_{CE} at three values of Φ_{BE} marked in (a).

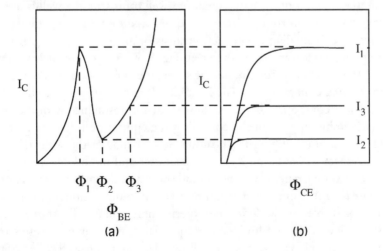

the emitter–base voltage provides the basis for effective control of the collector current.

Another way to design a resonant-tunneling transistor is to make direct contact with the quantum well of the double-barrier resonant-tunneling structure. Such a three-terminal device would have a three-dimensional emitter and a collector and a two-dimensional base – the quantum well. Both unipolar and BTs of this type are possible. These transistors are referred to as gated quantum-well transistors.

Unipolar Resonant-Tunneling Transistor

We now consider the unipolar resonant-tunneling transistor. In such a device, the quantum-well base has to be conducting, i.e., at least the lowest state of the quantum well should be occupied by electrons in equilibrium. As we know, electrons typically tunnel through the lowest quasi-bound state of the well. It is easy to understand that in such a case there is no separation between the controlling and the current-carrying electrons, although the separation is required for transistor action. This case is illustrated by Fig. 9.56(a). If the region under base contact B is supposed to be conductive up to quantum well QW, it follows that only one barrier exists between the well and collector; this is depicted in the lower energy diagram. In the region under the emitter, we have a typical double-barrier structure, as shown in the upper energy diagram. For this case, effective transistor action is not possible.

To separate the controlling and the current-carrying electrons, one can use tunneling through the second, excited state of the well. For this case, a sketch of the three-terminal structure and the energy diagram are presented in Fig. 9.56(b). Resonant tunneling occurs when the excited state aligns with the occupied energy states of the emitter. This alignment is controlled by the gated quantum well. The development of such a transistor is in progress in several laboratories. The major technological problem is the difficulty in making contact with a thin quantum-well base whose width is only ~100 Å.

In another variant of the gated quantum-well transistor, the well is used as a collector, while the controlling electrode – called a gate – is placed behind the collector layer. The band diagram of such a transistor under an applied gate voltage and an emitter–collector bias is shown in Fig. 9.57. The collector is grounded, that is, in terms of transistor circuits the schematic corresponds to the common-collector biasing configuration. The Fermi levels in the emitter and the gate are marked. The second quasi-bound state is shown in resonance with the emitter electrons. The mechanism of the control of the emitter–collector current is the following. The gate modifies the position of subbands in the well (the collector) with respect to the Fermi level of the emitter and modulates the tunneling emitter–collector current. The latter should show a peak when the resonant tunneling from the three-dimensional emitter to the two-dimensional quantum-well collector is possible. In Fig. 9.58, the experimental data are illustrated for such a transistor operating at a low temperature. The particular parameters of the structure presented in Fig. 9.58 are as follows: The quantum-well collector made

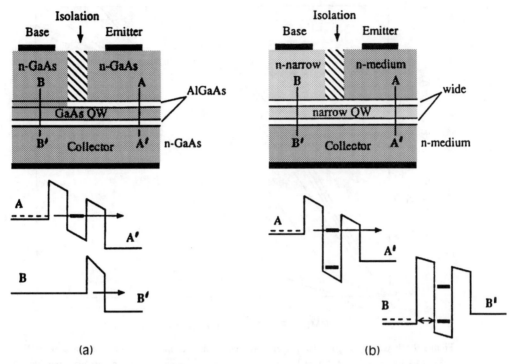

(a)   (b)

Figure 9.56. The principle of operation of a unipolar tunneling transistor in the common-collector configuration. Schematic cross-sections and energy-band diagrams of such devices are presented. The top doped layer is divided into two electrically insulated parts, the emitter and the base contacts. For cases (a) and (b) the emitter is separated from the quantum well by a barrier as for a conventional RTD. The base region and the well layer are slightly doped. [After A. C. Seabaugh and M. A. Reed, "Resonant-tunneling transistors," in *Heterostructures and Quantum Devices*, N. G. Einspruch and W. R. Frensley, eds. (Academic, San Diego, 1994), pp. 352–383.]

Figure 9.57. Band diagram of the gated quantum-well resonant-tunneling transistor with the collector at reference and the biases $\Phi_G > 0$ and $\Phi_E < 0$ corresponding to peak resonant tunneling of emitter electrons into the second subband of the well.

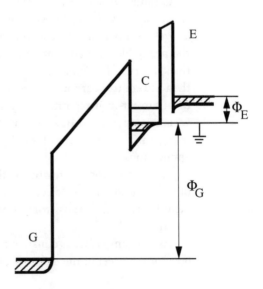

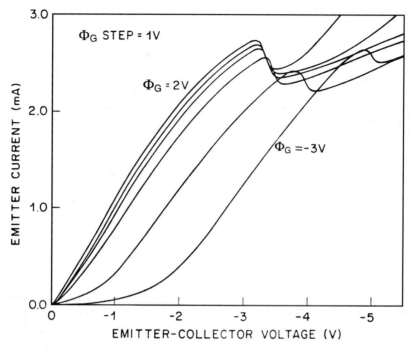

Figure 9.58. Common-collector characteristics of the resonant-tunneling transistor of Fig. 9.57 at various gate biases $\Phi_G = 2, 1, 0, -1, -2,$ and -3 V. The measurements were performed at 7 K. [F. Capasso, *et al.*, "Quantum functional devices: resonant-tunneling transistors, circuits with reduced complexity, and multiple-valued logic," IEEE Trans. Electron Devices **36**, pp. 2065–2082 (1989).]

of GaAs is 120 Å thick and the layer is provided by the collector contact; the n^+-GaAs doped emitter is separated from the well by a 40-Å-thick undoped AlAs tunneling barrier; and the second 1200 Å thick barrier of undoped AlAs is followed by an n^+-GaAs gate. The device possesses a NDR (see Fig. 9.58) as well as a negative transconductance. One of the advantages of the device is the negligible gate current since the gate and the collector are separated by a thick barrier. The gate current is always several orders of magnitude smaller than the emitter current. As we already know, a small base current corresponds to a current-transfer ratio close to unity and a large current gain. Thus this variant of the gated quantum-well devices possesses a NDR as well as a current gain. These features of the device make possible new multifunctional circuit applications.

As we mentioned previously, the RTD can be used as the emitter, the base, or the collector in hot-electron transistors. In all of these cases, the RTD element leads to new transistor properties and to possibilities for new operational functions.

As an example, let us consider a hot-electron transistor by using the RTD as an injector (emitter) of the electrons; this device is the so-called resonant-tunneling

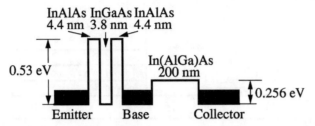

Figure 9.59. Band diagram of the InGaAs/InAlAs RHET under equilibrium. Widths of the InAlAs barriers, InGaAs quantum well, and In(AlGa)As collector barrier are indicated. The base width is 50 nm; its doping is 1×10^{18} cm^{-3}. [After N. Yokoyama, *et al.*, "Resonant-tunneling hot-electron transistors," in *Hot Carriers in Semiconductor Nanostructures: Physics and Applications*, J. Shah, ed. (AT&T and Academic, Boston, 1992), pp. 443–467.]

hot-electron transistor (RHET). In Fig. 9.59, the sketch of the energy-band diagram of a RHET is shown for the case of an InGaAs/InAlAs multilayer heterostructure. The widths of different layers are given in the figure. The structure is designed to direct the electrons into the base at an electron energy exceeding the base–collector junction height. In contrast to usual hot-electron-transistor structures, the resonant-tunneling structure injects electrons with a very narrow energy spread. This spreading can be of the order of 1 meV or even less, while this value for nominal hot-electron transistors is of the order of tens of milli-electron volts. The main feature of the RHET is the portion of the emitter–base current-voltage characteristic with negative slope. Figure 9.60 shows such characteristics. The collector current also has a negative-slope portion in a region of the emitter–collector voltage.

Currently RHETs operate primarily at low temperatures. At nitrogen temperature, InGaAs/InAlAs RHETs exhibit differential current gains up to 25, peak-to-valley ratios of 15, and collector-current densities of $\sim 3 \times 10^4$ A/cm^2.

RHETs have also been realized in Si/SiGe heterostructures. Because the valence-band offsets for this material system are larger than those of the conduction band, the resonant-tunneling effect and the transistors are based on p-doped heterostructures. In Fig. 9.61(a), the cross section of a pseudomorphic Si/SiGe RHET is presented. The structure is grown upon the p^+-Si substrate (collector)

Figure 9.60. Emitter (solid curve) and collector (dashed curve) currents of the RHET as functions of the base–emitter voltage. The area of the emitter and the temperature are indicated. [After N. Yokoyama, *et al.*, "Resonant-tunneling hot-electron transistors," in *Hot Carriers in Semiconductor Nanostructures: Physics and Applications*, J. Shah, ed. (AT&T and Academic, Boston, 1992), pp. 443–467.]

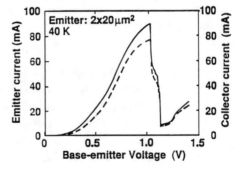

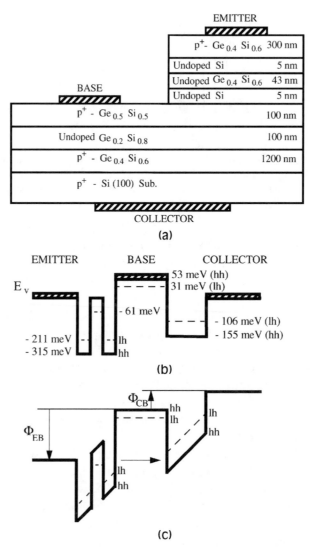

Figure 9.61. The Si/SiGe-based RHET: (a) the cross section of the device, (b) the energy-band diagram of the device under equilibrium, (c) the energy diagram under bias condition. The energies of the heavy and the light holes are different in the strained layers. The base–emitter and the collector–base biases are indicated for the common-base biasing configuration. [After S. S. Rhee, *et al.*, "SiGe resonant tunneling hot-carrier transistor," Appl. Phys. Lett. **56**, pp. 1061–1063 (1990).]

provided by the electrode. Then a conducting p^+-$Si_{0.6}Ge_{0.4}$ buffer layer is grown. An undoped, thick $Si_{0.8}Ge_{0.2}$ layer serves as the collector barrier. The base is made of a thick p^+-$Si_{0.5}Ge_{0.5}$ layer. The double-barrier structure consists of Si barriers and a narrow $Si_{0.6}Ge_{0.4}$ quantum-well layer. The emitter is made of p^+–$Si_{0.6}Ge_{0.4}$. The emitter contact is placed on the top of the structure. The dimensions of the layers are shown in Fig. 9.61(a). Such a combination of layers is used to obtain a pseudomorphic heterostructure with relatively thick base and collector barrier layers and simultaneously to provide a high barrier for the holes. Since in the strained Si and Si/Ge layers, the valence bands of the heavy and the light holes are split, the heights of the barriers and the bottoms of the wells are different for both kinds of carriers. For equilibrium conditions, they are given in Fig. 9.61(b). Figure 9.61(c) presents the energy-band diagram under bias conditions. The hot holes are injected from the emitter through the double-barrier structure and fly through the thin base almost without scattering and then enter the collector. At nitrogen temperature, the device exhibits a NDR and a negative differential transconductance. Although its characteristics are quite modest, the device is the first example of a functional device that can be developed on the basis of Si technology.

Bipolar Resonant-Tunneling Transistors

The double-barrier resonant-tunneling structure can be incorporated into a BT in different ways. In Fig. 9.62, energy-band diagrams are presented for different $n-p-n$ bipolar resonant-tunneling transistors at equilibrium when no voltage is applied. In Fig. 9.62(a) the resonant-tunneling structure is placed at the emitter–base junction. This case is similar to the previously studied RHET of Fig. 9.56. The double-barrier structure works as an injector. To achieve a resonant injection of electrons, the base–emitter bias has to be greater than the energy bandgap in the quantum-well layer. A sharp dependence of the injected current (i.e., the electron emitter current) on the base–emitter bias makes it possible to provide good control of the collector current and to realize effective operation of such a bipolar resonant-tunneling transistor up to room temperature.

Another design of the bipolar transistor with a resonant-tunneling structure as an injector is shown in Figs. 9.62(c) and 9.62(d). Here the resonant-tunneling structures are placed before the $n-p$–emitter–base junction. In the case of Fig. 9.62(d), a stack of two such structures is used. Compared with the case of Fig. 9.62(a), the designs of Fig 9.62(d) are characterized by larger peak-to-valley ratios in the collector current.

Figures 9.62(b) and 9.62(e) correspond to a double-barrier structure incorporated into the p base. Different emitters can be designed for this case. In Fig. 9.63, energy-band diagrams under bias conditions for the AlGaAs/GaAs bipolar resonant-tunneling transistor are shown for an abrupt emitter [Fig. 9.63(a)] and for tunneling emitters [Figs. 9.63(b) and 9.63(c)].

The design of the double-barrier structure determines the energy spacing of the levels in the quantum well. Of course, the spacing can be different. For example, Fig. 9.63(c) corresponds to a well with approximately equal spacing between the quasi-bound states. Several quasi-bound states in the well can cause more complex collector-current behavior. As the base–emitter voltage increases, the resonant-tunneling rate through each of the quasi-bound states of the well first reaches a maximum and then decreases when this state is lowered below the emitter conduction band. Such tunneling through the different subsequent states

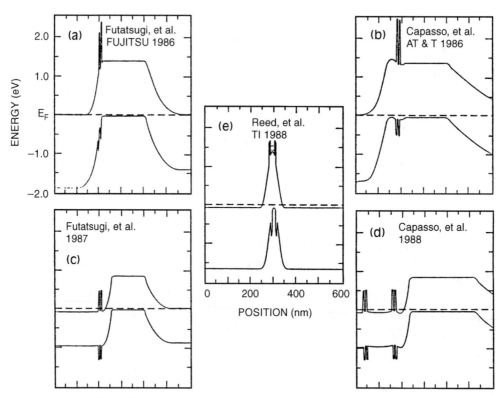

Figure 9.62. Energy-band diagrams for different designs of resonant-tunneling transistors. Profiles of both the conduction and the valence bands are depicted. Cases (a), (b), and (e) correspond to GaAs/AlGaAs structures grown on GaAs substrates; cases (c) and (d) correspond to InGaAs/InAlAs structures grown on InP. [After A. C. Seabaugh and M. A. Reed, "Resonant-tunneling transistors," in *Heterostructures and Quantum Devices*, N. G. Einspruch and W. R. Frensley, eds. (Academic, San Diego, 1994), pp. 352–383. The detailed references for papers indicated on Figure 9.62 are given at the end of Chapter 9.]

produces multiple peaks in the collector current, i.e., multiple regions with negative differential conductance. For the double-barrier structure design shown in Fig. 9.63(c), equally spaced peaks can be obtained in the collector current.

Depending on the structure design (dimensions, doping, etc.) electron transport in the base region of these transistors can be of different types: ballistic, quasi-ballistic, and thermionic injection through the base. The ballistic and the quasi-ballistic transport cases are sensitive to the quality of the structure, the doping, the dimensions, etc., as a result of the scattering in the base.

Let us consider a device that uses thermionic injection, which is much less sensitive to the structure parameters and can work up to room temperature. The band diagram of such a transistor is presented in Figs. 9.64(a), and 9.64(b) under conditions of operation. To line up the conduction band in the region adjacent to the emitter and the lowest quasi-bound state in the quantum well, one should choose a proper alloy composition in the region. For example, for a 75-Å-thick GaAs quantum well, the energy of the quasi-bound state is $E_1 = 65$ meV. The Al

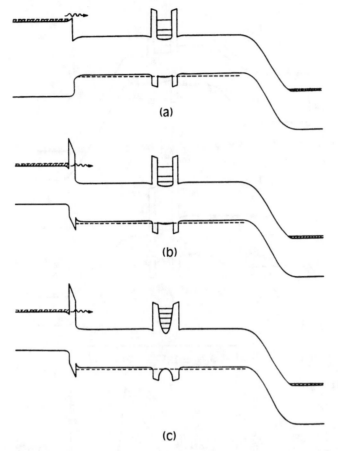

(a)

(b)

(c)

Figure 9.63. Energy-band diagrams of resonant-tunneling transistors with an RTD embedded into the doped base: the device with (a) an abrupt emitter, (b), (c) a tunneling emitter and different quantum wells. [After F. Capasso, *et al.*, "Quantum functional devices: resonant-tunneling transistors, circuits with reduced complexity, and multiple-valued logic," IEEE Trans. Electron Devices **36**, 2065–2082 (1989).]

mole fraction should be chosen as $x = 0.07$ for $\Delta E_c \approx E_1$. Thus the base consists of a wide-bandgap part ($Al_{0.07}Ga_{0.93}As$) and a narrow-bandgap part (GaAs). The quantum well is undoped, but a high two-dimensional hole concentration in the well is due to hole transfer from the wide-bandgap part of the base. This hole transfer partly eliminates the elastic scattering of the minority electrons by impurities: the irregular impurity potential is screened by the holes. Electrical contacts are made to both the well and the GaAs portion of the base. The $Al_{0.07}Ga_{0.93}As$ region is separated electrically from the GaAs part of the base by the first barrier adjacent to the AlGaAs. The emitter can be made of a wider bandgap AlGaAs alloy to prevent hole injection into the emitter.

The principle of operation of this transistor is the following. First, the collector–emitter voltage Φ_{CE} and the base current I_B are chosen so that the base–emitter junction is forward biased, while the base–collector junction is

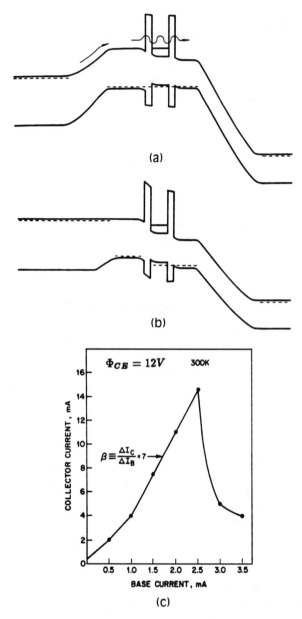

(a)

(b)

(c)

Figure 9.64. Energy-band diagrams of the resonant-tunneling BT with the thermal injection for different base currents I_B at a fixed collector–emitter voltage Φ_{CE} (not to scale). The base consists of two electrically insulated regions. As I_B is increased, the device first behaves as a conventional bipolar transistor (a) with a current gain until near-flat-band conditions in the emitter are achieved. For $I_B > I_{Bth}$ a potential difference develops across the AlAs barrier between the contacted and the uncontacted regions of the base. This raises the conduction-band edge in the emitter above the first resonance of the well, thus quenching resonant tunneling and the collector current (b). The collector current versus the base current in the common-emitter configuration at room temperature is shown in (c). [After F. Capasso, *et al.*, "Quantum functional devices: resonant-tunneling transistors, circuits with reduced complexity, and multiple-valued logic," IEEE Trans. Electron Devices **36**, 2065–2082 (1989).]

reverse biased, i.e., the active mode of operation is assumed. If one maintains Φ_{CE} at a constant value and increases the base current, the base potential also increases until a partial flat-band condition is reached, as shown in Fig. 9.64(a). In this regime, the collector current increases almost linearly with the base current, as shown in Fig. 9.64(c). The slope of this curve gives the current gain. In this regime, the electrons surmount the barrier of the emitter–base junction because of thermionic injection. There is a critical base current $I_{B,c}$ corresponding to the complete flat-band condition in which the conduction bands of the wide-bandgap emitter and the AlGaAs region of the base are lined up. A further increase in Φ_{EB} leads to an additional potential drop across the first AlAs barrier, which insulates both contacted and uncontacted parts of the base, as shown in Fig. 9.64(b). This breaks the resonant tunneling through the lowest quasi-bound state. As a result, the transfer characteristic and the current gain are sufficiently reduced, as presented in Fig. 9.64(c). It is obvious that this regime is characterized by negative transconductance. For this case, it is important that the base contact be made to the quantum well, which results in modulation of the energy difference between the state in the well and the emitter band. From Fig. 9.64, one can see that in the case of thermionic injection, the bipolar resonant-tunneling transistor exhibits a current gain as large as 7; large gains are obtained even at room temperature.

Let us discuss briefly the speed of operation of resonant-tunneling transistors. Incorporating the resonant-tunneling structure into the base or the emitter–collector junction, etc., should lead to an increase in emitter–collector delay time because of the finite tunneling time. Assume that for electrons injected into the structure, the energy distribution is wider than the width of the tunneling resonance, $\Gamma = 2\hbar/\tau$; here, τ is the tunneling time. Then one can write the total delay time as

$$t_{tr} = \frac{L_B}{v_B} + \frac{2\hbar}{\Gamma},$$

where the first term on the right-hand side is the classical transit time across the base and L_B and v_B are the base width and the electron velocity, respectively. The second term is associated with tunneling processes; see Subsection 8.2.4. The drift velocity in the base v_B can be estimated as $v_B \approx 10^7$ cm/s. Thus, for a base width of ~ 1000 Å, the first term is equal to or more than 10^{-12} s. The second term can be accounted for by the results presented in Fig. 9.49. They show that the second term can be of the order of 10^{-13}–10^{-12} s for a wide class of double-barrier structures. Combining these numbers, one can see that either the resonant-tunneling structure increases the transit time by a factor of 2 or it does not affect this time at all. An analogous conclusion can be made for the cutoff frequency of the transistor: $f_T = 1/(2\pi t_{tr})$. Thus incorporating a resonant-tunneling structure into a BT does not lead to a considerable decrease in the speed of operation. Indeed, it improves the device characteristics substantially.

9.8   Circuit Applications of Resonant-Tunneling Transistors

Among the possible applications of resonant-tunneling transistors, we consider those that are qualitatively novel and promise to reduce considerably the number of transistors and processing steps per operational function.

It is well known that a device with a NDR operating in the circuit as an active resistance can form a circuit with two stable states. The circuit is shown in Fig. 9.65(a). The load line

$$I = \frac{\Phi - \Phi_D}{R_L},$$

and the current-voltage characteristics are presented in Fig. 9.65(b) as functions of the voltage applied to the device. The intersections of these two curves determine two stable states of the circuit. A circuit possessing two or more stable states can be used as a memory cell or as a logic element. Thus the resonant-tunneling transistor with one or more peaks in the collector current can be used as a multifunctional element. On this basis it is possible to design different logical elements with fewer numbers of transistors. This leads to more densely packed devices, higher speed of operation, a decrease in a number of interdevice connections, etc.

An example of such a reduction in the number of components in the circuit is the exclusive NOR circuit that uses the resonant-tunneling transistor, which is shown in Fig. 9.66(a). The circuit includes a resistive summing network at the base terminal and exploits the negative-transconductance characteristic of the transistor. The RHET is chosen for this application. The circuit operates at room temperature. If both signals on inputs A and B are the same, they produce a low base–emitter voltage. This gives rise to a high resonant-tunneling current and consequently a low level of output C. If the signals on A and B are different, the base–emitter voltage is high and the resonant-tunneling current is switched off, which produces a high output signal on C. This is illustrated by the timing diagram in Fig. 9.66(b), in which different combinations of input signals, marked by A and B, and corresponding output signals, marked by C, are shown. As we

Figure 9.65. Simplest circuit (a) containing the device with NDR, (b) two stable values of the current in the circuit as the intersection of the load line and the current-voltage characteristics of the device.

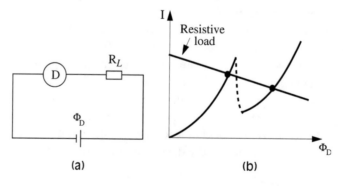

(a)   (b)

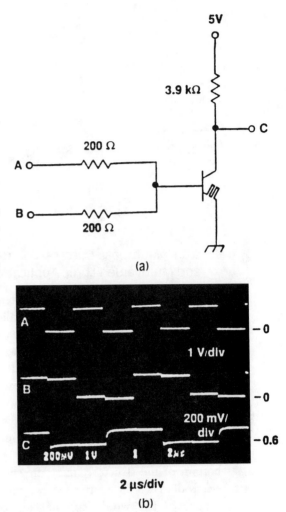

Figure 9.66. The RHET exclusive NOR gate operating at room temperature: (a) circuit, (b) inputs A and B and output C voltages. [After A. C. Seabaugh and M. A. Reed, "Resonant-tunneling transistors," in *Heterostructures and Quantum Devices*, N. G. Einspruch and W. R. Frensley, eds. (Academic, San Diego, 1994), pp. 352–383.]

see, such an operation becomes possible because of the negative transconductance of the RHET. The functional compression is approximately equal to 9 for this example, compared with the ordinary emitter coupled logic. However, at least one more transistor is required for increasing the output voltage for the integrated circuit.

Another example is the full-adder circuit presented in Fig. 9.67. This three-transistor adder uses one resonant-tunneling BT and two conventional BTs. The logic inputs are the addent (A), the augent (B), and the carry (C_n). The outputs are the sum and the carry (C_{n+1}). Comparing this circuit with that in Fig. 9.66, one can see that a resonant-tunneling BT with proper resistors constitutes an exclusive NOR function. The circuit operates at room temperature with a single 5-V supply. A conventional full adder uses approximately 42 transistors. Thus, by using the resonant-tunneling BT, one can reduce the number of transistors as many as 10 times and the number of other components as many as 3 times.

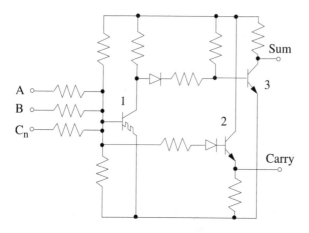

Figure 9.67. The full-adder circuit with single resonant-tunneling BT 1 and two conventional BTs 2 and 3. [After A. C. Seabaugh and M. A. Reed, "Resonant-tunneling transistors," in *Heterostructures and Quantum Devices*, N. G. Einspruch and W. R. Frensley, eds. (Academic, San Diego, 1994), pp. 352–383.]

9.8.1 Multipeak Current-Voltage Characteristics and Multivalued Logic Applications

Multipeak current-voltage characteristic dependences inherent for resonant-tunneling devices can be used in so-called multivalued logic applications. However, these circuit applications require characteristics with similar peak-to-valley ratios. This can be achieved by an integration of several RTDs. In this method, a single resonance of different but identical quantum wells produces the multipeak characteristic. There are two design approaches to achieving such integration: horizontal and vertical.

The case of horizontal integration is shown in Fig. 9.68. This device consists of a double-barrier and a modulation-doped AlGaAs/GaAs heterostructure with two electrically separated outputs denoted by 1 and 2; see Fig. 9.68(a). At the bottom of the structure, there are the common n^+-GaAs collector layer and electrode. Then the emitter spacer, double-barrier structure, and collector spacer are grown. The two-dimensional electron channel at the AlGaAs/GaAs interface is common for both outputs. The channel has a resistance of $\sim$10–15 Ω for real devices. The equivalent circuit of Fig. 9.68(b) clearly indicates that this device presents two parallel identical RTDs with output terminals connected by the electron channel. The principle of operation of the device is the following. The total current I_3 through common contact 3 is determined by the sum of two currents, I_1 and I_2, flowing through similar diodes 1 and 2. If the bias between contacts 1 and 2, Φ_{12}, is zero, these currents are equal and the total current I_3 usually exhibits one-peak–one-valley dependence on the bias $\Phi_{13} = \Phi_{23}$. (Since contact 1 is supposed to be grounded, a positive voltage is applied to contact 3.) If the voltage applied to contact 2 is negative, $\Phi_{12} < 0$, the resonance condition is reached first for electron tunneling in diode 2 and then in diode 1. As a result, resonant tunneling occurs separately for each diode and the total current I_3 exhibits two peaks; see Fig. 9.68(c). If $\Phi_{21} > 0$, the resonance is reached first for device 1 and then for device 2. Thus if there is some voltage between the two parallel RTDs Φ_{12}, one obtains a two-peak current-voltage characteristic with

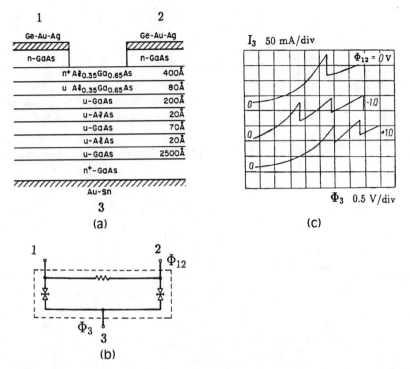

Figure 9.68. The multipeak resonant-tunneling device with horizontal integration of RTDs: (a) a sketch of the heterostructure, (b) the equivalent circuit, (c) the device current versus voltage at different biases between emitters of two integrated diodes. The biases are indicated at the curves. The data are measured at $T = 100$ K. [After S. Sen, *et al.*, "Resonant tunneling device with multiple negative differential mobility: digital and signal processing applications with reduced circuit complexity," IEEE Trans. Electron. Devices ED-34, pp. 2185–2191 (1987).]

the distance between the peaks being approximately equal Φ_{12}. It is clear that if the structure is composed of n such diodes, an n-peak characteristic can be achieved.

The vertical integration of RTDs can be achieved when a number of diodes are stacked in series. These diodes should be separated by heavily doped cladding layers in order to destroy quantum-mechanical coherence and decouple adjacent diodes from each other. Another feature of this structure is that each diode is designed so that quasi-bound states in the quantum wells lie substantially above the Fermi level in the adjacent cladding emitter layer. The principle of operation is as follows. Let an applied voltage correspond to resonant tunneling through both diodes. Increasing the bias leads to a higher electric field in the diode adjacent to the anode contact as a result of the accumulation of electric charges in the well. Thus the tunneling resonance in this diode will be quenched first. One obtains the first peak in the collector current. Further increase of the voltage quenches the resonance in another diode; this results in a second peak of the current, etc. By combining n tunnel diodes in series, one can reach n peaks for such a unipolar

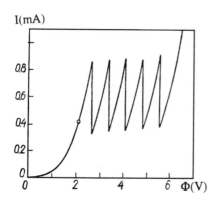

I(mA)

Figure 9.69. The current-voltage characteristic of five vertical integrated RTDs. [After A. A. Lakhani, *et al.*, "Eleven-bit parity generator with a single, vertically integrated resonant tunneling device," Electron. Lett. 24, pp. 681–682 (1988).]

structure. Figure 9.69 presents the current-voltage characteristic for the series of five vertically integrated RTDs. The five peaks and five valleys have almost the same magnitudes, which is necessary for circuit applications.

Analogous to Fig. 9.65, it is easy to see that a suitable load line in a combination with n peaks of the current-voltage characteristic of the resonant-tunneling device generates $(n+1)$ stable states of the circuit. Figure 9.70 illustrates this for the case of three peaks. The inset presents a circuit exhibiting three stable operating points. In contrast to the case of the circuit with the RTD, the latter multiple steady circuit can be controlled by a third terminal: the base–collector or the emitter–collector biases. This gives rise to the possibility of building memory elements for an $(n+1)$ stable logic system. Note that even for an ordinary binary computer, the storage system can be built on the basis of $(n+1)$ logic to increase packing density, etc.

Another potentially significant circuit application of multiple-state resonant-tunneling BTs is the analog-to-digital converter. In Fig. 9.71, an array of analog-to-digital circuits is shown. The analog input is applied to such an array of the circuits with different voltage scaling networks. Let us consider the simplest system that comprises only two transistors, Q_1 and Q_2. The analog input is Φ_i. The resistances R_0, R_1, and R_2 are chosen so that the biases of transistors Q_1 and Q_2, Φ_{B1} and Φ_{B2}, respectively, vary with the input voltage Φ_i according to linear curves Φ_{B1} and Φ_{B2}, as shown in the lower part of Fig. 9.72(a). For the input voltage Φ_1, let the outputs of both transistors be at operating point P_1 (the high output voltage), as shown in the upper part of Fig. 9.72(a), in which the collector voltage is presented as a function of the base voltage. If the input changes to Φ_2 the output of the transistor Q_1 should correspond to a new operating point P_2 (the low output voltage), while the output of the transistor Q_1 remains close to P_1. Applying this consideration to voltages Φ_3, Φ_4, etc., we find that the output follows the truth table presented in Fig. 9.72(b), that is, the outputs correspond sequentially to the high or the low voltages. Thus the circuit quantizes the input analog signal and converts it to a binary code. With a larger number of peaks in the current-voltage characteristic, the circuit can be extended to more bits. A NDR-based flash convertor requires only n transistors for an n-bit conversion. If a conventional convertor is used for this purpose, it requires 2^n devices. This reduces drastically the number of components of the convertor and enhances the speed of operation.

However, it is worth noting that in these circuits the minima of the collector voltage are determined by the maximum of the input signal applied to the base terminal, and they are actually higher than for normal BTs. The latter require proper circuit design in order to avoid effects such as breakdown.

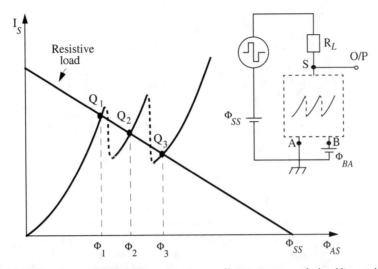

Figure 9.70. The schematic of a three-state memory cell. Intersections of a load line and the I–V characteristics determines three stable operating points Q_1, Q_2, and Q_3. The inset shows the equivalent circuit for this memory cell. [After F. Capasso, *et al.*, "Quantum functional devices: resonant-tunneling transistors, circuits with reduced complexity, and multiple-valued logic," IEEE Trans. Electron Devices **36**, pp. 2065–2082 (1989).]

In concluding this section, we emphasize that only a few examples of circuit applications of resonant-tunneling transistors are considered. There are other examples; for all of them the resonant-tunneling transistors behave as multifunctional devices, i.e., they can replace several conventional transistors. This reduces considerably the number of devices necessary for operation and, in turn, leads to the reduction of the number of other circuit components and connections. As a result, much higher packing density and speed of operation can be achieved.

Figure 9.71. Analog-to-digital converter circuit that uses multiple-state resonant-tunneling BTs. [After F. Capasso, *et al.*, "Quantum functional devices: resonant-tunneling transistors, circuits with reduced complexity, and multiple-valued logic," IEEE Trans. Electron Devices **36**, 2065–2082 (1989).]

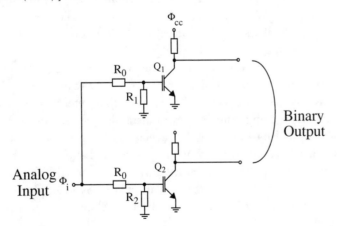

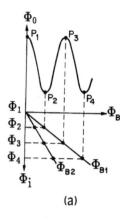

TRUTH TABLE		
INPUT	OUTPUT	
Φ_i	Q_2	Q_1
Φ_1	1	1
Φ_2	1	0
Φ_3	0	1
Φ_4	0	0

(a)   (b)

Figure 9.72. The principle of operation of the analog-to-digital converter circuit of Fig. 9.71, involving only two bits: (a) the voltages at different points of the circuit for various input voltages, (b) the truth table. [After F. Capasso, *et al.*, "Quantum functional devices: resonant-tunneling transistors, circuits with reduced complexity, and miltiple-valued logic," IEEE Trans. Electron Devices 36, pp. 2065–2082 (1989).]

9.9 Closing Remarks

In this chapter, we have studied electronic devices based on quantum heterostructures. We classified these devices as those that use classical regimes of electron transport and those that are based on the quantum-mechanical nature of electron transport. The devices that use classical transport have, as a rule, their homostructure prototypes and use known principles of operation. Examples are FETs and BTs. For such cases, we started with simple models of homostructure prototypes, which illustrate the basic properties and disadvantages of these devices. Then we analyzed the improvements brought about by using quantum heterostructures, selective doping, etc. We have studied heterostructure FETs and BTs for different combinations of materials and designs. We have shown that current technology achievements facilitate the fabrication of devices, taking into account a wide range of essential physical effects such as the quantization of electron spectra, high-field-mobility dependence, intervalley transitions, charge transfer in real space, thermionic overbarrier electron emission, etc. Each of these effects is useful, and artificial heterostructure devices provide the means to use them in the most efficient way. Other effects degrade device parameters, and, by optimizing heterostructures, one can either avoid undesirable effects or minimize them. As a result, superb device parameters can be realized by scaling down device dimensions. In particular, we have shown that heterostructure FETs and BTs based on III–V compounds achieve outstanding speeds of operation that are currently limited by the cutoff frequencies in the range of 200–300 GHz. We have studied Si-based heterostructure devices and shown that the application of the principles of heterostructure devices to Si-based FETs and BTs has already led to significant improvements. Particularly, Si/SiGe BTs with a cutoff frequency in the range of 75–95 GHz have been developed and further improvements are in progress.

We have studied heterostructure hot-electron transistors based on classical perpendicular electron transport. We have considered the basic principles of operation for ballistic-injection devices and for RST devices. We have found that only a heterostructure-based design of these devices can ensure their efficient operation.

We have studied devices based on quantum concepts, particularly RTDs and various resonant-tunneling transistors. We have shown that RTDs operate effectively in the frequency range of 500–700 GHz at room temperature. Exhibiting highly nonlinear current-voltage characteristics and a NDR, these diodes are used to generate microwave emission, to multiply frequency, and for different circuit applications. In particular, they can be used to design memory elements, etc. We have found that various heterostructures combining several RTDs or an RTD and a conventional transistor can be fabricated. Such novel devices possess characteristics that cannot be obtained either from a single RTD or from a conventional transistor. For example, these devices manifest multipeak current-voltage characteristics and multipeak transfer characteristics. In fact, these structures are multifunctional devices that facilitate the design of different circuits with considerable reduction in a number of devices, circuit components, and connections. This leads to higher density packing as well as increased speeds of operation of integration circuits.

Although in this chapter we have presented an analysis of conventional transistors, the reader can learn more about these devices in the following textbooks:

S. M. Sze, ed., *Physics of Semiconductor Devices* (Wiley, New York, 1981).

M. Shur, *Physics of Semiconductor Devices* (Prentice-Hall, Englewood Cliffs, NJ, 1990).

J. Singh, *Semiconductor Devices (An Introduction)* (McGraw-Hill, New York, 1994).

Additional materials on heterostructure devices based on III–V compounds, discussions of multifunctional devices, and new approaches to logic can be found in following publications:

W. Shockley, "The theory of *p–n* junctions in semiconductors and *p–n* junction transistors," Bell Sys. Tech. y., **28**, pp. 435–489 (1949).

H. Kroemer, "Theory of wide-gap emitter for transistors," Proc. IRE, **45**, 1535–1537 (1957).

F. Capasso, S. Sen, F. Beltram, *et al.*, "Quantum functional devices: resonant-tunneling transistors, circuits with reduced complexity, and multiple-valued logic," IEEE Trans. Electron Devices **36**, 2065–2082 (1989).

S. M. Sze, ed., *High-Speed Semiconductor Devices* (Wiley, New York, 1990).

M. Shur, *Physics of Semiconductor Devices* (Prentice-Hall, Englewood Cliffs, NJ, 1990).

M. Willander, *Heterojunction Transistors and Small Size Effects in Devices* (Studentlitteratur, Chartwell-Bratt, Linkoping, 1992).

Z. S. Gribnikov, K. Hess, and G. A. Kosinovsky, "Nonlocal and nonlinear transport in semiconductors: real-space transfer effects," J. Appl. Phys. 77, 1337–1373 (1995).

Si-based heterostructure transistors are described in the following review:

C. A. King, "Heterojunction bipolar transistors with $Si_{1-x}Ge_x$ alloys," in *Heterostructures and Quantum Devices*, N. G. Einspruch and W. R. Frensley, eds. (Academic, San Diego, 1994), pp. 157–187.

Recent reviews on resonant-tunneling diodes and resonant-tunneling transistors can be found in the following references:

M. A. Stroscio, "Quantum-mechanical corrections to classical transport in submicron/ultrasubmicron dimensions," in *Semiconductor Technology: GaAs and Related Compounds*, Cheng T. Wang, ed. (Wiley Interscience, New York, 1990).

T. C. L. G. Sollner, E.R. Brown, C. D. Parker, *et al.*, "High-frequency applications of resonant-tunneling devices," in *Electronic Properties of Multilayers and Low-Dimensional Structures*, J. M. Chamberlain, *et al.*, eds. (Plenum, New York, 1990), pp. 283–296.

G. I. Haddad and I. Mehdi, "Device applications of resonant-tunneling structures," in *Optoelectronic Materials and Device Concepts*, Manijeh Razeghi, ed., Vol. PMOS of the SPIE Press Monographs Series (The Society of Photo-Optical Instrumentation Engineers, Bellingham, WA, 1991), p. 57. J. Shah, ed., *Hot Carriers in Semiconductor Nanostructures (Physics and Applications)*, (AT&T and Academic, Boston, 1992).

E. R. Brown, "High-speed resonant-tunneling diodes," in *Heterostructures and Quantum Devices*, N. G. Einspruch and W. R. Frensley, eds. (Academic, San Diego, 1994), pp. 306–350.

A. C. Seabaugh and M. A. Reed, "Resonant-tunneling transistors," in *Heterostructures and Quantum Devices*, N. G. Einspruch and W. R. Frensley, eds. (Academic, San Diego, 1994), pp. 352–383.

The following references are indicated on Fig. 9.62:

F. Capasso, *et al.*, "Quantum-well resonant tunneling bipolar transistor operating at room temperature," IEEE Electr. Dev. Lett. EDL-7, pp. 573–576 (1986).

F. Capasso, *et al.*, "Multiple negative transconductance and differential conductance in a bipolar transistor by sequential quenching of resonant tunneling," Appl. Phys. Lett. 53, pp. 1056–1058 (1988).

M. A. Reed, *et al.*, "Realization of three terminal resonant tunneling device: the bipolar quantum resonant tunneling transistor," Appl. Phys. Lett. 54, pp. 1034–1036 (1989).

T. Futatsugi, *et al.*, "A resonant-tunneling bipolar transistor (RBT): a proposal and demonstration for new functional devices with high current gains," IEDM **86**, pp. 286–289 (1986).

T. Futatsugi, *et al.*, "InAlAs/InGaAs resonant tunneling bipolar transistors (RBTs) operating at room temperature with high current gains," IEDM **87**, pp. 877–878 (1987).

PROBLEMS

1. In terms of I_D, $l_d(x)$, μ_n, N_D, l, and L_z, derive an expression for the electric field, $F_x = -d\Phi/dx$, along the channel of an *n*-channel FET device. Show that the inversion charge per unit area is inversely proportional to F_x.

2. In deriving the excess minority carrier concentration of Eq. (9.54), we have assumed that the depletion region extends from $-l_{d,p}$ to $+l_{d,n}$ and that the diffusion region for positive z extends to $z \gg l_{d,n}$. If instead we assume that there is an ohmic contact at W_n, where $W_n > l_{d,n}$, show that

$$p'(z) = p_n - p_{n,0} = p_{n,0} \left[\exp\left(\frac{e\Phi}{k_B T} \right) - 1 \right] \frac{\sinh[(W_n - z)/L_p]}{\sinh[(W_n - l_{d,n})/L_p]}.$$

3. Consider a quantum-interference device such as those discussed in Subsection 9.3.2. Assume that the length of the device L is 4000 Å and the carrier velocity v_x is 2×10^7 cm/s. What is the characteristic transit time through the channel? What is the cutoff frequency of the device? What temperatures are required for destroying the coherence of this device?

4. What is the speed of operation of a GaAs/In$_{0.53}$Ga$_{0.47}$As/GaAs quantum-well transistor being switched through the lowest quasi-bound state? Assume that the GaAs barriers are 3 nm wide and that the In$_{0.53}$Ga$_{0.47}$As well is 4.6 nm wide. Assume that the carrier speed in the quantum well is 2×10^7 cm/s.

10

Optics of Quantum Structures

10.1 Introduction

Interactions of electromagnetic radiation with semiconductors include processes of emission, absorption, and scattering of light, as well as various nonlinear optical phenomena. These processes have always been major subjects of study in solid-state physics. The advent of semiconductor quantum structures with their unique properties makes it possible to observe many traditional optical effects in low-dimensional electron systems and to discover their novel features. Furthermore, heterostructures manifest a set of new effects that do not occur in the bulk materials. On the other hand, optical experiments are extremely powerful tools for the characterization of the heterostructures, including different electron properties, lattice parameters, surface and interface quality, etc. But the most important is that the unique optical and electrical properties of heterostructures have opened principally new avenues of development for optoelectronic and novel photonic applications.

All these facts make the optics of heterostructures an important topic in modern physics of solid-state devices. In this chapter we begin to study the optical properties of heterostructures. We start with a brief review of the basic concepts of electromagnetic fields. We define the classical characteristics of electromagnetic fields such as the energy, intensity, density of states, and we introduce the concept of quanta of these fields – the photons. Analyzing electromagnetic fields in free space and in optical resonators, we show that resonators drastically change the structure and the behavior of electromagnetic fields. These resonators facilitate the control of spatial distribution of the fields, i.e., carry out electromagnetic-field engineering. Next we study the interaction of light with matter and define three main optical processes: spontaneous emission, stimulated emission, and stimulated absorption. Calculations of probability amplitudes, total rates of phototransitions, and the absorption-amplification coefficient conclude the principal part of our treatment.

After this preparatory part, we review the optical properties of bulk semiconductors and define the major optical characteristics of semiconductors with emphasis on the specifics of III–V compounds. The next segments of this chapter are devoted to the optical properties of quantum structures. We study how electron confinement affects the optical spectra, stimulated emission, and other optical characteristics in quantum structures.

10.2 Electromagnetic Waves and Photons

10.2.1 Electromagnetic Fields, Modes, and Photons in Free Space

Electromagnetic waves are joint electric and magnetic fields that oscillate in both space and time. In the simplest homogeneous case, one can write the electric field in the form of a plane wave:

$$\vec{\mathcal{E}}(\vec{r}, t) = \vec{\xi} F_0 \cos(\vec{q}\vec{r} - \omega t), \tag{10.1}$$

where F_0 is the amplitude of the electric field, $\vec{\xi}$ is the polarization vector of the wave, $\vec{q}$ is the wave vector associated with the wavelength $\lambda = 2\pi/q$, and ω is the angular frequency of the wave. Alternatively, it is possible to use a complex form of the plane wave:

$$\vec{\mathcal{E}}(\vec{r}, t) = \vec{\xi} F_0 e^{-i(\vec{q}\vec{r} - \omega t)}, \tag{10.2}$$

but only the real part of Eq. (10.2) has physical meaning. This complex form is convenient for linear calculations. For example, for a uniform nonmagnetic dielectric, the magnetic field $\vec{\mathcal{H}}$ may be expressed in terms of $\vec{\mathcal{E}}$ as

$$\vec{\mathcal{H}}(\vec{r}, t) = \frac{c}{\omega} \vec{q} \times \vec{\mathcal{E}}, \tag{10.3}$$

where $\vec{q} \times \vec{\mathcal{E}}$ denotes the vector product and c is the speed of light in vacuum. For free space, the vector $\vec{\mathcal{E}}$ is always perpendicular to $\vec{q}$, so that if $\vec{q}$ is fixed, the electric field has, in general, two projections that correspond to the two possible polarizations of the electromagnetic wave.

The energy of the wave can be characterized by the density of the electromagnetic energy,

$$W = \frac{1}{4\pi} \kappa \overline{\mathcal{E}^2(t)} = \frac{1}{8\pi} \kappa F_0^2, \tag{10.4}$$

where $\overline{\mathcal{E}^2(t)}$ represents the time average of $\mathcal{E}^2(t)$. We can define the intensity of the wave as the energy flux through the unit area perpendicular to the wave vector $\vec{q}$:

$$\mathcal{I} = \frac{c}{8\pi} \kappa F_0^2. \tag{10.5}$$

In Eqs. (10.4) and (10.5) κ is the dielectric constant of the material. The frequency ω and the wave vector $\vec{q}$ are related through the dispersion relationship:

$$\omega = \omega_q \equiv \frac{c}{\sqrt{\kappa}} q. \tag{10.6}$$

In the simplest case, one can take κ to be independent of ω. It is assumed that the medium is uniform and isotropic and that the dissipation of the energy of the field is negligible in most cases of interest in this section. Equations (10.1)–(10.6) are associated with the classical description of the electromagnetic fields.

According to quantum physics, electromagnetic radiation consists of an infinite number of modes, each of which is characterized by a wave vector $\vec{q}$ and a specific polarization $\vec{\xi}$. Each mode $\vec{q}, \vec{\xi}$ may be described in terms of a harmonic oscillator of the frequency ω_q. Correspondingly, the energy separation between levels of this quantum-mechanical oscillator is

$$\hbar\omega_q = \frac{\hbar c}{\sqrt{\kappa}} q. \tag{10.7}$$

This oscillator can be in the nonexcited state, which manifests the so-called ground state or zero-point vibrations of the electromagnetic field. Existence of this zero-point energy is a purely quantum-mechanical phenomenon. The oscillator can be excited to some energy level. Let the integer $N_{\vec{q},\vec{\xi}}$ be a quantum number of this level; then the energy of the electromagnetic field associated with the oscillator in mode $\{\vec{q}, \vec{\xi}\}$ is

$$W_{\vec{q},\vec{\xi}} V = \left(N_{\vec{q},\vec{\xi}} + \frac{1}{2} \right) \hbar\omega_{\vec{q}}, \tag{10.8}$$

where $W_{\vec{q},\vec{\xi}}$ is the energy density of the mode and V is the volume of the system. One refers to the number of excited levels $N_{\vec{q},\vec{\xi}}$ as the number of quanta or number of photons in the mode under consideration. Thus any given electromagnetic field can be described by a set of the photon numbers.

One may recall that we have used a similar description for lattice vibrations in Chapter 5 and introduced lattice-vibration quanta – phonons. Indeed, the phonon formalism and the quantum description of electromagnetic waves have much in common. This fact has deep physical roots. In both cases, the quantum-mechanical principle of the second quantization and the same Bose statistics play a key role in the theory.

Because the quantum picture has to coincide with the classical picture for a large number of photons, i.e., when $N_{\vec{q},\vec{\xi}} \gg 1$, one can match Eqs. (10.4) and (10.8) in this limit. From this comparison, it is possible to find the relations of the classical amplitude of the electric field F_0 and of the intensity of the wave to the number of photons in the corresponding mode:

$$F_0 = \sqrt{8\pi\hbar\omega_q N_{\vec{q},\vec{\xi}} / \kappa V}, \tag{10.9}$$

$$\mathcal{I}_{\vec{q},\vec{\xi}} = c\hbar\omega_q N_{\vec{q},\vec{\xi}} / \sqrt{\kappa} V. \tag{10.10}$$

Moreover, the relation between electromagnetic waves and photons is an example of the wave–particle duality that is typical for quantum physics, as discussed in Chapter 2. A comparison of different characteristics of electromagnetic fields in the classical and quantum pictures is given in Table 10.1.

It is important that different modes of an electromagnetic field do not interact with each other as reflected by the linear character of the field in free space;

Table 10.1.
Comparison between Classical and Quantum Quantities

Classical quantity	Correspondent quantum quantity
Density of optical energy W	Photon number $N = WV/\hbar\omega$
Optical intensity $\mathcal{I}(\vec{r})$	Photon-flux density $\mathcal{I}(\vec{r})/\hbar\omega$
Total optical power P	Photon flux $P/\hbar\omega = Nc$

indeed, the equations of electromagnetism are linear. An interaction between these modes is possible only in special media. Such media is called nonlinear optical media. The peculiarities of nonlinear optical heterostructures are considered in Chapters 11 and 12.

Let us calculate the number of modes inside the frequency interval $d\omega$. We consider a uniform dielectric medium with dimensions much larger than the wavelength of the light λ. It is known that the number of wave vectors inside the elementary interval $(\vec{q}, \vec{q} + d\vec{q})$ is

$$V\frac{dq_x\,dq_y\,dq_z}{(2\pi)^3}.$$

For an isotropic medium, we can transform this expression to the spherical form

$$V\frac{4\pi q^2}{(2\pi)^3}dq = V\frac{\kappa^{3/2}\omega^2}{2\pi^2 c^3}d\omega,$$

where the first term indicates the number of the modes with the modulus of the wave vector in the interval $(q, q + dq)$. Taking into account two independent polarizations of the waves, we find that the density of electromagnetic modes expressed in terms of the number of all modes per unit interval of frequency is

$$\varrho(\omega) = V\frac{\kappa^{3/2}\omega^2}{\pi^2 c^3}. \tag{10.11}$$

In terms of the wavelengths, the mode number per unit interval of the wavelengths is

$$\varrho(\lambda) = V\frac{8\pi}{\lambda^4}. \tag{10.12}$$

Equations (10.11) and (10.12) show that the density of the modes increases rapidly when ω increases or, equivalently, as λ decreases. For example, for the same interval $d\lambda$ the number of modes in the middle of the infrared range ($\lambda \approx 5 \times 10^{-4}$ cm) differs by 4 orders of magnitude from that in the visible region ($\lambda \approx 5 \times 10^{-5}$ cm). Consider the numerical example for $\hbar\omega \approx 1$ eV, i.e., $\omega \approx 1.5 \times 10^{15}$ s^{-1} at $\kappa = 1$ and $V = 1$ cm^3. For this case one obtains $\varrho(\omega) = 7 \times 10^{-3}$ s. Increasing the density of states of the electromagnetic field has important consequences: a decrease in the radiative lifetimes and an increase in the scattering rates of light in the short-wavelength range.

10.2.2 Photons in Nonuniform Dielectric Media

In Subsection 10.2.1 we introduced the modes of the electromagnetic field, and, consequently, we described the photons for free space and for uniform dielectric media with dimensions much greater than the electromagnetic wavelengths. In fact, Eqs. (10.11) and (10.12) correspond to a very large boxlike resonator. In the general case, the modes of an electromagnetic field and photons may be introduced by the following method. Let a dielectric medium be characterized by a dielectric permittivity that is dependent on space coordinates $\kappa(\vec{r})$. Let us assume that the medium is embedded in some completely or partially reflective enclosure, which is called a resonator. Maxwell's wave equations determine all harmonic modes for electromagnetic fields, $\mathcal{E}_\nu$ and $\mathcal{H}_\nu$, in the system, where ν is the set of discrete or continuous parameters that classify the solutions. Each solution is characterized by a frequency ω_ν. In the absence of absorption of electromagnetic energy, all frequencies ω_ν are real quantities. One may consider any of these solutions as a mode of the field. Thus, in this way we obtain the total mode structure of the field. Note that, in the general case, the solutions of Maxwell's equations can differ considerably from the plane-wave solution given by Eq. (10.1). The mode structure, the frequency, the density of states, and other characteristics are strongly dependent on the type of dielectric medium, its geometry, and the properties of the optical resonator.

As an example of a spatially nonuniform system, let us consider a semiconductor structure with variations in the composition and bandgap in one direction, say, in the z direction, as illustrated in Fig. 10.1(a). This variation in the bandgap causes a change in the dielectric constant so that the narrow-bandgap layer of the structure has a larger optical density. For this case, the system of the modes is different from that of the plane waves. In particular, there are the modes – the waveguide modes – localized within the narrow-bandgap layer. Of course,

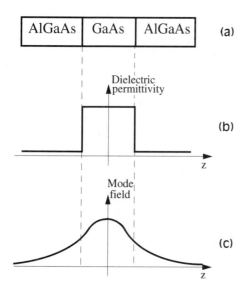

(a)

(b)

(c)

Figure 10.1. A double GaAs/AlGaAs heterostructure as a waveguide: (a) the scheme of the structure, (b) the coordinate dependence of the dielectric permittivity, (c) the sketch of a spatial profile of the fundamental optical mode localized near a GaAs narrow-bandgap layer.

the amplitudes of these modes decay far away from the device region. One can again recall an analogy with the lattice vibrations analyzed in Chapter 5: In semiconductor heterostructures, different vibrational modes can occur, including the localized modes. Of special significance is the fact that similar structures and the localized modes are used in the modern heterostructure laser.

We already know that according to the theory of quantization of the electromagnetic field, each of the previously considered modes represents a particular oscillator, or photon, with energy ω_ν. An excitation of this oscillator on the level N_ν corresponds to the number of photons equal to N_ν and the energy of the mode is equal to $W_\nu V = \hbar\omega_\nu(N_\nu + \frac{1}{2})$; see Eq. (10.8). The solutions $\mathcal{E}_\nu(\vec{r})$ and $\mathcal{H}_\nu(\vec{r})$, on the basis of which these photons are introduced, are frequently referred to as the form factors of the photon. The form factor determines the spatial distributions of the field for a fixed mode and represents an important characteristic of the mode. The probability $p(\vec{r})\mathrm{d}V$ of observing a photon at a point $\vec{r}$ is proportional to the density of electromagnetic energy $W(\vec{r})$, which can be calculated with the form factor. For example, let the form factor of a certain mode be a standing wave $\mathcal{E}(z) = F_0 \sin 2\pi z/\lambda$. Then at the points $z_n = \lambda n/2$ (nodal points), the probability of observing the photon is equal to zero; $p = 0$. An atom placed at a point where $\mathcal{E} = 0$ does not interact with this mode. At any point where the form factor does not vanish, the atom interacts with the field and can absorb or emit the photon corresponding to the form factor.

These properties of electromagnetic fields are important for tailoring the light–matter interaction because they facilitate the determination of the desired distributions of the fields and achieve efficient interaction with the electron subsystem. These principles work especially well in semiconductor lasers, photodetectors, waveguides, and other optoelectronics devices.

10.2.3 Optical Resonators

In the simplest case, the resonator consists of plane or curved mirrors that provide repeated reflections and some kind of trap of the light – cavity – in the region between the mirrors. The optical waves that can be trapped in this cavity comprise the resonator modes. A universal characteristic of the resonator mode is the quality factor Q, which can be defined as

$$Q = \omega \times \frac{\text{field energy stored in the cavity}}{\text{dissipated power in the resonator}}. \tag{10.13}$$

The dissipation of electromagnetic energy is caused by many factors: absorption by the mirrors or by matter inside the cavity, light transmission through the mirrors, light scattering, radiation out of the resonator as a result of light diffraction, etc. The dissipated power and the quality factor may be different for different modes.

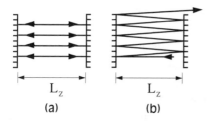

(a)   (b)

Figure 10.2. Longitudinal and transverse modes in the planar Fabry–Perot resonator. (a) Longitudinal modes. Strictly perpendicular light rays do not escape from the resonator. (b) Transverse modes. Slightly inclined light rays eventually escape from the resonator and have poor-quality factors.

Discussing Eq. (10.12) we have stressed one feature of electromagnetic waves: the very high density of modes in the optical region. This high mode density makes possible the emission of a large number of modes with different frequencies, polarizations, and all possible propagation directions. To avoid this type of extremely incoherent excitation of modes, one can use so-called open optical resonators. The simplest open resonator consists of two plane mirrors parallel to each other, which are finite in their transverse dimensions; an example is given by the well-known Fabry-Perot etalon, which is shown in Fig. 10.2. In this resonator, the majority of modes propagating through the cavity are lost in a single traversal of the cavity since the mirrors are inclined with respect to the mode-propagation directions. This implies that the majority of the modes – so-called transverse modes – have a low-quality factor. Only the waves propagating perpendicularly to the mirrors can be reflected back and travel from one mirror to another without escaping from the resonator. These waves correspond to the so-called longitudinal modes of the resonator. Thus, for finite dimensions of mirrors, only longitudinal modes can have a high-quality factor. Their loss and diffraction are caused by absorption by the mirrors, by transmission through the mirror, and by wave diffraction on the sides of the mirrors. The latter losses can be made much smaller than other loss mechanisms. Thus the open resonator provides a strong discrimination of modes. Relatively few of these modes have a high-quality factor Q. According to the definition of Eq. (10.13) it is possible to accumulate light energy in these high-quality modes.

For ideal mirrors with 100% reflection, the longitudinal modes are standing waves:

$$\vec{\mathcal{E}} = \vec{\xi} F_0 \sin qz \cos \omega t, \tag{10.14}$$

where $\vec{\xi}$ is perpendicular to the resonator axis z. Since the electric field should vanish at the surfaces of the mirrors, at $z = 0$ and at $z = L_z$, one obtains

$$q = \frac{\pi}{L_z} l, \quad l = 1, 2, 3, \ldots . \tag{10.15}$$

Thus, instead of a continuous spectrum of electromagnetic waves, as in the case of free space, one gets an infinite set of equally separated frequencies, as illustrated in Fig. 10.3(a):

$$\omega_{\text{long},l} = \frac{\pi c}{L_z \sqrt{\kappa}} l. \tag{10.16}$$

The separation between the frequencies of the longitudinal modes depends on the resonator length L_z.

It is easy to take into account finite reflection and transmission of the mirrors and losses in the cavity. Let us consider an external wave traveling through the resonator, say from left to right. Let its amplitude at a particular point of the resonator be F_0. After a double reflection and at the same point, we find a wave with the same propagation direction but with amplitude $re^{2i\delta}F_0$, where r is the attenuation factor due to the finite mirror transmission, absorption in the cavity, etc., and $\delta \equiv \omega\sqrt{\kappa}L_z/c$ is the change of the wave's phase after a double traversal of the resonator. We obtain a similar result after the next double traversal: $re^{2i\delta}(re^{2i\delta})F_0$. In fact, the amplitude of the electromagnetic field can be found as a superposition of these waves:

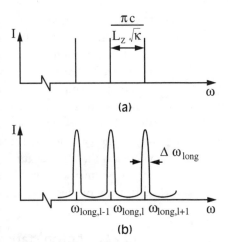

(a)

(b)

Figure 10.3. The schematics of spectral dependences of intensities of the resonator modes: (a) an ideal lossless resonator, (b) a resonator with finite losses.

$$F = F_0 + re^{2i\delta}F_0 + r^2 e^{4i\delta}F_0 + \cdots = \frac{F_0}{1 - re^{2i\delta}}.$$

Thus we can write the light intensity in the resonator as

$$\mathcal{I} = \frac{\mathcal{I}_0}{|1 - re^{2i\delta}|^2} = \frac{\mathcal{I}_0}{1 + (2\mathcal{F}/\pi)^2 \sin^2 \delta} \frac{1}{(1-r)^2}, \tag{10.17}$$

where $\mathcal{I}_0$ has a meaning of a maximal intensity inside the resonator and $\mathcal{F} \equiv \pi\sqrt{r}/(1-r)$ is a parameter, called the finesse or the contrast of the resonator. From Eq. (10.17) we see that the intensity is a periodic function of the frequency ω, and it has maxima when ω satisfies the resonant conditions of Eq. (10.16). The peaks have a full width at half-maximum equal to

$$\Delta\omega_{\text{long}} = \frac{\pi c}{L_z \mathcal{F}}.$$

Thus, instead of a set of infinitely narrow and high lines, as in Fig. 10.3(a), we obtain the mode structure presented in Fig. 10.3(b). In Eq. (10.17), the value $\mathcal{I}_0$ is still arbitrary. It is determined by the method of generation of the electromagnetic fields, which can be an external light accumulated in the cavity or light generated by the emission inside this cavity. The ratio between the maximum and the minimum of the intensity is independent of these methods:

$$\frac{\mathcal{I}_{\text{min}}}{\mathcal{I}_{\text{max}}} = \frac{1}{1 + (2\mathcal{F}/\pi)^2}.$$

Thus the Fabry–Perot etalon allows us to discriminate between transverse and longitudinal modes and provides high-quality factors for the latter.

The density of states for the longitudinal modes is

$$\varrho_{\text{long}}(\omega) = 2\frac{L_z}{\pi c}, \tag{10.18}$$

where we have taken into account the two possible polarizations of the wave. This density of states is independent of the light frequency. For $L_z = 1$ cm we find that the separation between the modes is 9.4×10^{10} s^{-1} and $\varrho_{\text{long}} = 2 \times 10^{-11}$ s. (Compare the latter value with the previously presented estimate for free space.)

Typical resonator lengths for semiconductor applications are a few hundreds of micrometers. For these resonators, the intermodal spacing is still less than the spectral range of emission. To decrease the number of excited modes it is necessary to increase the intermodal spacing, $2\pi c/\lambda$, to the extreme limit, $L_z = \lambda/2$, where λ is the wavelength in the resonator. Such a resonator is called a microresonator or microcavity. Modern semiconductor technology makes possible the fabrication of these resonators. We consider microresonators in Chapter 11.

10.2.4 Photon Statistics

For most cases of interest for optoelectronic applications, the electromagnetic fields are far from equilibrium. However, it is instructive to recall from physics that as a result of the interaction with a blackbody characterized by a temperature T, the electromagnetic radiation can also come into equilibrium and be characterized by the same temperature. We have already mentioned the analogy of the quantum description of lattice vibrations and electromagnetic waves. Both phonons and photons are bosons and obey the same laws of statistics. In Chapter 5, we introduced the boson equilibrium statistics and Planck's formula for phonons. Accordingly, the number of photons of some chosen mode in equilibrium is

$$N_{\vec{q},\xi} = \frac{1}{\exp(\hbar\omega_q/k_B T) - 1}. \tag{10.19}$$

Equations (10.4), (10.7), and (10.19) allow us to evaluate all the equilibrium properties of the radiation: spectral density of the energy, total energy, etc.

10.3 Light Interaction with Matter: Phototransitions

10.3.1 Photon Absorption and Emission

Among the different processes of the interaction of electromagnetic fields and matter, here we review briefly the three major processes: absorption, spontaneous emission, and stimulated emission.

To visualize these processes, let us consider a simple two-level system with energies E_1 and E_2, as depicted in Fig. 10.4. The different occupancies of the energy levels of this system correspond to particular states of a system of charged particles, say, the electrons. The charged particles interact with the electromagnetic field. This interaction is, of course, associated with the transitions between quantum states of the system. These transitions are frequently referred to as phototransitions. According to quantum theory, the system changes its energy

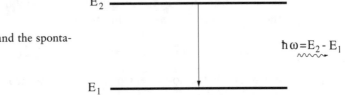

Figure 10.4. A two-level system and the spontaneous emission of a photon.

exclusively as a result of an interaction with electromagnetic waves with a frequency of

$$\omega = (E_2 - E_1)/\hbar. \tag{10.20}$$

If the lowest energy level E_1 is occupied, the wave can excite the system into upper level E_2 and the electromagnetic energy must decrease. One can describe this process as the absorption of one photon because the energy of the electromagnetic field decreases by $E_2 - E_1$. If the system occupies the upper level E_2, it can make a transition to level E_1 as a result of interaction with the electromagnetic field. Then the electromagnetic energy increases by $E_2 - E_1$. This process represents the emission of a photon with energy $\hbar\omega$. When activated by an external electromagnetic wave, the latter process is called stimulated emission. It is important that for stimulated emission, each emitted photon has the energy, direction, polarization, and even phase, coinciding precisely with those of the stimulating wave.

Both processes, absorption and stimulated emission, can be described in terms of an interaction with a classical electromagnetic wave. The rates of these processes are proportional to the intensity of the wave. Thus according to Eq. (10.10), these rates are proportional to the number of photons $N_{\vec{q},\vec{\xi}}$:

$$R_{\text{abs}} = B_{12} N_{\vec{q},\vec{\xi}} n_1, \tag{10.21}$$

$$R_{\text{st.em}} = B_{21} N_{\vec{q},\vec{\xi}} n_2, \tag{10.22}$$

where n_1 and n_2 are the numbers of particles in the system occupying levels 1 and 2, respectively; B_{12} and B_{21} are the kinetic coefficients describing these processes.

The two processes of absorption and emission are insufficient to describe the whole picture of the interaction between radiation and matter. For example, let us apply these two processes for only thermal-equilibrium conditions, under which the ratio of the populations of both levels is

$$\frac{n_2}{n_1} = e^{-(E_2 - E_1)/k_B T} = e^{-\hbar\omega/k_B T}; \tag{10.23}$$

for the sake of simplicity, we assume that the degeneracy equals 1 for each level. Using Eqs. (10.21)–(10.23), we can see that $R_{\text{abs}} \neq R_{\text{st.em}}$ at any temperature T. This result is in contradiction to the expected equilibrium between the system and the field. According to the Einstein theory, there is an additional quantum radiative transition in the system with the spontaneous emission of a photon of the same mode. The rate of this process is

$$R_{\text{sp.em}} = A_{21} n_2, \tag{10.24}$$

where A_{21} is the coefficient or rate of spontaneous emission. The spontaneous process does not depend on the intensity of the electromagnetic wave and takes place even in the absence of this wave. According to quantum electrodynamics, the excited material system spontaneously emits a photon because of interaction with zero-point vibrations of the electromagnetic fields. These zero-point vibrations are related to the fact that the ground-state energy of a harmonic oscillator of frequency ω is $\hbar\omega/2$ and not zero, as it would be for a quantum harmonic oscillator.

In contrast to the case of stimulated emission, a photon produced by the spontaneous process has an arbitrary phase. Moreover, this process produces photons with different directions of $\vec{q}$ and polarizations, but the energy is fixed, i.e., it produces photons of different modes.

Now we can apply the results of Eqs. (10.21), (10.22), and (10.24) to thermal equilibrium. Under equilibrium conditions the total rate of the photon emission has to be equal to the rate of the photon absorption; thus we will have

$$R_{\text{abs}} = R_{\text{sp.em}} + R_{\text{st.em}}. \tag{10.25}$$

Using the Planck formula of Eq. (10.19), the ratio of n_2/n_1 of Eq. (10.23), and substituting the expressions for R_{abs}, $R_{\text{sp.em}}$, and $R_{\text{st.em}}$ into Eq. (10.25), one can find the relation

$$B_{21} - A_{21} = (B_{12} - A_{21})e^{\hbar\omega/k_B T}. \tag{10.26}$$

Because this relation has to be satisfied at any arbitrary temperature T, one can obtain two equalities:

$$A_{21} = B_{21} = B_{12}. \tag{10.27}$$

Thus we have established the existence of three basic processes of resonant interaction of radiation and matter: absorption, stimulated emission, and spontaneous emission. Moreover, we have found the relations among the coefficients that determine the rates of these processes. It is worth emphasizing that all three processes are related to interactions with photons of the same mode.

As was shown in Subsection 10.2.3 the optical resonator effectively enhances the mode structure in the cavity. Combining this observation with the basic optical processes discussed previously, we conclude that these processes are not immutable properties of an atom, molecule, or another system. In fact, these processes, including spontaneous emission, are consequences of the matter–field coupling and they would disappear if there were no proper resonant states of the electromagnetic field. For example, in the mirror box considered previously, an atom does not emit or absorb photons if its frequency, as given by Eq. (10.20), differs from the resonant-mode frequencies. This result is not significant for resonators with dimensions much larger than the resonant wavelength but it becomes important for microresonators with large intermodal frequency differences.

The sum of the stimulated- and the spontaneous-emission rates, as determined by Eqs. (10.22) and (10.24), gives the total emission rate of photons for a fixed mode:

$$R_{em} = A_{21}(1 + N_{\vec{q},\vec{\xi}})n_2. \tag{10.28}$$

From Eq. (10.28) one can see that stimulated emission dominates over spontaneous emission for a fixed mode $\{\vec{q}, \vec{\xi}\}$ if the number of photons $N_{\vec{q},\vec{\xi}}$ is sufficiently larger than 1. However, there is a spontaneous emission of a great number of other modes with the same frequency but with different directions of $\vec{q}$ and different polarizations. This total spontaneous emission can be the dominant radiative process even if stimulated emission is the most important process for a particular mode $\{\vec{q}, \vec{\xi}\}$.

Now let us compare absorption and stimulated emission by calculating the rate of increase of the number of photons in some fixed modes:

$$\left(\frac{dN_{\vec{q},\vec{\xi}}}{dt}\right)_{st} \equiv R \equiv R_{st.em} - R_{abs} = B_{21} N_{\vec{q},\vec{\xi}}(n_2 - n_1). \tag{10.29}$$

This result shows that, if

$$n_2 - n_1 > 0, \tag{10.30}$$

stimulated emission dominates over absorption. Evidently, under equilibrium conditions, the opposite is true.

Inequality (10.30) is the criterion for population inversion. If a population inversion is achieved, electromagnetic waves with the resonant frequency can be amplified when passing through the material medium. This process of amplification of the radiation due to population inversion is the key mechanism underlying the operation of a laser (light amplification by stimulated emission of radiation).

In the previously obtained equations, the coefficients A_{21}, B_{21}, and B_{12} that describe the interaction of waves and matter are linked by the two equalities of Eq. (10.27), but at least one of these coefficients has to be calculated independently. At this point, we end our discussion of the simple two-level model, and we start to consider calculations for more realistic systems.

10.3.2 Calculation of Phototransition Probabilities

We assume that there are a number of electron states with energies E_i, where i can take on discrete as well as continuous values. For simple, but the most important cases, we can use the dipole approximation for the interaction energy of the light–matter interaction:

$$V_{\text{int}} = -\vec{\mathcal{D}}\vec{\mathcal{E}}, \tag{10.31}$$

where

$$\vec{\mathcal{D}} = \sum_n \vec{\mathcal{D}}_n \equiv e \sum_n \vec{r}_n Z_n \tag{10.32}$$

is the dipole moment of the system, $Z_n = 1$ for holes and -1 for electrons, and $\vec{\mathcal{E}}$ is the electric field of the wave in Eq. (10.1). In Eq. (10.32), the sum extends over all charges of the system whose coordinates are $\vec{r}_n$. We consider here only phototransitions involving electrons and do not analyze the interaction of light with the lattice.

For a bulk material, the photon wave vector $\vec{q}$ can be neglected in Eq. (10.1), because the wave amplitude varies over distances that usually are substantially larger than any characteristic scale of the electron system. In the framework of the semiclassical description of the electromagnetic waves, we can calculate the transition probability per unit time between any two states of the system, say, between the initial state i and the final state f. For such a process, the interaction energy of Eq. (10.31) acts as a perturbation of the system. According to Fermi's golden rule, derived in Section 2.4, the probability of transition of one of the electrons from state i to state f is

$$P_{i \to f} = \frac{2\pi}{\hbar} |\langle f \,|\, \bar{V} \,|\, i \rangle|^2 \delta(E_f - E_i - \hbar\omega), \tag{10.33}$$

where it is assumed that the perturbation has the form of Eq. (2.72):

$$V = \bar{V} e^{i\omega t} + \bar{V}^* e^{-i\omega t}.$$

In the case under consideration we have

$$\bar{V}^* = \bar{V} = -\frac{1}{2} F_0 \,\vec{\xi}\vec{\mathcal{D}}.$$

Now the transition probability is

$$P_{i \to f} = \frac{\pi}{2\hbar} |\langle f \,|\, F_0 \vec{\xi} \vec{\mathcal{D}} \,|\, i \rangle|^2 \delta(E_f - E_i - \hbar\omega), \tag{10.34}$$

where $\langle f \,|\, F_0 \vec{\xi} \vec{\mathcal{D}} \,|\, i \rangle$ is the matrix element calculated for the wave functions of states i and f. If $\mathcal{F}(E_\nu)$ denotes the average occupancy of some state ν of the system by electrons, one can write the rate of phototransitions from state i to state f in the form

$$R_{i \to f} = P_{i \to f} \mathcal{F}(E_i)[1 - \mathcal{F}(E_f)], \tag{10.35}$$

where the last multiplier is the number of empty places that can be populated as a result of the phototransition. Similarly, one can write the rate of the inverse process, $f \to i$, as

$$R_{f \to i} = P_{f \to i} \mathcal{F}(E_f)[1 - \mathcal{F}(E_i)], \tag{10.36}$$

with

$$P_{f \to i} = \frac{\pi}{2\hbar} e^2 |\langle i \, | \, F_0 \vec{\xi} \vec{r} \, | \, f \rangle|^2 \, \delta(E_f - E_i - \hbar\omega). \tag{10.37}$$

The δ functions in Eqs. (10.34) and (10.37) ensure that the initial and the final states are separated by the photon energy $\hbar\omega$. Finally, by summing Eqs. (10.35) and (10.36) over all possible initial and final states i and f, we can obtain the total rate of increase of the number of photons in the fixed mode:

$$
\begin{aligned}
R &= \sum_{i,f} (R_{f \to i} - R_{i \to f}) \\
&= \frac{\pi e^2}{2\hbar} \sum_{i,f} |\langle i \, | \, F_0 \vec{\xi} \vec{r} \, | \, f \rangle|^2 \, \delta(E_f - E_i - \hbar\omega)[\mathcal{F}(E_f) - \mathcal{F}(E_i)].
\end{aligned}
\tag{10.38}
$$

One can see that Eq. (10.38) is an analogy of Eq. (10.29) that describes the same quantity for the simple two-level model. Thus the rate of increase (decrease) of the number of the photons R is proportional to the difference of the occupancy of the electron states between which the phototransitions occur.

For a spatially homogeneous crystal, Eq. (10.38) allows one to calculate another important characteristic of the system, namely, the amplification (absorption) coefficient of the electromagnetic wave.

The change in the energy density of electromagnetic waves can be expressed through the rate of increase of the photon number R:

$$\frac{dW}{dt} = \frac{\hbar\omega}{V} R. \tag{10.39}$$

Let us assume that the wave propagates along the z axis. Then we can rewrite Eq. (10.39) in terms of the derivative with respect to z:

$$\frac{dW}{dt} = \frac{dW}{dz} \frac{c}{\sqrt{\kappa}}. \tag{10.40}$$

Eliminating F_0^2 from Eqs. (10.4) and (10.38), we get

$$\mathcal{I} \frac{1}{W} \frac{dW}{dz} \equiv \alpha. \tag{10.41}$$

According to Eq. (10.5), we can rewrite Eq. (10.41) as

$$\frac{d\mathcal{I}}{dz} = \alpha\mathcal{I}. \tag{10.42}$$

From Eq. (10.42) it is obvious that the coefficient α characterizes an amplification (absorption) of light. At $\alpha > 0$ this coefficient is called the amplification coefficient or the gain coefficient. At $\alpha < 0$ the value $-\alpha$ is called absorption coefficient. For the case of plane waves, this coefficient is

$$\alpha = \frac{4\pi^2 e^2 \omega}{c\sqrt{\kappa}\,V} \sum_{i,f} |\langle i\,|\,\vec{\xi}\vec{r}\,|\,f\rangle|^2 \,\delta(E_f - E_i - \hbar\omega)[\mathcal{F}(E_f) - \mathcal{F}(E_i)]. \tag{10.43}$$

Thus Eq. (10.43) gives the increment (decrement) of electromagnetic-wave intensity in space due to phototransitions in the system interacting with the wave:

$$\mathcal{I}(z) = \mathcal{I}_0 e^{\alpha z}, \tag{10.44}$$

where I_0 is the wave intensity at $z = 0$. Absorption of light occurs if $\alpha < 0$; amplification is possible if $\alpha > 0$.

It follows from these results that the properties of phototransitions depend on the wave functions and the energies of the electron states involved in the process. The electron occupancies of the states are also important. The initial and the final states can belong to different electron bands, or to impurities and a band, or to the same band. Depending on these factors, different mechanisms of absorption and emission of light are possible.

10.4 Optical Properties of Bulk Semiconductors

In this section we consider the mechanisms of absorption and emission of photons in semiconductors. Among them, the most important are the following mechanisms.

1. *Interband phototransitions (band to band)*: An absorbed photon can result in the creation of an electron–hole pair. An inverse process is radiative electron–hole recombination, resulting in the emission of a photon, as illustrated in Fig. 10.5(a).
2. *Impurity-to-band transitions*: In doped semiconductors, an absorbed photon can result in a transition between the bound state of an impurity – donor or acceptor – and the conduction or the valence band, as illustrated in Fig. 10.5(b) for the case of acceptors.

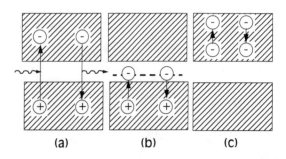

(a) (b) (c)

Figure 10.5. Schematics of absorption and emission of photons in a semiconductor: (a) band-to-band transitions, (b) the absorption of a photon leads to a valence-band to acceptor-level transition, (c) a free-carrier transition within the conduction band.

3. *Free-carrier transitions (intraband)*: An absorbed photon can transfer its energy to an electron or a hole, thereby increasing its energy within the same band, as illustrated in Fig. 10.5(c). It is important to note that free-carrier transitions require the participation of a third particle (another carrier, defect, phonon, etc). Inverse processes, such as emission of infrared photons, also occur.

4. *Excitonic transitions*: The absorption of a photon can lead to the formation of an electron and a hole in coupled states. These weakly bound electron–hole states are known as excitons, and they were studied in Section 3.7. Interband- and excitonic-absorption spectra of a semiconductor are illustrated in Fig. 3.22. An annihilating exciton can produce a photon as a result of electron–hole recombination.

5. *Phonon transitions*: Long-wavelength photons can be absorbed in the excitation of lattice vibrations, i.e., in the process of creating phonons. The last mechanism does not involve electrons, but it can result in an overlapping of this type of absorption with intraband phototransitions; for example, phonon absorption occurs for the regions from 0.02 to 0.07 eV for GaAs and from 0.1 to 0.2 eV for Si, while free-carrier transitions correspond to energies of less than 0.3 eV for both materials.

The above-listed phototransitions contribute to the overall absorption in a wide spectral region from far-infrared up to ultraviolet spectra. In the subsections below we consider in detail the phototransition mechanisms that have just been introduced.

10.4.1 Interband Emission and Absorption in Bulk Semiconductors

Figure 10.6 illustrates the dependence of the absorption coefficient on photon energy and wavelength for various semiconductors. From this figure, one can see that the absorption increases sharply in the short-wavelength region. Let E_g be the bandgap of the semiconductor. Then the material is relatively transparent for $\hbar\omega < E_g$. It changes properties to strong absorption for $\hbar\omega > E_g$, so that $\hbar\omega_g = E_g$ corresponds to the absorption edge. The shape of the absorption edge depends significantly on the structure of the electron bands. Direct-bandgap semiconductors such as GaAs have a more abrupt absorption edge and a larger absorption value than indirect-bandgap materials, of which Si provides an example. We can introduce the so-called bandgap wavelength or the cutoff wavelength $\lambda_g = 2\pi c\hbar/E_g$. If E_g is given in electron volts, the bandgap wavelength in micrometers is

$$\lambda_g = \frac{1.24}{E_g}. \tag{10.45}$$

The values of E_g and λ_g for various III–V semiconductor materials are apparent from the curves plotted in Fig. 10.6. One can see that interband transitions in

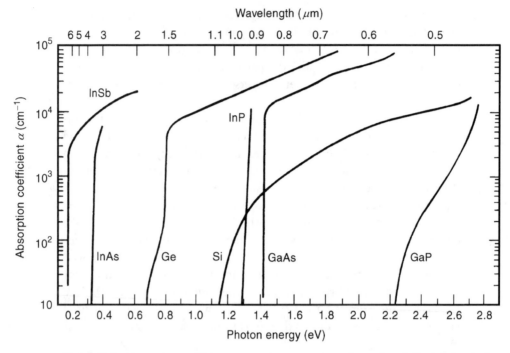

Figure 10.6. Absorption coefficient versus photon energy and wavelength for interband phototransitions in various semiconductors. [After G. E. Stillman, *et al.*, "III–V compound semiconductor devices: optical detectors," IEEE Trans. Electron Devices **ED-31**, pp. 1643–1655 (1984).]

III–V compounds cover a wide range from infrared to visible spectra. Optical activity in this spectral region is crucial for all optoelectronic applications of these materials.

A photon absorbed during an interband transition excites an electron from the valence band to the conduction band, i.e., it creates an electron–hole pair, as shown in Fig. 10.7(a). The inverse process – the phototransition of an electron

Figure 10.7. (a) The absorption of a photon results in electron–hole generation, (b) the radiative recombination, (c) stimulated emission.

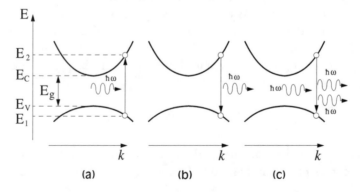

from the conduction band to the valence band – is referred to as the radiative recombination (annihilation) of an electron and a hole; this is depicted in Figs. 10.7(b) and 10.7(c). According to the general properties of phototransitions studied in Section 10.3 there are two such processes: spontaneous and stimulated emission, as illustrated by Figs. 10.7(b) and 10.7(c), respectively.

To analyze the interband transitions, we use the general expression for the rates of absorption and emission of Eq. (10.38). In the case under consideration, initial state i and final state f should be selected from the states of the conduction and the valence bands. For bulk materials, the set of quantum numbers depends on the type of band (conduction or valence), the wave vector of the carrier ($\vec{k}_c$ or $\vec{k}_v$, respectively), and the spin σ. Since we consider the simplest case, we neglect effects associated with the change of the spin states, so that spin conservation during phototransitions is assumed. That is why we should use $i = \{v, \vec{k}_v\}$ and $f = \{c, \vec{k}_c\}$ in Eq. (10.38) instead of the complete set of quantum numbers.

Before we can present further calculations, it is necessary to make a general remark concerning the electron wave functions and the interband transitions. So far, we have studied processes involving electron states of the same energy band, i.e., intraband processes. We have implicitly relied on the effective-mass method (see Section 2.6) and the corresponding Schrödinger equation and its solutions. It is now appropriate to recall that the true wave functions of electrons in a bulk crystal have the form

$$\Psi_{\vec{k}}(\vec{r}) = \frac{1}{\sqrt{V}} e^{i\vec{k}\vec{r}} u_{\vec{k}}(\vec{r}), \tag{10.46}$$

where V is the crystal volume, $\vec{k}$ is the electron wave vector, and $u_{\vec{k}}(\vec{r})$ is the Bloch function with the specific property that it is periodical with the period of the crystal lattice, and therefore $u_{\vec{k}}(\vec{r})$ repeats itself in each unit cell of the crystal. We normalize the Bloch functions by requiring that

$$\frac{1}{\Omega} \int_{\Omega} u_{\vec{k}}^*(\vec{r}) u_{\vec{k}}(\vec{r}) \, d\vec{r} = 1, \tag{10.47}$$

where Ω is the volume of the unit cell of the crystal. Any electron states not representable by the plane wave functions in Eq. (10.46), can be written as a linear combination of such plane waves:

$$\Psi(\vec{r}) = \int B(\vec{k}) e^{i\vec{k}\vec{r}} u_{\vec{k}}(\vec{r}) \, d\vec{k}. \tag{10.48}$$

For electron states that are smooth within each lattice period, one can approximate $\Psi(\vec{r})$ as

$$\Psi(\vec{r}) = u_{\vec{k}_0}(\vec{r}) \int B(\vec{k}) e^{i\vec{k}\vec{r}} \, d\vec{k} \equiv u(\vec{r}) F(\vec{r}), \tag{10.49}$$

where $u_{\vec{k}_0}(\vec{r}) \equiv u(\vec{r})$ is the Bloch function for the minimum (maximum) of the electron (hole) energy. The last transformation is based on the key assumption that within a given energy band, the Bloch function is a smooth function of $\vec{k}$ in the vicinity of the band edge. Therefore the actual wave function consists of two multipliers: The first is the Bloch function at the extremum of the corresponding energy band, and the second is the so-called envelope function, analyzed in Subsection 2.6.4.

The difference between intraband and interband processes is now clear. The former involves only one Bloch function $u_{\vec{k}_0}(\vec{r})$, and it does not appear explicitly in the final result. For example, let a potential Φ be responsible for the transitions between long-wavelength electron states in the same energy band, say b: $\{b, \nu\} \rightarrow \{b, \nu'\}$, where ν labels the set of intraband quantum numbers. Then the matrix element of the corresponding process is

$$
\begin{aligned}
\langle b, \nu \,|\, \Phi \,|\, b, \nu' \rangle &= \int u^*(\vec{r}) F_{\nu'}^*(\vec{r}) \Phi(\vec{r}) u(\vec{r}) F_\nu(\vec{r}) \, d\vec{r} \\
&= \int_\Omega u^*(\vec{r}) u(\vec{r}) \, d\vec{r} \sum_{\vec{R}_j} F_{\nu'}^*(\vec{R}_j) \Phi(\vec{R}_j) F_\nu(\vec{R}_j) \\
&= \Omega \sum_{\vec{R}_j} F_{\nu'}^*(\vec{R}_j) \Phi(\vec{R}_j) F_\nu(\vec{R}_j) = \int F_{\nu'}^*(\vec{r}) \Phi(\vec{r}) F_\nu(\vec{r}) \, d\vec{r}.
\end{aligned}
$$

Here we have used the normalization condition of Eq. (10.47), $\vec{R}_i$ is the coordinate vector at the center of the ith cell of the crystal, and the sum has been replaced by an integral. We see that for intraband processes, only the envelope functions are needed. A principally different situation occurs in the case of interband transitions, which are studied here.

We now return to an analysis of Eqs. (10.33)–(10.38). If E_c and E_v are the energies of electrons and holes, the δ function in Eq. (10.37) dictates the energy-conservation law:

$$
E_c - E_v = \hbar\omega. \tag{10.50}
$$

In Eq. (10.34) for the phototransition rates, taking the spin into account, the matrix element is

$$
\langle i \,|\, \vec{\xi}\vec{D} \,|\, f \rangle \equiv \langle \nu, \vec{k}_v \,|\, \vec{\xi}\vec{D} \,|\, c, \vec{k}_c \rangle,
$$

with the wave functions

$$
\Psi_{v, \vec{k}_v} = \frac{1}{\sqrt{V}} u_{v, \vec{k}_{\max}}(\vec{r}) e^{i\vec{k}_v \vec{r}},
$$

$$
\Psi_{c, \vec{k}_c} = \frac{1}{\sqrt{V}} u_{c, \vec{k}_{\min}}(\vec{r}) e^{i\vec{k}_c \vec{r}},
$$

where $u_{v,\vec{k}_{max}}(\vec{r})$ and $u_{c,\vec{k}_{min}}(\vec{r})$ are the Bloch functions of the electrons in the valence and the conduction bands, respectively. Addressing the case of a direct-bandgap semiconductor, we let both the minimum of the conduction band and the maximum of the valence band be in the same point of $\vec{k}$ space, $\vec{k}_{min} = \vec{k}_{max} = \vec{k}_0$. In this case the matrix element can be written as

$$\langle v, \vec{k}_v \,|\, \vec{\xi}\vec{D} \,|\, c, \vec{k}_c\rangle \equiv \langle v, \vec{k}_v \,|\, \vec{\xi}\vec{D} \,|\, c, \vec{k}_c\rangle$$

$$= \frac{1}{V}\int u^*_{v,\vec{k}_0}\, e^{-i\vec{k}_v\vec{r}}(\vec{\xi}\vec{D})u_{c,\vec{k}_0}\,e^{i\vec{k}_c\vec{r}}\,d\vec{r}$$

$$= \frac{1}{V}\sum_{\vec{R}_i} e^{-i(\vec{k}_v-\vec{k}_c)\vec{R}_i}\int_\Omega u^*_{v,\vec{k}_0}(\vec{\xi}\vec{D})u_{c,\vec{k}_0}\,d\vec{r}. \qquad (10.51)$$

The last integral is taken over a unit cell of the crystal, and the sum runs over all cells. It is important that the integral does not depend on the cell number and is equal to $\Omega(\vec{\xi}\vec{D}_{vc})$. Here $\vec{D}_{vc}$ is the matrix element involving just the Bloch functions. This quantity provides the selection rules for phototransitions. It depends on the symmetries of the corresponding bands. The matrix element of Eq. (10.51) also depends on the polarization of the light with respect to the crystal axes.

In Eq. (10.51), the sum over all cells of the crystal gives, as usual, the Kronecker δ function:

$$\frac{\Omega}{V}\sum_{\vec{R}_i} e^{-(\vec{k}_v-\vec{k}_c)\vec{R}_i} = \delta_{\vec{k}_v,\vec{k}_c}.$$

Hence, for the matrix element in Eq. (10.51), we get

$$|\langle v, \vec{k}_v \,|\, \vec{\xi}\vec{D} \,|\, c, \vec{k}_c\rangle|^2 = |(\vec{\xi}\vec{D})_{c,v}|^2\delta_{\vec{k}_v,\vec{k}_c}. \qquad (10.52)$$

In the limit $V \to \infty$, we can replace $V\delta_{\vec{k}_v,\vec{k}_c}$ with $\delta(\vec{k}_c - \vec{k}_v)$. Thus we get an important result – the quasi-momentum law or the selection rule for the initial and the final momenta of electron and hole participating in the phototransition:

$$\vec{k}_v = \vec{k}_c. \qquad (10.53)$$

Transitions obeying this rule are presented in the E–$\vec{k}$ diagram in Fig. 10.7 by vertical lines. Consequently, changes in the wave vector $\vec{k}$ are neglected during phototransitions. Vertical transitions are possible only in direct-bandgap materials, in which the minimum of the conduction band and the maximum of the valence band are situated at same point of the $\vec{k}$ space, as we assumed in our derivation. In the case of indirect-bandgap material, phototransitions with the phonon energy equal to or slightly larger than the bandgap are forbidden without the participation phonons. A phonon carries momentum, and an adequate phonon momentum can make indirect phototransitions possible.

The criterion of (10.53) is approximate because we assume that the electric field of the electromagnetic wave is almost constant in space. If we use the exact formula for the electric field of the wave, as given by Eq. (10.1), we would obtain the exact selection rule:

$$\vec{k}_c - \vec{k}_v = \vec{q}, \qquad |\vec{q}| = \frac{2\pi}{\lambda}.$$

Expressing $k_{c,v}$ through the electron and the hole kinetic energies,

$$k_{c,v} = \sqrt{2m_{c,v} E_{c,v}}/\hbar,$$

one can estimate the ratio of the photon wave vector q to the particle wave vector $k_{c,v}$:

$$\frac{q}{k_{c,v}} = \frac{\hbar\omega}{E_{c,v}} \frac{v_{c,v}}{c} \sqrt{\kappa}.$$

Here $v_{c,v}$ are the velocities of the electron and the hole, respectively, and c is the velocity of light in vacuum. Because of the strong inequality $v_{c,v} \ll c$, we can conclude that this ratio is always very small. Consequently, the selection rule of Eq. (10.53) is sufficiently accurate. Using $k_v = k_c = k$ we can rewrite Eq. (10.50) in the form

$$E_c - E_v = E_c^0 + \frac{\hbar^2 k^2}{2m_e} - \left(E_v^0 - \frac{\hbar^2 k^2}{2m_h} \right)$$

$$= E_g + \frac{\hbar^2 k^2}{2m_r} = \hbar\omega, \tag{10.54}$$

where E_v^0 and E_c^0 are the top and the bottom of the valence and the conduction bands, respectively. In addition, m_r is the reduced mass of the electron–hole pair:

$$m_r = \frac{m_e m_h}{m_e + m_h}. \tag{10.55}$$

Note that the reduced mass of the pair appears in the conservation law, even though we are considering phototransitions involving unbound electrons and holes.

From Eq. (10.54), we can find the wave vector of the hole and the electron created during the phototransition:

$$k = \sqrt{\frac{2m_r}{\hbar^2}(\hbar\omega - E_g)}.$$

Taking into account $\delta(E_c - E_v - \hbar\omega)$ in Eq. (10.43) and $\delta_{\vec{k}_c,\vec{k}_v}$ in Eq. (10.52), we obtain the α coefficient in the explicit form:

$$
\begin{aligned}
\alpha &= \frac{4\pi^2\omega}{c\sqrt{\kappa}V}|(\vec{\xi}\vec{D})_{c,v}|^2 \sum_{\sigma,\sigma',\vec{k}_v,\vec{k}_c} \delta_{\sigma,\sigma'}\delta_{\vec{k}_v,\vec{k}_c}\,\delta[E_c(\vec{k}_c) - E_v(\vec{k}_v) - \hbar\omega] \\
&\quad \times \{\mathcal{F}[E_c(\vec{k}_v)] - \mathcal{F}[E_v(\vec{k}_c)]\} \\
&= \frac{4\pi^2\omega}{c\sqrt{\kappa}V}|(\vec{\xi}\vec{D})_{c,v}|^2 2 \sum_{\vec{k}} \delta[E_c(\vec{k}) - E_v(\vec{k}) - \hbar\omega]\{\mathcal{F}[E_c(\vec{k})] - \mathcal{F}[E_v(\vec{k})]\} \\
&= \frac{4\pi^2\omega}{c\sqrt{\kappa}V}|(\vec{\xi}\vec{D})_{c,v}|^2 2 \frac{V}{(2\pi)^3} \int d\vec{k}\,\delta[E_c(\vec{k}) - E_v(\vec{k}) - \hbar\omega] \\
&\quad \times \{\mathcal{F}[E_c(\vec{k})] - \mathcal{F}[E_v(\vec{k})]\}.
\end{aligned}
\tag{10.56}
$$

Here we have used the substitution

$$
\sum_{\vec{k}} \rightarrow \frac{V}{(2\pi)^3} \int d\vec{k}.
$$

In Eq. (10.56), functions $\mathcal{F}(E_c)$ and $\mathcal{F}(E_v)$ are the distribution functions of electrons and holes.

For further calculations, we use the formula of Eq. (10.54) and introduce an effective density of states $\varrho(\omega)$, which provides the number of states of the electron–hole system that can interact with the photons in the frequency range from ω to $\omega + d\omega$:

$$
\varrho(\omega) = \frac{(2m_r)^{3/2}}{2\pi^2\hbar^2}(\hbar\omega - E_g)^{1/2}, \quad \hbar\omega > E_g.
\tag{10.57}
$$

Expressing the α coefficient through Eq. (10.57), we get

$$
\alpha = \frac{4\pi^2\omega|(\vec{\xi}\vec{D})_{c,v}|^2}{c\sqrt{\kappa}\hbar}\varrho(\omega)\{\mathcal{F}[E_c(k)] - \mathcal{F}[E_v(k)]\}.
\tag{10.58}
$$

We see that the absorption (gain) coefficient is proportional to $\varrho(\omega)$. Therefore, in light of this proportionality and in view of the functional form of the curve in Fig. 10.8 (compare with Fig. 3.9), this quantity is often called the optical density of states. Evidently the optical density of states is an important characteristic because all rates describing basic optical processes depend on $\varrho(\omega)$.

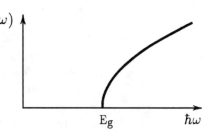

Figure 10.8. The optical density of states in a bulk semiconductor.

We can reveal the frequency dependence of the α coefficient of Eq. (10.58):

$$\alpha(\omega) = P(\hbar\omega - E_g)^{1/2}[\mathcal{F}(E_c) - \mathcal{F}(E_v)], \tag{10.59}$$

where P is the material constant, which can be expressed by the matrix element of the dipole $\mathcal{D}$ and the energies E_c and E_v, which satisfy Eq. (10.50). In some cases it is useful to rewrite the expression of Eq. (10.59) in terms of the contributions of both stimulated-emission and absorption processes:

$$\alpha(\omega) = \alpha_{\text{st.em}} - \alpha_{\text{abs}},$$

where

$$\alpha_{\text{abs}} = P(\hbar\omega - E_g)^{1/2}\mathcal{F}(E_v)[1 - \mathcal{F}(E_c)], \tag{10.60}$$

$$\alpha_{\text{st.em}} = P(\hbar\omega - E_g)^{1/2}\mathcal{F}(E_c)[1 - \mathcal{F}(E_v)], \tag{10.61}$$

have factors corresponding to those of Eqs. (10.35) and (10.36). It is also convenient to express P in terms of the radiative lifetime of electrons and holes τ_R:

$$P = \frac{\sqrt{2}c^2 m_r^{3/2}}{\kappa(\hbar\omega)^2} \frac{1}{\tau_R}. \tag{10.62}$$

Let us analyze Eq. (10.59) for thermal-equilibrium conditions. In this case, the occupancies of energy levels in both bands are described by the same Fermi function:

$$\mathcal{F}(E) = \frac{1}{e^{(E-E_F)/k_B T} + 1},$$

where E_F is the Fermi energy in the material. According to the latter formula,

$$\mathcal{F}(E_v) > \mathcal{F}(E_c),$$

since $E_v < E_c$. Hence the material absorbs light in accordance with the result $\alpha(\omega) > 0$. If E_F lies within the bandgap and

$$\left|E_c^0 - E_F\right|, \left|E_F - E_v^0\right| \gg k_B T,$$

we can set $\mathcal{F}(E_v) = 1$ and $\mathcal{F}(E_c) = 0$. The absorption coefficient then reduces to

$$-\alpha \equiv \alpha_{\text{abs}}(\omega) = \frac{\sqrt{2}c^2 m_r^{3/2}}{\kappa\tau_R} \frac{(\hbar\omega - E_g)^{1/2}}{(\hbar\omega)^2}. \tag{10.63}$$

Figure 10.9 depicts the absorption coefficient $\alpha_{\text{abs}}(\omega)$ for interband transitions in bulk GaAs at low temperatures; the excitonic absorption is not presented. We see that the interband contribution to the absorption increases sharply above the threshold photon energy ($\hbar\omega = E_g$) and reaches a value of the order of 10^4 cm^{-1}.

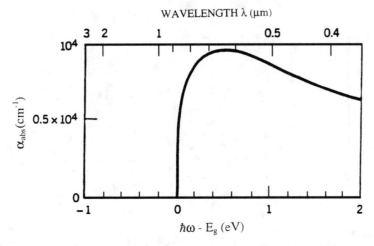

Figure 10.9. Calculated absorption coefficient in GaAs as a function of the photon energy (wavelength). Only interband phototransitions are taken into account. The exciton absorption is not shown. [After B. E. A. Saleh and M. C. Teich, *Fundamentals of Photonics* (Wiley, New York, 1991).]

The spectral region corresponding to Eq. (10.63) is frequently called the fundamental edge of absorption. According to Eq. (10.63), the maximum of $\alpha_{\text{abs}}(\omega)$ is reached at $\hbar\omega = 4E_g/3$. The high absorption due to interband phototransitions implies that in this spectral region the light is absorbed strongly in the micrometer material layer.

10.4.2 Spectral Density of Spontaneous Emission

From Eq. (10.27), we know that the Einstein coefficient A_{21} describing the spontaneous emission for some fixed mode can be expressed in terms of the coefficient of the stimulated process B_{21}. Using Eqs. (10.22) and (10.27), we can obtain the rates of spontaneous emission for the fixed mode from the rate of stimulated emission and Eqs. (10.39), (10.41), and (10.61) in the form

$$R_{\text{sp.em}}(\omega) = \frac{c}{\sqrt{\kappa}} P(\hbar\omega - E_g)^{1/2} \mathcal{F}(E_c)\left[1 - \mathcal{F}(E_v)\right].$$

To calculate the rate of spontaneous emission of all possible modes $R_{\text{sp.em}}(\omega)\,d\omega$ existing in the unit interval of frequency, we should multiply $R_{\text{sp.em}}$ by the number of these modes in this interval, i.e., by the density of states (10.11) per unit volume. Then we obtain

$$\mathcal{R}_{\text{sp.em}}(\omega)\,d\omega = R_{\text{sp.em}}(\omega)\varrho_{\text{ph}}(\omega)\,d\omega = \frac{\kappa\omega^2}{\pi^2 c^2} P(\hbar\omega - E_g)^{1/2}\mathcal{F}(E_c)[1 - \mathcal{F}(E_v)].$$

$$(10.64)$$

Assuming that the Fermi level lies in the bandgap, so that

$$\mathcal{F}(E_c)[1 - \mathcal{F}(E_v)] = e^{-(E_c - E_v)/k_B T} = e^{-\hbar\omega/k_B T},$$

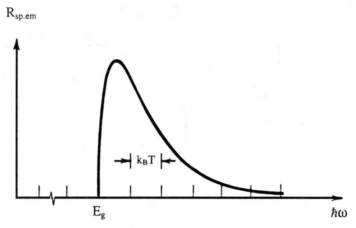

Figure 10.10. Spectral density of the interband spontaneous emission in thermal equilibrium. [After B. E. A. Saleh and M. C. Teich, *Fundamentals of Photonics* (Wiley, New York, 1991).]

we finally get the basic frequency dependence of the spectral density rate of spontaneous interband emission under equilibrium:

$$\mathcal{R}_{\text{sp.em}}(\omega) = D_0(\hbar\omega - E_g)^{1/2} e^{-\frac{\hbar\omega - E_g}{k_B T}}, \tag{10.65}$$

where D_0 is a constant that is independent of frequency:

$$D_0 = \frac{\sqrt{2}m_r^{3/2}}{\pi^2 \hbar^2 \tau_R} e^{-E_g/k_B T}.$$

$\mathcal{R}_{\text{sp.em}}(\omega)$ is shown in Fig. 10.10. The spectrum has a low-frequency edge at $\hbar\omega = E_g$ and extends over a width of the order of $2k_B T$.

10.4.3 Phototransitions in III–V Compounds

We now have equations describing absorption and emission in bulk materials. But we have not taken into account the actual complex structure of the valence band of III–V compounds, Si, Ge, etc. The valence-band structure was briefly discussed in Chapter 4.

To analyze the case, we should slightly generalize Eq. (10.56) by taking into account several valence bands v_j:

$$\alpha(\omega) = \sum_j \alpha_{c,v_j}(\omega), \tag{10.66}$$

$$\alpha_{c,v_j}(\omega) = 2\frac{4\pi^2\omega}{c\sqrt{\kappa}V}|(\vec{\xi}\vec{D})_{c,v_j}|^2$$

$$\times \sum_{\vec{k}_v, \vec{k}_c} \delta_{\vec{k}_v, \vec{k}_c} \delta[E_c(\vec{k}_c) - E_{v_j}(\vec{k}_v) - \hbar\omega]\{\mathcal{F}[E_{c_j}(\vec{k}_v)] - \mathcal{F}[E_{v_j}(\vec{k}_c)]\},$$

$$\tag{10.67}$$

where $(\vec{\xi}\vec{\mathcal{D}})_{c,v_j}$ is the dipole matrix element calculated by the Bloch functions of the conduction band c and the valence band v_j. The conduction and the valence bands are shown in Fig. 10.11, in which the three depicted valence bands are commonly known as the heavy-hole (hh), light-hole (lh), and split-off-hole (sh) bands. Each band originates from the energy levels of isolated atoms that form the crystal. In this sense, the conduction band originates from s atomic orbitals, while the three valence bands originate from three p atomic orbitals: p_x, p_y, p_z. Let the corresponding Bloch functions be u_s for the conduction band and u_x, u_y, and u_z for the valence bands. This association with the atomic orbitals is useful because it leads to the possibility of understanding the symmetry of the Bloch functions of the particular bands.

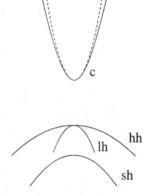

Figure 10.11. The sketch of a real band structure of III–V semiconductors. Parabolic and nonparabolic (dashed curve) conduction bands are depicted. Three valence bands are shown: heavy-hole, light-hole, and split-off-hole bands.

Thus the conduction-band Bloch function u_s has an even symmetry in all three directions, similar to the spherical symmetry of an s atomic orbital. In the same manner, u_z has odd symmetry along z, but even symmetry in the x and the y directions. These assignments are analogous to the case of p_z atomic orbitals. These observations allow us to conclude that the overlap integral vanishes, that is,

$$\langle u_s \,|\, u_z \rangle \equiv \int_\Omega u_s^*(\vec{r})u_z(\vec{r})\,\mathrm{d}\vec{r} = 0.$$

But the matrix element

$$\langle u_s \,|\, \mathcal{D}_z \,|\, u_z \rangle \equiv \int_\Omega u_s^*(\vec{r})\mathcal{D}_z u_z(\vec{r})\,\mathrm{d}\vec{r},$$

in the general case, is not zero, i.e.,

$$\langle u_s \,|\, \mathcal{D}_i \,|\, u_j \rangle = 0, \quad \text{if } i \neq j, \tag{10.68}$$
$$\langle u_s \,|\, \mathcal{D}_i \,|\, u_i \rangle \equiv \mathcal{D}, \tag{10.69}$$

where $i, j = x, y, z$. The spin does not change during dipole phototransitions, which is why

$$\langle u_s \,|\, \vec{\mathcal{D}} \,|\, \bar{u}_i \rangle = 0, \qquad \langle \bar{u}_s \,|\, \vec{\mathcal{D}} \,|\, u_i \rangle = 0,$$

where u_i, u_s and $\bar{u}_i$, $\bar{u}_s$ are the spin-up and the spin-down functions. Also, we define the constant $\mathcal{D}$ as the basic matrix element.

The electron wave functions in the valence bands should be constructed as certain linear combinations of the functions u_j. Let the electron wave vector be $\vec{k}$ and take the z axis along the vector $\vec{k}$. In this case, the linear combinations of the six Bloch functions, $u_i, \bar{u}_i$, corresponding to the hh, lh, and sh bands can be written, respectively, as

$$u_{hh} = -\frac{1}{\sqrt{2}}(u_x + iu_y), \qquad \bar{u}_{hh} = \frac{1}{\sqrt{2}}(\bar{u}_x - i\bar{u}_y),$$

$$u_{lh} = -\frac{1}{\sqrt{6}}(\bar{u}_x + i\bar{u}_y - 2u_z), \qquad \bar{u}_{lh} = \frac{1}{\sqrt{6}}(u_x - iu_y + 2\bar{u}_z),$$

$$u_{sh} = -\frac{1}{\sqrt{3}}(\bar{u}_x + i\bar{u}_y + u_z), \qquad \bar{u}_{sh} = -\frac{1}{\sqrt{3}}(u_x - iu_y - \bar{u}_z). \qquad (10.70)$$

Here, the prefactors are normalization constants. These linear combinations constitute a representation known as the angular-momentum representation. This representation is convenient for studying the spin-orbit interaction that arises from the coupling of the spin and the orbital angular momentum. This causes a splitting of the sh band.

Using Eqs. (10.68)–(10.70), one can obtain more complete and accurate descriptions of the matrix elements for phototransitions between the bands presented in Fig. 10.11. In these results, the constant $\mathcal{D}$ is the only parameter that remains to be either calculated or found from experiment. On the other hand, the angular-momentum representation of Eqs. (10.70) reveals the strong anisotropy of the electron–light interaction. For further calculations, we return to the general expressions for the absorption coefficient as given by Eqs. (10.66) and (10.67). The summation in Eq. (10.66) runs over all possible initial and final bands:

$$c \leftrightarrow hh, \qquad c \leftrightarrow lh, \qquad c \leftrightarrow sh.$$

In addition, the sum runs over both transitions: $u_c \leftrightarrow u_v$ and $\bar{u}_c \leftrightarrow \bar{u}_v$. Since the wave functions u_v and $\bar{u}_v$ in Eqs. (10.70) are states containing both spin-polarization orbitals, $u_i, \bar{u}_i$, the transitions $u_c \leftrightarrow \bar{u}_v$ and $\bar{u}_c \leftrightarrow u_v$ become possible. The complete scheme for phototransitions is shown in Fig. 10.12. Taking into account the relations of Eqs. (10.68)–(10.70), we obtain the intermediate result for the summation over all possible transitions:

$$|(\vec{\xi}\vec{\mathcal{D}})_{c,hh}|^2 = \frac{1}{2}|\mathcal{D}|^2(|\xi_x + i\xi_y|^2 + 0 + 0 + |\xi_x - i\xi_y|^2),$$

$$|(\vec{\xi}\vec{\mathcal{D}})_{c,lh}|^2 = \frac{1}{6}|\mathcal{D}|^2(|-2\xi_z|^2 + |\xi_x - i\xi_y|^2 + |\xi_x + i\xi_y|^2 + |2\xi_z|^2),$$

$$|(\vec{\xi}\vec{\mathcal{D}})_{c,sh}|^2 = \frac{1}{3}|\mathcal{D}|^2(|-\xi_z|^2 + |-\xi_x + i\xi_y|^2 + |-\xi_x - i\xi_y|^2 + |\xi_z|^2). \qquad (10.71)$$

Here ξ_i are the projections of the polarization vector of the electromagnetic wave; see Eq. (10.1). Each term in the parentheses corresponds to one of the four transitions shown in Fig. 10.12. Let us recall that the wave vector $\vec{k}$ is directed along the z axis. Then, from Eqs. (10.71), we get

$$\frac{|\vec{\xi}\vec{\mathcal{D}}|^2}{\mathcal{D}^2} = \begin{cases} 1 - \cos^2\theta & \text{for hh band} \\ 1/3 + \cos^2\theta & \text{for lh band} \\ 2/3 & \text{for sh band} \end{cases}$$

(10.72)

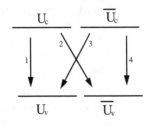

Figure 10.12. Scheme of interband photo-transitions: $u_c, u_v, \bar{u}_c,$ and $\bar{u}_v$ are the Bloch wave functions of the electrons and holes and their complex-conjugate values.

We see that transitions to (from) heavy- and light-hole bands depend on the angle between the wave polarization and the electron momentum. Figure 10.13 illustrates the corresponding transition strengths. These results indicate a strong polarization dependence of the electron–photon interaction. In contrast, the transition amplitude for the interactions of light with the split-off-hole-band electrons is completely isotropic.

In fact, in the three-dimensional case of bulk materials, electromagnetic waves interact with a great number of electrons, with all possible $\vec{k}$-vector directions. The average of these results over all possible $\vec{k}$ vectors is independent of the wave polarization:

$$\langle |\vec{\xi}\vec{\mathcal{D}}|^2 \rangle_{\text{average}} = \frac{2}{3}\mathcal{D}^2$$

for all three types of phototransitions. Thus, for bulk materials, one can use this result and Eq. (10.56) with the proper constant $\mathcal{D}$.

In conclusion, we emphasize that for low-dimensional heterostructures, the polarization dependence of the transition strength is much more important than for bulk crystals because the possible directions of $\vec{k}$ vectors are determined by the type of heterostructure.

Figure 10.13. Transition strength as a function of angle between an electron wave vector and the polarization vector of a light wave. The phototransitions $c \leftrightarrow$ hh and $c \leftrightarrow$ lh demonstrate different dependences. [After P. S. Zory, Jr., *Quantum Well Lasers* (Academic, Boston, 1993).]

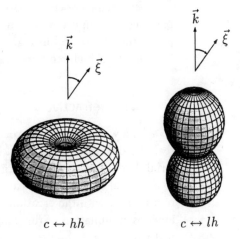

10.4.4 Excitonic Effects

In our previous discussions, we have considered interband phototransitions while ignoring the correlation of electron and hole motion due to the Coulomb interaction. In Chapter 3 we have studied the effects of this correlation, the main result of which is the formation of an exciton. It was described as a bound electron–hole pair. We have also considered the exciton-energy spectra in bulk materials as well as in low-dimensional structures. The existence of excitons results in the appearance of a series of the excitonic peaks in the optical spectra below the fundamental edge of interband transitions, that is, the peaks occur for $\hbar\omega < E_g$. These peaks schematically are shown in Fig. 3.22. In bulk materials, these peaks are usually observed at low temperatures and are washed out at room temperature. The overlapping of the exciton peaks forms an absorption tail below the fundamental edge that is observable at room temperature.

An important consequence of excitons is the fact that the matrix element of the optical transitions involving excitons is much greater than that of the interband transitions, because the electron and hole are close to each other and their wave functions overlap much more than those of the delocalized electron–hole pair.

The Coulomb correlation of unbound electron and hole reflects the fact that these particles generated by photon absorption in the conduction and the valence bands, respectively, move in the mutual Coulomb potential. This Coulomb potential causes corrections that may be described by a special multiplicative factor (the Sommerfeld factor) to the probability of the phototransition near the optical edge. As a result, the absorption becomes a sharper function of the photon energy. The spectral region, where Coulomb correlations are important, can be estimated as

$$0 < \hbar\omega - E_g < E_{\mathrm{ex}},$$

where E_{ex} is the exciton energy.

Phenomena associated with both excitons and Coulomb correlations of free particles are important at low temperatures. As a result, instead of the simple picture, presented in Fig. 10.10, the spectra become complex, as shown in Fig. 10.14. At high temperatures these effects are washed out.

10.4.5 Refractive Index

Along with the optical processes involving the absorption or the emission of real photons, there is another kind of electromagnetic-field interaction with material that occurs without energy exchange. These processes are usually referred to as virtual processes. We can imagine a virtual process such as the absorption of a photon with the creation of an electron–hole pair followed by the instantaneous annihilation of the pair by the emission of an equivalent photon. Thus real particles are absent, but the process results in a polarization of the material. Consequently, virtual processes contribute to the real part of the dielectric

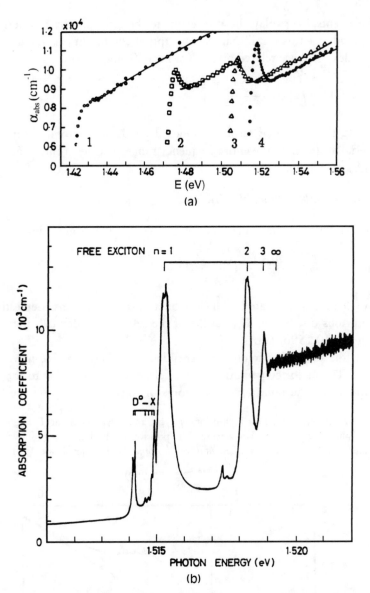

Figure 10.14. Exciton absorption in a bulk GaAs. (a) Evolution of the absorption coefficient with the temperature: 294 K (curve 1), 186 K (curve 2), 90 K (curve 3), and 21 K (curve 4). (b) The detailed structure of the absorption coefficient at a low temperature ($T = 1.2$ K). The peaks marked by 1, 2, and 3 correspond to an excitation of the three lowest excitonic states. The substructure below peak 1 (labeled as $D^0 - X$) is the absorption due to excitons localized on impurities.

constant κ or to the refractive index of the material:

$$n \equiv \sqrt{\kappa}.$$

These virtual processes result in a dispersion dependence of κ or n on the frequency ω. This dispersion relation is important for the design of photonic devices, particularly those that use optical microcavities, waveguides, and integrated optical elements. To explain how to estimate the electron contribution to the refractive index, let us recall that the absorption coefficient $\alpha_{abs}(\omega)$ and the refractive index, $n(\omega)$ are related by the Kramers–Krönig relation

$$n(\omega) = 1 + \frac{c}{\pi} \int_0^\infty d\omega' \frac{\alpha_{tot}(\omega')}{(\omega')^2 - \omega^2}. \tag{10.73}$$

In principle, the calculated refractive index depends on the total absorption of the material $\alpha_{tot}(\omega)$ over the entire frequency range. The electronic contribution to the refractive index is

$$\Delta n(\omega) \equiv n(\omega, N) - n(\omega, N = 0),$$

where N denotes the electron concentration. Now $\Delta n(\omega)$ is given by

$$\Delta n(\omega) = \frac{c}{\pi} \int_0^\infty d\omega' \frac{\alpha_{abs}(\omega', N)}{(\omega')^2 - \omega^2}, \tag{10.74}$$

where $\alpha_{abs}(\omega, N)$ is associated with electron absorption and has been estimated in our previous discussion. The results of the refractive-index calculations are shown in Fig. 10.15 for GaAs.

Briefly, the dispersive frequency dependence of the refractive index arises as follows. The major contribution to the refractive index is due to the lattice polarization, i.e., the polarization of atoms or ions that comprise the lattice.

Figure 10.15. Refractive index of high-purity, p-type, and n-type GaAs at $T = 300\,\text{K}$ as a function of photon energy and wavelength. The position of the bandgap is shown. [After B. E. A. Saleh and M. C. Teich, *Fundamentals of Photonics* (Wiley, New York, 1991).]

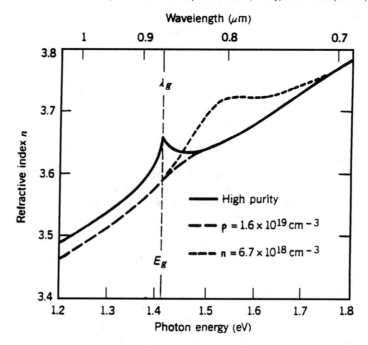

The contribution of free electrons is usually small, but in the frequency region corresponding to interband phototransitions, this contribution increases. In a high-purity sample, in the vicinity of the point where $\hbar\omega = E_g$, the contribution has almost a resonant character and can reach an absolute value of ~ 0.05. Figure 10.15 illustrates that in doped samples at high temperatures the resonant features are smeared out but the dependence on ω due to the electron contribution is still considerable.

Another noteworthy feature of the refractive index of the III–V compounds is its dependence on the alloy composition: an increase in a narrow-bandgap component in the alloy leads to a larger index of refraction at a fixed frequency. For example, for $Ga_xAl_{1-x}As$ alloys, increasing the Ga component decreases the bandgap and increases the refractive index: $n_{Ga_{1-x}Al_xAs} = n_{GaAs} + 0.62x$. This feature is of crucial importance for optoelectronic applications of heterostructures. It opens possibilities of creating optical waveguides on the basis of heterostructures and of combining electron confinement and optical confinement within the same narrow-bandgap layers, which are discussed in Sections 10.5 and 10.6.

10.5 Optical Properties of Quantum Structures

After consideration of the basic interband optical processes in bulk materials, we are ready to study these processes in quantum heterostructures. Specific features of optical processes originate from two basic physical peculiarities. First, spatial nonuniformity causes specific characteristics of the interaction of light with matter, including light propagation, absorption, etc. Second, electrons in quantum structures have energy spectra different from those of electrons in bulk materials. Both factors are analyzed in the following discussions. We introduce several parameters that characterize the interaction of light with matter for different cases of light propagation. The parameters for light absorption are calculated for type-I heterostructure quantum wells. Various factors affecting the optical properties of low-dimensional electrons, such as the broadening of spectra that is due to intraband-scattering processes, excitonic effects, etc., are analyzed in this section.

10.5.1 Electrodynamics of Heterostructures

Let us start by studying the electrodynamic features of heterostructures in the spectral region that corresponds to band-to-band phototransitions. In bulk materials, one can use the electrodynamics of uniform media and consider the interaction of light with matter as homogeneous in space. In heterostructures, both the refractive index and the bandgap, i.e., the fundamental edge of absorption, vary spatially. This changes the light propagation and the character of the interaction of light with matter.

For example, a layer with a larger refractive index causes (1) partial reflection of electromagnetic waves propagating through the layer and (2) localization of electromagnetic modes propagating along the layer. As a result, the electromagnetic fields (modes) in quantum structures are substantially different from plane waves. On the other hand, spatial modulation of the bandgap leads to nonuniform absorption and emission of light. In particular, in some spectral regions, narrow-bandgap layers interact with the light, while wide-bandgap layers are transparent to it. Consequently, only certain layer(s) of heterostructures can absorb and emit photons in such a spectral region, i.e., below the fundamental edge of wide-bandgap layers and above that of narrow-bandgap layers.

To take into account the above-mentioned features of electrodynamics, we return to the general results of Eqs. (10.34)–(10.38). These equations allow us to consider arbitrary electromagnetic fields, including those different from the plane waves of Eq. (10.1). We exploit the fact that sizes of quantum structures are always much less than the wavelength λ of the light in the spectral region of interest ($L = 100$–$200\,\text{Å}, \lambda > 1000\,\text{Å}$). Equations (10.34)–(10.38) therefore can be simplified if we assume a particular shape of electromagnetic fields. Let us consider two cases of light interaction with a quantum-well layer: (1) light propagation perpendicularly to the layers and (2) propagation along the layers. Both cases are presented in Figs. 10.16(a) and 10.16(b). In the first case, the electromagnetic field depends on the z coordinate only, and we can express the electric field as

$$\mathcal{E}(z, t) = \frac{1}{2}[e^{i\omega t} F(z) + e^{-i\omega t} F^*(z)].$$

Introduction of a local absorption coefficient is meaningless in this case. Loss or gain of the light energy can be characterized by the attenuation, which is defined as the change of the light intensity after the light passes through the layer to the initial intensity:

$$\beta \equiv \frac{\mathcal{I}_{\text{in}} - \mathcal{I}_{\text{out}}}{\mathcal{I}_{\text{out}}}. \tag{10.75}$$

Figure 10.16. Two main types of electromagnetic modes in quantum structures: (a) standing waves, (b) localized waves.

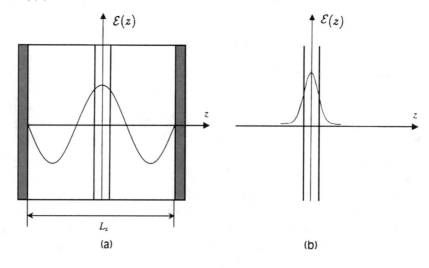

(a)

(b)

Another useful characteristic of the interaction of light with matter is the decrement (increment) or decay (gain) of the energy of standing waves. The standing waves are formed by the optical resonator in which a heterostructure is embedded. A simple resonator with two plane mirrors is sketched in Fig. 10.16(a). The decrement (increment) of the mode is

$$\gamma \equiv \frac{1}{N}\frac{dN}{dt} = \frac{1}{N}R, \tag{10.76}$$

where N is the number of photons of the mode under consideration and R is the total rate of photon absorption (emission) given by Eq. (10.38). Since $\lambda \gg L$, as mentioned above, we can rewrite R in the form

$$R = \frac{\pi}{2\hbar}|F(z_w)|^2 \sum_{i,f}|\langle i\,|\,\vec{\xi}\vec{D}\,|\,f\rangle|^2\delta(E_i - E_f - \hbar\omega)[\mathcal{F}(E_i) - \mathcal{F}(E_f)], \tag{10.77}$$

where z_w is the position of the center of the quantum-well layer and $\vec{D}$ is the dipole vector. The number of photons N of the fixed mode can be calculated as the total energy of this mode divided by $\hbar\omega$:

$$N = \frac{1}{8\pi\hbar\omega}\int \kappa(z)|F(z)|^2(z)\,d\vec{r} = \frac{S}{8\pi\hbar\omega}\int dz\kappa(z)|F(z)|^2, \tag{10.78}$$

where S is the cross section of the resonator. Thus the γ coefficient can be written as

$$\gamma = 4\pi^2\omega\frac{|F(z_w)|^2}{\int \kappa(z)|F(z)|^2\,dz}\frac{1}{S}\sum_{i,f}|\langle i\,|\,\vec{\xi}\vec{D}\,|\,f\rangle|^2\delta(E_i - E_f - \hbar\omega)[\mathcal{F}(E_i) - \mathcal{F}(E_f)], \tag{10.79}$$

where the integral in the denominator extends over the entire resonator of L_z size. For a plane wave of light coming through the quantum-well layer, the attenuation of Eq. (10.75) can be expressed by γ as

$$\beta = \gamma\frac{L_z}{c}\sqrt{\kappa}, \tag{10.80}$$

where γ can be calculated from Eq. (10.79) with $|F|^2 = 1$.

In the simplest case of a narrow quantum well, the changes in the refractive index can be neglected and the standing wave is

$$\vec{\mathcal{E}}(z) = \vec{\xi}\,F_0\cos qz\cos\omega t, \tag{10.81}$$

where $\vec{\xi}$ is the vector of light polarization, which is in the x, y plane, q is a discrete number that depends on the resonator length L_z ($q = \pi n/L_z$), and n is an integer. The form factor of this mode is $\vec{F}(z) \equiv \vec{\xi}\,F_0\cos qz$. For the standing wave of Eq. (10.81), the γ coefficient takes the form

$$\gamma = \frac{8\pi^2\omega}{\kappa}\cos^2 qz_w\frac{1}{SL_z}\sum_{i,f}|\langle i\,|\,\vec{\xi}\vec{D}\,|\,f\rangle|^2\delta(E_i - E_f - \hbar\omega)[\mathcal{F}(E_i) - \mathcal{F}(E_f)]. \tag{10.82}$$

As expected, this result depends on the position of the quantum-well layer. Accordingly, if the position coincides with a nodal point of the standing wave, the decrement is zero. At an antinodal point, the decrement reaches its maximum.

Another example is presented in Fig. 10.16(b), in which light waves propagate along a heterostructure layer. Basically, there are three types of such designs. Among them, the most interesting design is that in which a narrow-bandgap layer, possibly a graded layer, localizes the electromagnetic modes. The amplitudes of these modes depend on z and decay far away from the layer. These modes are called the waveguide modes. One of them is depicted in Fig. 10.16(b). Let the electromagnetic field of this mode be

$$\vec{\mathcal{E}} = \vec{\xi} F(z) \cos(\vec{q}\vec{r} - \omega t), \tag{10.83}$$

where $F(z)$ is the form factor of the mode. The wave vector $\vec{q}$ is in the x, y plane. For a waveguide mode, one can define the electromagnetic energy per unit area of the waveguide plane as

$$w = \frac{1}{4\pi} \int \kappa(z)\overline{\mathcal{E}^2}(z)\,dz = \frac{1}{8\pi} \int \kappa F(z)^2\,dz, \tag{10.84}$$

and the number of photons of this mode is $N_w = wS/(\hbar\omega)$ [compare with the bulk case of Eq. (10.8)]. The rate of change of the number of photons in this waveguide mode R is given by Eq. (10.76). One can introduce the total intensity of the waveguide mode: $I = wc_{\text{eff}}$, where c_{eff} is the group velocity of the mode under consideration. Within the simplest approach we can set $c_{\text{eff}} \approx c/\sqrt{\kappa}$, neglecting the dependence of κ on the z coordinate. Now we can define the gain (absorption) coefficient for the waveguide mode:

$$\frac{1}{I}\frac{dI}{dx} = \alpha. \tag{10.85}$$

Then α can be expressed as

$$\alpha = \frac{4\pi^2 \omega \sqrt{\kappa}}{c} \Gamma \frac{1}{SL} \sum_{i,f} |\langle i | \vec{\xi}\vec{\mathcal{D}} | f \rangle|^2 \delta(E_i - E_f - \hbar\omega)[\mathcal{F}(E_i) - \mathcal{F}(E_f)], \tag{10.86}$$

where

$$\Gamma \equiv \frac{F^2(z_w)L}{\int \kappa(z)F^2(z)\,dz} \tag{10.87}$$

is the so-called optical-confinement factor and L is the width of the quantum well. The optical-confinement factor characterizes a portion of the light energy accumulated within the active layer where the phototransitions take place. Note that Γ is always less than 1. The better the optical confinement, the larger the light absorption or gain.

10.5.2 Light Absorption by Confined Electrons

Now let us consider the optical properties of confined electrons. The energy spectrum of a confined system is different from that of a bulk material because of the additional quantization of both electrons and holes. In general, the equations that we obtained previously should be used in conjunction with the appropriate quantized energy levels as the initial and the final states, taking into account also the densities of states of these levels. Another important factor is the particular shape of electron and hole wave functions. These wave functions determine the matrix element and not only change the magnitudes of the absorption and emission rates, but also lead to new selection rules for the phototransitions.

Let us consider the simplest case of a three-layered semiconductor structure of type I. Type-I structures have band-edge discontinuities such that the same embedded layer provides quantum wells for electrons and holes; see Fig. 10.17(a). Figure 10.17(a) shows the energy levels of both types of carriers. The energy dependence of the in-plane wave vector is shown in Fig. 10.17(b) for each low-dimensional subband. The first obvious conclusion can be drawn, i.e., that there must be a shift of the interband spectra toward a short-wavelength region since the threshold energy of interband phototransitions can be estimated as

$$\hbar\omega \geq \hbar\omega_0^{(1)} \equiv E_{c,1} - E_{v,1} = E_g^{QW,1} > E_g. \tag{10.88}$$

Here $E_{c,j}$ and $E_{v,i}$ are the positions of the electron and the hole two-dimensional subbands, respectively. The indices i and j correspond to the subband numbers. Thus the lowest subbands determine the threshold given by Eq. (10.88). Another conclusion is that the same transition energy $\hbar\omega$ may correspond to different combinations of hole and electron subbands involved in the phototransitions.

In Eqs. (10.77), (10.79), and (10.82), the matrix elements should be estimated from the true wave functions containing the envelope functions different from those of bulk electrons:

$$\Psi_{v,\vec{k}_v,i} = \frac{1}{\sqrt{S}} u_{v,\vec{k}_0}(r) e^{i\vec{k}_v \vec{r}} \chi_{v,i}(z),$$

$$\Psi_{c,\vec{k}_c,j} = \frac{1}{\sqrt{S}} u_{c,\vec{k}_0}(r) e^{i\vec{k}_c \vec{r}} \chi_{c,j}(z),$$

Figure 10.17. (a) Band diagram, (b) dispersion curves of type-I quantum wells. Parabolic dispersion relations are assumed for both the conduction and the simple valence bands. Two subbands are shown for both bands. [After B. E. A. Saleh, and M. C. Teich, *Fundamentals of Photonics* (Wiley, New York, 1991).]

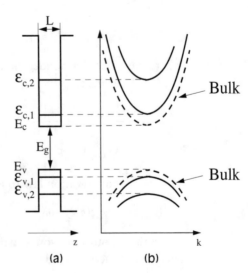

where $u_{v,\vec{k}_0}(r)$ and $u_{c,\vec{k}_0}(r)$ are the Bloch functions. Here and in the following discussion, $\vec{k}_v$ and $\vec{k}_c$ are the two-dimensional in-plane wave vectors, S is the area of the quantum-well layer, and $\chi_{v,i}(z)$ and $\chi_{c,j}(z)$ are the envelope wave functions for the quantized transverse motion, as discussed in Chapter 3. Thus the matrix element is

$$\langle v, i, \vec{k}_v | \vec{\xi}\vec{D} | c, j, \vec{k}_c \rangle = |(\vec{\xi}\vec{D})_{cv}|^2 \frac{\Omega}{S} \sum_{\vec{R}_n} e^{i(\vec{k}_c - \vec{k}_v)\vec{R}_n} \chi^*_{v,i}(Z_n) \chi_{c,j}(Z_n)$$

$$= \delta_{\vec{k}_v, \vec{k}_c} |(\vec{\xi}\vec{D})_{cv}|^2 \int dz\, \chi^*_{v,i}(z) \chi_{c,j}(z), \qquad (10.89)$$

where $\vec{D}_{cv}$ is the interband dipole matrix element introduced in Subsection 10.4.1, Ω is the unit cell volume of the crystal, and $\vec{R}_n = \{X_n, Y_n, Z_n\}$ is the position vector of the nth cell. The summation runs over all crystal cells. Equation (10.89) shows that the overlap integral of the envelope functions depends on the quantum numbers of the initial and the final states of transitions. From Eq. (10.89), the selection rule for two-dimensional wave vectors can be derived:

$$\vec{k}_v = \vec{k}_c. \qquad (10.90)$$

This selection rule differs from Eq. (10.53) for bulk crystals as a result of the lack of translational symmetry in the z direction. Note that, compared with the bulk case, a new factor appears in the matrix element of Eq. (10.89):

$$\int dz\, \chi^*_{v,i}(z) \chi_{c,j}(z) \equiv \langle v, i \,|\, c, j \rangle,$$

which is the overlap integral of the envelope functions from different bands.

Using the matrix element of Eq. (10.89), one can calculate all characteristics, β, γ, and α, of Eqs. (10.75), (10.77), and (10.85) for the geometries presented in Fig. 10.16. For example, from Eq. (10.86) we get

$$\alpha = \frac{4\pi^2 \omega \sqrt{\kappa}}{c} \Gamma \frac{1}{SL} \sum_{i,\vec{k}_c; j, \vec{k}_v} \delta_{\sigma,\sigma'} |(\vec{\xi}\vec{D})_{cv}|^2 \delta_{\vec{k}_c, \vec{k}_v} |\langle v, j \,|\, c, i \rangle|^2 \delta(E_{c,i} - E_{v,j} - \hbar\omega)$$

$$\times [\mathcal{F}(E_{c,j}) - \mathcal{F}(E_{v,i})], \qquad (10.91)$$

where we assume a nondegenerate valence band. A further transformation of Eq. (10.91) leads to

$$\alpha = \frac{4\pi^2 \omega \sqrt{\kappa}}{c} \Gamma \frac{|(\vec{\xi}\vec{D})_{cv}|^2}{2\pi^2 L} \sum_{i,j} |\langle v, j \,|\, c, i \rangle|^2 \int d\vec{k}\, \delta[E_{c,i}(\vec{k}) - E_{v,j}(\vec{k}) - \hbar\omega]$$

$$\times [\mathcal{F}(E_{c,i}) - \mathcal{F}(E_{v,j})]. \qquad (10.92)$$

The energy-conservation law following from the δ function,

$$E_{c,i}(\vec{k}) - E_{v,j}(\vec{k}) = \hbar\omega, \qquad (10.93)$$

shows that phototransitions can involve different subbands from both the valence and the conduction bands.

In the case of parabolic subbands, i.e.,

$$E_{c,i}(\vec{k}) = E^0_{c,i} + \frac{\hbar^2 \vec{k}^2}{2m^*_e}, \qquad E_{v,j}(\vec{k}) = E^0_{v,j} - \frac{\hbar^2 \vec{k}^2}{2m^*_h}, \tag{10.94}$$

Eq. (10.93) gives the magnitude of the two-dimensional vector $\vec{k}$, corresponding to vertical transitions between the v, j and c, i subbands:

$$k = \sqrt{\frac{2m_r}{\hbar^2}\left(\hbar\omega - E^0_{c,i} + E^0_{v,j}\right)}. \tag{10.95}$$

Transitions are possible when $\hbar\omega > E^0_{c,i} + E^0_{v,j}$. Finally, we get

$$\alpha = \frac{4\pi^2\omega\sqrt{\kappa}}{c}\Gamma|(\vec{\xi}\vec{\mathcal{D}})_{cv}|^2 \sum_{i,j} |\langle v, j \,|\, c, i\rangle|^2$$

$$\times \frac{m_r}{\pi\hbar^2 L}\Theta\left(\hbar\omega - E^0_{c,i} + E^0_{v,j}\right)[\mathcal{F}(E_{c,i}) - \mathcal{F}(E_{v,j})]. \tag{10.96}$$

Here, m_r is the reduced effective mass of the electron–hole pair. The optical density of states is represented by the factors

$$\varrho^{opt}_{i,j}(\omega) \equiv \frac{m_r}{\pi\hbar^2}\Theta\left(\hbar\omega - E^0_{c,i} + E^0_{v,j}\right), \tag{10.97}$$

which appear in Eq. (10.96); this optical density of states corresponds to phototransitions between subbands: $v, j \leftrightarrow c, i$. The energies $E_{c,i}$ and $E_{v,j}$ in the argument of Fermi function in Eq. (10.96) must be calculated for the wave vector given by Eq. (10.95). The quantities $\varrho^{opt}_{i,j}(\omega)$ are the steplike functions, which are consistent with the results on the density of states of two-dimensional electrons (holes) studied in Chapter 3. The function $\alpha(\omega)$ also has a steplike shape; see Fig. 10.18. When the photon energy exceeds the threshold energy corresponding to the transitions between a new pair of electron and hole subbands, a new step in the absorption should be observed. As a result, $\alpha(\omega)$ can be represented as a sum over all pairs of the subbands involved in phototransitions:

$$\alpha(\omega) = \sum_{i,j} \alpha_{sub}(\omega, i, j), \tag{10.98}$$

$$\alpha_{sub}(\omega, i, j) \equiv \frac{4\pi^2\omega\sqrt{\kappa}}{c\hbar}\Gamma|(\vec{\xi}\vec{\mathcal{D}})_{cv}|^2|\langle v, j \,|\, c, i\rangle|^2\varrho^{opt}_{i,j}(\omega)[\mathcal{F}(E_{c,i}) - \mathcal{F}(E_{v,j})]. \tag{10.99}$$

Schematically $\alpha_{abs}(\omega) = -\alpha(\omega)$ is shown in Fig. 10.18 for two multiple-quantum-well heterostructures with different well widths.

Each contribution, $\alpha_{sub}(\omega, i, j)$, is proportional to the overlap integral of the envelope functions $\chi_{c,i}(z)$ and $\chi_{v,j}(z)$. These overlap integrals result in new selection rules. Let us examine the case of infinitely deep quantum structures for electrons and holes. In such quantum structures, the eigenfunctions are independent of the effective mass and form an orthogonal set of functions. The set

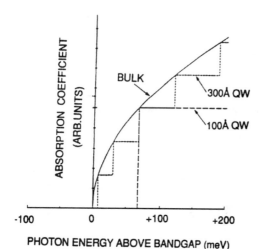

ABSORPTION COEFFICIENT (ARB.UNITS)

BULK

300Å QW

100Å QW

-100   0   +100   +200

PHOTON ENERGY ABOVE BANDGAP (meV)

Figure 10.18. The sketch of absorption by multiple subbands of an ideal quantum well. The results for two GaAs quantum wells with well widths equal to 100 Å and 300 Å are presented. For the narrower well, only one step in the absorption coefficient is shown, while for wider well five steps are seen for the same frequency range. The absorption coefficient for bulk-like material is depicted for comparison. [After S. Schmitt-Rink, *et al.*, "Linear and nonlinear optical properties of semiconductor quantum wells," Adv. Phys. 38, pp. 89–188 (1989).]

depends parametrically on the size of the quantum-well layer, as analyzed in Chapter 3. If we choose the subband numbers $i = j$, the wave functions $\chi_{v,i}(z)$ and $\chi_{c,i}(z)$ are identical and the overlap integral

$$|\langle v, i \,|\, c, i \rangle|^2 = 1,$$

and phototransitions between these subbands are allowed. If $i \neq j$, the envelope functions are orthogonal, $\langle v, i \,|\, c, j \rangle = 0$ and the transitions are forbidden. Figure 10.19 shows allowed and forbidden transitions.

For quantum wells with finite barrier heights, the wave functions depend on the effective mass, the barrier parameters, etc., i.e., the two sets of the wave functions are different. Thus, when $i \neq j$, orthogonality between the electron and the hole functions no longer holds. However, $|\langle v, i \,|\, c, i \rangle|^2$ is still of the order of 1, while $|\langle v, i \,|\, c, j \rangle|^2 \ll 1, i \neq j$. This means that the transition probability is typically very small for $i \neq j$.

Thus we can conclude that the quantization of the electron energy leads to major changes in the selection rules and in the intensity of the band-to-band transitions.

10.5.3 Effects of Complex Valence Bands of III–V Compounds

Above we discussed the interband transitions in quantum wells with simple structures of both the valence and the conduction bands; see Eqs. (10.94). In fact, for heterostructures based on III–V compounds, we should take into account the complex structure of the valence band, which was analyzed in Subsection 10.4.3.

The energies of the valence subbands are complex functions of the wave vector and depend on the orientation of the quantum-well layer with respect to the crystal axes. Figure 10.20(a) shows these subband energies for a particular

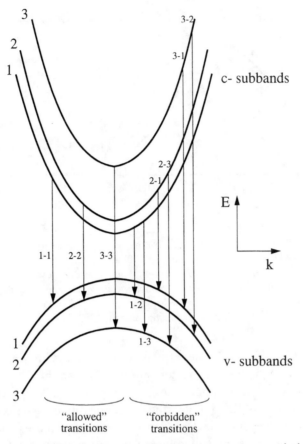

Figure 10.19. Allowed and forbidden interband phototransitions in an ideal quantum well. Vertical and horizontal axes are energy and wave vector, respectively. Three subbands for electrons and holes are shown. The vertical lines connecting different subbands indicate all possible transitions. The overlap integrals of subband wave functions of the electrons and holes suppress some of the transitions.

AlGaAs/GaAs quantum well for two directions: $\langle 100 \rangle$ and $\langle 110 \rangle$. Comparing these results with Fig. 4.11, given for a bulk GaAs material, one can see that the quantization of holes in a quantum well leads to a splitting of the heavy- and the light-hole bands as well as more complex dispersion dependences, giving rise to $\vec{k}$ intervals with negative effective masses $[d^2 E(k)/dk^2 < 0]$. Given the $E_{c,i}(\vec{k})$ and $E_{v,j}(\vec{k})$ dependences, we can apply Eq. (10.92). For the integral over $\vec{k}$ in Eq. (10.92) one has

$$\frac{1}{2\pi^2 L} \int d\vec{k}\,\delta[E_{c,i}(\vec{k}) - E_{v,j}(\vec{k}) - \hbar\omega][\mathcal{F}(E_{v,i}) - \mathcal{F}(E_{c,j})]$$

$$= \frac{1}{\pi L} \frac{k_t}{|d[E_{c,i}(k) - E_{v,j}(k)]/dk|_{k_t}} \Theta[\hbar\omega - E_{c,i}(k_t) + E_{v,j}(k_t)]$$

$$\times \{\mathcal{F}[E_{c,i}(k_t)] - \mathcal{F}[E_{v,j}(k_t)]\}, \tag{10.100}$$

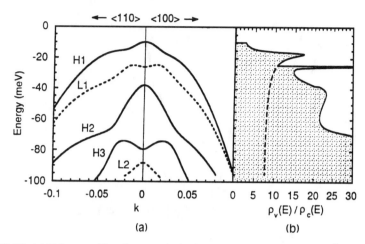

Figure 10.20. (a) Valence subband energies, (b) the density of states in an 80-Å $Al_{0.2}Ga_{0.8}As/$ GaAs quantum well for two possible directions, $\langle 110 \rangle$ and $\langle 100 \rangle$, of the in-plane wave vector, $\vec{k}$. The quantization of the holes in the well of 95-meV height leads to a strong coupling between the heavy and the light holes. It results in complex energy-subband dependences. The total density of states and a contribution of the heavy holes (dashed curves) are presented. [After P. S. Zory, Jr., *Quantum Well Lasers* (Academic, Boston, 1993).]

where k_t is the absolute value of the wave vector satisfying the energy-conservation law. The first two factors are proportional to the optical density of states corresponding to phototransitions $c, i \leftrightarrow v, j$. Within the effective-mass approximation of Eqs. (10.94), the optical density of states coincides with Eq. (10.96). The nonparabolic dependences $E_{v,j}(\vec{k})$ yield more complex optical densities of states and essentially different transition strengths. The density of states of the valence band is presented in Fig. 10.20(b). These data correspond to the energy dependences presented in Fig. 10.20(a). The density of states apparently reflects the $E(\vec{k})$ behavior and, in particular, it causes a nonmonotonous dependence on the hole energy. Peaks in the density of states correspond to the points with zero derivatives, $d^2 E(k)/dk^2$. Thus the real band structure results in more complex dependences of interband spectra compared with the two parabolic bands of Eq. (10.54). These dependences are especially important for laser generation, as they affect the light gain and the spectrum of laser emission.

10.5.4 Other Factors Affecting the Interband Optical Spectra

There are different physical factors that have not been taken into account in the previous discussion. However, these factors also affect the interband optical spectra and cause them to differ from the ideal steplike shape shown in Fig. 10.18. One of them is the broadening of the spectra due to intraband-scattering processes. Indeed, intraband-scattering processes lead to uncertainties of the electron and the hole energies. According to the general uncertainty relations of Chapter 2, one can estimate energy uncertainties for conduction and valence bands as $\hbar/\tau_{v,c}$, where $\tau_{v,c}$ is the intraband-scattering time in the valence (v)

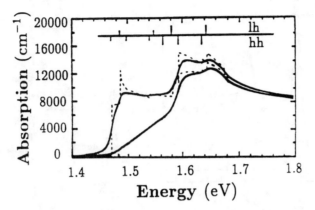

Figure 10.21. Light absorption coefficient as a function of the photon energy in an 80-Å $Al_{0.2}Ga_{0.8}As/GaAs$ quantum well. The dashed curves correspond to calculations based on the equations discussed in the text. The solid curves are obtained for a broadening due to scattering processes; see Subsection 10.5.4. At the top of the figure, markers indicate the $c \leftrightarrow$ hh and $c \leftrightarrow$ lh transitions between the first electron subband and the three lowest hole subbands. Lower and upper curves are for the carrier density $2 \cdot 10^{18}$ cm^{-3} and 0, respectively. [After P. S. Zory, Jr., *Quantum Well Lasers* (Academic, Boston, 1993).]

and the conduction (c) bands. The energy uncertainties or, in other words, the broadening of the energy levels result in the broadening of the optical spectra. It is possible to take this effect into consideration by means of a simple procedure: replace the δ function in Eqs. (10.82) and (10.86) with the broadening function $\Delta(E_c - E_v - \hbar\omega)$. An example of the broadening function is the Lorentzian:

$$\Delta(E_c - E_v - \hbar\omega) = \frac{1}{\pi} \frac{\hbar/\tau_c + \hbar/\tau_v}{(E_c - E_v - \hbar\omega)^2 + (\hbar/\tau_c + \hbar/\tau_v)^2}. \quad (10.101)$$

Note that $\Delta(E_c - E_v - \hbar\omega)$ has the same dimensionality as the δ function, i.e., energy in the power minus one. The results presented in Fig. 10.21 show that the spectral shape of the intensities of interband transitions becomes smooth and broad in spite of the sharp steplike density of states.

It is known that the scattering rates increase as the temperature increases. Hence the optical spectra become broader at higher temperatures. Besides the broadening effects, an essential temperature dependence of the spectra is caused by the temperature dependence of the distribution functions, $\mathcal{F}_c$ and $\mathcal{F}_v$.

The optical density of states of the two-dimensional systems of Eq. (10.97) is small. To increase it, one can use multiple-quantum-well structures with 20–100 quantum-well layers separated by relatively thick barriers.

Another factor that affects optical spectra is the presence of excitonic effects. Bulk-material excitonic spectra were studied in Section 10.4. As we know from Chapter 3, the exciton energy in quantum wells is greater than that in bulk material. This leads to an enhancement of the excitonic spectral lines, so that they are clearly observed up to room temperature. The absorption spectrum at low temperature is presented in Fig. 10.22 for AlAs/GaAs multiple-quantum-well

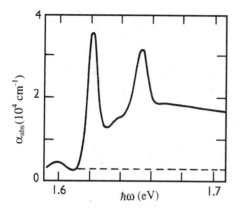

Figure 10.22. Measured absorption spectra of the AlAs/GaAs multiple-quantum-well structure. [After G. Bastard, *Wave Mechanics Applied to Semiconductor Heterostructures* (Halsted, New York, 1988).]

heterostructures near the fundamental absorption edge. The two pronounced peaks in absorption are due to the excitation of heavy- and the light-hole excitons in 76-Å GaAs wells. In Fig. 10.23, the spectrum of another AlGaAs/GaAs heterostructure is shown for room temperature. The structure consists of 65 periods with 96-Å GaAs wells. The exciton resonances with quantum numbers $n = 1$, 2, and 3 are pronounced. The heavy- and the light-hole transitions are resolved. The large resonant effects related to excitons are used in many applications, as will be discussed in Chapters 11 and 12.

Figure 10.23. Absorption spectrum measured at room temperature for a $Al_{0.3}Ga_{0.7}As$/GaAs multiple-quantum-well structure. Arrows show the positions of the three lowest exciton transitions with $n = 1, 2, 3$. [After J. Singh, *Physics of Semiconductors and Their Heterostructures* (McGraw-Hill, New York, 1993).]

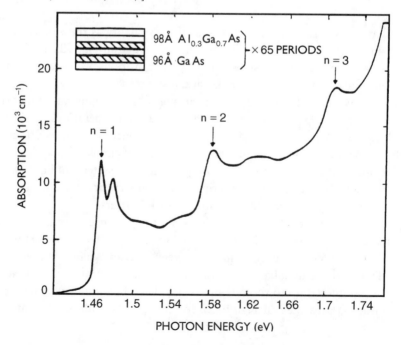

So far we have considered type-I quantum structures. In type-II semiconductor heterostructures, the interband phototransitions are indirect in real space. This leads to a smoothing of the absorption (emission) spectra, similar to the indirect transitions in bulk materials. To measure the absorption spectra in type-II quantum structures, one has to have a multilayer structure with more than 100 layers.

10.5.5 Polarization Effects

In low-dimensional electron systems, the effects of light polarization on the optical spectra are much more pronounced than in bulk materials. It has been shown in Section 10.4 that for a fixed wave vector $\vec{k}$, the interaction with photons is strongly polarization dependent. However, after the averaging over of all possible directions in three-dimensional space, this dependence disappears in bulk materials, such as Si, Ge, and III–V compounds. In fact, it is true for any cubic crystal. A completely different physical situation takes place in quantum heterostructures, in which electrons and holes are confined to two, one, or zero directions. Therefore one can expect strong polarization effects in the optical spectra of these structures.

Let us consider a quantum well in a type-I heterostructure. The confined states can be treated as a sum of two plane waves that result in a standing wave with wave vectors parallel and antiparallel to the z axis. Thus we can approximately treat any electron state as a combination of states characterized by the wave vector $\vec{k} = \{\vec{k}_\parallel, k_z\}$. Within this approach, the polarization dependence is similar to that of the bulk case of Fig. 10.12. The only difference is that we must average over all possible directions of $\vec{k}_\parallel$ in the plane of a quantum well. As a result, we have an average wave vector $\vec{k}_{av}$ that coincides with k_z. This means that the polarization dependences of the transition strengths are proportional to those shown in Fig. 10.13. We assume that the phototransitions are near the band edge, where the $\vec{k}$ vector mainly consists of the quantized component k_z. Otherwise, we should take into account additional mixing of the valence bands in the quantum well. If we set $\vec{k} \approx \vec{k}_{av} \approx \{0, k_z\}$, we can use Eq. (10.72). For example, light with polarization perpendicular to the quantum-well layer does not cause phototransitions between the conduction band and the heavy-hole band, i.e., $(\vec{\xi}\vec{D})_{c\text{-hh}} = 0$. In contrast, the interaction of this light with the light-hole band reaches a maximum just for this perpendicular polarization and one has $(\vec{\xi}\vec{D})_{c\text{-lh}} = 4\mathcal{D}/3$. On the other hand, for light with polarization parallel to the quantum-well layer, the phototransition strength to the heavy-hole band increases to the maximum and $(\vec{\xi}\vec{D})_{c\text{-hh}} = \mathcal{D}$, while the phototransition strength to the light-hole band decreases to the minimum with $(\vec{\xi}\vec{D})_{c\text{-lh}} = \mathcal{D}/3$. These results are illustrated by the diagram in Fig. 10.24, in which the bulk case is presented as well for comparison. In these diagrams, the relative intensities of phototransitions for different polarizations and types of valence bands are shown.

Consequently, there is a strong polarization dependence of the interaction of light with carriers in confined quantum structures. These effects are important for light-emitting diodes and laser diodes based on quantum structures.

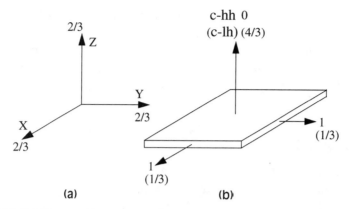

Figure 10.24. Relative transition intensities for various valence bands for (a) bulk, (b) quantum-well III–V compound structures. In case (b) the upper figures correspond to conduction-band–heavy-hole transitions, while the lower figures correspond to conduction-band–light-hole transitions.

10.6 Intraband Transitions in Quantum Structures

In Section 10.5 we studied the interaction of electrons and holes with light, which results in band-to-band phototransitions. In this section we analyze intraband phototransitions. For this case, free carriers absorb or emit photons while remaining in the same energy band. We are going to study two types of phototransitions: between subbands in a quantum well and from subbands to extended electron states. For both cases, we calculate the attenuation of light passing across the quantum-well layer and the absorption coefficient for the light propagating along this layer. We show that intraband phototransitions are strongly polarization dependent. We present several experimental and theoretical results for intraband phototransitions in the more complex heterostructure systems with two coupled quantum wells as well as those with three coupled quantum wells.

10.6.1 Intraband Absorption and Conservation Laws

We start by recalling that in an ideal bulk crystal, intraband phototransitions are impossible because of the energy- and the momentum-conservation laws

$$E(\vec{k}) - E(\vec{k}') = \hbar\omega, \qquad \vec{k} - \vec{k}' = \vec{q}(\approx 0) \tag{10.102}$$

cannot be satisfied simultaneously; here, $E(\vec{k}) = \hbar^2\vec{k}^2/(2m^*)$ is the electron kinetic energy, $\hbar\vec{k}$ is the three-dimensional momentum, $\omega = cq/\sqrt{\kappa}$ is the light frequency, and $\vec{q}$ is the wave vector of the light. In fact, as shown in Sections 10.4 and 10.5, the change in momentum is negligible. Intraband phototransitions can be induced only by phonons, impurities, and other crystal imperfections.

In contrast to bulk materials, intraband phototransitions occur in semiconductor heterostructures. The prediction based on the conservation laws of Eq. (10.102) needs to be modified since the correct conservation laws do not forbid these types of transitions. Consider a quantum well. Let the x, y axes be in the plane of the quantum-well layer. In systems with heterojunctions, a carrier is not characterized by a z component of the momentum. In a proper description, one introduces quantum numbers: subband number i, in-plane wave vector $\vec{k}_{\|} \equiv \{k_x, k_y\}$, and spin number σ. The subband-energy dispersion is given by $E_i(\vec{k}_{\|})$. Now, instead of Eq. (10.102), for i–j intersubband phototransitions, one can write

$$E_i(\vec{k}_{\|}) - E_j(\vec{k}'_{\|}) = \hbar\omega, \qquad \vec{k}_{\|} - \vec{k}'_{\|} \approx 0, \tag{10.103}$$

$$E_i(\vec{k}_{\|}) = \epsilon_i + \frac{\hbar^2 \vec{k}_{\|}^2}{2m^*}. \tag{10.104}$$

Consequently, if $\hbar\omega$ coincides with intersubband distances $\epsilon_i - \epsilon_j$, Eq. (10.103) is satisfied. If one of the electron states involved in a phototransition is in an unconfined extended state, the energy of the transverse motion E_z is a true quantum number and the conservation laws take the form

$$E(E_z, \vec{k}_{\|}) - E_j(\vec{k}'_{\|}) = \hbar\omega, \qquad \vec{k}_{\|} - \vec{k}'_{\|} \approx 0, \tag{10.105}$$

$$E(E_z, \vec{k}_{\|}) = V_b + E_z + \frac{\hbar^2 \vec{k}_{\|}^2}{2m^*}, \tag{10.106}$$

where V_b is the depth of the quantum well. Equation (10.105) is satisfied at $E_z = \epsilon_j + \hbar\omega - V_b$.

We see that both Eqs. (10.103) and (10.105) can be satisfied for certain photon energies. Consequently, intraband phototransitions are allowed and can occur without the presence of phonons, defects, etc. Two types of processes corresponding to the cases of Eqs. (10.103) and (10.105) are shown in Figs. 10.25(a) and 10.25(b), respectively. Figure 10.25(a) corresponds to the so-called intersubband phototransition. The second type of process may be described as photoionization of a quantum well and as photocapture into the well. Note that for the intrasubband case, when $i = j$, we return to the conditions given by Eq. (10.102), which forbid intrasubband phototransitions.

Taking into account that the energy distances between subbands are tens to hundreds of millielectron volts and the depths of quantum wells are of the same

Figure 10.25. The schematics of intraband phototransitions: (a) intersubband transitions, (b) confined-to-extended-state transitions.

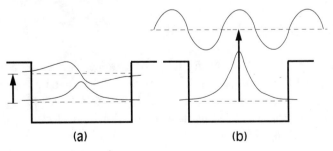

(a)   (b)

order of magnitude, one can see from Eqs. (10.103) and (10.105) that intraband phototransitions correspond to absorption or emission of infrared light.

Thus we may conclude that intraband phototransitions are allowed in heterostructures. The intensities of these phototransitions depend strongly on heterostructure parameters and may be varied artificially over a wide range. This makes quantum structures attractive for applications in different infrared devices such as photodetectors, infrared emitters, and lasers, etc.

10.6.2 Probability of Intersubband Phototransitions

For intraband processes between two different subbands, we can apply general equations for phototransition rates with initial and final quantum number sets denoted by

$$\nu = \{b, i, \vec{k}_\parallel, \sigma\}, \qquad \nu' = \{b, j, \vec{k}'_\parallel, \sigma'\},$$

where b is the label of the band ($b = v, c$). Label b does not change for intraband transitions. This is why we omit it in further equations. We also omit the label of the spin σ because of spin conservation. Now the phototransition probability can be written as

$$P_{i,\vec{k}_\parallel; j, \vec{k}'_\parallel} = \frac{2\pi}{\hbar} |\langle i, \vec{k}_\parallel | \vec{\mathcal{E}} \vec{D} | j, \vec{k}'_\parallel \rangle|^2 \delta[E_j(\vec{k}'_\parallel) - E_i(\vec{k}_\parallel) - \hbar\omega], \tag{10.107}$$

where $\vec{\mathcal{E}}$ is the electric field of the light wave and $\vec{D} \equiv e\vec{r}$ is the dipole moment of the electron. To be specific, we suppose that $E_j(\vec{k}'_\parallel) > E_i(\vec{k}_\parallel)$, i.e., we consider absorption of light. The electric field of Eq. (10.1) is a function of coordinates x, y, and z. If the quantum-well thickness L is substantially smaller than the wavelength of light, i.e., $Lq \ll 1$, we can rewrite Eq. (10.1) in the form

$$\vec{\mathcal{E}}(\vec{r}) = \vec{\xi} F(z_w) \exp[i(q_x x + q_y y)],$$

where $F(z_w) = F_0 \cos(q_z z_w)$ is the magnitude of the electric field in a quantum-well layer situated at $z \approx z_w$. As we found in Section 10.5, for interband phototransitions only the envelope wave functions define the matrix element in Eq. (10.107). According to Chapter 3, the envelope wave functions of confined electrons have the form

$$\psi_{b,l,\vec{k}_\parallel}(\vec{r}) = \frac{1}{\sqrt{S}} \chi_l(z) \exp[i(k_x x + k_y y)], \tag{10.108}$$

where S is the area of the quantum-well layer and $\chi_l(z)$ is the confined wave function for the lth subband; $l = 1, 2, \ldots$. Since the electric-field dependence on position $F(\vec{r})$ is smooth compared with both the quantum-well thickness L and the electron wavelength $2\pi/k$, this dependence can be neglected in the matrix element. Then, taking into account the conservation of spin in electric dipole transitions, we can calculate the matrix element:

$$\langle i, \vec{k}_\| \,|\, \vec{\xi}\vec{r} \,|\, j, \vec{k}'_\| \rangle = \langle i, \vec{k}_\| \,|\, \vec{\xi}\vec{r} \,|\, j, \vec{k}'_\| \rangle$$

$$= \frac{1}{S} \int d^3 r\, \chi_j^*(z) e^{-i\vec{k}'_\| \vec{\rho}} (\xi_x x + \xi_y y + \xi_z z) \chi_i(z) e^{i\vec{k}_\| \vec{\rho}}$$

$$= \xi_z \delta_{\vec{k}_\|, \vec{k}'_\|} \int dz\, \chi_j^*(z) z \chi_i(z) \equiv \xi_z \delta_{\vec{k}_\|, \vec{k}'_\|} \langle \chi_j \,|\, z \,|\, \chi_i \rangle. \qquad (10.109)$$

Here we take into account that for $i \neq j$ the integral

$$\int dz\, \chi_j^*(z) \chi_i(z) = 0,$$

because the wave functions of different subbands are orthogonal.

From Eq. (10.109), one can see that if the vector of the polarization of light $\vec{\xi}$ lies in plane of the quantum-well layer (x, y plane), the matrix element is zero and phototransitions are impossible. If the vector $\vec{\xi}$ has a z component, intersubband processes take place. The probability of phototransitions can be written as

$$P_{i,\vec{k}_\|,\sigma;j,\vec{k}'_\|,\sigma'} = \frac{2\pi}{\hbar} e^2 |\xi_z|^2 |F(z_w)|^2 \delta_{\vec{k}_\|, \vec{k}'_\|} \delta_{\sigma,\sigma'} |\langle \chi_j(z) \,|\, z \,|\, \chi_i(z) \rangle|^2$$

$$\times \delta[E_j(\vec{k}'_\|) - E_i(\vec{k}_\|) - \hbar\omega]. \qquad (10.110)$$

Equation (10.110) gives both conservation laws as well as an explicit polarization dependence of the phototransitions. For the case of an infinitely deep well that has wave functions

$$\chi_l(z) = \sqrt{\frac{2}{L}} \sin\frac{\pi l}{L} z, \quad l = 1, 2, \ldots, \quad (0 < z < L), \qquad (10.111)$$

we get the overlap integral

$$\langle \chi_j(z) \,|\, z \,|\, \chi_i(z) \rangle = \frac{L}{\pi^2} \left\{ \frac{\cos[\pi(j-i)]-1}{(j-i)^2} - \frac{\cos[\pi(j+i)]-1}{(j+i)^2} \right\}. \qquad (10.112)$$

Thus we find the expected result: the matrix element is proportional to the thickness of the quantum well L. Equation (10.112) implies some selection rules: for transitions between two subbands with the same parity, the matrix element is zero, i.e., these transitions are forbidden. The transitions are possible only between states with different parity. Another result that is evident from Eq. (10.112) is that the matrix element decreases when the subband numbers i, j increase or their difference $|i - j|$ increases. The maximum absolute value of the matrix element corresponds to transitions between the lowest subbands:

$$|\langle \chi_2(z) \,|\, z \,|\, \chi_1(z) \rangle| = \frac{16}{9\pi^2} L.$$

For a quantum well with a finite depth V_b, the corrections to Eq. (10.112) are of the order of $\hbar/L\sqrt{2m^* V_b}$. The inverse of this quantity determines the number of confined states in the well (see Subsection 3.2). Consequently the above analysis is correct if the well contains at least several subbands.

From these results, it is straightforward to calculate the absorption rate for the intraband transitions:

$$R_{\text{abs}}(\omega) = \sum_{i,\vec{k}_\parallel;j,\vec{k}'_\parallel} P_{i,\vec{k}_\parallel;j,\vec{k}'_\parallel}[\mathcal{F}_i(\vec{k}_\parallel) - \mathcal{F}_j(\vec{k}'_\parallel)], \tag{10.113}$$

where $\mathcal{F}_i(\vec{k}_\parallel)$ is the occupation number of the state $\{i, \vec{k}\}$. For the parabolic subbands of Eq. (10.104), the δ function in Eq. (10.113) is independent of the wave vector and the sum over $\vec{k}$ can be calculated in an explicit form:

$$R_{\text{abs}}(\omega) = \frac{2\pi}{\hbar} e^2 |\xi_z|^2 F^2(z_w) S \sum_{i,j} |\langle \chi_j \,|\, z \,|\, \chi_i \rangle|^2 \, \delta(\epsilon_j - \epsilon_i - \hbar\omega)(n_i - n_j). \tag{10.114}$$

Here $n_l \equiv \sum_{\vec{k}_\parallel,\sigma} \mathcal{F}_{l,\sigma}(\vec{k}_\parallel)$ represents the total population of the lth subband, i.e., the concentration of electrons per unit area in that subband. From Eq. (10.114), it is seen that the absorption spectrum consists of a series of infinitely high and narrow peaks at photon energies corresponding to the intersubband separation energies. This result is a consequence of our assumption of exactly the same dispersion relations $E_l(\vec{k}_\parallel)$ of all subbands involved in the transition. Allowance for energy-broadening mechanisms would change these δ peaks into lines with finite heights and widths. Equation (10.101) presents a possible broadening function $\Delta(\omega)$, which leads to finite peaks of $\alpha(\omega)$. The parameters $\tau_{c,v}$ in Eq. (10.101) should be replaced by the intraband-scattering time τ_{int}. Another cause of the finite peaks is a deviation of $E_i(\vec{k}_\parallel)$ from parabolic dependence. A nonparabolicity always exists in the electron and hole spectra because the kinetic energy with effective mass is only the second term of an infinite series expansion in the electron energy, as shown in Chapter 2.

We see that the absorption rate depends directly on the subband population of the quantum wells. A population of the lowest subband can be achieved by means of doping. For one of the most important applications – photodetectors – a special type of selective doping is used: quantum wells are doped while barriers remain undoped and ensure high mobility for the overbarrier motion of electrons.

In Section 10.5, we studied some features of the electrodynamics of nonuniform media, particularly of heterostructures. We introduced the main parameters that characterize the interaction of light with matter. All those considerations remain valid for intraband absorption. Thus we can introduce parameters like α, β, and γ of Eqs. (10.85), (10.75), and (10.76) that can describe light absorption (amplification). We consider two possible physical situations, as illustrated in Fig. 10.26.

The first is light propagation along a quantum-well layer, as shown in Fig. 10.26(a), but the light wave is not localized within the layer (compare with Fig. (10.26). If the transverse dimensions of the light beam are L_y and L_z, we get

$$\frac{1}{\mathcal{I}_x} \frac{d\mathcal{I}_x}{dx} = -\alpha_{\text{abs}} = -\frac{\beta_o}{L_z}, \tag{10.115}$$

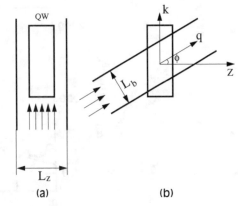

Figure 10.26. Two geometries for intraband absorption of the light beam in a quantum well: (a) the beam with the aperture $L_y \times L_z$ propagates along the well layer, (b) the beam axis is tilted with respect to the well layer; the area of the aperture is $L_y \times L_b$.

where $\mathcal{I}_x$ is the total power of the light beam propagating along the x axis; obviously the light intensity is $\mathcal{I}_x/L_y L_z$. Next, β_o can be written as

$$\beta_o = \frac{16\pi^2 e^2 \omega}{c\sqrt{\kappa}} |\xi_z|^2 \sum_{i,j} |\langle \chi_j | z | \chi_i \rangle|^2 \Delta(\epsilon_j - \epsilon_i - \hbar\omega)(n_i - n_j). \tag{10.116}$$

From Eq. (10.116), one can see that β_o is independent of the aperture of the beam $L_y L_z$, while the absorption coefficient α is inversely proportional to L_z.

The second situation corresponds to having the light beam tilted with respect to the layer and having a finite cross section $L_b L_y$, as shown in Fig. 10.26(b). For this case, taking into account a small absorption by one layer, we find the light attenuation per transit through the quantum-well layer:

$$\frac{\mathcal{I}_{\text{in}} - \mathcal{I}_{\text{out}}}{\mathcal{I}_{\text{in}}} = \frac{\beta_o}{\cos\phi} \equiv \beta, \tag{10.117}$$

where ϕ is the angle between the light-propagation direction and the z axis. Equation (10.117) is valid if the beam aperture L_b is less than the length of the layer L_x: $L_b < L_x \cos\phi$.

According to these results, light with polarization parallel to the heterointerface of the quantum-well layer, $\xi_z = 0$, does not interact with electrons in the heterostructure. When light propagates along the layer, as in Fig. 10.26(a), the light polarization can be either in the x, y plane, i.e., $\xi_y = 1$, $\xi_z = 0$, so that the light does not interact with the heterostructure, or perpendicular to it, i.e., $\xi_y = 0$, $\xi_z = 1$, so that the light is absorbed. In the case of a tilted light beam, we may neglect the light reflection from the quantum-well layer and set $|\xi_z|^2 \approx \sin^2\phi$. Thus the angular dependence of attenuation β is

$$\beta \propto \frac{\sin^2\phi}{\cos\phi}. \tag{10.118}$$

The magnitude of light attenuation by one quantum well is typically small and is typically ~0.01 or less; thus intersubband phototransitions should be measured in multiple-quantum-well structures with the number of wells ranging from 50

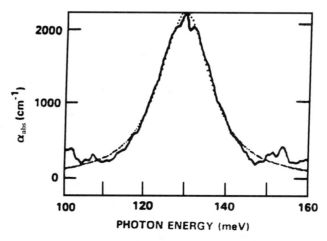

Figure 10.27. Intersubband absorption coefficient of a multiple-quantum-well heterostructure. The solid curve represents experimental data; the dashed curve exhibits the Lorentzian distribution. [After Z. C. Feng, ed., *Semiconductor Interfaces, Microstructures, and Devices. Properties and Applications* (Institute of Physics, Bristol and Philadelphia, 1993).]

to 100. Thick barriers between quantum wells suppress the tunneling effect as well as the formation of minibands. Figure 10.27 shows an example of this intersubband absorption. The GaAs wells are 70 Å wide and the $Al_{0.27}Ga_{0.63}As$ barriers are 150 Å thick. The doping density of the well equals 2×10^{18} cm^{-3}. The calculated subband energies are $\epsilon_1 = 47$ meV and $\epsilon_2 = 171$ meV for a barrier height of 202 meV. The results are given in terms of the absorption coefficient α_{abs} of Eq. (10.115). The attenuation β_o, which is due to a single quantum well, can be estimated by the multiplication of α_{abs} by the thickness of one period of the structure, i.e., 2.2×10^{-6} cm. Thus the maximum attenuation equals 0.0044. The line shape of the absorption coefficient has the Lorentzian shape of Eq. (10.101) with a lifetime broadening of ~0.18 ps. The latter is of the order of the intraband-relaxation time at room temperature. The result of calculations of the absorption coefficient is presented in Fig. 10.27 for comparison.

Above, we have derived the equations for a single-quantum-well or a multiple-well structure with identical wells. Advanced technology facilitates the fabrication of periodic structures with a complex design for each period. For example, a period can consist of two asymmetric coupled quantum wells or of three coupled wells, etc. The results obtained for a single quantum well can be used for the calculation of light absorption in complex structures. One must only substitute the proper envelope functions $\chi_i(z)$ into the matrix element of Eq. (10.109). Then, on the basis of these matrix elements, one can calculate the parameters of light absorption. Figure 10.28(a) shows the asymmetric two-coupled-well and three-coupled-well structures based on $Al_{0.48}In_{0.52}As/Ga_{0.47}In_{0.53}As$ heterostructures; the conduction-band offset is 510 meV. The calculated energy levels and wave functions are also presented. The double-coupled-well structure consists of wells that are 59 Å and 24 Å wide and a coupling barrier with a width of

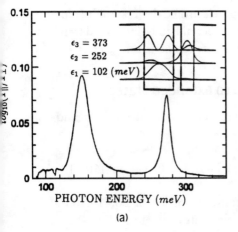

(a)

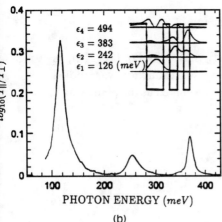

(b)

Figure 10.28. Measured intersubband absorption as a function of photon energy in (a) asymmetric two-coupled-quantum-well structure, (b) three-coupled-quantum-well structure. Parameters of the structures are given in the text. Experiments have been carried out at $T = 10\,\text{K}$. Insets show subband energies and corresponding wave functions. [After C. Sirtori, *et al.*, "Nonparabolicity and a sum rule associated with bound-to-bound and bound-to-continuum intersubband transitions in quantum wells," Phys. Rev. B 50, pp. 8663–8674 (1994).]

13 Å. There are three confined electron states, $\epsilon_1 = 102\,\text{meV}$, $\epsilon_2 = 252\,\text{meV}$, and $\epsilon_3 = 373\,\text{meV}$, in this structure. The experimental results for light absorption are obtained for a 20-period structure. The areal doping concentration in the well equals $1.2 \times 10^{11}\,\text{cm}^{-2}$. Only the lowest level is populated for this doping concentration. The experiment is performed at a temperature of $10\,\text{K}$ with eight passes of the light beam through the heterostructure. The two absorption peaks correspond to intersubband phototransitions: $\epsilon_1 \rightarrow \epsilon_2$ (the first peak) and $\epsilon_1 \rightarrow \epsilon_3$ (the second peak).

The triple-coupled-well structure of Fig. 10.28(b) consists of wells with widths of $46, 20$, and 19 Å, from left to right respectively, and 10 Å thick coupling barriers. The wells contain four confined states, $\epsilon_1 = 126\,\text{meV}$, $\epsilon_2 = 242\,\text{meV}$, $\epsilon_3 = 383\,\text{meV}$, and $\epsilon_4 = 494\,\text{meV}$. The experiments are carried out with the 40-period structure at an areal doping concentration equal to $3.2 \times 10^{11}\,\text{cm}^{-2}$. Three absorption peaks correspond to intersubband transitions from the lowest energy level: $\epsilon_1 \rightarrow \epsilon_2$, $\epsilon_1 \rightarrow \epsilon_3$, and $\epsilon_1 \rightarrow \epsilon_4$.

The results shown in Fig. 10.28 ensure that desirable absorption spectra in a infrared region can be tailored with intersubband phototransitions. Structures with complex periods can be used in various optoelectronic devices: multicolor photodetectors for light detection in two or more spectral bands, intraband transition lasers, generators of microwave radiation in the terahertz range, etc.

10.6.3 Probability of Phototransitions to Extended States

Let us now consider phototransitions between confined states $i, \vec{k}_\parallel, \sigma$ and extended states $E_z, \vec{k}_\parallel', \sigma'$. The transition probability is

$$P_{i,\vec{k}_\parallel,\sigma;E_z\vec{k}_\parallel',\sigma'} = \frac{2\pi}{\hbar}\delta_{\sigma,\sigma'}|\langle E_z, \vec{k}_\parallel' \mid \vec{\mathcal{E}}\vec{D} \mid i, \vec{k}_\parallel\rangle|^2\delta(V_b + E_z - \epsilon_i - \hbar\omega). \tag{10.119}$$

The δ function in Eq. (10.119) gives the energy-conservation law of Eq. (10.105). In the same manner as before, one finds matrix elements of the form

$$\langle E_z, \vec{k}_\parallel' \mid \vec{\mathcal{E}}\vec{D} \mid i, \vec{k}_\parallel\rangle = F(z_w)\xi_z\delta_{\vec{k}_\parallel,\vec{k}_\parallel'}\int dz\,\chi_{E_z}^*(z)z\chi_i(z).$$

For confined states, we can use the wave function χ_i given by Eq. (10.111), as was discussed in Subsection 3.2.2. The extended-state wave function was studied in Subsection 3.2.2. The matrix element can be approximated as the plane wave:

$$\chi_{E_z}(z) \equiv \chi_K(z) = \frac{1}{\sqrt{L_z}}e^{iKz}, \tag{10.120}$$

where

$$K = \frac{\sqrt{2m^*E_z}}{\hbar} = \frac{2\pi n}{L_z}, \quad n = 0, 1, 2, \ldots,$$

and L_z is the size of the system in the z direction; L_z does not appear in the final, physically significant results. Here we have neglected electron reflection from the discontinuities in the potential at $z = 0$ and L. Now the absorption rate for the case under consideration can be written as

$$\mathcal{R}_{\mathrm{abs}}(\omega) = \frac{2\pi}{\hbar}e^2|\xi_z|^2F^2(z_w)\sum_{i,K}|\langle\chi_K \mid z \mid \chi_i\rangle|^2\delta(V_b + E_z - \epsilon_i - \hbar\omega)$$
$$\times [\mathcal{F}_i(\vec{k}_\parallel) - \mathcal{F}_K(\vec{k}_\parallel)]. \tag{10.121}$$

The last term in the brackets, $\mathcal{F}_K$, can be omitted if most of the electrons are inside the well. Finally, summation over K gives

$$\mathcal{R}_{\mathrm{abs}}(\omega) = \frac{2m^*e^2}{\hbar^3}|\xi_z|^2F^2(z_w)\sum_i\frac{L_z}{K_i(\omega)}|\langle\chi_{K_i} \mid z \mid \chi_i\rangle|^2Sn_i, \tag{10.122}$$

where

$$K_i(\omega) \equiv \frac{\sqrt{2m^*E_{z,i}(\hbar\omega)}}{\hbar},$$
$$E_{z,i} = \epsilon_i + \hbar\omega - V_b > 0.$$

To avoid cumbersome equations, we present the matrix element for the most interesting case of phototransitions involving the lowest subband $j = 1$:

$$|\langle \chi_K \,|\, z \,|\, \chi_1 \rangle| = \pi \sqrt{\frac{2L}{L_z}} L \left| \frac{4KL\cos(KL/2) - (\pi^2 - K^2L^2)\sin(KL/2)}{(\pi^2 - K^2L^2)^2} \right|.$$

$$(10.123)$$

This matrix element is a nonmonotonous function of E_z and vanishes in the two limits of $KL \to 0$ and $KL \gg 1$. It reaches a maximum at the resonant condition, $KL = \pi$:

$$|\langle \chi_K \,|\, z \,|\, \chi_1 \rangle|_{\max} = \frac{L}{\pi} \sqrt{\frac{L}{2L_z}}.$$

It is instructive to compare this result with Eq. (10.112). There is an important difference between this case and the case studied in Section 10.5. For confined-to-confined-state transitions, the absorption spectrum consists of a series of peaks; for confined-to-extended-state transitions the absorption spectrum is rather a wideband situated at $\hbar\omega > V_b - \epsilon_1$, with its maximum at $\hbar\omega \approx V_b$ and a half-width of $\sim\epsilon_1$.

Now we can use Eqs. (10.122) and (10.123) for the calculation of optical characteristics due to phototransitions to extended states. The attenuation β_o is

$$\beta_o = \frac{16\pi^2 e^2 m^* \omega}{\hbar^2 c \sqrt{\kappa}} |\xi_z|^2 L_z \sum_i \frac{1}{K_i(\omega)} |\langle \chi_{K_i} \,|\, z \,|\, \chi_1 \rangle|^2 n_i. \qquad (10.124)$$

Since the square of the matrix element of Eq. (10.123) is proportional to $1/L_z$, the attenuation β_o, as given by Eq. (10.124), is independent of L_z. This result is consistent with the meaning of the quantity β_o: it determines the loss of the light energy during a single passage through the quantum-well layer and it should not depend on the structure size in the z direction.

An example of photoabsorption to extended states is presented in Fig. 10.29. The experimental and the calculated results are shown for a $Al_{0.2}Ga_{0.8}As/GaAs$ multiple-quantum-well structure. The well width is $46\,\text{Å}$; the barrier height is $145\,\text{meV}$. The narrow wells are used to provide only one two-dimensional subband with energy $\epsilon_1 = 68\,\text{meV}$. One can see from Fig. 10.29 that there is a wide absorption band with a maximum at the resonant condition $KL \approx \pi$, which corresponds to the energy $E_z = 15\,\text{meV}$ above the barriers. Considerable sensitivity of the spectrum to the well thickness is demonstrated by curve (b), which is calculated with the same parameters except for a thickness of $40\,\text{Å}$. For this case, only level ϵ_1 has a larger energy, which leads to a blue shift of the absorption spectrum. This result shows that the fabrication of structures with different layer thicknesses provides the possibility of controlling the absorption spectrum.

Note that intraband phototransitions to extended states correspond to the photoionization of a quantum well, i.e., photogeneration of the electrons moving

Table 10.2.
The Selection Rules for Intersubband Phototransitions
in Quantum Wells

Polarization of the wave	ξ_x	ξ_y	ξ_z
Propagation in the			
z direction	forbidden	forbidden	
x direction		forbidden	allowed
y direction	forbidden		allowed

over the barriers. This type of photoabsorption is similar to the usual photoionization of impurities.

Let us stress once again that intraband absorption is strongly polarization dependent. Table 10.2 briefly summarizes the selection rules for intraband transitions with different polarizations and propagation directions of light. For each direction of propagation, there are only two possible polarizations of the transverse electromagnetic wave. Quantum wells are assumed to be perpendicular to the z direction.

Figure 10.29. Experimental results and calculations for the absorption coefficient of confined-to-extended-state phototransitions. The multiple-quantum-well heterostructure is described in the text. Calculated curves (a) and (b) correspond to well widths of 46 and 40 Å, respectively. [After Z. C. Feng, ed., *Semiconductor Interfaces, Microstructures, and Devices. Properties and Applications* (Institute of Physics, Bristol and Philadelphia, 1993).]

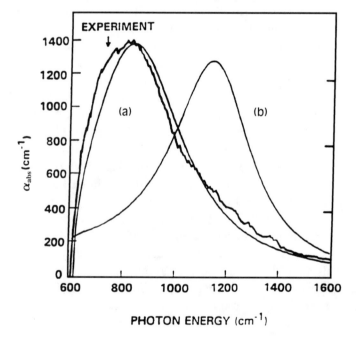

In conclusion, the absorption spectra corresponding to intraband transitions in quantum heterostructures are sensitive to the parameters of the structures. This flexibility of tuning to necessary absorption spectra has many potential applications for photodetectors. These applications will be analyzed in Chapter 12.

10.7 Closing Remarks

In this chapter we reviewed the basic results for the optics of semiconductors. We started with a consideration of electromagnetic fields and introduced the quanta of these fields – photons. It has been emphasized that in optical resonators and nonuniform dielectric media, the structures of electromagnetic fields differ considerably from those of free space. As a result, the properties of photons, their form factors, polarization, frequencies, and densities of states are also different and can be controlled through the use of proper resonators, dielectric waveguides, etc.

We reviewed interband optics of bulklike, direct-bandgap semiconductors, calculated absorption, and spontaneous and stimulated emission. Excitonic effects and phototransitions for a real band structure of III–V compounds were analyzed. We briefly considered the refractive index of the materials and the electron contribution to this quantity.

Then we studied optical interband transitions in quantum heterostructures. We derived basic formulas for optical effects in these systems and found that optical spectra of quantum heterostructures have more complex forms and depend substantially on the materials and the parameters of the structures. In particular, the optical density of states exhibits a sharper dependence on the frequency compared with that of bulk materials. The excitonic absorption is much more pronounced in quantum heterostructures and is observable even at room temperature. All these results are important for optoelectronics and particularly for lasers and nonlinear optical devices based on quantum heterostructures.

We studied intraband phototransitions. In a bulk material, these types of optical transitions are possible only if a third particle – phonon, defect, interface, etc. – is involved in the process. In heterostructures, these optical transitions occur even for ideal structures. We calculated the phototransitions between confined states of electrons and holes and absorption from confined to extended states of carriers. Such intersubband optical spectra can be tailored for a desired spectral region. In many cases, such tailoring techniques are applied in the infrared regions of the electromagnetic spectrum. It is worth mentioning that intersubband phototransitions exist in heterostructures of any kind of materials: direct- and indirect-bandgap materials can be used. In Chapters 11 and 12 we will use these results to analyze photodetectors and lasers based on intersubband transitions.

More details on the optics of solid-state structures, including semiconductor structures, can be found in the following textbooks:

C. Kittel, *Quantum Theory of Solids* (Wiley, New York, London, 1963).
D. K. Ferry, *Semiconductors* (Macmillan, New York, 1991).

I. Ipatova and V. Mitin, *Introduction to Solid-State Electronics* (Addison-Wesley, Reading, MA, 1996).

Additional discussions on the interband optics of heterostructures based on III–V compounds are presented in the following books:

G. Bastard, *Wave Mechanics Applied to Semiconductor Heterostructures* (Halsted, New York, 1988).

C. Weisbuch and B. Vinter, *Quantum Semiconductor Structures* (Academic, San Diego, 1991).

Intraband phototransitions are analyzed in detail in the following references:

G. Bastard, *Wave Mechanics Applied to Semiconductor Heterostructures* (Halsted, New York, 1988).

M. O. Manasreh, ed., *Semiconductor Quantum Wells and Superlattices for Long-Wavelength Infrared Detectors* (Artech House, Boston, 1993).

PROBLEMS

1. How many photons per unit volume are there in GaAs when an electromagnetic wave with $\lambda = 1\,\mu\text{m}$ and $F_0 = 10^4$ V/cm propagates through the crystal?

2. Quantum-well infrared photodetectors incorporate multiple-quantum-well or superlattice structures since the absorption in a given quantum well is relatively low as a result of the small number of electrons in a given quantum well. Furthermore, it is observed that in order to produce significant absorption, the polarization of the infrared electric field must have a component perpendicular to the plane of the well. In other words, for the case of normal illumination, the polarization is nearly in plane and the coupling of radiation to electrons is low. Explain why the absorption in a quantum-well infrared photodetector is larger for the case in which there is an appreciable component of electric field perpendicular to the plane of the quantum well than for the case in which the infrared electric field is parallel to the quantum-well interface.

3. In resonant-tunneling devices exhibiting negative differential resistance, the off state is usually associated with a nonzero minimum current that is frequently referred to as the valley current. It is known that phonon-assisted tunneling transitions contribute significantly to this valley current. Consider two cases: (a) polar-optical–phonon-assisted tunneling between bands of the same symmetry and (b) polar-optical–phonon-assisted tunneling between bands of different symmetry. By separating the Bloch states into periodic and envelope functions, argue that the matrix element for such transitions should be smaller for transitions between bands of different symmetry.

11

Electro-Optics and Nonlinear Optics

11.1 Introduction

The effect of an external electric field on the refractive index or the absorption coefficient, i.e., its effect on the propagation of light through a material or on the reflection of the light is known as the electro-optical effect. Because of the high electric-field sensitivity of the electron-energy spectra and the feasibility of controlling carrier concentrations in quantum heterostructures, the electro-optical behavior of such heterostructure devices is unique.

In this chapter we study the electro-optical effect for quantum heterostructures including quantum wells, double-quantum-well, and multiple-quantum-well (MQW) structures, and superlattices. We consider the quantum-confined Stark effect, the Burstein–Moss effect, and the effect of destroying excitons in gated heterostructures. In all three of these cases, the absorption edge is controllable over a wide range of external voltages. In addition, we examine the electro-optics of superlattices as well as evidence of Stark ladder effects. For double quantum wells, we study coherent oscillations of an electron wave packet and terahertz microwave emission controlled by an electric field.

We show that electro-optical effects have great potential for the optoelectronic applications of using a field to control light and of realizing the tunable generation of microwave emission.

The second part of this chapter is devoted to the consideration of nonlinear optical effects in quantum heterostructures. Such effects occur at high intensities of light, when the optical characteristics of a heterostructure become dependent on the amplitude of the light wave or the light intensity. Semiconductor heterostructures lead to enhancements of the nonlinear effects observed in bulk materials, and they also exhibit unique nonlinear optical behavior that is peculiar to only low-dimensional systems. This class of effects provides the possibility of controlling light by light and creates the basis for new devices with all-optical addresses.

11.2 Electro-Optics in Semiconductors

11.2.1 Electro-Optical Effect in Conventional Materials

We start with a brief review of the electro-optical effect in commonly used materials.

The effect of an external electric field $\vec{F}$ on the refractive index $n(\omega)$ of a transparent dielectric medium can be introduced by a function $n(\omega, \vec{F})$. In fact, the field dependence is relatively weak, and we can restrict ourselves to several terms of the expansion of $n(\omega, \vec{F})$ in a series with respect to the field $\vec{F}$. Since n is a scalar value and $\vec{F}$ is a vector, the expansion depends on the symmetry of the medium under consideration.

In noncentrally symmetric crystals, the first term in the expansion is linearly dependent on the magnitude of the field. Thus for the linear electro-optical effect we can write

$$n(F) = n - \frac{1}{2} a_P n^3 F, \tag{11.1}$$

where n is the refractive index at $F = 0$ and a_P is the linear electro-optical co-efficient. This case is known as the linear electro-optical effect or the Pockels mechanism.

In centrally symmetric media, the quadratic electro-optical effect or the Kerr mechanism occurs. Changes in the refractive index are proportional to the square of the magnitude of the field:

$$n(F) = n - \frac{1}{2} b_K n^3 F^2, \tag{11.2}$$

where b_K is the quadratic electro-optical coefficient. Generally, both coefficients a_P and b_K depend on light frequency. Typical values of a_P lie in the range of 10^{-10}–10^{-8} cm/V. For a field F of $\sim 10^4$ V/cm, such a material undergoes changes in refractive index from 10^{-6} to 10^{-4}. Typical values for b_K range from 10^{-14} to 10^{-10} cm^2/V^2, that is, in a field of $F = 10^4$ V/cm, changes in the refractive index are in the range from 10^{-6} to 10^{-2}.

From a practical point of view, the main operational function that can be provided by the electro-optical effect is the modification of the optical properties of a material as a result of an applied electric field. While acoustic waves and a magnetic field could be used for the same purposes by means of acousto-optical effects and magneto-optical effects, the electro-optical effect is of special importance because it is widely used for high-speed applications and it is compatible, in principle, with modern electronics.

To gain more insight, let us briefly sketch some applications of the effect under consideration. We know that a light wave can be characterized by three parameters: a frequency ω, an amplitude F_0 (or an intensity $\mathcal{I}$), and a phase ϕ. The electro-optical effect may be used to control two of these parameters – phase and intensity. For a plane wave, we can write

$$\vec{\mathcal{E}}(\vec{r}) = \vec{F}_0 \cos[\omega t - \phi(\vec{r})], \tag{11.3}$$

where the phase ϕ is position dependent. For a homogeneous medium, $\phi = n\omega z/c$, where c is the velocity of light in free space and z is the direction of light propagation.

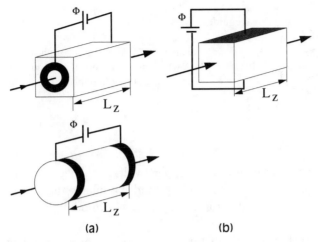

Figure 11.1. Electro-optical modulators: (a) a longitudinal modulator: an electric field is applied along the optical path; (b) a transverse modulator: an electric field is perpendicular to the optical path.

Consider first the phase modulation. If, by external means, one changes the refractive index n of a medium along the optical path L_z, the phase varies as

$$\phi = \phi_0 - \frac{\omega}{2c} a_P n^3 F L_z \equiv \phi_0 - \pi \frac{\Phi}{\Phi_\pi}, \tag{11.4}$$

where ϕ_0 is the phase without an electric field, and we have introduced the voltage $\Phi = Fd$, the dimension d of the region with the field F, and a characteristic value $\Phi_\pi = 2\pi dc/a_P n^3 \omega L_z$, which is appropriate to the case of the Pockels mechanism. According to Eq. (11.4), a variation of the field or, equivalently, the voltage, leads to a modulation of the wave's phase.

In Fig. 11.1, two schemes of phase modulators are shown. The case of Fig. 11.1(a) corresponds to the so-called longitudinal modulators, in which the direction of the electric field coincides with the direction of light propagation (here the optical path L_z coincides with d). The case of Fig. 11.1(b) corresponds to the transverse modulator. The characteristic times of nonstationary processes required for reaching the refractive index changes are small because both the Pockels and the Kerr mechanisms are extremely short (10^{-13} s or so), for both Figs. 11.1(a) and 11.1(b) the modulation speed is limited only by the transition time of the light through the crystal: $T = L_z n/c$. Thus the transit-time-limited modulation bandwidth is $1/T$. If the electric wave traveling through the line has a velocity matched to that of the light, the transition delay can be eliminated. Although commonly known modulators operate at modulation frequencies of roughly hundreds of megahertz, modulation speeds of up to tens of gigahertz can be achieved in the latter scheme. Integrated optical phase modulators working on the electro-optical effect have also been developed. Figure 11.2 presents the schematic of such a modulator. Light propagates through a waveguide that is

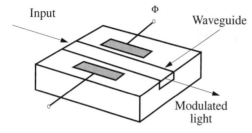

Input Φ Waveguide

Modulated light

Figure 11.2. An integrated optical phase modulator. A waveguide and controlling electrodes are presented. Input and output (modulated) light beams are shown.

placed between two electrodes. An applied voltage changes the refractive index of the waveguide and its surroundings and modulates the phase of the output light signal. The modulator can operate with modulation speeds of up to 100 GHz.

There are different methods for electric modulation of the intensity of light. For example, one can modulate the wave intensity if a phase modulator is placed in one branch of an interferometer. Such a case, the Mach–Zehnder interferometer, is shown in Fig. 11.3. Here, the light wave with intensity $\mathcal{I}_{in}$ is split into two beams, say, with equal intensity $\mathcal{I}_{in}/2$. As a result of interference of the two

Figure 11.3. The Mach–Zehnder interferometer with a phase modulator placed in one branch. The lower part of the figure shows the interferometer transmission T as a function of an applied voltage Φ. The dependence $T(\Phi)$ is periodic, with a period equal to $2\Phi_\pi$. Near point B the device operates as an almost linear modulator of the light intensity, as illustrated. Voltage switching between points A and C provides optical switching between zero and total transmission.

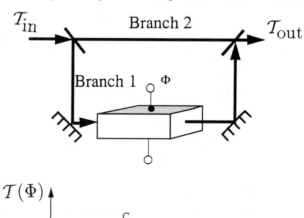

$\mathcal{I}_{in}$ Branch 2 $\mathcal{I}_{out}$

Branch 1 Φ

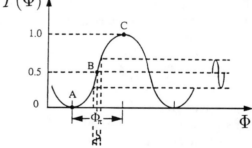

$T(\Phi)$

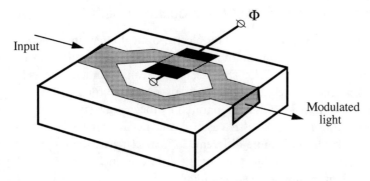

Figure 11.4. An integrated optical intensity modulator or optical switch. The device consists of an integrated Mach–Zehnder waveguide interferometer with electro-optical modulation in one of its branches.

beams, the output intensity $\mathcal{I}_{\text{out}}$ is

$$\mathcal{I}_{\text{out}} = \frac{1}{2}\mathcal{I}_{\text{in}} + \frac{1}{2}\mathcal{I}_{\text{in}} \cos\phi = \mathcal{I}_{\text{in}} \cos^2 \frac{\phi}{2}, \tag{11.5}$$

where ϕ is the difference between the phase shifts in the both branches of the interferometer. Let this difference be ϕ_0 at zero electric field. Then, in accordance with Eq. (11.4), the transmission coefficient of the device is

$$T(\Phi) = \frac{\mathcal{I}_{\text{out}}}{\mathcal{I}_{\text{in}}} = \cos^2\left(\frac{\phi_0}{2} - \frac{\pi}{2}\frac{\Phi}{\Phi_\pi}\right). \tag{11.6}$$

The phase ϕ_0 can be controlled, for example, by the length of the branches. Thus different regimes of operation are possible. If $\phi_0 = \pi/2$, the light intensity can be modulated almost linearly around $\mathcal{I}_{\text{out}} = \mathcal{I}_{\text{in}}/2$. If $\phi_0 = 2\pi m$, where m is an integer, the device provides almost 100% modulation of the output intensity over the range of voltages from 0 to Φ_π.

This principal idea can be used to realize an integrated optical intensity modulator, as illustrated in Fig. 11.4. In this device, an input signal is split and propagates through two waveguide branches; the refractive index for one of the branches is controlled by an applied voltage, similarly to the case of Fig. 11.2. Modulation of the light intensity at a few gigahertz can be achieved for such devices.

The most common modulators are based on LiNbO$_3$ crystals. Unfortunately, this material is not directly compatible with modern optoelectronic and electronic devices based on heterostructures of III–V and group IV materials.

Recently it was verified that the electro-optical effect is strong in quantum heterostructures. In addition to changing the refractive index, the electric field strongly affects light absorption in these structures. It is expected that quantum-based modulator structures – working on the basis of the principles just discussed – will play an increasingly important role in optical modulation as well as in other operations for controlling and processing light signals.

11.2.2 Electro-Optical Effect in Quantum Wells

The most known and important electro-optical effect in semiconductors is the Franz–Keldysh mechanism of modifying the fundamental edge of interband absorption in an electric field. This mechanism can be understood as follows. In Fig. 11.5, valence and conduction bands are presented for $F = 0$ and $F \neq 0$. At zero electric field, the bands are separated by the energy gap E_g and absorption starts at photon energies $\hbar\omega > E_g$, as discussed in Chapter 10. For a finite electric field F, the situation is quite different. In a strict sense, there is no longer any bandgap, since the wave functions of the electrons and the holes overlap under the influence of the applied field and interband phototransitions become possible for $\hbar\omega < E_g$, as shown in Fig. 11.5. Thus the Franz–Keldysh mechanism leads to the electro-optical effect in absorption and refraction.

As a result of the field, the fundamental absorption edge is modified and an absorption tail appears in the long-wavelength spectral region:

$$\alpha(\omega) \propto \exp\left[-\left(\frac{E_g - \hbar\omega}{\hbar\omega_F}\right)^{3/2}\right], \quad \hbar\omega < E_g,$$

where the factor ω_F depends on the electric field:

$$\omega_F = \frac{e^2}{2\,\hbar}\left(\frac{m_e + m_h}{m_e m_h}\right)F^{2/3},$$

m_e and m_h are the effective masses of the electrons and the holes, respectively. The presence of Planck's constant $\hbar$ provides further insight into the quantum origin of the Franz–Keldysh mechanism.

Although the Franz–Keldysh effect is the most important electro-optical mechanism within the fundamental absorption edge, its use is quite limited in bulk crystals. Furthermore, this mechanism washes out excitonic effects, because relatively weak electric fields dissociate the excitons. For III–V compounds, the destruction of excitons begins in electric fields of the order of 10^3 V/cm.

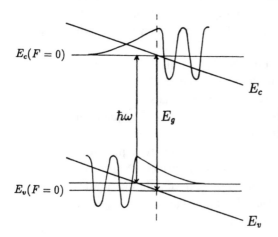

Figure 11.5. An illustration of the Franz–Keldysh effect.

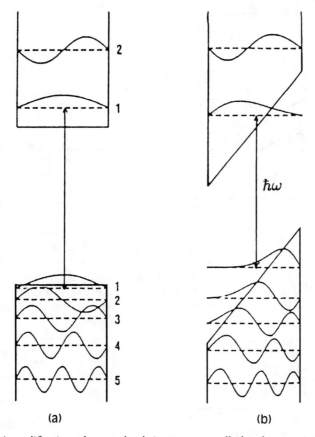

(a)   (b)

Figure 11.6. A modification of energy levels in quantum wells for electrons and holes, when an electric field is applied: (a) wells and levels without an electric field; electron, hole subbands, and their wave functions are shown, (b) the same in an electric field. This field always leads to a decreasing energy separation of electron and hole subbands and a red shift of interband optical spectra. [After D. A. B. Miller, *et al.*, "Relation between electroabsorption in bulk semiconductors and in quantum wells: The quantum-confined Franz–Keldysh effect," Phys. Rev. B 33, pp. 6976–6982 (1986).]

Let us consider the case of a quantum well placed in an electric field perpendicularly to the quantum-well layer. Figure 11.6(a) depicts energy levels in quantum wells for both electrons and holes. These levels are modified by the field, as shown in Fig. 11.6(b). One can see that a shift of energies occurs in the field; this leads to the so-called quantum-confined Franz–Keldysh (or Stark) effect. For gas media, a shift of atomic levels by an electric field is known as the Stark effect. So the effect under consideration is sometimes called the quantum-confined Stark effect (QCSE). In contrast to the case of bulk systems, in quantum-well structures, the gap between the conduction and the valence bands still remains, but the separation between the lowest electron and hole subbands decreases in high electric fields. The upper subbands are also modified. This leads to a red shift of interband optical spectra. In addition, excitonic effects are preserved up to the very high fields of 10^5 V/cm and more because the confining well potential

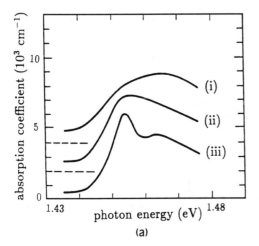

(a)

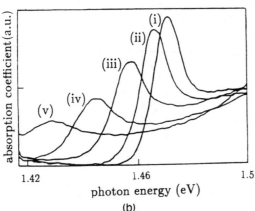

(b)

Figure 11.7. The electric-field effect on the absorption coefficient of a 94-Å-thick GaAs quantum well for (a) parallel, (b) perpendicular field configurations. For the parallel configuration, curves (iii), (ii), and (i) correspond to fields equal to 0, 1.6×10^4, and 4.8×10^4 V/cm, respectively. For the perpendicular configuration, curves (i)–(v) correspond to fields equal to $0, 6 \times 10^4, 1.1 \times 10^5, 1.5 \times 10^5$, and 2×10^5 V/cm, respectively. [After S. Schmitt-Rink, *et al.*, "Linear and nonlinear properties of semiconductor quantum wells," Adv. Phys. 38, pp. 89–188 (1989).]

suppresses dissociation processes. Indeed, the exciton level is shifted and follows the bottom of the subband.

A comparison of the bulklike Franz–Keldysh effect with its confined analog can be performed with the same quantum-well structure for two electric-field configurations: parallel and perpendicular to layers of the structure. In Fig. 11.7, the effect of an electric field on the absorption coefficient is presented for fields parallel and perpendicular to a quantum-well layer. The experiments were performed with 94-Å GaAs quantum wells. In accordance with the previous discussion, the parallel field configuration is rather similar to the Franz–Keldysh effect in a bulk crystal since quantum confinement is not of major importance. The field magnitude is varied from 0 to 4.8×10^4 V/cm. The results presented for three magnitudes of the field show a red shift in the interband absorption tail and a suppression of the excitonic effect. In the perpendicular field configuration, the field is varied in a wider range, from 0 to 2×10^5 V/cm. The results demonstrate a large red shift of light absorption and existence of the excitonic effects up to high fields.

Note that, in contrast to the Pockels and the Kerr mechanisms, electro-optical effects in quantum structures are neither linear nor quadratic functions of the electric field. The effects are rather characterized by complex field and frequency dependences. To enhance the electroabsorption effect, MQW structures may be used. The use of MQW structures facilitates reaching almost 100% modulation of the absorption. Thus quantum structures can be used for direct modulation of the light transmission (without an interferometer) because of the large electroabsorption effect.

As we discussed in Subsection 10.4.5, changes in the absorption coefficient are closely related to the electro-optical effect in the refractive index. The large magnitude of this effect in quantum wells leads to its use in controlling light by an electric field in devices based on the Mach–Zehnder interferometers. Accordingly, the sizes of the necessary heterostructures are much smaller than conventional light modulators based on better materials, such as $LiNbO_3$.

For quantum wells, there are other natural mechanisms in addition to the electro-optical effect. In Chapter 4 we studied the control of the carrier concentration in gated heterostructures by applying voltage to the gate. It was shown that one can deplete quantum wells or increase the carrier concentration up to sheet densities of $\sim 10^{12}$ cm^{-2}. Variations of the carrier concentration induce significant changes in the absorption coefficient as a result of the following two main mechanisms.

The first is caused by the filling factor: An increased population of the electron or the hole subband blocks the absorption; for bulk crystals, this effect is known as the Burstein–Moss effect. Electrons and holes populate almost all energy states near the top of the valence band and the bottom of the conduction band, which makes phototransitions impossible in some spectral range near the fundamental absorption edge. If the quasi-Fermi levels for both types of carriers are E_{Fp} and E_{Fn}, the phototransitions are allowed for photon energies $\hbar\omega > E_g + E_{Fp} + E_{Fn}$. In other words, an increase in population of the bands leads to a blue shift in absorption. Since the density of states of two-dimensional subbands in quantum wells is quite small, as discussed in Chapter 3, the filling factor can be more important: The blue shift of the absorption coefficient is reached easily.

The second mechanism is related to the suppression of the excitonic effect: An increase in the carrier concentration provides more effective screening of the electron–hole Coulomb interaction, which leads to the destruction of excitons. This causes an additional blue shift in the absorption.

Both phenomena contribute to the electro-optical effect in quantum wells. They are illustrated in Fig. 11.8. In this figure, the absorption coefficient α and its changes $\Delta\alpha$ are presented as functions of the photon energy for a GaAs well. The three curves in the left part of figure correspond to the following cases: Both the concentration per unit area and field are zero ($N_s = 0$, $F = 0$), there are a finite electric field and zero concentration (F up), and there are a finite concentration and zero field (N_s up). In the right part of figure, changes in the absorption coefficient are shown: One is due to the QCSE, the other one is the carrier-induced effect. A comparison shows that both effects are of the same order of magnitude.

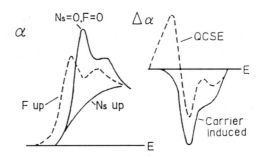

Figure 11.8. Schematic illustrations of changes in absorption spectra of quantum wells. The electric-field effect (dashed curves) and the carrier-induced effect (solid curve) on absorption are shown. For comparison the absorption coefficient at zero electric field and in the absence of the carrier concentration is presented. The left part is the interband-absorption coefficient α as a function of the photon energy E. The right part shows partial contributions of both effects. [After C. Weisbuch and B. Vinter, *Quantum Semiconductor Structures* (Academic, San Diego, 1991).

Within the absorption edge, carrier-induced changes in the refractive index can be large. The effect is strongly dependent on the light frequency and can reach values up to 0.01. Thus, by controlling the carrier concentration by the gate voltage, one can realize effective electro-optical effects that vary either the absorption coefficient or the refractive index.

Note, however, that the quantum-confined Franz–Keldysh (Stark) effect and the carrier-induced effect are characterized by different response times. The first effect is manipulated with the carrier energy states and their wave functions. So the response time is limited by a value of the order of $\hbar/\epsilon_i$, where ϵ_i are subband energies; this follows from the uncertainty relations. The carrier-induced effect is limited by the lifetime of the carriers, i.e., a value that is much larger and is on the nanosecond scale of time or even longer. Figure 11.9(a) is a schematic of an integrated optical modulator based on both effects. The effective waveguide of an optical mode has a 5000-Å core of $Al_yGa_{1-y}As$, with $y = 0.24$. Beneath and over this core, there are cladding layers of $Al_xGa_{1-x}As$, with $x = 0.28$. Wave confinement in the lateral direction is reached by etching, for example. A heterostructure modulator is embedded in one of the branches of the interferometer. The heterostructure consists of 90-Å GaAs MQWs separated by $Al_{0.24}Ga_{0.76}As$ barriers, as shown in Fig. 11.9(c). The modulator length is $L = \sim230\,\mu m$. To realize the QCSE and the carrier-induced, effect, a field-effect-transistor configuration is used, as shown in Fig. 11.9(b). By applying a gate and a source–drain voltage, one controls the electric field and the carrier concentration in the region of the wells. Let us estimate the characteristic parameters of this modulator. Let both effects produce a change in the refractive index equal to Δn. If the confined factor of the optical mode is Γ (this factor was introduced in Subsection 10.5.1), then an effective change of the guided-mode refractive index is estimated as $\Gamma\Delta n$. For the above waveguide parameters, the confined factor Γ is ~0.12. The maximum change in the refractive index for the case presented in Figure 11.8 corresponds

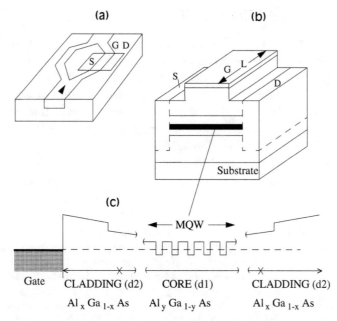

Figure 11.9. The schematics of a Mach–Zehnder interferometer with the quantum-well modulator in one branch: (a) the principal scheme of the interferometer, (b) the MQW modulator with control of the electric field and the carrier concentration, (c) the energy diagram of the heterostructure (the conduction-band profile is shown).

to the photon energy $\hbar\omega_{max} = 1.55\,\mathrm{eV}$ and equals $\Delta n \approx 0.012$. To achieve a phase shift of π [i.e., 100% modulation according to Eq. (11.6)], the required length of the controlled part of branch should be only $33\,\mu\mathrm{m}$. Thus the device can effectively operate not only with a photon energy of $\hbar\omega_{max}$, but also over a wide range of the energies.

It is important that the switching speed of such a modulator be essentially the same as that of the field-effect transistor. If the thickness of the electron channel of the transistor (lateral thickness of the waveguide) is $\sim$1–2 μm, the cutoff frequency for operation of the device is $\sim$15 GHz or higher. In this example, the integrated optical modulator operates with light propagating parallel to the quantum-well layers.

For the perpendicular geometry (the direction of the light propagation is normal to layers comprising a heterostructure), one uses transparent electrodes. In such a case, modulators based on MQW structures are efficient at thicknesses of roughly a few micrometers. Two improvements can be made. The first is to use an integrated multilayer reflector in a double-pass optical system. The second comes about when two such integrated reflectors are used to form a Fabry–Perot resonator. These devices can increase the contrast ratio of the modulation by as much as a factor of 100. Modern technology facilitates the fabrication of arrays of these devices for electrically addressed spatial light modulators. Such modulators are key components of the future optical computing system that utilizes parallel processing of the signals.

11.2.3 Electro-Optical Effect in Superlattices

As we learned in Section 3.6, systems with a large number of MQWs and with thin barrier layers manifest novel characteristics, including the formation of minibands with wave functions that extend throughout the structure. In Section 3.6, such structures are known as superlattices. The electro-optics of superlattices is a more complex subject than that of isolated quantum wells. It involves reconstructions of the energy spectra in the electric field such as the formation of the Stark ladder discussed in Subsection 8.3.3.

To examine the electro-optical effect in superlattices, we sketch in Fig. 11.10(a) electron and hole MQW potentials and minibands for a type-I semiconductor

Figure 11.10. The diagram of interband phototransitions in a superlattice. (a) The potential profile for electrons and holes in a type-I superlattice. One electron miniband and two hole minibands are shown. Transitions T_1 and T_2 originate from heavy- and light-hole minibands, respectively. (b) Interband phototransitions under a moderate electric field: 0 labels vertical transitions, ± 1 and ± 2 label transitions into neighboring wells. (c) Transitions at a high electric field. The miniband structure is destroyed, the carriers are localized in the wells, and dominant transitions are vertical. [After E. E. Mendez and F. Agullo-Rueda, "Optical properties of quantum wells and superlattices under electric fields," J. Luminesc. **44**, pp. 223–231 (1989).]

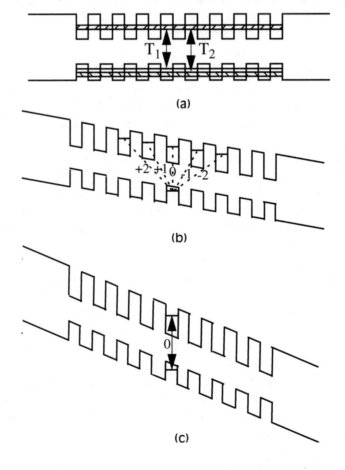

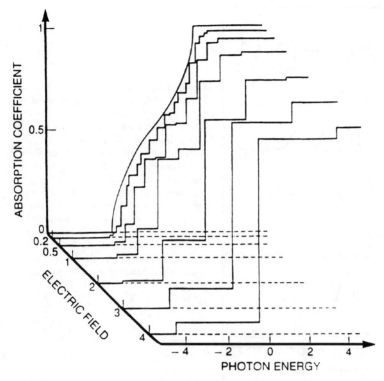

Figure 11.11. Calculated interband-absorption spectra of a superlattice for different electric fields. The absorption coefficient is given in arbitrary units; the electric field (dimensionless) and the photon energy are dimensionless.

material without an electric field. The diagrams depict a GaAs/Al$_{0.35}$Ga$_{0.65}$As superlattice with well and barrier thicknesses equal to 30 and 35 Å, respectively. The fields are 2×10^4 V/cm for Fig. 11.10(b) and 1×10^5 V/cm for Fig. 11.10(c). In Fig. 11.10(a), only the lowest miniband is presented for the conduction band, while for the valence band, two minibands are plotted; one is associated with heavy holes and the other with light holes. In the absorption spectrum of such a superlattice, two types of interband phototransitions should be observed: heavy-hole miniband $\leftrightarrow$ electron miniband (T_1) and light-hole miniband $\leftrightarrow$ electron miniband (T_2), as indicated in Fig. 11.10(a). Consistent with the calculations of the absorption spectra given in Chapter 10, each type of interband phototransition is proportional to the optical density of states of the superlattice. In Fig. 11.11, the absorption coefficient is presented for a model with two simple minibands occurring for the case of parabolic electron and hole dispersion relations, $E_{c,v}(\vec{k})$, given by Eq. (10.54). In the figure, the curve corresponding to the zero electric field has a shape similar to that of the density of states of the superlattice presented in Fig. 3.21. In realistic situations, both types of phototransitions, T_1 and T_2, contribute to the spectrum; these contributions are overlapping and the spectrum becomes more complicated. One such example is shown in Fig. 11.12. This spectrum has been measured for a superlattice with 40-Å GaAs

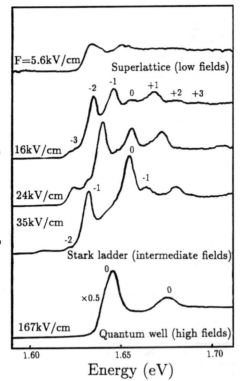

Figure 11.12. Absorption spectra of the superlattice with the parameters discussed in Fig. 11.10 for different electric fields. Magnitudes of the fields are indicated on the curves. At low fields the results correspond to absorption spectra of superlattices and show two phototransitions, T_1 and T_2. At intermediate fields the peaks correspond to the transitions from fully localized hole states to different states of the electron Stark ladder. At high fields the spectra correspond to interband phototransitions in a single-quantum-well structure. [After E. E. Mendez and F. Agullo-Rueda, "Optical properties of quantum wells and superlattices under electric fields," J. Luminesc. **44**, pp. 223–231 (1989).]

wells and 20-Å $Al_{0.35}Ga_{0.65}As$ barriers at $T = 5$ K. The upper curve is presented for a weak electric field of $F = 5.6$ kV/cm. For this case, contributions originating from both the heavy- and the light-hole minibands are resolved.

As discussed in Subsection 8.3.3, an electric field F, applied to a superlattice with the period d, causes a localization of carriers and transforms the minibands into a set of equidistant energy levels,

$$\epsilon_n = \epsilon_0 + eFdn, \quad n = \pm 1, \pm 2, \pm 3, \ldots, \tag{11.7}$$

i.e., a Stark ladder of energy levels. In small fields, the localization of carriers is weak, that is, their wave function extends over several superlattice periods. As the field increases, the localization increases too.

The holes with larger effective masses are localized first. Thus, in intermediate fields, wave functions of both types of holes are confined in one of the identical quantum wells, while the electron wave functions are still delocalized over a few periods of superlattice. As a result, besides strictly vertical interband transitions, which are labeled by 0, there are several well-pronounced transitions, labeled as $\pm 1, \pm 2$, etc., into neighboring wells, that is, the transitions involve several energy levels of the electron Stark ladder of Eq. (11.7). The calculation for the simple model of Fig. 11.11 clearly demonstrates the appearance of these levels in the absorption spectrum at intermediate electric fields, as shown in this figure by the curves for dimensionless fields f of ~1, where $f = eFd/\Delta$ and Δ is the width of the miniband under consideration. For the superlattice under consideration, the experimental evidence of these levels is given in Fig. 11.12 for fields of 16–35 kV/cm.

A further increase in the field causes the localization of electrons in the same manner that it occurred for the holes. Figure 11.10(c) shows the interband transitions for this case. Now the phototransitions and the absorption spectrum are those of a single quantum well. The calculation for such a limiting case is plotted in Fig. 11.11 for $f \geq 3$. The absorption coefficient has a simple steplike shape. Experimental results are shown in Fig. 11.12 for $F = 167$ kV/cm, in which both types of transitions, hh $\leftrightarrow c$ and lh $\leftrightarrow c$, are marked. The evolution of the spectrum is dramatic: from two partially overlapping bands through the multiband behavior and back again to two more peaked and more resolved bands.

Plotting the positions of the observed peaks of the absorption coefficient as functions of the electric field, one can obtain a field dependence of the Stark ladder of Eq. (11.7). In this manner, the results obtained for the superlattice under consideration are collected in Fig. 11.13, which apparently shows the splitting of the electron miniband into a discrete-level series and a linear displacement of the levels with increasing field (the width of the hole miniband is negligible). The

Figure 11.13. Interband-transition energies as functions of the electric field for a superlattice with the parameters discussed in Fig. 11.10. The numbers at the curves correspond to different energy states of the electron Stark ladder. [After E. E. Mendez and F. Agullo-Rueda, "Optical properties of quantum wells and superlattices under electric fields," J. Luminesc. 44, pp. 223–231 (1989).]

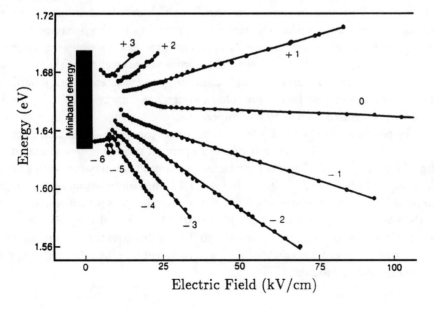

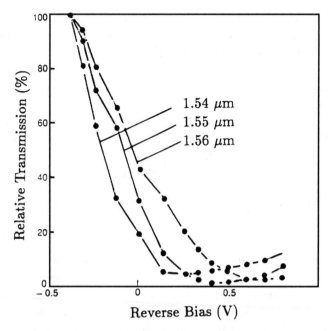

Figure 11.14. Optical transmission through a InGaAs/InAlAs superlattice as a function of the applied voltage. For all three wavelengths the large decrease in the transmission is due to resonant absorption corresponding to the −1 transition of the Stark ladder.

numbers at the curves indicate different steps in the Stark ladder. Juxtaposing these results with those of Fig. 11.10(b), one can conclude that transitions up to the right over six quantum wells are present at low electric fields. The left transitions are less observable for this particular heterostructure; however, three of them are clearly seen in Fig. 11.13. These results demonstrate that the electron wave functions have a spatial coherence of not less than 10–13 periods of the superlattice. With an increasing electric field, the high-order transitions disappear and only the vertical transition, 0, remains. The reasons for this behavior include the fact that there is a suppression of coherent transverse motion as a result of a progressive localization of the electrons within several wells and the fact that there are low rates of tunneling and scattering processes, etc. In superlattices grown by the most advanced technology, coherence lengths above 10 periods have been measured even at room temperature.

The physical processes that we have just discussed result in a strong electro-optical effect. A typical blue shift of the absorption spectra is demonstrated in Figs. 11.11 and 11.12. The effect can be used in intensity modulators with a large contrast between on and off states. In Fig. 11.14, the light transmission through an InGaAs/InAlAs superlattice embedded in a *p–i–n* diode is shown as a function of the voltage bias for three wavelengths: $\lambda = 1.54$, 1.55, and $1.56\,\mu$m. One can see that a moderate voltage bias of -0.5 to 0.5 V modulates the transmission by a factor of 10 or more because of the electroabsorption effect. No resonator is needed for such a modulator.

11.2.4 Terahertz Coherent Oscillations of Electrons in an Electric Field

In Chapter 8 we studied Bloch oscillations of the electrons moving in super-lattice minibands without scattering. These oscillations were deduced from the classical Newton equation. Classical vibrations of charged electrons can generate microwave emission. This electro-optical effect has attracted much attention owing to its important applications. In the quantum picture, the frequency of these oscillations correspond to phototransitions between the above-studied Stark ladder of the electron-energy levels. Semiclassical oscillations of the electric charge can be realized by the creation of an electron wave packet as a superposition of several Stark states; this wave packet oscillates in real space.

Despite the experimental observation of the Stark ladder and confirmation of a high degree of coherence in electron propagation across barriers in a super-lattice, which was discussed in Subsection 11.2.3, until now, coherent Bloch oscillations have not been excited by an external field for such a superlattice and microwave emission has not been detected directly.

Oscillations and microwave emission have been verified for less complex heterostructures that contain two quantum wells. Let us examine the case of coupled double quantum wells. The insets of Fig. 11.15 show the energy diagram of a double-well structure. According to the results of Chapter 3, in such a system of coupled wells, any electron state can be considered as a superposition of two states that originate from each of the wells. The coupling is due to the tunneling

Figure 11.15. Absorption spectra of the double-well heterostructure for different electric fields. The heterostructure, electron- and hole-energy levels, and wave functions are shown in the insets.

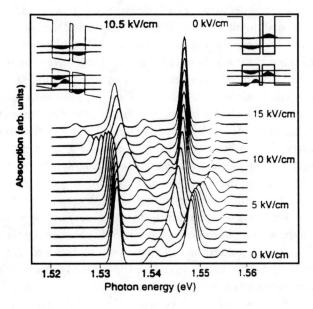

mechanism. For example, in the case of symmetric wells we have the splitting of initially degenerate states $\chi_{1,2}(z)$ to form the following two states:

$$\psi^+(t, z) = \frac{1}{\sqrt{2}}(\chi_1 + \chi_2)e^{it\epsilon_+/\hbar}, \qquad \psi^-(t, z) = \frac{1}{\sqrt{2}}(\chi_1 - \chi_2)e^{it\epsilon_-/\hbar}, \qquad (11.8)$$

with the energy difference $\Delta\epsilon = \epsilon_- - \epsilon_+ \equiv 2|T_t|$, where T_t is the tunneling matrix element.

Now let us suppose that at $t = 0$ we put the electrons inside the left well, that is, we populate an electron state with the wave function

$$\Psi(t = 0, z) = \chi_1(z).$$

Such a state is not an eigenstate of our problem, and it should change in time. If, as we assumed previously, other energy levels are situated quite far from ϵ_- and ϵ_+, we can write the time-dependent wave function $\Psi(t, z)$ as a superposition of the eigenstates of Eqs. (11.8):

$$\Psi(t, z) = A\psi^+(t, z) + B\psi^-(t, z). \qquad (11.9)$$

The coefficients A and B can be found from the boundary condition at $t = 0$ and the normalization condition. As a result of such a coherent superposition of the states, the probability of finding the electrons at some point z, as determined by $|\Psi(t, z)|^2$, oscillates in space, that is, at some point z we get

$$|\Psi(t, z)|^2 = \frac{1}{2}\left[|\chi_1(z)|^2\left(1 + \cos\frac{\Delta\epsilon}{\hbar}t\right) + |\chi_2(z)|^2\left(1 - \cos\frac{\Delta\epsilon}{\hbar}t\right)\right]. \qquad (11.10)$$

Thus we see that the electric charge oscillates coherently in space with a frequency of $\Omega = \Delta\epsilon/\hbar$. For example, while at $t = 0$ we have $|\Psi(z)|^2 = |\chi_1(z)|^2$, i.e., the electrons occupy the first quantum well. At $t = \pi/2\,\Omega$ we find that $|\Psi(z)|^2 = |\chi_2(z)|^2$, i.e., the electrons have transferred to the second well, and so on. These oscillations of the charged electrons should give rise to microwave emission.

In the case of asymmetric quantum wells, by applying the same procedure again, we can obtain two states $\psi^-(t, z)$ and $\psi^+(t, z)$, but since an asymmetric structure is assumed, the total probability of finding the electrons in the left and the right wells are different for these wave functions and $\Delta\epsilon$ depends on both the tunneling matrix element and the splitting due to the asymmetry. The latter is important because an applied electric field can control this splitting and therefore controls the frequency of the charge oscillations.

To detect these oscillations, asymmetric double-quantum-well structures, which manifest the energy-level splitting in the absorption spectra, are used. One chooses light with a photon energy between the energies of the split levels. A short light pulse with a duration of less than the inverse frequency of oscillations, Ω^{-1}, predominantly populates one of the wells. By this means, an initial coherent superposition of electron states – a wave packet – is created. This wave packet then oscillates between the two wells. The oscillations persist after the

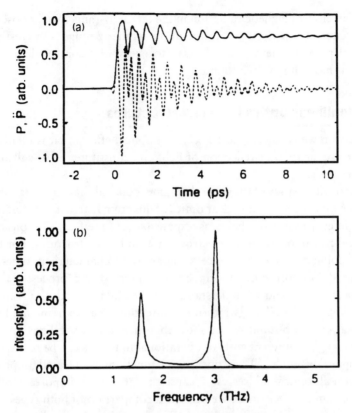

Figure 11.16. Coherent oscillations in the double-well structure presented in Fig. 11.15. (a) Time-dependent dipole moment (solid curve) of oscillating carriers and its second derivative (dashed curve). The results are obtained for a 150-ps pumping light pulse centered near the photon energy 1.54 eV. (b) Intensity of emitted radiation as a function of the frequency of the emitted radiation.

initial pulse for a time scale of the order of a relaxation time. In fact, the existence of the holes generated during the pulse complicates the whole picture since a hole wave packet can also oscillate but with another frequency and another decay time. Figure 11.15 depicts the particular asymmetric double-quantum-well system as well as its absorption spectra as functions of an applied electric field. The excitation pulse has a duration of 150 fs and the photon energy is centered at ~1.54 eV. The time dependence of the electric dipole moments induced by the oscillations is presented in Fig. 11.16(a). The oscillations are not exactly harmonic, and there are at least two types of oscillations, one for electrons and the other for holes, that decay in time. Figure 11.16(b) presents the power spectrum of emitted microwave radiation as a function of frequency; it shows typical microwave oscillations in the terahertz range. The peak with the larger frequency corresponds to coherent oscillations of the electron, and the other is associated with the hole.

In conclusion, through the example of a double-barrier structure, we have considered the effect of coherent oscillations of the electron wave packet that generate an electromagnetic radiation in the terahertz range. Similar oscillations can be excited in superlattices. Although the terahertz range is interesting from a practical point of view, a simple way to excite and control these coherent oscillations has still not been found.

11.3 Nonlinear Optics in Heterostructures

In this section we consider another class of optical effects that is related to the nonlinear propagation or reflection of light and constitute the so-called field of nonlinear optics.

We start with a study of the general nonlinear optical characteristics of materials. Then we consider the nonlinear optics of quantum heterostructures, focusing on the different physical mechanisms responsible for the effects. We consider both virtual and real populations of electron states in heterostructures. For the case of a real population, we introduce a hierarchy of characteristic times, which affords us the opportunities to classify a number of the processes that lead to nonlinear effects and to understand their advantages and disadvantages in potential applications. Then we turn our attention to an extreme nonlinear effect such as optical bistability and show that there are possibilities for controlling light by light. Different sorts of operational functions can be realized on the basis of optical bistability. These operational functions can be used for processing signals with all-optical addresses. Because the latter is an important subject in conventional microelectronics, we provide a comparison of both types of signal processing – all optical and electronic. We show that the nonlinear optics of quantum heterostructures has great potential in signal-processing applications.

We begin by introducing the basic ideas and results in this branch of optics known for conventional nonlinear optical materials.

11.3.1 Linear and Nonlinear Optics

Throughout the previous sections, we considered all optical phenomena as linear phenomena, that is, previously we assumed that (1) the refractive index and the absorption (amplification) coefficients are independent of light intensity, (2) the light frequency is not altered during its passage through the medium, (3) the principle of superposition of any light waves holds, and (4) light does not interact with any other light waves and therefore light cannot control light. These assumptions are valid and can be applied for a range of low-amplitude light waves.

In the opposite case, in which the amplitude of the wave or the light intensity is large, light propagation through an optical medium can exhibit nonlinear behavior: (1) the refractive index and consequently the speed of light are changed with intensity, (2) light can alter its frequency, (3) the superposition principle fails, and (4) light can control light, i.e., photons can interact with photons. Such a medium, which changes the behavior of light from linear to nonlinear, is known as a nonlinear optical medium.

To introduce parameters characterizing nonlinear optical properties, let us consider the displacement vector $\vec{\mathcal{D}}$, which can be represented as

$$\vec{\mathcal{D}} = \vec{\mathcal{E}} + 4\pi \vec{\mathcal{P}}, \tag{11.11}$$

where $\vec{\mathcal{E}}$ is the electric field of the light wave and $\vec{\mathcal{P}}$ is the polarization of a medium. In linear optical media, $\vec{\mathcal{P}}$ is simply proportional to $\vec{\mathcal{E}}$. For example, for an isotropic medium, we can write

$$\vec{\mathcal{P}} = \chi(\omega)\vec{\mathcal{E}}, \qquad \vec{\mathcal{D}} = \kappa\vec{\mathcal{E}}, \tag{11.12}$$

where $\kappa = 1 + 4\pi\chi(\omega)$ is the dielectric constant and $\chi(\omega)$ is the electric susceptibility. These linear relations were used widely in our previous discussions.

In the case in which a material exhibits nonlinear properties, the latter can be accounted for with a nonlinear polarization $\vec{\mathcal{P}}$, which, in general, is a functional of the electric field $\vec{\mathcal{E}}$. For example, if it is possible to expand $\vec{\mathcal{P}}$ in a series with respect to $\vec{\mathcal{E}}$, one gets for its projections

$$\mathcal{P}_i = \chi_{ij}\mathcal{E}_j + \chi_{ijl}^{(2)}\mathcal{E}_j\mathcal{E}_l + \chi_{ijlk}^{(3)}\mathcal{E}_j\mathcal{E}_l\mathcal{E}_k + \cdots, \tag{11.13}$$

where $\chi_{ijl}^{(2)}$ and $\chi_{ijlk}^{(3)}$ are coefficients describing second- and third-order nonlinear effects.

When the optical medium is nonlinear, the linear equation for light waves is no longer valid. If we rewrite $\vec{\mathcal{P}}$ as a sum of a linear part and a nonlinear contribution,

$$\vec{\mathcal{P}} = \chi\vec{\mathcal{E}} + \vec{\mathcal{P}}_{\mathrm{NL}}, \tag{11.14}$$

the wave equation for $\mathcal{E}$ can be represented in the form

$$\vec{\nabla}^2\vec{\mathcal{E}} - \frac{\kappa}{c^2}\frac{\partial^2\vec{\mathcal{E}}}{\partial t^2} = -\vec{\mathcal{S}}, \tag{11.15}$$

where

$$\vec{\mathcal{S}} \equiv -\frac{4\pi}{c^2}\frac{\partial^2\vec{\mathcal{P}}_{\mathrm{NL}}}{\partial t^2}.$$

In this way, we keep the left-hand side of the wave equation linear, while the term $\vec{\mathcal{S}}$ is nonlinear. This term can be regarded as a source for the generation of fields different from the incident wave, including those with different frequencies, wave vectors, etc.

In contrast to the linear wave equation, Eq. (11.15) cannot be solved analytically in the general case. For each particular case it is necessary to develop a proper approach for analysis and solve the nonlinear equation of Eq. (11.15). The most common method in use is the so-called Born approximation, which is similar to that used in perturbation theory in quantum mechanics. This approximation is an iterative procedure in which, for the first step, the incident wave is used for calculation of the right-hand side of Eq. (11.15), i.e., the source $\vec{S}$. Thus the first nonlinear correction to the incident wave is estimated as a solution of the linear wave equation with a given right-hand side $\vec{S}$, which is a nonlinear function of the amplitude of the incident wave. After this first-order correction is found, the procedure should be iterated until the accuracy necessary for the description of the nonlinear effect under consideration is achieved.

If this method cannot be applied, an analysis of the particular physical situation can sometimes lead to an approach for solving the specific problem at hand. For example, such an analysis can be based on a comparison of characteristic times. If a physical process in a material is described by a relaxation time τ_{ch} that is larger than the characteristic time for changes in the light-wave amplitude t_{lw} and its inverse frequency ω,

$$\tau_{ch} > t_{lw}, \omega^{-1}, \tag{11.16}$$

a set of susceptibilities of different orders can be introduced in accordance with Eq. (11.13) and the Born approximation can be used.

Another limiting case corresponds to the situation in which the relaxation time is less than t_{lw}:

$$\omega^{-1} \ll \tau_{ch} < t_{lw}. \tag{11.17}$$

For this case, time-averaged parameters of the wave should specify the process, that is, the intensity of the wave governs the nonlinear effect. It can be taken into account by the introduction of a refractive index and an absorption coefficient that depend on the intensity of wave $\mathcal{I}$:

$$n = n(\omega, \mathcal{I}), \qquad \alpha = \alpha(\omega, \mathcal{I}). \tag{11.18}$$

Since $\mathcal{I} \sim \bar{\mathcal{E}}^2$, Eq. (11.15) can be transformed into an equation for the wave amplitude with the dielectric function $\kappa \equiv n^2$, which depends on the square of the amplitude. Thus one obtains another kind of nonlinear equation for $\mathcal{E}$; that is, $\kappa(\bar{\mathcal{E}}^2)$ provides a new nonlinear relation.

11.3.2 Optical Nonlinearities in Quantum Wells

Although there are a number of different mechanisms that cause optical nonlinearities in semiconductors and semiconductor heterostructures, a nonlinear optical response depends on whether a virtual or a real excited-state population is involved in the process.

If the light frequency is well below the absorption edge, the light wave induces virtual excitations: the electric field of the light brings about a coherent polarization of the material that persists only as long as the field is applied. The foregoing analysis is valid independently of whether the wave amplitude is large or small. But if the wave amplitude is large, the induced polarization can couple various optical fields, providing a photon–photon interaction mediated by means of the material, etc. During the action of the field, the state of the crystal is a coherent superposition of its excited states. Such a coherence exists until relaxation processes are not active. In the opposite case, the quantum-mechanical phase of the excitation will be lost, i.e., there is absorption of a real photon. One can perform a simple analysis based on the uncertainty principle. In a coupled system of an excited material and a light wave, there is an oscillation of energy between these two parts. The frequency of the oscillation is determined by the detuning parameters,

$$\Delta\omega = (E - \hbar\omega)/\hbar,$$

where E is the excited energy level of the material. As long as this detuning is much greater than the inverse relaxation time $1/\tau_{ch}$, the mechanism of the nonlinearity is predominantly virtual. Thus a large detuning below the band edge is favored for the virtual picture of the interaction of light and a semiconductor. According to this analysis, the virtual mechanism corresponds rather to inequality (11.16). In such a case, field-induced nonlinearities usually correspond to the small off-resonance case but they are extremely fast; their characteristic times can be estimated as $(\Delta\omega)^{-1}$.

For a photon energy above the fundamental absorption edge, there is a predominant generation of a real population of excited states. The mechanisms change from virtual to real as a result of the fact that within the detuning interval $\hbar\Delta\omega = \hbar\omega - E_g$, there are many electron states that absorb the energy of the light and then scatter it. This strong absorption produces large populations, which leads, in turn, to changes in optical properties. However, these changes are incoherent and depend on only the intensity of the wave, as for the case of inequality (11.17). The populations of excited states have a finite lifetime and suffer various relaxation processes. Changes in optical spectra induced by such an excitation persist as long as the populations exist themselves; in particular, they can exist even when the light pulse is switched off rapidly. For the particular cases of semiconductors or their heterostructures, the indicated excitations are the excitons and the electron–hole pairs that form an electron–hole plasma.

11.3.3 Virtual Field-Induced Mechanism of Nonlinear Optical Effects

Let us consider light with a frequency well below the absorption edge so that we have the off-resonance case. As was discussed in Subsection 11.3.2, in this case one can expect a coherent nonlinear response. Since exciton levels also lie below the fundamental absorption gap, the contribution of the excitons to such an effect should be dominant. The simplest way to explain the nature of this virtual mechanism is to recall the close analogy between excitons and the hydrogen atom, which was emphasized in Chapter 3. From quantum mechanics, we know that a hydrogen atom placed in an ac electric field F_0 manifests shifts of its energy levels, known as the ac Stark effect. This effect is quadratic in the amplitude of the electric field. The shift and its sign depend strongly on the frequency of the light. The effect can be calculated in the second order by the quantum-mechanical perturbation theory. Thus, in the case of the exciton, we can also use the second-order perturbation theory and obtain the shift of the exciton energy:

$$\delta E_{ex} = A \frac{|\mathcal{D}F_0|^2}{E_{ex} - \hbar\omega}, \tag{11.19}$$

where $\mathcal{D}$ is the interband matrix element (see Section 10.4) and F_0 is the amplitude of the light wave. The multiplier A is positive and dependent on the properties of the material, the heterostructure, etc. This result demonstrates that if $\hbar\omega$ lies below the exciton energy E_{ex}, the energy shift is positive, i.e., a blue shift appears.

To examine this phenomenon further, the following experiment has been performed. A short, high-power light beam (pumping beam) with the frequency below the exciton line illuminates an AlGaAs/GaAs quantum well. Transient changes in the light transmission are measured with another relatively weak probe beam. The latter can be detuned either slightly below or above the exciton line. The results are presented in Figs. 11.17(a) and 11.17(b). The positions of two exciton lines associated with heavy holes, hh, and light holes, lh, and also the frequencies of two probe beams, $\omega_{probe} = \omega_a, \omega_b$, are identified in the inset. The main changes that can be seen in the transmission persist during the action of the pump; there is also a much weaker, long-time effect due to the photogeneration of an electron–hole plasma. In accordance with Eq. (11.19), in the region slightly below the exciton, as in Fig. 11.17(a), the transmission increases, while in the region slightly above the exciton, the transmission decreases, as in Fig. 11.17(b). The results clearly demonstrate a blue shift of the exciton spectrum during the illumination by the pump beam. In principle, these changes can be expressed in terms of a third-order susceptibility, $\chi^{(3)}$, but one can see that this quantity has complex dependences on both the frequencies, ω_{pump} and ω_{probe}. Figure 11.18(a) depicts the transmission as a function of the detuning frequency, while Fig. 11.18(b) shows clearly that the nonlinear effect is third order and, accordingly, changes in the transmission are proportional to F_0^2 and they increase near the resonance condition $\hbar\omega = E_{ex}$ in accordance with Eq. (11.19).

11.3.4 Nonlinear Optical Effects due to Generation of Excitons and Electron–Hole Plasma

If the light frequency is above the absorption edge, other mechanisms of nonlinear effects come into play. These mechanisms are due to the generation of real excitons and electron–hole pairs that form a plasma. Before considering these mechanisms of nonlinear response, let us classify the possible relaxation processes that can occur when real electrons and holes are generated optically.

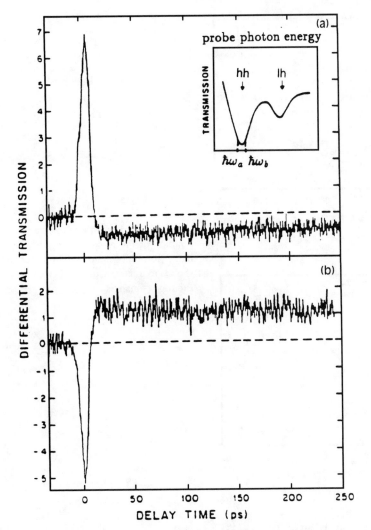

Figure 11.17. The differential transmission of an AlGaAs/GaAs quantum well as a function of the delay time between the pump and the probe beams. The pump light has photon energy approximately 25 meV below the hh exciton resonance. The photon energies of two probe beams, *a* and *b*, are indicated in the inset, where the transmission under equilibrium is plotted. Both measurements show a blue shift of the exciton spectrum under such a virtual excitation.

Possible processes have different time scales, as illustrated by Fig. 11.19. The left part of this figure shows a time axis with more than 10 decades. The right part shows the distributions of photoexcited carriers over energy for five different stages of the relaxation process. The real particles can be excited by a light pulse with a duration ranging from 10 to 100 fs in the continuous-wave regime. The lifetime of the excitations is of the order of 100 ps–100 ns. Thus there is a very wide time range, which should be analyzed. Let us suppose that the pulse is ultrashort. Since the time between electron-scattering events can be estimated

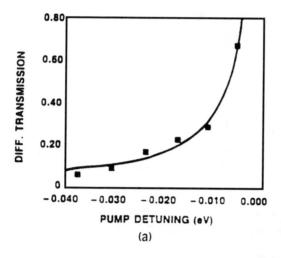

PUMP DETUNING (eV)

(a)

Figure 11.18. The differential transmission as a function of the photon energy of the (a) pump light, (b) pump intensity. The latter case corresponds to a fixed detuning 18 meV below the hh exciton resonance.

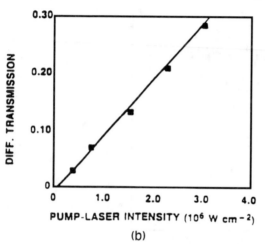

PUMP-LASER INTENSITY (10^6 W cm^{-2})

(b)

as $\tau_{dph} = 100$–1000 fs, for $t < \tau_{dph}$ the behavior of the system is more like the virtual, or coherent, case which was considered in Subsection 11.3.3. Thus this first stage can be thought as a coherent stage. In Fig. 11.19, this stage is marked by 1. During this coherent stage the carrier distribution is determined mainly by the spectrum of the pump beam. Then a dephasing of the initial excitation occurs ($t > \tau_{dph}$), the system absorbs photons, and we get real particles in both of the bands with almost classical distribution functions centered around energies, corresponding to those consistent with the conservation laws,

$$E_c(\vec{k}_c) - E_v(\vec{k}_v) = \hbar\omega, \quad k_c = k_v,$$

as shown by case 2 of Fig. 11.19. In the next stage, the particles exchange their momenta and energies and their distribution functions relax to Fermi functions, but their effective temperature T_e is far from that of the crystal; this situation defines the third relaxation stage. Both the second and the third phases correspond to so-called nonthermalized plasma. At longer times, the electrons and the

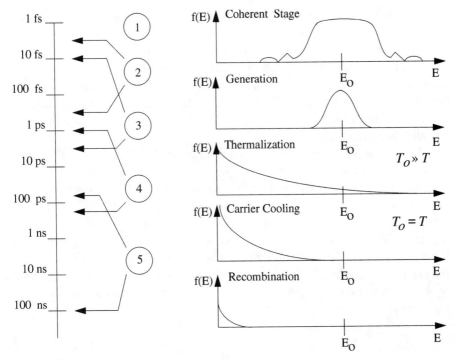

Figure 11.19. Time scales (left) and carrier distributions at different stages of the relaxation of photoexcited electron–hole plasma (right).

holes lose excess energy and relax to distributions with the crystal temperature; at the end of this cooling stage we have a thermalized electron–hole plasma, and the electron–hole pairs begin to form the excitons; this sequence defines the fourth stage. During the fifth and last stage, both the free electron–hole pairs and excitons recombine toward the true equilibrium. The left part of Fig. 11.19 shows typical time scales for each of the relaxation stages.

Now we consider the simplest case. This occurs when the excitons and the electron–hole plasma are thermalized, i.e., equilibrium within the conduction and the valence bands is assumed to be established. (In terms of the above analysis, this corresponds to the recombination stage.) In this case, only concentrations of excitons, n_{ex}, and the electron–hole plasma, $n = n_c = n_v$, characterize the nonequilibrium state of the system. Optical parameters of the system can be regarded as functions of these concentrations that, in turn, are dependent on the light intensity. Thus we get optical nonlinearity in form of Eqs. (11.18) by means of the generation of the concentrations of real electrons and holes.

In fact, the effect of the electrons and holes on the optical spectra was discussed previously in Chapter 10 and Section 11.2, when we studied the electro-optical effect; see Fig. 11.8. Optically generated carriers affect the spectra in the same manner. Both previously discussed phenomena, the filling factor and the screening of the electron–hole interaction, are important physical reasons of such an effect.

The filling factor reflects only that the absorption coefficient is proportional to the difference in the distribution functions of generated holes and electrons: $\mathcal{F}_v - \mathcal{F}_c$. The populations of some states in the conduction and the valence bands decrease this difference and lead to bleaching of the spectral regions, which corresponds to the transitions between populated states. When nonequilibrium carriers are thermalized within both of the bands, the mechanism leads to bleaching of the absorption edge and effectively increasing the optical bandgap; this is the so-called Burstein–Moss effect.

The mechanism related to the exciton screening has a more complex explanation. It depends on whether the excitons or the electron–hole pairs are excited. For the two-dimensional case we can assign an effective disk with radius a_{ex}, where a_{ex} coincides with the exciton radius. If exciton states are excited directly – a resonant absorption of the pump beam – and there is no plasma, we can imagine that those disks are hard, so two disks cannot occupy the same space. If $\pi a_{\text{ex}}^2 n_{\text{ex}} \ll 1$, excitons do not interfere with one another. In the opposite case, the excitons are destroyed and the exciton peaks in the spectra are washed out. More accurate estimates give a threshold exciton concentration, above which the excitons disappear:

$$n_{\text{ex,th}} = \frac{0.117}{\pi a_{\text{ex}}^2}. \tag{11.20}$$

Thus, if one resonantly generates excitons, there is a threshold light intensity, $\mathcal{I}_{\text{th}} \approx n_{\text{ex,th}}/\tau_{\text{ex}}$, ($\tau_{\text{ex}}$ is the lifetime of the exciton) beyond which the exciton absorption disappears as a result of a bleaching effect.

If free-electron–hole pairs are generated, one can use for estimates the screening length, which was introduced in Chapter 6. The meaning of this length is as follows. As the distance from an electric charge exceeds this length, the electric field of this charge is suppressed. Therefore, if the screening length becomes less than the exciton radius, the interaction between the electron and the hole decreases, and that brings about the destruction of the exciton. For two-dimensional carriers and a low temperature T, the screening length coincides with an exciton radius a_{ex} at a high temperature, which is proportional to $\sqrt{T}$. Calculations give the criteria for destroying the excitons:

$$n\pi a_{\text{ex}}^2 = 0.056, \quad \text{if } k_B T \ll Ry_{\text{ex}}^*, \tag{11.21}$$

$$n\pi a_{\text{ex}}^2 = 0.042 \frac{k_B T}{Ry_{\text{ex}}^*}, \quad \text{if } k_B T \gg Ry_{\text{ex}}^*, \tag{11.22}$$

where n is the concentration of free-electron-hole pairs, Ry_{ex}^* is the coupling energy of the exciton.

Using these results, one can obtain a phase diagram for the existence of an exciton. Because three essential parameters, n_{ex}, n, and T, control the effects, this phase diagram, in principle, should be a three-dimensional plot. We present this diagram as projections on an n–T plane, as shown in Fig. 11.20. The results

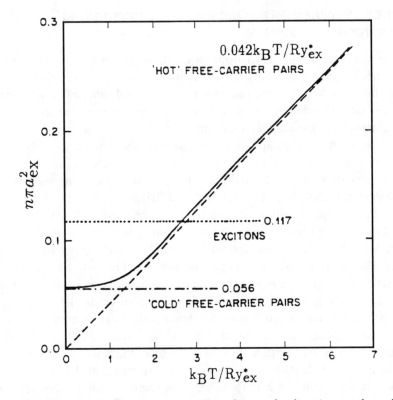

Figure 11.20. The concentration–temperature phase diagram for the existence of two-dimensional excitons. The concentration is given in units of πa_{ex}^2, the temperature in Ry_{ex}^*. The dotted line shows the limitation associated with the criterion of Eq. (11.20). The solid curve presents limitations due to plasma screening. The low-temperature and the high-temperature limits correspond to Eqs. (11.21) and (11.22), respectively.

can be understood as follows. If n is the plasma concentration, the excitons exist only below the solid curve. This curve represents the screening effect with the limiting cases of Eqs. (11.21) and (11.22) marked as contributions of cold and hot free-carrier pairs. But the exciton concentration n_{ex} cannot exceed the threshold of Eq. (11.20). The latter is represented by the dotted line in the figure for comparison.

Thus the solid curve divides the n–T plane into two regions. Above the curve, the excitons and their features in the optical spectra vanish. Below the curve, there are almost free excitons and exciton features in the optical spectra, if the exciton subsystem is weakly excited. At high excitation, the excitons exist below the dotted line. A similar diagram can be plotted for the intensity–temperature variables, that is, there is a similar diagram in the $\mathcal{I}$–T plane.

In conclusion, in the case of thermalized photoexcitations, nonlinear optical effects manifest themselves as vanishing excitons in the spectra for high light intensities.

11.3.5 Nonlinear Effects Induced by Nonthermalized Electron–Hole Plasma

The processes of the thermalization of photoexcited plasma and excitons take times ranging above tens of picoseconds. Thus the applications of these effects are restricted by subnanosecond or nanosecond time scales. Meanwhile, advanced optical techniques make it possible to generate light beams with temporal durations in the femtosecond range. Therefore this technology facilitates the investigation and exploitation of femtosecond phenomena.

In the femtosecond regime, all physical processes, including nonlinear optical processes, differ greatly from those of the thermalized case. As was discussed in Subsection 11.3.4 the reason for such a difference is that the electron–hole pairs are generated initially in the continuum of the conduction band as well as that of the valence band; subbands serve as bands for the case of quantum-well structures. In both cases, the states are populated in relatively narrow regions of energy. The electrons and the holes then begin to relax toward the bottom of the conduction band and the top of the valence band, respectively. The next stage of the relaxation process is the formation of coupled electron–hole pairs, i.e., the excitons, which now can be screened dynamically by the relaxing plasma. As the electrons undergo transitions to lower energies and the holes make similar transitions as they are transferred in the direction of the band edge, there are significant time-dependent changes in the optical properties of the semiconductor medium. In sharp contrast with the other cases, there is a pronounced delay with respect to the initial, short femtosecond optical pulse. For a structure with multiple GaAs quantum wells, these fast changes have been examined on the basis of the pump–probe method in which two light beams are used. One such example is illustrated in Fig. 11.21. The first pulse – the higher-power pump pulse, which had an energy 20 meV above the ground state of the two-dimensional exciton but below the excited state associated with the next subband – had the spectrum shown in Fig. 11.21. A probe beam with a broad spectral band served for recording the light transmission with a controlled delay Δt between the two pulses. The differential transmission is plotted for several values of Δt. One can see that for $\Delta t \leq -50$ fs (just at the front of the pump) a spectral hole in the spectral region of the pump is formed in the transmission, i.e., an increase in the transmission. This spectral hole corresponds to the population of some states within the subbands. The created electron–hole pairs immediately screen the exciton lines belonging to the first and the second subbands; for this time interval in the transmission there is no exciton absorption. In this phase, there are only small changes in the transmission. In the next interval of time, the spectral hole shifts down to lower energies, which correspond to a progressive relaxation of the electron–hole pairs. Within ~ 200 fs, the transmission spectrum that exhibits bleaching near the fundamental absorption edge is formed. This corresponds to the case in which the carriers occupy the lowest energy states in both bands. Thus the short excitation results in a strong suppression of the exciton absorption and a bleaching of the band-to-band absorption edge. Thus

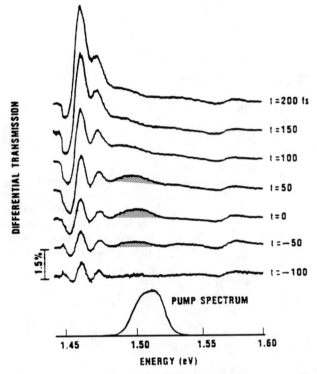

Figure 11.21. The differential transmission of excited AlGaAs/GaAs quantum wells as a function of the photon energy. The transmission has been measured with a broad-spectral-band probe pulse of 50-fs duration. The temporal dependence of the spectra have been measured with 50-fs intervals before and after the excitation. At the bottom of the figure the spectrum of the pump light is shown. The bleaching of optical spectra near the fundamental absorption edge ($\hbar\omega \approx 1.455\,\text{eV}$) demonstrates a dynamic of the thermalization of the photoexcited plasma. [After W. H. Knox, "Optical studies of femtosecond carrier thermalization in GaAs," in *Hot Carriers in Semiconductor Nanostructures: Physics and Applications*, J. Shah, ed. (AT & T and Academic, Boston, 1992), pp. 313–344.]

this experiment supports the existence of a strong and fast nonlinear dynamic effect in the absorption spectra in the femtosecond time domain.

11.3.6 Optical Bistability

Now we consider extremely large nonlinear phenomena such as optical bistability. An optically bistable system is a system with two stable states that has an output optical signal that can take on one of two quite different stable values, no matter what input signal is arranged. The sketch of a bistable output–input characteristic with a hysteresis loop is presented in Fig. 11.22(a). If a system exhibits optical bistability, its parameters can be adjusted so that the hysteresis is eliminated, but a sharp dependence $\mathcal{I}_{out}(\mathcal{I}_{in})$ still remains, as shown in Fig. 11.22(b). The figures correspond to a bistability with the so-called bleaching

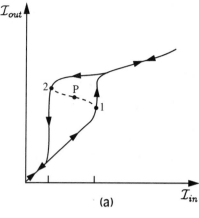

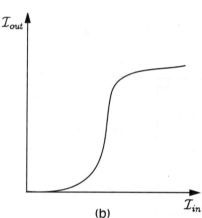

(a)

Figure 11.22. The output intensity $\mathcal{I}_{out}$ versus input intensity $\mathcal{I}_{in}$ for an optically bistable device. (a) The case of an established hysteresis (solid curve). The middle branch (dashed curve) is unstable. (b) The case in which the bistability transforms into a sharp dependence, $\mathcal{I}_{out}(\mathcal{I}_{in})$.

(b)

of the system (an increase in the light intensity leads to a decrease in absorption). There are other examples of bistability that demonstrate increasing absorption (an increase in the intensity leads to an increase in absorption).

For a bistable optical device or one with sharp output–input dependences, $\mathcal{I}_{out}(\mathcal{I}_{in})$, large changes in the output signal should be reachable. From the previous considerations, we know that, in fact, nonlinear optical effects are quite modest in absolute value – less than a percent in the refractive index and not more that a few tenths of a percent in the absorption coefficient. To reach large variations of the signals in devices with small dimensions, we should introduce some kind of feedback, as shown schematically in Fig. 11.23. There are different ways to realize such a feedback in optical systems. A straightforward way is to place a nonlinear element in an optical resonator. Then, by use of the resonator

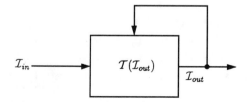

Figure 11.23. The scheme of an optical feedback: the transmittance of a device is a function of the output signal.

to reach multiple passages of light through this nonlinear element, it is possible to increase the nonlinear effects strongly and bring about optical bistability.

To obtain a bistable characteristic for a system with feedback, we introduce the light transmission T of this system. Since feedback is assumed, the transmission depends on the output intensity $\mathcal{I}_{out}$. Then we can rewrite the definition of the transmission as

$$\mathcal{I}_{in} = \frac{\mathcal{I}_{out}}{T(\mathcal{I}_{out})}. \tag{11.23}$$

Let the dependence $T(\mathcal{I}_{out})$ be a sharp function, as shown in Fig. 11.24(a). In accordance with Eq. (11.23), we can plot the dependence $\mathcal{I}_{in}(\mathcal{I}_{out})$. Because the output is controlled by the input intensity, from Fig. 11.24(b) we can see that for some range of the input $\mathcal{I}_{in}$ there are three possible values of the output $\mathcal{I}_{out}$. The intermediate branch of the dependence $\mathcal{I}_{in}(\mathcal{I}_{out})$ between points 1 and 2 is unstable. We just obtain the bistable output–input characteristic shown in Fig. 11.24(c). So, to get optical bistability, it is necessary to have a sharp enough dependence of the transmission on the intensity.

We can assume a nonlinear effect in the refractive index that may be characterized by

$$n = n_0 + n_2 \mathcal{I}_{out}, \tag{11.24}$$

where n_0 and n_2 are regarded as constants. From the considerations of Section 11.1 for a Mach–Zehnder interferometer, the transmission given by Eq. (11.6) can be rewritten in the form,

$$T \equiv \frac{\mathcal{I}_{out}}{\mathcal{I}_{in}} = \cos^2 \frac{\phi}{2} = \cos^2 \left(\frac{\omega}{2c} L_z n_2 \mathcal{I}_{out} + \phi_1 \right), \tag{11.25}$$

where ϕ_1 is another constant phase. This dependence can provide a bistable regime of light transmission.

Another important example is given by the Fabry–Perot etalon that comprises two plane mirrors separated by some distance L_z; see Subsection 10.2.3. According to Eq. (10.17), for this case the transmission can be represented as

$$T = \frac{T_{max}}{1 + (\mathcal{F}/2\pi)^2 \sin^2(\omega L_z n_2 \mathcal{I}_{out}/c + \phi_1)} \frac{1}{(1-r)^2}. \tag{11.26}$$

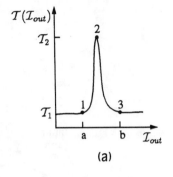

(a)

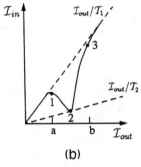

(b)

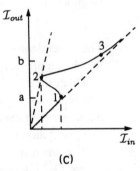

(c)

Figure 11.24. The reconstruction of $\mathcal{I}_{out}(\mathcal{I}_{in})$ for the case of bistability. (a) A sharp dependence of the transmission on the output intensity in an interval $\{a, b\}$. The transmissions in points marked by 1 and 3 equal T_1 and those marked by 2 equal T_2. (b) The input intensity versus the output according to Eq. (11.23). The dashed lines restrict the variation of $\mathcal{I}_{in}(\mathcal{I}_{out})$. (c) The bistable characteristic. Points 1, 2, 3 in (b) and (c) correspond to the indicated points in (a).

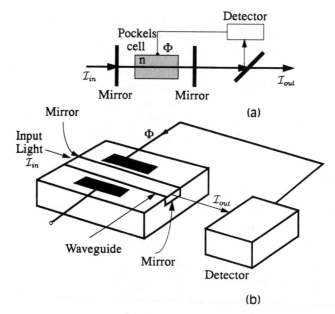

Figure 11.25. The example of an hybrid bistable optical device. A Fabry–Perot etalon contains an electro-optical element (the Pockels cell). The light beam is split and controls the voltage on the element through a photodetector (an external feedback).

The dependence of Eq. (11.26) is a periodic sharply peaked function of $\mathcal{I}_{out}$, which can provide not only bistability, but also multistability. In both examples in which the bistability results from the dependence of n on $\mathcal{I}$, the bistability is referred to as a dispersive bistability. In a similar manner, it can be shown that nonlinearity in the absorption coefficient of a medium placed in a resonator leads to a bistability; this case is referred to as absorption, or dissipative, bistability.

Optical bistability can also be reached by means of an external feedback. Figure 11.25 depicts such a hybrid electro-optical bistable system. It uses a split-off beam light to generate an electrical signal in a photodetector that, in turn, controls an optical element (Pockels cell) by means of the electro-optical effect.

Both systems based on an all-optically addressed bistability and on an electro-optical hybrid can use quantum heterostructures. Two variants of all-optical systems, for which the Fabry–Perot etalon is used, are shown in Fig. 11.26. For these cases, multilayer dielectric mirrors are placed on both sides of the quantum heterostructure to create the etalon. In the case of Fig. 11.26(a), a bistability of the transmitted beam is used, while the reflected light is used in the case of Fig. 11.26(b).

The hybrid type of an optically bistable device has been worked out on the basis of MQW structures. It is called the self-electro-optic-effect device (SEED) and is shown in the left part of Fig. 11.27. For this device, a MQW structure is embedded in the i region of a p–n junction and simultaneously serves as a photodetector and an optical modulator; compare this case with that of Fig. 11.25. Consider the principle of the operation of this SEED. In the right

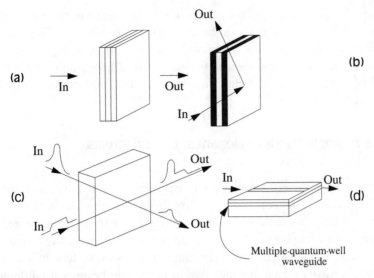

Figure 11.26. Possible geometries for applications of a MQW structure in a resonator bistable device. The structure is placed between mirrors. (a), (c), (d) The geometry for the bistability in the transmitted signal, (b) the geometry for the bistability in the reflected signal.

part of Fig. 11.27, the absorption coefficient as a function of the photon energy is presented for different voltages across the device. Let the voltage be chosen so that the absorption of the light is small. This choice corresponds to a voltage of 10 V for the example of absorption presented in Fig. 11.27(b). For reference, the photon energy of the input light is marked by the vertical dashed line. If light illuminates the device, a voltage drop across the device changes because of a small photocurrent generated by this light. This change in the voltage drop

Figure 11.27. The schematics of (a) a SEED, (b) the absorption coefficient versus the photon energy for different voltages at the device. [After C. Weisbuch and B. Vinter, *Quantum Semiconductor Structures* (Academic, San Diego, 1991).]

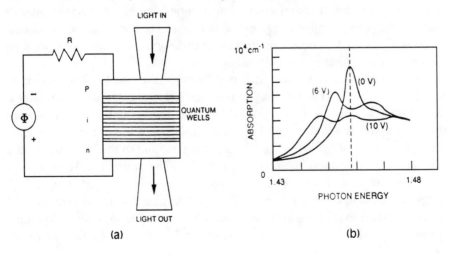

increases the absorption coefficient because of the electro-optic effect while the bleaching, which we studied in Subsection 11.2.2, is canceled. The absorption coefficient, in turn, induces a larger current and a further voltage drop, etc. As a result of such an external electric feedback, the SEED exhibits either bistable optical behavior or a sharp dependence $\mathcal{I}_{out}(\mathcal{I}_{in})$. In Chapter 12 we will analyze particular SEEDs in details.

11.3.7 Applications of Nonlinear Optical Effects in Quantum Wells

There are a number of well-known applications of conventional nonlinear optical materials. Here we mention some of them: (1) high-order harmonic generation, which arises when the incident intense light with the frequency ω generates light beams with frequencies $2\omega, 3\omega, \ldots$, in a nonlinear medium; (2) optical rectification resulting from intense light that generates dc voltage in a nonlinear medium; (3) wave mixing, in which two or more light beams with different frequencies $\omega_1, \omega_2, \ldots$, generate light beams with frequencies $\omega_1 + \omega_2, \omega_1 - \omega_2$, etc., by means of a nonlinear medium – this is also possible for three-wave mixing, four-wave mixing, etc.; and (4) optical phase conjugation, which is a particular case of four-wave mixing (degenerate case) in which all the frequencies coincide. This case is especially interesting because the nonlinear medium generates a light beam with a phase that is exactly the same as that of the incident wave, but it moves strictly in the opposite direction. This list of nonlinear optics applications can be expanded. For many of these applications, the quantum heterostructures studied herein are used.

In this section we concentrate on the discussion of applications of nonlinear optical effects supporting the main goal of microelectronics and optoelectronics, i.e., processing of information. Such processing requires a few typical operations with signals. One important operation is switching. We need such an operation for communication networks, processing systems, and digital computing. The latter also requires fast gates and memory devices such as flip–flops. Thus, in a system with all-optical addresses, nonlinear optics should, in principle, provide the following series of operations: optical switching, fast optical gates, and optical memory elements. Consider a realization of these and some other functions for the example of a bistable optical system.

Figure 11.28 illustrates how it is possible to realize the flip–flop operation in a bistable system. The bistable output intensity versus the input intensity is sketched in the upper left portion of this figure. The lower left portion depicts the input signal as a function of time. The output signal versus time is shown in the right portion of the figure. Let the input correspond to the low-output state for times such that $t < 2$ (point 1 in the $\mathcal{I}_{out}$–$\mathcal{I}_{in}$ characteristic). At time $t = 2$, an additional positive-input pulse switches the system to the state with a high output (point 2 in the $\mathcal{I}_{out}$–$\mathcal{I}_{in}$ characteristic). At time $t = 3$, a decrease of the input signal to a level below a threshold value (a negative pulse) leads to switching the system down to the low-output state (point 3 in the $\mathcal{I}_{out}$–$\mathcal{I}_{in}$ characteristic).

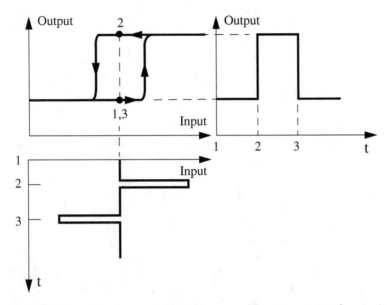

Figure 11.28. Flip–flopping of an optical bistable system. The output–input characteristic, the time-dependent input, and the resultant output are presented.

The resultant output signal $\mathcal{I}_{out}(t)$ is controlled by relatively small changes in the input signal.

Bistable device parameters can be adjusted so that the hysteresis loop disappears and only the sharp dependence $\mathcal{I}_{out}(\mathcal{I}_{in})$ still remains, as shown in Fig. 11.29. In this case, a bistable system can serve as an amplifier, a pulse sharper, or a limiter. Consider Fig. 11.29(a). If the $\mathcal{I}_{out}/(\mathcal{I}_{in})$ dependence is quite sharp,

Figure 11.29. Bistable device as (a) an amplifier of the input modulated signal, (b) an optical pulse sharper or limiter. The input and the output signals are shown. The output–input characteristic has to have a sharp (threshold) dependence.

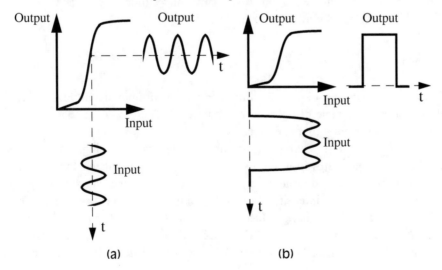

(a)   (b)

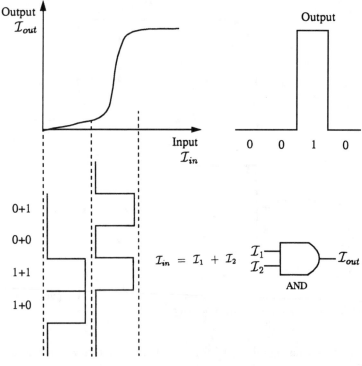

Figure 11.30. An AND logic gate based on an optical bistable device. The output–input characteristic is presented. Two time-dependent input signals and corresponding output are shown. The symbol of this device is presented in the right portion of this figure.

the stationary input intensity can be chosen so that

$$d\mathcal{I}_{out}/d\mathcal{I}_{in} > 1.$$

Let the input be modulated in time with a small magnitude $\Delta\mathcal{I}_{in}$. Then the output should be modulated with a larger magnitude:

$$\Delta\mathcal{I}_{out} = \frac{d\mathcal{I}_{out}}{d\mathcal{I}_{in}}\Delta\mathcal{I}_{in} > \Delta\mathcal{I}_{in}.$$

This condition corresponds to an amplification of an input signal. Figure 11.29 illustrates the principle of operation of a pulse sharper or limiter based on the same type of output–input characteristics.

This output–input characteristic can also be used for logic operations. Figure 11.30 illustrates how an AND logic gate can be realized for this case. Consider the two time-dependent signals, $\mathcal{I}_1(t)$ and $\mathcal{I}_2(t)$, shown in the lower left part of the figure. The intensity of the first signal is less than the threshold $\mathcal{I}_{th}$ of the device. Thus this signal produces only a small output. The second signal is also less than $\mathcal{I}_{th}$, but the sum $\mathcal{I}_1 + \mathcal{I}_2$ can exceed the threshold $\mathcal{I}_{th}$ and switch the device to the state with a large output. As is seen from Fig. 11.30, this device can operate as an all-optically addressed AND logic gate (the right part of the figure shows the symbol of such an optical device). There are a number of other particular devices designed on the basis of semiconductors and heterostructures that have been proposed and realized for different operations with optical signals.

Thus, for many envisioned applications, it is possible to identify a complete set of the devices required for all-optical signal processing; these devices include laser sources of light, amplifiers of optical signals, nonlinear optical switchers, elements for different logic operations, memory elements, etc.

Since all-optical processing systems are an alternative to conventional microelectronics that nowadays successfully serves for nearly all tasks in signal processing, one must compare these two approaches to understand what advantages and disadvantages are expected in optical processing. The main parameters for which new systems should be competitive are

- the switching time (time necessary to switch from one state to another)
- the switching energy (energy necessary to activate or deactivate the switch)
- the power dissipation (energy dissipated per unit time in the switch process).

There are other characteristics that are more dependent on the strategy of processing, the device design, etc. For example, the propagation delay time (time taken by crossing the switch), the throughput (maximum data rate that can flow through the switch), the cross talk (undesired power leakage to other lines), physical dimensions, etc.

Although it is difficult to predict precise limits for microelectronics, which continues to advance rapidly, some approximate limits in the main parameters are presented in Fig. 11.31 for different microelectronic devices. Thus these restrictions are a minimum switching time of $\sim$10–20 ps, a minimum energy per operation of $\sim$10–20 fJ, and a minimum switching power of $\sim$1 μW.

As for optical operations, there are two types of limitations: fundamental limitations and practical limitations. One of the fundamental limits in switching energy is related to fluctuations of light. The simplest way to estimate this limit is to assume that the number of the photons corresponding to an input light signal fluctuates according to Poisson statistics, that is, if $\bar{N}$ is an average number of photons in a pulse, a random signal will consist of N photons with the probability

$$p(N) = \bar{N}^N e^{-\bar{N}} / \bar{N}!.$$

For example, let us fix the average number of the photons such that $\bar{N} = 21$; then there is a finite probability that no photons will enter device: $p(0) = 10^{-9}$. If we construct a processing system, which allows an average of 1 error per 10^9 trials, we should conclude that the light energy of the signal,

$$E_{\text{signal}} = 21\,\hbar\omega,$$

is just the limit for an operation (ω is the light frequency). For light with a wavelength $\lambda = 1\,\mu$m and $\hbar\omega = 1.24$ eV, we obtain $E_{\text{signal}} = 26$ eV $= 4.2 \times 10^{-18}$ J.

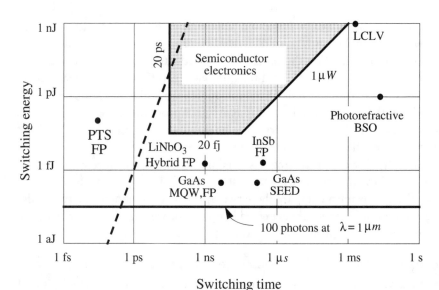

Figure 11.31. The switching energy versus switching time for several optical bistable devices. The following abbreviations are used: LCLV, liquid-crystal light valve; FP, Fabry–Perot resonator; PTS, polymerized diacetylene, an organic nonlinear material; BSO, bismuth silicon oxide. The photon-fluctuation limit on the switching energy is indicated (100 photons of $\lambda = 1\,\mu$m). For comparison, the island of these ultimate parameters for electron devices is presented. [After B. E. A. Saleh and M. C. Teich, *Fundamentals of Photonics* (Wiley, New York, 1991).]

Since the photon statistics can be different from Poisson statistics, we can accept 100 photons for each light signal as the approximate energy limit for an operation with this signal. (Note that this energy is much greater than the characteristic thermal energy $k_B T$: for room temperature $k_B T = 0.026$ eV.) Such a choice, $\bar{N} \gg 1$, also implies that the well-known energy–time uncertainty relation does not limit the switching time, say T, since the inequality $E_{signal} \times T > \hbar$ is satisfied for any time T greater than one time period of the light wave. It is worth noting that optical pulses of $\sim$10 fs have been generated already and, as shown above, subpicosecond switching has been demonstrated for semiconductor heterostructures.

The next fundamental limit is related to the size of light switches. As a result of diffraction effects, the size of optical devices cannot be less than the wavelength of light.

In fact, limitations of systems with all-optical addresses are dictated by the amplitudes of nonlinear effects in currently available materials and structures. These practical limitations are far from the fundamental limits. Another practical limitation is related to the heating of optical devices in a regime with a high rate of signals. The power dissipated in the system can be estimated by $P \approx E_{signal}/T$. As a result, heat removal restricts both the switching energy and the time. In Fig. 11.31 the switching energy versus switching time is presented for different optical devices. The fundamental limit of a signal of 100 photons at $\lambda = 1\,\mu$m is indicated. The island of similar parameters for electron devices is shown for comparison. One can see that optical devices based on heterostructures (nonlinear GaAs MQW elements, GaAs SEEDs, etc.) have parameters that are competitive with discrete electronic devices.

The state of the art of particular optical processing devices is much less developed than that for electronics. But fundamental advantages of light – such as propagation of the different signals without interaction through linear optical media (through fibers, for example), extremely large bandwidths for telecommunications (up to 100 THz), direct connection with optical-processing devices, a high degree of parallelism, etc. – continually lead to new approaches to the optical processing of signals (for example, the parallel-processing strategy).

11.4 Closing Remarks

In this chapter we have studied two classes of optical phenomena: electro-optical effects and nonlinear effects. These effects are well known for conventional bulklike materials, and they manifest remarkable features for the case of quantum heterostructures.

Large electro-optical effects in quantum structures occur as a result of two major mechanisms. The first is an analog of the Franz–Keldysh or Stark effect, i.e., a reconstruction of energy spectra of two-dimensional free carriers and excitons in an electric field. This spectrum reconstruction results in large changes in optical absorption near the fundamental absorption edge. The second mechanism is due to the filling factor's being controlled by an external field, i.e., a field-dependent population of energy states in quantum structures. An electric field affects both the absorption and the refractive index of the structures; it also provides a way to achieve effective control of optical properties.

We considered electrically biased superlattices and double-quantum-well structures and found that high-frequency microwave emission (up to the terahertz range) can be generated by such devices.

As for the nonlinear optical properties of quantum structures, there is a hierarchy of different mechanisms responsible for these effects. The fastest mechanisms are due to the virtual processes used for the excitation of the structures. In this case, an off-resonance short light pulse creates a coherent superposition of excited states – the polarization of a quantum structure – providing optical nonlinearity (photon–photon interaction). This mechanism occurs for light frequencies below the lowest exciton energy. As discussed in Chapters 3 and 10, in low-dimensional systems, exciton features are much more pronounced in optical spectra; thus the virtual mechanism leads to larger optical nonlinearities. At higher resonant frequencies, when real excitons and electron–hole pairs are generated, nonlinear mechanisms and their magnitudes depend significantly on the time scale of the light pulse. The fastest nonlinearity (the subpicosecond range) arises as a result of the filling factor and the screening of the Coulomb interaction. This occurs in a nonthermalized photoexcited plasma. On a longer time scale (the nanosecond range), the electron–hole plasma is thermalized, and the screening effect and bleaching near the absorption edge determine the nonlinearity. The required intensities of light are sufficiently lower than those for bulklike crystals.

We discussed how a nonlinear optical system with feedback can produce bistable optical behavior, i.e., for the same optical input two different outputs can be observed. Optical bistability, as an extreme nonlinear effect, provides the means to carry out different operations with light beams. We discussed how such operational functions, elements for optical switching, amplification of a modulated signal, memory elements, and different logic elements can be implemented in a bistable system. Quantum heterostructures together with other optical elements such as waveguides, multilayered resonators, etc., and lasers create the basis for the all-optical signal processing, which can be a new alternative to conventional electronics.

More details about electro-optics and nonlinear optics can be found in the following books:

N. Bloembergen, *Nonlinear Optics* (Benjamin, New York, 1965).

R. Loundon, *The Quantum Theory of Light* (Clarendon, Oxford, 1973).

A. Yariv, *Quantum Electronics* (Wiley, New York, 1975).

B. E. A. Saleh and M. C. Teich, *Fundamentals of Photonics* (Wiley, New York, 1991).

Further discussions of electro-optics and nonlinear optics of quantum heterostructures are presented in the following reviews:

D. A. B. Miller, D. S. Chemla, and S. Schmitt-Rink, "Electric field dependence of optical properties of semiconductor quantum wells," in *Optical Nonlinearities and Instabilities in Semiconductors*, H. Haug, ed. (Academic, Boston, 1988), p. 325.

D. S. Chemla, D. A. B. Miller, and S. Schmitt-Rink, "Nonlinear optical properties of semiconductor quantum wells," in *Optical Nonlinearities and Instabilities in Semiconductors*, H. Haug, ed. (Academic, Boston, 1988) p. 83.

Optical bistability and perspectives of its applications are discussed in the following references:

H. M. Gibbs, *Optical Bistability: Controlling Light by Light* (Academic, Orlando, FL, 1985).

N. Peyghambarian and H. M. Gibbs, "Semiconductor optical nonlinearities and applications to optical devices and bistability," in *Optical Nonlinearities and Instabilities in Semiconductors*, H. Haug, ed. (Academic, Boston, 1988), p. 295.

PROBLEMS

1. What is the electro-optic energy $\hbar\omega_F$ associated with the Franz–Keldysh effect for an $In_{0.15}G_{0.85}As$ layer grown on an unstrained GaAs [111] surface? Assume that the strain-induced piezoelectric field in the $In_{0.15}G_{0.85}As$ layer is equal to 2.1×10^5 V/cm and is directed along the (111) direction. Assume that the dominant Franz–Keldysh effect occurs as a result of the interaction of the electron wave function and the heavy-hole wave function; take $m_e(111) = 0.067\,m$ and $m_{hh}(111) = 0.65\,m$. Further assume that the InGaAs may be treated as a region of bulk material and that the only nonzero electric field in this material is the piezoelectric field. [Readers interested in learning more about the use of the Franz–Keldysh effect to measure the strain-induced piezoelectric field in InGaAs may refer to H. Shen, M. Dutta, W. Chang, R. Moerkirk, D. M. Kim, K. W. Chung, P. P. Ruden, M. I. Nathan, and M. A. Stroscio, "Direct measurement of piezoelectric field in a [111] B grown InGaAs/GaAs heterostructure by Franz–Keldysh oscillations," Appl. Phys. Lett. **60**, 2400–2402 (1992).]

2. Consider the two-well system of Subsection 11.2.4 that has been used to observe coherent oscillations of electrons. Derive a relationship between the energy separation $\Delta\epsilon$ and the period of oscillation T. Is this result consistent with the Heisenberg uncertainty relation?

3. Suppose that the exciton radius is 50 Å for a system with $k_B T / Ry_{ex}^* = 4$. What value of the exciton density n corresponds to the density at which these excitons will be washed out from exciton–exciton interactions?

4. It is known that Ti_2O_3 is an indirect narrow-bandgap semiconductor with a bandgap of only ~ 0.1 eV at 300 K. At 600 K the bandgap closes completely and the Ti_2O_3 becomes a semimetal. It is known further that coherent optical phonons with a frequency of 7 THz may be excited in Ti_2O_3 by an appropriate laser field. Find the temperature of Ti_2O_3 crystal for this 7 THz laser field excitation to produce semiconductor-to-semimetal transition. [If interested in the experimental details, see H. J. Zeiger, T. K. Cheng, E. P. Ippen J. Vidal, G. Dresselhaus, M. S. Dresselhaus, and "Femtosecond studies of the phase transition in Ti_2O_3," Phys. Rev. B **54**, pp. 105–123 (1996).]

12

Optical Devices Based on Quantum Structures

12.1 Introduction

In Chapters 10 and 11 we studied the optical properties of quantum heterostructures. We found that these properties are frequently significantly different from those of bulk materials and that they depend on the parameters of the heterostructure. The novel optical properties of heterostructures open new possibilities for the application of quantum heterostructures in various optical devices. In this chapter we analyze such optical applications of quantum heterostructures.

We start with a study of population inversion and light amplification due to interband phototransitions. As the major method of pumping such systems, we consider the injection of electrons and holes into the active region of a p–n junction with embedded quantum structures. Since the width of such an active region is small, specially designed heterostructure waveguides are necessary for the confinement of light waves in this region. Accordingly, our discussion includes an analysis of the effect of optical confinement. Then we show that the amplification coefficient of the confined optical mode can be large enough to produce laser oscillations in an optical cavity. Our discussion focuses on heterostructures that facilitate the simultaneous formation of optical modes and realization of laser oscillations that convert electric power into laser emission. A variety of different laser designs are analyzed; indeed, we consider a set of particular quantum-well and quantum-wire lasers including devices operating in the near-infrared and visible optical regions. Furthermore, our discussion includes an analysis of a new type of laser based on intraband phototransitions: the quantum-cascade laser.

Two other types of optical devices that use quantum heterostructures are also analyzed in this chapter: self-electro-optic-effect devices (SEEDs) and quantum-well photodetectors.

12.2 Light Amplification in Semiconductors

This section is devoted to a subject of prime importance: the phenomenon of light amplification in bulk semiconductors and semiconductor heterostructures. In Section 10.3, we established the existence of two fundamental processes: light

556

absorption and the spontaneous and stimulated emission of light. Stimulated emission can dominate over absorption when there is a population inversion for two energy levels separated by the energy of the light, $\hbar\omega$. Such a physical system can serve as an optical amplifier. Here we study the criteria for the formation of a population inversion between the conduction and the valence bands as well as how these criteria depend on the light frequency, pumping rate, temperature, etc. In addition, we calculate the gain and discuss the methods to create population inversion. Then we briefly consider double heterostructures from the point of view of both light amplification and gain in low-dimensional systems.

12.2.1 Criteria for Light Amplification

To examine light amplification in the case of interband phototransitions in semiconductors and semiconductor structures, let us return to the results obtained in Chapter 10 for the light amplification and absorption coefficients; see Eqs. (10.43), (10.56), and (10.58). Note that the analysis of population inversion does not depend on the dimensionality of an electron system. We already know that the absorption rate is proportional to

$$\mathcal{F}[E_v(\vec{k})] - \mathcal{F}[E_c(\vec{k})],$$

where $\mathcal{F}[E_v(\vec{k})] \equiv \mathcal{F}_p$ and $\mathcal{F}[E_c(\vec{k})] \equiv \mathcal{F}_n$ are the population numbers in the valence and the conduction bands, respectively. Hence the condition for negative absorption, or light amplification, is

$$\mathcal{F}(E_c) - \mathcal{F}(E_v) > 0, \tag{12.1}$$

where the energies $E_v(\vec{k})$ and $E_c(\vec{k})$ satisfy the condition

$$E_c(\vec{k}) - E_v(\vec{k}) = \hbar\omega. \tag{12.2}$$

Here ω is the light frequency. Obviously, inequality (12.1) imposes the population inversion of the energy states $E_v(\vec{k})$ and $E_c(\vec{k})$.

Population inversion can be obtained under only strongly nonequilibrium conditions. In general, functions $\mathcal{F}_n$ and $\mathcal{F}_p$ may take a variety of forms, depending on the conditions. Two types of processes – intraband and interband – have special importance in establishing the forms of these functions. As shown in Section 11.3, intraband processes are much faster than interband processes. In the most interesting case of high carrier concentrations, the carrier–carrier interaction defines the shape of the functions $\mathcal{F}_n$ and $\mathcal{F}_p$; this interaction is also responsible for the evolution of these functions to forms close to those of the Fermi distributions. This is similar to the case analyzed in Section 2.9 for nonequilibrium electron kinetics in an electric field. As a consequence of electron–electron interactions, quasi-equilibrium distribution functions can be assumed for each band. Thus

$$\mathcal{F}_{p,n} = \frac{1}{\exp[(E - E_{F_{p,n}})/k_B T] + 1},$$
(12.3)

where E_{Fn} and E_{Fp} are quasi-Fermi levels that depend on the nonequilibrium electron and hole concentrations. Within this approximation, the population numbers depend on two parameters: the lattice temperature T and the quasi-Fermi levels $E_{F_{p,n}}$; in other words, they depend on the concentrations of electrons and holes.

Now one can easily deduce that inequality (12.1) is equivalent to

$$\hbar\omega < E_{Fn} - E_{Fp}.$$

Thus, to achieve light amplification, the photon energy must be smaller than the separation between the quasi-Fermi levels of the electrons and the holes. Taking into account the fact that interband transitions are possible only for $\hbar\omega > E_g$, where E_g is the bandgap, one can obtain the criteria for light amplification for interband phototransitions:

$$E_g < \hbar\omega < E_{Fn} - E_{Fp}.$$
(12.4)

The two-band semiconductor system under consideration is transparent for light in the frequency range $\omega < E_g/\hbar$, amplifies light in the frequency range given by inequality (12.4), and absorbs light in the range $\hbar\omega > E_{Fn} - E_{Fp}$. Under equilibrium conditions, $E_{Fn} = E_{Fp}$, absorption is possible if $\hbar\omega > E_g$.

Both quasi-Fermi levels are functions of concentrations and therefore functions of the pumping rates of the electrons and holes. The difference in the right-hand side of inequality (12.4) increases as the rate of pumping increases, i.e., as the rates of both the carrier injection and the optical generation of the electrons and holes increase. Inequality (12.4) indicates the existence of a threshold level of pumping that corresponds to the very appearance of the population inversion for a fixed light frequency ω. For the case in which population inversion is driven by an injection current, there should be a threshold current that leads to a population inversion and to light amplification at frequency ω. When the pumping rate increases, the spectral bandwidth over which the amplification exists expands as well. In this respect, a semiconductor laser amplifier is unlike an atomic or molecular laser amplifier, for which the bandwidth is relatively independent of the pumping parameters.

Let us consider the spectral dependence of the amplification coefficient for interband transitions in bulk semiconductors. According to Eq. (10.58), the amplification coefficient is

$$\alpha = \frac{4\pi^2 \omega |(\vec{\xi}\vec{\mathcal{D}})_{c,v}|^2}{c\sqrt{\kappa}\hbar} \varrho(\omega) \left[\mathcal{F}(E_c) - \mathcal{F}(E_v) \right].$$

The prefactor is a slowly varying, linear function of frequency and can be ignored; thus the main frequency dependence comes from both the optical density of states $\varrho(\omega)$ and the population inversion $\mathcal{F}_n - \mathcal{F}_p$. At low temperatures, we set $\mathcal{F}(E_c) = 1$ for $E_c < E_{Fn}$ and 0 otherwise; as well, $\mathcal{F}(E_v) = 1$ for $E_v > E_{Fp}$ and 0 otherwise. Thus,

$$\mathcal{F}(E_c) - \mathcal{F}(E_v) = \begin{cases} +1, & \hbar\omega < E_{Fn} - E_{Fp} \\ -1, & \hbar\omega > E_{Fn} - E_{Fp} \end{cases}.$$

For the case of a bulk crystal, both frequency factors are presented in Figs. 12.1(a) and 12.1(b). As a result, the frequency dependence of $\alpha(\omega)$, almost coincides

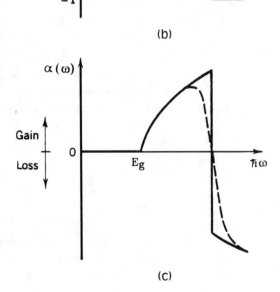

(a)

Figure 12.1. Different frequency factors that contribute to the amplification coefficient α. Schematics of (a) the energy dependence of the optical density of states ρ as a function of the photon energy $\hbar\omega$, (b) the population factor $\mathcal{F}(E_c) - \mathcal{F}(E_v)$ (the solid line corresponds to $T = 0$, and the dashed curve corresponds to a finite temperature), (c) the amplification coefficient. Light amplification and absorption correspond to $\alpha > 0$ and $\alpha < 0$, respectively. [After B. E. Saleh and M. C. Teich, *Fundamentals of Photonics* (Wiley, New York, 1991).]

(b)

(c)

with the optical density of states and changes its sign at $\hbar\omega = E_{Fn} - E_{Fp}$; see Fig. 12.1(c). Finite temperatures result in a flattening of the functions $\mathcal{F}(E_v)$, $\mathcal{F}(E_c)$, and $\alpha(\omega)$, as shown in Fig. 12.1.

12.2.2 Estimates of Light Gain

As the pumping rate grows, we see that the amplification spectrum broadens and the amplification coefficient increases. At low temperatures, when electrons and holes can be considered as completely degenerate gases, we can easily calculate the gain bandwidth $\Delta\omega$ and the maximum gain $\alpha_{\max}$ as functions of the carrier concentrations n. Using Eq. (2.81) for E_{Fn} and E_{Fp} in the three-dimensional case, we find

$$\Delta\hbar\omega = (3\pi^2)^{2/3} \frac{\hbar^2}{2m_r} n^{2/3},$$

$$\alpha_{\max} = P(3\pi^2)^{1/3} \sqrt{\frac{\hbar^2}{2m_r}} n^{1/3},$$

where the factor P is determined by Eq. (10.62) and m_r is the reduced electron-hole mass. The concentrations of electrons and holes are assumed to be equal. The maximum of α is reached at the edge of the spectral gain band, as shown in Fig. 12.1. Because the carrier concentration n is proportional to the pumping rate, the last equations show how the gain spectral bandwidth and the gain coefficient increase with increased pumping. More accurate calculations of α for finite temperatures are shown in Fig. 12.2(a) for an InGaAsP alloy for different electron and hole concentrations and for $T = 300$ K. This compound supports light amplification in the near-infrared spectral range from 1.1 to 1.7μm. The maximum values of α are collected in Fig. 12.2(b), which clearly demonstrates the threshold character of the population inversion. It is worth noting that the absolute values of the amplification coefficients are large; typically they are of the order of several hundreds of inverse centimeters.

In accordance with inequality (12.4), at a finite temperature there is a minimum threshold value of the carrier concentration n_{th} below which there is no amplification. Near the threshold it is possible to approximate the amplification coefficient by

$$\alpha = \gamma \left(\frac{n}{n_{\mathrm{th}}} - 1 \right). \tag{12.5}$$

Equation (12.5) is useful for making rough estimates of the gain coefficient. For example, in the case of InGaAsP, the parameters needed to estimate α are given in Fig. 12.2 as $\gamma = 600$ cm^{-1} and $n_{\mathrm{th}} = 1.1 \times 10^{18}$ cm^{-3}; thus, for $n = 1.4\, n_{\mathrm{th}}$, one obtains $\alpha = 180$ cm^{-1}. If we choose the length of the active region of the amplifier in the direction of light propagation to be $L_y = 300\ \mu$m, we find that

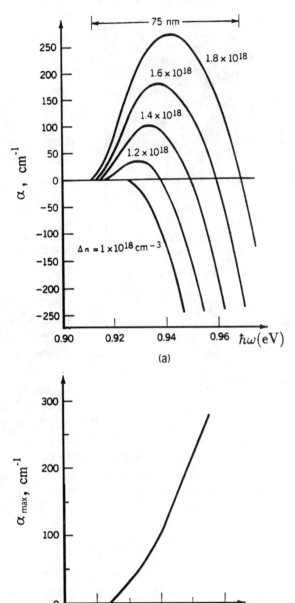

Figure 12.2. Characteristics of an InGaAsP light amplifier. (a) Calculated light-amplification coefficient as a function of the photon energy. Different curves correspond to different injected carrier concentrations indicated at the curves. The threshold injected concentration is ~1×10^{18} cm^{-3}. The maximum of light amplification is centered near the photon energy corresponding to 1.3 μm. For particular carrier concentrations the width of the spectral region of light amplification is indicated. (b) The maximum amplification coefficient as a function of the carrier concentration. After [N. K. Dutta, "Calculated absorption, emission and gain in In$_{0.72}$Ga$_{0.28}$As$_{0.6}$P$_{0.4}$," J. Appl. Phys. **51**, pp. 6095–6100 (1980)], [N. K. Dutta and R. Nelson, "The case for Auger recombination in In$_{1-x}$Ga$_x$As$_y$P$_{1-y}$," J. Appl. Phys. **53**, pp. 74–92 (1982).]

the total gain

$$G \equiv \frac{\mathcal{I}_{\text{out}}}{\mathcal{I}_{\text{in}}} = e^{\alpha L_y}$$

is equal to $e^{5.4} = 221 = 2.3$ dB. As a result of facet reflections from facet structure imperfections as well as other loss mechanisms, the losses of light in optical

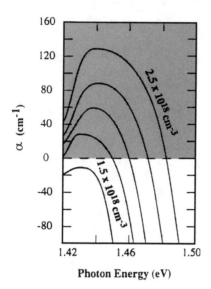

Figure 12.3. The amplification coefficient versus the photon energy in a GaAs light amplifier for different carrier concentrations. The temperature is 300 K. The results are given for five values of the concentration with equal steps of 0.25×10^{18} cm^{-3}. [After J. Singh, *Semiconductor Devices: An Introduction* (McGraw-Hill, New York, 1994).]

cavities are typically of the order of 3–5 dB/path. Such losses must of course be overcome by gain if the medium is to function as an amplifier. Figure 12.3 gives another example of a calculation of $\alpha(\hbar\omega)$ for GaAs at $T = 300$ K and for different carrier concentrations. Despite the different photon-energy range (1.42–1.48 eV), the order of magnitude of the concentrations required for obtaining population inversion remains almost the same.

12.2.3 Methods of Pumping

From the previous numerical examples, one can see that to create population inversion for a bulk crystal one should generate a density of nonequilibrium carriers as high as 10^{18} cm^{-3}. Light gain comparable with the optical losses requires even more intensive pumping. In the active region of a device, where the electron and the hole concentrations are highly nonequilibrium, the characteristic lifetimes of these excess carriers are small. Figure 12.4 shows the radiative lifetime for electrons and holes with different concentrations in direct-bandgap GaAs at 300 K. For concentrations of $\sim 10^{18}$ cm^{-3}, which are encountered in practice, this time is less than 10 ns.

Let us introduce the density of the pumping rate, $\mathcal{R}_{\text{pump}}$, that represents the number of electron–hole pairs excited in a unit volume per unit time. For the case in which the radiation mechanism is the decay of electron–hole pairs, we can write

$$\mathcal{R}_{\text{pump}} = Bn^2 = \frac{n}{\tau_R(n)}, \tag{12.6}$$

where B is a parameter and $\tau_R = 1/(Bn)$ is the radiation lifetime. The previously discussed numerical values allow us to estimate the pumping rate $\mathcal{R}_{\text{pump}}$ necessary to induce a population inversion:

$$\mathcal{R}_{\text{pump}} \geq 10^{26} \text{ cm}^{-3}\text{s}^{-1}. \tag{12.7}$$

If the bandgap E_g is supposed to be ~ 1 eV, the pumping rate of relation (12.7) corresponding to the density of the pumping power given in the previous example is

$$E_g \mathcal{R}_{\text{pump}} \geq 16 \text{ MW/cm}^3,$$

that is, a population inversion can be created at only an extremely high density of excitation power.

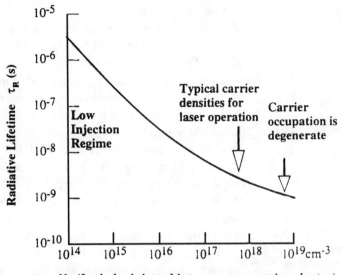

Figure 12.4. The radiative lifetime for GaAs as a function of electron and hole concentrations at room temperature. The low injection regime, the carrier concentrations necessary for the lasing, and the regime of strong degeneracy of the electron–hole gas are indicated. [After J. Singh, *Semiconductor Devices: An Introduction* (McGraw-Hill, New York, 1994).]

At very high levels of the excitation of electron–hole pairs, another mechanism competitive with radiative recombination comes into play; it is the Auger mechanism. Three carriers participate in Auger processes – two electrons and one hole or one electron and two holes. The energy released by the recombining electron and hole is absorbed by the third particle and then dissipated by intraband-relaxation processes. The rate of Auger recombination can be written as

$$\mathcal{R}_{\text{aug}} = C_a n^3. \tag{12.8}$$

A typical order of magnitude for C_a is 5×10^{-30} cm^6/s for bulk GaAs at room temperature.

In the following discussion, we consider the conventional methods of pumping semiconductor materials to establish population inversion and light gain due to interband phototransitions.

The first method is optical pumping, in which external light, often incoherent, with a photon energy larger than the bandgap is absorbed and creates nonequilibrium electron–hole pairs. Then, because of the short time associated with the intraband relaxation, the electrons and the holes relax to the bottom of the conduction band and the top of the valence band, respectively, where they are accumulated. If the rate of optical pumping is sufficiently large, population inversion can be induced between the bands and light amplification becomes possible, as illustrated in Fig. 12.5.

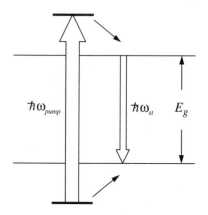

Figure 12.5. The schematic of the optical-pumping process. A pump light with photon energy, $\hbar\omega_{pump}$, creates electron–hole pairs with the excess energy. The electrons relax to the conduction-band bottom, while the holes relax to the top of the valence band. Stimulated emission occurs with the energy $\hbar\omega_{st}$ near the bandgap E_g.

Optical pumping corresponds to a conversion of one kind of radiation, not necessarily coherent, into coherent radiation with a lower photon frequency. One uses optical pumping in the cases in which electrical-current pumping is either not possible or is ineffective. This pumping method is often used to test prototypical laser structures before the design of the current pumping system.

A much more convenient method for pumping semiconductor lasers is electron and hole injection into a depletion region of a p–n junction. The physics of a p–n junction was studied in Subsection 9.4.1. Let us show that such a junction can be used to obtain population inversion. A p–n junction is shown in Fig. 9.23(b) in an unbiased state in which electrons and holes have the same equilibrium Fermi level E_F and are separated in space; see also Fig. 9.24(a). If a voltage bias is applied, the electrons are injected into the region with the existing hole population and vice versa. Nonequilibrium electron and hole concentrations are determined by the quasi-Fermi levels E_{Fn} and E_{Fp}, as shown in Fig. 9.24(b). As a result of the carrier injection, a space region is formed where both electrons and holes exist simultaneously and interband phototransitions occur. An increase in the voltage bias leads to the separation of the quasi-Fermi levels, as shown in Fig. 9.24(b). When these levels are well separated, light amplification becomes possible. In accordance with inequality (12.4), the applied voltage should exceed Φ_{th}, which is given by

$$e\Phi_{th} \geq E_g - E_{Fn} + E_{Fp}.$$

Note that the region of population inversion – the so-called active region – is quite narrow. Its typical thickness d for GaAs and InGaAsP is 1–3 μm.

Let us introduce the cross-sectional area $A = L_x \times L_y$, through which the current I is injected into the p–n region of a diode; L_x and L_y are the linear sizes of this area. Then, in steady state, the rate of injection of electrons and holes into a unit volume per unit time can be expressed as

$$\mathcal{R} = \frac{I}{eAd} = \frac{j}{ed}, \tag{12.9}$$

where j is the injection current density. Injection leads to the accumulation of nonequilibrium electron–hole pairs with concentration

$$n = \tau\mathcal{R} = \frac{\tau}{ed}j, \tag{12.10}$$

where τ is the total lifetime of the nonequilibrium pairs in the active region.

Because the injected concentration is proportional to the current density, we can rewrite Eq. (12.5) for light gain in another form:

$$\alpha(\omega) = \gamma \left(\frac{j}{j_{th}} - 1 \right). \tag{12.11}$$

Here we introduce the threshold current density j_{th}, given by

$$j_{th} = \frac{ed}{\eta_i \tau_R} n_{th}. \tag{12.12}$$

The ratio

$$\eta_i \equiv \frac{\tau}{\tau_R} \tag{12.13}$$

characterizes the internal efficiency of the emission in the p–n junction. Note that at $j = j_{th}$ the absorption (amplification) coefficient $\alpha = 0$; this condition corresponds to the case in which there is no absorption or amplification. One can say that the threshold current corresponds to transparent conditions.

From Eq. (12.12), we can draw some important conclusions. First, we can see that j_{th} is proportional to the thickness of the active region of the p–n junction, d. This means that a lower threshold current can be achieved in a narrow p–n junction. The current density j_{th} is inversely proportional to the internal efficiency; therefore a material with high internal efficiency (large probability of radiative recombination) is preferable.

Let us make some numerical estimates of the characteristic parameters for a semiconductor diode light amplifier. An InGaAsP system can function at room temperature and has the following typical parameters: $\tau_R = 2.5$ ns, $\eta_i = 0.5$, $n_{th} = 1.1 \times 10^{18}$ cm^{-3}, and $\gamma = 600$ cm^{-1}. Let the junction thickness and the transverse sizes of the diode be $d = 2$ μm, $L_x = 10$ μm, and $L_y = 200$ μm; then the threshold current density is $j_{th} = 4.6 \times 10^4$ A/cm^2. The current density $j = 5.4 \times 10^4$ A/cm^2 results in a light gain coefficient α of 100 cm^{-1} or a total gain (in the y direction) G of magnitude $G = e^{\alpha L_y} = 7.3$. The total current flowing through the area $A = 2 \times 10^{-5}$ cm^2 is $\sim$1 A. These numbers show that laser amplifiers usually exploit very high electric currents and electric powers in accordance with our previous estimates.

So far we have considered that the electron–hole concentrations are fixed and that the amplification coefficient does not depend on the light intensity. However, if the light intensity $\mathcal{I}$ is large, the phototransitions induced by the light wave change the balance between the generation and the recombination of the carriers; in turn, this results in a decrease in both the carrier concentration and the population inversion. Accordingly, the amplification coefficient decreases. The concentration balance is given by

$$\frac{dn}{dt} = \mathcal{R}_{pump} - \frac{n}{\tau} - \mathcal{R}_{st}, \quad n = p, \tag{12.14}$$

where $\mathcal{R}_{\text{pump}}$ is the rate of the electron–hole generation in the active region and $\mathcal{R}_{\text{st}}$ is the rate of the stimulated emission. According to Section 10.3, we find

$$\mathcal{R}_{\text{st}} = \alpha \frac{c}{\sqrt{\kappa}} \mathcal{I}. \tag{12.15}$$

Here we take into account that the generation of one photon corresponds to one event of the radiative electron–hole recombination. Thus the amplification becomes a function of the light intensity. In the limit of very high intensity, the population inversion goes to zero,

$$\mathcal{F}_n - \mathcal{F}_p \to +0,$$

and light amplification saturates. This saturation effect in the light amplification is important for laser operation because it is responsible for steady-state laser performance.

12.2.4 Motivations for Using Heterostructures for Light Amplifications

From the previous considerations, one can see that if the thickness of the active region decreases, the electric current required for obtaining population inversion and light amplification also decreases. Thus the length of the active region of the p–n junction is one of the critical parameters for injection pumping of the population inversion. As we already know, an excess extrinsic carrier concentration in some spatial region always leads to carrier diffusion out of this region. This leakage of the carriers is prevented by their finite lifetime; for the case under consideration, this lifetime is the recombination time τ. If the diffusion coefficient is D, the characteristic width, d, of the region with the excess concentration cannot be less than diffusion length $L_D = \sqrt{D\tau}$, which is the average distance of electron (hole) transfer before recombination, that is,

$$d > L_D.$$

Since the diffusion length of electrons and holes in III–V compounds is of the order of a few micrometers, it is not possible to make the active region less than the diffusion length in typical homostructure p–n junctions.

To localize nonequilibrium electrons and holes in a smaller active region, one can use two heterojunctions. The basic idea of using a double heterostructure is to design potential barriers on both sides of the p–n junction – this protects electrons and holes from diffusion. The potential profile of a double heterostructure is sketched in Fig. 12.6(a). The heterostructure consists of three materials with the bandgaps E_{g1}, E_{g2}, and E_{g3}. The band offsets are chosen so as to design a structure with a barrier for electrons in the left part of the structure before the p region and a barrier for holes in the right part before the n region. The middle region with the bandgap E_{g2} is accessible by both types of carriers, and it serves as the active region. Figure 12.6(a) corresponds to the case of a flat-band condition

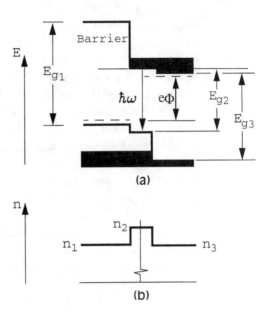

Figure 12.6. (a) The energy-band diagram of a double heterostructure for the light amplifier. The applied voltage Φ induces the flat-band condition; E_{g1}, E_{g2}, and E_{g3} are the bandgaps in different regions of the structure. (b) The schematic of the refractive index corresponding to a double heterostructure; n_1, n_2, and n_3 are the values of the refractive index in the regions with different bandgaps.

of a p–n junction with a double heterostructure embedded in the depletion region. In this case, it is not the diffusion length but the distance between the barriers that determines the size of the active region. As a result, the size can be as small as 0.1 μm and the critical electric current decreases by 1 order of magnitude or more compared with that of a conventional homostructure p–n junction.

Another reason for using heterostructures for light amplification arises as a result of the electrodynamic peculiarities of light propagation in a narrow active region. In Section 10.5, we discussed such peculiarities. We came to the conclusion that for spatially inhomogeneous media, the concept of a simple plane light wave is no longer valid. We found that propagation and amplification are strongly dependent on both the magnitude and the geometry of the inhomogeneity as well as on the wave direction. We showed that in nonuniform media, light modes are formed because of the inhomogeneity of the refractive index. On the basis of these modes, we can introduce confined photons into these media and then consider the amplification of these modes. If the volume of the active region is less than the volume confining the light mode, the amplification of the mode decreases roughly by the ratio of these volumes. In our previous discussions, this reduction was described in terms of the so-called optical confinement factor; see Eq. (10.87). For the case in which the active region is in a plane layer perpendicular to the z axis, the confinement factor is

$$\Gamma \approx \frac{\int_{\text{Over act. region}} dz |F(z)|^2}{\int_{\text{Over sample}} dz |F(z)|^2}, \tag{12.16}$$

where $F(z)$ is the distribution of the wave amplitude as a function of z. For plane waves, we can easily approximate this factor by $\Gamma \approx d/L_z$, where L_z represents a light confinement region that for homojunction devices may coincide with the size of the sample in the z direction. The factor Γ is, in fact, the portion of the electromagnetic energy of the wave that propagates through the active region. We can easily understand that for a narrow active region, the factor Γ is small and a decrease of the amplification by this factor can be crucial.

The second disadvantage associated with the use of a narrow active layer is that there is absorption of the light in the nonamplifying regions adjacent to the active layer of the device; indeed, some optical losses always exist there. Even if these losses are small, they can result in a significant reduction in or even elimination of the total optical gain if the passive region is wide. Thus there are potential problems associated with the small localization of light within the active region.

With double heterostructures, it is possible to avoid these disadvantages. Let us return to the simplest double-heterostructure design, which requires three basic layers of different materials. Figures 12.6(a) and 12.6(b) show the schematics of a double heterostructure, the associated band profiles, and the refractive index. Layer 1 is p doped, and it has a bandgap E_{g1} and a refractive index n_1. The second layer has a bandgap E_{g2} and a refractive index n_2; usually this layer is undoped. The third layer is n doped, and it has a bandgap E_{g3} and a refractive index n_3. If the bandgaps satisfy the conditions

$$E_{g2} < E_{g1},\ E_{g3},$$

the refractive index of the middle layer, n_2, is greater than that of the surrounding layers. This structure is designed for two goals:

1. The band discontinuities in the conduction and the valence bands lead to the confinement of the nonequilibrium electrons and the holes within the active layer. Compared with the usual homogeneous p–n junction, this double heterostructure efficiently accumulates both types of carriers in a thinner region.

2. The larger refractive index in the second narrow-bandgap layer causes optical confinement within this layer. When solving the wave equation for light, we obtain solutions localized within the layer. The lowest mode $\vec{F}_0(z)$ that has no nodal points is presented in Fig. 12.7. There are other solutions with transverse distributions of the electric field $\vec{F}_1(z)$, $\vec{F}_2(z)$, ..., which have 1, 2, ..., nodal points and weaker confinement. These waves are the transverse modes. In Fig. 12.7, the lowest modes with wavelength $\lambda = 0.9\,\mu$m are presented for three different active-layer thicknesses in the range from 0.1 to 0.3 μm; the discontinuity in n equals 0.1. This discontinuity leads to the localization of the modes with characteristic extensions of $\sim$0.5 to 1 μm.

Thus, in a double heterostructure, one can assume proper localization of photon modes for which the factor Γ is much greater than that for unconfined, extended waves. For the confined modes, factor Γ can be of the order of 1.

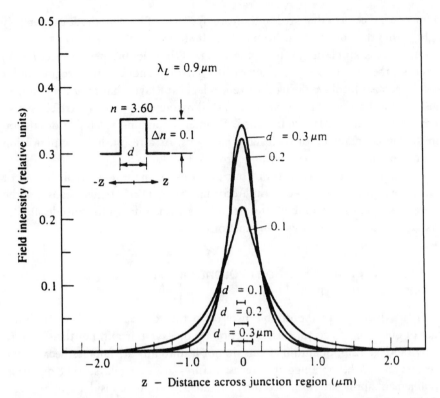

Figure 12.7. Distributions of the intensities in the lowest optical modes confined by the step discontinuities in the refractive index. The inset shows the refractive index across the heterostructure. The results are given for the wavelength $\lambda = 0.9$ μm and three widths of the dielectric waveguide d. [After H. Kressel and J. K. Butler, *Semiconductor Lasers and Heterojunction LEDs* (Academic, New York, 1977).]

A complication can arise from the mechanism of electron–hole recombination at the interfaces of the heterostructure. Within the context of a phenomenological model, we can describe this mechanism by introducing the interface recombination velocity s_{if}, so that the rate of this additional recombination is

$$R_{if} = s_{if}\, n, \tag{12.17}$$

where n is the three-dimensional carrier concentration in the barriers of the heterostructure. Typical values of s_{if} are of the order of 100–1000 cm/s. At modest levels of pumping, this channel of carrier losses can be important. At higher injection levels, radiative recombination dominates over interface recombination because the rate of the former mechanism is proportional to the cube of the carrier concentration; see Eq. (12.6).

As a particular example of light localization in a double heterostructure, we consider a GaAs/AlGaAs diode with an active layer fabricated from GaAs with $E_{g2} = 1.42$ eV and $n_2 = 3.6$. The surrounding layers can be made of $Al_xGa_{1-x}As$, with x in the range from 0.35 to 0.5 and $E_{g1,3} > 1.42$ eV; such layers have the refractive indices $n_{1,3}$ that are less than 3.6 by 5% to 10%. This diode can amplify light within the optical band from 0.82 to 0.88 μm.

Another example is given by the InGaAsP/InP double heterostructure. In this case, the $In_{1-x}Ga_xAs_{1-y}P_y$ layer is the narrow-bandgap active region and it simultaneously localizes the light wave, while the InP layers have a wider bandgap and serve as barriers for the carriers. The alloy fractions, x and y, are chosen so that all materials are lattice matched; Chapter 4 contains a detailed discussion on the lattice matching of such material layers. As a result, light amplification can be obtained in the spectral range from 1.1 to 1.7 μm.

The double heterostructures of our previous examples have typical sizes of the active region d of $\sim$0.2–0.3 μm. According to Eq. (12.12), this means that the threshold currents for double heterostructures are 1 order of magnitude smaller than those for homogeneous p–n junctions.

12.2.5   Light Amplification in Quantum Wells and Quantum Wires

If the thickness of the double heterostructure decreases further, the influence of quantum effects on carrier motion becomes important. For the structure shown in Fig. 12.6, quantum effects do not lead to any advantages because the advantages can be obtained if heterostructures are designed so that quantum confinement applies to both electrons and holes. Here we consider only the case of light propagation along the active layer. In Section 12.3 we will study particular structures in more details. The simplest case of quantum confinement can be achieved if a quantum-well layer is embedded in an active region of a type-I heterostructure. Three possible designs of the active regions exhibiting quantum confinement are sketched in Fig. 12.8. Figures 12.8(a) and 12.8(b) correspond to the resultant confinement in a single quantum well, while Figure 12.8(c) corresponds to the confinement in multiple quantum wells. For these designs, electrons

Figure 12.8. Composition profiles of $(Al_xGa_{1-x})_{0.5}In_{0.5}P$ heterostructures providing simultaneously a quantum confinement of the carriers and an optical confinement: (a) a single quantum well and a steplike refractive-index heterostructure, (b) a single quantum well and a graded-index optical confinement structure, (c) a multiple quantum well and a steplike index structure. [After P. S. Zory, Jr., *Quantum Well Lasers* (Academic, Boston, 1993).]

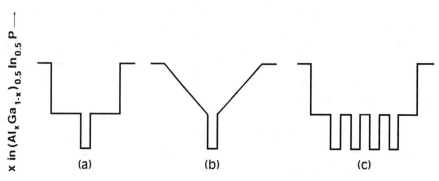

and holes that are either generated by external light or injected from p and n regions move in barrier layers and then are captured into the active region and quantum wells. The characteristic time of this capture is less than 1 ps. Escape processes require additional energy and have low relative probabilities of occurrence. Carriers in the quantum wells relax to the lowest energy states available. This results in the accumulation of both types of carriers in an extremely narrow active region, which is typically 100 Å wide or even less. A similar situation can be realized if quantum wires or dots are embedded in the active region.

From the analysis of interband phototransitions of confined electrons and holes given in Section 10.5, it is clear that the criterion for population inversion given by inequality (12.1) can be presented in the form of inequality (12.4) with only one modification: the bandgap of the quantum-confining layer, E_g, should be replaced by the optical bandgap,

$$E_g + \epsilon_{c1} + \epsilon_{v1},$$

where ϵ_{c1} and ϵ_{v1} are the bottom and the top of the lowest electron- and hole-energy subbands, respectively. For low-dimensional cases, the difference between the quasi-Fermi levels determining population inversion increases with increasing carrier concentrations faster than in a bulk crystal. E_F is proportional to n_{2D} in quantum wells and to n_{1D}^2 in quantum wires, where n_{2D} and n_{1D} are the surface and the linear carrier concentrations, respectively; see Eqs. (2.83) and (2.84). These scaling results imply that the lower is the dimension of the structure, the easier to satisfy the criterion of inequality (12.4) and create population inversion at lower pumping rates.

As discussed in Chapter 10, two factors – the optical density of states $\varrho(\omega)$ and the value of population inversion $\mathcal{F}_n - \mathcal{F}_p$ – determine the frequency dependence of light amplification as described by Eqs. (10.99) and (10.100). As shown in Chapter 3, quantum confinement modifies the density of states significantly. In turn, this causes changes of the spectral dependence of the amplification coefficient α. Figure 12.9 illustrates both factors for a quantum-well layer. A comparison with the bulk case of Fig. 12.1 shows the differences in the frequency dependences of light amplification. The results shown in Fig. 12.9 are obtained for a pumping rate such that only the lowest electron and hole subbands exhibit population inversion. From Fig. 12.9, it is seen that the steplike density of states causes a narrower spectral range of light amplification. This fact is used in quantum-well lasers for the generation of a smaller number of optical modes. Thus, using quantum wells in many cases is advantageous for improving the spectral characteristics of light amplification.

For light amplification in quantum wells, optical-confinement factor Γ is a critical parameter. It is convenient to present the amplification coefficient of Eq. (10.99) by isolating the Γ factor, as follows,

$$\alpha(\omega) = \Gamma \alpha_0(\omega). \tag{12.18}$$

The narrow-bandgap quantum-well layer can serve simultaneously as a waveguide for the light. From Fig. 12.7 it is seen, however, that the degree of optical

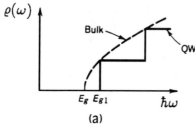

(a)

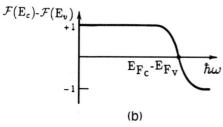

(b)

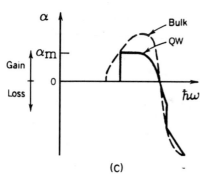

(c)

Figure 12.9. Different frequency factors that contribute to the amplification coefficient in a quantum-well layer (for comparison these factors are also shown for a bulk material by dashed curves). (a) Schematic of the energy dependence of the optical density of states ρ, (b) the population factor $\mathcal{F}(E_c) - \mathcal{F}(E_v)$ (a finite temperature is assumed), (c) the amplification coefficient. [After B. E. Saleh and M. C. Teich, *Fundamentals of Photonics* (Wiley, New York, 1991).]

localization decreases in thinner confining layers. Thus a single quantum well does not lead to good localization of the light wave. Accordingly, factor Γ is usually small.

To localize the light within a quantum well and to realize simultaneously a high rate of the capture into the well, one designs special heterostructure wave guides in the active region. Two such designs of the active region with variable refractive indices are shown in Fig. 12.10 for InGaAsP/InGaP systems. These designs correspond to the structure shown in Figs. 12.8(a) and 12.8(b). The structure provides for a steplike or gradual change in the refractive index and confines light modes within the active region. The dependence of confinement factor Γ on the half-thickness $d/2$ of this structure is presented in Fig. 12.10 for the same 100-Å-wide quantum well. Nonmonotonous dependencies $\Gamma(d/2)$ can be understood as follows. The limits $d \to 0$ and $d \to \infty$ both correspond to the confinement of the mode in a single-quantum-well layer. In the case of the first limit, this layer provides the step in the refractive index corresponding to the composition with $x = 1$; for the case of the second limit, the step corresponds to the composition with $x = 0.6$. Both quantum-well layers lead to a relatively weak localization of light modes, with Γ being ~1%. If d is finite, the modes are more localized. The figure shows that if the half-thickness of the structure

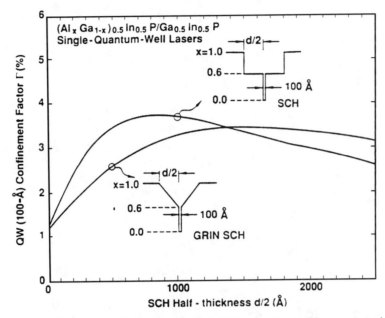

Figure 12.10. The optical-confinement factor for $(Al_xGa_{1-x})_{0.5}In_{0.5}P/Ga_{0.5}In_{0.5}P$ single-quantum-well lasers. Results are given for two designs of separate confinement heterostructures (SCH) shown in the inset. [After P. S. Zory, Jr., *Quantum well Lasers* (Academic, Boston, 1993).]

changes from 0 to 0.25 μm, the factor Γ changes from 1% to 3.7%, that is, specially designed heterostructures improve the confinement factor and the gain coefficient in accordance with Eq. (12.18).

In the case of injection current pumping, the quantum-well layer and the light-confining waveguide structure should be embedded in the region of the p–n junction. For such a case, results of calculations of the parameter $\alpha_0(\omega)$ in Eq. (12.18) are shown in Fig. 12.11(a) as a function of the current density j. The calculations are given for InGaAsP/InP and AlGaAs/GaAs heterostructures with the 50-Å quantum wells. The results demonstrate low values of zero-gain (threshold) currents for both heterostructures: 20 and 60 A/cm^2 for InGaAs/InP and AlGaAs/GaAs, respectively. Both $\alpha_0(I)$ curves increase sharply with the current density. For comparison, we also present in Fig. 12.11(b) the theoretical and the experimental results for double-heterostructure diodes based on the same materials, which contain bulklike active regions where the carriers are not quantized. As one can see, the threshold currents are approximately 100 times less in structures with quantum wells than those in double heterostructures with bulklike carriers.

Taking into account the small value of the confinement factor (see Fig. 12.10), one can estimate that for optimally designed quantum structures it is possible to obtain a threshold current for laser oscillations 5 to 10 times less than that of classical double heterostructures. The threshold can be lowered further if instead of a single quantum well one uses a multiple-quantum-well structure embedded in

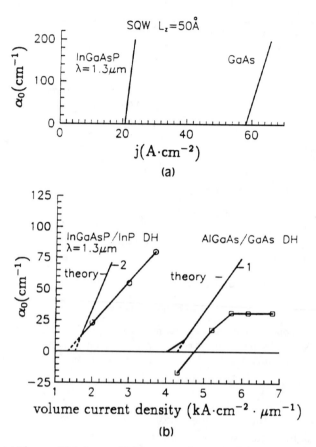

Figure 12.11. (a) The amplification coefficient α_0 as a function of the pumping current density for single InGaAsP and GaAs quantum wells of 50-Å width, (b) the amplification coefficient α_0 versus the current density for InGaAsP/InP and AlGaAs/GaAs double heterostructures with bulklike carriers. Theoretical and experimental results are presented. The dashed lines show the zero-amplification current density. [After P. S. Zory, Jr., *Quantum Well Lasers* (Academic, Boston, 1993).]

the active region; see Fig. 12.8. In a multiple-quantum-well amplifier, the capture of carriers into the wells is more effective; moreover, all the wells interact with the same optical mode and emit the same photons. As a result, both the optical-confinement factor and the light amplification increase.

12.3 Light-Emitting Diodes and Lasers

12.3.1 Light-Emitting Diodes

In Section 12.2 we stressed that carrier injection into the active region of a semiconductor diode is the most applicable method for the generation of nonequilibrium electrons and holes and for the creation of population inversion and stimulated emission. Although stimulated emission from the injection diode is important practically, subthreshold operation of the diode – when only spontaneous light is emitted – is in many cases advantageous. It does not require feedback for controlling the power output, it facilitates operation over a wide range of temperatures, and it is reliable and inexpensive. Such diodes operating on spontaneous light emission are called light-emitting diodes.

Let us consider these light-emitting diodes. We base our considerations on the results obtained in Section 12.2. For a forward-biased p–n diode, we can write the rate of spontaneous emission or the total photon flux as

$$\Upsilon = \frac{n}{\tau_R} V_0, \tag{12.19}$$

where n is the electron–hole concentration in the active region of the diode, V_0 is the volume of this region, and τ_R is the radiative lifetime. Using Eq. (12.13), we obtain

$$\Upsilon = \eta_i \frac{I}{e}, \tag{12.20}$$

where η_i is internal quantum efficiency of the diode. In fact, not all photons emitted in the active region contribute to the output light flux. This is so because photons are lost as a result of internal absorption in n and p regions, reflections from air–semiconductor boundaries, etc. To evaluate these losses, one introduces the overall transmission efficiency η_t so that the product $\eta_{ext} = \eta_t \eta_i$ characterizes the external efficiency of the diode and the output photon flux can be written as

$$\Upsilon_{out} = \eta_{ext} \frac{I}{e}. \tag{12.21}$$

The quantity η_{ext} is the ratio of the externally produced photon flux to the injected electron flux. Obviously, η_{ext} depends not only on the physical mechanisms but also on the design of the diode. The optical power of the light-emitting diode can be calculated from Eq. (12.21) as

$$\mathcal{P} = \hbar\omega \Upsilon_{out} = \eta_{ext} \hbar\omega \frac{I}{e}, \tag{12.22}$$

where $\hbar\omega$ is the photon energy. Practically, the linear dependence of $\mathcal{P}(I)$ is valid within only a limited current range. At high injection currents, the light output starts to saturate for a variety of reasons: a decrease of the radiative lifetime, heating of the device, etc. Figure 12.12 illustrates typical light-current characteristics of the diode.

The spectral distribution of the diode emission is an important characteristic. For a homogeneous p–n junction, the spectral density of spontaneous emission is given by Eq. (10.65); see also Fig. 10.10. For relatively weak pumping, when the quasi-Fermi levels lie within the bandgap, the peak value of the spectral distribution is

$$\hbar\omega_m = E_g + \frac{1}{2} k_B T.$$

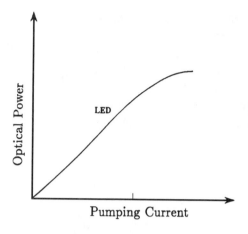

Figure 12.12. The output of a light-emitting diode (LED) as a function of the pumping current.

The full width at half-maximum of the distribution is $\Delta\omega \approx 2k_B T/\hbar$, and it is independent of ω. In terms of the wavelength λ, we obtain $\Delta\lambda = (\lambda_m^2/2\pi c)\Delta\omega$ or

$$\Delta\lambda = 1.45\lambda_m^2 k_B T, \tag{12.23}$$

where λ_m corresponds to the maximum of the spectral distribution $\Delta\lambda$; λ_m are expressed in micrometers, and $k_B T$ is expressed in electron volts. Figure 12.13 shows the spectral density as a function of the wavelength of light-emitting diodes based on various materials. The spectral density is normalized so that its maximum equals 1 for all samples. For different materials, the spectral linewidths increase in proportion to λ^2, in accordance with Eq. (12.23). From Fig. 12.13 one can see that light-emitting diodes cover a wide spectral region, from the infrared – ~8 μm for InGaAsP alloys – to near ultraviolet – 0.4 μm for GaN. They are indeed universal light sources.

Light-emitting diodes may be designed in either a surface-emitting configuration or in an edge-emitting configuration. These configurations are illustrated in

Figure 12.13. The spectra of light-emitting semiconductor diodes with different bandgaps. [After S. M. Sze, ed., *Physics of Semiconductor Devices* (Wiley, New York, 1981).]

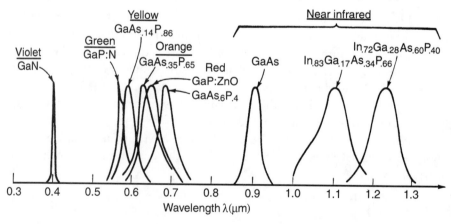

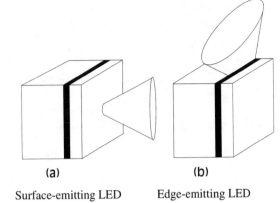

Figure 12.14. Two possible outputs of light-emitting diodes: (a) a surface-emitting diode, (b) an edge-emitting diode.

(a) (b)

Surface-emitting LED   Edge-emitting LED

Figs. 12.14(a) and 12.14(b), respectively. Surface-emitting diodes radiate from the face parallel to the *p–n* junction plane. The light emitted in the opposite direction is either absorbed by a substrate or reflected by metallic contacts. The edge-emitting diodes radiate from the edge of the junction region. Usually surface-emitting diodes are more efficient.

Since the diodes under consideration radiate through spontaneous emission, the spatial patterns of the emitted light depend on only the geometries of the devices. Figure 12.15(a) illustrates emission patterns for a surface-emitting diode in the absence of a lens, when the intensity varies as $\cos \theta$, where θ is the angle from the normal to the face. Different lenses can improve the emission pattern, as shown in Figs. 12.15(b) and 12.15(c) for hemispherical and parabolic lenses, respectively. Usually the edge-emitting diodes have narrower emission patterns.

Light-emitting diodes find many applications in the fields of signal processing and communications. Therefore an important figure of merit is their response time. To estimate this time, we can consider a small injection level, so that all processes are described by linear differential equations. Then we can assume an injection current in the form

$$I = I_0 + I_1 \cos \Omega t, \quad I_1 \ll I_0. \tag{12.24}$$

Let us use this expression to calculate the injection rate of Eq. (12.9). Then, from

Figure 12.15. Patterns of output emission for surface-emitting diodes: (a) the pattern in the absence of a lens, (b) the pattern of a diode with a hemispherical lens, (c) the pattern of a diode with a parabolic lens. [After B. E. Saleh and M. C. Teich, *Fundamentals of Photonics* (Wiley, New York, 1991).]

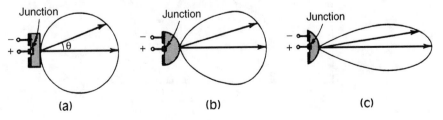

(a) (b) (c)

Eq. (12.14) we can find the electron and hole concentrations and calculate the optical power of Eq. (12.22) $\mathcal{R}_{st} = 0$. The result is

$$\mathcal{P} = \mathcal{P}_0 + \mathcal{P}_1 \cos(\Omega t + \phi),$$

where ϕ is a phase shift. For the modulated output $\mathcal{P}_1$, we get

$$\mathcal{P}_1 \sim \frac{I_1}{\sqrt{1 + (\Omega\tau)^2}}. \tag{12.25}$$

Here τ is the lifetime of the injected carriers in the active region. Therefore this time limits the device speed. As a modulation characteristic, we introduce the modulation bandwidth:

$$f \equiv \frac{1}{2\pi\tau}. \tag{12.26}$$

The bandwidth values are presented in Fig. 12.16 for different injected concentrations. We can consider that $\tau \approx \tau_R \approx 1/(Bn)$ [see Eq. (12.6)]. Then, using Eq. (12.26), we obtain an expression for the modulation bandwidth as a function of the current:

$$f = \frac{1}{2\pi}\sqrt{\frac{BI}{eAd}}. \tag{12.27}$$

This square-root dependence on the current agrees with the experimental data presented in Fig. 12.16 for InGaAsP light-emitting diodes. Results are given for diodes with different thicknesses of the active region d and cross sections A. The typical bandwidth of such devices is of the order of hundreds of megahertz, although a bandwidth above 1 GHz has also been demonstrated.

Figure 12.16. The modulation bandwidth as a function of the pumping-current-to-active-region width ratio, I/d for InGaAsP light-emitting diodes. [After O. Wada, *et al.*, "High radiance InGaAsP/InP lensed LED's for optical communication systems at 1.2–1.3 μm," IEEE J. Quantum Electron. QE-17, pp. 174–178 (1981).]

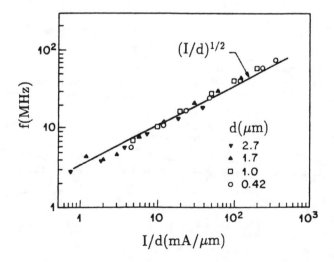

12.3.2 Amplification, Feedback, and Laser Oscillations

So far we have considered only one effect related to the stimulated emission, namely, light amplification. Lasing can be achieved if a light amplifier is supplied with a path for optical feedback. A straightforward way to provide optical feedback is to place an amplifier between two mirrors of an optical resonator, as discussed in Subsection 10.3. On passing through the amplifier, the light wave gains energy; then it is reflected back with some energy losses, and on passing through the amplifier again, it gains the energy once again, and so forth. If the amplification exceeds the optical losses in the system, the light energy inside the resonator increases. This increase may not be infinite because of the saturation effect discussed in Subsection 12.2.3; specifically, an increase in the light energy leads to an additional electron–hole recombination through the radiative channel and consequently to decreasing population inversion and amplification. Steady-state laser operation will be reached when the amplification becomes equal to the optical losses. This criterion defines the steady-state amplitude of a light wave in the optical resonator. These considerations are valid for any type of laser.

Let us consider a semiconductor injection laser. Our goal is to find the threshold current for lasing. For an injection laser, the feedback is usually obtained by cleaving the crystal planes normal to the plane of the p–n junction. In Fig. 12.17, a device with two cleaved surfaces forming an optical resonator is shown. For the light reflected from the crystal boundaries, we define the reflection coefficient as

$$r = \frac{\mathcal{I}_r}{\mathcal{I}_{\text{in}}},$$

where $\mathcal{I}_{\text{in}}$ and $\mathcal{I}_r$ are the intensities of incident and reflected light, respectively.

Figure 12.17. The schematic of an injection laser with two cleaved facets that act as reflectors. [After B. E. Saleh and M. C. Teich, *Fundamentals of Photonics* (Wiley, New York, 1991).]

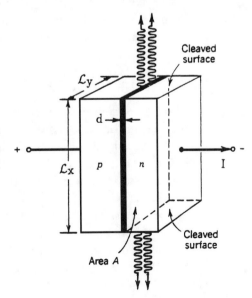

The reflection coefficient for an air–semiconductor boundary is

$$r = \left(\frac{n_{\mathrm{ri}} - 1}{n_{\mathrm{ri}} + 1}\right)^2.$$

Since semiconductor materials usually have large refractive indices n_{ri}, the coefficients r are large enough. The intensity of the light transmitted through this mirror is

$$\mathcal{I}_{\mathrm{out}} = (1 - r)\mathcal{I}_{\mathrm{in}}.$$

Let two cleaved surfaces be characterized by two coefficients, r_1 and r_2. After two passages through the device, the light intensity is attenuated by the factor $r_1 \times r_2$. We can define an effective overall distributed coefficient of optical losses:

$$\alpha_r \equiv \frac{1}{2\mathcal{L}_x} \ln \frac{1}{r_1 r_2}.$$

Here $\mathcal{L}_x$ is the distance between the cleaved surfaces. Note that the optical cavity area, $\mathcal{L}_x \times \mathcal{L}_y$, may be larger than the active area, $L_x \times L_y$, of the device. In principle, there can be other sources of optical losses in the resonator. Let they be characterized by the absorption coefficient α_s. Then the total loss coefficient is

$$\alpha_{ls} = \alpha_r + \alpha_s.$$

If α is the amplification coefficient of some light mode in this resonator, we can write the criterion for laser oscillations as

$$\alpha \geq \alpha_{ls} = \alpha_r + \alpha_s. \tag{12.28}$$

For injection lasers, the criterion of relation (12.28) is a condition imposed on the magnitude of the injection current density j. Using Eq. (12.11) for $\alpha(j)$, we find the threshold injection current density, j_{th}^l, necessary for laser oscillations:

$$j_{\mathrm{th}}^l = \frac{\alpha_{ls} + \gamma}{\gamma} j_{\mathrm{th}}, \tag{12.29}$$

where j_{th} is the threshold current density for population inversion. If the loss coefficient is small, $\alpha_{ls} \ll \gamma$, the threshold currents nearly coincide.

The threshold current j_{th}^l is the key parameter characterizing laser diodes. Equation (12.29) shows that in order to minimize j_{th}^l, one needs to minimize the optical losses α_{ls} and optimize structure parameters. For numerical estimates, we can use the example of the InGaAsP amplifier with a homogeneous p–n junction that was considered in Subsection 12.2.3. We have $j_{\mathrm{th}} = 4.6 \times 10^4$ A/cm^2 for a diode with sizes $d \times L_x \times L_y = 2 \times 10 \times 200\ \mu\mathrm{m}^3$ and $\gamma = 600\ \mathrm{cm}^{-1}$ at $T = 300$ K. The reflection coefficient is $r = 0.3$, and the corresponding effective absorption coefficient is $\alpha_r = 60\ \mathrm{cm}^{-1}$ for $\mathcal{L}_x = L_x$. We assume that other intracavity losses give the same absorption $\alpha_s = \alpha_r$; thus $\alpha_{ls} = 120$ cm^{-1}. According to Eq. (12.29), we find the laser threshold current density $j_{th}^l = 1.2 \times j_{th} = 5.6 \times 10^4$ A/cm^2 and the total current $I = 1.1$ A. The current through a device with a homogeneous p–n junction is so high that, for continuous-wave operation of this laser diode, it is necessary to provide for effective heat removal from the device. In contrast, for a double-heterostructure device, in which characteristic currents are 1 order of magnitude lower, laser operation at room temperature is possible. For example, if we assume that the size of the active region of the double heterostructure is $d = 0.1$ μm and the optical-confinement factor is $\Gamma = 1$, we obtain the threshold current density $j_{th} = 2.3 \times 10^3$ A/cm^2 and the total threshold current of the laser operation $I_{th}^l = 55$ mA.

12.3.3 Laser Output Power and Emission Spectra

According to Eq. (12.28), laser oscillations are possible for those resonator modes that have the lowest optical losses. As we have seen in Chapter 10, the optical resonator strongly discriminates different optical modes, and a relatively small number of them enjoy low losses as a result of their high-quality factors. Below threshold, $j < j_{th}^l$ and there is no stimulated emission so that the intracavity photon flux Υ of these high-quality modes is almost zero. Above threshold, this flux grows with time. According to Eqs. (10.40) and (12.15), we can rewrite the differential equation for the photon flux, taking into account the total losses coefficient α_{ls}:

$$\frac{d\Upsilon}{dt} = \frac{c}{\sqrt{\kappa}} \Upsilon \left[\alpha(\Upsilon) - \alpha_{ls} \right]. \tag{12.30}$$

It is important that the amplification coefficient is now a function of Υ. Above threshold when $\Upsilon \neq 0$, in the steady state the amplification coefficient is equal to the total loss coefficient:

$$\alpha(\Upsilon_0) = \alpha_{ls}, \tag{12.31}$$

where Υ_0 is the photon flux in the steady state. In order to calculate $\alpha(\Upsilon_0)$ at fixed injection current density j, we should return to the balance equation for the carrier concentration; see Eq. (12.14). Now, we rewrite Eq. (12.14) as

$$\frac{dn}{dt} = \mathcal{R} - \frac{n}{\tau} - \alpha(\Upsilon) \frac{\Upsilon}{\mathcal{L}_y d} = 0, \tag{12.32}$$

where $\mathcal{R} = j/ed$ is the injection rate density and τ is the electron lifetime. According to Eq. (10.40), the last term in Eq. (12.32) represents the stimulated-emission rate per unit volume; see also Eq. (12.15). Thus, instead of Eq. (12.10), we obtain

$$n_0 = \frac{j}{ed} \tau - \frac{\alpha_{ls} \tau}{\mathcal{L}_y d} \Upsilon_0. \tag{12.33}$$

Let us use Eq. (12.5) for the calculation of $\alpha(n)$. Now we have three algebraic equations, (12.31), (12.5), and (12.33), for the three variables α, n_0, and Υ_0, which can be easily solved. For the intracavity photon flux we obtain

$$\Upsilon_0 = \begin{cases} 0, & j < j^l_{\text{th}} \\ \eta_i (I - I^l_{\text{th}})/e, & j > j^l_{\text{th}} \end{cases}. \tag{12.34}$$

Here we have introduced the internal efficiency of the stimulated emission as $\eta_i \equiv \mathcal{L}_y/(\alpha_{ls} A) \equiv 1/(\alpha_{ls} \mathcal{L}_x)$ and the current $I = j\mathcal{L}_x\mathcal{L}_y$.

Since the light transmission through the mirrors is $(1 - r_{1,2})$, the laser output flux has the form

$$\Upsilon_{\text{out}} = (1 - r)\eta_i \frac{I - I^l_{\text{th}}}{e} \equiv \eta_{\text{ext}} \frac{I - I^l_{\text{th}}}{e}, \tag{12.35}$$

where we have assumed that $r_1 = r_2 = r$ and have introduced the external quantum efficiency of the laser diode,

$$\eta_{\text{ext}} = (1 - r)\,\eta_i.$$

Finally, above the laser threshold the laser output power is

$$\mathcal{P}_{\text{out}} = \hbar\omega\,\eta_{\text{ext}} \frac{I - I^l_{\text{th}}}{e}. \tag{12.36}$$

In Fig. 12.18, the light-current characteristic of Eq. (12.36) is shown for an ideal laser diode and is compared with that of a real InGaAsP injection laser emitting 1.3-μm light. For the latter, the characteristic saturates at currents greater than 75 mA; this is not shown in the figure.

To illustrate a typical laser output power, let us use our previous numerical example of the InGaAsP laser. For the double-heterostructure design, when the laser threshold current is 55 mA and when $\eta_{\text{ext}} = 0.25$, we obtain an output power $\mathcal{P}_{\text{out}} = 4.9$ mW.

The spectral distribution of the output light is one of the optical characteristics that can be improved considerably by stimulated emission. In general, several factors govern the spectral distribution. One such factor is that stimulated emission is possible only in the spectral region where the amplification exceeds the optical losses: $\alpha(\omega) > \alpha_{ls}$. Then the oscillations are possible only for the resonator modes with high-quality factors. In the case of a resonator with plane mirrors, these modes are separated by the spacing $\Delta\omega = \pi c/(\mathcal{L}_x\sqrt{\kappa})$; see Eq. (10.15). For the InGaAsP laser with a central

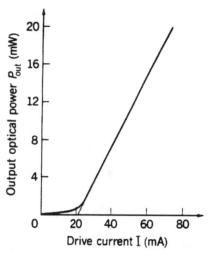

Figure 12.18. The light-current characteristics for an ideal (straight line) and real (solid curve) InGaAsP injection laser operating at $\lambda = 1.3$ μm. The laser uses an index-guided heterostructure. [After B. E. Saleh and M. C. Teich, *Fundamentals of Photonics* (Wiley, New York, 1991).]

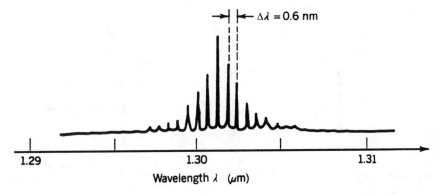

Figure 12.19. The spectral distribution of the emission from a 1.3-μm InGaAsP index-guided heterostructure laser. [After R. J. Nelson, et al., "CW electrooptical properties of InGaAsP ($\lambda = 1.3 \, \mu$m) buried-heterostructure lasers," IEEE J. Quantum Electron. **QE-17**, pp. 202–207 (1981).]

wavelength $\lambda = 1.3 \, \mu$m and $\mathcal{L}_x = 300 \, \mu$m, the intermode spacing is $\Delta\lambda = 0.9$ nm while the bandwidth of the amplification coefficient approximately equals that of the spontaneous emission of approximation (12.25) and is $\sim$7 nm. Thus approximately eight longitudinal modes can be generated. Figure 12.19 shows a specific example of the spectral distribution of a 1.3-μm InGaAsP index-guided heterostructure laser, in which light confinement is provided by variation of the refractive index. Comparing this figure with Fig. 12.13, in which the spontaneous spectra are presented, one can conclude that the laser spectra are much narrower. By reducing the resonator length to $\sim$30–40 μm, one can obtain only one longitudinal mode within the amplification bandwidth; in this case, single-mode generation is possible with the linewidth of the generated mode approximately equal to 0.01 nm.

Another advantage of stimulated emission is the sharp spatial distribution of the output power. Let us consider an injection current laser. The spatial distribution of the laser emission is determined by the number and by the type of the resonator modes that are excited under lasing conditions. Suppose we have the index-guided double-heterostructure laser. The dielectric waveguide has a cross-sectional area $d \times \mathcal{L}_y$. For simplicity we can treat this waveguide as a dielectric slab with the same rectangular cross section. This waveguide has modes with different transverse structures. Since the ratio d/λ is small, the waveguide maintains only a single mode in the transverse direction perpendicular to the plane of the p–n junction. But since $\mathcal{L}_y$ is much greater then λ, the waveguide maintains several modes, having spatial structure in the direction parallel to the junction; these modes are called lateral modes. Fortunately only a few of these lateral modes enjoy low optical losses and can be excited during laser operation. Note that the number of excited lateral modes can be reduced by a reduction in the transverse device scale $\mathcal{L}_y$. Figure 12.20 illustrates the generation of several transverse modes. We can estimate the far-field angular divergence of the laser emission. In the plane perpendicular to the junction, the angular divergence is

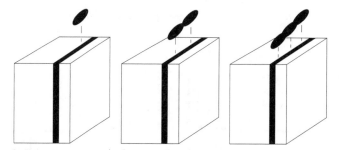

Figure 12.20. Schematic illustration of spatial distributions of the light output for the lowest waveguide modes.

approximately λ/d, while in the plane parallel to the junction it is approximately $\lambda/\mathcal{L}_y$. For example, if the cross section of the waveguide is $2 \times 10 \ \mu\mathrm{m}^2$, the two angles are $40°$ and $8.5°$. Thus the laser diode has a significantly sharper spatial distribution of emission than light-emitting diodes operating in spontaneous regime; see Fig. 12.15.

12.3.4 Modulation of the Laser Output

As was stressed previously, a major application of advanced emitters is optical communication and processing systems. This application requires high-speed modulation of the light output. Unlike other lasers that are modulated externally by means of changing the mirror transmission, semiconductor lasers can be modulated internally by modulating the injection current. In Subsection 12.3.1, we considered the modulation of the light-emitting diodes. This is an easy task because the number of radiatively recombining electron–hole pairs is directly governed by the injection current. In estimating the laser modulation one should take into account that, in general, the lasing process is nonlinear. Besides pumping, it involves stimulated emission. The steady state has been obtained as a result of a self-consistent solution of nonlinear Eqs. (12.30)–(12.32) and (12.5). Now we again suppose that the current consists of dc and ac components, as in Eq. (12.24). Then we define the small-signal modulation response of the electron–hole concentration and flux:

$$n - n_0 = n_1 = \delta n e^{i\Omega t} + \delta n^* e^{-i\Omega t},$$
$$\Upsilon - \Upsilon_0 = \Upsilon_1 = \delta \Upsilon e^{i\Omega t} + \delta \Upsilon^* e^{-i\Omega t}.$$

We rewrite Eqs. (12.30) and (12.32) as

$$\frac{d\Upsilon_1}{dt} = \frac{c}{\sqrt{\kappa}} \Upsilon_0 \frac{\gamma n_1}{n_{\mathrm{th}}}$$
$$\frac{dn_1}{dt} = \frac{I_1 \cos(\Omega t)}{ed\mathcal{L}_x\mathcal{L}_y} - n_1 \left(\frac{1}{\tau} + \Upsilon_0 \frac{\gamma}{n_{\mathrm{th}} \mathcal{L}_y d} \right) - \Upsilon_1 \frac{\alpha_{ls}}{\mathcal{L}_y d}.$$

The last two terms in the second equation are related to the self-consistent action of the pumping current and the stimulated emission flux. The system of

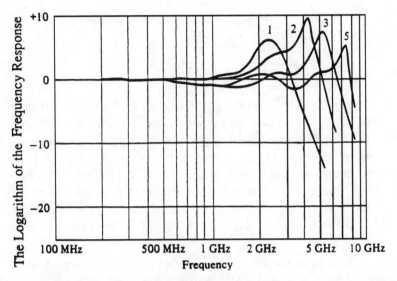

Figure 12.21. The modulated light output as a function of the frequency of the current modulation at different dc currents. The curves correspond to different dc currents, whose magnitudes increase with numbers at the curves. [After K. Y. Lau, *et al.*, "Superluminescent damping of relaxation resonance in the modulation response of GaAs lasers," Appl. Phys. Lett. 43, pp. 329–331 (1983).]

these equations can be solved easily. Since our main interest is in the modulation response, we present here the ratio of the complex amplitude $\delta\Upsilon$ and the amplitude of the current variation I_1:

$$\frac{\delta\Upsilon}{I_1} = -\frac{c\gamma}{2e\sqrt{\kappa}n_{\text{th}}\,\mathcal{L}_x\mathcal{L}_yd}\frac{\Upsilon_0}{\Omega^2 - i\Omega[1/\tau + \Upsilon_0\gamma/(\mathcal{L}_ydn_{\text{th}})] - \frac{c\Upsilon_0\gamma\alpha_{ls}}{\sqrt{\kappa}n_{\text{th}}\mathcal{L}_yd}}.$$

$$(12.37)$$

A typical response,

$$\frac{\Upsilon_1}{I_1} \equiv \frac{|\delta\Upsilon|}{I_1},$$

is shown in Fig. 12.21 for different dc currents. The curves labeled 1 to 5 correspond to an increase in the bias current. The response curve is flat at low frequencies as in the case of light-emitting diodes described by approximation (12.25), but unlike that case it has a peak at the frequency

$$\Omega_m = \sqrt{\frac{\gamma\Upsilon_0\alpha_{ls}c}{n_{\text{th}}\mathcal{L}_yd\sqrt{\kappa}} - \frac{1}{2}\left(\frac{1}{\tau} + \frac{\gamma\Upsilon_0}{n_{\text{th}}\mathcal{L}_yd}\right)^2}.$$

$$(12.38)$$

For a typical semiconductor laser, only the first term in Ω_m is important so that, to a good degree of accuracy,

$$\Omega_m = \sqrt{\frac{\gamma\Upsilon_0\alpha_{ls}c}{n_{\text{th}}\mathcal{L}_yd\sqrt{\kappa}}}.$$

$$(12.39)$$

If Ω exceeds the value of Eq. (12.39) the response decreases quickly. We can introduce the modulation bandwidth as

$$f = \frac{1}{2\pi}\Omega_m = D\sqrt{P_{\text{out}}},\qquad(12.40)$$

where D is some constant. [Compare with Eq. (12.26) for light-emitting diodes.] Equations (12.39) and (12.40) show that in order to increase the modulation response, one needs to increase the light amplification, to decrease the optical losses, and to operate at as high a photon flux as possible. For advanced laser structures, modulation bandwidths as high as tens of gigahertz are achieved.

In closing this section, it is worth stressing that, like light-emitting diodes, semiconductor lasers operate over a very wide spectral region: from the middle-infrared to the near-ultraviolet region. In Fig. 12.22, data on spectral regions

Figure 12.22. Spectral regions of semiconductor lasers with different compounds. The lasers emitting in the region $\lambda > 3$ μm usually operate at low temperatures. [After P. L. Derry, *et al.*, "Ultralow-threshold graded-index separate-confinement single quantum well buried heterostructure (Al, Ga)As lasers with high reflectivity coatings," Appl. Phys. Lett. 50, pp. 1773–1775 (1987).]

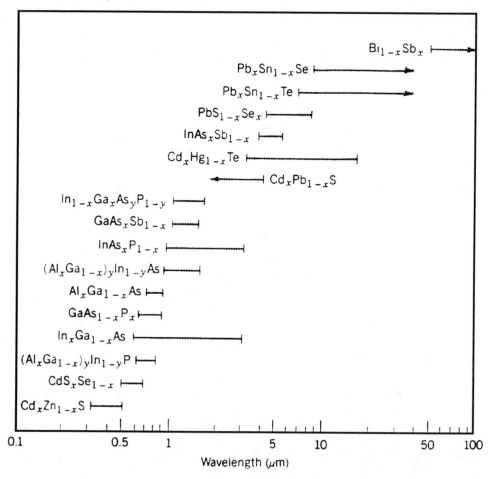

of different semiconductor lasers are presented. Most of these laser systems operate with injection current pumping. Some of them require optical excitation. Semiconductor lasers operating at $\lambda > 3 \ \mu$m usually require cooling below room temperature.

12.3.5 Quantum-Well Lasers

In Subsection 12.2.5 we analyzed light amplification in heterostructures with quantum wells. Now, using the results of the analysis of the resonator feedback and optical losses given in Subsection 12.3.3, we can study lasers based on light amplification in quantum wells. We start with a consideration of a single quantum well and rewrite Eq. (12.18) in terms of the optical-confinement factor of this well Γ_0 and the coefficient α_0 calculated for a single well:

$$\alpha(\omega) = \Gamma_0 \alpha_0(\omega).$$

The threshold condition for laser generation as given by relation (12.28) now takes the form

$$\Gamma_0 \alpha_0(\omega) > \alpha_s + \frac{1}{2\mathcal{L}_x} \ln \frac{1}{r_1 r_2}. \tag{12.41}$$

The intracavity optical losses α_s include light scattering out of the waveguide α_{sc}, losses in the passive region of the guide α_{pp}, as well as those of the active region in the quantum well α_{qw} . Thus one can write these losses as $\alpha_s = \alpha_{sc} + \Gamma_0 \alpha_{qw} + (1 - \Gamma_0)\alpha_{pp}$. Note that one of the main mechanisms of loss in the passive region of the guide is absorption by free carriers. This mechanism becomes important at high injection levels when a large number of carriers move over the barrier layers in the guiding region and contribute to α_{pp}.

Since the electric current governs the injection and pumping processes, let us rewrite the criterion of inequality (12.41) as a condition on the current. Following Eq. (12.11), we can represent $\alpha_0(\omega)$ as

$$\alpha_0 = \gamma \left(\frac{j}{j_{th}} - 1 \right), \tag{12.42}$$

where j_{th} is the threshold current density for the establishment of population inversion in a quantum well. Then the criterion of inequality (12.41) can be rewritten as

$$j > j_{th} \left(1 + \frac{\alpha_{sc}}{\gamma \Gamma_0} + \frac{1}{2\gamma \Gamma_0 \mathcal{L}_x} \ln \frac{1}{r_1 r_2} \right). \tag{12.43}$$

We see the critical role of the optical-confinement factor on the injection current needed for laser action.

To estimate the parameters j_{th} and γ we can use, for example, the data of Fig. 12.11(b) for the InGaAsP quantum-well laser. According to these data we can set $j_{th} \approx 20$ A/cm^2 and $\gamma = 10^3$ cm^{-1}. We can estimate the optical-confinement factor by using the results of Fig. 12.10: $\Gamma_0 = 0.03$. A typical value of α_{sc} is

3 cm^{-1}. Thus the criterion is

$$j \, (\text{A/cm}^2) > 20 + 2 + \frac{1}{3\mathcal{L}_x} \ln \frac{1}{r_1 r_2}.$$

Let the length of the optical cavity, $\mathcal{L}_x$, be $200 \, \mu\text{m}$; for cleaved surfaces we can estimate the reflection coefficients as $r_1 = r_2 = 0.31$. Finally, we obtain

$$j \, (\text{A/cm}^2) > 20 + 2 + 38 = 60,$$

where we intentionally present all three contributions in inequality (12.43) in explicit form. Since the first contribution corresponds to the threshold for population inversion, we can conclude that for this typical example the lasing threshold current is as many as three times higher than that of population inversion. To lower the laser threshold we can increase the mirror reflectivity or increase the cavity length $\mathcal{L}_x$.

The laser output power can be estimated with Eq. (12.36), for which the external efficiency η_{ext} typically can be taken to be ~ 0.5 for the case of quantum-well lasers. In Fig. 12.23, the output power of a single-quantum-well AlGaAs

Figure 12.23. The laser output versus the pumping current for a GaAs single-quantum-well laser with an optical-confinement structure. Curve (a) is for a high facet reflection ($r_1 = r_2 = 0.8$) due to the coated end facets. Curve (b) is for uncoated end facets ($r_1 = r_2 = 0.3$). [After B. E. Saleh and M. C. Teich, *Fundamentals of Photonics* (Wiley, New York, 1991).]

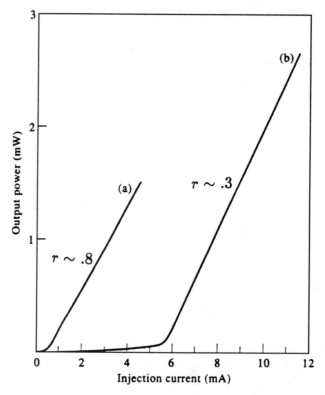

laser is shown as a function of the current. The GaAs quantum well has a width of 100 Å. The well is embedded in a symmetric graded-refractive-index waveguide structure with a thickness of ~0.4 μm and length $\mathcal{L}_x = 120$ μm. The results are presented for two different reflection coefficients: $r_1 = r_2 = 0.31$ (uncoated end facets) and $r_1 = r_2 = 0.8$ (highly reflective coated end facets). Each dependence shows a sharp bend corresponding to the laser threshold. Beyond the threshold, the output power increases linearly in accordance with Eq. (12.36). A typical power that can be obtained for a single-quantum-well laser is approximately a few milliwatts.

Because of the inherently small transverse size (the width) of quantum wells, the optical-confinement factor Γ_0 is always small and can be increased only when more quantum wells are added to the guiding region. If ν_{QW} is the number of the quantum wells in the region, for estimates of the threshold current one can use inequalities (12.41) and (12.43) with

$$\Gamma = \Gamma_0 \nu_{QW}. \tag{12.44}$$

It is easy to see that for values of ν_{QW} of 10 to 30, the laser threshold current can be decreased practically to the value of the current for population inversion j_{th}, which means that the confinement factor is of the order of 1.

Modulation bandwidth: Important applications of light-emitting diodes and conventional laser diodes in communication systems, optical signal processing, etc., require high-speed modulation of the light output. Hence the modulation bandwidth is a crucial parameter of these devices. Considering the modulation speed, one must take into account two additional relaxation stages in the process of quantum-well pumping. In Fig. 12.24, these two processes are shown schematically: carrier diffusion (or drift) through the region of the optical confinement and the subsequent capture processes into the wells. In the optical-confinement region, charge carriers are in extended states. In a quantum-well, they are in confined states. The transition between these two types of states can be characterized by a time that usually exceeds the intraband-relaxation times. One can introduce the surface concentrations of carriers in the extended states, n, and the surface concentrations of carriers in the confined states, n_{QW}. Instead of a single equation for the carriers, as in Eq. (12.32), we should now write two

Figure 12.24. Schematic illustration of the main processes that determine the rate of population of the quantum-well states.

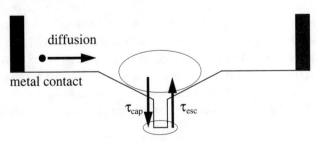

equations:

$$\frac{dn}{dt} = \mathcal{R} - \frac{n}{\tau} - \left(\frac{n}{\tau_{\text{cap}}} - \frac{n_{\text{QW}}}{\tau_{\text{esc}}}\right), \tag{12.45}$$

$$\frac{dn_{\text{QW}}}{dt} = \left(\frac{n}{\tau_{\text{cap}}} - \frac{n_{\text{QW}}}{\tau_{\text{esc}}}\right) - \frac{n_{\text{QW}}}{\tau^*} - \alpha\frac{\Upsilon}{\mathcal{L}_y d}, \tag{12.46}$$

where $\mathcal{R}$ is the injection rate density, τ and τ^* are the lifetimes due to interband recombination in extended and confined states, respectively, and τ_{cap} and τ_{esc} are the effective times of the capture and the escape processes. The last term in Eq. (12.46) is the stimulated-emission rate; see Eq. (12.32). τ_{cap} includes processes of the delivery of carriers to the quantum-well layer and subsequent quantum transitions of carriers from extended to confined states. If the delivery process is the carrier diffusion, one can estimate these parameters as

$$\tau_{\text{cap}} = \frac{d^2}{2D} + \tau_{\text{cap}}^q\frac{d}{L}, \tag{12.47}$$

where D is the diffusion coefficient and τ_{cap}^q is the average lifetime of the carriers in the region of the quantum well due to quantum transitions into the well. The first term is equal to the time of carrier diffusion over a distance d. The intrinsic quantum-capture time, τ_{cap}^q, is scaled up by a factor d/L. Recall that L is the width of the well. For typical values of $\tau_{\text{cap}}^q \approx 0.5$ ps and $d/L = 10$, the maximum modulation bandwidth due to quantum capture is $1/(2\pi\tau_{\text{cap}} \approx 30$ GHz). The diffusion contribution in Eq. (12.47) dominates for $d \geq 0.2$ μm; it increases τ_{cap} and limits the modulation width. Structures with graded optical-confinement regions, as in Fig. 12.10, have built-in forces that can produce driftlike carrier motion and decrease considerably the delivery time. (Such a drift effect in an built-up effective field has been discussed in detail in Sections 9.4 and 9.5.) In this case, the only intrinsic quantum capture restricts τ_{cap}. Therefore we can conclude that with proper design of the narrow optical-confinement region, quantum-well lasers can be modulated by an injection current with frequencies up to 30 GHz and higher.

Strained-layer quantum-well lasers: So far we considered examples of AlGaAs and InGaAsP quantum-well lasers that are based on lattice-matched heterostructures. In some cases, lasers based on lattice-mismatched heterostructures demonstrate superior properties. These heterostructures were analyzed in Chapter 4. Lasers using lattice-mismatched heterostructures can operate in spectral regions different from those of lasers based on lattice-matched structures, and they can provide lower threshold currents or higher amplification. Let us consider briefly a strained-layer GaAs/InGaAs/GaAs laser. The lattice constants of both materials are significantly different: $a = 6.0585$ Å (InAs) and $a = 5.6535$ Å (GaAs). If a thin layer of InGaAs is embedded between two thick GaAs layers, the InGaAs layer experiences a considerable in-plane compression of the layer. This compressive strain changes the band structure of the

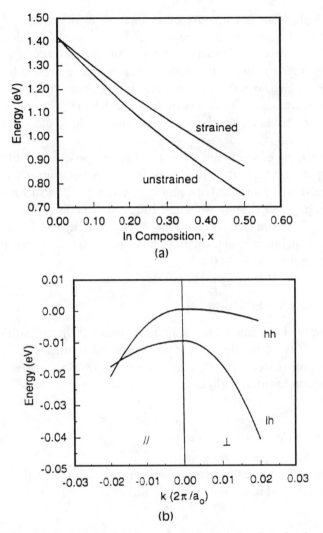

Figure 12.25. The bandgap and energy dispersion in the hole bands under the strain: (a) the bandgaps for strained and unstrained $In_xGa_{1-x}As$ as functions of the indium component, (b) the heavy-hole and the light-hole valence-band structure resulting from a biaxial compression for $In_xGa_{1-x}As$. [After P. S. Zory, Jr., *Quantum Well Lasers* (Academic, Boston, 1993).]

layer in three ways: (1) it increases the bandgap, as shown in Fig. 12.25(a); (2) it removes the degeneracy at $\vec{k}=0$ between the light and the heavy holes, as illustrated in Fig. 12.25(b); (3) it makes the valence band strongly anisotropic. In Fig. 12.25(b), the negative and the positive wave vectors are presented for different directions. As can be seen from Fig. 12.25(b), in the directions parallel to the layer plane, the topmost band has a light effective mass while in the direction perpendicular to the plane, the same band has a heavy effective mass. These changes in the energy spectrum of the valence band improve the performance of strained-layer lasers. First, the peak emission of the laser is shifted to higher energies because of an increase in the energy of the lowest hole subband of the strained quantum-well layer. Second, the decrease in the in-plane effective mass leads to a reduction of the threshold current. This result follows from the criterion of the population inversion expressed by Eq. (12.4). Since in the strained layer the hole effective mass decreases, the quasi-Fermi level of the holes increases at the same carrier concentration. Thus the difference in the quasi-Fermi levels increases under strain and the threshold injection current decreases.

As an example of the family of strained-layer lasers, we mention InGaAsP devices operating within the broad wavelength region from 0.9 to 1.55 μm. One particular laser fabricated of a multiple-quantum-well structure with 20-Å-thick $In_{0.78}Ga_{0.22}As$ quantum wells and 200-Å barriers operates at $\lambda = 1.55$ μm with a submilliampere threshold current. Another example is a GaInP/InGaAlP strained-layer quantum-well laser that emits light with $\lambda = 634$ nm and achieves an output power in excess of 0.5 W.

12.3.6 Surface-Emitting Lasers

Thus far, we have studied the amplification of light propagating along the quantum-well layers. In Chapter 10, we mentioned another possible design in which light propagates perpendicularly to the layers in a so-called vertical geometry. The amplification of light passing through a quantum-well layer can be defined as

$$\frac{\mathcal{I}_{out} - \mathcal{I}_{in}}{\mathcal{I}_{in}} \equiv \beta.$$

Here $\mathcal{I}_{in}$ and $\mathcal{I}_{out}$ are input and output light intensities, respectively. The quantity β can be expressed through $\alpha_0(\omega)$ of Eq. (12.18) as

$$\beta(\omega) = \alpha_0(\omega)L,$$

where L is the quantum-well width. It is easy to see that β is typically very small. For example, if $\alpha_0 = 100$ cm^{-1} and $L = 100$ Å, we get $\beta = 10^{-4}$. To obtain laser oscillations in a vertical-geometry structure, one should use a multiple-quantum-well structure and provide near-perfect mirrors with extremely high reflection. Figure 12.26(a) shows the schematics of a surface-emitting laser. The laser design includes an active region providing for high light gain, dielectric multilayers, metallic contacts, and implanted regions that form the light output. Layered dielectric mirrors give high reflection while the active region contains a multiple-quantum-well structure. The lateral sizes of this laser can be reduced to the 1–10-μm range. A decrease in the surface area of the diode leads to a considerable decrease in the magnitude of the threshold current. In the case of the quantum-well structures just considered, we can assume a characteristic current density of $\sim$100 A/cm^2; accordingly, if the lateral sizes are each 5 μm,

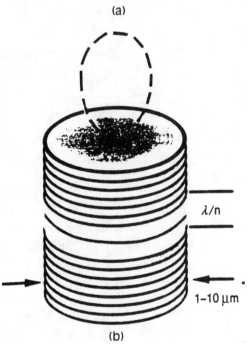

light output

| | metallization | | active region |
| | multilayer | | implanted region |

(a)

Figure 12.26. Surface-emitting lasers: (a) schematic diagram of a surface-emitting laser, (b) a surface-emitting microlaser. Dielectric mirrors, active region (with the width of λ/n), and output pattern (dashed curve) of the lowest mode are indicated.

λ/n

1–10 μm

(b)

the pumping surface area is 2.5×10^{-7} cm^2 and the threshold current equals 25 μA.

Surface-emitting quantum-well lasers offer a new advantage of high packing density. Nowadays, technology allows one to fabricate an array of $\sim 10^6$ surface-emitting electrically pumped lasers. The structure of such a single microlaser is shown in Fig. 12.26(b). The cylindrical boundary is formed by an etching procedure that results in lateral dimensions of less than 10 μm. The dominant mode pattern emitted from this microlaser is shown schematically. Microlasers can operate at room temperature and have threshold currents of the order of 0.1 mA.

12.3.7 Quantum-Wire Lasers

As we have learned from Chapter 3, additional carrier confinement leads to modifications of the energy spectra and the density of states. These modifications may significantly improve laser performance as the structures under consideration change from quantum-well structures to quantum-wire and quantum-dot structures. Let us recall the density of states per unit volume for electron systems with different dimensionalities:

$$\varrho_{3D} = \frac{(2m^*/\hbar^2)^{3/2}}{2\pi^2}\sqrt{E},$$

$$\varrho_{2D} = \frac{m^*}{\pi\hbar^2 L_z}\sum_i \Theta(E - \epsilon_i),$$

$$\varrho_{1D} = \frac{(2m^*)^{1/2}}{\pi\hbar L_y L_z}\sum_{i,j} \frac{1}{\sqrt{E - \epsilon_{i,j}}}\,\Theta(E - \epsilon_{i,j}),$$

$$\varrho_{0D} = \frac{2}{L_x L_y L_z}\sum_{i,j,k} \delta(E - \epsilon_{i,j,k}),$$

where L_x, L_y, and L_z are the sizes of the structures in the confined dimensions, E is the kinetic energy, and ϵ_i, $\epsilon_{i,j}$, and $\epsilon_{i,j,k}$ are the energy levels in quantum wells, quantum wires, and quantum dots, respectively. A comparison of the densities of states for different dimensionalities is presented in Fig. 12.27. The density of states acquires the sharper features as the carrier dimensionality decreases. Comparing these results, one can obtain several general conclusions related to the transition from a quantum well (confinement in one direction) to a quantum wire and dot (multidimensional confinement): a sharper density of states leads to a narrower spectral linewidth of emission and to an increase in the quasi-Fermi levels E_{Fn} and E_{Fp}. According to the criterion for population inversion as expressed by inequality (12.4), an increase of E_F leads to a reduction in the threshold carrier concentration and, hence, to a lowering of the threshold current. Other advantages of a multidimensional confinement are related to the small sizes of these structures, the possibility of densely packed laser arrays, the potential for monolithic integration with low-power electronics, etc.

Let us consider briefly existing results for quantum-wire lasers. For the parallel laser configuration, in which the light wave propagates along the wires, we can use Eq. (12.18), in which the optical-confinement factor now is the ratio of the area of the carrier's confinement and that of the optical confinement:

$$\Gamma \approx \frac{S_{QWR}}{S_{mode}}.$$

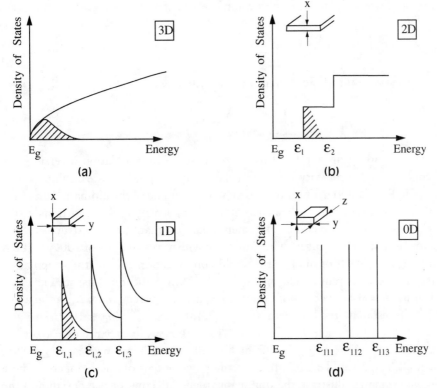

Figure 12.27. Schematics of the energy dependences of the density of states and occupations for systems with different dimensionalities. The hatched areas indicate occupied states: (a) a three-dimensional system, (b) a quantum well, (c) a quantum wire, (d) a quantum dot.

Here S_{QWR} and S_{mode} are the areal cross sections of the quantum wire and the waveguide region, respectively. We assume a linear dependence of the amplification coefficient on the carrier concentration n:

$$\alpha_0 = \gamma^{1D}\left(\frac{n}{n_{\mathrm{th}}} - 1\right),$$

where γ^{1D} is to be calculated for phototransitions between one-dimensional subbands of the electrons and holes. To pump the quantum wire, one needs to apply the following electric current per unit length of the wire,

$$I = \frac{e S_{\mathrm{QWR}}}{\eta_i \tau} n,$$

where η_i is the internal efficiency and τ is the recombination time. Hence the threshold transparency current is

$$I_{\mathrm{th}} = \frac{e S_{\mathrm{QWR}}}{\eta_i \tau} n_{\mathrm{th}}.$$

One can rewrite the local amplification coefficient as

$$\alpha_0 = \gamma^{1D}\left(\frac{I}{I_{th}} - 1\right).$$

Now the lasing threshold current can be written as

$$I > I_{th} + \frac{I_{th}}{\gamma^{1D}\Gamma}\left(\alpha_s + \frac{1}{2\mathcal{L}_x}\ln\frac{1}{r_1 r_2}\right) \equiv I_{th} + I_{cav}, \tag{12.48}$$

where the first term is related to the appearance of population inversion; the second depends on cavity losses and the cavity design. Let us examine various terms in Eq. (12.48) in order to gain an understanding of the ultimate limit of the laser threshold current. The transparency current density I_{th} is proportional to the wire cross section and can be extremely small for quantum-wire structures. For example, assume a quantum wire fabricated with a cross section of 100 Å × 100 Å and a length of 100 μm. The threshold concentration for creating population inversion (transparent condition) is $n_{th} \approx 10^{18}$ cm^{-3} and $\tau = 3$ ns; see Fig. 12.4. Then for $\eta_i = 0.5$, we obtain a total transparency current, $I_{th} \times \mathcal{L}_x$, as small as 1 μA. To estimate the last term in Eq. (12.48) we set $\gamma^{1D}/n_{th} \approx 10^{-15}$ cm^2 and $S_{mode} = 1 \times 1$ μm^2; then $I_{th}/(\gamma^{1D}\Gamma) = 1.1$ mA. If $r_1 = r_2 = 0.9$ and $\alpha_s = 10$ cm^{-1}, we find $I_{cav} \times \mathcal{L}_x = 110$ μA. For highly reflective mirrors with $r_1 = r_2 \approx 0.99$, we find the threshold current to be in the range of tens of microamperes. These simple estimates illustrate the dominant trends affecting device design and the achievement of the lowest current thresholds.

In Fig. 12.28, a calculation of the local amplification coefficient α_0 is presented as a function of the wavelength for the GaInAs/GaInAsP rectangular quantum wire with the dimensions 120 Å × 200 Å at different carrier densities n. The calculations take into account the broadening of the subbands due to the intrasubband relaxation with characteristic time $\tau_{in} = 0.1$ ps; see Subsection 10.5.4.

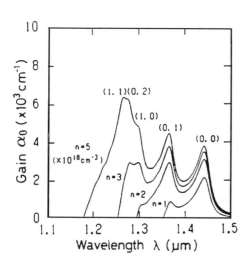

Figure 12.28. The amplification coefficient as a function of the light wavelength for the GaInAs/GaInAsP quantum-wire laser with a 120 Å × 200 Å cross section. Results are calculated for the set of carrier concentrations n and $T = 77$ K. Indices under the curves indicate contributions of different one-dimensional subbands (n_1, n_2). [After P. S. Zory, Jr., *Quantum Well Lasers* (Academic, Boston, 1993).]

The carrier density changes from 1×10^{18} cm^{-3} to 5×10^{18} cm^{-3}. The complex structure of $\alpha_0(\lambda)$ is due to phototransitions from the upper subbands. According to Subsection 10.5.2, phototransitions between different electron and hole subbands contribute to the spectra. Each subband is characterized by two quantum numbers $\{n_1, n_2\}$. The contributions of electron and hole subbands with the same pair $\{n_1, n_2\}$ are the most important; they are marked in Fig. 12.28. This figure shows that the local amplification coefficient reaches a value of the order of 10^3 cm^{-1}. The dependence of these results on the concentration supports the previously assumed estimate for γ/n_{th}.

There are different technological approaches for the fabrication of quantum-wire lasers. The most direct approach is fabrication by the etching and regrowth technique. In Fig. 12.29, such a device is presented schematically. The structure is fabricated by growing a conventional quantum-well structure first. Next, a wire pattern is formed by etching the quantum-well active region with an electron-beam-written grating mask. Then the upper part of the structure is formed in the second growth step. The wires are 100 Å × 300 Å. The distance between the wires is 400 Å. The lower part of the structure is n-doped InP, the upper part is

Figure 12.29. The schematic of the GaInAs/GaInAsP quantum-wire laser fabricated by etching and regrowth. The lower part shows the optical-confinement GaInAsP region with embedded quantum wires. [After P. S. Zory, Jr., *Quantum Well Lasers* (Academic, Boston, 1993).]

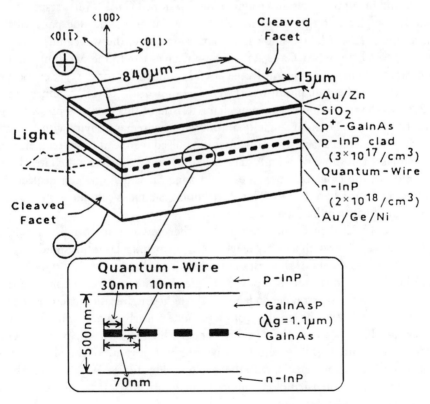

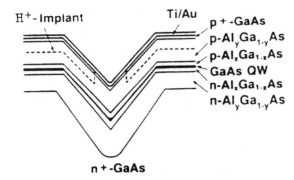

Figure 12.30. The cross section of an AlGaAs/GaAs single-quantum-well laser grown on a V-grooved substrate. [After E. Kapon, *et al.*, "Single quantum wire semiconductor lasers," Appl. Phys. Lett. 55, pp. 2715–2717 (1989).]

p-doped InP. They are accessed by electrical contacts. The optical-confinement region is made of a GaInAsP layer with a width of $\sim$5000 Å. The cleaved facets serve as light reflectors. Other details of the structure are given in the figure. A similar procedure is used for the formation of an array of quantum dots with lateral dimensions of less than 1000 Å.

Another method uses growth on a patterned nonplanar substrate. Cross sections of a single AlGaAs/GaAs quantum-wire laser grown on a V-grooved GaAs substrate are shown in Fig. 12.30. The figure shows the sequence of layers and their doping. The quantum-well layer is situated in an undoped *i* region. The lower and the upper parts are heavily *n*- and *p*-doped GaAs layers, respectively. The H$^+$ implantation provides for electric insulation of the side parts of the structure. It is worth stressing that, for this structure, the waveguide layer of $Al_xGa_{1-x}As$ between $Al_yGa_{1-y}As$ with $x < y$, gives rise to the two-dimensional optical confinement. Analogously, the lateral confinement for the carriers occurs because of the bent quantum-well layer. The dimension of this confinement is $\sim$200 Å. The output-power-current characteristic of this device is presented in Fig. 12.31. This particular single-quantum-wire device has a cavity with a length of 350 μm, uncoated facets, and operates in a pulsed regime at room temperature with a laser threshold of $\sim$4 mA. The inset in Fig. 12.31 shows the lateral distribution of the emission in the near-field zone (in which one can neglect the diffraction of the light beam). This distribution is almost circular with roughly a 1-μm full width at half-maximum.

Figure 12.32 depicts the evolution of the emission spectra for a single-quantum-wire laser with a cavity length of 270 μm and a laser threshold current of 4.3 mA. The design of this laser corresponds to that of Fig. 12.30. Near the threshold many longitudinal modes oscillate. It is instructive to compare this behavior with the spectral dependence of the light-amplification coefficient presented in Fig. 12.28. The envelope of these modes clearly demonstrates that two different subbands are involved in the lasing. At high levels of the pumping current, the oscillations shift to shorter wavelengths as a result of the filling of the upper subbands in agreement with the results of Fig. 12.28. With an increase in the current, the intensity of stimulated emission increases, while the number of oscillating modes decreases.

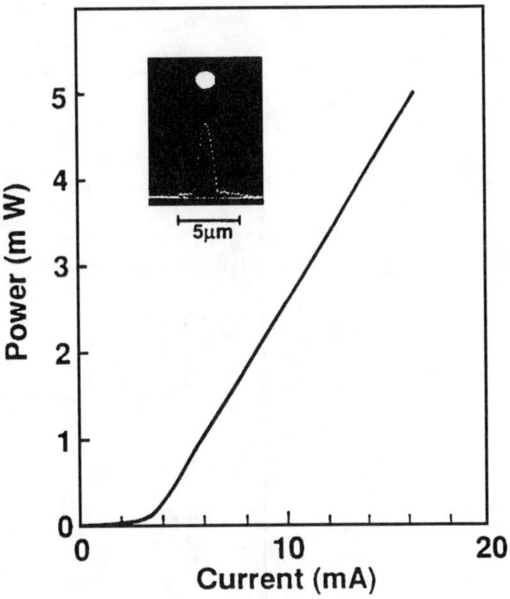

Figure 12.31. The light output versus the pumping current of the single-quantum-wire laser with the structure presented in Fig. 12.30. The inset shows the lateral distribution of the output light in the near-field zone. [After E. Kapon, *et al.*, "Single quantum wire semiconductor lasers," Appl. Phys. Lett. 55, pp. 2715–2717 (1989).]

In concluding this section, we summarize the previous discussion in Fig. 12.33, which illustrates decreasing threshold currents for semiconductor lasers with different types of heterostructures and designs. One can see that the transition from homostructure semiconductor lasers to quantum heterostructure lasers reduces the threshold current by 4 orders of magnitude. Further progress in the field of quantum-wire lasers will require considerable improvement in the technology of these structures. Small sizes, better interfaces, improved uniformity of the quantum wires, cavities with small optical losses, and improved fabrication of arrays of wires are all necessary for achieving high performance from this type of low-threshold lasers.

12.3.8 Blue Quantum-Well Lasers

From Fig. 12.22, one can see the lasers based on III–V compounds cover the wavelength range approximately from 0.6 to 3 μm. In the interesting range of shorter wavelengths, this family of injection lasers is still not developed. It is necessary, though, to note that a III–V compound light-emitting diode with a blue output is available now and has excellent characteristics.

There is another family of injection lasers based on II–VI materials that operate in the short-wavelength range. For this family, all necessary elements – quantum wells and a waveguide layer embedded in a p–n junction – have been developed. Figure 12.34 depicts a particular example of such a ZnCdSe/ZnSSe/ZnMgSSe structure. The quantum-well layer is made of narrow-bandgap $Zn_{1-x}Cd_xSe$ material with a thickness of $\sim$75 Å. The optical-confinement layer is ZnSSe with a thickness of $\sim$1000 Å. The left and the right parts of the structure are made of $Zn_{1-z}Mg_zS_ySe_{1-y}$ and are p and n doped, respectively. The

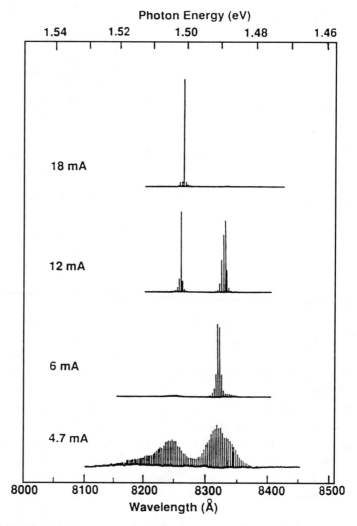

Figure 12.32. The emission spectra of a single-quantum-wire laser with the structure presented in Fig. 12.30 for different pumping currents. [After E. Kapon, *et al.*, "Single quantum wire semiconductor lasers," Appl. Phys. Lett. 55, pp. 2715–2717 (1989).]

quantum-well layer is pseudomorphic with a lattice mismatch greater than 1%. This produces strain and a deformation of the valence bands. The anisotropy of the lowest hole subband is shown in the inset. This effect has been discussed in Subsection 12.3.5, see also Fig. 12.25(b). The depths of the quantum wells are 180 meV for the electrons and 70 meV for the holes.

Unlike the III–V compounds, in II–VI-based quantum wells, the exciton binding energy is large. Such binding energies were discussed in Subsection 3.7.2. For ZnCdSe and ZnSe quantum wells, this energy is ~40 meV and exceeds $k_B T$ – even at the room temperature. These factors make electron–hole correlations and exciton effects even more important than those of the III–V materials. The

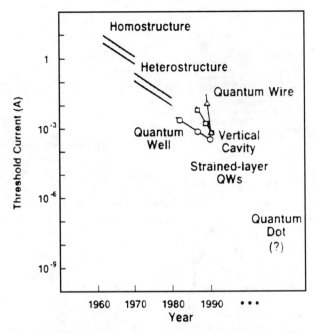

Figure 12.33. The threshold currents for semiconductor lasers with different designs. The evolution of the threshold current is plotted as a function of the year. Circles, squares, and triangles correspond to quantum-well lasers, surface-emitting lasers, and quantum-wire lasers, respectively. [After P. S. Zory, Jr., *Quantum Well Lasers* (Academic, Boston, 1993).]

exciton is stable against thermal dissociation, screening, and other many-body effects. In accordance with the general considerations of Subsection 10.5.4, exciton effects enhance the strength of the radiative transitions and, as a result, they dominate in the optical spectra of these structures. Another important consequence of excitonic effects is that radiative recombination through two-dimensional excitons dominates over possible nonradiative paths, including the recombination of free electron–hole pairs. These dominant radiative transitions result in increased internal emission efficiency for the devices fabricated of these materials. The stability of the excitons against screening facilitates the realization of high exciton concentrations. All these factors result in a peak amplification coefficient below the edge of the band-to-band transitions. This is in strong contrast to the case of III–V compound lasers, in which, under the condition of population inversion, the

Figure 12.34. Schematic of a ZnCdSe/ZnSSe/ZnMgSSe heterostructure for a blue laser. The energy diagram of the device, the 75-Å ZnCdSe quantum well, and the 1000-Å ZnSSe optical-confinement region are indicated. The inset illustrates the anisotropy of the lowest hole band in the ZnCdSe strained layer. [After A. V. Nurmikko and R. L. Gunshor, "II–VI heterostructures: from quantum well physics to blue-green lasers," in *Proceedings of the International Conference on Semiconductor Physics*, pp. 27–34 (World Scientific, Singapore, 1995).]

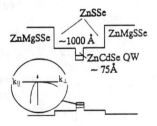

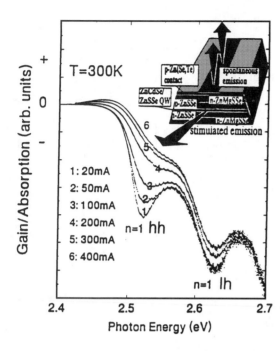

Figure 12.35. The amplification (absorption) coefficient as a function of the photon energy for the heterostructure presented in Fig. 12.34. The curves correspond to different pumping currents. [After J. Ding, *et al.*, "Role of Coulomb-correlated electron-hole pairs in ZnSe-based quantum-well diode lasers," Phys. Rev. B **50**, pp. 5787–5790 (1994).]

exciton peaks completely disappear. In Fig. 12.35, the amplification–absorption spectra are shown for the structure under consideration at different injection currents and at room temperature. The transparency current – corresponding to the threshold for population inversion – is equal to approximately $300\,\mu$A. At higher currents, positive amplification occurs at energies below the $n = 1$ heavy-hole resonance, i.e., in the exciton region for $\hbar\omega < 2.5$ eV. For ZnCdSe/ZnMgSSe structures, an amplification coefficient of the order of 1500 cm^{-1} is reached at room temperature. The inset of the figure shows a sketch of this laser and the outputs of the stimulated and spontaneous emissions. In Fig. 12.36, the current-voltage and light power-current characteristics are shown for such a blue laser diode. The light power-current characteristics clearly demonstrate the threshold behavior of the stimulated emission. The light output is roughly several milliwatts for pumping currents above 10 mA.

The major problems associated with this class of quantum-well lasers are device degradation and short device lifetimes. The electric power dissipated in present II–VI diodes is still high and causes rapid degradation as a result of the generation of intrinsic defects. Much improvement must be realized in this field in order to have wide practical application of these devices.

12.3.9 Unipolar Intersubband Quantum-Cascade Laser

So far in this chapter, we considered light amplification and laser action based on interband phototransitions involving both electrons and holes. Another type of phototransitions – intraband absorption of infrared light – has been studied in

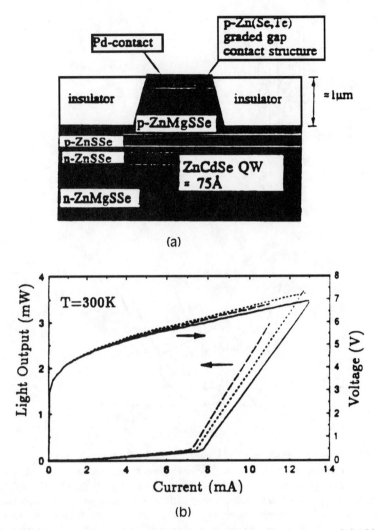

Figure 12.36. The current-voltage and light output-current characteristics of the blue laser presented in Figs. 12.34 and 12.35.

Section 10.6. This process involves only one type of carriers and is allowed in an ideal crystal system with heterojunctions. In accordance with the general properties on light–matter interactions, there should also be an emission of infrared light from intraband phototransitions. Since these phototransitions are drastically different from intersubband transitions, laser action associated with these intraband transitions should differ in a fundamental way from that studied for the laser schemes considered previously in this text. First, intraband transition lasers should use only one type of carrier, i.e., it is a unipolar device. Second, it should be based on electron transitions between confined states arising from the quantization in semiconductor heterostructures. To create a population inversion between two confined states, one needs to provide for (1) electron injection into

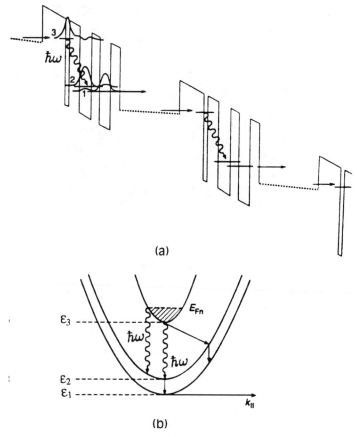

(a)

(b)

Figure 12.37. (a) Two periods of the 25-stage staircase coupled-well region of a quantum-cascade laser under operation condition. The laser phototransitions are indicated by arrows. They occur between levels (subbands) 3 and 2 with the photon energy of 295 meV. Level 2 depopulates through level 1 and subsequent tunneling. The energy separation between levels 2 and 1 is ~30 meV. (b) Energy dispersion for subbands 1, 2, and 3, phototransitions, and interband-scattering processes (straight lines). [After F. Capasso, *et al.*, "Quantum cascade laser," Science **264**, pp. 553–556 (1994).]

a higher lasing state and (2) depletion of a lower lasing state. For this purpose, a vertical scheme of the electron transport has been proposed. This scheme is illustrated in Fig. 12.37(a). The proposed heterostructure is a superlattice with a complex design for each period. Each of the periods consists of four AlInAs barriers forming three GaInAs quantum wells and a digitally graded AlInGaAs region that is doped. Under zero-bias conditions, the overall band diagram appears like a sawtooth structure. Under an applied electric field, the band diagram takes on a staircase structure, as shown in Fig. 12.37(a). The barriers form three coupled quantum wells with three quasi-bound levels. These three levels are marked in the figure by 1, 2, and 3. Each of the confined states originates from one of the wells.

The structure is chosen so that there is a considerable overlap between the wave functions of upper state 3 and intermediate state 2 as a result of tunneling processes. The same is valid for wave functions of states 2 and 1. Under a voltage bias, the potential in the doped regions is almost flat, as shown in Fig. 12.37(a). The electrons are injected from the doped regions through the barrier in confined state 3 of the first quantum well. From this state, they relax primarily to state 2. There are two processes of relaxation: phonon emission and photon emission. In Fig. 12.37(b), these processes are shown for electrons with different values of the in-plane wave vector $\vec{k}$. The three indicated subbands, $\epsilon_{1,2,3}(\vec{k})$, correspond to the three confined states. The straight arrows represent intersubband phonon relaxation. The third confined state, 1, is selected to provide depletion of state 2 as fast as possible. Thus in this manner we have a three-level scheme in which the upper level is pumped by the direct injection of electrons from the doped region. The second level is depleted because of strong coupling with the lowest level 1. From level 1, electrons escape to the next doped region. Then the processes are repeated in each subsequent period of the superlattice. One can say that the carriers make transitions down through such a cascade structure.

We can assume that intraband-relaxation processes are faster than the interband processes. Then we can introduce the distribution functions for the three subbands: $\mathcal{F}_3(\vec{k})$, $\mathcal{F}_2(\vec{k})$, and $\mathcal{F}_1(\vec{k})$. We assume that quasi-Fermi distributions hold for each of the confined states. In this case, we need to define only the numbers of electrons in these states: n_3, n_2, and n_1. The criterion for population inversion between levels 2 and 3 should be

$$n_3 > n_2.$$

We can write simple balance equations for n_3 and n_2:

$$\frac{dn_3}{dt} = -\frac{1}{e}j - \frac{n_3}{\tau_{32}}, \tag{12.49}$$

$$\frac{dn_2}{dt} = \frac{n_3}{\tau_{32}} - \frac{n_2}{\tau_{21}}, \tag{12.50}$$

where j is the density of the injection current and τ_{32} and τ_{21} are relaxation times between the states 3 and 2 and 2 and 1, respectively. In Eq. (12.50), we neglect the inverse $1 \rightarrow 2$ process since state 1 can be regarded as almost empty as a result of fast electron escape to the doped region. For steady-state conditions, we obtain the concentrations

$$n_3 = -\frac{1}{e}j\tau_{32}, \qquad n_2 = n_3 \frac{\tau_{21}}{\tau_{32}},$$

and population inversion

$$\Delta n \equiv n_3 - n_2 = -\frac{1}{e}j\tau_{32}\left(1 - \frac{\tau_{21}}{\tau_{32}}\right). \tag{12.51}$$

Material	n	Thickness (nm)	Region
GaInAs Sn doped	$n \approx 2.0 \times 10^{20}\,\mathrm{cm^{-3}}$	20.0 nm	Contact layer
GaInAs	1.0×10^{18}	670.0	Contact layer
AlGaInAs Graded	1.0×10^{18}	30.0	Contact layer
AlInAs	5.0×10^{17}	1500.0	Waveguide cladding
AlInAs	1.5×10^{17}	1000.0	Waveguide cladding
AlGaInAs Digitally graded	1.5×10^{17}	18.6	Waveguide core
Active region	undoped	21.1	Waveguide core
GaInAs	1.0×10^{17}	300.0	Waveguide core
AlGaInAs Digitally graded	1.5×10^{17}	14.6	Waveguide core
AlGaInAs Digitally graded	1.5×10^{17}	18.6	Waveguide core (x 25)
Active region	undoped	21.1	Waveguide core (x 25)
GaInAs	1.0×10^{17}	300.0	Waveguide core
AlGaIn Digitally graded	1.5×10^{17}	33.2	Waveguide core
AlInAs	1.5×10^{17}	500.0	Waveguide cladding
Doped n^+ InP substrate			

Figure 12.38. Schematic cross section of the cascade laser structure. The whole structure consists of 500 layers. [After F. Capasso, *et al.*, "Quantum cascade laser," Science **264**, pp. 553–556 (1994).]

Thus, to create a population inversion, one should design the laser so that

$$\tau_{21} < \tau_{32}. \tag{12.52}$$

To fabricate such unipolar laser structures with vertical electron transport, precise and sophisticated semiconductor technology is necessary. In Fig. 12.38, a schematic cross section of the unipolar n-type $\mathrm{Al_{0.48}In_{0.52}As/Ga_{0.47}In_{0.53}As}$ laser structure is shown in detail. The structure is grown upon an n^+-InP substrate that serves as a contact. Then a waveguide cladding of a 500-Å-thick AlInAs-doped layer follows. The additional cladding layers are designed at the bottom and the top of the structure to increase the refractive-index step between the core and the cladding in order to enhance the optical confinement. The waveguide core is grown upon the waveguide cladding layer. It consists of AlGaInAs, and GaInAs layers that provide the optical confinement and optical amplification. For the structure presented in Fig. 12.38, the active region and the doped layer are repeated alternately 25 times. The active region contains the barriers and the quantum wells discussed previously. The waveguide provides a confinement factor for a single-mode waveguide of about $\Gamma \approx 0.46$ (see Section 10.5). For this structure, the necessary cascade regime is reached at an electric field of $\sim 10^5$ V/cm. The characteristic times in Eqs. (12.49) and (12.50) are $\tau_{32} \approx 4.3$ ps and $\tau_{21} \approx 0.6$ ps, and the time of escape from level 1 is $\tau_{1,esc} < 0.5$ ps, that is, the criterion for population inversion as given by inequality (12.52) is fulfilled. For this structure, it follows that the spontaneous radiative processes are negligible compared with phonon relaxation: $\tau_R \approx 10$ ns and $\tau_{32}/\tau_R \approx 4.3 \times 10^{-4}$. For a particular device with an optical path of ~ 700 μm and mirror reflectivity $r_1 = r_2 = 0.27$, the laser output-current characteristics are shown in Fig. 12.39 for different temperatures. The insets of Fig. 12.39 show the current-voltage characteristics and temperature dependence of the laser threshold current. The

Figure 12.39. The measured optical power P from a single facet of the quantum-cascade laser with the structure presented in Fig. 12.38 and an optical cavity length of 1.2 μm. The results are given for different temperatures. The insets show the dependence of the laser threshold current as a function of temperature and the current-voltage characteristics of the device. [After F. Capasso, *et al.*, "Quantum cascade lasers: a unipolar intersubband semiconductor laser," in *Proceedings of the International Conference on Semiconductor Physics*, pp. 1636–1640 (World Scientific, Singapore, 1995).]

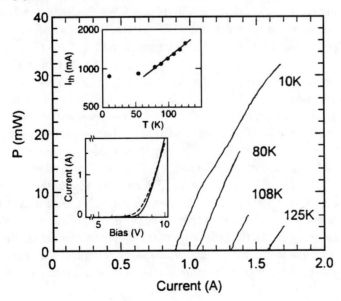

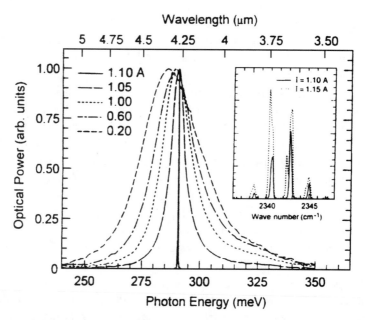

Figure 12.40. The spectra for the emission of the quantum-cascade laser presented in Figs. 12.38 and 12.39 at $T = 80$ K. Results are given for different pumping currents. The inset shows the high-resolution spectra for the cascade laser with a shorter cavity (750 μm) at two pumping currents. [After F. Capasso, *et al.*, "Quantum cascade lasers: a unipolar intersubband semiconductor laser," in *Proceeding of International Conference on Semiconductor Physics*, pp. 1636–1640 (World Scientific, Singapore, 1995).]

laser threshold current can be approximated by $I_{th} = C \exp(T/112)$, where the constant A is ~900 mA and T is measured in degrees Kelvin. Using typical values of the electric current, one can estimate the values of the population inversion as given by Eq. (12.51): $\Delta n \approx 10^{11}$ cm^{-2} per each period of the cascade structure. From the figure, it follows that the output power reaches tens of milliwatts.

The emission energy is in the range 275–310 meV. Spectra of the laser output are presented in Fig. 12.40 for different currents and for $T = 80$ K. For this case, according to Fig. 12.39, the threshold current is ~1.06 A. Figure 12.40 clearly demonstrates a sharp narrowing of the emission spectra above the laser threshold: the spectra reduce to a sharp peak at $I = 1.1$ A $> I_{th}$. The inset shows details of the laser emission spectra for particular values of the pumping currents, which are greater than the threshold current; in particular, the inset shows equally spaced longitudinal modes and a higher-order transverse mode for the higher current of 1.15 A. In accordance with the selection rules for intersubband phototransitions given in Table 10.2, the laser emission is polarized normal to the layers.

In conclusion, the unipolar cascade laser is drastically different from the lasers based on intersubband phototransitions. The properties of interband phototransitions and, consequently, the properties of the unipolar laser are determined to a large degree by quantum confinement; accordingly, this novel laser can be tailored for operation in the spectral region from the middle infrared to submillimeter waves.

12.4 Self-Electro-Optic-Effect Devices

In previous sections of this chapter, we found that laser sources of light based on quantum heterostructures have been developed extensively. They are characterized by low-threshold currents, high efficiencies, narrow spectra, and they can be modulated in the high-frequency region. Contemporary technology facilitates the fabrication of large arrays of such lasers with very small lateral dimensions (of the order of the laser wavelength); such lasers are frequently referred to as microlasers. This generates a practical interest in developing systems for optical signal processing, computing, telecommunications switching, etc. As is clear from Chapter 11, for this purpose, the lasers should be used in combination with optical bistable devices. In Subsection 11.3.6, we introduced two kinds of optical bistable devices: devices that use an intrinsic bistability with an internal feedback and hybrid devices that use an external feedback. Hybrid devices are exploited widely. This section is devoted to the studying of such devices.

A representative schematic of such a bistable hybrid device is shown in Fig. 11.25. One embodiment of such a device that uses a quantum structure – a SEED structure – is presented in Fig. 11.27. As discussed in Subsection 11.3.6, the SEED exploits the quantum-confined Stark effect, which is a strong electroabsorptional mechanism; in particular, changes in the voltage across the quantum-well layers produce significant changes in optical interband absorption. The absorption either increases or decreases, depending on the wavelength of the light. The changes in the optical transmission can be large enough for device applications, that is, they can be of factor of 2 or 3 for a 1-μm-thick multiple-quantum-well stack. Such a controllable absorption allows one to use these devices for processing perpendicular light beams.

12.4.1 Resistor SEED

The device illustrated by Fig. 11.27 and Fig. 12.41(a) is the simplest, so-called resistor SEED. It requires a load resistor R_L. The multiple-quantum-well structure is embedded in the i region of a p–i–n diode. The diode is supposed to be reversed biased. The structure acts as a light modulator and as a photodetector. The load resistance is chosen so that if no light is incident upon the diode, essentially all voltage drops across the diode. As we increase the optical input power, photoexcited carriers cause an increase in the photocurrent I_{photo} and a voltage drop across the resistor. If we operate the device in the spectral region where a decrease in the voltage causes an increase in absorption, then this increase produces a further increase in the photocurrent and a redistribution of the voltage drop between the resistor and the device. This will continue until the quantum efficiency of the device – acting as a photodiode – drops off as its voltage approaches the forward-bias regime, i.e., the condition of zero voltage applied to the device. The net result is that the device switches abruptly from a high- to a low-voltage state, which corresponds to the switching in the optical output from a transparent (high) to a nontransparent (low) optical state.

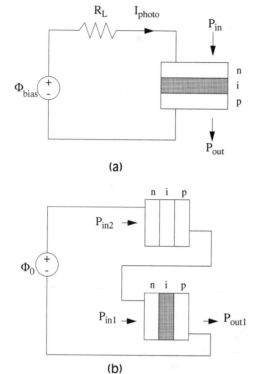

(a)

(b)

Figure 12.41. Schematics of the simplest SEEDs: (a) resistor-biased SEED, (b) photodiode-biased SEED.

In Fig. 12.42 the output–input light characteristics are presented for a SEED made of AlGaAs/GaAs. The multiple-quantum-well region of this device contains fifty 95-Å GaAs wells separated by 98-Å AlGaAs barriers. This region is undoped and embedded between p and n regions made of AlGaAs. The measurements are done for the photon energy equal to 1.456 eV, which corresponds to $\lambda = 851.7$ nm. The load resistance is $R_L = 1$ MΩ and the applied voltage equals 20 V. The figure illustrates the characteristic hysteresis for the input power in the range ∼45–50 μW. The switching directions – from a high to a low optical state with an increase in the input – are in agreement with the previously discussed qualitative picture. In terms of output power, the contrast between the high and the low optical states is factor of ∼2. For the same device, pronounced hysteresis

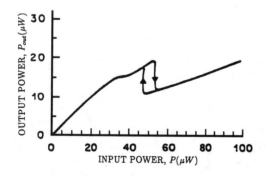

Figure 12.42. The light output–input power characteristics of the SEED. [After D. A. B. Miller, *et al.*, "Novel hybrid optically bistable switch: the quantum well self-electro-optic effect device," Appl. Phys. Lett. **45**, pp. 13–15 (1984).]

and the switching effect are observable over a wide range of parameters: the applied voltage from 15 to 40 V, the spectral region from 851 to 862 nm, and the light spot diameter from 10 to 100 μm. The optical and electrical switching energies per unit area are approximately 4 and 14 fJ/μm^2, respectively. These parameters can be reduced considerably: the physical limit of the light spot diameter is determined by the diffraction of light and equals λ^2/n^2, where n is the refractive index. In the diffraction limit the total energy required for the switching is estimated to be as low as 1 fJ.

According to the analysis of applications of bistable effects given in Chapter 11, this example of a SEED illustrates the simplest optical logic gate controlled by an external electric circuit. The switching energy of a few femtojoules is comparable with that of advanced electronic devices and the best high-finesse resonance-cavity bistable optical devices; data relevant to this discussion are presented in Fig. 11.31.

The next step in developing SEEDs is the use of a photodiode instead of a resistor, as shown in Fig. 12.41(b). In this so-called diode-biased SEED, the signal P_{in1} and the optical bias P_{in2} light beams are incident upon the quantum well of p–i–n diode and are absorbed in the load photodiode, respectively. The load photodiode can be thought of as a resistor whose value is governed by a bias beam power. The use of the load photodiode instead of a fixed resistor makes it easier to integrate the necessary load. It also makes it possible to obtain a smaller device size and to avoid stray capacitance. This step leads to an integrated optoelectronic device with good switching energy. Another important point is that, unlike other bistable devices, the operating light power is not fixed during device fabrication. Actually, the operating range of the devices could be scaled by adjusting the optical bias to the photodiode. The use of a load photodiode also allows new functions to be performed. The device can act as a dynamic memory, a spatial light modulator, etc. Furthermore, arrays of diode-biased SEEDs can be fabricated.

12.4.2 Symmetric SEED

As we mentioned in Subsection 12.4.1, the use of SEEDs is similar to that of most intrinsic bistable devices; see Subsection 11.3.7 and Figs. 11.28–11.30. Their operation is based on adjusting the input-beam energy to a value close to the switching threshold so that a small incident input signal provides just enough energy for switching, signal amplification, etc. In other words, a critical biasing of the device is required. It means that the power of the input beam should be controlled quite precisely. This causes additional difficulties. For example, because the devices make no difference between the input and the output ports, they can be uncontrollably switched by any reflection back into the output device port, etc. In general, such a poor input–output isolation is known as the classical problem of two-terminal devices.

The next development of SEEDs is a device that includes all necessary attributes of three-terminal devices. This so-called symmetric SEED is presented

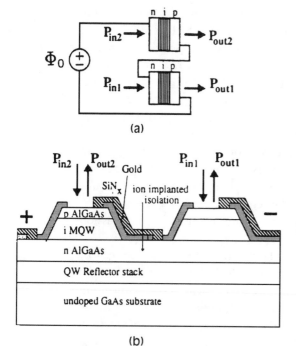

(a)

(b)

Figure 12.43. Symmetric SEED: (a) the circuit with two p–i–n photodiodes, (b) the cross section of the particular AlGaAs/GaAs SEED consisting of two diodes in series. The multiple-quantum-well region (MQW) and reflector stack are indicated. [After A. L. Lentine, *et al.*, "Evolution of the SEED technology: bistable logic gates to optoelectronic smart pixels," IEEE J. Quantum Electron. **29**, pp. 655–669 (1993).]

schematically in Fig. 12.43. A symmetric SEED consists of two identical SEEDs in series; the output of one SEED is electrically connected to the input of another, as shown in Fig. 12.43(a). No external load is necessary: one of the diodes behaves as the load for another and vice versa. The cross section of one of the possible variants of a AlGaAs device is shown in Fig. 12.43(b). This device uses the reflected light as the output. The device is grown upon an undoped GaAs substrate. Adjacent to the substrate is a multiple-layer stack for light reflection. The n and the p regions are made of AlGaAs, and the i region contains a multiple-quantum-well AlGaAs/GaAs structure. Two separated p–i–n diodes are designed by proper processing of an initially layered heterostructure. Typical dimensions of each diode mesa are $200\,\mu m \times 200\,\mu m$ with optical windows of approximately $100\,\mu m \times 150\,\mu m$. Electrical contacts, insulation layers, and other details are shown in the figure.

The operation of a symmetric SEED can be understood through the use of load lines, as shown in Fig. 12.44. Let the supply voltage be fixed. Then one can plot the photocurrents for both the upper and the lower diodes as functions of the voltage across one of them, say across the lower diode. Let a pair of unequal power beams set the states of the device, as shown in Fig. 12.43(a). For example, the incident optical power into the lower diode, P_{in1}, is twice of that of the upper diode, P_{in2}. Under these conditions, photocurrents as functions of the voltage are plotted in Fig. 12.44(a) for each diode. In fact, the photocurrent of the upper diode versus Φ_1 is the load line. The intersection of both photocurrent curves defines the stationary state of the device. In Fig. 12.44, this state is indicated

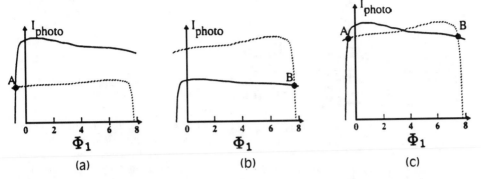

Figure 12.44. Different states of a symmetric SEED. The photocurrents for both diodes are presented as functions of the voltage applied to one of the diodes. The intersections of curves correspond to states of the SEED. (a) A larger input optical power is on one of the diodes. The stable state of a SEED is marked by A. (b) A larger input power is on another diode. The stable state is marked by B. (c) An equal optical power is on both diodes. The system has two stable states, A and B. [After A. L. Lentine, *et al.*, "Evolution of the SEED technology: bistable logic gates to optoelectronic smart pixels," IEEE J. Quantum Electron. **29**, pp. 655–669 (1993).]

as A. Likewise, we can plot similar curves if P_{in2} is twice P_{in1}, as shown in Fig. 12.44(b). Point B corresponds to the stationary state of the device in the latter case. The situation is exactly reversed, corresponding to the previous case. These two figures illustrate that the ratio of the input powers I_{in1} and I_{in2} determines the states of the device. It is important that a variation of the incident powers does not cause switching between the states. If we apply two equal input power beams, $P_{in1} = P_{in2}$, we get two possible states, A and B, as shown in Fig. 12.44(c). For this case, the system obviously displays a bistable behavior. Thus the device is bistable only when the optical input powers on the diodes are comparable; the device has only a single state when the power on one diode significantly exceeds the power on the other.

In operation, two sets of two beams are used, as shown in Fig. 12.45. First, a set of unequal-power beams (signal beams) sets the state of the device. The second set of beams is made up of equal-power clock beams. They are used to read the state of the device. Since the state of the device is determined by the ratio of the total power incident upon each of the diodes, the clock power has to be small when the device state is set. Then the clock beams should have approximately equal optical powers to ensure that either state can be read. For example, let assume that the signal beams have already produced state A, as shown in Fig. 12.44(a), i.e., $P_{in1} > P_{in2}$. Then the lower diode is in an absorptive state and the output of the clock beam will be greater for the upper diode ($P_{out1} < P_{out2}$), as shown in Fig. 12.45(b) for the first period. If during a clock-beam pause the signal beams set another state of the device, B, one gets $P_{out1} > P_{out2}$. This switching is shown in Fig. 12.45(b) by the trace s. The trace r illustrates the reset of the device. It is important that the outputs of the device are dependent on only the signal beams acting during the clock-beam pause. This means that the clock beams can be chosen to be much larger than the signal beams, that is, some kind

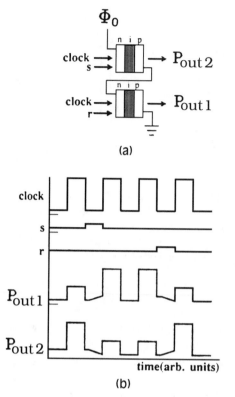

(a)

(b)

time(arb. units)

Figure 12.45. The schematics of one mode of operation of a symmetric SEED. (a) Input and output signals. The clock inputs, set, s, and reset, r, inputs, and outputs P_{out1} and P_{out2}, are indicated. (b) The timing diagram illustrating a time-sequential optical gain. [After A. L. Lentine, *et al.*, "Evolution of the SEED technology: bistable logic gates to optoelectronic smart pixels," IEEE J. Quantum Electron. **29**, pp. 655–669 (1993).]

of optical gain occurs. This gain is known as time-sequential gain. This is not the optical gain in the sense of an optical amplifier studied in Subsection 11.3.7; compare Figs. 12.45(b) and 11.29. In other words, the outputs do not coincide in time with the application of the inputs. This provides effective input–output isolation: a reflection of the output signals back to the device input will not occur at a time when the device is most sensitive to the input. Therefore one can avoid the difficulties with the critical biasing of the device that were discussed previously for resistor-biased SEEDs and diode-biased SEEDs.

In conclusion, the symmetric SEED has many desirable qualities, such as insensitivity to optical supply fluctuations, good input–output isolation like a three-terminal device, and time-sequential gain. Using these devices, one can perform memory functions and all logic operations, including NOR, OR, NAND, and AND. Arrays and cascades of symmetric SEEDs have been fabricated. All of this makes the devices flexible, powerful, and potentially useful for optical signal processing.

12.5 Photodetectors on Intraband Phototransitions

In general, the detection of light in semiconductor materials exploits changes in conductive processes. In bulk materials, there are three main processes activated by light. The intrinsic photoconductivity involves the excitation of electrons from the valence band to the conduction band. The conductivity of the sample increases because additional concentrations of free electrons and holes are generated. This process occurs when the energy of incident photons exceeds the bandgap, i.e., the process has a long-wavelength cutoff, as determined by Eq. (10.45). The second photoconductive mechanism in bulk materials is the extrinsic photoconductivity that occurs in doped semiconductors. At low temperatures, most of the impurities are not ionized (neutral), and photoconductivity results from the excitation of an electron (hole) bound to a neutral donor (acceptor) into the conduction (valence) band. Such a process occurs at long wavelengths corresponding to the ionization energy of impurities. This type of photoconductivity is used for the detection of infrared light signals. The third process that affects the conductivity involves absorption of light by free carriers. This absorption process heats up electrons, and detectors that use this type of photoconductivity are referred to as free-electron bolometers. These bolometers can work at such long wavelengths that they detect far-infrared signals. However, in bulk materials, intraband absorption is too weak and does not lead to the effective detection of light.

Although quantum heterostructures can be used in photodetectors operating in different spectral regions, the most interesting applications of these structures are in infrared technology in which high sensitivity, spectral tuning, and other desirable properties are the challenging problems. According to Section 10.6, in quantum heterostructures with electron confinement, infrared absorption corresponds to intraband (intersubband) processes. These processes can be much more pronounced than those in bulk semiconductors, and they can be controlled through the structure design. In Subsection 12.5.2 we show that intraband phototransitions can be used for the effective detection of long-wavelength light.

12.5.1 Photoconductive Detectors

We start with a brief review of the principles of operation of photodetectors and their basic characteristics. The simplest photodetector – a photoconductive device – is illustrated in Fig. 12.46. The schematic of this photodetector is presented in Fig. 12.46(a): a photosensitive sample of length L and cross-sectional area S is under a voltage bias Φ. When light with a proper wavelength illuminates the device, the electric current increases. The changes in the current induced by light can be detected by a circuit, as illustrated in Fig. 12.46(b). The circuit includes a load resistor R_L and a capacitance C. The latter is necessary if only the ac signal is to be detected.

An important characteristic of photoconductive detectors is the gain of the device. The gain indicates how many electrons can be collected for each absorbed photon. Let g_l be the rate of photogeneration of free carriers; more specifically, let g_l be the number of carriers generated by photon absorption in a unit volume per unit time. If the lifetime of the free carriers is τ, the concentration of these carriers is $n_l = \tau g_l$. The total current I consists of two contributions. The first is

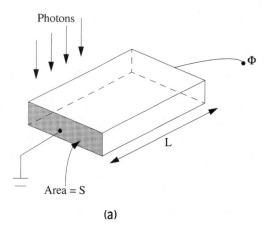

(a)

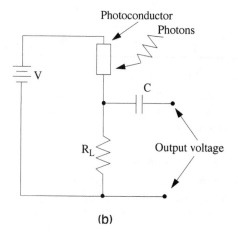

(b)

Figure 12.46. The schematics of a photoconductor detector: (a) the photoconductor geometry, (b) a typical photodetector circuit.

the dark current I_d, and the second is the photocurrent I_l:

$$I = I_d + I_l, \qquad I_d = e\mu n_d F S, \qquad I_l = e\mu n_l F S,$$

where μ is the mobility, n_d is the carrier concentration contributing to the current in the absence of illumination, and F is the electric field. If we define the transit time of the electrons in the device by $t_{\mathrm{tr}} = L/\mu F$, we can express the photocurrent in terms of τ and t_{tr}:

$$I_l = e\frac{\tau}{t_{\mathrm{tr}}} g_l L S. \tag{12.53}$$

This is the current generated in the circuit by light. We can introduce the primary photocurrent $I_{l,p} = e g_l L S$, which is the photocurrent in the case in which each generated carrier transfers one electron charge e from one contact to the other contact. Now we can define the gain of the detector as

$$G = \frac{I_l}{I_{l,p}} = \frac{\tau}{t_{\mathrm{tr}}}. \tag{12.54}$$

Values of gain larger than one, $G \geq 1$, arise if a photogenerated electron can go around the circuit several times during its lifetime. It is easy to see that the gain increases with the applied electric field.

Another important figure of merit of a photodetector is the so-called responsivity r, which can be defined as

$$r = \frac{I_l}{P_l},$$

where P_l is the light power illuminating the detector. Since the rate of the photogeneration of carriers is $g_l = \eta P_l / (\hbar \omega S L)$, where η is the internal absorption efficiency of the detector, the responsivity has the form

$$r = e \frac{\eta G}{\hbar \omega}. \tag{12.55}$$

Thus the responsivity is determined by the internal efficiency and by the gain of the device.

Both figures of merit, the gain and the responsivity, characterize the photocurrent and the efficiency of the device. Despite the existence of a dark current, the photocurrent can certainly be distinguished from the dc dark current by use of, for example, a time modulation of the photosignal with a frequency f and by detection of the electric current in a frequency band Δf around f.

However, there are always fluctuations of the dark current that constitute the so-called current noise. Specifically, there are small time-dependent deviations of the instantaneous current from its mean value, $\delta I(t) = I(t) - I$. These fluctuations can be caused by different random processes: shot noise, Nyquist noise, generation–recombination noise, etc. These fluctuations make a contribution to the electric signal in any frequency range, including the one in which the measurements are made. As a result, the fluctuations limit significantly the intensity of a detectable signal. Let the Fourier transform of the current fluctuations be ΔI_f. We can introduce the spectral density of current fluctuations ΔI_f^2 through the following equation:

$$\langle \Delta I_f \Delta I_{f'} \rangle = \Delta I_f^2 \delta(f - f'), \tag{12.56}$$

where the angle brackets $\langle \; \rangle$ denote a statistical average and $\delta(f - f')$ is the δ function. It is easy to check that the average current fluctuations, $\langle [\Delta I(t)]^2 \rangle$, are expressed by the spectral density ΔI_f^2. Now we can characterize the current noise by the parameter

$$I_{ns} = \sqrt{\Delta I_f^2 \Delta f}. \tag{12.57}$$

The signal-to-noise ratio,

$$\mathcal{N}_{(sgn/ns)} = I_l^2 / I_{ns}^2,$$

determines the detection limit of the device.

Often we use the noise-equivalent optical power, i.e., the power of the optical signal at which the photocurrent is equal to I_{ns}:

$$P_{NEP} = \frac{I_{ns}}{eG\eta} \hbar\omega. \tag{12.58}$$

In this section, we defined three quantities: the gain G, the responsivity r, and the noise-equivalent power P_{NEP} that characterize the utility of any photodetector in converting an optical signal into an electrical signal.

12.5.2 Intraband Phototransitions and Electron Transport in Multiple-Quantum-Well structures

Let us turn to an analysis of intraband absorption, electron transport, and current noise in multiple-quantum-well photodetectors.

In Section 10.6, we studied intraband phototransitions in a quantum well. We found that light absorption in a single well is relatively small. Since the main application of intraband phototransitions is the detection of weak infrared light signals, one needs to obtain an absorption as large as possible. For this purpose, multiple-quantum-well structures are used. Figure 12.47 sketches the potential profiles and energy spectra of some photodetecting systems. In all cases, the barriers that separate the wells are assumed to be thick enough to prevent the formation of minibands due to electron tunneling. Figure 12.47(a) shows wells that are designed so that two energy levels exist in each well. Intersubband absorption can activate an electric current through this system when a voltage bias is applied. Absorption by such a structure is presented in Fig. 10.27. For this particular case, the maximum absorption coefficient reaches 2000 cm^{-1}. The conductivity is due to either the tunneling between ϵ_2 states in different wells or thermally activated overbarrier transport. Figure 12.47(b) shows multiple-well structures with a single bound energy level. Two different cases can occur in such structures. If the barriers are of sufficient thickness, a modification of the electron spectrum in the energy region above the barrier is negligible and can be considered as a continuum of the extended states studied in Section 10.6. The absorption coefficient for such a case, presented in Fig. 10.29, is $\sim$1000–1400 cm^{-1}. Conductivity occurs as a result of photogenerated free carriers. In

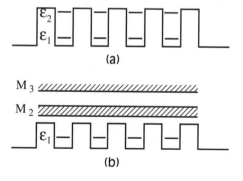

(a)

(b)

Figure 12.47. Two examples of possible energy spectra of multiple-quantum-well structures with different device parameters: (a) ϵ_1 and ϵ_2 indicate the quantized levels in the wells, (b) wells with a single quantized level ϵ_1, while overbarrier energies are split into minibands because of multiple electron reflection from the relatively narrow barriers. M_2 and M_3 indicate these minibands.

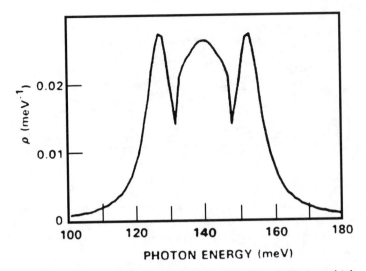

Figure 12.48. Calculated optical density of states of an $Al_{0.5}Ga_{0.5}As/GaAs$ multiple-quantum-well structure with 40-Å wells and 200-Å barriers. [After Z. C. Feng, ed., *Semiconductor Interfaces, Microstructures, and Devices. Properties and Applications* (Institute of Physics, Bristol and Philadelphia, 1993).]

Fig. 12.48, the optical density of states for such an electron-energy spectrum is shown as a function of the photon energy for an ideal $Al_{0.25}Ga_{0.75}As$ multiple-quantum-well structure. The well width is 40 Å and the period is $d = 240$ Å. Three peaks in the density of states are due to the formation of three minibands above the barriers. For such a structure, the absorption coefficient α can reach magnitudes of 200–400 cm^{-1}. Thus different designs of multiple-quantum-well structures result in different absorption spectra and different magnitudes of absorption.

Consider next the important factors determining the detection of photosignals, i.e., the dark currents and the photocurrents through the structure. For photodetection, the wells are usually doped; thus there are always carriers at least in the lowest subband. If a voltage is applied to a multiple-quantum-well structure, a dark current flows through the structure, even in the absence of illumination. The dark current is caused by different mechanisms: (1) sequential tunneling from one well to a neighboring well, which may be assisted by phonon absorption or emission; (2) excitation to the second subband and subsequent tunneling; and (3) thermoionization to the continuum and subsequent free propagation for energies above the barrier potential. According to the analysis given in Chapters 7 and 8, only coherent tunneling is nearly independent of the temperature of the system. Other previously mentioned mechanisms in multiple-quantum-well structures with thick barriers are sharply activated with increasing temperature. For example, the thermionic current is an exponential function of the temperature, as indicated by Eq. (9.85).

The identification of the different mechanisms contributing to the dark electron transport can be accomplished straightforwardly by use of the hot-electron

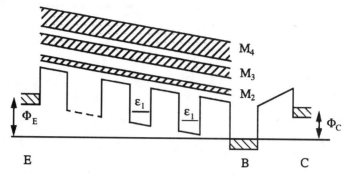

Figure 12.49. The band diagram of an electron-energy spectrometer: E, B, and C denote the emitter, base, and collector, respectively. A single energy level is supposed to be in the wells; M_2, M_3, and M_4 are minibands formed above the energies of the potential barriers; Φ_E and Φ_C are the emitter and the collector biases, respectively.

spectroscopy method that we studied in Chapter 9. With this purpose, a multiple-quantum-well photodetector structure should serve as the emitter part of the hot-electron spectrometer. Then an additional doped layer should be grown to create a base. This base is separated from the collector by a barrier. The energy diagram of such a spectrometer is shown in Fig. 12.49. The diagram corresponds to the case in which the wells have a single bound level, ϵ_1. The energy spectrum of the extended states can consist of several minibands, M_2, M_3, ... The voltage drop on the multiple-quantum-well part of the device is Φ_E, and the base–collector bias is Φ_C. By varying the emitter–base bias Φ_E and the base–collector bias Φ_C, one can separate electrons with different energies coming into the base from the multiple-quantum-well part of the structure. The electron distribution function $\mathcal{F}(E)$ can be found in the form of $\mathcal{F}(E_0 - e\Phi_E)$, where E_0 is a constant; that is, the electrons with small energies are detected at high voltage Φ_E, while decreasing the voltage facilitates the detection of electrons with higher energy. Experimental results obtained by this method are shown in Fig. 12.50. Parameters of the multiple-quantum-well structure are those identified in the caption to Fig. 12.48. The structure consists of 50 periods of GaAs wells and $Al_{0.25}Ga_{0.75}As$ barriers. A thin, 300-Å pseudomorphic $In_{0.25}Ga_{0.85}As$ layer is grown on the top of the multiple-quantum-well structure as a base layer. A 2000-Å layer of $Al_{0.25}Ga_{0.75}As$ separates the base from the GaAs collector. All the layers except the barriers are doped to $n_D = 1.2 \times 10^{18}$ cm^{-3}. The distribution of hot electrons entering the collector is presented as a function of emitter–base bias. The position of the single subband ϵ_1 is indicated by the arrow. Another arrow corresponds to energy ϵ_2, which has a value of 225 meV. The results are given for different temperatures. It is clearly seen that at low temperatures the distribution function is centered around ϵ_1, i.e., the major mechanism of the electron transport is the tunneling through the lowest energy level ϵ_1. With increasing temperature, the high-energy tail grows. Electrons with energies between ϵ_1 and ϵ_2 are associated with thermally assisted tunneling. Such thermally assisted tunneling prevails at temperatures near 80 K. Electrons coming into the base with energies close to ϵ_2

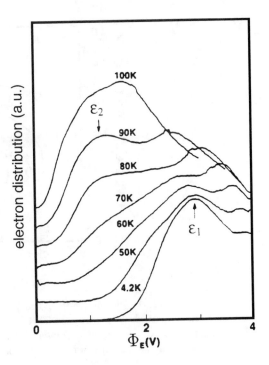

Figure 12.50. Results of the measurement of the electron distributions for the multiple-quantum-well structure as functions of the bias at different temperatures. The thicknesses of the wells and the barriers are 40 Å and 200 Å, respectively; ϵ_1 indicates the position of the lowest subband; ϵ_2 corresponds to the energy 225 meV. The origins of the curves are shifted up for clarity. [After Z. C. Feng, ed., *Semiconductor Interfaces, Microstructures, and Devices. Properties and Applications* (Institute of Physics, Bristol and Philadelphia, 1993).]

propagate as a result of thermionic processes. These processes become dominant at temperatures above 90–100 K. Figure 12.51 shows the total dark (emitter) current and the collector current as functions of the emitter–base voltage at $T = 77$ K.

Consider another example of the multiple-quantum-well structure with GaAs well layer thicknesses of 70 Å and $Al_{0.36}Ga_{0.64}As$ barrier widths of 140 Å. The doping concentration of the wells is 1.4×10^{18} cm^{-3}. The barriers prevent large tunneling currents, so the broadening of the lowest energy level in the wells $\Delta\epsilon_1$ is ~7 meV. The applied voltage is such that the voltage drop over one period of the structure is approximately $\Phi_p = 1$ meV. In Fig. 12.52, the relative contributions from sequential tunneling I_{st}, phonon-assisted tunneling I_{pt}, and thermionic emission I_{th} to the dark current are presented as functions of the temperature. From these results a general conclusion can be drawn: at low temperatures the tunneling contribution into the dark current dominates. This process is possible until the energy-level width exceeds the voltage drop, $\Delta\epsilon_1 > \Phi_p$. Phonon-assisted tunneling contributes over a relatively narrow temperature region, while the thermionic current dominates at higher temperatures above 100 K.

As we now know the key characteristics of the dark current, let us compare it with the photocurrent. We use a multiple-quantum-well structure with the parameters defined in Fig. 12.50. The geometry and the principal components of the photodetector are shown in the inset of Fig. 12.51: the coupler with light, emitter E, the quantum-well infrared photodetective (QWIP) region, base B, the barrier for filtering electrons with different energies, and collector C. The light is

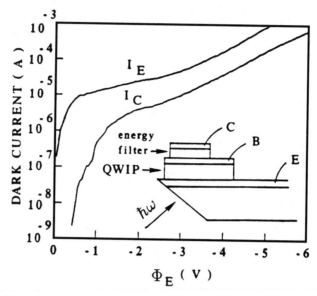

Figure 12.51. The emitter I_E and collector I_C dark currents for the device with the multiple-quantum-well structure at $T = 77$ K. The inset shows the detector and the coupler with light. [After Z. C. Feng, ed., *Semiconductor Interfaces, Microstructures, and Devices. Properties and Applications* (Institute of Physics, Bristol and Philadelphia, 1993).]

coupled into the device through a 45° angle to provide for the interaction of light with the quantum-wells (see Table 10.2). This three-terminal device facilitates the investigation of both the dark current and the photocurrent, as well as the control of the dark current. The thermostimulated collector current is shown in Fig. 12.53 as a function of the emitter–base voltage Φ_E at three different temperatures from 50 to 100 K. The photocurrent is presented at $T = 90$ K. As one can see, this current corresponds to electron propagation at energies greater than the barrier potential. All currents are in arbitrary units, but in the same scale. We can conclude that the photocurrent dominates over the dark current at temperatures below 100 K and the voltage bias on the photodetective part of the structure is $\sim$0.5 V.

The measurement of the energy distribution of photoexcited electrons leads to conclusions that allow us to formulate the model of electron transport in multiple-quantum-well infrared photodetectors. In Fig. 12.53, the arrows marked by ϵ_1 and ϵ_2 have the same meaning as in Fig. 12.50. Thus we deduce that the photocurrent corresponds to electron propagation at energies greater than the barrier potential. In particular, Fig. 12.53 clearly indicates that there are no ballistic electrons. The energy width of the photogenerated electrons is approximately several tens of millielectron volts and is comparable with the device temperature of 90 K, that is, the distribution of photocarriers is consistent with that of the usual electron drift over barriers, despite large applied electric fields. Note that for the superlattice with N periods, the electric field can be estimated as $F = |\Phi_E|/(Nd) \approx 5 \times 10^3 - 10^4$ V/cm. These conclusions allow us to develop a simple model of the processes taking place in such devices.

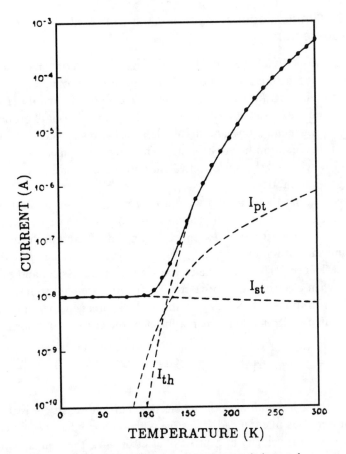

Figure 12.52. The experimental temperature dependence of the total current through a multiple-quantum-well structure at a low bias (shown by the filled circles). The dashed curves show the contributions to the current from different partial mechanisms: sequential tunneling (I_{st}), phonon-assisted tunneling (I_{pt}), and the thermionic emission (I_{th}). [After Z. C. Feng, ed., *Semiconductor Interfaces, Microstructures, and Devices. Properties and Applications* (Institute of Physics, Bristol and Philadelphia, 1993).]

Figure 12.53. Hot-electron spectra at different temperatures. Curves (a), (b), and (c) correspond to dark (thermal stimulated) currents at $T = 100$, 77, and 50 K, respectively. Curve (d) corresponds to the spectra under illumination at $T = 90$ K. The curves are shifted up for clarity. [After Z. C. Feng, ed., *Semiconductor Interfaces, Microstructures, and Devices. Properties and Applications* (Institute of Physics, Bristol and Philadelphia, 1993).]

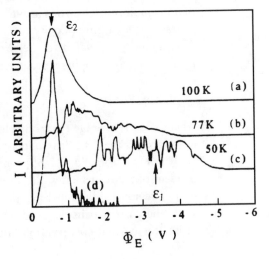

12.5.3 Simple Model of Multiple-Quantum-Well Photodetectors

In Subsection 12.5.1 we introduced the figures of merit for photodetectors by using the characteristics of bulklike materials: the lifetime of photoexcited electrons τ, their generation rate g_l, mobility μ, etc. Our goals in this section are to formulate a simple model describing a multiple-quantum-well photodetector, to write down proper equations, and to calculate the figures of merit of these devices in terms of the quantum heterostructure. These figures of merit are the gain G, the responsivity r, and the noise-equivalent optical power P_{NEP} or, equivalently, the signal-to-noise ratio.

We assume that the quantum wells are narrow, contain only one subband, and have the same width L. The barriers are supposed to be thick so that the spatial period of the structure d is much larger than L, and both the tunneling current and the formation of minibands above the barrier energy are entirely suppressed. Hence perpendicular electron transport is classical and takes place primarily when electrons are above the barriers in extended states. The quantum wells are supposed to be doped and to be sources of thermal (dark) and photogenerated electrons. In such an approach, the current equations are

$$\frac{1}{e}\frac{d\,j_z(z)}{dz} = \sum_{i=1}^{N}\{g_i - s[n(z) - n_0]\}\,\delta(z - z_i), \qquad j_z = en\mu F + eD\frac{dn}{dz}, \quad (12.59)$$

where j_z is the zth component of the electron current density, n is the volume density of electrons in extended states, n_0 is the volume density of electrons at equilibrium, s is the velocity of capture of electrons into the well, and $\delta(z - z_i)$ is the δ function that reflects the fact that the wells are assumed to be very narrow and located at coordinates $z = z_i$. The summation extends over all wells; N is the total number of the wells in the structure, F is the electric field that can depend on z, and μ and $\mathcal{D}$ are the electron mobility and the diffusion coefficient, respectively. The quantities $g_i\delta(z - z_i)$ describe the photogeneration rate of electrons into extended states. The quantity s can be formally expressed through the lifetime τ of nonequilibrium electrons in extended states: $s = d/\tau$. It is worth noting that the characteristic capture velocity s acts as the recombination rate for charge carriers. This quantity is related to the probability p_c of capturing electrons that are propagating over the barriers. The relation between s and p_c can be estimated as

$$p_c = (s/v),$$

where v is the average electron velocity. As discussed in Chapter 7, usually v is much greater than the drift velocity $v_{dr} = \mu F$.

It follows from Eq. (12.59) that the current density in each barrier is independent of the coordinates, i.e., $j_z = j_i$, where j_i is the current density in the ith barrier region, which separates the $(i + 1)$th and ith quantum wells. In steady state, the current densities in different barriers should be the same since they satisfy the current-conservation law, $j_{i+1} = j_i$. After integration of Eq. (12.59) over a slab that includes the ith quantum well, we obtain

$$j_{i+1} - j_i = eg_i - es\,(n_i - n_0) = 0, \tag{12.60}$$

which leads to $j_i = j = \text{const.}$ and $n_i = n_0$.

We now consider the case of high photogeneration rates, i.e., $g_i \gg n_0 s$. Then, from Eq. (12.60), we obtain $n_i = g_i/s$. Because only strong electric fields are interesting for photodetector applications, we can assume that the drift term in Eq. (12.59), $en\mu F$, is dominant and therefore we omit the diffusion term, $eD dn/dz$. Now, after the integration of Eq. (12.59) with boundary conditions for $n(z)$ imposed by Eq. (12.60), we obtain an expression for the current density in the ith barrier:

$$\frac{1}{e}\, I_i = \frac{g_i v_i}{s}, \tag{12.61}$$

where $v_i = \mu F_i$. Note that we assume the simplest case, i.e., when the mobility is independent of the field. The result can be easily generalized for any dependence $\mu(F)$. At the end of this section, we present the case in which the velocity-saturation effect is taken into account.

The photogeneration rate g_i equals $\beta \Upsilon_i$, where Υ_i is the photon-flux density through the ith well; it can be simply expressed in terms of the light power $\mathcal{I}$: $\Upsilon_i \equiv \mathcal{I}_i/S\hbar\omega$. Here S is the cross section of the device and β represents the light attenuation during one transit through the well. It can also be interpreted as the quantum efficiency of absorption by a single quantum well.

Photoconductive gain in a multiple-quantum-well structure: According to the definition of Eq. (12.54), the gain can be written as

$$G_p = \frac{I}{e \Upsilon_0 \eta} = \frac{I}{e \sum_{i=1}^{N} g_i}. \tag{12.62}$$

Equating the current densities in the two-barrier regions adjacent to the ith quantum well, we obtain a recurrent equation for electric fields in the ith and $(i+1)$th barrier layers: $v_i g_i = v_{i+1} g_{i+1}$, or

$$F_{i+1} = \frac{g_i}{g_{i+1}}\, F_i.$$

Because the photon flux changes while passing through the well, the rate of photogeneration g_i depends on the well number i: $g_i \neq g_{i+1}$. In the presence of such nonuniform absorption, the electric fields in different barriers are different. This occurs primarily because the quantum wells are charged until the conservation of the photocurrent is satisfied. It follows from Poisson's equation that the charge of the ith well is

$$N_i^+ = \frac{\kappa}{4\pi e} \frac{(g_i - g_{i+1})}{g_{i+1}}\, F_i.$$

Usually, $g_i > g_{i+1}$, i.e., the wells are charged positively.

We can introduce the average electric field $\bar{F}$ in the detector active region as

$$\bar{F} = \frac{1}{N} \sum_{i=1}^{N} F_i = \frac{g_1 F_1}{N} \sum_{i=1}^{N} g_i^{-1}.$$

Then, from Eq. (12.62), we can rewrite the photoconductive gain as

$$G_p = G_0 K_p, \tag{12.63}$$

where G_0 is the gain of the detector with a uniform electric field $F = \bar{F}$:

$$G_0 = \frac{\bar{v}}{s N}. \tag{12.64}$$

Here, $\bar{v}$ is the drift velocity in the uniform electric field $\bar{F}$. Keep in mind that, in the previous expressions, we introduced an effective lifetime $\tau = d/s$ and a total length for the device active region Nd. Thus it is easy to see that in these terms, Eq. (12.64) represents the gain of an ordinary photodetector given by Eq. (12.54). In fact, the multiplier K_p describes the deviations from a uniform case:

$$K_p = \frac{1}{(1/N) \sum_{i=1}^{N} g_i \, (1/N) \sum_{i=1}^{N} g_i^{-1}}. \tag{12.65}$$

In general, we can find that

$$K_p \leq 1.$$

It follows from Eq. (12.65) that $K_p = 1$ for the case of uniform generation.

For a single-pass detector without backreflecting mirrors, we can take into account absorption by the quantum wells ranging from 1 to $i-1$ to obtain the photogeneration rate in the ith well as

$$g_i = \Upsilon_0 (1 - \beta)^{i-1} \beta,$$

where Υ_0 is the photon flux entering the first quantum well with $i = 1$. The quantum efficiency of absorption for a single-pass device is

$$\eta = \frac{\sum_{i=1}^{N} g_i}{\Upsilon_0}. \tag{12.66}$$

The sums in Eqs. (12.62), (12.65), and (12.66) can be calculated approximately for $N \gg 1$ when the summation can be approximated by an integration over i. Then the photoconductive gain of the detector with single-pass geometry is

$$G_p = G_0 \frac{\ln^2(1 - \beta)}{2} \frac{N^2}{\cosh[N \ln(1 - \beta)] - 1}. \tag{12.67}$$

For almost uniform absorption when $\beta N \ll 1$, we get results that do not differ from those found in Eq. (12.64), i.e., $G_p = G_0$. In the opposite case of strong absorption, $\beta N \gg 1$, we obtain

$$G_p \simeq G_0 (N\beta)^2 \exp(-\beta N).$$

Numerical calculations for a structure with a one-pass geometry and with 50 quantum wells yield $K_p = 0.92$ for $\beta N = 1$ and $K_p = 0.72$ for $\beta N = 2$. Corresponding figures for a double-pass photodetector with a backreflecting mirror are $K_p = 0.98$ and $K_p = 0.84$. Thus nonuniform absorption leads to a decrease of the gain from G_0 to G_p.

Using Eqs. (12.63) and (12.66), one can calculate the responsivity of the detector as

$$r_p = e \frac{\sum_i g_i}{\Upsilon_0 \hbar \omega} G_0 K_p. \tag{12.68}$$

Generation–recombination noise: As was mentioned in Subsection 12.5.1, the important figure of merit for a photodetector is the signal-to-noise ratio. For the device under consideration, the major source of the noise is the generation–recombination noise, i.e., the noise originating from the randomness of processes of electron capture into the wells and electron escape from the wells. Calculating such a noise, we can estimate the signal-to-noise ratio and therefore the limitation of the device sensitivity.

The simplest way of doing so is by introducing random sources into Eq. (12.59). These random sources, $\delta g_{i,\text{th}}(x, y, t)$ and $\delta r_{i,\text{th}}(x, y, t)$, describe both thermal processes: electron escape from the well (thermal generation) and electron capture (thermal recombination) into the well. In general, the latter quantities depend on the in-plane coordinates x and y. This means that the fluctuation δn is a function of all coordinates, x, y, z, and time t. Since we are interested in the calculation of fluctuations of the total current through the device, we introduce the averaged in-plane fluctuation of the concentration, $\delta n(z, t)$, and the averaged z component of the current density fluctuation, $\delta j(z, t)$:

$$\delta j(z, t) = e\mu F_i \, \delta n_i + e\mu n_i \, \delta F(z, t); \quad z_i \leq z \leq z_{i+1}, \tag{12.69}$$

where $\delta F(z, t)$ is the fluctuation of the electric field in the ith barrier region $z_i < z < z_{i+1}$. Then the equation for a fluctuation of the electron concentration $\delta n(z, t)$ can be written in the so-called Langevin equation form. For this type of equation one should assume that the evolution of $\delta n(z, t)$ is governed by the linearized, nonstationary basic equation, Eq. (12.59), with additional random sources:

$$\frac{\partial \delta n(z, t)}{\partial t} + \frac{1}{e} \frac{\partial \delta j(z, t)}{\partial z} = -s \sum_{i=1}^{N} \delta n(z, t) \, \delta(z - z_i)$$

$$+ \sum_{i=1}^{N} \left[\delta g_{i,\text{th}}(t) - \delta r_{i,\text{th}}(t) \right] \delta(z - z_i), \tag{12.70}$$

where $\delta g_{i,\text{th}}(t)$, and $\delta r_{i,\text{th}}(t)$ are random terms describing the rate of random capture and escape processes per unit area of the device cross section.

Estimates show that the main reason for electric-field fluctuations is the charge fluctuations in the wells, i.e., the contribution of fluctuations of the electron concentration in the barrier regions can be neglected. Thus the fluctuative field is independent of coordinates within each of the barriers:

$$\delta F(z, t) = \delta F_i(t) \quad \text{for } z_i \leq z \leq z_{i+1}.$$

The time Fourier transformation of Eq. (12.70) gives

$$i[f\delta n(z, f)] + \frac{1}{e}\frac{d\delta j(z, f)}{dz} = -s\sum_{l=1}^{N} \delta n(z, f)\delta(z - z_l)$$

$$+ \sum_{l=1}^{N}[\delta g_{l,\text{th}}(f) - \delta r_{l,\text{th}}(f)]\delta(z - z_l), \qquad (12.71)$$

where $\delta n(z, f)$, $\delta j(z, f)$, $\delta g_i(t)$, and $\delta r_i(t)$ are the Fourier transforms of $\delta n(z, t)$, $\delta j(z, t)$, and the random terms, respectively. For low-frequency fluctuations one can neglect the first term on the left-hand side of Eq. (12.71). After integration of Eq. (12.71) over $z_{i-1} < z < z_{i+1}$, we obtain

$$\delta j_i(f) - \delta j_{i-1}(f) = -s\delta n_i(z) + \delta g_i(f) - \delta r_i(f) = 0, \quad i = 1, 2, \ldots, N.$$
$$(12.72)$$

This result shows that the fluctuating current through the device is independent of z and the fluctuation of the concentration is

$$\delta n_i(f) = \frac{\delta g_i(f) - \delta r_i(f)}{s}.$$

The electric-field fluctuation in the ith barrier can be determined from Eqs. (12.71) and (12.72):

$$\delta F_i(f) = \frac{\delta j(f) - e\mu F_i \, \delta n_i(f)}{e\mu n_i}, \qquad (12.73)$$

where it has been taken into account that the low-frequency fluctuations of the current, $\delta j(f)$, are constant along the device, so that $\delta j_i(f) = \delta j(f)$. For a stable fixed voltage bias across the device, we obtain

$$\sum_{i=1}^{N} \delta F_i(f) = 0,$$

that is, for the current fluctuations $\delta j(f)$ after integration of Eq. (12.73) along the detector we get

$$\delta j(f) = e\mu \frac{\sum_{i=1}^{N} F_i n_i^{-1} \, \delta n_i(f)}{\sum_{i=1}^{N} n_i^{-1}}. \qquad (12.74)$$

We assume that the processes of random thermal generation and recombination in different wells are not correlated. For the ith narrow well, intensities are proportional to the electron concentration n_i and the recombination velocity s:

$$\langle \delta g_i(f) \, \delta g_k(f) \rangle = \langle \delta r_i(f) \, \delta r_k(f) \rangle = 4n_i \frac{s}{A} \delta_{ik}, \quad \langle \delta g_i(f) \, \delta r_k(f) \rangle = 0, \quad (12.75)$$

where $\langle \cdots \rangle$ denotes a statistical average and A is the detector cross-sectional area. In Eq. (12.75), the stationary concentration is $n_i = g_i s^{-1} + n_0$. According to the definition, the mean square total current noise in the frequency bandwidth Δf is

$$\Delta I_f^2 = \delta I_f^2 A^2 \Delta f,$$

where δI_f^2 should be calculated with the procedure given by Eq. (12.56). Finally, we obtain ΔI_f^2 in the form

$$\Delta I_f^2 = 4 \frac{(e\mu)^2 A}{s} \frac{\sum_{i=1}^{N} F_i^2 \, n_i^{-1}}{\left(\sum_{i=1}^{N} n_i^{-1} \right)^2} \Delta f. \quad (12.76)$$

For the case of homogeneous thermal generation, we have $n_i = n_0$ and from Eq. (12.76) we obtain

$$\Delta I_f^2 = 4e G_0 I_d \Delta f,$$

where $I_d = e \, \bar{v} \, n_0 A$ is the dark thermal current. Equation (12.76) exhibits a current (voltage) dependence typical for the generation–recombination noise:

$$\Delta I^2 \propto I^2 \propto F^2.$$

Under a high level of photogeneration, using Eq. (12.63) for the current noise, we obtain

$$\Delta I_f^2 = 4e I G_p K_{ns} \Delta f, \quad (12.77)$$

where

$$K_{ns} = \frac{\sum_{i=1}^{N} g_i \sum_{i=1}^{N} g_i^{-3}}{\left(\sum_{i=1}^{N} g_i^{-1} \right)^2}.$$

Generally one can show that

$$K_{ns} \geq 1.$$

For the single-pass geometry and in the limit of a small attenuation, $\beta \ll 1$, the summations can be estimated, and we obtain

$$K_{ns} = \frac{(1 - \beta)^{2N}[(1 - \beta)^{-3N} - 1]}{3[1 - (1 - \beta)^N]}. \quad (12.78)$$

We can further simplify this expression for the two limits: $K_{ns} \approx 1$ for $\beta N \ll 1$, and $K_{ns} \simeq \frac{1}{3} \exp(\beta N)$ if $\beta N \gg 1$. Numerical calculations performed for a structure with $N = 50$ wells show that $K_{ns} = 1.72$ for $\beta N = 1$ and $K_{ns} = 3.23$ for $\beta N = 2$ in the single-pass geometry.

Now we have specified all the major figures of merit for photodetectors in terms of the following quantum structure parameters: light attenuation for a single well β, rate of electron capture s, number of wells N, and mobility in barrier regions μ. Of some importance is the fact that the first two parameters depend on the quantum-well depth and width as well as the doping level of the wells. They can be designed to achieve a large attenuation and a small capture rate. The mobility is determined by processes in the barrier regions. These regions can be taken as undoped to achieve a high value of the mobility. These results reveal one of the main advantages of using multiple quantum-well structures for the detection of light: light absorption and transport of photoexcited carriers can be controlled almost independently. In comparison, in bulklike extrinsic photodetectors, increasing the doping leads to increased absorption but degrades the transport properties.

According to these results, the photoconductive gain and responsivity increase in proportion to the electric field (the applied voltage), while the efficiency and the noise-equivalent power remain almost constant. This is valid at relatively low voltage biases, when the electron transport can be thought of as linear. At high biases, one needs to take into account nonlinear transport effects. The model we have just formulated can be extended to such a case if we assume an approximation for the field dependence of the electron mobility $\mu(F)$. For example, to include the effect of velocity saturation, as described in Chapter 7, we can assume a drift velocity v_{dr} in the form

$$v_{dr}(F) = \frac{\mu F}{\sqrt{1 + (F/F_{st})^2}},$$

where F_{st} is the characteristic field of the saturation. Let us consider a photodetector based on AlGaAs/GaAs multiple quantum wells with a long-wavelength cutoff of ~ 10 μm. We can set $N = 50$ and $d = 500$ Å. For a single quantum well, we estimate the attenuation as $\beta = 0.02$; see Subsection 10.6.2. For transport parameters, we assume $\mu = 2 \times 10^3$ cm^2/V s, $F_s = 5 \times 10^3$ V/cm, and $s = 10^6$ cm/s, which corresponds to an effective capture time of approximately $\tau = d/s = 5 \times 10^{-12}$ s. For these parameters of the multiple-quantum-well photodetector, the responsivity, efficiency, and photocurrent gain are presented in Fig. 12.54 as functions of the voltage bias. One can see that the quantum efficiency is almost independent of the bias, while both the responsivity and the gain increase with a tendency to be saturated at high biases.

In conclusion, intraband absorption in multiple-quantum-well structures allows one to design principally new devices to detect weak infrared signals. They can be designed in selected spectral regions to yield greater sensitivity than existing infrared technology based on doped bulklike materials.

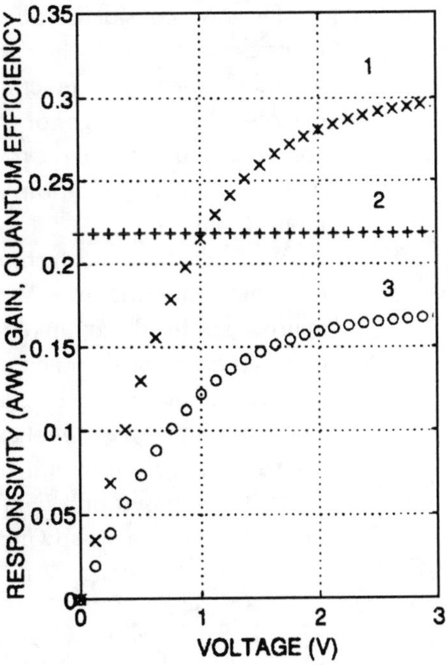

Figure 12.54. Voltage dependences of the responsivity 1, quantum efficiency 2, and photoconductive gain 3 calculated for the simple model of a multiple-quantum-well photodetector and the drift-velocity approximation discussed in the text. [After V. D. Shadrin, *et al.*, "Photoconductive gain and generation–recombination noise in quantum-well photodetectors biased to strong electric field," J. Appl. Phys. 78, pp. 5765–5774 (1995).]

12.6 Closing Remarks

In this chapter, we applied the results obtained in Chapters 10 and 11 to the optics of quantum heterostructures. We studied applications of these quantum heterostructures to optical devices. It was shown that these heterostructures make possible principal improvements in major elements of optoelectronic technologies; such elements include sources of spontaneous and stimulated emission, devices for processing optical signals, and infrared photodetectors.

We started with semiconductor injection lasers based on the homostructure $p-n$ junction, and then we considered the development of advanced heterojunction lasers. We traced the steps critical to the improvement of injection lasers: from a double heterojunction to single- and multiple-quantum-well lasers and quantum-wire lasers. Each of these steps leads to an additional confinement of injected carriers and, as a consequence, to a considerable decrease in the threshold pumping current. It is important that the confinement of both carriers and light occurs in these heterostructures. This increases significantly the gain of confined optical modes and lowers the injection currents necessary for laser oscillations. It makes possible the design of microlasers excited by low currents as well as the design of arrays of such devices. Quantum-heterostructure lasers operate from the infrared to the blue spectral region.

We analyzed a new type of experimentally realized heterostructure lasers – quantum-cascade lasers. These laser sources are unipolar injection devices. They can be tailored for operation in the wave region from the middle infrared to the submillimeter.

Then we considered SEED heterostructure devices, with bistable output–input characteristics. They are used for processing optical signals. Components with different logical functions and memory elements are designed with these devices. The advantages of these devices are high speed of operation, small energy consumption, and the possibility of parallel signal processing. At the present time, their main applications appear to be information technologies and telecommunications.

We studied photodetectors based on intraband phototransitions in multiple-quantum-well structures. We have found that this technology provides a new approach to the detection of weak infrared signals. Multiple-quantum-well photodetectors can be designed to operate in the infrared spectral region with great sensitivity. They have considerable potential for further development and will compete with conventional extrinsic infrared photodetectors.

Additional information on semiconductor lasers including heterostructure lasers, lasers based on quantum wells and wires, and quantum-cascade lasers can be found in the following books and reviews:

A. Yariv, *Quantum Electronics* (Wiley, New York, 1991).

B. E. Saleh and M. C. Teich, *Fundamentals of Photonics* (Wiley, New York, 1991).

Y. Suematsu, K. Iga, and S. Arai, "Advanced semiconductor lasers," Proc. IEEE 80, 383–397 (1992).

E. Kapon, "Quantum wire lasers," Proc. IEEE 80, 398–410 (1992).

P. S. Zory, Jr., *Quantum Well Lasers* (Academic, Boston, 1993).

J. Faist, F. Capasso, D. L. Sivco, C. Sirtori, A. L. Hutchinson, and A. Y. Cho, "Quantum-cascade laser," Science 264, 553–556 (1994).

A. L. Lentine, *et al.*, "Evolution of the SEED technology: bistable logic gates to optoelectronic smart pixels," IEEE J. Quant. Electr. 29, pp. 665–669 (1993).

SEEDs and the latest related developments are discussed in these references:

D. A. B. Miller, *et al.*, "Nobel hybrid optically bistable switch: the quantum well self-electro-optic effect device," Appl. Phys. Lett. 45, pp. 13–15 (1984).

H. M. Gibbs, *Optical Bistability: Controlling Light by Light* (Academic, Orlando, FL, 1985).

A. L. Lentine, *et al.*, "Evolution of the SEED technology: bistable logic gates to optoelectronic smart pixels," IEEE J. Quant. Electr. 29, pp. 665–669 (1993).

Analyses of different heterostructure applications in infrared photodetectors can be found in the following books:

M. O. Manashreh, *Semiconductor Quantum Wells and Superlattices for Long-Wavelength Infrared Detectors* (Artech House, Boston, 1993).

K. L. Wang and R. P. G. Karunasiri, "Infrared detectors using SiGe/Si quantum-well structures," in *Semiconductor Interfaces, Microstructures, and Devices. Properties and Applications*, Z. C. Feng, ed. (Institute of Physics Publishing, Bristol and Philadelphia, 1993).

W. W. Chow, S. W. Koch, and M. Sargent, *Semiconductor Laser Physics* (Springer-Verlag, Berlin, New York, 1994).

PROBLEMS

1. Photodetector arrays operating in the infrared region of the spectrum have proved to be difficult to develop. Amazing progress has been made in this regard by advanced fabrication techniques for producing photodetector arrays with N separate detector elements with $Hg_xCd_{1-x}Te$ as the active detection medium. The narrow-bandgap $Hg_xCd_{1-x}Te$ must be grown with great care to prevent variations in the material properties from rendering particular elements of the array inoperable in the desired wavelength range. Suppose the array has N elements. Assume that the failures of the individual elements are uncorrelated and that probability for the failure of an individual element is f. Show that the probability P of obtaining an array with no bad elements is $P = (1 - f)^N$. What is P for a 1×128 array, assuming that only 0.01% of the elements fails? What is P for a 128×128 array assuming a failure probability of 0.01%? (Note: A 128×128 array is now well within the size of available arrays.)

2. Intersubband quantum-well lasers made of III–V semiconductors have been designed to emit radiation over a wide variety of frequencies. When these lasers are designed to operate in the current injection mode, carriers are injected over the laser quantum-well barrier with an energy E_{inj} that is greater than the energy ϵ_2 of the upper laser state. The electrons injected over the quantum-well barrier must lose an amount of energy $E_{inj} - \epsilon_2$ in order to thermalize to the energy of the upper level ϵ_2 of the laser transition $\epsilon_2 \to \epsilon_1$. This thermalization process involves the loss of energy, and in III–V semiconductor lasers this loss process is dominated by the emission of LO phonons. Would you expect LO phonons to be reabsorbed by the thermalizing electrons? What effect would this reabsorption process have on the electron-thermalization time? Would this effect limit the response of the laser output to a temporal change in the number of injected carriers? This temporal response is characterized in terms of a modulation bandwidth that indicates how quickly such a laser may be switched from the on state to the off state. For conventional quantum-well lasers this modulation bandwidth is limited to $\sim$20 GHz. In GaAs, a zone-center LO phonon decays into a pair of LA phonons with a lifetime in the range of 4–10 ps, depending on the temperature of the lattice. Would you expect this LO phonon lifetime to influence the modulation bandwidth of such a laser? By the mid-1990s, engineers had discovered ways to increase the modulation bandwidth by ensuring that $E_{inj} - \epsilon_2 = n\hbar\omega_{LO}$, where n is restricted to be a small integer such as 1 or 2; see for example, H. Yoon, X. Zhang, A. Gutierrez-Aitken, and Y. Lam, "Tunneling injection lasers: a new class of lasers with reduced hot carrier effects," IEEE J.

Quantum Electron. **32**, 1620 (1996). Why should limiting n to 1 or 2 lead to increased modulation bandwidth?

3. Intersubband quantum-well lasers based on III–V semiconductors have been made to operate at infrared wavelengths through advanced bandgap engineering. One technique used to produce infrared lasing is to design a system with lasing between energy levels ϵ_3 and ϵ_2, $\epsilon_3 - \epsilon_2 = \hbar\omega_{laser}$, and with $\epsilon_2 - \epsilon_1 = \hbar\omega_{LO}$ so that there is rapid depopulation of ϵ_2 through phonon-assisted transitions between ϵ_2 and ϵ_1. In many such intersubband quantum-well lasers, the quantum wells have dimensions of the order of 30 Å. Assuming that such a laser is fabricated from $Al_xGa_{1-x}As$ material, is it clear that $\hbar\omega_{LO}$ should always be near 36 meV? Is it possible that $\hbar\omega_{LO}$ should be approximately equal to 50 meV? [This problem has been considered in M. Stroscio, J. Appl. Phys. **80**, 6864 (1996).]

Index

Airy, equation, 79
Airy, function, 79, 162
All-optical signal processing, 551
Amplifier, 372
Angular momentum representation, 482
Annihilation, 475
Approximation, Born, 533
Approximation, Hartree, 44
Approximation, adiabatic, 86
Approximation, dipole, 468
Approximation, drift-diffusion, 70
Approximation, electron temperature, 67
Approximation, hydrodynamic, 69
Approximation, tight-binding, 101
Approximations, local, 64
Attenuation, 488

BT, 366
Balance principle, 68
Band, bending, 153
Band, conduction, 48, 107, 129, 357, 481
Band, heavy-hole, 273, 481, 499
Band, light-hole, 481, 499
Band, offset, 110
Band, split-off, 481
Band, split-off, valence, 131
Band, splitting, valence, 131
Band, valence, 48, 107, 129, 173, 470
Bandgap, 47, 106
Bandgap, narrowing, 406
Bandgap, wavelength, 471
Barrier height, 120
Basic vectors of the Bravais lattice, 45
Bias, forward, 396
Bias, reverse, 396
Biaxial strain, 144
Bipolar case, 397
Bistability, dispersive, 546
Bistability, dissipative, 556

Bleaching of the system, 543
Bloch, function, 45, 46
Bloch, oscillations, 104, 351
Bloch, theorem, 101
Bohr radius, 43, 112, 180, 245
Boltzmann, distribution, 39
Boltzmann, transport equation, 61, 62
Bosons, 198
Boundary conditions, cyclic, 46
Boundary conditions, electrostatic, 226
Boundary conditions, mechanical, 221
Built-in, Schottky potential, 369
Built-in, Schottky voltage, 368
Built-in, potential, 157

CHINT, 420
CMOS, 363
Capacitively coupled, 362
Carriers, intrinsic, 151
Carriers, majority, 397
Carriers, minority, 397
Channel, 368
Charge, accumulation, 371
Charge, transfer, 154
Coefficient, amplification, 470
Coefficient, diffusion, 18, 66
Coefficient, force, quasi-elastic, 186
Coefficient, reflection, 58
Coefficient, transmission, 58
Common base, 346
Concentration, intrinsic, 151, 406
Concentration, surface, electrons, 80, 162
Condition, flat-band, 154, 566
Condition, quasi-neutrality, 395
Conditions, cyclic, 188
Conditions, transparent, 565
Conductance, 60, 167, 357, 392, 428
Conductance, completely opened channel, 370

Conductance, quantum, 61, 357
Confined states, 114
Confinement, 213
Conservation laws, 264
Constant, Boltzmann's, 15
Constant, Frölich, coupling, 331
Constant, Planck's, 13, 518
Constant, Planck's, reduced, 13
Constant, Rydberg, 110
Constant, Rydberg, effective, 116
Constant, lattice, 138, 186, 189
Contrast of the resonator, 463
Coordinates, normal, 193, 274
Coordinates, spherical, 92, 266
Correlator, fluctuations, 241
Coulomb, blockade, 358
Coulomb, interaction, 40
Criterion for laser oscillations, 580
Cross-section area, 60, 341, 564
Cross-section, resonator, 489
Crystalline potential, 45
Crystalline potential, ideal periodic, 253
Cubic symmetry, 132
Current, amplification, base-collector, 400
Current, density, 60
Current, diffusion, 65
Current, displacement, 69
Current, drift, 65
Current, emitter, crowding, 407
Current, thermionic, 422
Current, transfer ratio, 400
Cutoff, frequency, 344, 365, 374, 445
Cutoff, mode, 403

DBRTD, 7
DCFL, 375
DRAM, 3
De Broglie wavelength, 15
Decrement, 489
Defect, pill-box, 121
Defect, semi-Gaussian, 121
Defect, shoe-box, 121
Defect, trench-like, 122
Defects, interfacial, 119
Defects, misfit, 140, 413
Deformation potentials, 255, 256
Density of particle flow, 29
Density of states, 192

Density of states, effective, 477
Density of states, effective, electrons and
  holes, 151
Density of states, optical, 477
Depth of well, 75
Devices, ballistic-injection, 367, 416
Devices, depletion mode, 369
Devices, enhancement mode, 369
Devices, homostructure, 73
Devices, normally-off, 369
Devices, normally-on, 369
Devices, single-electron, 357
Devices, voltage-controlled, 362
Devices, with a doped base, 419
Dielectric permittivity, medium, 41
Dielectric permittivity, two-dimensional
  gas, 245
Diodes, blue-green laser, 145
Diodes, light-emitting, 575
Dipole matrix element, 481
Dipole moment, 223, 531
Dipole moment, electron, 223, 502, 531
Dipole moment, system of electrons, 468
Distribution function, Fermi, 38, 48, 60
Distribution function, electron, 48, 69,
  159, 236
Distribution, Bose-Einstein, 198
Distribution, Fermi, 38
Distribution, needle-like, electron, 305
Doping, δ, 180
Doping, selective, heterostructures, 152
Double-well system, 96
Drain, 362
Drift velocity, average, 355
Drift velocity, saturation, 307

Edge, absorption, 471
Edge, absorption, fundamental, 106, 479
Edge-emitting configuration, 576
Effect, Burstein-Moss, 521
Effect, Gunn, 314
Effect, ac Stark, 535
Effect, electro-optical, 513
Effect, quantum confined Franz-Keldysh,
  519
Effect, quantum confined Stark, 519
Effect, saturation in the light amplification,
  566

Effect, stationary overshoot, 312
Effect, transient overshoot, 311, 312
Effective mass, 15
Effective mass, electron, 51
Effective mass, exciton, 109
Effective mass, heavy hole, 131
Effective mass, hole, 52
Effective mass, light hole, 131
Effective mass, longitudinal, 131
Effective mass, negative, 104
Effective mass, transverse, 131
Effects, linear optical, 514
Effects, nonlinear optical, 513
Effects, quadratic electro-optical, 514
Effects, quantum size, 17
Effects, real-space transfer, 326
Effects, transverse classical size, 21
Efficiency, emitter, 400
Efficiency, external, diode, 575
Efficiency, external, quantum, 582
Efficiency, internal, absorption, 617
Efficiency, internal, emission, 565
Eigenvalue, 33, 117
Eigenvalue, equation, 26
Eigenvalue, problem, 117, 214
Elastic, collision, 18
Elastic, energy per unit volume, 139
Elastic, energy, 140, 147, 200
Elastic, moduli of the crystal, 139, 200
Elastic, modulus, 186
Electric field, 62, 162, 223, 255
Electromagnetic cavity, 426
Electron affinity, 135
Electron affinity, rule, 136
Electron concentration, 68
Electron concentration, per unit area, 39
Electron concentration, per unit length, 39
Electron concentration, per unit volume, 39
Electron gas, one-dimensional, 39
Electron gas, three-dimensional, 39
Electron gas, two-dimensional, 39, 81
Electron, effective temperature, 304
Electron, energy band, 46
Electron, energy, 22, 49, 88
Electron, mean free path, 18
Electron, mobility, 64, 66
Electron, temperature, 68

Electron-hole pair, center of masses, 108
Electron-hole pair, reduced mass, 108, 109
Electron-phonon, coupling factor, 258
Electron-phonon, interaction, 253
Electron-phonon, interactions, piezoelectric, 255
Electron-phonon, interactions, polar, 255
Emission, spontaneous, 466
Emission, stimulated, 465
Emitter spacer, 448
Energy region, active, 309
Energy region, passive, 308
Energy, Fermi, 38
Energy, band, 100
Energy, dissipation, 69
Energy, exchange-correlation, 53
Energy, exciton, 106, 109, 484, 536
Energy, exciton, internal interaction, 109
Energy, extra elastic, 140
Energy, hole, 49
Energy, ionization, 110, 161
Energy, ionization, hydrogen atom, 110
Energy, kinetic, exciton, 109
Energy, of a transverse motion, 75
Energy, phonon, 218, 271
Energy, potential, 24, 56
Energy, subbands, 81
Energy-band, engineering, 141
Energy-band, tailoring, 141
Envelope phonon wave function method, 274
Equation, Langevin, 627
Equation, Poisson, 44
Equation, continuity, 68
Equation, neutrality, 160
Equation, two-particle Schrödinger, 108
Exact selection rule, 476
Exciton, 106
Exciton, interface, 111
Exciton, lifetime, 540
Exciton, radius, bulk, 109
Exclusive NOR-circuit, 446
Expectation value, 29
Extended state, 501

FET, 362
Fabry-Perot etalon, 462
Factor, Sommerfeld, 484

Factor, base transport, 400
Factor, optical confinement, 490
Factor, quality, 461
Fermi, golden rule, 35, 37, 61, 254
Fermi, level, 38, 49, 136, 151, 245
Fermi, wave vector, 40
Finesse of resonator, 463
First Brillouin zone, 46, 101
Fock equations, 42
Folded zone representation, 150
Folding procedure, 150
Forbidden gap, 191
Force, external, 52, 140, 308
Force, restoring, 186
Form factor, 247
Fourier, coefficients, 193
Fourier, series, 53
Free electron, mass, 15
Free-electron, bolometers, 615
Free-standing semiconductor slab, 213
Frequencies, high, 23
Frequencies, longitudinal optical phonon, 226
Frequencies, low, 23
Frequencies, transverse optical phonon, 226
Frequencies, ultra-high, 22
Frequency, 13
Function, delta, Dirac, 36
Function, envelope, 25, 53, 55, 254
Function, probe, 121
Function, radial, 92
Function, step, Heaviside, 82, 157, 427
Function, trial, 113
Function, work, 135
Functional, 32, 44, 63
Functional, variational, 122
Functions, Bessel, 94
Functions, spherical, 92

Gain, coefficient, 470
Gain, current, 401
Gain, power, 375
Gain, time-sequential, 614
Gate, 362, 436

HBT, 367
HEMT, 152, 365

HFET, 152, 365
Hamiltonian of a system, 23, 34
Harmonic oscillator, 193
Heat, dissipation, 327
Heat, removal, 327
Hermite polynomials, 90, 197
Heterojunction, 73
Heterostructure device, 73
Heterostructures, gated, 156
Heterostructures, pseudomorphic, 140
Heterostructures, ungated, 156
High-order harmonic generation, 548
Homojunctions, 73
Hot-electron, phenomena, 306
Hot-electron, problem, 304
Hot-phonon, problem, 315
Hydrogen-atom, problem, 109
Hydrogen-like impurity, 117

Increment, 489
Infrared light, 502
Integral, collision, 62, 236
Integral, overlap, 98
Integral, shift, 98
Integral, transfer, 98
Interaction, long-range, 261
Interaction, spin-orbital, 38, 131
Internal pinch-off voltage, 369
Islands, 120

JBT, 367
JFET, 365
Joule power, 68

Kronecker delta, 200, 276
Kronig-Penney model, 100

LED, 9
Lame constants, 200
Landauer formula, 55, 60
Laser, 467
Laser, output flux, 582
Lattice-matched, 138
Lattice-mismatched, 139
Layer, depletion, 161
Layer, inversion, 154
Layer, spacer, 163
Length, ambipolar, 20
Length, coherence, 17, 19

Length, diffusion, thermal, 19
Length, penetration, 393
Length, recombination, electron, 398
Length, recombination, hole, 398
Length, scattering, inelastic, 18
Length, screening, 245
Lifetime, of a carrier, 20
Lifetime, radiative, 478
Light-current characteristics, 575
Linear density of the string, 186

MBT, 367
MESFET, 365
MISFET, 365
MODFET, 152, 365
MOSFET, 362
Mach-Zehnder interferometer, 516
Macroatom, 91
Macroscopic polarization vector, 223
Mean free path, 17
Mechanism, Auger, 563
Mechanism, Franz-Keldysh, 518
Mechanism, Kerr, 514
Mechanism, Pockels, 514
Microcavity, 464
Microresonator, 464
Minibands, 102
Misfit, dislocations, 140
Misfit, imperfections, 182
Mode, active, operation, 403
Mode, saturation, 403
Modes, acoustic, 187
Modes, dilatational, 208
Modes, flexural, 205, 208
Modes, interface, 220, 227, 230
Modes, lateral, 583
Modes, longitudinal, 462
Modes, optical, 187
Modes, torsional, 208
Modes, transverse, 462
Modes, waveguide, 460, 490
Modulation bandwidth, 578, 589
Momentum vector, 13
Multivalued logic applications, 448
Mutual drag, 327

NERFET, 420
NMOS, 363

Negative dispersion, 218
Negative transconductance, 424
Newton's equations, 223
Noise, current, 617
Noise, equivalent optical power, 618
Noise, generation-recombination, 627
Noise, shot, 357
Nonequilibrium-phonon problem, 315
Nonlinear, optical media, 459
Nonlinear, optical medium, 532
Nonlocalized wavefunctions, 114
Number, base Gummel, 413
Number, of atoms per primitive crystal cell, 193
Number, of photons, 13, 458
Number, spin, 37
Number, total, electrons, 38

One-dimensional subbands, 88
Open optical resonators, 462
Operator, momentum, internal, 107
Operator, of momentum, 107
Operator, phonon annihilation, 198
Operator, phonon creation, 198
Operator, translation, 45
Optical, phase conjugation, 548
Optical, rectification, 548

PBT, 365
PMOS, 363
Partial factorization, 112
Pauli exclusion principle, 37
Peak-to-valley ratio, 344
Permittivity, optical, 224
Permittivity, static, 225
Perturbation method, 33
Phonon, 197
Phonon, occupation numbers, 198
Phototransitions, 464
Phototransitions, indirect, 475
Phototransitions, interband, 470
Phototransitions, intersubband, 501
Phototransitions, intraband, 500
Phototransitions, rate, 468
Photoconductivity, extrinsic, 615
Photoconductivity, intrinsic, 614
Photon, absorption, 465
Photon, emission, 465

Photon, flux in steady state, 581
Photon, flux, intracavity, 582
Photon, flux, total, 575
Photon, form factors, 461
Photons, 456
Polaron, 254
Population inversion, 467
Potential, Hartree, 43
Potential, deformation, tensor, 143, 261
Potential, effective, 93
Potential, self-consistent, 40
Potential, triangular, 162
Power dissipation, dynamic, 376
Power dissipation, static, 376
Primitive cell, 46, 53
Probability, of electron transition, 37
Probability, of finding the electron in the
     barrier layer, 96
Probability, of transition, 37
Probe-particle approach, 236
Processes, intraband, 473
Processes, intraband, scattering, 496
Processes, virtual, 484
Pumping, power, 562
Pumping, threshold level, 558
Punch-through, 407

QUIT, 366
Quantization of electron motion, 17
Quantum number, magnetic, 119
Quantum number, momentum, angular, 92
Quantum number, principal, 93
Quantum wires, 88
Quasi-Fermi levels, 397, 521, 558, 592
Quasi-ballistic motion, 16
Quasi-momentum law, 475

RHET, 439
RST, 420
Random sources, 627
Real-space transfer, 324
Recombination, Auger, 563
Recombination, electron-hole, 397
Recombination, radiative, 473
Recombination, velocity, interface, 569
Refractive index, 485
Regime, classical, ballistic, 21
Regime, classical, transport, 20

Regime, mesoscopic, transport, 20
Regime, streaming, 308, 322
Region, Schottky depletion, 168
Region, population inversion, 564
Relation, Einstein, 66, 331
Relation, Kramers-Krönig, 486
Relation, Lyddane-Sachs-Teller, 225
Relation, dispersion, 187
Remote impurities, 239
Resistance, base spreading, 407
Resistance, differential, 425
Resistance, multiplier, 432
Resistively coupled, 366
Resonant-tunneling process, 335
Responsivity, 617

SDHT, 152
SEED, 546
SOI, 363
Scattering by interface roughness, 249
Scattering, backward, 281
Scattering, duration, 21
Scattering, forward, 281
Scattering, quasi-elastic, 265
Scattering, in-elastic, 18
Scattering, rate, total, 237
Semiconductors, direct-bandgap, 129
Semiconductors, indirect-bandgap, 129
Semiconductors, many-valley, 128
Separation of variables, 30
Sheet density, 159
Short-range coupling, 260
Signal-to-noise ratio, 617
Source, 362
Spacer, 163, 168, 181, 240
Spherical average, 263
Split-gate technique, 89
Stark, ladder, 104
Stark, splitting, 353
Stationary states, 22
Statistics, Bose-Einstein, 198
Statistics, Fermi, 38
Statistics, Poisson, 551
Streaming, 305
Structure, Schottky gate, 167
Structure, cascade, 605
Structure, double quantum-well, 96
Structure, layered, semiconductor, 202

Structure, monopolar, 348

Structure, normally off, 168

Structure, normally on, 168

Superlattice, period, 101, 149, 211, 297, 353

Superlattice, sawtooth-doped, 181

Superlattices, 100

Surface-emitting, configuration, 576

Surface-emitting, laser, 592

Systems, many-particle, 37

Systems, mesoscopic, 20

Systems, zero-dimensional, 91

TEGFET, 152

Tensor, effective mass, reciprocal, 51, 130

Tensor, piezoelectric, 262

Tensor, strain, 139

Tensor, stress, 139

Time, characteristic, changes in the light wave amplitude, 534

Time, mean free flight, 19

Time, mean, two elastic scattering events, 18

Time, mean, two inelastic collisions, 18

Time, of electron interaction with lattice, 67

Time, of spreading, 19

Time, phase, 344

Time, phonon-decay, 315

Time, recombination, electron-hole, 397

Time, relaxation, energy, 69, 238

Time, relaxation, momentum, 65, 237

Time, scattering, electron-electron interaction, 67

Time, thermalization, 315

Time, transit, 22

Time, transit, semiclassical, 344

Transconductance, 346, 348, 372, 401, 423

Transistor, induced-base, 419

Transistor, quantum-interference, 389

Transistor, velocity-modulation, 385

Transistors, field-effect, 362

Transistors, hot-electron, 367

Transistors, quantum-well, gated, 436

Transitions, between the stationary states, 35

Transitions, excitonic, 471

Transitions, free-carrier, 471

Transitions, impurity-to-band, 470

Transitions, intersubband, 86

Transitions, phonon, 471

Transitions, vertical, 493

Transport, electron, semiclassical, 62

Transport, parallel or horizontal, 295

Transport, perpendicular or vertical, 295

Transport, quantum ballistic, 20

Transport, quasi-ballistic, 21

Tunneling, Zener, 352

Tunneling, coherent, processes, 336

Tunneling, matrix element, 530

Tunneling, sequential, 340

Turned off, 369

Turned on, 369

Two-dimensional, Fourier transformation, 241

Two-dimensional, optical confinement, 598

Two-dimensional, phonons, 284

Two-level system, 464

Unipolar case, 397

VMT, 366

Vacuum level, 135

Valleys, 128

Variational principle, 32

Vector of relative displacement, 139

Velocity, electron, 18

Velocity, group, 18, 26

Velocity, of light, 13

Velocity, overshoot, 311

Velocity, phase, 26

Velocity, saturation, 310

Velocity, sound, 187

Vibrations, collective, 187

Vibrations, zero-point, 458

Virtual excitations, 534

Volume, of the primitive cell, 256

Volume, of the system, 39

Wannier-Stark ladder, 353

Warm electrons, 304

Wave, intensity, 457

Wave, mixing, 548

Wave, phase, 13
Wave, polarization, 457
Wave function, many-electron, 40
Wave function, one-electron, 41
Wave function, time-dependent, 530
Wave function, trial, 163
Wave function, true, 33, 473
Wave number, 13

Waves, dilatational, 205
Waves, shear horizontal, 205
Waves, shear vertical, 205
Waves, torsional, 200
Wave vector, 46

f-process, 272
g-process, 272